Introduction to Computational Proteomics

CHAPMAN & HALL/CRC
Mathematical and Computational Biology Series

Aims and scope:
This series aims to capture new developments and summarize what is known over the entire spectrum of mathematical and computational biology and medicine. It seeks to encourage the integration of mathematical, statistical, and computational methods into biology by publishing a broad range of textbooks, reference works, and handbooks. The titles included in the series are meant to appeal to students, researchers, and professionals in the mathematical, statistical and computational sciences, fundamental biology and bioengineering, as well as interdisciplinary researchers involved in the field. The inclusion of concrete examples and applications, and programming techniques and examples, is highly encouraged.

Proposals for the series should be submitted to one of the series editors above or directly to:
CRC Press, Taylor & Francis Group
4th, Floor, Albert House
1-4 Singer Street
London EC2A 4BQ
UK

Published Titles

Algorithms in Bioinformatics: A Practical Introduction
Wing-Kin Sung

Bioinformatics: A Practical Approach
Shui Qing Ye

Biological Sequence Analysis Using the SeqAn C++ Library
Andreas Gogol-Döring and Knut Reinert

Cancer Modelling and Simulation
Luigi Preziosi

Cancer Systems Biology
Edwin Wang

Cell Mechanics: From Single Scale-Based Models to Multiscale Modeling
Arnaud Chauvière, Luigi Preziosi, and Claude Verdier

Combinatorial Pattern Matching Algorithms in Computational Biology Using Perl and R
Gabriel Valiente

Computational Biology: A Statistical Mechanics Perspective
Ralf Blossey

Computational Hydrodynamics of Capsules and Biological Cells
C. Pozrikidis

Computational Neuroscience: A Comprehensive Approach
Jianfeng Feng

Data Analysis Tools for DNA Microarrays
Sorin Draghici

Differential Equations and Mathematical Biology, Second Edition
D.S. Jones, M.J. Plank, and B.D. Sleeman

Engineering Genetic Circuits
Chris J. Myers

Exactly Solvable Models of Biological Invasion
Sergei V. Petrovskii and Bai-Lian Li

Gene Expression Studies Using Affymetrix Microarrays
Hinrich Göhlmann and Willem Talloen

Glycome Informatics: Methods and Applications
Kiyoko F. Aoki-Kinoshita

Handbook of Hidden Markov Models in Bioinformatics
Martin Gollery

Introduction to Bioinformatics
Anna Tramontano

Introduction to Computational Proteomics
Golan Yona

An Introduction to Systems Biology: Design Principles of Biological Circuits
Uri Alon

Kinetic Modelling in Systems Biology
Oleg Demin and Igor Goryanin

Knowledge Discovery in Proteomics
Igor Jurisica and Dennis Wigle

Meta-analysis and Combining Information in Genetics and Genomics
Rudy Guerra and Darlene R. Goldstein

Modeling and Simulation of Capsules and Biological Cells
C. Pozrikidis

Niche Modeling: Predictions from Statistical Distributions
David Stockwell

Normal Mode Analysis: Theory and Applications to Biological and Chemical Systems
Qiang Cui and Ivet Bahar

Optimal Control Applied to Biological Models
Suzanne Lenhart and John T. Workman

Pattern Discovery in Bioinformatics: Theory & Algorithms
Laxmi Parida

Python for Bioinformatics
Sebastian Bassi

Spatial Ecology
Stephen Cantrell, Chris Cosner, and Shigui Ruan

Spatiotemporal Patterns in Ecology and Epidemiology: Theory, Models, and Simulation
Horst Malchow, Sergei V. Petrovskii, and Ezio Venturino

Stochastic Modelling for Systems Biology
Darren J. Wilkinson

Structural Bioinformatics: An Algorithmic Approach
Forbes J. Burkowski

The Ten Most Wanted Solutions in Protein Bioinformatics
Anna Tramontano

Chapman & Hall/CRC Mathematical and Computational Biology Series

Introduction to Computational Proteomics

Golan Yona

CRC Press is an imprint of the
Taylor & Francis Group an **informa** business
A CHAPMAN & HALL BOOK

Cover design created by Jawahar Sudhamsu.

First published 2011 by Chapman & Hall

Published 2019 by CRC Press
Taylor & Francis Group
6000 Broken Sound Parkway NW, Suite 300
Boca Raton, FL 33487-2742

First issued in paperback 2019

ISBN-13: 978-0-367-45228-5 (pbk)
ISBN-13: 978-1-58488-555-9 (hbk)

Visit the Taylor & Francis Web site at
http://www.taylorandfrancis.com

and the CRC Press Web site at
http://www.crcpress.com

Library of Congress Cataloging-in-Publication Data

Yona, Golan, author.
Introduction to computational proteomics / Golan Yona.
p. ; cm. -- (Chapman & Hall/CRC mathematical & computational biology)
"A Chapman & Hall book."
Includes bibliographical references and index.
Summary: "This book tackles the steps and problems involved with protein analysis, classification, and meta-organization. It starts with the analysis of individual entities and proceeds to the analysis of more complex entities, from protein families to interactions, cellular pathways, and gene networks. The first part of the book presents methods for identifying the building blocks of the protein space, algorithms for assessing similarity between proteins, and mathematical models for representing protein families and classifying new instances. The second part covers methods that investigate higher order structure in the protein space through the application of unsupervised learning algorithms"--Provided by publisher.
ISBN 978-1-58488-555-9 (hardcover : alkaline paper)
1. Proteomics--Mathematical models. I. Title. II. Series: Chapman & Hall/CRC mathematical and computational biology series (Unnumbered)
[DNLM: 1. Proteomics--methods. QU 58.5]

QP551.Y66 2011
572'.6--dc22 2010042932

Acknowledgments

I would like to thank the following people for their help in reviewing chapters of this book and providing useful comments: Ran El-Yaniv, Nir Kalisman, Klara Kedem, Danny Barash, Yoram Gdalyahu, Sergio Moreno, Peter Mirani, Dahlia Weiss, Adelene Sim, Assaf Oron and Fengzhu Sun.

Certain sections of this book are based on joint research with my former students, and I would like to thank them for their contributions: Itai Sharon, Liviu Popescu, Chin-Jen Ku, Niranjan Nagarajan, Helgi Ingolfsson, Umar Syed, Michael Quist, Richard Chung, Shafquat Rahman, William Dirks, Aaron Birkland, Paul Shafer and Timothy Isganitis.

To Michael Levitt, for his continuous encouragement and support over the years. To Jawahar Sudhamsu, for his kind help with many figures and his invaluable friendship.

Contents

I The Basics **1**

1 What Is Computational Proteomics? **3**
1.1 The complexity of living organisms 3
1.2 Proteomics in the modern era 4
1.3 The main challenges in computational proteomics 5
1.3.1 Analysis of individual molecules 5
1.3.1.1 Sequence analysis 5
1.3.1.2 Structure analysis 6
1.3.2 From individual proteins to protein families 6
1.3.3 Protein classification, clustering and embedding . . . 7
1.3.4 Interactions, pathways and gene networks 7

2 Basic Notions in Molecular Biology **9**
2.1 The cell structure of organisms 9
2.2 It all starts from the DNA 10
2.3 Proteins 12
2.4 From DNA to proteins 15
2.5 Protein folding - from sequence to structure 18
2.6 Evolution and relational classes in the protein space 20
2.7 Problems 22

3 Sequence Comparison **23**
3.1 Introduction 23
3.2 Alignment of sequences 24
3.2.1 Global sequence similarity 25
3.2.1.1 Calculating the global similarity score . . 26
3.2.2 Penalties for gaps 27
3.2.2.1 Linear gap functions 28
3.2.3 Local alignments 29
3.2.3.1 Calculating the local similarity score . . . 30
3.3 Heuristic algorithms for sequence comparison 31
3.4 Probability and statistics of sequence alignments 32
3.4.1 Basic random model 33
3.4.2 Statistics of global alignment 33

3.4.2.1 Fixed alignment - global alignment without gaps 34
3.4.2.2 Optimal alignment 34
3.4.2.3 The zscore approach 35
3.4.3 Statistics of local alignments without gaps 37
3.4.3.1 Fixed alignment 37
3.4.3.2 Optimal alignment 38
3.4.4 Local alignments with gaps 43
3.4.5 Handling low-complexity sequences 45
3.4.6 Sequence identity and statistical significance 48
3.4.7 Similarity, homology and transitivity 49
3.5 Scoring matrices and gap penalties 50
3.5.1 Scoring matrices for nucleic acids 50
3.5.2 Scoring matrices for amino acids 50
3.5.2.1 The PAM family of scoring matrices . . . 52
3.5.2.2 The BLOSUM family of scoring matrices 57
3.5.3 Information content of scoring matrices 58
3.5.3.1 Choosing the scoring matrix 60
3.5.4 Gap penalties 61
3.6 Distance and pseudo-distance functions for proteins 62
3.7 Further reading 66
3.8 Conclusions 68
3.9 Appendix - non-linear gap penalty functions 69
3.10 Appendix - implementation of BLAST and FASTA 72
3.10.1 FASTA 72
3.10.2 BLAST 72
3.11 Appendix - performance evaluation 75
3.11.1 Accuracy, sensitivity and selectivity 76
3.11.2 ROC 79
3.11.3 Setup and normalization 80
3.11.4 Reference datasets, negatives and positives 82
3.11.5 Training and testing algorithms 83
3.12 Appendix - basic concepts in probability 85
3.12.1 Probability mass and probability density 85
3.12.2 Moments 86
3.12.3 Conditional probability and Bayes' formula 87
3.12.4 Common probability distributions 88
3.12.5 The entropy function 91
3.12.6 Relative entropy and mutual information 92
3.12.7 Prior and posterior, ML and MAP estimators 93
3.12.8 Decision rules and hypothesis testing 95
3.13 Appendix - metrics and real normed spaces 98
3.14 Problems 100

4 Multiple Sequence Alignment, Profiles and Partial Order Graphs **105**
4.1 Dynamic programming in N dimensions 106
4.1.1 Scoring functions 107
4.2 Classical heuristic methods 108
4.2.1 Star alignment 109
4.2.2 Tree alignment 110
4.3 MSA representation and scoring 113
4.3.1 The consensus sequence of an MSA 113
4.3.2 Regular expressions 114
4.3.3 Profiles and position-dependent scores 116
4.3.3.1 Generating a profile 116
4.3.3.2 Pseudo-counts 117
4.3.3.3 Weighting sequences 122
4.3.4 Position-specific scoring matrices 126
4.3.4.1 Using PSSMs with the dynamic programming algorithm 128
4.3.5 Profile-profile comparison 128
4.4 Iterative and progressive alignment 132
4.4.1 PSI-BLAST - iterative profile search algorithm . . . 132
4.4.2 Progressive star alignment 136
4.4.3 Progressive profile alignment 137
4.5 Transitive alignment 138
4.5.1 T-coffee 139
4.6 Partial order alignment 141
4.6.1 The partial order MSA model 142
4.6.2 The partial order alignment algorithm 144
4.7 Further reading 148
4.8 Conclusions 149
4.9 Problems 150

5 Motif Discovery **155**
5.1 Introduction 155
5.2 Model-based algorithms 156
5.2.1 The basic model 157
5.2.2 Model quality 158
5.2.2.1 Case 1: model unknown, patterns are given 159
5.2.2.2 Case 2: model is given, patterns are unknown 160
5.3 Searching for good models 160
5.3.1 The Gibbs sampling algorithm 161
5.3.1.1 Improvements 162
5.3.2 The MEME algorithm 162

5.3.2.1 E-step 164
5.3.2.2 M-step 165
5.3.2.3 The iterative procedure 166
5.4 Combinatorial approaches 167
5.4.1 Clique elimination 167
5.4.2 Random projections 170
5.5 Further reading 173
5.6 Conclusions 175
5.7 Appendix - the Expectation-Maximization algorithm 176
5.8 Problems 180

6 Markov Models of Protein Families 183

6.1 Introduction 183
6.2 Markov models 184
6.2.1 Gene prediction 184
6.2.2 Formal definition 188
6.2.2.1 Visible symbols and hidden Markov models 190
6.2.2.2 The model's components 190
6.3 Main applications of hidden Markov models 191
6.3.1 The evaluation problem 192
6.3.1.1 The HMM forward algorithm 194
6.3.1.2 The HMM backward algorithm 194
6.3.1.3 Using HMMs for classification 196
6.3.2 The decoding problem 196
6.3.3 The learning problem 198
6.3.3.1 The forward-backward algorithm 198
6.3.3.2 Learning from multiple training sequences 200
6.3.4 Handling machine precision limitations 201
6.3.5 Constructing a model 202
6.3.5.1 General model topology 202
6.3.5.2 Model architecture 202
6.3.5.3 Hidden Markov models for protein families 204
6.3.5.4 Handling silent states 206
6.3.5.5 Building a model from an MSA 206
6.3.5.6 Single model vs. mixtures of multiple models 209
6.4 Higher order models, codes and compression 210
6.4.1 Fixed order models 211
6.4.2 Variable-order Markov models 213
6.4.2.1 Codes and compression 214
6.4.2.2 Compression and prediction 217
6.4.2.3 Lempel-Ziv compression and extensions . 218
6.4.2.4 Probabilistic suffix trees 219

6.4.2.5 Sparse Markov transducers 227
6.4.2.6 Prediction by partial matches 229
6.5 Further reading 231
6.6 Conclusions 232
6.7 Problems 233

7 Classifiers and Kernels **235**
7.1 Generative models vs. discriminative models 235
7.2 Classifiers and discriminant functions 237
7.2.1 Linear classifiers 238
7.2.2 Linearly separable case 241
7.2.3 Maximizing the margin 244
7.2.4 The non-separable case - soft margin 246
7.2.5 Non-linear discriminant functions 249
7.2.5.1 Mercer kernels 253
7.3 Applying SVMs to protein classification 255
7.3.1 String kernels 256
7.3.1.1 Simple string kernel - the spectrum kernel 256
7.3.1.2 The mismatch spectrum kernel 257
7.3.2 The pairwise kernel 257
7.3.3 The Fischer kernel 258
7.3.4 Mutual information kernels 259
7.4 Decision trees 262
7.4.1 The basic decision tree model 263
7.4.2 Training decision trees 264
7.4.2.1 Impurity measures for multi-valued attributes 267
7.4.2.2 Missing attributes 268
7.4.2.3 Tree pruning 268
7.4.3 Stochastic trees and mixture models 270
7.4.4 Evaluation of decision trees 272
7.4.4.1 Handling skewed distributions 274
7.4.5 Representation and feature extraction 275
7.4.5.1 Feature processing 276
7.4.5.2 Dynamic attribute filtering 277
7.4.5.3 Binary splitting 278
7.5 Further reading 279
7.6 Conclusions 280
7.7 Appendix - estimating the significance of a split 281
7.8 Problems 288

8 Protein Structure Analysis **291**
8.1 Introduction 291
8.2 Structure prediction - the protein folding problem 293
8.2.1 Protein secondary structure prediction 296
8.2.1.1 Secondary structure assignment 297
8.2.1.2 Secondary structure prediction 299
8.2.1.3 Accuracy of secondary structure prediction 301
8.3 Structure comparison 303
8.3.1 Algorithms based on inter-atomic distances 305
8.3.1.1 The RMSd measure 305
8.3.1.2 The structal algorithm 309
8.3.1.3 The URMS distance 311
8.3.1.4 The URMS-RMS algorithm 312
8.3.2 Distance matrix based algorithms 318
8.3.2.1 Dali 319
8.3.2.2 CE 322
8.3.3 Geometric hashing 324
8.3.4 Statistical significance of structural matches 327
8.3.5 Evaluation of structure comparison 330
8.4 Generalized sequence profiles - integrating secondary structure with sequence information 332
8.5 Further reading 336
8.6 Conclusions 339
8.7 Appendix - minimizing RMSd 340
8.8 Problems 342

9 Protein Domains **345**
9.1 Introduction 345
9.2 Domain detection 348
9.2.1 Domain prediction from 3D structure 349
9.2.2 Domain analysis based on predicted measures of structural stability 351
9.2.3 Domain prediction based on sequence similarity search 355
9.2.4 Domain prediction based on multiple sequence alignments 361
9.3 Learning domain boundaries from multiple features 364
9.3.1 Feature optimization 365
9.3.2 Scaling features 366
9.3.3 Post-processing predictions 366
9.3.4 Training and evaluation of models 369
9.4 Testing domain predictions 370
9.4.1 Selecting more likely partitions 373
9.4.1.1 Computing the prior $P(D)$ 375

9.4.1.2 Computing the likelihood $P(S|D)$ 376
9.4.2 The distribution of domain lengths 378
9.5 Multi-domain architectures 380
9.5.1 Hierarchies of multi-domain proteins 380
9.5.2 Relationships between domain architectures 381
9.5.3 Semantically significant domain architectures 385
9.6 Further reading 387
9.7 Conclusions 389
9.8 Appendix - domain databases 390
9.9 Problems 393

II Putting All the Pieces Together **395**

10 Clustering and Classification **397**
10.1 Introduction 397
10.2 Clustering methods 399
10.3 Vector-space clustering algorithms 401
10.3.1 The k-means algorithm 402
10.3.2 Fuzzy clustering 404
10.3.3 Hierarchical algorithms 408
10.3.3.1 Hierarchical k-means 409
10.3.3.2 The statistical mechanics approach 409
10.4 Graph-based clustering algorithms 410
10.4.1 Pairwise clustering algorithms 411
10.4.1.1 The single linkage algorithm 412
10.4.1.2 The complete linkage algorithm 414
10.4.1.3 The average linkage algorithm 414
10.4.2 Collaborative clustering 415
10.4.3 Spectral clustering algorithms 421
10.4.4 Markovian clustering algorithms 425
10.4.5 Super-paramagnetic clustering 427
10.5 Cluster validation and assessment 428
10.5.1 External indices of validity 430
10.5.1.1 The case of known classification 430
10.5.1.2 The case of known relations 433
10.5.2 Internal indices of validity 434
10.5.2.1 The MDL principle 434
10.5.2.2 Cross-validation 439
10.6 Clustering proteins 440
10.6.1 Domains vs. complete proteins 440
10.6.2 Graph representation 441
10.6.3 Graph-based protein clustering 442
10.6.4 Integrating multiple similarity measures 444

10.7 Further reading 448
10.8 Conclusions 450
10.9 Appendix - cross-validation tests 451
10.10 Problems 457

11 Embedding Algorithms and Vectorial Representations 459
11.1 Introduction 459
11.2 Structure preserving embedding 461
11.2.1 Maximal variance embeddings 461
11.2.1.1 Principal component analysis 462
11.2.1.2 Singular value decomposition 467
11.2.2 Distance preserving embeddings 467
11.2.2.1 Multidimensional scaling 468
11.2.2.2 Embedding through random projections . 474
11.2.3 Manifold learning - topological embeddings 478
11.2.3.1 Embedding with geodesic distances 479
11.2.3.2 Preserving local neighborhoods 482
11.2.3.3 Distributional scaling 484
11.3 Setting the dimension of the host space 488
11.4 Vectorial representations 490
11.4.1 Internal representations 492
11.4.2 Collective and external representations 493
11.4.2.1 Choosing a reference set and an association measure 494
11.4.2.2 Transformations and normalizations . . . 495
11.4.2.3 Noise reduction 495
11.4.2.4 Comparing distance profiles 496
11.4.2.5 Distance profiles and mixture models . . . 500
11.5 Further reading 502
11.6 Conclusions 503
11.7 Problems 504

12 Analysis of Gene Expression Data 505
12.1 Introduction 505
12.2 Microarrays 509
12.2.1 Datasets 512
12.3 Analysis of individual genes 513
12.4 Pairwise analysis 515
12.4.1 Measures of expression similarity 517
12.4.1.1 Shifts 520
12.4.2 Missing data 521
12.4.3 Correlation vs. anti-correlation 523
12.4.4 Statistical significance of expression similarity 524

12.4.5 Evaluating similarity measures 527
12.4.5.1 Estimating baseline performance 528
12.5 Cluster analysis and class discovery 529
12.5.1 Validating clustering results 534
12.5.2 Assessing individual clusters 536
12.5.3 Enrichment analysis 538
12.5.3.1 The gene ontology 538
12.5.3.2 Gene set enrichment 541
12.5.4 Limitations of mRNA arrays 544
12.6 Protein arrays 545
12.6.1 Mass-spectra data 546
12.7 Further reading 548
12.8 Conclusions 550
12.9 Problems 551

13 Protein-Protein Interactions **553**
13.1 Introduction 553
13.2 Experimental detection of protein interactions 556
13.2.1 Traditional methods 557
13.2.1.1 Affinity chromatography 557
13.2.1.2 Co-immunoprecipitation 558
13.2.2 High-throughput methods 558
13.2.2.1 The two-hybrid system 558
13.2.2.2 Tandem affinity purification 560
13.2.2.3 Protein arrays 561
13.3 Prediction of protein-protein interactions 561
13.3.1 Structure-based prediction of interactions 562
13.3.1.1 Protein docking and prediction of interaction sites 563
13.3.1.2 Extensions to sequences of unknown structures 567
13.3.2 Sequence-based inference 568
13.3.2.1 Gene preservation and locality 568
13.3.2.2 Co-evolution analysis 571
13.3.2.3 Predicting the interaction interface 578
13.3.2.4 Sequence signatures and domain-based prediction 582
13.3.3 Gene co-expression 589
13.3.4 Hybrid methods 589
13.3.5 Training and testing models on interaction data . . . 591
13.4 Interaction networks 592
13.4.1 Topological properties of interaction networks 593
13.4.2 Applications 601

13.4.3 Network motifs and the modular organization of networks . . . 603
13.5 Further reading . . . 606
13.6 Conclusions . . . 607
13.7 Appendix - DNA amplification and protein expression . . . 608
13.7.1 Plasmids . . . 608
13.7.2 SDS-PAGE . . . 608
13.8 Appendix - the Pearson correlation . . . 610
13.8.1 Uneven divergence rates . . . 610
13.8.2 Insensitivity to the size of the dataset . . . 610
13.8.3 The effect of outliers . . . 611
13.9 Problems . . . 613

14 Cellular Pathways **615**
14.1 Introduction . . . 615
14.2 Metabolic pathways . . . 618
14.3 Pathway prediction . . . 621
14.3.1 Metabolic pathway prediction . . . 621
14.3.2 Pathway prediction from blueprints . . . 623
14.3.2.1 The problem of pathway holes . . . 623
14.3.2.2 The problem of ambiguity . . . 623
14.3.3 Expression data and pathway analysis . . . 624
14.3.3.1 Deterministic gene assignments . . . 626
14.3.3.2 Fuzzy assignments . . . 629
14.3.4 From model to practice . . . 632
14.4 Regulatory networks: modules and regulation programs . . . 635
14.5 Pathway networks and the minimal cell . . . 640
14.6 Further reading . . . 642
14.7 Conclusions . . . 645
14.8 Problems . . . 646

15 Learning Gene Networks with Bayesian Networks **649**
15.1 Introduction . . . 649
15.1.1 The basics of Bayesian networks . . . 650
15.2 Computing the likelihood of observations . . . 654
15.3 Probabilistic inference . . . 655
15.3.1 Inferring the values of variables in a network . . . 656
15.3.2 Inference of multiple unknown variables . . . 660
15.4 Learning the parameters of a Bayesian network . . . 661
15.4.1 Computing the probability of new instances . . . 666
15.4.2 Learning from incomplete data . . . 667
15.5 Learning the structure of a Bayesian network . . . 669
15.5.1 Alternative score functions . . . 672

15.5.2 Searching for optimal structures 674
15.5.2.1 Greedy search 675
15.5.2.2 Sampling techniques 675
15.5.2.3 Model averaging 676
15.5.3 Computing the probability of new instances 678
15.6 Learning Bayesian networks from microarray data 678
15.7 Further reading 682
15.8 Conclusions 683
15.9 Problems 684

References **687**

Conference Abbreviations **735**

Acronyms **737**

Index **739**

Preface

Computational molecular biology, or simply **computational biology**, is a term generally used to describe a broad set of techniques, models and algorithms that are applied to problems in biology. This is a relatively new discipline that is rooted in two different disciplines: computer science and molecular biology. Being on the border line between the two disciplines, it is related to fields of intensive research in both. The goal of this book is to introduce the field of computational biology through a focused approach that tackles the different steps and problems involved with protein analysis, classification and meta-organization. Of special interest are problems related to the study of protein-based cellular networks. All these tasks constitute what is referred to as **computational proteomics**.

This is a broad goal, and indeed the book covers a variety of topics. The first part covers methods to identify the building blocks of the protein space, such as motifs and domains, and algorithms to assess similarity between proteins. This includes sequence and structure analysis, and mathematical models (such as hidden Markov models and support vector machines) that are used to represent protein families and classify new instances. The second part covers methods that explore higher order structure in the protein space, through the application of unsupervised learning algorithms, such as clustering and embedding. The third part discusses methods that explore and unravel the broader context of proteins, such as prediction of interactions with other molecules, transcriptional regulation and reconstruction of cellular pathways and gene networks.

The book is structured also based on the type of the biological data analyzed. It starts with the analysis of individual entities, and works its way up through the analysis of more complex entities. The first chapters provide a brief introduction to the molecular biology of the main entities that are of interest when studying the protein space, and an overview of the main problems we will focus on. These are followed by a chapter on pairwise sequence alignment, including rigorous and heuristic algorithms, and statistical assessment of sequence similarity. Next we discuss algorithms for multiple sequence alignment, as well as generative and discriminative models of protein families. We proceed to discuss motif detection, domain prediction and protein structure analysis. All these algorithms and models are elemental to the methods that are discussed in the next couple of chapters on clustering, embedding and protein classification. The last several chapters are devoted to the analysis of the broader biological context of proteins, which is essential to fully and accurately characterize proteins and their cellular counterparts. This includes gene expression analysis, prediction and analysis of protein-protein interactions, and the application of probabilistic models to study pathways, gene networks and causality in cells.

The book is intended for computer scientists, statisticians, mathematicians and biologists. The goal of this book is to provide a coherent view of the field and the main problems involved with the analysis of complex biological systems and specifically the protein space. It offers rigorous and formal descriptions, when possible, with detailed algorithmic solutions and models. Each chapter is followed by problem sets from courses the author has taught at Cornell University and at the Technion, with emphasis on a practical approach. Basic background in probability and statistics is assumed, but is also provided in an appendix to Chapter 3. Knowledge of molecular biology is not required, but we highly recommend referring to a specialized book in molecular biology or biochemistry for further information (for a list of recommended books, see the book's website at `biozon.org/proteomics/`)

It should be noted that the interaction of computer science and molecular biology as embodied in computational biology is not a one way street. In this book we focus on algorithms and models and their application to biological problems. The opposite scenario, where biological systems are used to solve mathematical problems (as in DNA computing), is also of interest; however it is outside the scope of this book. Nevertheless, it is fascinating to see how biology affects the way we think, by introducing new concepts and new models of computation (well known examples include neural networks and genetic algorithms). This interaction invigorates fields like statistics and computer science and triggers the development of new models and algorithms that have a great impact on other fields of science as well.

Before we start, we should mention the term **Bioinformatics**, which is equivalent to computational biology. Some make a distinction and use the term computational biology to refer to the *development* of novel algorithms and models to solve biological problems, while Bioinformatics is used to refer to the *application* of these algorithms to biological data. However, this difference in semantics is somewhat fuzzy, and practically the terms are used interchangeably.

Part I

The Basics

Chapter 1

What Is Computational Proteomics?

1.1 The complexity of living organisms

Living organisms are very complex. In a way, they are similar to complex computer programs. The basic identity of an organism is defined by its DNA, its genetic blueprint, which is referred to as the genome. One can perceive the DNA as a master code that basically codes for everything else that is going on in the cell. The DNA codes for the basic functional elements like proteins and RNA molecules. In computer science lingo we can refer to these molecules as functional procedures. Proteins, RNA and DNA can interact and form "functional modules". These modules and the basic functional elements are put together to form pathways which are the cellular counterpart of programs, with input and output. All these programs talk to each other, in what might be considered equivalent to an operating system[1].

Although we know much about the way cells function, we really know only very little. In order to better understand their complexity we first need to decipher their genome. Not surprisingly, when the human genome project was completed in 2002, it garnered a lot of excitement. However, although it was a great achievement, it was really just the first step toward understanding our own code at multiple levels. The outcome of the human genome project was only the master code, the long sequence of nucleotides that encodes for all the functional elements and indirectly for the more complex constructs. Hence, upon its completion the main interest in genomics has shifted to studying the biological meaning of the entities that are encoded within.

The main entities of interest in this book are proteins, since they are the main machinery of cells and are at the core of almost all cellular processes. The term **proteome** refers to the set of all proteins of an organism[2]. Char-

[1]The analogy between living cells and computers or computer programs seems natural, but cells are structured according to different principles, with information transfer back and forth between the master code and the procedures it codes for, which can result in "reprogramming" of the master code.

[2]Sometimes the term is used to refer only to the set of *expressed* (active) proteins, which may differ from one cell type to another. Even the same cell might have different proteomes under different environmental conditions, such as different temperatures or growth mediums.

acterizing the structure and function of proteins has been one of the main goals of molecular biology. Knowledge of the biological function of proteins is essential for the understanding of fundamental biochemical processes, for drug design and for genetic engineering, to name a few reasons.

Proteomics is concerned with the large-scale study of proteins[3]. Beyond structural and functional analysis of individual proteins, proteomics research seeks to characterize the proteome of an organism (the set of expressed proteins) in different cells and under different conditions and identify the molecular counterparts each protein interacts with.

At the next level, proteomics research aims to study the organization of proteins into cell-wide networks and characterize the properties of these networks. Reconstruction of gene networks is important for understanding causality in biological systems. It is also important for the identification of cellular procedures with specific biological functions (cellular pathways) and can shed light on how they communicate with each other.

1.2 Proteomics in the modern era

While it is possible to answer the questions listed above through experimentation, this has become exceedingly difficult in view of the massive flow of new biological data. Modern technologies keep generating a vast amount of data on various aspects of biological systems. For example, large-scale sequencing projects throughout the world turn out new sequences whose function is unknown. These ongoing sequencing efforts have already uncovered the sequences of over 10,000,000 proteins (as of January 2010), the majority of which have not been characterized yet. In parallel, new technologies such as RNA and protein microarrays enable researchers to take snapshots of cell activity by measuring RNA and protein expression on a cellular level. Other technologies, such as yeast two-hybrid assays, enable researchers to detect interactions on a genomic scale. This multitude of new biological data exposes new processes and phenomena in biological systems. Consequently, there is a growing interest in exploring modules and pathways involving multiple proteins and other subcellular agents. By studying the organization of macromolecules into cellular processes we wish to better understand the role of these procedures, and by studying cellular procedures, we hope to better understand cellular "computations" and to gain insight into the functions of the component molecules.

[3]The term proteomics was introduced in the 1990s [1–4], referring mostly to cell-wide measurements of protein abundance levels, but it later evolved to encompass protein analysis in the broader sense.

This sheer volume of new biological data rules out directed research of all known protein sequences, interactions and pathways and consequently some domains of the proteome will remain untouched by experimentation in the near future. **Computational proteomics** uses computational tools to address the questions proteomics is concerned with: from sequence or structure analysis that aims to characterize proteins based on their similarity with proteins of known function, to algorithms that seek to identify recurring patterns in proteins and classify proteins into families that share a common function, and algorithms that predict interactions and pathways. These are the main goals of computational proteomics.

1.3 The main challenges in computational proteomics

1.3.1 Analysis of individual molecules

A vast majority of the proteins that were sequenced already were never studied in vivo. Many of them have an unknown function, or contain segments (domains) of unknown function. Deciphering the rules that define protein functionality, and characterizing computationally the biological role of these molecules is one of the main research problems in computational biology.

Driven by the principle of evolution, many biological discoveries are based on inference through induction, rather than through experimentation. If entity A has a property X and entity B is similar to A, then it is likely that B has property X. Hence, the similarity relation is fundamental to biological data analysis. Reliable and effective methods for comparison of biological entities are essential not only for the deduction of relationships between objects, they are also needed for the meta-organization of biological entities (be it proteins, DNA sequences, pathways or interactions) into relational classes (see Section 1.3.3).

When it comes to the analysis of proteins, there are multiple attributes we need to consider to establish their similarity. This includes (but is not limited to) their sequence, their structure and their domain architecture. Furthermore, a proper analysis should employ multiple metrics and representations that exploit different aspects of the data. Throughout this book we will discuss the topic extensively, characterizing these metrics, their advantages and their limitations.

1.3.1.1 Sequence analysis

Sequence data is the most abundant type of data on proteins. Sequence analysis considers the basic properties of individual amino acids, as well as their combination. The goal is to predict, based only on their sequence, the biochemical properties of proteins and their functional role in living cells. Significant sequence similarity entails similar or related functions. Therefore, detecting similarities between protein sequences can help to reveal the biological function of new protein sequences, by extrapolating the properties of their "neighbors" which have been assigned a function. Thus, biological knowledge accumulated over the years can be transfered to new genes.

Sequence similarity is not always easily detectable. During evolution sequences have changed by insertions, deletions and mutations. These evolutionary events may be traced today by applying algorithms for sequence comparison. New protein sequences are routinely compared with the sequences in one or more of the main sequence databases. Such methods were applied during the last four decades with considerable success and helped to characterize many protein sequences, as well as to reveal many distant and interesting relationships between protein families. A major part of this book will cover different aspects of sequence analysis that are geared toward function prediction.

1.3.1.2 Structure analysis

In many cases sequences have diverged to the extent that their common origin is untraceable by current means of sequence comparison. In such cases sequence analysis fails to provide clues about the functionality of the protein in question. However, proteins that are related through evolution (homologous) often have similar structures even if their sequences are no longer similar. Hence, detecting structural similarity can help to identify possible functional relationships. Structure analysis is also the key to understanding the mechanisms of protein-protein and protein-ligand interactions, which are of special interest in areas such as immunology. We will discuss structural aspects of proteins in Chapters 8, 9 and 13.

1.3.2 From individual proteins to protein families

Functional analysis of proteins is greatly enhanced by applying advanced algorithms that utilize the information in groups of homologous proteins (protein families). Over the years several different mathematical representations were developed, including statistical generative models such as position-specific scoring matrices (Chapter 4) and hidden Markov models (Chapter 6) and discriminative models such as support vector machines (Chapter 7). Comparing a new sequence with a database of protein families using these representations is more effective than a standard database search in which each sequence is compared with only one library sequence at a time. Such models can capture subtle features of proteins that belong to those families and therefore are more sensitive in detecting remote homologies. These models also play an essential role in the study of the global organization of the protein space, since the application of pattern-recognition algorithms requires models that can represent protein families efficiently and effectively.

1.3.3 Protein classification, clustering and embedding

The goal of protein classification is to organize proteins into functional classes (families). Detecting classes in biological data can help to better understand and characterize the entities in each class and discover global principles that underlie biological systems, similar to the periodic table in chemistry. This is true not only for proteins but for other entities as well, such as interactions or pathways.

Protein classification is extremely useful for function and structure analysis and is especially important for large-scale annotation efforts such as those that follow the large-scale sequencing projects. Classification is usually achieved by means of clustering: grouping entities into clusters based on their similarity (Chapter 10). Beyond classification, clustering helps to study meta-organization and organize proteins into a hierarchy of classes. Protein sequences are usually classified into a hierarchy of sub-families, families and super-families. If structural information is available, they can be further classified into fold families and classes. Some of these relational classes are based on clear evolutionary relationships, such as families and super-families. The definition of other classes is subjective and depends on our current knowledge of the protein structure repertoire.

Higher level organization of proteins can also be studied by applying embedding techniques and mapping the protein space into real-normed spaces. This mapping not only facilitates the application of advanced data analysis tools. It can also provide a global view of the protein space, which may expose subtle relationships between protein families and add insights about the function of a protein or a protein family from its relative position in the space, and from the protein families in its close neighborhood. This is especially useful when currently available means of comparison fail to give satisfactory information about the biological function of a protein. In such cases, a global view may add the missing clues.

The task of finding high-order organization in the protein space entails the solution of several fundamental problems, some of which are inherent to the analysis of protein sequences. For example, proteins can be analyzed at the level of their building blocks (domains) or as complete protein chains. They can be compared based on their sequence, structure or other attributes. How should we assess their similarity while accounting for their multiple facets? At the next level, we have to explore the most effective ways to represent and cluster proteins and protein families. These issues pertain to open problems in computational proteomics, as mentioned above, and in search of meta-organization of the protein space we must address all of them.

1.3.4 Interactions, pathways and gene networks

Biological entities are tightly related and mutually dependent on each other, and to fully understand the role of an individual entity one has to know which

molecular components with different roles (Figure 2.1). These components move around compartments, communicate with each other, act upon other molecules or serve as the building blocks of more complex molecules.

The functioning of a cell is an orchestrated sequence of events that are carried out to sustain life, respond to surrounding stimuli and communicate with other cells. These include energy metabolism, metabolism and degradation of molecules (recycling), signal transduction, duplication, cell division and more. A general distinction is made between organisms whose cells contain a nucleus (**eukaryotes**) and those that do not (**prokaryotes**). Eukaryotes are often multi-cellular organisms (a notable exception is yeast), while almost all prokaryotes are unicellular organisms (although some can form colonies composed of multiple cells).

2.2 It all starts from the DNA

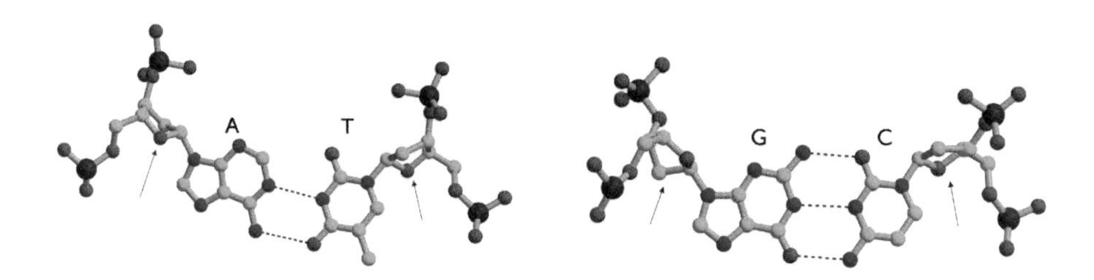

FIGURE 2.2: A ball-and-stick model of nucleotides in base pairs. A and T on the left, and G and C on the right. The hexagonal and pentagonal shapes are the rings that form the bases. The nucleotides A, G are called **purines** (characterized by a two ring structure), while C, T are called **pyrimidines**. The arrows point to the ribose sugar rings. The green balls represent carbon atoms, the red oxygen and the blue nitrogen. The larger black balls at the periphery represent phosphorus atoms. The dashed lines indicate hydrogen bonds between nucleotides. C forms three hydrogen bonds with G, while A forms two with T; therefore the GC pairing is stronger.

The **DNA molecule** is the main carrier of genetic information in every cell of almost every organism[1]. The DNA is a macromolecule, one of a class of molecules called **nucleic acids**, which is composed of four small monomer

[1]There are some exceptions, such as RNA viruses.

molecules called **nucleotides**[2]: **adenine** (A), **guanine** (G), **cytosine** (C) and **thymine** (T). The four molecules are similar in structure, but they have different bases, attached to a sugar ring, called **ribose**.

The DNA sequence is formed by concatenating nucleotides one after the other in a strand. The nature of the bond between consecutive nucleotides (called phosphodiester bond) is such that the resulting sequence is directed, with one side called 5' and the other side called 3'. The four nucleotides can also interact with each other across different DNA strands to create **base pairs**, by forming another type of bond called a hydrogen bond, as shown in Figure 2.2. Two nucleotides that can bind to each other are called complementary. The famous double helix structure is composed of two complementary DNA strands (Figure 2.3). The double helix structure increases the stability of the DNA and provides means for easy replication (mainly during cell division) and translation as well as self-correction mechanisms. The DNA can be linear, or circular as in some prokaryotes. It should be noted though that not all organisms have a double helical DNA. Certain organisms have linear single-strand DNA (such as some viral genomes).

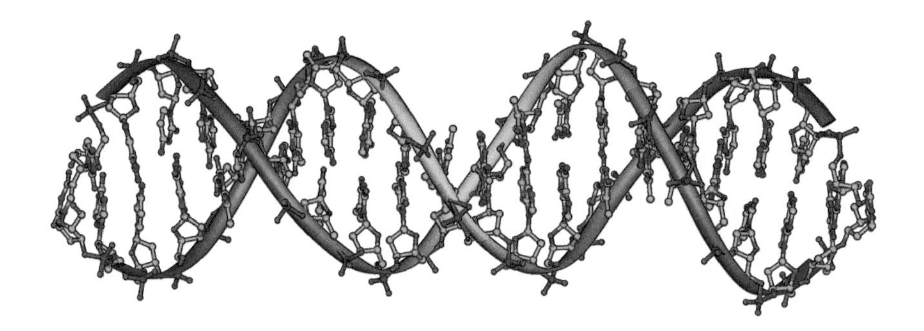

FIGURE 2.3: **The DNA double helix.** (Figure reproduced from Wikipedia [5]. Original figure created by Michael Ströck.)

The whole DNA sequence of an organism comprises the **genome** of that organism. Almost every cell of an organism has a copy of its genome (mammalian red blood cells, for example, do not). Inside the cell, the genome is divided and stored in multiple **chromosomes**. The total number of nucleotides and the total number of chromosomes in a genome varies from one

[2]The name nucleotides refers not only to the four bases that make up DNA, but to other molecules as well. Some are involved in metabolism or energy production and storage cycles (e.g., ATP), others serve as intermediates in multiple cellular processes.

organism to another (see Table 2.1). For example, the human genome contains about 3 billion base pairs stored in 46 chromosomes, while the bacteria *Escherichia coli* (E. coli) contains about 5 million stored in a single chromosome. In fact, all known prokaryotes contain a single chromosome, stored in the cell's cytoplasm (the liquid medium contained within the cell membrane). In eukaryotes, most of the DNA is stored in the cell's nucleus and the rest is stored in other compartments, such as the mitochondria (see Figure 2.1).

TABLE 2.1: The number of chromosomes and nucleotides in the genomes of several well-studied organisms. Information was retrieved from the ORNL website at `www.ornl.gov`

Organism	No. of Chromosomes	Genome Size (Mega bases)	Estimated No. of Genes
Homo sapiens (human)	46	3038	~30,000
Rattus norvegicus (rat)	42	2750	~30,000
Drosophila melanogaster (fruit fly)	8	180	13,600
Arabidopsis thaliana (plant)	10	125	25,500
Saccharomyces cerevisiae (yeast)	32	12	6,300
Escherichia coli (bacteria)	1	5.1	3,200
Haemophilus influenzae (bacteria)	1	1.8	1,700

The DNA molecule carries information about almost all functional elements of the cell. Most notable are regions along the DNA that are referred to as protein coding regions or simply **coding regions**[3]. Each coding region typically encodes for a single protein and is called a **gene**. To understand how the DNA codes for proteins we first need to understand what proteins are.

2.3 Proteins

Proteins are the molecules that form much of the functional machinery of each cell in every organism. They make a large part of the structural framework of cells and tissues (such as collagen in skin and bones), they carry out the transport and storage of small molecules, and they are involved with the transmission of biological signals. Proteins include the enzymes that catalyze chemical reactions and regulate a variety of biochemical processes in the cell, as well as antibodies that defend the body against infections.

[3]This term is a bit misleading, since the segments between protein coding regions can code for other molecular entities, including RNA molecules, regulatory elements, patterns that control DNA folding and others. In fact, the coding regions are only a small percentage of the genomes of higher organisms.

Depending on the organism, each type of cell has up to several thousand kinds of proteins that play a primary role in determining the characteristic of the cell and how it functions.

TABLE 2.2: Properties of amino acids. Based on RasMol's classification of amino acids, courtesy of the RasMol open source team at `www.rasmol.org`.

Residue	A	R	N	D	C	E	Q	G	H	I	L	K	M	F	P	S	T	W	Y	V
acidic				*		*														
acyclic	*	*	*	*	*	*	*	*		*	*	*	*			*	*			*
aliphatic	*							*		*	*									*
aromatic									*					*				*	*	
basic		*							*			*								
buried	*				*					*	*		*	*				*		*
charged		*		*		*			*			*								
cyclic									*					*	*			*	*	
hydrophobic	*							*		*	*		*	*	*			*	*	*
large		*				*	*		*	*	*	*	*	*				*	*	
medium			*	*	*										*		*			*
negative				*		*														
neutral	*		*		*		*	*	*	*	*		*	*	*	*	*	*	*	*
polar		*	*	*	*	*	*		*			*				*	*			
positive		*							*			*								
small	*							*								*				
surface		*	*	*		*	*	*	*			*			*	*	*		*	

Proteins are complex molecules. They are assembled from 20 different **amino acids**, which have diverse chemical properties. The amino acids form the vocabulary that allows proteins to exhibit a great variety of structures and functions. All amino acids have the same base structure, but different side chains (Figure 2.4). In general, amino acids can be divided into several categories, based on their physical and chemical properties (see Table 2.2 and Figure 2.4).

	Second				
First	T	C	A	G	Last
T	Phe	Ser	Tyr	Cys	T
	Phe	Ser	Tyr	Cys	C
	Leu	Ser	**Stop**	**Stop**	A
	Leu	Ser	**Stop**	Trp	G
C	Leu	Pro	His	Arg	T
	Leu	Pro	His	Arg	C
	Leu	Pro	Gln	Arg	A
	Leu	Pro	Gln	Arg	G
A	Ile	Thr	Asn	Ser	T
	Ile	Thr	Asn	Ser	C
	Ile	Thr	Lys	Arg	A
	Met (**Start**)	Thr	Lys	Arg	G
G	Val	Ala	Asp	Gly	T
	Val	Ala	Asp	Gly	C
	Val	Ala	Glu	Gly	A
	Val	Ala	Glu	Gly	G

TABLE 2.3: Codon Table. The 64 possible combinations of nucleotides that make up codons. Note that most amino acids are encoded by more than one codon. Although all coding sequences must start with a start codon, not all proteins start with a methionine because of post transcription and translation modifications of the gene sequence. It also has been observed that some prokaryotic genes start with TTG (leucine) or GTG (valine) codons.

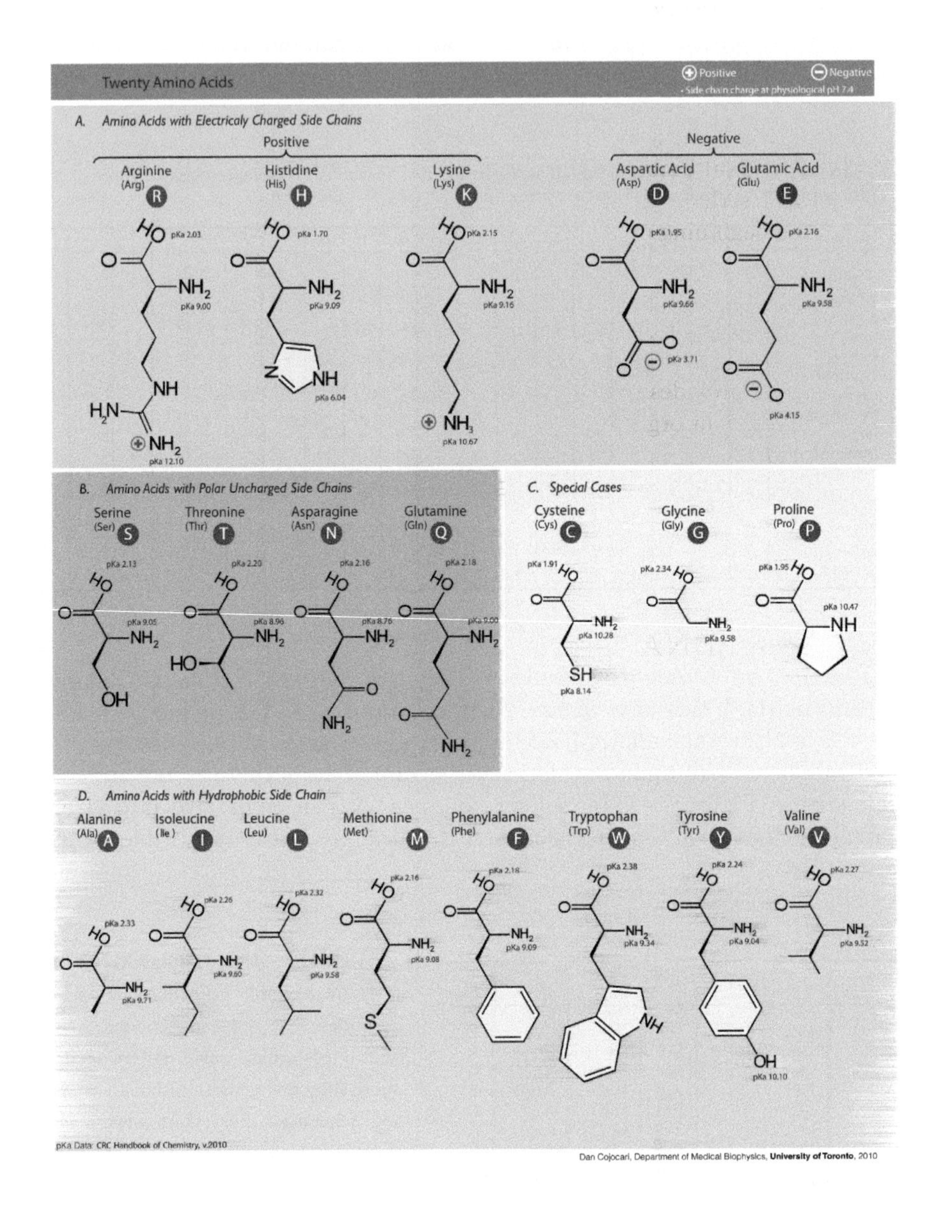

FIGURE 2.4: Amino acids. The molecular structure of the 20 naturally occurring amino acids that make up proteins. Next to the name of each amino acid appear the three-letter code and the one-letter code (in red circle). (Figure modified from Wikipedia [6]. Original figure created by Dan Cojocari.)

What is the relation between DNA and proteins? How can an alphabet of four bases code for sequences over an alphabet of 20 amino acids? Clearly at least three nucleotides are needed to span the 20 possible amino acids, since two nucleotides allow for only 16 possible combinations. In the DNA gene sequence of a protein every block of three nucleic acids (**codon**) corresponds to an individual amino acid. For example, the codon AGC encodes for the amino acid serine. The **start codon** (which also encodes for methionine) marks the beginning of a coding sequence and the **stop codon** marks its end. The set of rules that specify which amino acid is encoded in each codon is called the **genetic code** (Table 2.3). With 64 possible combinations there is redundancy, and some amino acids are coded by more than one codon, up to six different codons (arginine, leucine, serine). The redundancy of the genetic code provides robustness to random mutations. The code is universal, although different organisms might use different codons that code for the same amino acid.

2.4 From DNA to proteins

How is the information being propagated from the DNA forward? While this is a complicated puzzle, of which some pieces are still unknown, we do know quite a bit already, and according to the **central dogma of molecular biology**, the information is being transfered in several main steps as depicted in Figure 2.5.

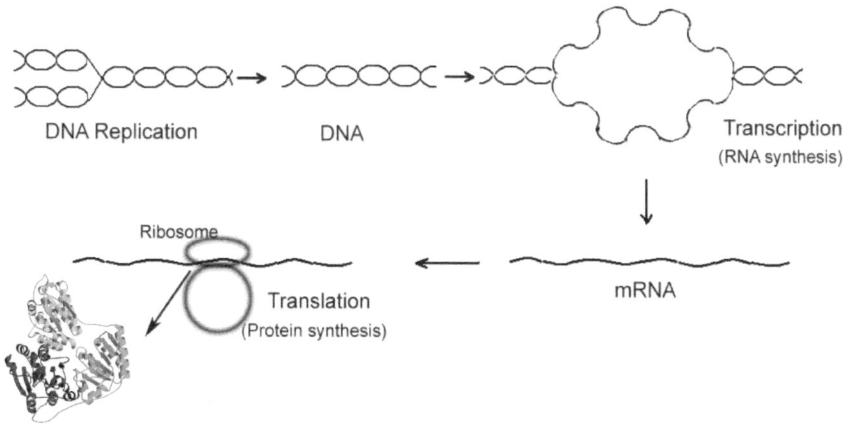

FIGURE 2.5: The central dogma of molecular biology.

The first step is **transcription**, the conversion of coding regions along the DNA to a molecular template called **messenger RNA (mRNA)** that is later translated to a protein. Transcription is carried by a protein molecule called **RNA polymerase**. Like DNA, **RNA** is a nucleic acid built from four bases, three of which are identical to the bases that make up DNA. The fourth one is **uracil** (U), which replaces thymine (T). Both have very similar chemical structures.

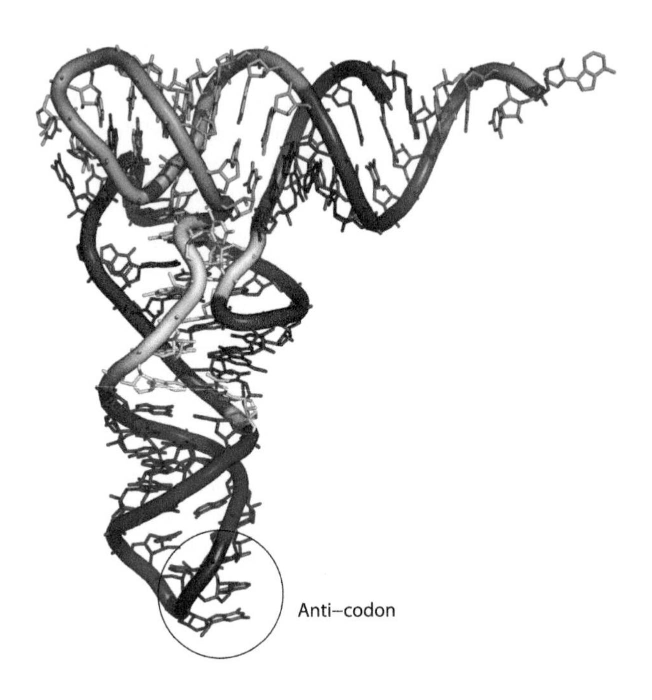

FIGURE 2.6: **Transfer RNA molecule (tRNA).** (Figure modified from Wikipedia [7].) Note the secondary structures formed by the RNA molecule through hydrogen bonds between A and U or C and G (see Figure 2.2). For each amino acid there is a different tRNA molecule, with a different anti-codon sub-structure (circled). The amino acid binds to the other end of the tRNA molecule. The anti-codon mechanism is essential for the recognition of the relevant codon locations along a gene sequence. Helped by the ribosome (a large complex of proteins and RNA molecules), tRNA molecules attach to the mRNA gene template (by forming bonds between the codons and the corresponding anti-codons) and "decipher" the instructions encoded within. The amino acids attached to the other end of the tRNA molecules form **peptide bonds**, and gradually the complete protein sequence is built.

In addition to their role as an intermediate step toward the assembly of proteins, RNA molecules have other important functions, including protein translation (see below), catalyzing enzymatic reactions and the control of other processes (such as replication and expression). These functions are prescribed in the RNA three-dimensional structure that is formed through base pairing of nucleotides (Figure 2.6). However, unlike DNA, RNA molecules are single stranded and the base pairing occurs between nucleotides along the single-stranded molecule.

The next step is **translation**, the decoding of the mRNA templates into proteins. A special machinery in the cell, called the **ribosome**, is responsible for the decoding process, i.e., the translation and execution of a gene's "instructions" resulting in the formation of a protein sequence. This is a fascinating process during which the ribosome reads the mRNA template sequence, one codon at a time, and guides **transfer RNA (tRNA)** molecules to bind to the template using the **anti-codon** mechanism, while forming the amino acid sequence on the other end (see Figure 2.6). Once the translation is done, the mRNA is degraded by an enzyme called RNase, and the nucleotides are recycled for future uses. Computer scientists and mathematicians might notice the resemblance between the translation process and a Turing machine. The difference is that this "Turing machine" is not allowed to write to its tape, at least not directly.

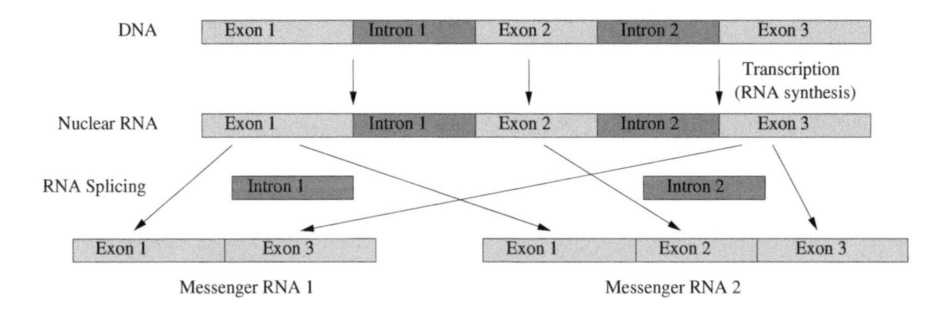

FIGURE 2.7: Alternative splicing. In this example there are seven possible protein products, consisting of the exon combinations 1, 2, 3, 12, 13, 23 and 123 (although not all necessarily exist). Up to an additional eight products may exist if shuffling occurs (corresponding to the combinations 21, 31, 32, 132, 213, 231, 312 and 321).

In prokaryotes each gene usually consists of a single coding sequence. However, in eukaryotes, a gene sequence often contains non-coding segments that are skipped during translation. The coding segments are called **exons** and the non-coding ones are called **introns**. After transcription and before translation the template RNA is processed and the introns are spliced out. The remaining exons make up the actual gene sequence that is being translated to a protein. The mechanism of **splicing** is actually more complex. Sometimes, only some of the exons are used. This results in what is called **alternative splic-**

ing, where different combinations of exons result in different protein products (Figure 2.7). Thus, the same DNA gene sequence may encode for multiple proteins, diversifying the proteins encoded in the genomes of higher organisms (without substantially increasing the genome size). With alternative splicing and **post-translational modifications (PTMs)**[4] of the protein sequence, the number of different protein products has been estimated to be as high as 10 times the number of genes.

Interestingly, although the process of translation suggests that DNA predates RNA and RNA predates proteins, it is generally accepted that RNA was actually present in ancient organisms before the evolution of DNA and proteins (the **RNA World** hypothesis), in light of its versatility and ability to carry functions that are characteristic of both DNA and proteins [8]. However, it has not been established yet whether DNA predates proteins or vice versa.

2.5 Protein folding - from sequence to structure

Each protein is created in the cell first as a well-defined chain of amino acids, whose sequence is called the **primary structure** of a protein. The average length of a protein sequence is 350 amino acids, but it can be as short as a few amino acids, and as long as a few thousand (the longest protein known, the muscle protein titin, is over 30,000 amino acids long). Schematically, each amino acid may be represented by one letter, and proteins can be viewed as long "words" over an alphabet of 20 letters.

According to the central dogma of protein folding, the protein sequence (the primary structure) dictates how the protein folds in three dimensions. It is the specific three-dimensional structure that enables the protein to fulfill its particular biological role. It prescribes the function of the protein, and the way it interacts with other molecules. However, to date the relation between the sequence, structure and the function of a protein is not yet fully understood.

Different terms are used to describe the structural hierarchy in proteins: **secondary structures** are local sequence elements (3-40 amino acids long) that have a well-determined regular shape, such as a **alpha helix**, or a **beta strand** (see Figure 2.8). Other local sequence elements, which are neither helices nor strands, are usually called **loops** or **coils**, and they may adopt a large variety of shapes. Secondary structures are packed into what is called the

[4]PTMs are modifications of the protein sequence or the protein structure, after it is translated from the template mRNA sequence. These could be a modification of certain amino acids or the attachment of other small molecules (see page 554).

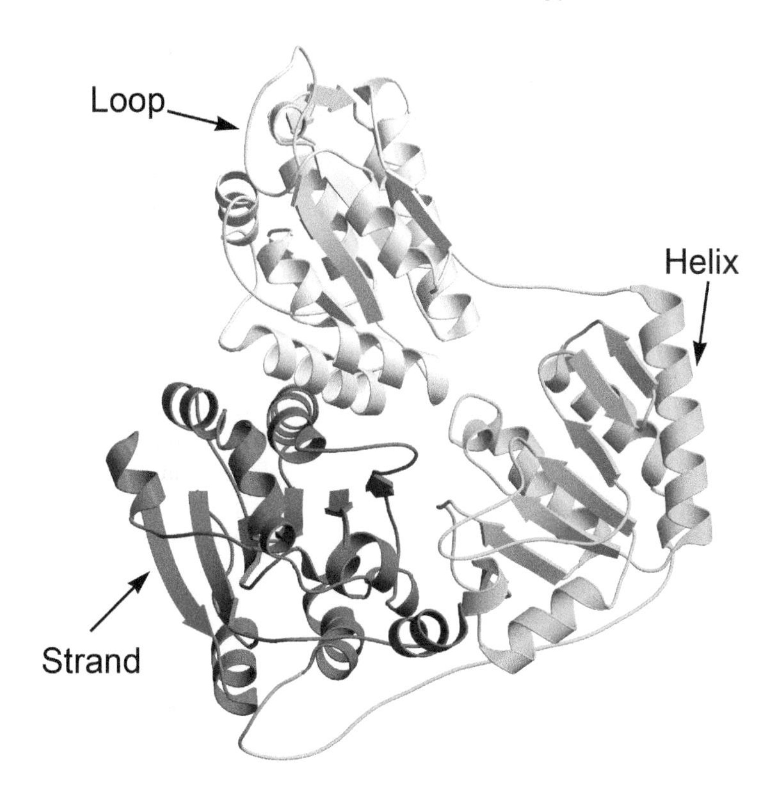

FIGURE 2.8: **Ribbon diagram of the crystal structure of the enzyme pyruvate decarboxylase (PDB 1qpb).** The protein is composed of three domains, colored black, light gray and dark gray.

tertiary structure - the three-dimensional (3D) structure of the whole sequence. The process during which a protein acquires its 3D structure is called **protein folding**. This is a spontaneous process, and although searching the space of all possible 3D conformations could take enormous time, the protein folds in a matter of milliseconds or microseconds (the so-called **Levinthal's paradox**). In some cases the process is aided by helper molecules called **chaperones**, also proteins, that guide the protein into its native structure.

In general, helices and strands form the **core** of the protein structure, whereas loops are more likely to be found on the surface of the molecule. The shape (the architecture) of the 3D structure is called the **fold**. Each protein sequence has a unique 3D structure, but several different proteins may adopt the same fold, i.e., the shape of their 3D structure is similar. Finally, several protein sequences may arrange themselves together to form a complex, which is called the **quaternary structure**.

The terms "motifs" and "domains" are also very common when describing protein structure/function. A **motif** is a simple combination of a few consecutive secondary structure elements with a specific geometric arrangement (e.g., helix-loop-helix). Some, but not all motifs are associated with a specific biological function. A **domain** is considered the fundamental unit of structure. It combines several secondary elements and motifs, not necessarily contiguous, which are usually packed in a compact globular structure. A protein may be comprised of a single domain or several different domains, or several copies of the same domain (Figure 2.8).

Not every possible combination of amino acids can form a stable protein sequence that folds and functions properly. Evolution has "selected" only some of those sequences that could fold into a stable functional structure. The term **sequence space** is used to describe the set of all currently known protein sequences and those expected to exist. The term **structure space** refers to the set of all folds adopted by protein sequences. Again, not every shape is possible (because of physical constraints), and not every possible shape is expected to exist.

2.6 Evolution and relational classes in the protein space

The well-established theory of evolution assumes that all organisms evolved from an ancient common ancestor through many events of speciation. These events can be summarized in what is called the **tree of life**, which describes the relations between organisms and their ancestors. Evolution manifests itself in relations between biological entities. For example, proteins are called **homologous** if they share a common ancestor (homologous proteins are referred to as **homologs**). When protein sequences exhibit significant sequence similarity they are assumed to be homologous. Proteins that are evolutionarily related usually have the same core structure, but they may have very different loops. A **protein family** is a group of homologous proteins, and proteins that belong to the same family have the same or very similar biological function. Different families have different **molecular clocks**, meaning they evolve at different rates, which depend on their function and their molecular counterparts. Proteins involved in transcription and translation (such as some of those that make up the ribosome) are essential for reproduction and normal functioning of cells and tend to be highly conserved, while proteins that are part of the immune system tend to evolve rapidly, adapting to foreign agents such as viruses and bacteria.

Protein families that are distantly related through evolution are usually grouped into a **protein superfamily**. Proteins that belong to the same superfamily have close or related biological functions and are usually assumed to have evolved from the same ancestor protein, but sequence similarity is not always detectable between members of the same superfamily. Different protein superfamilies may adopt the same **fold**, meaning their 3D structures are similar. This relation might not be due to common evolutionary ancestry, but could be the result of **convergent evolution**, a term used to describe the scenario in which evolution rediscovers the same "solution" (e.g., fold) starting from different seeds and exploring different paths.

Throughout this book we will discuss algorithms that attempt to detect relationships between proteins, based on their sequence and structure, and classify them into families.

2.7 Problems

For updates, additional problems, files and datasets check the book's website at `biozon.org/proteomics/`.

1. **DNA, RNA, proteins and the genetic code**

   (a) Explain why there are six possible Open Reading Frames (ORFs) in a gene DNA sequence of a double-stranded DNA.

   (b) Derive the six possible protein sequences encoded in this sequence:

   $$TATGACCAGGGAGCTCTGGAT$$

   (c) Derive the sequence of the mRNA transcribed from the following gene sequence:

   $5'-AACTGATGGTACCCTACAGGAATTGTACCCGATTTTGAGGTC-3'$

   (d) Translate the mRNA sequence to a protein (identify the correct open reading frame first).

2. **Alternative splicing**

   Introns are recognized and spliced out at specific splicing sites. The most common splicing sites are GT at the 5' end of an intron and AG at the 3' end of an intron.

   (a) How many introns are there in the following DNA sequence?
   $TATGACCAGGGAGCTCGTGGATAGCCTGACATCTGCGTAACGAC$
   $AGAGACCCGCATAATTCA$

   (b) Derive all possible protein products coded in this sequence.

Chapter 3

Sequence Comparison

3.1 Introduction

Sequence analysis is the main source of information for most new genes. Evolution has preserved sequences that have proved to perform correctly and beneficially, and proteins with similar sequences are found in different organisms. Proteins that evolved from the same ancestor protein are called **homologous proteins**. Similar proteins can also be found within the same organism as a result of duplication and shuffling of coding segments in the DNA during evolution, either before or after speciation. These proteins are referred to as **paralogous proteins**. Proteins can also be similar as a result of convergent evolution and evolve to have similar folds or similar functions, even if they do not have a common ancestor and do not exhibit sequence similarity. Such proteins are called **analogous proteins**.

Over the last four decades a considerable effort has been made to develop algorithms that compare sequences of macromolecules (proteins, DNA). The purpose of these algorithms is to detect evolutionary, structural and functional relations among sequences. Successful sequence comparison would allow one to infer the biological properties of new sequences from data accumulated on related genes. For example, sequence comparison can help to detect genes in a newly sequenced genome, since a similarity between a translated nucleotide sequence and a known protein sequence suggests a homologous coding region in the corresponding nucleotide sequence. Significant sequence similarity among proteins may imply that the proteins share the same secondary and tertiary structure and have close biological functions. Prediction of the structure of a protein sequence is often based on the similarity with homologous proteins whose structure has been solved.

This chapter discusses the basics of sequence comparison, scoring schemes, and the statistics of sequence alignments, which is essential for distinguishing true relations among proteins from chance similarities[1].

[1]Several sections of this chapter overlap slightly with [9], contributed by the author, and were included with permission by OUP.

3.2 Alignment of sequences

When comparing sequences of macromolecules, we wish to determine whether they evolved from a common ancestor sequence and estimate their evolutionary distance. The more similar the sequences are, the more likely they are to be related and have similar functions.

Evolution manifests itself through random changes in the DNA, and to assess the evolutionary distance between sequences we need to trace these changes. The basic events of molecular evolution are the substitution of one base by another (**point mutations**) and the insertion or deletion of a base. Other events could be duplication of a segment within the same gene, transposition of a segment and reversal. However, these events are relatively rare, and since they complicate the mathematical analysis of sequence similarity they are usually ignored or simply treated as insertions/deletions.

Suppose that the sequence $\mathbf{b}$ is obtained from the sequence $\mathbf{a}$ by substitutions, insertions and deletions. It is customary and useful to represent the transformation by an **alignment** where $\mathbf{a}$ is written above $\mathbf{b}$ with the common (conserved) bases aligned appropriately. For example, say that $\mathbf{a} = ACTTGA$ and $\mathbf{b}$ is obtained by substituting the second letter from C to G, inserting an A between the second and the third letters, and by deleting the fifth base (G). The corresponding alignment would be

$$\begin{aligned}\mathbf{a} &= A\ C\ -\ T\ T\ G\ A \\ \mathbf{b} &= A\ G\ A\ T\ T\ -\ A\end{aligned}$$

We usually do not know which sequence evolved from the other. Therefore the events are not directional and insertion of A in $\mathbf{b}$ is equivalent to a deletion of A in $\mathbf{a}$.

In a typical application we are given two possibly related sequences and we wish to recover the evolutionary events that transformed one to the other. The goal of sequence alignment is to find the *correct* alignment that summarizes the series of evolutionary events that have occurred. The alignment can be assigned a score that accounts for the number of identities (a match of two identical bases), the number of substitutions (a match of two different bases) and the number of gaps (insertions/deletions). With high scores for identities and low scores for substitutions and gaps (see Table 3.1), the basic strategy toward recovering the correct alignment seeks the alignment that scores best, as described next.

The algorithms described below may be applied to the comparison of protein sequences as well as to DNA sequences (coding or non-coding regions). Though the evolutionary events occur at the DNA level, the main genetic pressure is on the protein sequence. Consequently, the comparison of protein sequences has proved to be much more effective [10]. Mutations at the DNA

	A	C	G	T
A	1	-1	-1	-1
C	-1	1	-1	-1
G	-1	-1	1	-1
T	-1	-1	-1	1

TABLE 3.1: A simple scoring function for nucleotides. Scoring functions for amino acids are more complicated and are discussed in Section 3.5.2.

level do not necessarily change the encoded amino acid because of the redundancy of the genetic code, or they often result in conservative substitutions at the protein level, namely, replacement of an amino acid by another amino acid with similar biochemical properties. Such changes tend to have only a minor effect on the protein's functionality. With this in mind and in light of the scope of this book, this chapter focuses on comparison of protein sequences, although most of the discussion applies also to comparison of nucleic acid sequences.

3.2.1 Global sequence similarity

We are given two sequences $\mathbf{a} = a_1 a_2 ... a_n$ and $\mathbf{b} = b_1 b_2 ... b_m$, where $a_i, b_j \in \mathcal{A}$, and $\mathcal{A}$ is the alphabet of amino acids. A **global alignment** of the two sequences is an alignment where all letters of **a** and **b** are accounted for.

Let $s(a, b)$ denote the similarity score of amino acids of type a and b (for example, arginine and leucine) and $\alpha > 0$ be the penalty for deleting/inserting one amino acid. The score of an alignment in which the number of matches between amino acid a and amino acid b is N_{ab} and the number of insertions/deletions is N_{gap} is defined as

$$\sum_{a \in \mathcal{A}} \sum_{b \in \mathcal{A}} N_{ab} \cdot s(a, b) - N_{gap} \cdot \alpha$$

The **global similarity** of sequences **a** and **b** is defined as the highest score of any alignment of sequences **a** and **b**, i.e.,

$$S(\mathbf{a}, \mathbf{b}) = \max_{alignments} \left\{ \sum_{a \in \mathcal{A}} \sum_{b \in \mathcal{A}} N_{ab} \cdot s(a, b) - N_{gap} \cdot \alpha \right\}$$

As we will see later (Section 3.5.2), the entries in amino acid scoring matrices are usually defined such that the score of an alignment has a statistical meaning, reflecting the likelihood that the sequences are evolutionary related. The similarity score is thus essentially a maximum likelihood measure.

How to find the best alignment? Since there is an exponentially large number of possible alignments (see Problem 3.1) it is impossible to perform an exhaustive search. For example, the number of possible alignments of two sequences of length 350 (which is the average length of protein sequences in protein sequence databases) exceeds 10^{250}. However, using a **Dynamic Programming (DP)** algorithm we can find the optimal alignment efficiently without checking all possible alignments, but only a very small portion of the search space. Dynamic programming algorithms were already introduced in the late 1950's [11]. However, the first authors to propose a dynamic programming algorithm for *global* comparison of macromolecules were Needleman and Wunsch [12].

3.2.1.1 Calculating the global similarity score

Denote by $S_{i,j}$ the score of the best alignment of the substring $a_1a_2...a_i$ with the substring $b_1b_2...b_j$, i.e.,

$$S_{i,j} = S(a_1a_2...a_i \ , \ b_1b_2...b_j)$$

The *optimal* alignment of $a_1a_2...a_i$ with the substring $b_1b_2...b_j$ can end only in one of following three ways:

$$\begin{matrix} a_i \\ b_j \end{matrix} \quad \textbf{or} \quad \begin{matrix} a_i \\ - \end{matrix} \quad \textbf{or} \quad \begin{matrix} - \\ b_j \end{matrix}$$

Also, each sub-alignment must be optimal as well (otherwise, the alignment won't be optimal). Therefore, the score $S(\mathbf{a}, \mathbf{b})$ can be calculated recursively, considering these three options:

$$S_{i,j} = \max\{S_{i-1,j-1} + s(a_i, b_j),\ S_{i,j-1} - \alpha,\ S_{i-1,j} - \alpha\}$$

after initialization

$$S_{0,0} = 0 \qquad S_{i,0} = -i \cdot \alpha \quad \text{for } i = 1..n \qquad S_{0,j} = -j \cdot \alpha \quad \text{for } j = 1..m$$

In practice, the scores are stored in a two-dimensional array of size $(n+1) \cdot (m+1)$ (called the **DP matrix**). The initialization sets the values at the zero row and the zero column. The computation proceeds row by row, exhausting every row before proceeding to the next one. Each matrix cell stores the maximal possible score for aligning the corresponding subsequences, which is calculated from entries that were already calculated. Specifically, we need the three matrix cells: on the left, on the top and the one diagonal to the current cell (see Figure 3.1). The global similarity score is stored in the last cell of the matrix, $S_{n,m}$. The time complexity of this algorithm is $O(n \cdot m)$. Note however that the space complexity is $O(n)$, since it is enough to keep just two rows at a time.

The algorithm described above computes the similarity score of the best alignment. But how can we recover the alignment itself? Since we do not commit to a specific alignment until the very end, we have to keep track for each cell how we got there (that is, which of the previous cells was used to maximize the alignment score of the current cell). This is done by using an array of pointer variables (called the **trace matrix**) and filling in the values as we compute the similarity score in each cell of the DP matrix. Once we completed the computation of the DP matrix, we use the pointers of the trace matrix to recover the optimal alignment, aligning characters if we used a diagonal move and inserting gaps in one of the sequences otherwise, depending on whether we move along the column or the row.

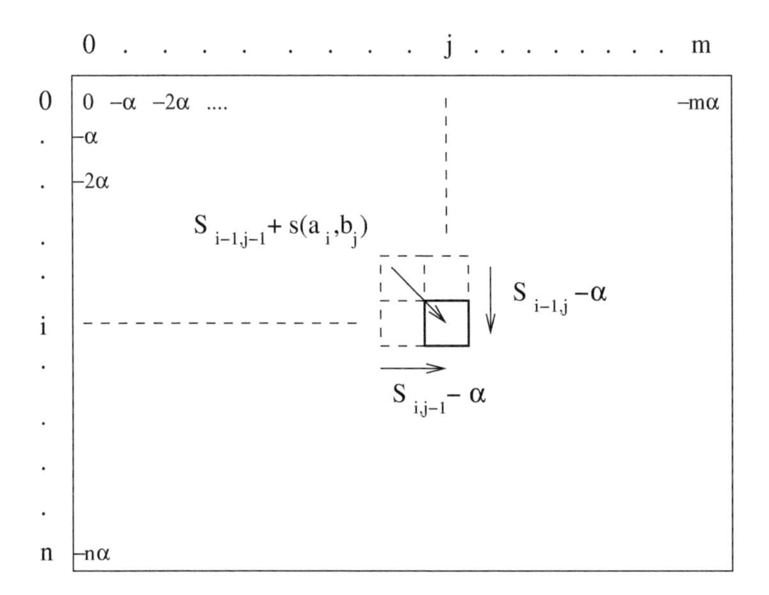

FIGURE 3.1: **Calculating the global similarity score.** (Figure modified from [9] with permission.) The score of the (i, j) entry in the matrix is calculated from three matrix cells: the one on the left (corresponding to the scenario where b_j is aligned with a gap), the one on the top (aligning a_i with a gap) and the one located at the top left corner of the current cell (aligning a_i with b_j). In the case of a non-constant gap penalty we need also to check all the cells in the same row and all the cells in the same column (along the dashed lines).

3.2.2 Penalties for gaps

The event of an insertion or deletion often affects a segment of a protein, consisting of several adjacent amino acids. In other words, a gap of several amino acids is more likely to occur as a single event rather than several consecutive events of deletions or insertions of individual amino acids. In this view, opening a gap is considered to be a more significant event than extending a gap, and most computational models for sequence comparison assign a penalty for a gap of length k that is smaller than the sum of k independent gaps of length 1. Denote the penalty for a gap of length k as $\alpha(k)$, then we are interested in **sub-additive** functions such that $\alpha(x+y) \leq \alpha(x) + \alpha(y)$.

Let $N_{k\text{-}gap}$ be the number of gaps of length k in a given alignment. Then the score of the alignment is now defined as

$$\sum_{a \in \mathcal{A}} \sum_{b \in \mathcal{A}} N_{ab} \cdot s(a,b) - N_{k\text{-}gap} \cdot \alpha(k)$$

With a non-constant gap penalty the alignment can end in a gap of arbitrary length, and therefore $S_{i,j}$ is now defined as follows:
Initialize

$$S_{0,0} = 0 \qquad S_{i,0} = -\alpha(i) \qquad S_{0,j} = -\alpha(j)$$

then

$$S_{i,j} = \max \left\{ S_{i-1,j-1} + s(a_i, b_j),\ \max_{1 \leq k \leq j} \{S_{i,j-k} - \alpha(k)\},\ \max_{1 \leq l \leq i} \{S_{i-l,j} - \alpha(l)\} \right\}$$

Note that when the gap penalty is fixed and set to α then all we need to check when computing the score of a certain matrix cell is the three adjacent cells (diagonal, left, top), even if the gap length is $k > 1$. However, if the gap penalty is an arbitrary function $\alpha(k)$ then it is necessary to consider all the cells in the current column and the current row. In other words, the adjacent cells do not have all the information necessary to compute $S_{i,j}$ (for a concrete example, see Appendix 3.9 of this chapter). For arbitrary function $\alpha(k)$, the time complexity of the DP algorithm is $\sum_{i,j}(i+j+1) = O(n^2 \cdot m + n \cdot m^2 + n \cdot m)$ plus $n+m$ calculations of the function $\alpha(k)$. For $n = m$ we get $O(n^3)$.

3.2.2.1 Linear gap functions

A better time complexity than $O(n^3)$ can be obtained for a linear gap function

$$\alpha(k) = \alpha_0 + \alpha_1 \cdot (k-1)$$

where α_0 is the penalty for opening a gap, and α_1 is the penalty for extending the gap.

Set $R_{i,j} = \max_{1 \leq k \leq j}\{S_{i,j-k} - \alpha(k)\}$. This is the maximum value over all the matrix cells in the same row, where a gap is tested with respect to each one of them. Note that

$$\begin{aligned}
R_{i,j} &= \max\left\{S_{i,j-1} - \alpha(1),\ \max_{2\le k\le j}\{S_{i,j-k} - \alpha(k)\}\right\} \\
&= \max\left\{S_{i,j-1} - \alpha_0,\ \max_{1\le k\le j-1}\{S_{i,j-(k+1)} - \alpha(k+1)\}\right\} \\
&= \max\left\{S_{i,j-1} - \alpha_0,\ \max_{1\le k\le j-1}\{S_{i,(j-1)-k} - \alpha(k)\} - \alpha_1\right\} \\
&= \max\{S_{i,j-1} - \alpha_0,\ R_{i,j-1} - \alpha_1\}
\end{aligned}$$

Therefore, only two matrix cells from the same row need to be checked. These cells correspond to the two options we need to consider, i.e., opening a new gap at this position (first term) or extending a previously opened gap (second term). The sub-linearity of α implies that it is never beneficial to concatenate gaps.

Similarly, for a column define $C_{i,j} = \max_{1\le l\le j}\{S_{i-l,j} - \alpha(l)\}$ and in the same way obtain

$$C_{i,j} = \max\{S_{i-1,j} - \alpha_0, C_{i-1,j} - \alpha_1\}$$

With these two new arrays, we can now compute the similarity score more efficiently. We initialize

$$C_{0,0} = R_{0,0} = S_{0,0} = 0$$

$$R_{i,0} = S_{i,0} = -\alpha(i) \qquad C_{0,j} = S_{0,j} = -\alpha(j)$$

then

$$S_{i,j} = \max\{S_{i-1,j-1} + s(a_i, b_j),\ R_{i,j},\ C_{i,j}\}$$

and the time complexity is $O(m \cdot n)$

Gonnet et al. [13] proposed a model for gaps that is based on gaps observed in pairwise alignments of related proteins. The model suggests an exponentially decreasing gap penalty function (see Section 3.5.4). However, a linear penalty function has the advantage of better time complexity, and in most cases the results are satisfactory. Therefore the use of linear gap functions is very common.

3.2.3 Local alignments

In many cases the similarity of two sequences is limited to a specific motif or domain, the detection of which may yield valuable structural and functional insights, while outside of this motif/domain the sequences may be essentially unrelated. In such cases global alignment may not be the right tool. In search for an optimal global alignment, local similarities may be masked by long unrelated regions (see Figure 3.2). Consequently, the score of such an alignment can be as low as for totally unrelated sequences. One possible solution is to use **semi-global alignment** (see Problem 3.3), which does not penalize gaps at the ends of the alignment. However, if the similarity is confined to a short domain, then it might be still dominated by the unrelated segments. Moreover, the algorithm may misalign the common region. A minor modification of the previous algorithm solves this problem. In the literature, this algorithm is often called the Smith-Waterman (SW) algorithm, after those who introduced this modification [14].

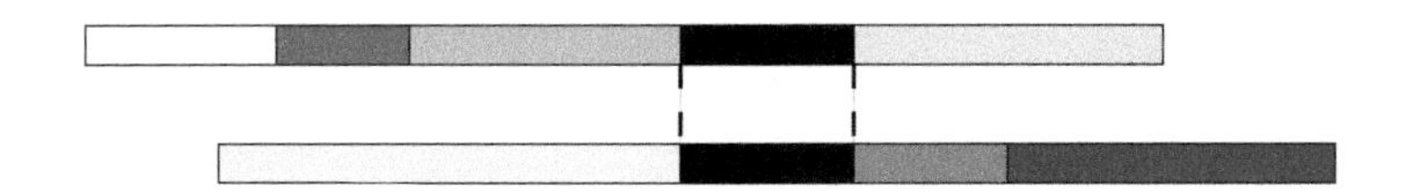

FIGURE 3.2: **Local alignments.** Proteins are composed of domains. If two proteins share in common only one domain, then it might be overlooked with a global alignment. However, if two proteins are similar globally then local alignment will most likely find the right solution. Therefore, it is almost always recommended to use local alignment instead of global alignment.

3.2.3.1 Calculating the local similarity score

A **local alignment** of **a** and **b** is defined as an alignment between a substring of **a** and a substring of **b**. The **local similarity** of sequences **a** and **b** is defined as the maximal score over all possible local alignments. Define $S^{local}_{i,j}$ to be the maximum local similarity of two segments ending at a_i and b_j

$$S^{local}_{i,j} = \max\{0,\ S(a_x a_{x+1} ... a_i,\ b_y b_{y+1} ... b_j) :\ 1 \le x \le i,\ 1 \le y \le j\}$$

where the score is set to zero if no positive match exists. This similarity is calculated recursively as before. Initialize

$$S^{local}_{i,0} = S^{local}_{0,j} = 0 \quad 1 \le i \le n,\ 1 \le j \le m$$

then

$$S^{local}_{i,j} = \max\left\{0,\ S^{local}_{i-1,j-1} + s(a_i, b_j),\ \max_{1 \le k \le i}\{S^{local}_{i-k,j} - \alpha(k)\},\ \max_{1 \le l \le j}\{S^{local}_{i,j-l} - \alpha(l)\}\right\}$$

and the local similarity of sequences **a** and **b** is obtained by maximizing over all possible segments

$$S^{local}(\mathbf{a}, \mathbf{b}) = \max\{S^{local}_{k,l} \quad 1 \le k \le n,\ 1 \le l \le m\}$$

That is, the best local alignment can end anywhere in the matrix, not necessarily the last cell. Clearly, the scoring matrix $s(a, b)$ must have positive and negative entries, otherwise extending an alignment will always increase its score.

To recover the alignment we start from the best cell and trace back until we get to a cell with zero score. By the definition of the local similarity score, it is guaranteed that any alignment of the subsequences that precede this cell score negatively, and therefore they are excluded from the optimal local alignment. For linear gap functions (which improve computational complexity) the formulation is done as before, with initialization $S_{i,j}^{local} = R_{i,j} = C_{i,j} = 0$ for $i \cdot j = 0$.

3.3 Heuristic algorithms for sequence comparison

During the last two decades sequencing techniques have greatly improved. Many large scale sequencing projects of various organisms are carried throughout the world, and as a result the number of new sequences that are stored in sequence databases (such as GenBank [15], SwissProt [16] and others) is increasing rapidly. In a typical application a new protein sequence is compared with all sequences in a database, in search of related proteins.

The DP algorithms described above may not be suitable for this purpose. The complexity of the DP algorithm is quadratic in the sequence length (when using a linear gap function), and the comparison of an average-length sequence against a typical protein database that contains several million sequences may take a couple of days on a current standard PC.

Several algorithms have been developed to speed up the search. The two main algorithms are **FASTA** [17] and **BLAST** [18]. These are heuristic algorithms that are not guaranteed to find the optimal alignment. However, they proved to be very effective for sequence comparison, and they are significantly faster than the DP algorithm. It is also worth mentioning hardware-based solutions. For example, biotechnology companies such as Compugen and Paracel developed special purpose hardware that accelerates the dynamic programming algorithm [19]. This hardware has again made the dynamic programming algorithm competitive with FASTA and BLAST, both in speed and in simplicity of use, but was short lived. Another interesting development emerged from the field of computer graphics. Graphic cards can be programmed and modified to perform various computational tasks including sequence comparison [20, 21]. However, FASTA and especially BLAST have become standard in this field and are being used extensively by biologists all over the world. Both algorithms are fast and effective and do not require the purchase of additional hardware. The two algorithms are based on a similar approach. They start by pre-processing the query to create a lookup table of short k-mers that appear in the query. This table facilitates the detection of "seeds" of potential sequence similarity with database sequences.

For example, when comparing two sequences BLAST (Basic Local Alignment Search Tool) starts by locating all pairs of similar ungapped segments, whose similarity score exceeds a certain threshold. These pairs of segments are called high-scoring segment pairs (**HSPs**) and the segment pair with the highest score is called the maximum segment pair (**MSP**). It is possible to find the MSP with the dynamic programming algorithm, by setting the gap penalty to ∞. However, BLAST finds it much faster because of its efficient implementation. A more detailed description of FASTA and BLAST appears in Appendix 3.10.

3.4 Probability and statistics of sequence alignments

When comparing protein sequences our goal is to determine whether they are truly related. If we detect strong sequence similarity between the two proteins we can infer homology, meaning a common evolutionary origin. Homologous proteins almost always have similar structures [10, 22, 23] and in many cases close biological functions.

The algorithms that were described in the previous sections can be used to identify such similarities. However, these algorithms will produce an alignment for any two input protein sequences, even if totally unrelated. Hence, sequence similarity does not necessarily imply homology. For unrelated sequences this similarity is essentially random and meaningless. As the length of the sequences compared increases, this random similarity may increase as well. Naturally, we would like to identify those similarities that are genuine and biologically meaningful, and the raw similarity score may not be appropriate for this purpose.

When can we infer homology from similarity? In order to answer this question we need a statistical measure that would assess the significance of a similarity score. The **statistical significance** is a measure of the "surprise" in observing a certain event; it is the likelihood that the event could have happened by chance. The smaller the likelihood, the less likely it is that the event happened by chance and the higher the statistical significance.

Though statistically significant similarity is neither necessary nor sufficient for a biological relationship, it may give us a good indication of such a relationship. When comparing a new sequence against a database in search of close relatives, this is extremely useful since we are interested in reporting only significant hits, and sorting the results according to statistical significance seems reasonable. When the sequence similarity is statistically significant, then the likelihood to observe it by chance is small, and we can deduce with a high confidence level that the sequences are related (two exceptions are segments with unusual amino acid composition and similarity that is due to convergent evolution). The reverse implication is not always true. There are many examples of low sequence similarity despite common ancestry and structural and functional similarity.

In order to determine whether a similarity score is unlikely to occur by chance it is important to know what score to expect simply by chance. The statistical significance of similarity scores for "real" sequences is estimated by computing the probability that the same score could have been obtained for *random* sequences. To compute this probability we need to establish the statistical theory of sequence matches. A great effort was made in the last two decades to establish such statistical theory. Currently, there is no complete theory, though some important results were obtained. Although most of the theoretical results were obtained for ungapped alignments, these results have created a framework for assessing the statistical significance of various similarity scores, including gapped sequence alignments, structural alignments [24, 25], profile-profile alignments [26, 27] and alignments of gene expression profiles [28, 29].

3.4.1 Basic random model

In the basic random model, the sequences are random sequences of characters where the characters are drawn independently and identically (i.i.d.) from a certain distribution over the alphabet $\mathcal{A}$.

Each sequence is thus viewed as a sequence of i.i.d. *random variables* drawn from the distribution $\mathcal{P}$ over the alphabet $\mathcal{A}$. In what follows, uppercase letters (A_i) denote random variables and lowercase letters (a_i, $a_i \in \mathcal{A}$) indicate a specific value of the random variable. Uppercase bold letters denote sequences of random variables, and lowercase bold letters denote sequences of amino acids.

For two random sequences $\mathbf{A}$ and $\mathbf{B}$, the scores $S(\mathbf{A}, \mathbf{B})$ (the global similarity score) and $S^{local}(\mathbf{A}, \mathbf{B})$ (the local similarity score) are functions of random variables, and therefore are also random variables. The distributions of these scores for randomly drawn sequences differ, and the next two sections summarize the main properties known about these distributions.

3.4.2 Statistics of global alignment

Though the distribution of global similarity scores of random sequences has not been fully characterized yet, some important properties of this distribution were partly determined. The main characteristic of this distribution is the linear growth with the sequence length. It is easy to see that this is the case under a simplified scenario, and then generalize it as was outlined in [30].

3.4.2.1 Fixed alignment - global alignment without gaps

This is a very simplistic case and the results are straightforward. Let $\mathbf{A} = A_1 A_2 ... A_n$ and $\mathbf{B} = B_1 B_2 ... B_n$ where A_i, b_j are i.i.d. random variables as defined above[2]. For the following alignment with no gaps

$$\begin{array}{l} \mathbf{A} = A_1 \ A_2 \ \ldots \ A_n \\ \mathbf{B} = B_1 \ B_2 \ \ldots \ B_n \end{array}$$

the score is simply defined as $S = \sum_{i=1}^{n} s(A_i, B_i)$.

S is the sum of i.i.d. random variables, and therefore for large n it is distributed approximately normally with expectation value $E(S) = n \cdot E(s(A, B)) = n\mu$ and $Var(S) = n \cdot Var(s(A, B)) = n\sigma^2$, where μ and σ are the mean and standard deviation of the scoring matrix $s(a, b)$

$$\mu = \sum_{a \in \mathcal{A}} \sum_{b \in \mathcal{A}} p_a p_b s(a, b) \qquad \sigma^2 = \sum_{a \in \mathcal{A}} \sum_{b \in \mathcal{A}} p_a p_b [s(a, b) - \mu]^2$$

Hence, under this setup the score S increases (or decreases, depending on the mean of the scoring matrix) with the sequence length. Surprisingly, perhaps, this characteristic holds for the general case as well.

3.4.2.2 Optimal alignment

In the general case, gaps are allowed in the alignment, and the similarity score is defined as the maximum over all possible alignments,

$$S = S(\mathbf{A}, \mathbf{B}) = \max_{alignments} \left\{ \sum_{a \in \mathcal{A}} \sum_{b \in \mathcal{A}} N_{ab} \cdot s(a, b) - N_{k\text{-}gap} \cdot \alpha(k) \right\} \quad (3.1)$$

where $\alpha(k)$ (the penalty for a gap of length k) is a general non-negative subadditive function, i.e., $\alpha(x + y) \leq \alpha(x) + \alpha(y)$.

The normal distribution limit law no longer holds because of the optimization over all possible alignments. However, it has been shown [30] that for random sequences the *expected* global similarity score S (as defined in Equation 3.1) grows **linearly** with the sequence length such that

$$\lim_{n \to \infty} S = \rho \cdot n$$

where the growth rate ρ is a constant. Intuitively, it is clear why the score is at most linear in n, since there are only n characters to align. It is also easy to see that $\rho \geq E(s(A, B))$, since the fixed alignment discussed above is one of the alignments we theoretically scan through when optimizing the

[2]When aligning sequences globally it is desirable that the sequences are of comparable length. To simplify the analysis we assume the sequences are of identical lengths.

global similarity score, and the score of the fixed alignment converges to $n \cdot E(s(A, B))$ as was shown above. However, the exact growth rate ρ has not been determined yet.

The statistical significance of a similarity score obtained for "*real*" sequences, which exceeds the expected score by a certain amount, is estimated by the probability that the similarity score of *random* sequences would exceed the expected mean by the same amount. However, since the distribution of scores is unknown, the available estimates give only a rough bound on that probability. For example, the Azuma-Hoeffding lemma gives a bound on the probability that a random variable exceeds its mean by a certain margin. It provides a measure of concentration for a broad class of random variables, which includes this case[3] such that

$$\text{Prob}(S - E(S) \geq \gamma \cdot n) \;\leq\; e^{-\frac{\gamma^2 n}{2c^2}}$$

Though the linear growth of the global similarity score is an important feature, both results are theoretical and have little practical use. Since ρ is unknown, the linear growth remains only a qualitative result. Moreover, the bound obtained by the Azuma-Hoeffding lemma is not very useful for a typical protein where n is of the order of 300. For example, for a typical scoring matrix such as the BLOSUM 62 matrix (see Section 3.5.2.2), and a gap opening penalty of 12, the constant c equals 30. If a global similarity score exceeds the expected mean by $3 \cdot n$ (which, for a global similarity score is usually significant), then the corresponding bound would be $\text{Prob}(S - E(S) \geq 3 \cdot n) \leq 0.223$, which is not very significant (suggesting that the bound is not tight). The variance of the global similarity score has not been determined either, and the best results give only an upper bound.

While it has not proved to be the case, in practice the scores tend to be distributed normally, and the significance of a score S can be approximated by computing the total probability to observe higher scores

$$\text{Prob}(S' \geq S) = \int_S^{\infty} P(x)dx \qquad (3.2)$$

where the parameters of the normal distribution $P(x)$ can be derived from the empirical distribution of global similarity scores over a large dataset.

3.4.2.3 The zscore approach

The zscore (z-score) method associates a significance measure with each similarity score based on the empirical distribution of similarity scores (Fig-

---

[3]The Azuma-Hoeffding lemma holds for random variables that are defined as partial sums of i.i.d. random variables $S(n) = X_1 + X_2 + ... + X_n$, where the difference between consecutive random variables is bounded by a constant c. In our case the individual variables X_i are the scores of matching two amino acids or introducing a gap, and the values of consecutive variables are bounded, with the bound c depending on the scoring matrix and the gap penalty.

ure 3.3). For a given combination of scoring matrix and a gap penalty function, the distribution can be simulated by shuffling the sequences and comparing the shuffled sequences. The shuffled sequences can be considered as random sequences, and by repeating this procedure many times we practically sample the distribution of similarity scores for random sequences.

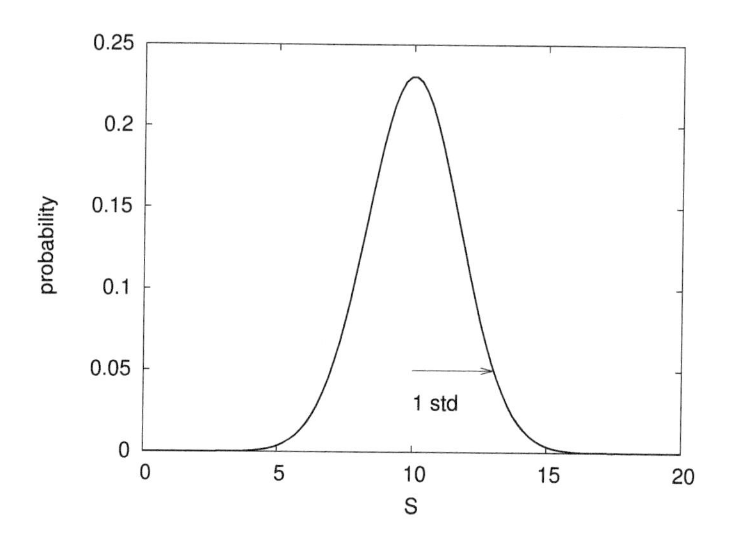

FIGURE 3.3: **Computing the zscore.** The distribution of global similarity scores can be approximated with a normal distribution (here plotted for $\mu = 10$ and $\sigma = 3$). The zscore of a certain score S is defined as its distance from the mean in units of the standard deviation (std). In the example above, a score of 13 corresponds to a zscore of 1.

Let μ and σ^2 be the mean and the variance of the empirical distribution of scores. The zscore associated with the global similarity score S is defined as

$$zscore(S) = \frac{S - \mu}{\sigma}$$

The zscore measures how many units of standard deviation apart the score S is from the mean of the distribution. The larger it is, the more significant is the score S. While the zscore is not a probability measure per-se, it can be associated with one if the underlying distribution can be well approximated with a normal distribution, as in Equation 3.2. The advantage of the zscore approach is that it can be applied in other situations as well, even if the distribution is not exactly normal. This approach can quite effectively detect statistically significant alignments; however, it is computationally intensive since it requires computing the alignment of many shuffled sequences.

The zscore transformation is a popular standardization procedure in general, as we will see in subsequent chapters. It works well in situations where the background distribution is uni-modal and normal-like. It is especially useful when comparing scores based on different attributes, since it converts scores to a uniform scale. It is also invariant to different scalings of attributes. Moreover, since the transformation is defined based on the empirical distribution of scores, the zscore self-calibrates itself to the characteristics of the data. Hence it is often referred to as the **whitening transformation**.

3.4.3 Statistics of local alignments without gaps

The statistics of local similarity scores, and especially ungapped alignments, was studied extensively in the early 1990s. The exclusion of gaps allowed a rigorous mathematical treatment, and several important results were obtained. As with global alignments, interesting results for local alignments are obtained already under a very simplistic model. However, it is the work of Dembo, Karlin, Altschul and colleagues on the more general case that really changed the field of sequence analysis.

3.4.3.1 Fixed alignment

The following simple case intuitively gives away one of the main statistical results regarding local similarity scores. Consider the asymptotic behavior of the longest perfect match between two random sequences of length n, when the alignment is given and fixed. The probability that two aligned letters are identical is given by

$$\lambda = \text{Prob(match)} = \sum_{a \in \mathcal{A}} p_a^2$$

where p_a is the probability of the letter a. The problem of finding the longest match can be rephrased in terms of the length S_n of the longest streak of heads in n coin tosses, when the probability of head is λ.

According to Erdös and Renyi [31] $S_n \to \log_{1/\lambda} n$. An intuitive proof is given in Waterman [30]. The intuition is that the probability of a streak of m heads is λ^m, and for $m << n$ there are about n possible streaks (one for each possible starting point). Note that this setup is similar to the standard coin-toss problem, where the probability for a "success" (i.e., heads) is $p = \lambda^m$. The expected number of successes is distributed as the binomial distribution and therefore

$$E(\text{number of perfect matches of length } m) \simeq n \cdot \lambda^m$$

If the longest streak is unique (i.e., it occurs only once) then S_n should satisfy $1 = n \cdot \lambda^{S_n}$ and hence

$$S_n = \log_{1/\lambda} n$$

Clearly, this is an approximation since the overlapping streaks are not independent of each other. Yet, it produces a useful result. This result holds for exact matches, which start at the same position in both sequences. Allowing shifts makes the problem more interesting in the context of sequence comparison. The length of the longest match in this case is actually the score of the best local alignment $\mathcal{S} = S^{local}(\mathbf{A}, \mathbf{B})$ given that the score for a match is 1, it is $-\infty$ for a mismatch, and ∞ for opening a gap.

When shifts are allowed, there are n^2 possible starting positions (i, j) for the match (the number of possible starting positions of a match defines the size of the **search space**). Therefore, following the intuition of the previous case we expect that

$$\mathcal{S} \rightarrow \log_{1/\lambda} n^2 = 2 \cdot \log_{1/\lambda} n$$

and the logarithmic characteristic is preserved. However, these results assume perfect matches, while the original problem allows mismatches as well. Surprisingly, the results hold even when mismatches are allowed (see next section).

3.4.3.2 Optimal alignment

This section proceeds to the case of a general scoring scheme $s(a, b)$ for $a, b \in \mathcal{A}$, with the constraints (i) $E(s(a, b)) < 0$ and (ii) $s^* = \max\{s(a, b)\} > 0$. The first requirement implies that the average score of a random match will be negative (otherwise, extending a match would tend to increase its score, and this contradicts the idea of local similarity). The second condition implies that a match with a positive score is possible (otherwise a match would always consist of a single pair of residues).

The following theorem concerns local matches (as defined in Section 3.2.3) with a general scoring scheme but *without gaps* (i.e., the penalty for opening a gap is ∞). It characterizes the maximal score of a local alignment $\mathcal{S} = S^{local}(\mathbf{A}, \mathbf{B})$ of two random sequences, and the distribution of amino acids in the maximal scoring segments.

Theorem 1 (the asymptotic properties of maximally scoring segment pairs [32]): Let $\mathbf{A} = A_1 A_2 ... A_n$ and $\mathbf{B} = B_1 B_2 / dos B_n$ where A_i, B_j are i.i.d. and sampled from the same distribution $\mathcal{P}$ over an alphabet $\mathcal{A}$. Assume that $s(a, b)$ satisfies $E(s(A, B)) < 0$ and $s^* = \max\{s(a, b) : a, b \in \mathcal{A}\} > 0$. Let $\lambda > 0$ be the largest real root of the equation $E(e^{\lambda \cdot s(A,B)}) = 1$. Then

$$\lim_{n \to \infty} \mathcal{S} = \frac{\ln n^2}{\lambda} \tag{3.3}$$

and the proportion of letter a aligned with the letter b in the best scoring match converges to $q_{a,b} = p_a p_b e^{\lambda \cdot s(a,b)}$.

The direct implication of this theorem is that for two random sequences of length n and m, the score of the best local alignment (referred to as the the MSP score) is centered around $\frac{\ln(n \cdot m)}{\lambda}$. That is, the local similarity score grows **logarithmically** with the length of the sequences and with the size of the search space $n \cdot m$. Practically, given the distribution $\mathcal{P}$ (for example, the overall **background distribution** of amino acids in the database), λ is obtained by solving the equation

$$\sum_{a,b} p_a p_b e^{\lambda \cdot s(a,b)} = 1 \tag{3.4}$$

using a method such as Newton's method.

This result in itself is still not enough to obtain a measure of statistical significance for local similarity scores. This can be done only once a measure of concentration is obtained or the distribution of similarity scores is defined. Indeed, one of the most important results in this field is the characterization of the distribution of local similarity scores without gaps. This distribution was shown to follow the extreme value distribution [32–34].

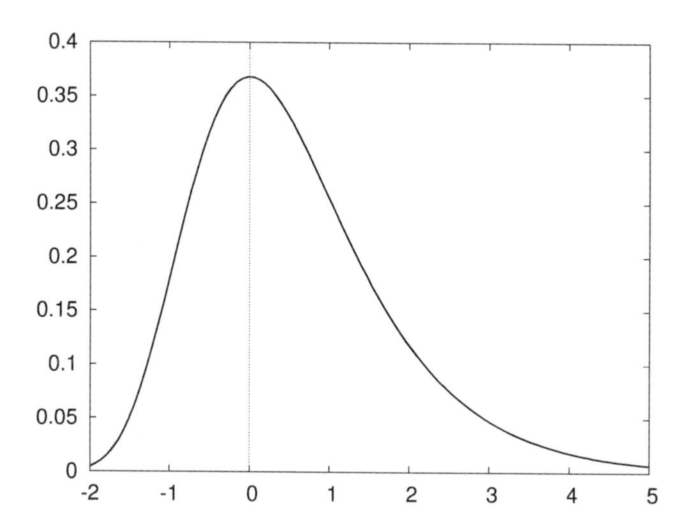

FIGURE 3.4: Probability density function for the extreme value distribution with $u = 0$ and $\lambda = 1$.

It is well known the **sum** of many i.i.d. random variables is distributed **normally**. In a similar manner, it has been shown that the **maximum** of many i.i.d. random variables is distributed as an **extreme value distribution** (EVD) [35]. The EVD looks like a normal distribution with a long tail and is characterized by two parameters: the index value u and the decay constant λ (for $u = 0$ and $\lambda = 1$, the distribution is plotted in Figure 3.4). The distribution is not symmetric. It is positive definite and unimodal with one peak at u. Practically, the score of the best local alignment (the MSP score) is the maximum of the scores of many different and essentially random alignments (although not necessarily independent), which explains the observed distribution. This is summarized in the next theorem, which concerns the distribution of local similarity scores for random sequences.

Theorem 2 (the significance of local similarity scores [32–34]): Let $\mathcal{S}$ be a random variable whose value is the local similarity score for two random sequences of length n and m, as defined above. $\mathcal{S}$ is distributed as an extreme value distribution and

$$\text{Prob}(\mathcal{S} \geq x) \sim 1 - \exp(-e^{-\lambda \cdot (x-u)}) \tag{3.5}$$

where λ is the root of Equation 3.4 and $u = \frac{\ln Kmn}{\lambda}$, where K is a constant that can be estimated from first-order statistics of the sequences (i.e., the background distribution $\mathcal{P} = \{p_a\}$) and the scoring matrix $\mathbf{s} = \{s_{ab}\}$ [32]. Denote $f(x) = e^{-\lambda \cdot (x-u)}$, then $\text{Prob}(\mathcal{S} \geq x) \sim 1 - \exp(-f(x))$ is the complementary cumulative density function (Figure 3.5). The term $\exp(-f(x))$ is the cumulative density function, and its derivative, the probability density function (Figure 3.4), is given by $\lambda f(x) \exp(-f(x))$.

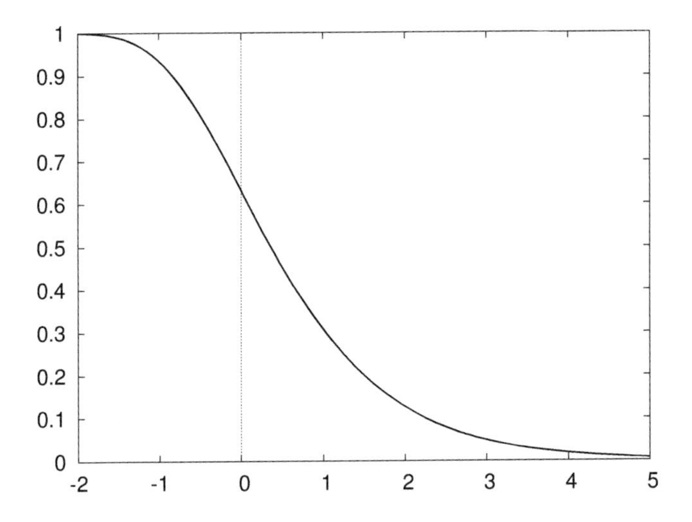

FIGURE 3.5: **The complementary cumulative density function for the extreme value distribution with $u = 0$ and $\lambda = 1$.**

Theorem 2 holds, subject to the restriction that the amino acid composition of the two sequences that are compared are not too dissimilar [32]. Assuming that both sequences are drawn from the background distribution, the amino acid composition of both should resemble the background distribution. Without this restriction theorem 2 overestimates the probability of similarity scores, and indeed, this is observed in protein sequences with unusual compositions (we will get back to this issue in Section 3.4.5).

For a small a we can use the approximation $1 - \exp(-a) \sim a$. Therefore, for a large x

$$\begin{aligned}\text{Prob}(\mathcal{S} \geq x) &\sim 1 - \exp(-e^{-\lambda \cdot (x-u)}) \sim e^{-\lambda(x-u)} \\ &= e^{-\lambda \cdot x} e^{\lambda \cdot u} = Kmne^{-\lambda \cdot x}\end{aligned} \tag{3.6}$$

and note the dependency on the size of the search space mn. This approximation helps to calculate the probability that a given MSP score could have been obtained by chance. The score will be statistically significant at the 1% level if $\mathcal{S} \geq x_0$, where x_0 is determined by the equation $Kmne^{-\lambda \cdot x_0} = 0.01$. Hence, a pairwise alignment with a score $\mathcal{S}$ has a **pvalue** of p where $p = Kmne^{-\lambda \cdot \mathcal{S}}$. I.e., there is a probability p that this score could have happened by chance.

In general, the same two sequences may have more than one high-scoring pair of segments. This may happen whenever insertions/deletions are introduced to align the sequences properly. The combined assessment of scores from several ungapped alignments can be evaluated by applying Poisson statistics with the parameter $p = e^{-\lambda(x-u)}$. Given the k highest scoring HSPs, among which the lowest HSP score is x, the Poisson distribution can be used to calculate the probability that at least k segments pairs, all with a score of at least x, would appear by chance in one pairwise comparison. This approach has the disadvantage of being dependent on the lowest score among the k highest scores. Another alternative is to calculate the sum S_k of the highest k scores. The distribution of such sums has been derived [36], and the probability of a given sum is calculated (numerically) by a double integral on the tail of the distribution. In either case, the HSPs should first satisfy a consistency test before the joint assessment is made. This approach of combining segments was later replaced with direct assessment of gapped alignments, as discussed in the next section.

The discussion so far was in the context of a pairwise comparison. The probability p is the probability that a similarity score $\mathcal{S}$ could have been obtained simply by chance when comparing two random sequences. This probability should be adjusted when multiple comparisons are performed, for example, when a sequence is compared with each of the sequences in a database of size D. Denote by **p-match** a match between two sequences that has a pvalue $\leq p$ (i.e., its score $\geq \mathcal{S}$). We will refer to a p-match as a "success". Assuming independence between the database sequences, then multiple comparisons can be considered as multiple "experiments" (tests), similar to multiple coin tosses, where the probability of success in each test is p. Therefore, the number of p-matches in a database search is expected to follow the binomial distribution with a parameter p (see Section 3.12.4). If the database has D sequences, then the probability P of getting by chance m ($m \geq 1$) p-matches is

$$P = \text{Prob}(m \geq 1) = 1 - \text{Prob}(m = 0) = 1 - (1-p)^D \tag{3.7}$$

A simpler term can be obtained by using the Poisson distribution with parameter $\delta = Dp$ to approximate the binomial distribution $Bin(D,p)$. This approximation holds for D large and $\delta = Dp$ small. With the Poisson distribution, the probability of m successes is $P(m) \simeq \frac{\delta^m}{m!}e^{-\delta}$. Therefore, the probability P of getting by chance at least one p-match when searching a database containing D sequences is

$$P = \text{Prob}(m \geq 1) = 1 - \text{Prob}(m = 0) = 1 - e^{-\delta} = 1 - e^{-Dp} \tag{3.8}$$

and for $P < 0.1$ this is well approximated by Dp (using again the approximation $1 - \exp(-a) \sim a$). Thus,

$$P = 1 - e^{-Dp} \simeq Dp = KDmne^{-\lambda \cdot \mathcal{S}} \tag{3.9}$$

This discussion assumes that all sequences in the database (**library sequences**) have the same probability of sharing a similar region with the **query sequence**. However, it is more appropriate to assume that all *equal-length* protein segments in the database have an equal a priori probability to be related to the query sequence (since similarity usually follows domains). Therefore, if the query sequence is of length n, the (pairwise) alignment of interest involves a library segment of length m and the database has a total of N amino acids, then D should be replaced with N/m. Hence, Equation 3.9 is corrected to obtain

$$P \simeq KNne^{-\lambda \cdot \mathcal{S}} \tag{3.10}$$

where the term Nn can be viewed as the "effective size" of the search space.

Usually it is the **expectation value** (evalue) that is used as a measure of statistical significance, rather than the pvalue. The expectation value is the first moment (the mean) of a random variable (see Section 3.12.2). In our case we are interested in the expected number of p-matches. The parameter $\delta = Dp$ of the Poisson distribution is also the expectation value of this distribution (and Dp is the expectation value of the original binomial distribution, which we approximated with the Poisson distribution). Therefore,

$$E = E(\text{number of } p\text{-matches}) = \delta = Dp \tag{3.11}$$

and as discussed above, D should be replaced with N/m. Hence

$$E = KNne^{-\lambda \cdot \mathcal{S}} \tag{3.12}$$

This is the expected number of distinct matches (segment pairs) that would obtain a score $\geq \mathcal{S}$ by chance in a database search, with a database of size N (amino acids) and composition $\mathcal{P}$ (the background distribution of amino acids). If $E = 0.01$, then the expected number of random hits with a score $\geq \mathcal{S}$ is 0.01. In other words, we may expect a random hit once in 100 independent searches. If $E = 10$, then we should expect 10 hits with a score $\geq \mathcal{S}$ by chance, in a single database search. This means that such a hit is not significant. Note that $E \simeq P$ for $P < 0.1$ (Equation 3.10 and Equation 3.12), and consequently in many publications in this field there is no clear distinction between E and P. However, for $P > 0.1$ they differ, where $P = 1 - \exp(-E)$ (Equation 3.8 and Equation 3.11).

Finally, by setting a value for E and solving the equation above for $\mathcal{S}$, it is possible to define a threshold score, above which hits are reported. This is the score above which the number of hits that are expected to occur at random is $< E$. Therefore, we can deduce that a match with this score or above reflects true biological relationship, but we should expect up to E errors per search. The specific value of E affects both the sensitivity of a search (a function of the number of true relationships detected) and its selectivity (a function of the number of errors). A lower value of E would decrease the error rate. However, it would decrease the sensitivity as well[4].

It is difficult to set a clear dividing line between true homology and chance similarity. An expectation value below 10^{-5} indicates that a false match would occur once in 100,000 searches and can be safely considered significant[5]. On the other hand, an expectation value above 10 reflects mostly pure chance similarities. However, the midrange is more difficult to characterize, and homologous proteins can have expectation values around 1 or even higher. Unfortunately, it is hard to assess similarities between remote homologs that are borderline significant (the so-called **twilight zone** of sequence similarity). A reasonable choice for E is between 0.1 and 0.001, which usually ensures that most reported similarities are due to a true common evolutionary origin, without missing too many important similarities. In subsequent chapters we will discuss other methods for sequence analysis that attempt to enhance homology detection in the twilight zone.

3.4.4 Local alignments with gaps

Though local alignments without gaps may detect most similarities between related proteins and give a good estimate of the true similarity of the two sequences, it is clear that gaps in local alignments are crucial in order to obtain the correct alignment and a more accurate measure of similarity. However, introducing gaps in alignments greatly complicates the analysis of their statistical properties.

Several studies showed that the score of local gapped alignments can be characterized in the same manner as the score of local ungapped alignments. According to theorem 1 the local *ungapped* similarity score grows logarithmically with the sequence's length and the size of the search space. Arratia and Waterman [37] have shown that for a range of substitution matrices and gap

[4]We will revisit the problem of hypothesis evaluation and multiple testing in Section 3.12.8.
[5]This is not necessarily true for low-complexity sequences, see Section 3.4.5.

penalties, local *gapped* similarity scores have the same asymptotic characteristic. Furthermore, empirical studies [38,39] strongly suggest that local gapped similarity scores are distributed according to the extreme value distribution, though some correction factors may apply [40]. More recently, the statistical properties of the score distribution were characterized theoretically indicating a phase transition between a linear asymptotic phase and a logarithmic asymptotic phase, depending on the gap penalties [41].

Multiple approximations and practical solutions were suggested over the years. For example, in [40], the parameters of the EVD were estimated empirically for several popular scoring matrices and a range of gap penalties based on a large set of sequence alignments. Other studies use simulations over random sequences (e.g., [42]). The approach we will describe here briefly is the one suggested in [43], where statistical estimates for local alignments with gaps are derived based on scores obtained from a database search. A database search provides tens of thousands of scores from sequences that are unrelated to the query sequence, and therefore are effectively random. As discussed above, these scores are expected to follow the extreme value distribution. This is true as long as the gap penalties are not too low. Otherwise the alignments shift from local to global and the extreme value distribution no longer applies.

Since the logarithmic growth in the sequence length holds in this case, scores are corrected first for the expected effect of sequence length. The correction is done by calculating the regression line $S = a+b \cdot \ln n$ for the scores obtained in a database search, after removing very high scoring sequences (probably related sequences). The process is repeated as many as five times. The regression line and the average variance of the normalized scores are used to define a zscore:

$$zscore = \frac{S - (a + b \cdot \ln n)}{var}$$

and the distribution of zscores is approximated by the extreme value distribution

$$p = \text{Prob}(zscore > x) = 1 - \exp(-e^{c_1 \cdot x - c_2})$$

where c_1 and c_2 are constants. The expectation value is defined as before $E(zscore > x) = N \cdot p$, where N is the number of sequences in the database (the number of tests).

This empirical approach has the advantage that it internally calibrates the estimates to the specific properties of the query sequence and the database, which is most effective when the query contains low-complexity segments and has proved to be very accurate in estimating the statistical significance of gapped similarity scores (see also [44,45]). Other approaches are described in [46–49].

3.4.5 Handling low-complexity sequences

Low-complexity sequences, also known as "simple sequences", are abundant in proteins. These compositionally biased sequences are frequent in structural proteins such as collagens (which are found in connective tissues such as skin and bones) and cell-wall proteins. They are usually rich with amino acids such as alanine (A), serine (S), proline (P) and glycine (G), while cysteine (C) and tryptophan (W) are very rare [50–52]. Almost 30% of the known sequences as of 2008 contain at least one low-complexity segment. However, not much is known about the potential function of these sequences.

Low-complexity sequences pose a problem for sequence homology searches. Because of the repetitive nature of these sequences, they often result in high-scoring similarities that are biologically meaningless. However, the statistical theory that was developed for estimating the significance of sequence matches fails to provide accurate estimates in this case. Recall that Equation 3.5, which characterizes the significance of local similarity scores, holds only if the compositions of the compared sequences are similar to the overall composition of amino acids in the database. The parameters λ and K determine the exact characteristic of the extreme-value distribution for a given population of amino acid sequences. Both depend on the scoring matrix and the background distribution of amino acids in the database. For example, λ is obtained by solving the equation

$$\sum_{a,b} p_a p_b e^{\lambda \cdot s(a,b)} = 1 \tag{3.13}$$

where p_a and p_b are the background probabilities of amino acids a and b. Low-complexity sequences however have unusual sequence compositions and therefore their true λ differs markedly from the one used in BLAST. Because of the exponential dependency on λ in Equation 3.5, a change in the value of that parameter results in many orders of magnitudes change in the significance of a given similarity score. As a result, the evalue of matches with proteins of unusual amino acid composition is often overestimated, and biologically meaningless similarities might be reported as significant.

There are several methods for handling low-complexity sequences. Some algorithms work by masking low-complexity segments before alignment [53,54]. For example, SEG [53] uses entropy to measure sequence complexity along a sliding window and segments of low entropy are marked and filtered out (by replacing them with X). However, not all low-complexity segments are filtered with this method, and the method is somewhat sensitive to the choice of the parameters, such as the window size and the low-entropy threshold. On the other hand, potentially useful information is lost because of this masking process. These low-complexity fragments may play an important role in determining the structure and function of proteins [55], and many relationships of biological significance can be missed if only sequences that pass the filter are to be considered [56]. Other algorithms attempt to reduce information loss by keeping the sequence intact and using permutation methods to reassess the significance [52, 57]. For example, the *zscore* based approach generates a background population of "random" alignments by repeatedly permuting the database sequence and aligning the query sequence with each one of the permuted sequences. The significance of the alignment between the non-permuted sequences is estimated from the mean μ and the standard deviation σ of the background distribution in terms of the zscore $\frac{S-\mu}{\sigma}$. Permutation methods have the advantage that they preserve the context without making any assumptions on the source. However, they are very slow since they require tens of samples to generate the background distribution from which a zscore can be estimated. Some programs use the distribution of similarity scores with respect to the database sequences to derive a zscore and assess its statistical significance, as described in the previous section. These self-normalized scores are more efficiently computable and are quite reliable.

In theory, one way to correct the evalue in the presence of low-complexity sequences would be to recompute the parameters based on the compositions of the specific sequences compared. However this simple solution cannot be applied in practice since the analytical solutions for λ and K were established only for ungapped alignments, while BLAST in its current format generates gapped alignments. To overcome this problem, gapped BLAST uses pre-computed parameters that are derived from distributions of similarity scores of large sequence datasets. The parameters can be estimated empirically, by fitting an extreme-value distribution to the empirical distributions. The parameters are computed with a variety of scoring matrices and gap penalties but for a fixed sequence composition (the background distribution of amino acids in the database) [40]. As noted above, using the pre-computed parameters leads to substantially inaccurate estimates when the query or library sequence has unusual sequence compositions. Recomputing the parameters for each pair of sequences is currently impractical, but a few approaches were developed for rapid correction of the statistical estimates for gapped alignments [42, 47].

One of the recent solutions is the one employed by the newer versions of BLAST. This method, referred to as **composition based statistics**, rescales the scoring matrix for *gapped* alignments based on a factor that depends on the compositions of the sequences compared and is computed from the statistical parameters of *ungapped* alignments. Specifically, given a query sequence and a database sequence, the program computes the parameters λ_u (for ungapped alignments with the background distribution) and λ'_u (for ungapped alignments with the specific compositions of the query and database sequences, solving analytically Equation 3.13), to derive the factor $r = \lambda'_u/\lambda_u$. This factor is used to re-scale the substitution scores and the sequences are re-aligned using the re-scaled scoring matrix. The parameter λ_g (for gapped alignments with the background distribution) is then used to estimate the significance of the new similarity scores. Note that this provides an approximate solution for Equation 3.13, using gapped alignments and the sequence-specific

Query:

```
>swissprot: (P40273) Histone H1.M6.1.
MSDAAVPPKKASPKKAAAKKASPKKSAARKTAAKKTAKKPAVRKPAAKKRAAPKKKPAAA
KKPAAKKAPKKAVKKAPKKK
```

Top match (composition-based statistics):

```
>trembl: (Q9RU01) Glucose-6-phosphate 1-dehydrogenase (EC 1.1.1.49) (G6PD).
          Length = 590

 Score = 39.7 bits (91), Expect = 0.008
 Identities = 31/74 (41%), Positives = 41/74 (54%), Gaps = 3/74 (4%)

Query: 8  PKKASPKKAAAKKASPKKSAARKTAAKKTAKKPAVRKPAAKKRAAPKKKPAAAKKPAAK- 66
          PKK+SPKK+   +KA  K+SAA+  AAK T +       + A K   A   PA ++K A K
Sbjct: 7  PKKSSPKKSGPEKALAKESAAQGEAAKATRQATQQTEAAKKVGVAQPGAPAQSRKAARKS 66

Query: 67 --KAPKKAVKKAPK 78
            + PK A   APK
Sbjct: 67 RQRVPKHAGDNAPK 80
```

FIGURE 3.6: **Effect of SEG and composition-based statistics on a BLAST search.** Composition-based statistics can help to eliminate spurious matches with low-complexity sequences. However, in some cases it may produce unintuitive results. For example, a histone protein (SwissProt P40273) was compared against a large sequence database that also contains the query sequence. When composition-based statistics was disabled, about 450 matches were detected with evalue < 0.001 of which more than half were Histones or Histone-like proteins. When repeated with the SEG flag on, the search reported 0 matches. When composition-based statistics was enabled, no matches were detected with evalue < 0.001 and only two were detected with evalue < 0.1 (the top scoring match is displayed above). The query sequence was missing from the list and only one Histone was reported (with evalue $= 1.4$). (Figure reproduced from [58] with permission.)

composition. Thus, composition-based statistics effectively changes the scoring matrix and the gap penalties for each pair of sequences compared. To save time, the new evalues are computed only for matches that score higher than a certain threshold with the original parameters.

The composition-based statistics method is quite effective in eliminating many chance similarities that are due to unusual sequence compositions [59]. However, the method does not work well in all cases. It might affect the sensitivity of the algorithm, and in some cases significant matches are eliminated. It also introduces some unusual artifacts: most notably the ranking of the query sequence in cases when the query sequence is also part of the database searched. In some cases the query sequence is no longer the most significant match, or it is eliminated altogether from the match list (see Figure 3.6). This occurs because the re-scaling method essentially uses different parameters to estimate the significance of each pair, resulting in a nonmonotonic transformation of the raw similarity scores. The problem is especially pronounced when the iterative PSI-BLAST algorithm is used (where the query sequence might be eliminated from the list after a few iterations). Nevertheless, it is one of the most effective methods for handling low-complexity sequences. For further reading see [58, 60], where alternative methods are described.

3.4.6 Sequence identity and statistical significance

When comparing two sequences, a popular alternative to similarity scores is the **Percent Identity** (PI). That is, the percentage of identical pairs of residues in the optimal alignment. There are several variations on this measure, depending on whether the alignment is local or global and whether the gaps are counted or not. For example, consider the alignment

$$\begin{array}{l} \mathbf{a} = T\ A\ G\ A\ C - T\ T\ G\ A\ - \\ \mathbf{b} = - - - A\ G\ A\ T\ T - A\ C \end{array}$$

The PI can be defined as the number of identities (4) divided by the length of the longer sequence (9), shorter sequence (7) or the average length (8). The more common approach is to normalize the number of identities by the length of the local alignment. The percent identity in the local alignment is $\frac{4}{7} * 100$ if gaps are included and $\frac{4}{5} * 100$ if not. Since the similarity might be limited to a local region with long gaps due to loops that were inserted or deleted, it is best to measure the percent identity with respect to overlapping (aligned) residues in the best local alignment (in the example above, this amounts to 80% identity).

The higher the percent identity, the more likely it is that the sequences are related and the percentage can be used as a coarse similarity/distance measure (Figure 3.7). As a rule of thumb, many use 30% as the threshold above which two proteins are assumed to be related [44, 61–63]. However, if the local alignment involves only a fragment of the sequences compared, then one should keep in mind that any hypothesis on the relation between the proteins compared applies only to that fragment. Moreover, there are many examples of evolutionary related sequences with lower than 30% sequence identity, as well as of short (less than 80 residues) unrelated sequences with higher sequence identity [44, 64, 65].

The degree of sequence conservation varies among protein families, and some can diverge greatly resulting in borderline significant or insignificant similarities, with less than 30% identity (often referred to as the **twilight zone** of sequence similarity [66]). These similarities are harder to detect with standard sequence comparison algorithms. In chapters 4-8 we will discuss more advanced methods for detection of subtle similarities between remotely related sequences.

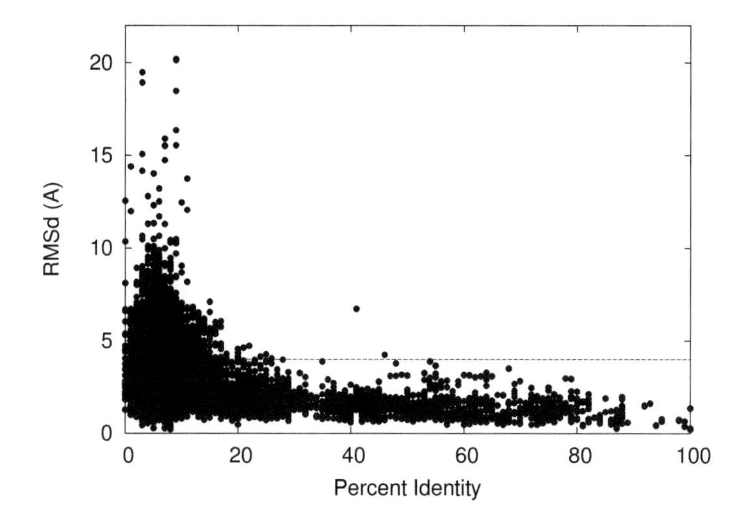

FIGURE 3.7: **Structural similarity vs. sequence identity.** A scatter-plot showing the structural distance between pairs of homologous proteins vs. the percent identity in their respective sequence alignment. Structural distance is measured in terms of the root mean square distance (RMSd; Section 8.3.1.1) under structural alignment with the RMS-URMS algorithm (Section 8.3.1.4). Structural similarity with RMSd < 4 is usually considered significant, often suggesting common evolutionary ancestry (we will discuss the problem of structure comparison in Chapter 8). As the graph indicates, sequence alignments with 30% identity or more almost always correspond to significant structural alignments with RMSd < 4.

3.4.7 Similarity, homology and transitivity

As discussed in previous sections, establishing homology can tremendously help in determining the function of new proteins. By definition, homologous proteins have evolved from the same ancestor protein, and if they have not diverged much, then their sequences are similar. We can deduce homology from statistically significant sequence similarity.

Homology is, by definition, a **transitive relation**: if protein A is homologous to B, and B is homologous to C, then A is homologous to C. This simple observation can be very effective in discovering homology and is particularly useful in the so called "twilight zone". In such cases transitivity can be used to detect distantly related proteins beyond the power of a direct search, as has been demonstrated in [13, 65, 67–72].

Though transitivity is an attractive concept, it has its perils, and when applied simple mindedly, it can lead to many pitfalls. First, note that similarity is not transitive. Similarity is a quantitative feature, whereas homology is a relation that either holds or does not hold. We can infer homology from significant sequence similarity. The more significant the similarity is, the more confident we can be in the inference. However, in general, similarity does not necessarily imply homology, and one should be especially cautious when dealing with weak similarities or similarities involving low-complexity sequences. In some cases, transitivity can lead to completely false conclusions, for example, when two sequences are connected through a third, as illustrated later in Figure 3.8b, although they are unrelated evolutionarily. Therefore, transitivity should be applied very carefully to infer homology. We will get back to these issues in Section 10.6, when discussing clustering algorithms and their application to proteins.

3.5 Scoring matrices and gap penalties

3.5.1 Scoring matrices for nucleic acids

Scoring matrices for nucleic acids are usually quite simple. For example, some applications use the identity matrix (where $s(a_i, b_j) = 1$ if $a_i = b_j$ and 0 otherwise). BLAST uses a similarity matrix where $s(a_i, b_j) = 5$ if $a_i = b_j$ and -4 otherwise. A more refined approach is to consider the chemical structure of nucleotides when scoring substitutions. Nucleotides can be classified into two categories: **purines** (A,G) and **pyrimidines** (C,T). Purines have two-ring base, while pyrimidines have a single-ring base (see Chapter 2). A mutation that conserves the number of rings is called a transition ($A \leftrightarrow G$ or $C \leftrightarrow T$), otherwise it is called a transversion. Although there are more ways to create transversions than transitions, the majority of accepted mutations are transitions. Respectively, the transition/transversion matrix (Table 3.2) scores substitutions differently, depending on their nature. This scoring matrix can potentially generate more accurate alignments [73].

	A	C	G	T
A	1	-4	-4	0
c	-4	1	0	-4
G	-4	0	1	-4
T	0	-4	-4	1

TABLE 3.2: **The transition/transversion scoring matrix for nucleotides.**

3.5.2 Scoring matrices for amino acids

Scoring matrices for protein sequences are more complicated. Since the late 1960s, several different approaches were taken to derive reliable and effective scoring matrices, some of which are summarized next. For more details, and comparison of different scoring methods see [74, 75].

The genetic code matrix [76] measures amino acid similarity by the number of common nucleotides in their codons. This quantity is maximized by considering the closest matching representative codons. Identical amino acids share three bases, while non-identical share two or fewer. Indeed, there is some correlation between the codons of different amino acids and their biochemical properties (see Problem 3.8) and it has been shown [13] that for proteins that diverged slightly the structure of the code influences the distribution of accepted point mutations. However, despite this nice rationale, which goes back to the very basic evolutionary events at the DNA level, most of the genetic pressure is on the protein sequence, and the correlation with the codon structure is not strong enough to detect weak homologies [77].

Some matrices are based directly on similarities between physico-chemical properties of amino acids [78, 79]. However, these matrices do not perform as well as other matrices when used for sequence comparison. One of the reasons is that there is no single property of amino acids that accounts for the conservation of proteins' structure and function.

The most effective matrices are those that are based on actual frequencies of mutations that are observed in closely related proteins. Substitutions occur more frequently among amino acids that share certain properties. Indeed, these matrices reflect the biochemical properties of amino acids, which influence the probability of mutual substitution, and amino acids with similar properties have a high pairwise score. Matrices that are based on sequence alignments include the family of PAM matrices [80] (and their improvement by [81]), the BLOSUM matrices [82], Gonnet matrix [13] and more (e.g., [83]).

Some matrices were extracted from the secondary structure propensities of amino acids [84, 85]. Other matrices, which proved to be very effective for protein sequence comparison, are those that are based on structural principles [75, 86]. These matrices reflect the statistics of pairwise substitutions that are observed at structurally equivalent positions in aligned structures of proteins from the same family and perform especially well in detecting remote homologs. These matrices can be a good choice when aligning relatively mutable regions (sequence derived matrices are based on conserved regions, and therefore cannot accurately model more mutable regions that diverged significantly). In such regions gaps are much more frequent, and to obtain accurate alignments it is necessary to unify gaps and substitutions in a single model. However, existing models for gaps are based on different principles than substitution matrices (see Section 3.5.4), and it is not clear a priori how to select the parameters for gap penalties relative to the substitution matrix.

The two most extensively used families of scoring matrices are the PAM matrices and the BLOSUM matrices. The underlying principle behind both is similar. The score of aligning residue a with residue b is defined as the log-odds of the pair occurrence under two different hypotheses: the first is that the sequences are related and a mutation happened, while the second is that the occurrence of the pair happened by chance as a result of independent selection (random match). If p_{ab} is the probability of mutation $a \leftrightarrow b$ and p_a, p_b are the background probabilities of a and b, then the ratio $\frac{p_{ab}}{p_a p_b}$ is called a **likelihood ratio** or **odds ratio.** If $p_{ab} > p_a p_b$ then the ratio > 1 and it is more probable that the joint occurrence of a and b is due to a mutation, while if the ratio < 1 the more likely hypothesis is that of independent selection. Taking the logarithm of the ratio $\log \frac{p_{ab}}{p_a p_b}$ we get a scoring function that assigns positive scores to favorable mutations and negative scores otherwise. These scores are referred to as **log-odds** scores. With the logarithm transformation, the score of an ungapped alignment **A** can be re-written as

$$S(\mathbf{A}) = \sum_{i=1}^{N} s(a_i, b_i) = \sum_i \log \frac{p_{a_i b_i}}{p_{a_i} p_{b_i}} = \log \prod_i \frac{p_{a_i b_i}}{p_{a_i} p_{b_i}} \tag{3.14}$$

and thus the definition of the alignment score as the sum of amino acid pairwise scores may be interpreted as the log-odds of the sequences being similar under the two hypotheses outlined above. The DP algorithm that maximizes the similarity score over all possible alignments essentially results in a maximum likelihood measure.

A detailed description of the PAM and BLOSUM matrices is given in the next two sections. The difference between these matrices is the way the probabilities are calculated.

3.5.2.1 The PAM family of scoring matrices

PAM matrices were proposed by Dayhoff et al. in 1978 [80] based on observations of hundreds of alignments of closely related proteins. The frequencies of substitution for each pair of amino acids were extracted from alignments of proteins of small evolutionary distance, below 1% divergence (i.e., at most one mutation per 100 amino acids, on average). These frequencies, normalized to account for the frequencies of single amino acids, resulted in the PAM-1 matrix. The PAM-1 matrix corresponds to substitutions observed in one unit of evolution yielding on average one mutation per 100 amino acids. Accordingly, it is suitable for comparison of proteins that have diverged by 1% or less. The PAM-1 matrix is then extrapolated to yield the family of PAM-k matrices. Each PAM-k matrix is obtained from PAM-1 by k consecutive multiplication and is suitable for comparison of sequences that have diverged k%, or are k evolutionary units apart. For example, PAM-250 = $(\text{PAM-1})^{250}$ reflects the frequencies of mutations for proteins that have diverged 250% or 250 mutations per 100 amino acids (though the definition of PAM-250 seems odd, it still make sense, as is subsequently explained). These matrices were later refined by [81] based on a much larger dataset. The main differences were detected for substitutions that were hardly observed in the original dataset of [80].

Computing the PAM-1 matrix. The basic PAM-1 matrix is a transition probability matrix that corresponds to mutations that occur in "one unit of evolution". There is really no such thing as a unit of evolution, but in order to assess the similarity of sequences at different evolutionary distances it is convenient to define a certain distance unit that corresponds to a small evolutionary interval. The distance could be defined, for example, as the average number of mutations per 100 residues, with 1 PAM corresponding to evolutionary change that resulted in one point mutation per 100 residues. The name PAM stands for point accepted mutations (and hence the distance in number of mutations per 100 amino acids), but it could also mean percentage of accepted mutations (and hence the distance in percentages). Note that this definition of evolutionary distance does not necessarily correspond to a certain time scale, since different proteins from various protein families have evolved at different rates.

To characterize the transition probabilities in 1 PAM, Dayhoff and colleagues analyzed alignments of close homologs. The choice of closely related homologous proteins almost ensures that the correct alignments are obvious. Selecting closely related proteins also minimizes the probability of indirect mutations ($a \rightarrow b \rightarrow c$), which are inappropriate for the computation of PAM-1. The mutations that occur in these alignments are referred to as "accepted mutations", emphasizing that these mutations were accepted by natural selection. In other words, they did not destroy the protein's functionality and the properties of the protein were largly preserved (otherwise, in most cases the protein is degraded and broken into amino acids).

Given the alignments, we can compute the frequencies f_{ab} of mutations ($a \leftrightarrow b$). The events are assumed to be undirected and symmetric, meaning $f_{ab} = f_{ba}$. Every mutation $a \leftrightarrow b$ is counted once when considering mutations that started from occurrences of a (termed **a-mutations**) and once when considering mutations from b (b-mutations). The total number of a-mutations is

$$f_a = \sum_{b \neq a} f_{ab}$$

and

$$f = \sum_{a} f_a \tag{3.15}$$

is the total number of amino acids involved in mutations (note that f equals twice the number of mutations). It should be noted that in their original work, Dayhoff and colleagues determined the mutations based on phylogenetic trees and comparison of sequences with the inferred ancestor proteins, rather than by means of direct comparison between the observed sequences. This step (of building phylogenetic trees) was skipped in later studies that used the PAM model.

The transition matrix M consists of the non-diagonal elements M_{ab}, which denote the probability that amino acid a mutates into amino acid b, and the diagonal elements M_{aa}, which denote the probability that the amino acid a

remains unchanged during the corresponding evolutionary interval. Consider the diagonal elements first.

To compute the probability M_{aa} that an amino acid a remains unchanged we need to know the complementary probability, denoted m_a, which is the probability of amino acid a to change. The term m_a is referred to as the relative **mutability** of amino acid a. Once we determined m_a we can compute the diagonal probabilities

$$M_{aa} = 1 - m_a$$

To determine m_a, we define the following events

- The event A: amino acid a occurred.
- The event B: a mutation happened (any mutation).
- The event C: amino acid a occurred *and* mutated (i.e., $C = A$ and B).
- The event D: amino acid a changed (i.e., $D = B|A$).

The mutability of a is the probability of the event D. Note that the event C corresponds to all a-mutations that are observed in the dataset

$$P(C) = P(a\text{-mutations}) = \frac{f_a}{100 \cdot f}$$

The division by 100 normalizes this quantity to correspond to the specific evolutionary time interval and ensures that

$$P(\text{mutation}) = \sum_a P(a\text{-mutations}) = \sum_a \frac{f_a}{100 \cdot f} = 0.01$$

i.e., the total amount of possible change sums to 1 mutation per 100 amino acids[6].

On the other hand,

$$P(C) = P(A, B) = P(A)P(B|A) = P(A)P(D) = P(A)P(a \text{ changed})$$

Therefore

$$m_a = P(a \text{ changed}) = P(D) = \frac{P(C)}{P(A)} = \frac{f_a}{100 \cdot f \cdot p_a} \tag{3.16}$$

[6]Note that if f is defined as the total number of residue pairs in the dataset (and not just those that are involved in mutations), then there is no need to normalize by 100. However, the total number of mutations might not sum exactly to 1% of all residue pairs, and by defining f as in Equation 3.15 and introducing the factor 100 we ensure that the total change is exactly 1 point mutation per 100 residues on average.

where p_a is the probability of occurrence of amino acid a, which can be estimated from the relative frequencies of amino acids in the dataset[7]. From the mutability we can compute the probability that amino acid a remains unchanged as $M_{aa} = 1 - m_a$.

Computing the non-diagonal transition probabilities is easy, once we computed the mutability, since the probability that a mutates into b is given by

$$M_{ab} = P(a \text{ mutated into } b) =$$

$$P(\text{mutation into } b|\text{a changed}) \cdot P(\text{a changed}) = \frac{f_{ab}}{f_a} \cdot m_a$$

The matrix is a proper transition probability matrix, since the probability that amino acid a mutates into any amino acid b including itself (the probability that it remains unchanged) sums to 1, as required:

$$\sum_b M_{ab} = \sum_{b \neq a} \frac{f_{ab}}{f_a} \cdot m_a + M_{aa} = \frac{f_a}{f_a} m_a + M_{aa} = m_a + 1 - m_a = 1$$

Also, the number of amino acids that remain unchanged per 100 residues is

$$\sum_a 100 \cdot p_a M_{aa} = 100 \sum_a p_a (1 - m_a) =$$

$$100 \left(\sum_a p_a - \sum_a p_a \frac{f_a}{100 \cdot f \cdot p_a} \right) = 100 \left(1 - \frac{1}{100} \right) = 99 \qquad (3.17)$$

Therefore, the expected rate of mutations is 1 per 100 amino acids, and the PAM-1 matrix corresponds to one unit of evolution, as required (this explains the normalization in Equation 3.16).

Computing the PAM-k matrices. Given the basic matrix M, it is now possible to get the transition probabilities for larger evolutionary distances by extrapolation. For example, the probability that a mutates to b in two PAM units of evolution is given by

$$P_2(a \rightarrow b) = \sum_c P(a \rightarrow c) P(c \rightarrow b) = \sum_c M_{ac} M_{cb}$$

Therefore, the matrix that corresponds to two units of evolution is given by PAM-2 $= M \cdot M$, and for k units of evolution PAM-k $= M^k$. Already in their original paper, Dayhoff et al. observed that the application of PAM-1 to

[7]The original paper on PAM is a bit vague, but a good interpretation appears in Setubal and Meidanis [87]. Here I use a similar formalism. The definition of mutability according to Dayhoff is based on the ratio of changes to occurrences of a. Note that this is consistent with Equation 3.16 above, since f_a is the frequency of changes involving a and $100 f p_a$ is the extrapolated frequency of occurrence of a (although it can be computed directly).

a sequence of average composition does not change much the composition of the sequence, and application of PAM-k to sequences of compositions other than the average composition tend to change the composition toward the average composition. Interestingly, for large k (on the order of a thousand) M^k converges to a steady-state matrix with identical rows, where in each row $M^k_{ab} = p_b$ as in the average composition. Therefore, for large evolutionary distances, the distribution of mutations converges to the background distribution of amino acids, regardless of the ancestor amino acid.

Computing the scoring matrices. The scoring matrix S is obtained from the transition probability matrix M. The entries are defined based on the likelihood ratio of two events: a pair is a mutation versus a random occurrence. Consider the transition from a to b. The probability of the first event is the mutation probability M_{ab} that a changes into b. The probability of the second event is the probability p_b that we encounter b in the second sequence, just by chance. The score is defined in proportion to the likelihood ratio (also called odds ratio)

$$S_{ab} \sim \frac{M_{ab}}{p_b}$$

If the odds ratio is larger than 1 then mutation is the more likely event. Otherwise, random occurrence is the more likely event. Given an alignment we can assess the total odds ratio by multiplying the ratios of the individual aligned pairs of residues. If the total odds ratio is larger than 1 then the alignment reflects a sequence of mutations (and hence, a true evolutionary relation), otherwise it can be considered as a random or chance alignment.

It is more convenient to define the score S_{ab} as the logarithm of the odds ratio

$$S_{ab} = 10 \cdot \log_{10} \frac{M_{ab}}{p_b}$$

With the logarithm transformation the alignment score, which is defined as the sum of pairwise scores, corresponds to the logarithm of the product of the ratios (log-likelihood), as in Equation 3.14. The factor 10 is introduced because the similarity scores are often rounded to integers to speed up the calculations, and this reduces the numerical error. For distance of k evolutionary units

$$S^k_{ab} = 10 \cdot \log_{10} \frac{(M^k)_{ab}}{p_b}$$

and the matrix is symmetric $S^k_{ab} = S^k_{ba}$ (recall that the accepted mutations used for defining the scores are undirected events).

The PAM-250 matrix. The PAM-250 matrix is one of the most extensively used matrices in this field. This matrix corresponds to a divergence of 250 mutations per 100 amino acids. Naturally one may ask whether it makes sense to compare sequences that have diverged this much. Surprising as it may seem, when calculating the probability that a sequence remains unchanged after 250 PAMs (applying Equation 3.17 for PAM-250), the outcome is that such sequences are expected to share about 20% of their amino acids. For reference, note that the expected percentage of identity in a random match is $100 \cdot \sum_a p_a^2$. Therefore, for a typical distribution of amino acids (in a large ensemble of protein sequences), we should expect less than 6% identities.

3.5.2.2 The BLOSUM family of scoring matrices

Unlike PAM matrices, which are extrapolated from a single matrix PAM-1, the BLOSUM series of matrices was constructed by direct observation of sequence alignments of related proteins, at different levels of sequence divergence. The matrices are based on **blocks** - a collection of multiple sequence alignments (MSAs) of similar segments without gaps [88], each block representing a conserved region of a protein family (we will discuss MSAs in the next chapter). A dataset of accepted substitutions is collected from these blocks by counting over all pairs of residues in each column of the MSAs. As with the PAM model, the pairs are unordered and b, a is considered the same as a, b. Let f_{ab} denote the number of pairs a, b or b, a in the dataset and

$$f = \sum_{a=1}^{20} \sum_{b \leq a} f_{ab}$$

denote the total number of pairs, then the observed relative frequency of the pair $a \leftrightarrow b$ given by

$$q_{ab} = \frac{f_{ab}}{f}$$

and $\sum_{a=1}^{20} \sum_{b \leq a} q_{ab} = 1$ as required[8]. Note that the events are undirected and $q_{ab} = q_{ba}$.

The BLOSUM log-odds scoring matrix is defined from the ratio between the observed and the expected frequencies

$$s_{ab} = \log \frac{q_{ab}}{e_{ab}}$$

The expected probability of the pair, e_{ab}, is estimated from the dataset by computing the probability of observing the pair under independent selection. The probability of the occurrence of amino acid a *in a pair* is

$$p_a = q_{aa} + \sum_{b \neq a} \frac{q_{ab}}{2}$$

where q_{ab} is divided by two since it contributes both to p_a and p_b. With this definition, $\{p_a\}$ is a proper probability distribution and $\sum_{a=1}^{20} p_a = 1$.

[8]The probability space in the case of the BLOSUM matrices is defined over the top (or bottom) half of the transition matrix (including the diagonal).

The expected probability of the pair $a \leftrightarrow b$ is simply the product of the amino acid occurrence probabilities. If $a = b$ then $e_{aa} = p_a p_a$, but if $a \neq b$ then $e_{ab} = p_a p_b + p_b p_a = 2p_a p_b$ since we either select a first and then b or vice versa. Under this definition, $\sum_{a=1}^{20} \sum_{b \leq a} e_{ab} = 1$ as with the transition probabilities.

The main difference between PAM and BLOSUM is in the way the frequencies are computed. To reduce the bias in the amino acid pair frequencies caused by multiple counts from closely related sequences, segments in a block with at least $x\%$ identity are clustered and pairs are counted *between* clusters, i.e., pairs are counted only between segments less than $x\%$ identical. When counting pair frequencies between clusters, the contributions of all segments within a cluster are averaged, so that each cluster is weighted as a single sequence. Varying the percentage of identity x within clusters results in a family of matrices BLOSUM-x, where x ranges from 30 to 100. For example, BLOSUM-62 is based on pairs that were counted only between segments less than 62% identical.

3.5.3 Information content of scoring matrices

Theorem 1 has a direct bearing on the question of how to choose the appropriate substitution matrix. It states that the frequency of a letter a aligned with the letter b in the best scoring match of two random sequences converges to

$$q_{ab} = p_a p_b \exp^{\lambda \cdot s(a,b)} \tag{3.18}$$

as the length of the compared sequences grows without bound. These frequencies are called **target frequencies**. Hence, any substitution matrix has an implicit target distribution for aligned pairs of amino acids, which can be easily calculated from the scores $s(a, b)$. According to Equation 3.13 these frequencies sum to 1. The implicit target frequencies of a matrix characterize the highest scoring alignments, i.e., the alignments this matrix is *optimized* to find. In other words, only if the frequencies of aligned pairs in a match resemble the target frequencies will the corresponding match have a high score. Therefore, it is claimed [32,89] that a matrix is optimal for distinguishing true distant homologies from chance similarities, if the matrix's target frequencies correspond to the real frequencies of paired amino acids in the alignment of distantly related proteins.

Equation 3.18 can be restated as

$$s_{ab} = \frac{1}{\lambda} \left(\ln \frac{q_{ab}}{p_a p_b} \right) \tag{3.19}$$

i.e., the score for an amino acid pair can be written as the logarithm (to some base) of the pair's target frequency divided by the product of their probability of occurrence under independent selection. This ratio thus compares the probability of an event under two alternative hypotheses (as in the PAM model, Section 3.5.2.1). Therefore, each scoring matrix is implicitly a log-odds matrix, even if the underlying model is not based on observed substitutions. The PAM and the BLOSUM matrices are explicitly of this form.

By Equation 3.13 it follows that multiplying a substitution matrix by a constant factor α is equivalent to dividing λ by α but does not alter the implicit target frequencies, nor the implicit form of log-odds matrix[9]. Such scaling merely corresponds to a different base for the logarithm in Equation 3.19. If λ is chosen to be $\lambda = \ln 2$, the base for the logarithm is 2, and scores can be viewed as bits of information.

How many bits of information are needed to deduce a reliable relationship? If the expected number of MSPs with a score S or higher (Equation 3.12) is set to E and the equation is solved for S, then

$$S = \log_2 \frac{K}{E} + \log_2 Nn$$

Recall that this is the score above which the number of hits that are expected to occur at random is $< E$. If a very low value is set for E, then the match is very significant and probably biologically meaningful. Therefore, this score, expressed in bits, can be viewed as the (minimal) number of bits needed to distinguish an MSP from chance (with error rate $< E$). For a typical substitution matrix K is of the order of 0.1, and for an alignment to be considered significant, E should be 0.1 or less [40,89]. Therefore the dominant term is the $\log_2 Nn$. In other words, to distinguish an MSP from chance, the number of bits needed is roughly the number of bits needed to specify where the MSP starts among Nn possible positions [89]. For example, for an average protein length of 350, and the SwissProt+TrEMBL database with over 2×10^9 amino acids (as of January 2009), at least 39 bits are needed.

With this interpretation, it is possible to get a coarse measure of which matrix is appropriate for a search. Matrices can be evaluated by their information content, i.e., the average information per position,

$$H = \sum_{a,b} q_{ab} s_{ab} = \sum_{a,b} q_{ab} \log_2 \frac{q_{ab}}{p_a p_b}$$

which is the relative entropy of the target and background distributions. The higher the value, the better the distributions are distinguished, and the shorter the length of an alignment *with the target distribution* that can be distinguished from a chance similarity is (in other words, the **minimum significant length** is shorter).

[9]Theorem 1 and its implications for scoring matrices hold for local alignments. For global alignments, multiplying all scores by a constant has no effect on the relative scores of different alignments (as for local alignments). However, the same is true when adding a constant a to the score of each aligned pair and $a/2$ to a gap. Such transformation entails that no unique log-odds interpretation of global substitutions matrices is possible, and probably no theorem about target frequencies can be proved [18].

For PAM matrices the information content decreases as the PAM distance increases. For example, PAM-120 has an information content of 0.98 bits per position. Assuming that at least 38 bits are needed to distinguish a true relationship from a chance similarity (see above), the minimum significant length is 39. PAM-250 has an information content of 0.36 bits per position and the minimum significant length is 106. This is much longer than many domains or motifs. Therefore, short motifs can be detected by PAM matrices only if they have diverged a small PAM distance. For BLOSUM matrices the information content is higher when the index of the matrix is higher (e.g., BLOSUM-100 has an information content of 1.45 bits per position, while BLOSUM-45 has an information content of 0.38 bits per position). It is important to note that higher information content does not signify a better performance in terms of detecting distant homologies. It is the target distribution of a matrix that determines whether the matrix is optimal for a specific search. Therefore, a matrix like the BLOSUM-100, which reflects substitutions between closely related proteins, is not appropriate for the purpose of detecting distant homologies, despite its high information content. A commonly used matrix is the BLOSUM-62, which has an information content of 0.7 bits per position (the same as PAM-160). This matrix is appropriate for comparison of moderately diverged sequences, and it is considered to be one of the best matrices for database searches because of its overall performance, which is superior to all PAM matrices (see next section).

3.5.3.1 Choosing the scoring matrix

When comparing two sequences, the most effective matrix to use is the one that corresponds to the evolutionary distance between them (see previous section). However, we usually do not know this distance. Therefore, it is recommended to use several scoring matrices that cover a range of evolutionary distances, for example PAM-40, PAM-120 and PAM-250. In general, low PAM matrices are well suited to finding short but strong similarities, while high PAM matrices are best for finding long regions of weak similarity.

Several studies compared the performance of different scoring matrices [90, 91]. To evaluate each matrix a few hundred protein families are used as a benchmark, and the quality of a matrix is assessed by means of its ability to detect family members in a database search. Specifically, a query sequence is chosen from each family, and a database search is performed, each time with a different scoring matrix. All family members with a score above a certain threshold are considered to have been detected. The threshold can be chosen, for example, as the score at 1% error rate, or the score above which the number of related sequences equals the number of unrelated sequences (for discussion on performance evaluation, see Section 3.11). The matrix that detects the maximum number of members from the family is considered optimal for this family. The best matrix may vary among different families; therefore the quality is averaged over all families in the reference set.

These studies show that log-odds matrices derived directly from alignments of highly conserved regions of proteins (such as BLOSUM matrices or the Overington matrix, which is based on structural alignment [75]) outperform extrapolated log-odds matrices based on an evolutionary model, such as PAM matrices. Moreover, the accuracy of alignments based on extrapolated matrices decreases as the evolutionary distance increases. This suggests that extrapolation cannot accurately model distant relationships and that the PAM evolutionary model is inadequate. BLOSUM matrices were shown to be more effective in detecting homologous proteins. Specifically, BLOSUM-62 and BLOSUM-50 gave superior performance in detecting weak homologies. These matrices offer good overall performance in searching the databases. According to these tests, the best hybrid of matrices for searching in different evolutionary ranges is either BLOSUM 45/62/100 or BLOSUM 45/100 plus the Overington matrix.

3.5.4 Gap penalties

Gap penalties can greatly affect the performance of sequence comparison algorithms and the alignment accuracy. Currently, there is no widely accepted mathematical model to explain the evolution of gaps. However, there is evidence that the distribution of gaps is quite different from the one implied by the linear model we discussed in Section 3.2.2.1. By observing alignments of related proteins, Gonnet et al. [13] have empirically shown that the probability for opening a gap increases linearly with the PAM distance of the two sequences. However, the probability of observing a gap of length k decreases as $k^{-3/2}$, *independent* of the PAM distance of the two sequences. That is

$$\text{Prob(gap of length } k) \sim \frac{d}{k^{3/2}}$$

where d is the PAM distance between the sequences. This offers further evidence to the hypothesis that gaps are not created by consecutive events of insertions/deletions. As with scoring matrices that are derived from substitution probability matrices, the penalty for a gap of length k is defined proportional the the log probability of observing the gap, and the proposed gap penalty function is of the form

$$\alpha(k) = \alpha_0 + \alpha_1 \cdot \log k + \alpha_2 \log d$$

where α_0 and α_1 are positive constants and α_2 is a negative constant, such that the gap penalty decreases with the PAM distance (recall that our definition of the alignment score, as in Section 3.2, subtracts gap penalties from the total score).

Gonnet et al. suggested an explanation for this functional dependency. Given an accepted gap (inserted/deleted chain), it is reasonable to assume that the two ends of the extracted/inserted chain lie structurally close to each other, so that the chain's insertion/deletion does not affect much the global 3D structure of the protein, and hence its functionality. The probability that the two ends of a randomly coiled chain are placed spatially close is inversely proportional to the mean volume it occupies [92]. This volume increases as $k^{3/2}$ for random chains of length k, which may explain the dependency of $k^{-3/2}$. The statistical interpretation of [93, 94] further provides support for this approach, but practical considerations (the need for a simple mathematical model, time complexity) have led to the broad use of linear gap functions.

Since the linear model for gaps is independent of the models used to derive substitution matrices, it is unclear how to select the gap penalties relative to the substitution matrix (see discussion in [95]). The exact parameters are usually determined based on performance evaluation studies (e.g., [91]), and the parameters can change from one matrix to another. For example, the parameters 12 (for opening a gap) and 2 (for extending a gap) were proposed as a good set of parameters for a wide range of scoring matrices. The default set in BLAST is 11,1 with the BLOSUM62 matrix.

3.6 Distance and pseudo-distance functions for proteins

All the algorithms we discussed in the previous sections for the analysis of proteins were similarity-based algorithms. But in some cases it is useful to define a distance function among proteins, for example, when constructing evolutionary trees, or when investigating higher order organization in the protein space (as will be discussed in Chapter 10 and Chapter 11). A distance function would induce a metric over the protein space and will enable us to explore properties and aspects of the protein space that are otherwise not accessible or hidden.

In contrast to similarity, where the alignment score is a measure of how much the strings are alike, a distance function measures the *dissimilarity* between proteins. A proper distance function should be non-negative and symmetric and should satisfy the triangle inequality (see Section 3.13). One possible approach to distance-based alignment is to assign a cost to elementary edit operations (evolutionary events) and seek a series of operations that transforms one string to another, with the minimal cost. This measure is called the **edit distance**.

To determine the alignment and compute the total distance we can use a dynamic programming algorithm very similar to the one we discussed in Section 3.2, as was proposed by Sellers [96]. The definition of distance resembles the definition of similarity, except that $s(a_i, b_j)$ is replaced with $d(a_i, b_j)$, which reflects the *distance* between amino acids a_i and b_j, and the gap penalty $\alpha(k)$ now adds to the total distance instead of decreasing the similarity. To produce a metric over sequences, the function $d(\cdot,\cdot)$ must be a metric on $\mathcal{A}$, the alphabet of amino acids, and satisfy the three metric requirements. Also $\alpha(k)$ has to be positive. The minimum distance is obtained by minimizing the sum of matches/mismatches costs and the penalties for gap

$$D(\mathbf{a},\mathbf{b}) = \min_{alignments} \left\{ \sum_{i,j} N_{ij} \cdot d(a_i,b_j) + \sum_k N_{k-gap} \cdot \alpha(k) \right\}$$

This algorithm results in a **global distance alignment**. Note that the global distance $D(\cdot,\cdot)$ is zero only if the two sequences are identical.

As was shown by [30, 96], distance and similarity measures are related by a simple formula in some cases. Let $s(a_i,b_j)$ be a similarity measure over $\mathcal{A}$ and $\alpha(k)$ the penalty for gap of length k. Let $d(a_i,b_j)$ be a metric on $\mathcal{A}$ and $\hat{\alpha}(k)$ be a corresponding cost for gaps of length k. If there is a constant c such that

$$d(a_i,b_j) = c - s(a_i,b_j) \qquad \forall a_i,b_j \in \mathcal{A}$$

and

$$\hat{\alpha}(k) = \alpha(k) + \frac{kc}{2}$$

then each alignment A_l satisfies

$$D(A_l) = \frac{c(n+m)}{2} - S(A_l)$$

where n (m) is the length of the sequence $\mathbf{a}$ $(\mathbf{b})$. In particular,

$$\begin{aligned} D(\mathbf{a},\mathbf{b}) &= \min_{alignments\ A_l} D(A_l) \\ &= \frac{c(n+m)}{2} - \max_{alignments\ A_l} S(A_l) \\ &= \frac{c(n+m)}{2} - S(\mathbf{a},\mathbf{b}) \end{aligned}$$

i.e., an alignment is similarity optimal if and only if it is distance optimal (the proof of this claim is based on the observation that $n + m = 2 \cdot \sum_{i,j} N_{ij} + \sum_k N_{k-gap}$).

Though the formula suggests a simple transformation from a similarity measure to a distance measure, it should be noted that the transformation from $s(a_i,b_j)$ to $d(a_i,b_j)$ does not yield a metric when applied to common scoring matrices such as BLOSUM or PAM. These matrices were designed using a log-odds approach, and the value $s(a_i,a_i)$ varies among different amino acids (see Section 3.5.2). Consequently, no c exists such that $d(a_i,a_i) = c - s(a_i,a_i) = 0$ for all $a_i \in \mathcal{A}$, as needed. Distance-based scoring matrices for amino acids were proposed in [76, 97]. While not as effective as the log-odds matrices, they provide an alternative that can be used directly with the global distance alignment algorithm.

The major culprit with distance measures is that they are limited to global alignments, but proteins often share only local similarities. Since global distance measures are strongly affected by the length differences among sequences, they are likely to miss weak local similarities, even if they are significant, and they often produce incorrect alignments. Hence, they are not really suited for the comparison of proteins.

On the other hand, local similarity measures are not suitable for defining distance measures. Whereas global similarity measures can sometime be transformed into global distance measures as described above, no such transformation is known for local similarity measures. A case that seems to rule out the possibility of defining a local distance measure is depicted in Figure 3.8. In principle, local similarities, such as those observed in multi-domain proteins, make the problem of defining distances among protein sequences ill-posed.

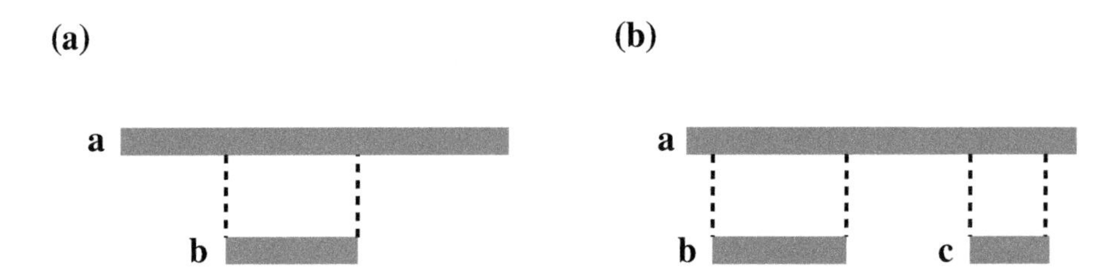

FIGURE 3.8: **Local similarities cannot be easily transformed to a distance function.** Two cases to keep in mind are (a) If **b** is a subsequence of **a**, the sequences are obviously related. However, a distance measure should account for those parts of the sequence **a** that are not matched with **b**. Consequently, the distance $D(\mathbf{a}, \mathbf{b})$ may be as high as for totally unrelated sequences. (b) Multi-domain proteins: if **b** and **c** are unrelated sequences (i.e., $D(\mathbf{b}, \mathbf{c}) >> 1$), then assigning a low distance for $D(\mathbf{a}, \mathbf{b})$ and $D(\mathbf{a}, \mathbf{c})$ will violate the triangle inequality.

While it is hard or even impossible to define a sensitive distance function for complete protein sequences, we could define a fairly sensitive **pseudo-distance** measure that would enable us to work with the many learning algorithms that were designed for metric spaces. Formally, the term pseudo-distance is used in the literature to describe distance functions where the distance between two distinct points is not necessarily positive and can be zero. Here we will use the term pseudo-distance in a broader sense to refer to any function that measures the dissimilarity value between instances and has some of the properties of a distance function (symmetric and non-negative) but is not necessarily monotonic and is not guaranteed to satisfy the triangle inequality.

For example, the following transformation [98] defines a pseudo-distance between two sequences from their local similarity scores

$$d(x,y) = -\log \frac{s(x,y) - s_{rand}}{s_{ave}(x,y) - s_{rand}} \tag{3.20}$$

where $s_{ave}(x,y) = \frac{s(x,x)+s(y,y)}{2}$ and $s(x,x)$ is the self-similarity score of x. The score s_{rand} is the similarity score of the shuffled sequences. Another possible transformation which results in a pseudo-distance function is

$$d(x,y) = s(x,x) + s(y,y) - 2 \cdot s(x,y) \tag{3.21}$$

In [99] this transformation was applied to segments of fixed length, and it has been observed that failures rarely occur. However, this pseudo-distance function has similar properties to global distance measures (see Figure 3.9) implying that the global alignments of the self-similarities $s(x,x)$ and $s(y,y)$ dominate over the local similarity $s(x,y)$.

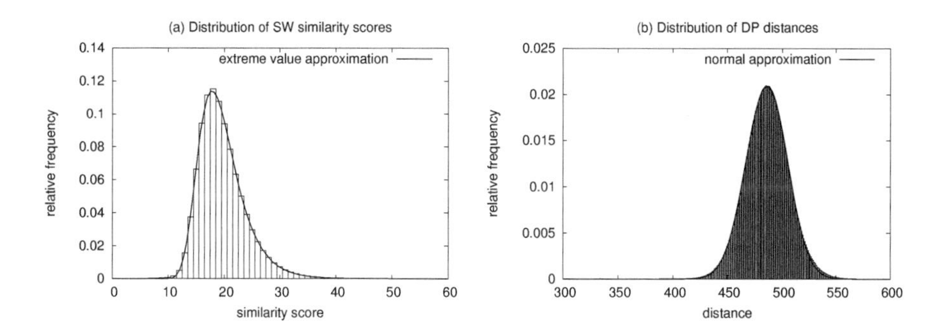

FIGURE 3.9: **(a) Distribution of local similarity scores $s(x,y)$. (b) Distribution of distances $d(x,y) = s(x,x) + s(y,y) - 2 \cdot s(x,y)$.** Recall that scores for global alignments without gaps are distributed normally (see Section 3.4.2.1). Therefore, it seems that though pairwise comparisons originally yielded a measure of local similarity, with a clear extreme value distribution of scores (left), the transformation to pseudo-distance (right) effectively changed the nature of the similarity from local to global. The distributions were obtained based on all pairwise similarities among 5000 randomly chosen protein segments of length 50. The distribution of similarity scores is approximated by the extreme value distribution $\lambda \cdot \exp -e^{-\lambda\cdot(x-u)} \cdot e^{-\lambda\cdot(x-u)}$ where $\lambda = 0.31$ and $u = 17.8$, while the distribution of distances is approximated by the normal distribution (see Section 3.12.4) where $\mu = 486$ and $\sigma = 19$.

A different, more sensitive pseudo-metric is the expectation value (evalue) of the similarity score. As pointed out in Section 3.4, statistical measures that assess the significance of a match are much more effective than the raw similarity scores in discerning true homologies from chance similarities. Therefore, whenever possible we should use the pvalue or the evalue instead of the original similarity score. Recall that the evalue of a similarity score S is the number of occurrences that a match with the same score or higher could have been obtained by chance in a database search.

The nice thing about the evalue is that it has the properties of a dissimilarity measure; the less similar two proteins are the higher their evalue. Although the evalue of the self-similarity score is not zero by definition, it often is because of machine precision limitations. However, since expectation values increase as the length of the similar region decreases, the evalue of the self-similarity is likely to be > 0 for short sequences. Therefore, when a distance function between protein sequences is sought, the expectation values may need to be corrected for the query length. One possible solution is to normalize all similarities with a short sequence by its self-similarity evalue.

Another issue with the evalue measure is that it is not necessarily symmetric, although the similarity score of comparing x with y and y with x is the same. Evalues are often computed empirically based on a database search (see Section 3.4.4) or normalized taking into account the length of the query sequence and the distribution of hits with respect to one protein might differ from the distribution of hits with respect to another. This can be solved by defining the evalue of two proteins as the maximum of the two evalues (the less significant) or the minimum (the more significant).

Finally, for practical reasons, it is easier to work with the $\log(evalue)$ instead of directly with the evalue because significant evalues are $<< 1$ and the range of meaningful evalues is only $[0, 1]$, while after the logarithm transformation the effective range is much larger and is usually between $[-\text{minfloat}, 0]$ (this range can change depending on the implementation and the architecture). By adding minfloat we can shift the range and obtain distance-like values in the range $[0, \text{minfloat}]$. Note that we set the evalue cutoff evalue to 1, as similarities with evalue> 1 are usually meaningless (see discussion in Section 3.4.3). As such, this pseudo-metric is not effective for unrelated sequences and tends to violate the triangle inequality frequently. However, it is effective for related proteins, where it matters more (see Problem 3.13).

3.7 Further reading

There is an extensive literature on sequence analysis and comparison. A good start is the seminal papers on global sequence comparison [12], local sequence comparison [14], FASTA [17], BLAST [18] and gapped BLAST [100]. Many review papers, surveys and book chapters were published over the years (e.g. [10, 40, 95, 101, 102]). Of special note are the books by Waterman [30] and Setubal & Meidanis [87], which were the first of many to follow. The book by Gusfield [103] studies thoroughly various techniques of sequence analysis and representation. Other notable books are [104–106]. Analysis of statistical significance in the context of sequence comparison is reviewed in [60, 107]. For updates and additional references see the book's website at `biozon.org/proteomics/`.

3.8 Conclusions

- Sequence comparison is the main method of analysis of new genes.
- The function of most new genes can be determined based on extrapolation of the properties of homologous genes with similar sequences.
- Sequences can be compared globally or locally. Local alignments are better suited for the comparison of proteins, which often share only one motif or domain in common.
- Sequence alignment can be computed rigorously with a dynamic programming (DP) algorithm. Heuristic algorithms are almost as sensitive as the DP algorithm while being much faster.
- To infer possible homology between sequences their similarity score needs to be statistically significant. Significance can be estimated based on the distribution of matches between random sequences.
- The distribution of local similarity scores was characterized and shown to follow the extreme value distribution.
- Sequence comparison algorithms use scoring matrices for pairs of amino acids, reflecting their substitution rates. Log-odds matrices, such as BLOSUM and PAM, are among the most successful ones.
- Sequences can be evolutionarily related even if their sequence similarity is insignificant. These similarities are referred to as remote homologies and are harder to detect.

3.9 Appendix - non-linear gap penalty functions

In Section 3.2.2 we noted that the dynamic programming (DP) matrix can be computed efficiently, if the gap penalty function is constant or linear. For example, when the gap penalty is constant and set to α then the score of each cell in the DP matrix depends only on the diagonal cell and the last cells in the same row and the same column, even if the gap length is > 1. This can be proved by contradiction. Assume the optimal alignment $\mathbf{A}$ of $a_1a_2...a_i$ and $b_1b_2...b_j$ ends in a gap of length $k > 1$. Denote the sub-alignment of $a_1a_2...a_i$ and $b_1b_2...b_{j-1}$ that is induced by $\mathbf{A}$ and ends with a gap of length $k-1$ as $\mathbf{B}$. Note $S(\mathbf{A}) = S(\mathbf{B}) - \alpha$. Now assume the optimal alignment $\mathbf{B}'$ of $a_1a_2...a_i$ and $b_1b_2...b_{j-1}$ does not end in a gap of length $k-1$. Then $S(\mathbf{B}') > S(\mathbf{B})$. However, if this is the case, then there is an alignment $\mathbf{A}'$ that is better than $\mathbf{A}$, which is formed from $\mathbf{B}'$ by aligning b_j with a gap, because

$$S(\mathbf{A}') = S(\mathbf{B}') - \alpha > S(\mathbf{B}) - \alpha = S(\mathbf{A})$$

which contradicts our assumption that $\mathbf{A}$ is optimal.

However, in the general case of sub-additive gap functions, which are not linear, there is a possibility that the score for some cell will be set according to another cell that is neither its neighbor nor the cell from which the neighboring cell's score was obtained. The following example is artificial but demonstrates this scenario.

Assume the following function $\alpha(k)$ where k is the gap length:

- $\alpha(k) = 6k$ for $k \leq 3$
- $\alpha(k) = 2k + 10$ for $k > 3$

It is easy to verify that this function is sub-additive such that $\alpha(x + y) \leq \alpha(x) + \alpha(y)$. Note that $\alpha(1) = 6$, $\alpha(2) = 12$, $\alpha(3) = 18$, $\alpha(4) = 18$, $\alpha(5) = 20$, $\alpha(6) = 22$, $\alpha(7) = 24$ and so on.

Assume that the scoring scheme for identities and substitutions is the following:

- score$(x, x) = 2$
- for $(x, y) \in \{(C, G), (G, C)\}$ score$(x, y) = -5$
- for any other pair (x, y) such that $x \neq y$ score$(x, y) = -20$

Suppose we are given the following sequences $\mathbf{s} = ACTTGTT$ and $\mathbf{t} = AG$. The dynamic programming matrix for $ACTTG$ against AG will look as follows:

	*	A	C	T	T	G	T	T
*	0	-6	-12	-18	-18	-20		
A	-6	2	-4	-10	-16	-16		
G	-12	-4	-3	-9	-15	-14		

Note that the score for the bottom-right cell was obtained by aligning **s**'s G against **t**'s G, using the diagonal move. This results in the following subalignment:

$$\mathbf{s} = A\ C\ T\ T\ G$$
$$\mathbf{t} = A\ -\ -\ -\ G$$

The cell that contains the score for the alignment of $ACTTGT$ against AG gets its score from its neighboring cell on the left:

	*	A	C	T	T	G	T	T
*	0	-6	-12	-18	-18	-20	-22	
A	-6	2	-4	-10	-16	-16	-18	
G	-12	-4	-3	-9	-15	-14	-20	

This corresponds to the following alignment:

$$\mathbf{s} = A\ C\ T\ T\ G\ T$$
$$\mathbf{t} = A\ -\ -\ -\ G\ -$$

Last, we add the scores in the last column. If we are to look only at the previous cell and the cell from which the previous cell got its score, then the bottom-right cell gets a score of -26; however, if we take into account the sequence of gaps that begins at the third character of **s** we get -23:

	*	A	C	T	T	G	T	T
*	0	-6	-12	-18	-18	-20	-22	-24
A	-6	2	-4	-10	-16	-16	-18	-20
G	-12	-4	-3	-9	-15	-14	-20	-23

This corresponds to the following alignment:

$$\mathbf{s} = A\ C\ T\ T\ G\ T\ T$$
$$\mathbf{t} = A\ G\ -\ -\ -\ -\ -$$

Note that the final alignment is not based on the optimal subalignment of the cell on the left. The structure of the alignment changed completely, and the previous cell does not (and cannot) contain the information needed to deduce that.

As this example shows, in the general case when the function is not linear, we have to look at all the cells in the current cell's row and column in order to get its score right.

3.10 Appendix - implementation of BLAST and FASTA

3.10.1 FASTA

FASTA performs fast sequence comparison by searching for short segments with many identities or conservative substitutions. The process is composed of the following steps

- Create a hash table of all k-mers in the query sequence using a sliding window. Usually, $k = 1$ or 2 for protein sequences, where $k = 1$ gives higher sensitivity. For each k-mer there is an index vector with all positions of the k-mer in the query sequence. For example, for the query sequence $ATTCGCG$ and $k = 2$ the following index vectors are created:

AA	1
TT	2
TC	3
CG	4,6

- Scan a library sequence (Figure 3.10): Initialize an offset vector with zeros. Each offset value corresponds to one diagonal in the DP matrix. Look up each k-mer of the library sequence in the query sequence using the hash table. If it exists, then for each appearance in the query sequence compute the offset (the relative displacement of the k-mer) and increase the offset vector in the corresponding entry by one. If the k-mer appears in position i in the query sequence and position j in the library sequence then the offset is $i - j$.

- Rescan the top diagonals. Extend and combine segments that are on the same diagonal and are not too far from each other by allowing substitutions. Then join nearby segments even if they are not on the same diagonal. Finally, run a bounded DP around these regions.

The algorithm has several parameters (such as the maximal distance between nearby segments, or the number of diagonals scanned), which are set heuristically.

3.10.2 BLAST

BLAST is an outgrowth of the statistical theory for local alignments without gaps (see Section 3.4). This theory gives a framework for assessing the probability that the similarity score of two protein sequences could have emerged by chance.

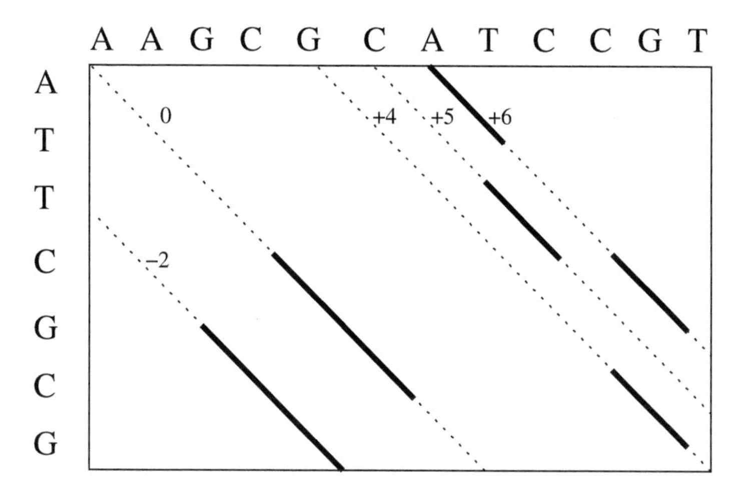

FIGURE 3.10: **Scanning a library sequence.** In this example $k = 2$. Note that in some cases k-mers that have the same displacement (on the same diagonal) overlap. For instance, GCG on the diagonal with offset -2 consists of two hits: the k-mers GC and CG. The values of the offset vector o in this example are $o(-2) = 2,\ \ o(0) = 2,\ \ o(4) = 1,\ \ o(5) = 1,\ \ o(6) = 2$.

Algorithmically, BLAST is similar to FASTA, although it is even faster since it pre-processes the database as well, not just the query. Another major difference is that BLAST uses a scoring matrix already in the first step when defining the set of possible seeds (k-mers) in the query. This set is represented using a hash table or a deterministic finite automaton.

The original BLAST algorithm [18] consisted of four main steps:

- Compile a list of words of length w (usually $w = 3$ or 4 for protein sequences) that score at least T with some substring of length w of the query sequence. For example, if the query sequence is $MKCLLI$... then in addition to the k-mer MKC that appears in the query we will also add the k-mer LKC if $S(MKC, LKC) > T$. Changes in the threshold T permit a trade-off between speed and sensitivity. A higher value of T yields greater speed, but also an increased probability of missing weak similarities.

- Scan database sequences in search for perfect matches with words in the list from the first step. Each match is a seed for a potential alignment.

- Extend each seed in both directions, without gaps, until the maximum possible score for the extension is reached. The resulting HSP (high scoring segment pair) is recorded. If the score of the extension falls below a certain threshold then the process stops. Therefore, there is a chance (usually small, depending on the threshold) for the algorithm to miss a possible good extension.

- Attempt to combine multiple HSP regions. For each consistent combination, calculate the probability of this combination using the Poisson or sum statistics [108] and report the most significant one (lowest probability). If the probability is very low, then the similarity is statistically significant and the algorithm reports the similarity along with its statistical significance.

In the latest version of BLAST the criterion for extending seeds has been modified to save processing time [100]. The new version requires the existence of two non-overlapping seeds on the same diagonal (i.e., the seeds are at the same distance apart in both sequences), and within a certain distance (typically 40) of one another, before an extension is invoked. To achieve comparable sensitivity, the threshold T is lowered, yielding more hits than previously. However, only a small fraction of these hits are extended, and the overall speed increases. Though there are no analytic bounds on the time complexity of this algorithm, in practice, the run time is proportional to the product of the lengths of the query sequence and the database size.

Although BLAST's similarity score is only an approximate measure for the similarity of the two sequences, since gaps are ruled out, it also has an advantage over the DP algorithm in some cases, since it may reveal similarities that are missed by the DP algorithm, for example when two similar regions are separated by a long dissimilar region. Moreover, the statistical theory of alignments without gaps provided a reliable and efficient way of distinguishing true homologies from chance similarities, thus making this algorithm an important tool for molecular biologists.

Nevertheless, the original BLAST algorithm did not always detect complex similarities that included gaps. Gaps have been introduced in the second generation of BLAST, called **gapped BLAST**. Gapped BLAST allows for gapped alignments by using dynamic programming to extend a central seed in both directions [100]. This is complemented by PSI-BLAST, an iterative version of BLAST, with a position-specific score matrix (see Chapter 4) that is generated from significant alignments found in round i and used in round $i+1$. The latter may better detect weak similarities that are missed in database searches with a simple sequence query.

3.11 Appendix - performance evaluation

Performance evaluation is critical and integral to any learning task in computational biology, from homology detection, to structure prediction, domain prediction, clustering and more. Here we will focus on performance evaluation in the context of homology detection, where the goal is to assess the success of a certain algorithm in detecting true relationships between proteins and distinguishing them from false or chance similarities. However, the methodology is more general and applies to other problems as well.

A typical setup consists of a labeled set **F** with n sequences (proteins that belong to a specific protein family), which is a subset of a larger database **D** with N sequences (that belong to many different protein families). The database **D**, including the family **F**, is called the **reference dataset**. We refer to the members of **F** as **positives** and the members of $\mathbf{D} \setminus \mathbf{F}$ as **negatives** (Figure 3.11).

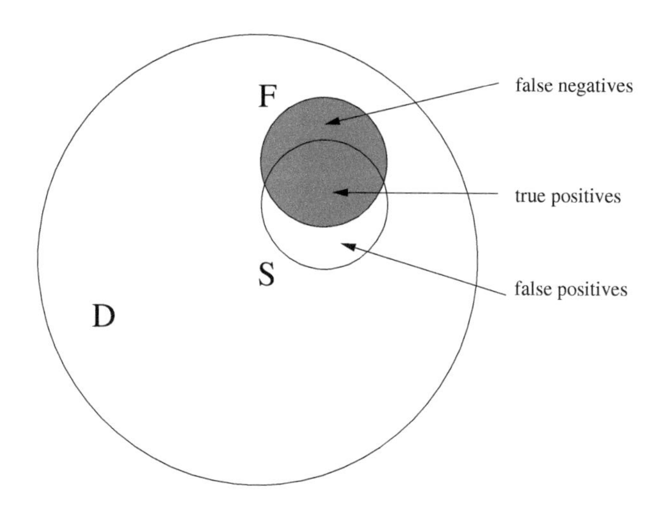

FIGURE 3.11: **Positives and negatives.** Note that $\mathbf{D} = tn + tp + fn + fp$.

Given a query q from **F** and an algorithm M that compares q with sequences in **D**, our goal is to detect all other members of **F** (the homologs of q). That is, in the sorted list of results, we hope that the related sequences score higher than all other sequences in **D**. In the perfect scenario all homologs are at the top of the list, but this is almost never the case, and we usually have to sift through a list that is a mix of related and unrelated sequences. Assume for a moment that we stop after observing the top k scoring sequences, and denote that set by **S**. The set **S** is referred to as the set of **detected** sequences. The performance of the algorithm depends on the exact mix of sequences in **S** and the number of homologs in the set. The subset of sequences in **S** that are also in the set **F** is referred to as **true positives (tp)** indextrue positive

$$tp = \mathbf{S} \cap \mathbf{F}$$

These are the sequences that were successfully detected. The subset of sequences in **S** that are not in the set **F** is referred to as **false positives (fp)**

$$fp = \mathbf{S} \setminus \mathbf{F}$$

These are the sequences that were falsely marked as related. The subset of sequences in **F** that are not in the set **S** is referred to as **false negatives (fn)**

$$fn = \mathbf{F} \setminus \mathbf{S}$$

These are the sequences that were falsely rejected and marked as unrelated. The last set in Figure 3.11 is the set of **true negatives**, which is defined as

$$tn = \mathbf{D} \setminus (\mathbf{F} \cup \mathbf{S})$$

This is the set of unrelated sequences that are correctly rejected. However, since the size of the database **D** is typically much larger than **F**, this set is usually of less interest. In the next sections we will discusses several indices to assess the quality of the set **S**.

3.11.1 Accuracy, sensitivity and selectivity

In statistical analysis the three most basic measures of performance are accuracy, sensitivity and selectivity. The **accuracy** measures the relative number of samples (sequences) that are labeled correctly. Namely, it is the fraction of true positives *and* true negatives out of the whole database **D**

$$\text{accuracy} = \frac{tp + tn}{|D|} = \frac{tp + tn}{tp + tn + fp + fn}$$

However, when the target set **F** is small compared to **D**, this measure can be misleading since it may be dominated by the number of true negatives. The sensitivity and selectivity are usually more informative since they focus on the sets **F** and **S**.

The **sensitivity** (also called **recall** in the context of information retrieval) is the fraction of the set **F** that we detected successfully. The more sensitive the algorithm, the higher the number of true positives and so is the ratio

$$\text{sensitivity} = \frac{tp}{|F|} = \frac{tp}{tp + fn}$$

The **selectivity** (also called **precision**) is concerned with the overall makeup of the set **S** and the fraction of true positives in the set **S**

$$\text{selectivity} = \frac{tp}{|S|} = \frac{tp}{tp + fp}$$

If the algorithm is not very selective, then **S** will contain many unrelated sequences (false positives) and the precision decreases. Another common measure that is complementary to precision is the **error rate**. This is the fraction of false positives in the set **S**

$$\text{error} = \frac{fp}{|S|} = \frac{fp}{tp + fp} = 1 - \text{precision}$$

Note that an algorithm can be very sensitive (detecting all members of **F**) but highly non-selective (detecting also many unrelated sequences). Or, it can be very selective (detecting only members of **F**), but very insensitive (detecting only a small fraction of **F**).

These two measures are a function of the set **S** of detected sequences, and the definition of this set depends on our stopping criterion or the threshold score. A simple criterion is to stop once we observed $k = |\mathbf{F}|$ sequences (the **top-k index**). That is, we focus on a neighborhood of the query that is as big as the family the query belongs to.

Considering the top k matches is a simple criterion, but it often results in poor selectivity, especially for large and diverse protein families. If the algorithm is associated with a measure of statistical significance (such as evalue), then another natural choice would be to stop once the evalue is higher than a certain threshold (say evalue of 0.1). However, not all algorithms are associated with a statistical significance measure.

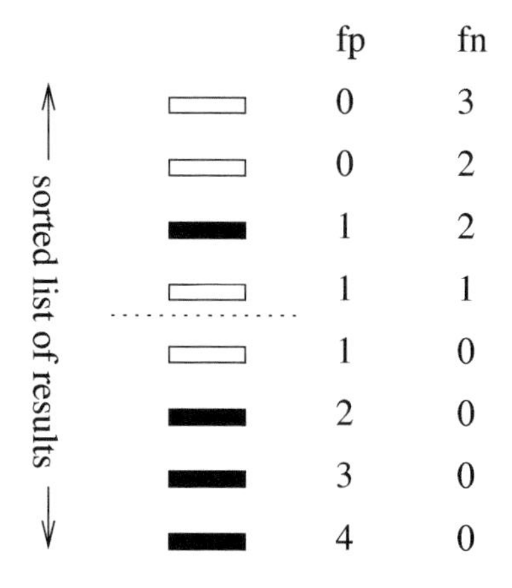

FIGURE 3.12: The equivalence number criterion. A sorted list of results with positives (white bars) and negatives (black bars). The numbers of false positives and false negatives at each point in the list are given on the right. The equivalence point is the point where the number of false positives equals the number of false negatives (marked with a dashed line). The set of detected sequences is defined as the set of hits above the equivalence point.

Sensitivity and selectivity are clearly related to each other. Hence we can measure the performance of one while monitoring or fixing the other. For example, we can measure the sensitivity of an algorithm at a certain error rate, of say 0.1 (which means 90% accuracy or selectivity). With this criterion, the set **S** is defined by scanning the list of top scoring matches until the ratio of false positives fp/k reaches the preset threshold. Similarly, we can measure the accuracy at different sensitivity levels. A variation of this method is the **equivalence number** criterion [91]. The equivalence number criterion sets a threshold at the point where the number of false positives (the number of unrelated sequences with a score above the threshold) equals the number of false negatives (the number of homologs with a score below the threshold), i.e., it is the point of balance between selectivity and sensitivity (Figure 3.12). The set **S** is defined as the set of all sequences above the threshold, and a homolog that scores above this threshold is considered successfully detected (true positive). The quality of the algorithm is measured in terms of the sensitivity of the set **S**, which in this case equals the selectivity.

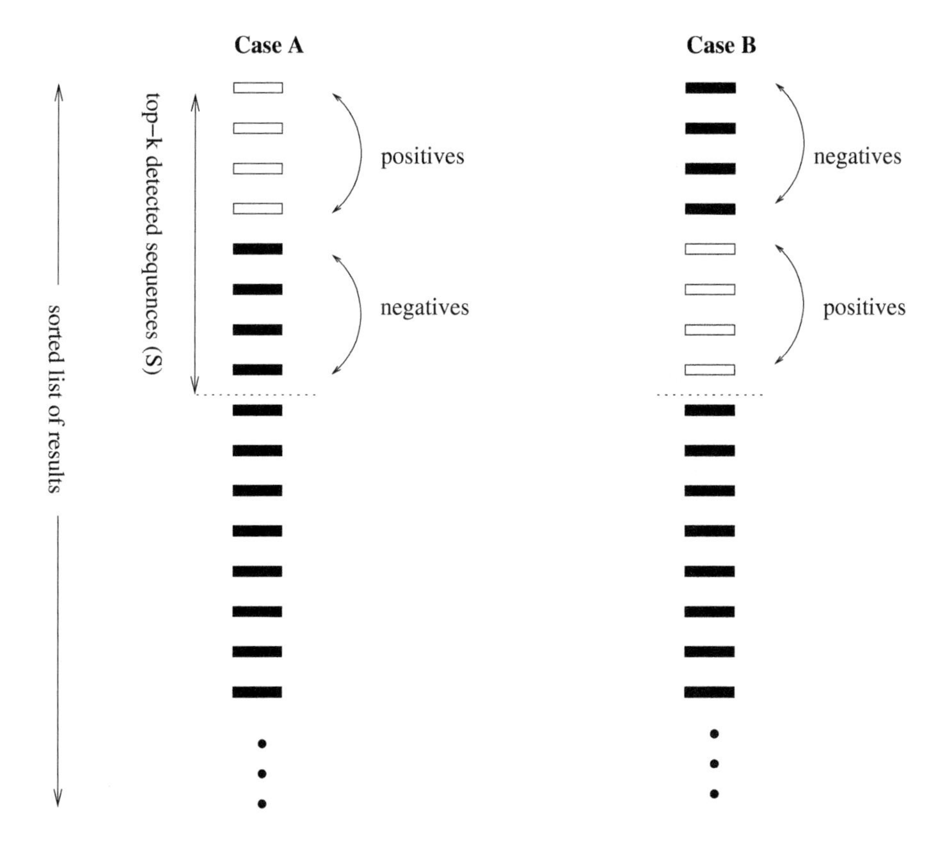

FIGURE 3.13: Accuracy and sensitivity. In both cases the accuracy and sensitivity of the set S are the same, although the left one separates the positives and negatives while the right one does not.

3.11.2 ROC

The coupling of sensitivity and selectivity makes it difficult to get a true sense of an algorithm's performance. If we set a high threshold, the set of detected sequences **S** will be inevitably small and the algorithm is likely to be highly selective, yet insensitive. If we lower the threshold many more sequences will be detected, thus increasing the sensitivity, but at the potential cost of decreasing the selectivity. In this view, a single measure that takes both sensitivity and selectivity into account is desirable.

Another issue with accuracy and sensitivity is that in some cases they can be misleading, since they do not depend on the order of positives and negatives in the set **S** but just on the composition of the mix. Consider for example the scenarios in Figure 3.13. Both have the same accuracy and sensitivity, but the left one is clearly a more favorable situation.

The **Receiver Operating Characteristic (ROC)** [109,110] is a popular quality measure, typically used in signal detection and classification, which summarizes the relation between sensitivity and selectivity in a single curve. Given a sorted list of hits, the ROC curve plots the number of positives as a function of the number of negatives in the list. The ROC index measures the area under the curve, and a larger index implies better performance. Thus it can be used to compare the overall performance of different methods. Maximal performance translates to a perfect separation between true positives and false positives, and a maximal normalized ROC score of 1 (Figure 3.14).

The ROC-N measure is a variation over the ROC index, where the plot is truncated at N negatives. In other words, the ROC-N measure is the number of true positives detected at the point where N false positives were reported. N should be chosen based on the task and the setup (see Section 3.11.3). For example, if the size of **F** is small (hence also the number of true relationships) then ROC-1 would be a good choice (= the number of true positives detected before the *first* false positive), while if assessing a large dataset we can increase N. The idea behind this measure is that in scanning the results of a database search one may be willing to overlook few errors, if additional meaningful similarities can be detected. Furthermore, in many cases supposedly false positives with respect to the reference database are actually true positives [45,111] (see also the discussion in Section 10.5.1.1 of Chapter 10).

The definition of positives and negatives depends on the dataset and the task. Take the SCOP hierarchy for example (Section 9.8). A true positive can be defined as a match between proteins within the same family, casting all other types of connections as false positives. This definition of false positives is very strict, and there are other approaches. Therefore, it is usually beneficial to repeat this procedure with different definitions of true and false positives as discussed in Section 3.11.4.

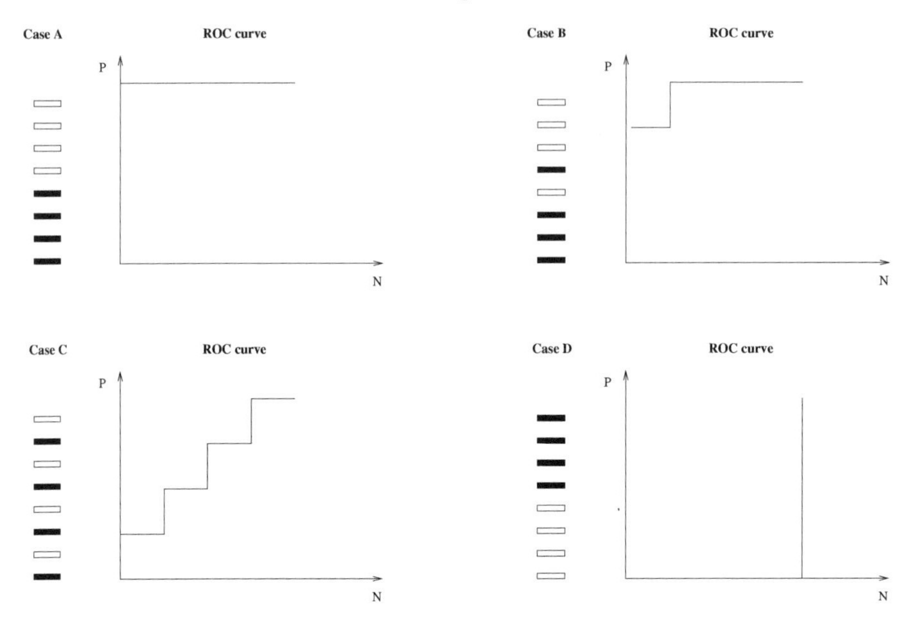

FIGURE 3.14: ROC curves of positives vs. negatives in different scenarios. Positives are denoted by white bars and negatives by black bars. To generate the curve for a specific method sort the results based on the score, scan the list from the top and plot the number of positives vs. the number of negatives at each position in the list. To compute ROC-N count the number of true positives that are detected until N errors occur.

3.11.3 Setup and normalization

There are two common setups for assessing the performance of an algorithm. One is the assessment of **individual queries**, while the second is a **collective assessment**. Consider for example the following task: we are given a set of k protein families $\{\mathbf{F}_1, \mathbf{F}_2, ..., \mathbf{F}_k\}$, and every member of each family is compared with all sequences in the dataset, using a method M. Our goal is to detect the homologs of each protein.

In the individual setup, the performance of M is assessed by analyzing the results for each query sequence separately. This has to be repeated for many queries and averaged per family. That is, the overall performance of a method M with respect to a set of queries $\mathbf{Q} = \{q_1, ..., q_n\}$, using index I is assessed as the sum

$$I(M, \mathbf{Q}) = \sum_{i=1}^{k} \frac{1}{n_i} \sum_{q_i \in \mathbf{F}_i} I(M, q_i)$$

where $n_i = |\mathbf{F}_i|$ and $I(M, q_i)$ is the performance of the method M on query q_i. For example, ROC-1(BLAST, q_i) is the ROC-1 performance of BLAST on protein q_i. Note that self-similarities should be ignored with all performance indices.

The overall performance indices might be affected by the specific makeup of the query set. A given algorithm may perform well for specific queries that belong to large families possessing many members in the database. The indices associated with these elements alone may generate large global performance indices while potentially masking a poor performance of the algorithm on other queries that belong to small families.

To eliminate the bias toward larger families, the performance should be evaluated with normalized counts. Specifically, the counts (number of true positives detected) should be divided by the total number of positives in the dataset (the total number of proteins in the same family, sans the query). That is, the normalized performance of M on query q that belong to family $\mathbf{F}_i$ is

$$I^N(M,q) = \frac{I(M,q)}{n_i - 1}$$

The overall normalized performance is given by the sum

$$I^N(M,\mathbf{Q}) = \sum_{i=1}^{k} \frac{1}{n_i} \sum_{q_i \in \mathbf{F}_i} I^N(M,q_i)$$

and the average performance index per protein family is bounded between 0 and 1. With this normalization each family with n members can add 1 at the most to the overall performance measure, while without normalization each family can add up to n to the total counts. Note that without averaging, each family can add up to $n(n-1)$ to the total counts.

The individual setup is useful when assessing the overall performance of different algorithms, as well when comparing the performance of an algorithm over different families. An algorithm might work well on some families while performing less well on others. By studying the cases on which the algorithm performs poorly we can better understand the properties of the data and the algorithm and come up with a better algorithm.

However, the individual setup might overlook variations in performance that are detected only when considering all the results. For example, it might happen that a match between a query $q_1 \in \mathbf{F}_1$ and one of its negatives scores lower than all homologs of q_1 but higher than a match between another query $q_2 \in \mathbf{F}_2$ and some of its homologs. If the score used by M is a measure of statistical significance, then we are supposed to believe the false match between q_1 and its negative more than the true matches between q_2 and its homologs. Thus, although the performance on q_1 might be perfect, the method might still lead to false predictions.

The **collective assessment** avoids such biases by compiling the results from all queries into one big list and sorting the list based on the scores. The list can then be assessed using indices such as the popular ROC-50, which is the number of true relationships that are detected before 50 false positives occur.

3.11.4 Reference datasets, negatives and positives

All performance measures depend on a labeled dataset (the reference dataset) and on the definition of positives and negatives. When evaluating algorithms for homology detection, the reference dataset can be one of the protein family and domain databases such as PROSITE, Pfam or SCOP (see Chapter 9)[10].

With established protein classifications, the definition of positives and negatives seems straightforward. Proteins that belong to the same family can be considered positives, while proteins that belong to different protein families can be considered negatives. However, in reality the situation is more complicated and the boundaries between negatives and positives are somewhat vague, since different protein families are often related to each other. Consider the SCOP hierarchy for example. This hierarchy is composed of families that are grouped into superfamilies, folds and classes (see Section 9.8). Usually a negative is defined as a relation between proteins from different families, but there is a structural and functional similarity between families that belong to the same superfamily and therefore detecting such relationships is not necessarily an error.

In fact, detecting this type of weak relationships is the goal of *remote homology* detection algorithms. Hence, the definition of positives and negatives also depends on the task. In the context of remote homology detection, a negative is usually defined as a relation between families that do not belong to the same superfamily. Even this definition is somewhat strict, since families that belong to the same fold can also be considered positives. In general, we can say that a relationship between two proteins is a *possible relationship* if both their families belong to the same SCOP fold, a *weak relationship* if they belong to the same class, *suspicious* if they belong to different classes (excluding the case of an all-alpha ↔ all-beta pair) and an *error* if one family is all-alpha and the second is all-beta.

The conclusion is that in order to measure the ability of a given algorithm to recognize different levels of structural and functional similarities between protein sequences, it is worthwhile to repeat the evaluation several times, each time using a different definition of a negative. Similarly, one is advised to inspect different neighborhoods of the query, of varying sizes, using variations on the top-k index. One such index which is applicable to hierarchal classifications such as SCOP is the top-x-y index. This index counts the total number of proteins sharing the same y SCOP denomination among the n_x closest sequences of the query, where n_x is the total number of sequences in the database that have the same x SCOP denomination as the query. For

[10]Another popular reference dataset is the Gene Ontology (GO) database. This database can be used to assess the biological validity and significance of results obtained in computational studies, as discussed in Section 12.5.3.1. However, GO analysis is not optimal for performance evaluation (see [58]) and is better suited for enrichment analysis, where the goal is to assess the coherence in a result set and identify the common or characteristic biological functions of the entities within.

example, to compute the top-fold-fold index for a query protein that belongs to a SCOP fold with n proteins of the same fold denomination, we need to look at the top n proteins in the sorted list of results and count how many of them are actually in the same fold as the query protein (for other variations, see [112]). Such indices can sometimes indicate significant differences in performance that might be overlooked when using indices such as ROC-1 (which can be overly sensitive to outliers).

Finally, it is important to note that the entities we test our algorithm on should be identical to the entities in the reference dataset. While this seems like an obvious requirement, it introduces a restriction on the type of sequence databases that can be used for performance evaluation. For the most part, established protein classifications are domain-based. Therefore, they are not suitable for evaluating sequence comparisons on databases such as SwissProt [16] and GenBank [15], which contain complete protein chains. To avoid problems due to multi-domain proteins it is advised to test an algorithm directly on the sequences of a reference database and use datasets that contain only single domains. We will discuss the protein-vs-domain problem in Chapters 9 and 10.

3.11.5 Training and testing algorithms

The performance evaluation procedures described above can also be used to optimize and tune the parameters of an algorithm, such as the scoring matrix or the gap penalties. However, to properly evaluate the performance of the tuned algorithm it is necessary to use a separate set from the one used for optimization. Otherwise, we are likely to obtain overly optimistic estimates of the performance that usually reflect the **overfitting** of the algorithm to the dataset used for optimization. A similar problem arises when we use a statistical model to describe the data, and we have to learn the parameters of the model from the data.

A proper procedure should divide the reference dataset into two disjoint subsets: a **training set** and a **test set**. As the names imply, the algorithm should be tuned or trained over the training set while tested on the test set. The reference dataset should be divided randomly into the two subsets, to ensure that they each resemble the original set as well as each other. With this procedure (often referred to as **cross-validation**), the independent test set provides an objective assessment of the algorithm's performance.

Usually the reference dataset is divided such that the majority of sequences is in the training set (two-thirds or three-quarters of the sequences). However, in some cases learning uses only the positive examples in the training set, and with small families every positive example matters. Many use the **leave-one-out** procedure (also called the **jackknife test**), where the algorithm is trained over all but one sequence and then tested over the excluded sequence. The procedure is repeated for each positive example in the dataset, and the overall performance is defined as the average over all excluded sequences. Despite theoretical results supporting the validity of this approach, performance evaluation with this procedure might still be overly optimistic, since in reality we are hardly ever in a situation where almost all data instances are known. Therefore, one is advised to use this procedure only if the labeled datasets available for training are very small and to use larger test sets when possible. Re-sampling techniques, such as bootstrapping, are another alternative (see Section 7.4.2.3).

3.12 Appendix - basic concepts in probability

The following is a brief overview of some basic notions in probability theory. A **random variable** X is a variable whose value is defined by an experiment (or process) with an outcome that cannot be predicted. There are two types of random variables: **discrete** and **continuous**. The result of rolling a die is a discrete random variable, while a random variable like height, weight or distance is a continuous variable. A discrete variable can take on a value out of a finite set of possible values, while for continuous variables, the set of possible values is infinite. The set of possible values is called the **sample space** $\mathcal{X}$ of the variable. For a discrete variable with m possible values the sample space is $\mathcal{X} = \{x_1, ..., x_m\}$. For example, the sample space of a die is $\{1, 2, 3, 4, 5, 6\}$. For a continuous variable (such as distance) the sample space is usually an interval of the real space $\mathcal{X} \subset \mathbb{R}$.

An **event** is a subset A of the sample space. It is a set of possible values of a certain experiment. For example, the event "even number" is one of the possible events when rolling a die.

3.12.1 Probability mass and probability density

Each possible value in the sample space is associated with a certain **probability value**, which is the **likelihood** of observing this value as an outcome of an experiment X. The total probability value over the whole sample space is 1, meaning that the experiment must take on one of the possible values in the sample space.

The set of all probability values over the sample space is called a **probability distribution**. For a discrete variable, the probability to observe a specific value x is called the probability mass of x and is denoted by $P(X = x)$ or simply $P(x)$. The **probability mass function** P (uppercase P) is the set of probabilities $\{P(x)\}_{x \in \mathcal{X}}$ such that

$$0 \leq P(x) \leq 1 \ \forall x \qquad \text{and} \qquad \sum_{x \in \mathcal{X}} P(x) = 1$$

It is sometimes convenient to refer to $P(x_i)$ as p_i and denote $\mathbf{p} = \{p_1, p_2, ..., p_m\}$, where m is the number of possible values for X. A simple probability distribution is the **uniform distribution** that associates a probability $1/m$ with every possible value in the sample space.

For a continuous variable, we cannot associate a finite mass with every possible value since there is an infinite number of possible values, and the total probability will end up being larger than 1. Instead, we define a **probability density function** $p(x)$ (lowercase p), which is the probability of observing an outcome in the range $(x - \Delta x/2, x + \Delta x/2)$ where $\Delta x \to 0$. The probability of a certain interval is given by the integral

$$p(x \in (a,b)) = \int_a^b p(x)dx$$

The density function $p(x)$ can get values larger than one, but the total integral over the sample space must satisfy

$$\int_{-\infty}^{\infty} p(x)dx = 1$$

The probability of an event is a measure of the likelihood of this event happening. If the variable is a discrete variable, then the probability of an event that corresponds to a subset A of values in the sample space is defined as

$$P(A) = \sum_{x \in A} P(x)$$

3.12.2 Moments

Moments are a useful way to characterize the properties of the probability distribution of a random variable. The n-th moment of a discrete random variable is defined as

$$M_n = E[X^n] = \sum_{x \in \mathcal{X}} x^n P(x)$$

and for a continuous random variable

$$M_n = E[X^n] = \int_{-\infty}^{\infty} x^n p(x)dx$$

where $E[Y]$ denotes the average value of Y. Two moments of special interest are the first and second moment. The first moment

$$M_1 = E[X] = \sum_x xP(x)$$

is called the **mean** or the **expectation value** of X and is often denoted by the symbol μ. The second moment

$$M_2 = E[X^2] = \sum_x x^2 P(x)$$

measures the scatter of a random variable. The **variance** and the **standard deviation** σ of a random variable are related to the second moment, where

$$Var_x = \sigma_x^2 = E[X^2] - E[X]^2$$

and

$$\sigma_x = \sqrt{Var_x}$$

The standard deviation is the typical amount one should expect a randomly drawn value for X to deviate from μ.

3.12.3 Conditional probability and Bayes' formula

The **joint probability distribution** of two variables X, Y over the combined sample sample $\mathcal{X} \times \mathcal{Y}$ characterizes the probability of both events $X = x$ and $Y = y$ occurring. It is denoted by $P(x, y)$ for discrete variables or $p(x, y)$ for continuous variables. As with a single variable distributions, the total probability must sum to 1

$$\sum_{x \in \mathcal{X}} \sum_{y \in \mathcal{Y}} P(x, y) = 1$$

The **marginal distributions** P_X, P_Y (or p_X, p_Y) characterize the probability distribution of one variable, regardless of the value of the second variable. For example,

$$P_X(x) = \sum_{y \in \mathcal{Y}} P(x, y)$$

The probability of an event A happening, given that event B happened is called the **conditional probability of** A **given** B and denoted by $P(A|B)$. If the events are **independent** then

$$P(A|B) = P(A)$$

For example, the probability to observe an even number when rolling a die is the same regardless of whether the last roll resulted in an odd number. This is because each roll is completely independent of the previous one.

If the events are dependent, then the probability of one happening is affected by the occurrence of the other (that is, whether the second happened or not). For example, the probability for rain (event A) is higher if the sky is cloudy (event B). Therefore,

$$P(A|B) = P(\text{rain}|\text{cloudy sky}) > P(\text{rain})$$

The joint probability $P(A \cap B)$ (also written $P(A, B)$) is the probability of both A and B happening. If A, B are independent then

$$P(A \cap B) = P(A) \cdot P(B)$$

otherwise we can express the joint probability in terms of the probability to observe A first, multiplied by the conditional probability to observe B

$$(1) \qquad P(A \cap B) = P(A)P(B|A)$$

Since the joint probability does not specify which event occurred first, then it also holds that

$$(2) \qquad P(A \cap B) = P(B)P(A|B)$$

Rewriting the joint probability in terms of the conditional probabilities is called **factorization**. From Equations (1) and (2) one can obtain the famous **Bayes' formula**

$$P(A|B) = \frac{P(B|A)P(A)}{P(B)}$$

The joint probability and Bayes' formula can be easily generalized to more than two variables. Factorization is generalized such that each variable in the multiplication is dependent on all previous variables in the expansion

$$P(A \cap B \cap C) = P(A)P(B|A)P(C|A \cap B)$$

When additional variables are introduced in Bayes' formula but ignored in the manipulation, then the simple rule is to add them to all the terms in the equation. That is, if we would like to add dependency on C, we need to change $P(A|B)$ to $P(A|B,C)$, change $P(B|A)$ to $P(B|A,C)$, change $P(A)$ to $P(A|C)$ and $P(B)$ to $P(B|C)$. Therefore

$$P(A|B,C) = \frac{P(B|A,C)P(A|C)}{P(B|C)} = \frac{P(A,B|C)}{P(B|C)}$$

3.12.4 Common probability distributions

Some experiments have a well-characterized probability distribution, knowledge of which can be very useful when assessing the likelihood of certain events happening. For example, say we run n independent experiments, and each experiment can either succeed with probability p or fail with probability $1-p$ (success can be, for example, getting "tails" in a coin toss, or getting a 6 when rolling a die or finding a sequence alignment with a certain score or higher). Define the discrete random variable X to be the number of successes, then X is distributed as the **binomial distribution** with **parameter** p, such that the probability to observe x successes is

$$P(X = x) = \text{Bin}(n,p) = \binom{n}{x} p^x (1-p)^{n-x}$$

where

$$\binom{n}{x} = \frac{n!}{(n-x)!x!}$$

denotes the number of possibilities to choose x successes out of n experiments. The binomial distribution is a probability mass function. The mean value of the distribution (which is the mean number of successes) is

$$E(X) = \mu = n \cdot p$$

and the standard deviation is

$$\sigma = \sqrt{np(1-p)}$$

The mean and standard deviation are called **statistics** of the data. Any function that is computed from the complete sample set and characterizes its properties in one way or another is a statistic. In addition to the sample mean and standard deviation these could be, for example, the minimum and maximum values or value histograms.

Another important distribution is the **normal distribution** (also referred to as the **Gaussian distribution**), which applies to continuous random variables. This distribution characterizes many natural phenomena, such as the distribution of height or weight over large populations, or the distribution of measurement values in many biological and medical experiments. The distribution has two **parameters**, the mean μ and the standard deviation σ, and is defined as

$$p(x) = N(\mu, \sigma) = \frac{1}{\sqrt{2\pi}\sigma} \exp\left(\frac{-(x-\mu)^2}{2\sigma^2}\right)$$

The normal distribution is a probability density function. It is perhaps the most popular and studied distribution, not only because of its mathematical tractability but also because of some fundamental properties it possesses. Interestingly, the normal distribution can approximate the binomial distribution for large n (typically, the approximation is applied when $np > 5$). Furthermore, according to the **central limit theorem**, the distribution of the **sum** of n i.i.d. (**i**ndependently **i**dentically **d**istributed) random variables approaches the normal distribution as n approaches ∞, regardless of the original underlying distribution from which the samples are drawn. This last property becomes especially useful when characterizing the distributions of various data features and measurements. Often we assume that a feature has a certain prototype, which in reality is affected and "corrupted" by a large number of random processes. For example, in handwriting recognition each character can be associated with a prototype. The observed character is a variation of the prototype, and a function of the personal style, the mood of the writer, the paper, the mode of writing etc. Each factor can be perceived as a random process, distributed perhaps normally, and the aggregate effect of all these factors is expected to be distributed normally by the central limit theorem.

The normal distribution can be generalized to a multivariate distribution in $\mathbb{R}^d$ over d random variables.

$$p(\vec{x}) = N(\vec{\mu}, \Sigma) = \frac{1}{(2\pi)^{d/2}|\Sigma|^{1/2}} \exp\left(-\frac{1}{2}(\vec{x}-\vec{\mu})^t\Sigma^{-1}(\vec{x}-\vec{\mu})\right)$$

where $\vec{\mu}$ is the mean vector, Σ is the covariance matrix and $|\Sigma|$ is the determinant of Σ. A normal distribution in $\mathbb{R}^d$ forms a hyper-ellipsoid. The sample mean is the center of gravity of the ellipsoid. It represents all the data in the sense that it minimizes the sum of squared distances from all examples. The sample covariance matrix describes how the data is scattered along different directions.

Other common distributions include the following:

- The **multinomial distribution**. This distribution is a generalization of the binomial distribution for $k > 2$ categories. The parameters of the multinomial distribution are the probabilities $p_1, p_2, ..., p_k$ of each category, such that $\sum_{i=1}^{k} p_i = 1$. The probability that n experiments will result in n_i instances in category i ($\sum_{i=1}^{k} n_i = n$) is

$$P(n_1, ..., n_k) = \binom{n}{n_1 ... n_k} p_1^{n_1} p_2^{n_2} \cdots p_k^{n_k}$$

  The mean of the distribution is the vector $\vec{\mu}$ such that $\mu_i = n \cdot p_i$.

- The **Poisson distribution**. This distribution characterizes the probability of observing an event k times in a certain period of time, given that each occurrence of the event is independent of the previous ones and the *average* rate of occurrences is known and equals λ. The distribution can also characterize events where location replaces the role of time, and the events are measured along a certain distance or volume. The process generating the events is called a **Poisson process**. For example, the event could be the radioactive decay of atoms. If the average number of decays per minute is known and equals p and we are interested in the number of decays in a time interval T, then $\lambda = T \cdot p$ and the probability of observing k decays is given by

$$P(k) = \text{Pois}(\lambda) = \frac{\lambda^k}{k!} e^{-\lambda}$$

  Another example is the number of typos in a book. If the error rate per page is p and we would like to characterize the number of typos k in n pages then $\lambda = n \cdot p$ and k is distributed as the Poisson distribution. The mean of the distribution is λ and the variance is λ as well. This distribution is often used as an approximation to the binomial distribution for large n and small p, with $\lambda = n \cdot p$.

- The **exponential distribution**. This distribution characterizes the times between consecutive events in a Poisson process. The probability that two occurrences are separated by time $t > 0$ is given by

$$p(t) = \text{Exp}(\lambda) = \lambda e^{-\lambda t}$$

  This probability density function has one parameter, λ, which is the rate parameter of the Poisson process. The mean of the distribution is $\frac{1}{\lambda}$ and the variance is $\frac{1}{\lambda^2}$.

- The **chi-square distribution**. This distribution characterizes the sum of squares of k independent random variables that are all distributed like the normal distribution with $\mu = 0$ and $\sigma = 1$. That is, if $X = \sum_{i=1}^{k} Y_i^2$

and $Y_i \sim N(0,1)$ for all i, then X is distributed like the chi-square distribution and the probability of $X = x$ $(x > 0)$ is given by

$$p(x) = \chi_k^2 = \frac{1}{2^{k/2}\Gamma(k/2)} x^{k/2-1} e^{-x/2}$$

where k is the parameter of the distribution, often referred to as the number of degrees of freedom. The function Γ is the gamma function (not to be confused with the gamma distribution), which for half integers $k/2$ is given by the term $\Gamma(k/2) = \sqrt{\pi}\frac{(k-2)!!}{2^{(k-1)/2}}$. The double factorial is defined as $n!! = n(n-2)(n-4)\cdots 6 \cdot 4 \cdot 2$ for even numbers and $n!! = n(n-2)(n-4)\cdots 5 \cdot 3 \cdot 1$ for odd numbers. The mean of the chi-square distribution is k and the variance is $2k$.

3.12.5 The entropy function

The entropy function is a measure of order or certainty associated with a random variable based on the probability distribution over the variable's sample space. The entropy $H(X)$ of a random variable X with probability distribution $\mathbf{p} = \{p_1, p_2, ..., p_m\}$ is defined as

$$H(X) = \text{Entropy}(\mathbf{p}) = -\sum_{i=1}^{m} p_i \log_2 p_i$$

The lower the entropy, the more ordered the distribution or the higher the **information content** of X is. The distribution with the lowest entropy over the sample space $\mathcal{X}$ is the distribution $\mathbf{p} = \{0, 0, ..., 1, 0, ..., 0\}$ where $P(x_i) = 1$ for some i and 0 otherwise. In other words, we are completely certain about the outcome of the corresponding experiment and the entropy is zero. On the other extreme, there is the uniform distribution $\mathbf{p} = \{\frac{1}{m}, \frac{1}{m}, ..., \frac{1}{m}\}$ where $P(x_i) = \frac{1}{m}$ $\forall i$, which has the highest entropy and maximum uncertainty about the outcome of the experiment.

For a continuous distribution, the entropy function is generalized to

$$\text{Entropy}(p) = -\int_{-\infty}^{\infty} p(x) \log_2 p(x) dx$$

Interestingly, the entropy of the normal distribution is the highest among all other distributions with the same mean and variance over the same sample space. In other words, it has maximal uncertainty or disorder. In physics, the entropy is related to the number of possible configurations of a system, which is also viewed as a measure of the total uncertainty (see Problem 3.15). According to the second law of thermodynamics, any physical system strives to the state of maximal entropy, spontaneously. Therefore, being the distribution with maximal entropy, we can speculate that without any restriction or pressure, any natural divergence will converge to a normal distribution.

Unless there is prior knowledge or reason to prefer another distribution, we can often assume (correctly) that the underlying data-generating distribution is a normal distribution.

One of the popular interpretations of the entropy function is as a measure of the average coding length of samples drawn according to the distribution $\mathbf{p}$. If each instance $x \in \mathcal{X}$ is encoded with a string of bits of length $l(x)$, then the average number of bits is

$$\sum_{x \in \mathcal{X}} p(x) l(x)$$

where every instance is weighted with its probability[11]. According to a famous theorem by Shannon, the optimal (i.e., shortest) encoding of a source that generates samples with distribution $\mathbf{p}$ uses $-\log_2 p(x)$ to encode x, with shorter code words (smaller number of bits) associated with more frequent instances. That is, $l(x) = -\log_2 p(x)$ and the average number of bits is

$$-\sum_{x \in \mathcal{X}} p(x) \log_2 p(x)$$

which is the entropy of $\mathbf{p}$. We will discuss codes in more detail in Chapter 6.

3.12.6 Relative entropy and mutual information

Often we need to estimate the statistical similarity or difference between two probability distributions over the same sample space. The **relative entropy**, also called the Kullback-Leibler (KL) divergence, measures the difference between a distribution $\mathbf{q}$ and a distribution $\mathbf{p}$ by measuring the difference in the average coding length, when using $\mathbf{q}$ to determine the code words rather than $\mathbf{p}$. That is,

$$D^{KL}[\mathbf{p}||\mathbf{q}] = -\sum_x p(x) \log_2 q(x) - \left(-\sum_x p(x) \log_2 p(x) \right) = \sum_x p(x) \log_2 \frac{p(x)}{q(x)}$$

This measure can be generalized to the continuous case by replacing the sum with an integral. As we shall see in Section 6.4.2.1, $D^{KL}[\mathbf{p}||\mathbf{q}]$ is always positive and is zero only if $\mathbf{p} = \mathbf{q}$. In Section 4.3.5 we will discuss other measures of statistical similarity between probability distributions.

The **mutual information** is a measure of statistical *dependence* between two variables X, Y. It is defined in terms of the relative entropy between

[11] Although $p(x)$ is usually reserved for continuous variables, from this point on will use $p(x)$ to denote either the probability mass of a discrete variable or the probability density function of a continuous variable. The nature of the variable will be clear from the context (e.g., the use of sums instead of integrals). Uppercase P will be usually reserved for a probability value.

two probability distributions: one being the joint distribution $p(x,y)$ and the second being the product of the marginal probability distributions $\hat{p}(x,y) = p_X(x)p_Y(y)$

$$I(X,Y) = D^{KL}[\mathbf{p}||\hat{\mathbf{p}}] = \sum_{x\in\mathcal{X}}\sum_{y\in\mathcal{Y}} p(x,y)\log_2 \frac{p(x,y)}{\hat{p}(x,y)} \tag{3.22}$$

If the two variables are independent, then $p(x,y) = \hat{p}(x,y)$, and $I(X,Y) = 0$ indicating independence. If the variables are dependent on each other then $p(x,y) \neq \hat{p}(x,y)$ and $I(X,Y) > 0$. When the two variables are identical, then $p(x,y) = p(x)$ and the mutual information equals the entropy of X.

3.12.7 Prior and posterior, ML and MAP estimators

In Section 3.12.4 we mentioned a few common probability distributions. Each one is characterized by a parameter θ or a parameter vector $\vec{\theta}$. For example, the probability of success p is the parameter θ of the binomial distribution. The mean μ and the standard deviation σ are the components of the parameter vector $\vec{\theta} = (\mu, \sigma)$ of the normal distribution.

Different experiments or processes are characterized by different distributions, with different parameters. To indicate that a certain random variable X is distributed following a certain probability distribution with parameter vector $\vec{\theta}$ (the **data-generating** distribution) we introduce the dependency on $\vec{\theta}$ explicitly in the probability function $P(x)$. That is, the probability to observe a certain value $X = x$ is denoted by

$$P(x|\vec{\theta})$$

where the type of the data-generating distribution is implicit. As before, this probability is referred to as the **likelihood** of $X = x$. Given a dataset $\mathbf{D} = \{x_1, ..., x_n\}$ with multiple instances of the random variable X, which are assumed to be independent of each other, then the likelihood of the dataset $\mathbf{D}$ is simply the product of the probabilities of the individual instances

$$P(\mathbf{D}|\vec{\theta}) = \prod_i P(x_i|\vec{\theta})$$

Often we do not know the underlying probability distribution and our goal is to find the distribution that best describes the data. Usually we know the type of the underlying distribution (meaning, its functional form), but we do not know the exact parameters. One approach to determine the parameters is to look for the parameters that maximize the likelihood of the data $P(\mathbf{D}|\vec{\theta})$. The parameter vector that maximizes the likelihood is called the **Maximum Likelihood (ML)** estimator of the probability distribution

$$\vec{\theta}_{ML} = \arg\max P(\mathbf{D}|\vec{\theta})$$

Another approach is to search for the parameters which maximize the **posterior probability** of $\vec{\theta}$ given the data

$$P(\vec{\theta}|\mathbf{D})$$

By Bayes' formula

$$P(\vec{\theta}|\mathbf{D}) = \frac{P(\mathbf{D}|\vec{\theta})P(\vec{\theta})}{P(\mathbf{D})}$$

The term $P(\theta)$ is called the **prior probability** and reflects our prior knowledge or belief about what should be the right parameter. As with any other probability distribution, the total prior probability must sum to 1 over the relevant sample space. The posterior probability $P(\vec{\theta}|\mathbf{D})$ updates our prior by considering also the likelihood of the data $P(\mathbf{D}|\vec{\theta})$. That is, the observations in $\mathbf{D}$ provide us with additional information that can change our prior belief. The term $P(\mathbf{D})$ is called the **marginal likelihood** of the data or the **evidence** and is given by integrating over all possible values of the parameter vector

$$P(\mathbf{D}) = \int_{\vec{\theta}} P(\mathbf{D}|\vec{\theta})d\vec{\theta}$$

The parameter vector which maximizes the posterior probability $P(\vec{\theta}|\mathbf{D})$ is called the **Maximum a Posteriori (MAP)** estimator of the distribution. If the prior $P(\vec{\theta})$ is uniform over the sample space of $\vec{\theta}$ then the ML and MAP estimators are identical (as the marginal likelihood is independent of $\vec{\theta}$).

Uniform priors are called uninformative priors. However, if we do have some knowledge on the underlying process, then it might be beneficial to incorporate it in the form of an informative prior. Priors are often chosen such that they are of the same functional form as the likelihood, to simplify the calculations of the posterior probability. Such priors are called **conjugate priors**. (For example, the conjugate prior of the normal distribution is also a normal distribution, while the conjugate prior of the multinomial distribution is the gamma distribution, as we will see in Section 15.4). The prior distribution itself can be characterized with a parameter vector $\vec{\alpha}$. That is,

$$P(\vec{\theta}) = P(\vec{\theta}|\vec{\alpha})$$

The vector $\vec{\alpha}$ is called **hyper-parameter** to distinguish it from the parameters $\vec{\theta}$ of the data generating distribution $P(x|\vec{\theta})$.

The notion of posterior probability is not limited to parameter vectors. For any two variables or events A, B we can say that $P(A)$ is the prior of event A happening, $P(B|A)$ is the likelihood of B happening given that A happened and $P(A|B)$ is the posterior probability that A happened given that B happened. The probability $P(B)$ is called the marginal likelihood (with the conditional probability $P(B|A)$ summed or integrated over all possible values of A). The four probabilities are related through Bayes' formula, as before

$$P(A|B) = \frac{P(B|A)P(A)}{P(B)}$$

If the two events are independent then the posterior of A is equal the prior since $P(B|A) = P(B)$.

3.12.8 Decision rules and hypothesis testing

Often we need to decide whether a given observation was generated by one process or another. For example, we might want to identify data instances that were generated by a certain process that characterizes a subset of all possible data instances, or distinguish a certain process from background noise. We refer to the process of interest as the **alternative process** and all other instances as the **background process**.

If A denotes the background process and B denotes the alternative process, then we can use **Bayes' decision rule** to decide if a measurement $X = x$ was generated by the background process or not, and choose B if

$$P(B|x) > P(A|x)$$

That is, we decide B if the posterior probability of B is higher than that of A. Using Bayes' formula

$$\frac{P(x|B)P(B)}{P(x)} > \frac{P(x|B)P(A)}{P(x)}$$

and since $P(x)$ is always positive this is reduced to

$$P(x|B)P(B) > P(x|B)P(A)$$

If the actual value of X is unknown, then the best we can do is to decide based on the prior probabilities $P(A)$ and $P(B)$. It can be shown that Bayes' decision rule minimizes the total probability of error (see Problem 3.16). However, the rule is impractical when we cannot characterize the underlying distribution of the process B.

Hypothesis testing is an alternative to the Bayesian approach in such cases. Usually, we can characterize quite easily the background process A. The **background distribution**, denoted here as P_0, is defined as the distribution of the random variable values over instances that were generated by the background process. The background distribution can be, for example, the distribution of scores of alignments between random sequences. The subpopulation we are interested in corresponds to pairs of sequences that are *related* in evolution. To assess whether an alignment score is due to a match between sequences in the subpopulation we ask whether the same score is likely to emerge from the background distribution. The assumption that the observed score is an outcome of the background process is called the **null hypothesis**. This is the essence of **hypothesis testing**, where the null hypothesis (also called the zero hypothesis) is tested against another alternative hypothesis. The definition of the null hypothesis depends on the problem.

The **statistical significance** of a measurement is the probability a measurement at least as extreme could have emerged as a result of the null hypothesis. This probability is called the **pvalue**.

$$pvalue(x) = \text{Prob}(X > x) = \int_{x'>x} P_0(x')dx'$$

If the pvalue is low, then we can reject the null hypothesis with confidence that depends on the exact pvalue. Rejection means that we believe the measurement is unlikely to emerge from the background distribution, and we accept the alternative hypothesis. If the null hypothesis is rejected, then the pvalue can also be viewed as a measure of the error probability, i.e., the probability that an instance was truly generated by the null hypothesis. Each such instance is called a **false positive** (see Section 3.11), and this type of error is referred to as **type I error**. For example, if the pvalue is 0.1 then the null hypothesis is not very likely and we might decide to reject it, but on average, the null hypothesis is expected to be true for one out of ten instances with the same score (or higher).

The second type of error occurs when we do not reject the null hypothesis for some instances that were actually generated by the alternative process, because their pvalues happen to be too high[12]. Each such instance is called a **false negative**, and this type of error is referred to as **type II error**. Characterizing the error probability in this case is harder, since we usually do not know the underlying distribution of B.

The discussion above assumed a single test. But what if instances are sampled again and again and a decision has to be made about each one? In this case each one is considered an individual test and the total experiment is referred to as **multiple testing**. In multiple testing the number of errors increases with the number of tests. What is the expected error rate for N tests? As mentioned above, the pvalue is a measure of the error probability in a single test. A popular approach for estimating the error rate under multiple testing is to apply the **Bonferroni correction**. This correction adjusts the pvalue of a certain experiment under multiple testing by multiplying the single experiment's pvalue by the number of tests. The justification is simple, as the process can be viewed as equivalent to that of tossing a coin where the probability p to get "tail" (error) equals the pvalue in an individual experiment. The number of tails obtained in N experiments is distributed like

[12]Although it is tempting to say that we "accept" the null hypothesis in such cases, the more common approach is to abstain from making a decision. If the pvalue is high then the null hypothesis is more plausible, but there is still a chance that the instance was generated by the alternative process B. Since we usually do not have enough information on how the instances are distributed in the alternative process, we do not accept the null hypothesis. We just do not reject it.

the binomial distribution (see Section 3.12.4) and the expected number is the mean which equals $N \cdot p$. This is the same logic that underlies the relation between the pvalue and the evalue in Equation 3.12 of Section 3.4.3.2. When the tests are independent, then the Bonferroni correction provides an accurate estimate of the false positive rate under multiple testing.

An alternative approach for estimating the significance under multiple testing is the **False Discovery Rate (FDR)** method. This method is more sensitive than the Bonferroni correction, in the sense that it can produce the same false positive rate (type I error) with fewer false negatives (type II error). The method is especially effective when the tests are not independent of each other. FDR controls the proportion of false positives out of those measurements that are deemed significant (e.g., scores that are higher than a certain preset threshold), as opposed to the rate of false positives out of all measurements.

With the Bonferroni correction the expected number of type I errors in N experiments is $N \cdot p$. By equating that number to a desired error rate α we can set a threshold on the pvalue of each test and reject the null hypothesis for all tests with pvalue $< \frac{1}{N}\alpha$. With this criterion we know that the type I error rate in N tests should not exceed α. However, this approach may be too conservative and often we may end up rejecting the alternative hypothesis for many samples that were actually generated by the alternative process, but whose pvalue is too high. This is especially true for datasets with positive correlation in their values (e.g., the set of database sequences similar to a query sequence).

The idea of the FDR approach is to control the rate of the false positives out of *a subset* of significant tests, and not all tests. The definition of the subset depends on the pvalues obtained in the tests, and hence the method adjusts to the distribution of observed pvalues. Specifically, the following procedure, suggested in [113], achieves the desired results. Given N tests, sort the tests in decreasing order of significance (increasing pvalue) and denote the pvalue of the i-th test by p_i. That is, $0 < p_1 < p_2 < ... < p_N < 1$. The correction factor for every test depends on its place in the sorted list and is set to $\frac{N}{i}$. To obtain a desired false positive rate α, find the largest i for which $p_i < \frac{i}{N}\alpha$ and mark the subset of tests $1..i$ as significant (i.e., reject the null hypothesis). As shown in [113], this procedure guarantees that the type I error rate within the subset $1..i$ of significant results (with pvalues $p_1, ..., p_i$) is bound by α. Note that with the Bonferroni correction we would have accepted only the tests for which $p_i < \frac{1}{N}\alpha$, while with FDR the test becomes more lenient as we scan through the list. Thus, if the tests are positively correlated (i.e., some of the instances generated by the alternative process are dependent on each other) we will tend to observe a cluster of tests with close pvalues and will accept some of them. This results in fewer false negatives (type II error). The method can also be extended to non independent tests with negative correlation. For details, see [114].

3.13 Appendix - metrics and real normed spaces

The notion of **metric space** is of importance when assessing relationships between protein sequences, as well as when discussing vectorial representations and embedding algorithms (Chapter 11). Central to its definition is the existence of a distance function (a metric) that satisfies certain requirements. A metric space does not need to be a vector space. Formally, a metric space is a pair $(\mathcal{X}, d)$, where $\mathcal{X}$ is a finite or infinite set, and $d: \mathcal{X} \times \mathcal{X} \rightarrow \mathbb{R}^+$ is a **distance function** over the set X that has the following properties

- Nonnegative: $d(x, y) \geq 0$ for all $x, y \in \mathcal{X}$ with equality if and only if $x = y$
- Symmetric: $d(x, y) = d(y, x)$
- Satisfies the **triangle inequality**: $d(x, y) \leq d(x, z) + d(z, y) \quad \forall x, y, z \in \mathcal{X}$

For example, the real space $\mathbb{R}^3$ with the Euclidean distance function

$$d(\vec{x}, \vec{y}) = \sqrt{(x_1 - y_1)^2 + (x_2 - y_2)^2 + (x_3 - y_3)^2}$$

is a metric space. So is the set of all strings over alphabet $\mathcal{A}$ with the edit distance metric.

In the real space $\mathbb{R}^d$, the distance between two vectors $\vec{x}, \vec{y} \in \mathbb{R}^d$ depends on the difference vector $\vec{x} - \vec{y}$. There are many distance functions we can define based on the difference vector, depending on the **norm** we associate with vectors. A norm $\|\cdot\|$ associates a nonnegative number $\|\vec{x}\|$ with every point $\vec{x}$ in the d-dimensional real space $\mathbb{R}^d$, where

- $\|\vec{x}\| \geq 0$ for all $\vec{x}$ and equality holds iff $\vec{x} = \vec{0}$
- $\|\lambda\vec{x}\| = |\lambda| \cdot \|\vec{x}\|$ for every $\vec{x} \in \mathbb{R}^d$ and every $\lambda \in \mathbb{R}$
- $\|\vec{x} + \vec{y}\| \leq \|\vec{x}\| + \|\vec{y}\|$ for every $\vec{x}, \vec{y} \in \mathbb{R}^d$

For example, the norm l_p applied on $\mathbb{R}^d$ is defined as

$$\|\vec{x}\|_p = \left(\sum_{i=1}^{d} |x_i|^p \right)^{1/p}$$

Three norms of main interest are the Manhattan distance l_1 (also called the city-block distance, after the street grid of Manhattan)

$$\|\vec{x}\|_1 = \sum_{i=1}^{d} |x_i|$$

the Euclidean norm l_2:

$$\|\vec{x}\|_2 = \sqrt{\sum_{i=1}^{d} x_i^2}$$

and the max or l_∞ norm:

$$\|\vec{x}\|_\infty = \max_i |x_i|$$

The space $\mathbb{R}^d$ equipped with the norm l_p is denoted by $\mathbb{R}^d_p$. Such spaces are called **real-normed spaces**. The metric associated with the norm $\|\cdot\|$ is $d(x, y) = \|\vec{x} - \vec{y}\|$. Unless specified otherwise, the default norm is usually the Euclidean norm, which is geometrically more intuitive than most other norms.

3.14 Problems

For updates, additional problems, files and datasets check the book's website at `biozon.org/proteomics/`.

1. **The complexity of the alignment problem**

   Given two sequences of length n, what is the number of possible pairwise alignments one can generate, assuming substitutions, insertions and deletions? Explain the derivation of that number. How many possible alignments do you get for $n = 350$ (the average length of a protein sequence)?

2. **Non-uniqueness of the optimal alignment**

   Assume the following scoring function: match scores $s(x, x) = 2$, substitution scores $s(x, y) = -2$, and a fixed gap penalty $\alpha = -2$. Show that the optimal global alignment of the sequences $\mathbf{a} = AGGCTT$ and $\mathbf{b} = AGCTCT$ is not unique. Perform the pairwise alignment algorithm by hand, filling the dynamic programming table with scores and paths.

3. **SemiGlobal alignment**

   Global alignment is effective for comparison of DNA or protein sequences that are fairly conserved across most of their sequence. However, sometimes the "tails" of the sequences are loosely or totally unrelated, and forcing the alignment to include these tails (by introducing gaps at the one or both ends) might result in a suboptimal or incorrect alignment. Propose a modification to the global alignment algorithm that does not penalize for end gaps (SemiGlobal alignment). Apply this algorithm to align AAAAAGCG and GCGAATTT. Compare to the global alignment you get with the scoring function of Problem 3.2.

4. **Local alignment with affine gap penalties**

   You are given the following scoring matrix and an affine gap penalty

	A	C	G	T
A	3	-3	2	2
C	-3	3	-4	-2
G	2	-4	3	-2
T	2	-2	-2	3

   function with $\alpha_0 = 4$ and $\alpha_1 = 1$. Compute the local alignment of the sequences $\mathbf{a} = ACAGTG$ and $\mathbf{b} = ATCG$. Show the four DP matrices for S_{ij}^{local}, R_{ij}, C_{ij} and the trace matrix $PATH_{ij}$.

5. **Circular permutations**

   Biological sequences can also change through circular permutations (for example `mentalin` is a circular permutation of `alignment`). Suggest a simple trick that would allow you to detect local similarities (including substitutions, deletions and insertions) between two sequences that are related through circular permutations.

6. **Blast search**

   You are given a hypothetical protein sequence for analysis (`sequence1`). Use BLAST to search the non-redundant (NR) protein sequence database at the NCBI website. What can you say about the possible function of this protein?

   You are given a second new protein (`sequence2`). Repeat the analysis with this protein. Can you determine the function of the protein? How is that protein different from the first one?

   Sequences and a link to the NCBI site are available at the book's website (`biozon.org/proteomics/`).

7. **The PAM model**

   (a) You are given the PAM-1 transition probability matrix M compiled from accepted mutations observed in highly similar proteins (see file `PAM1` in the book's website). Given the basic matrix M, it is now possible to obtain the transition probabilities for larger evolutionary distances. Compute the PAM-250 transition matrix.

   (b) Compute the probability that a sequence remains unchanged under PAM-250.

   (c) For the amino-acid glutamine (Q), compare the most frequent substitution in PAM-1 to that in PAM-250. Can you explain why they differ?

   (d) The scoring matrix S is obtained from the transition probability matrix M. The score S_{ab} is defined as $S_{ab} = 10 \cdot \log_{10}(M_{ab}/p_b)$. Rewrite S_{ab} using the basic frequencies f_{ab}, f and the background probabilities of occurrence p_a, p_b as computed from a large sequence database ($\sum_a p_a = 1$). Explain the term you got.

   (e) You are given the background probabilities (file: `background-prob`). Compute the PAM-250 *scoring* matrix.

   (f) For the amino-acid glutamine, compare the most frequent substitution in the PAM-250 *transition probability* matrix to the highest scoring replacement in the PAM-250 *scoring* matrix. Explain why they differ.

8. **The genetic-code scoring matrix**

   Derive a scoring matrix for amino acids based on the genetic code. Compare the most similar pairs of amino acids in this matrix to the same entries in PAM-250, and vice versa. What can you say about the different properties of these matrices?

9. **Implementing the dynamic programming algorithm**

   Implement the dynamic programming algorithm for global and local alignment of protein sequences with affine gap penalties. Compare `sequence3` and `sequence4` using the BLOSUM62, PAM-1 and PAM-250 scoring matrices. Try also different combinations of gap penalties. Input sequences, scoring matrices, gap parameters and output format are posted at `biozon.org/proteomics/`.

10. **The asymptotic behavior of local similarity scores**

    (a) You are given a set of protein sequences (see file `sequences`). Compute the all-vs-all local sequence similarities between the sequences using your program. Use the BLOSUM62 scoring matrix and gap penalties of 11 (opening) and 1 (extension).

    (b) For each pairwise comparison, compute the search space $n * m$ (where n and m are the lengths of the sequences compared) and collect data on the relation between similarity scores and the size of the search space (discretize the search space to bins of 10 or 100).

    (c) Show, based on the empirical data, that local similarities asymptotically behave like $a * ln(n * m)$.

11. **Significance of local similarity scores**

    (a) Using the data from Problem 3.10, generate the histogram of scores (use bins of size 1) and show that the empirical distribution follows the extreme value distribution. Estimate the parameters of this distribution (parameters can be estimated with curve fitting functions in gnuplot or matlab).

    (b) Given the parameters you just estimated, compute the significance (evalue) of the local similarity scores for the three pairs of sequences provided in the book's website. Compare the raw scores to the evalues. Which protein pairs can be assumed to be homologous?

12. **Assessing significance with zscores**

    (a) Using the dataset of Problem 3.10, run your program in global mode to compare all pairs of sequences and show that global similarity scores follow a normal distribution.

(b) Add a procedure to your program to shuffle a sequence. Given a pair of sequences (referred to as *query* sequence and *library* sequence), the zscore of their global similarity score can be computed by shuffling the query many times (e.g. 100), and aligning each shuffled sequence to the library sequence. The distribution of scores is the background distribution based on which the zscore of the similarity score between the unshuffled sequences can be computed. Compute the zscore of the global similarity score for the three pairs of sequences provided in the website.

(c) Repeat this procedure for local alignment. Compare the zscores you get to the evalues computed as in Problem 3.11. How useful is the zscore for assessing the significance of local similarity scores? Can you deduce homology with high confidence for the same pairs that you analyzed in Problem 3.11?

13. **Distance functions**

Given the set of sequences from Problem 3.10, transform the local similarity scores to a pseudo-distance measure using the following transformations:

(a) $d(x, y) = -\log\left[(s(x, y) - s_{rand})/(s_{ave}(x, y) - s_{rand})\right]$ as in Equation 3.20.

(b) $d(x, y) = s(x, x) + s(y, y) - 2 \cdot s(x, y)$ as in Equation 3.21.

(c) $d(x, y) = -\texttt{min\_evalue} - \log evalue[s(x, y)]$ where $\texttt{min\_evalue} = \min_{x,y} \log evalue[s(x, y)]$ over all possible pairs.

Compare the top scoring (with the evalue measure) to the closest pairs (distance-wise) with each method. Which one if any preserves the similarity relationships?

14. **Performance evaluation with ROC**

You are given a set of pairwise similarities between protein sequences that belong to different SCOP domain families (file `scores`). The similarities were computed using the SW local alignment algorithm. Assume the task is remote homology detection. Generate the ROC curve and compute the ROC, ROC1 and ROC10 scores for `query1` and `query2` (extract the list of pairwise similarities with respect to each and sort based on the similarity score). Repeat this procedure, now sorting results based on the evalue. Which one produces better ROC scores?

15. **Entropy as a measure of disorder/uncertainty**

Suppose you are given a random sample of N instances from two classes, with m instances from the first class, and $N - m$ instances from the second class. Assume also that N is large. We can define the probability for an instance to be in the first class as $p = \frac{m}{N}$.

(a) What is the number of possible configurations of the sample set, i.e., how many ways are there to divide the instances into two classes, so that m of them are in the first class and $N-m$ are in the second? This number is a measure of the disorder or uncertainty we have about the sample set.

(b) Show that this number can be approximated by $2^{N \cdot H(p)}$ where $H(p)$ is the entropy of the distribution. In other words, $N \cdot H(p)$ bits are required to code all possible configurations.

16. **Bayes' decision rule**

Denote by $\mathcal{X}_A$ the set of values $X = x$ for which $P(A|x) > P(B|x)$ (meaning, we decide A). Denote by $\mathcal{X}_B$ the set of values $X = x$ for which $P(B|x) > P(A|x)$. Write the total error probability over the sample space of X as the sum of type I error and type II error, and show that Bayes' decision rule minimizes the sum.

Chapter 4

Multiple Sequence Alignment, Profiles and Partial Order Graphs

The problem of Multiple Sequence Alignment (MSA) is an extension of the pairwise alignment problem discussed in Chapter 3. The input is multiple related sequences, and the goal is to recover the correct alignment that places together nucleotides or residues that evolved from the same ancestor nucleotide/residue. This problem received considerable attention for two main reasons. First, MSA-based algorithms are more powerful than pairwise alignment methods. By comparing a protein with an MSA of a protein family it is often possible to detect significant similarity that indicates remote homology, even when undetectable by means of pairwise sequence comparison. MSA-based algorithms usually generate more accurate alignments than pairwise methods and can help to discover important mutations (Figure 4.1) or highly conserved regions along sequences, with possible functional significance. MSAs also play an important role in reconstructing evolutionary relationships between sequences and deciphering the domain structure of proteins (see Chapter 9).

```
Q9Y6D9      MEDLGENTMVLSTLRSLNNFISQRVEGGS-GLDISTSAPGSLQMQYQQSMQLEERAEQI
AAD20359    MEDLGENTMVLSTLRSLNNFISQRVEGGS-GLDISTSAPGSLQMQYQQSMQLEERAEQI
Q9WTX8      MEDLGENTTVLSSLRSLNNFISQRMEGTS-GLDVSTSASGSLQKQYEYHMQLEERAEQI
Q13312      MEDLGENTMVLSTLRSLNNFISQRVEGGS-GLDISTSAPGSLQMQYQQSMQLEERAEQI
Q9W6G9      MDDSEDNTTVISTLRSFNKFLSQPLEGTApSLGTSTSTGASLQMQFQQRFLLEDQAAQI
Q9WTX9      MEDLGENTTVLSSLRSLNNFISQRMEGTS-GLDVSTSASGSLQKQYEYHMQLEERAEQI
```

FIGURE 4.1: Multiple sequence alignment of several mitotic checkpoint proteins. These proteins are part of a complex that prevents chromosome's separation in eukaryotic cells during meiosis or mitosis (cell division processes), until all chromosomes are properly aligned. Only the first 60 characters are shown, with the SwissProt [115,116] identifiers on the left. Note the mutation in the middle of the fifth protein (an insertion of a proline). The significance of this mutation might be overlooked in a pairwise alignment.

The second reason is the computational complexity of the problem. A straightforward algorithm for the alignment of N sequences is obtained by a generalization of the dynamic programming algorithm described in Chapter 3 to N sequences. However, the time complexity of this algorithm prevents it from being applied to more than a few sequences. To tackle this problem many heuristic algorithms were developed over the years.

There is an extensive literature on MSA algorithms. However, any survey of this field ought to mention some of the classical approaches to MSA, and they are reviewed briefly here for historical reasons and for the sake of completeness. We also discuss new approaches based on transitive relations and partial order graphs.

4.1 Dynamic programming in N dimensions

The DP algorithm in N dimensions is a generalization of the pairwise alignment algorithm described in Section 3.2, using an N-dimensional matrix instead of a two-dimensional matrix. Let $\mathbf{a}_1, \mathbf{a}_2, ..., \mathbf{a}_N$ be sequences over alphabet $\mathcal{A}$ of lengths $l_1, ..., l_N$. A multiple alignment is obtained by allowing substitutions and inserting gaps such that all sequences are of the same length. Once the leftmost characters are aligned, the procedure results in a sequence of columns, where each column contains N characters and gaps. The one exception is that columns with only gaps are not allowed.

Our goal is to recover the correct MSA where evolutionarily related characters are aligned. The underlying assumption is that evolutionarily related characters tend to be more similar to each other than unrelated characters, and as with pairwise alignment we attempt to recover the correct MSA by aligning sequences so as to maximize the total similarity of the aligned characters. There are different functions we can use to compute the score of each position (column), as will be discussed next. For now, we use the general notation $\text{score}(\sigma_1, ..., \sigma_N)$ to denote the score of a column that contains the characters $\sigma_1, ..., \sigma_N$, where $\sigma_i \in \mathcal{A}$.

Generalizing the function we used when comparing two sequences, we define the function $S(j_1, j_2, ..., j_N)$ as the maximal score of aligning the first j_i characters from the i-th sequence ($1 \leq i \leq N$, $1 \leq j_i \leq l_i$). To find the optimal MSA and compute its score, we first initialize $S(0, 0, ..., 0) = 0$. Then, we define S recursively

$$S(j_1, j_2, ..., j_N) = \max_{\vec{x}}\{S(j_1 - x_1, j_2 - x_2, ..., j_N - x_N) + \text{score}(\sigma_1, ..., \sigma_N)\}$$

where $\vec{x}$ is a binary vector $\neq \vec{0}$ indicating which of the previous cells we are considering and

$$\sigma_i = \begin{cases} \mathbf{a}_i(j_i) & \text{if } x_i = 1 \\ - & \text{if } x_i = 0 \end{cases}$$

For example, if $\vec{x} = \vec{1}$ then we are looking at the diagonal cell in the N-dimensional matrix. Note that the number of possible vectors $\vec{x}$ is $2^N - 1$. Therefore the computation of each cell in the N-dimensional matrix depends on $2^N - 1$ other cells. The score $S(\mathbf{A})$ of the best global MSA is stored in the last cell $S(l_1, l_2, ..., l_N)$. To generate a local MSA the algorithm is modified, similar to the modification introduced in the case of pairwise alignment, discussed in Section 3.2.3 (see problem Problem 4.1).

4.1.1 Scoring functions

When choosing a function to score columns of an MSA the general approach is to measure the mutual similarity (or conservation) of characters in the column. Possible scoring functions include:

1. **The sum-of-pairs score (SP-score)** [117]. The motivation behind the definition of this score is simple - a column should score high if there are many identical or similar residues. Given a column with N characters then its SP-score is defined as

$$\text{score}(\sigma_1, \sigma_2, ..., \sigma_N) = \sum_{i=1}^{N-1} \sum_{j=i+1}^{N} s(\sigma_i, \sigma_j)$$

   where $s(a, b)$ is an entry in a scoring matrix like BLOSUM or PAM. We can include spaces (gaps) in the scoring function and set $s(-, b) = \alpha$ where α is the gap penalty. However we define $s(-, -) = 0$ because the projection of the MSA on the two sequences the characters belong to creates a column that can be deleted.

   An interesting fact is that if $s(-, -) = 0$ then the SP-score of an MSA $\mathbf{A}$ can be decomposed into a sum of the projected pairwise alignment scores such that

$$S(\mathbf{A}) = \sum_{i \neq j} S(\mathbf{A}_{ij})$$

   where $\mathbf{A}_{ij}$ is the projection of the MSA on the i-th and j-th sequences. It is easy to see why this is true. In both cases we are adding the same pairwise scores, just in a different order. In the first case (MSA) we sum first over columns then rows. In the second case (sum of pairwise alignments) we sum first over rows and then over columns.

   A drawback of the SP-score is that it can harshly penalize mutations, especially when the number of aligned sequences is small. For example, consider a column with three A's and one C. If the scoring function is such that $s(A, C) = -s(A, A)$ then the score of this column would be zero, although it is highly conserved. If the column has only two A's and one C, then the score of this column would be negative if $s(A,C) > -\frac{1}{2}s(A,A)$. Common scoring matrices assign $s(a,a) > -s(a,b)\ \forall a,b \in \mathcal{A}$, so the problem is less pronounced in practice but it still exists for small datasets. This can be mitigated to some extent by weighting the sequences [118]. We will discuss weighting schemes in Section 4.3.3.3.

2. **The center of gravity measure (CG-score).** This measure scores a column with respect to the center of gravity (or the center of mass) of the set of residues in that column. That is, the amino acid that maximizes the similarity with all other residues in the column

$$\text{score}(\sigma_1, ..., \sigma_N) = \max_{a \in \mathcal{A}} \sum_{i=1}^{N} s(\sigma_i, a)$$

If the pairwise scoring matrix used is a log-odds matrix (see Section 3.5), and assuming all sequences are independent of each other, then this residue is also the residue the common ancestor most likely had at this position (Problem 4.2). However, the sequences in an MSA are related and hence the amino acid with the maximal score is not necessarily the most likely ancestor amino acid.

3. **The entropy measure.** Highly conserved columns are a strong indication of common ancestry. The entropy measure reflects the order observed within a set of measurements and is defined as

$$E(\sigma_1, ..., \sigma_N) = -\sum_{a \in \mathcal{A}} p(a) \log p(a)$$

where $p(a)$ is the empirical probability of the character a in the column[1]. Note that the entropy of a completely conserved column is zero. To use this measure with the dynamic programming algorithm described above, the entropy function has to be transformed first to a similarity function, for example by subtracting the entropy from the maximal possible value $\log |\mathcal{A}|$, obtained for a uniform distribution.

4.2 Classical heuristic methods

Despite its algorithmic simplicity, the dynamic programming algorithm in N dimensions suffers from severe computational issues. For N sequences of

[1]Notation-wise, we should use an uppercase letter $P(a)$ to denote a probability mass function over a discrete variable (as in Section 3.12), but in this and subsequent chapters we will use lowercase notation $p(a)$ to denote the probability mass function over amino acids (see also Footnote 3.11 on page 92.)

length n the total number of cells in the matrix is n^N and therefore the space complexity is n^N (although more efficient implementations exist [117]). Since the score of each cell depends on up to $2^N - 1$ cells, the time complexity is $O(n^N \cdot 2^N \cdot f(N))$ where $f(N)$ is the time needed to compute the score of a column. This depends on the scoring function, with $f(N) = N(N-1)$ for the SP-score and $f(N) = |\mathcal{A}| \cdot N$ for the CG-score. For 50 protein sequences of 300 residues each this means up to $O(300^{50} \cdot 2^{50} \cdot 50^2) = 10^{142}$ calculations. Therefore, this algorithm is impractical for more than a few sequences. There are more efficient implementations that can reduce the search space and the number of computations in some cases (e.g. [117, 119, 120]) However, it has been proved that finding the optimal MSA of N sequences with the SP-score is an NP-hard[2] problem [121].

Because of the computational issues, many heuristic algorithms were developed for the MSA problem. In this section we review some of the well-known ones. These algorithms are not guaranteed to find the optimal solution, per the definition of Section 4.1, but might produce a good approximation (in Section 4.6 we will see that the definition itself is worth revisiting).

4.2.1 Star alignment

This simple heuristic builds the MSA from pairwise alignments of a fixed sequence (termed the *center sequence* or the *star sequence*) and all the other input sequences. The algorithm identifies first the star sequence and computes the pairwise alignments of that sequence with all other sequences (either in local or global mode, depending on the goal). The MSA is created by aggregating the alignments one by one based on the similarity scores, starting from the most similar one, and padding with gaps where necessary (see Figure 4.2). The resulting alignment has the property that its *pairwise* projections $\mathbf{A}_{ij}$ are optimal if i or j is the center sequence.

To select the center sequence, we can try each one and pick the sequence that results in an MSA with the highest score using one of the scoring functions described above. Another possibility is to pick the sequence that is most similar to all other sequences in the group. That is, compute all pairwise scores and pick the sequence c that maximizes $\sum_{i \neq c} S(\mathbf{a}_i, \mathbf{a}_c)$.

The star alignment algorithm is much more efficient than the dynamic programming algorithm, and the main step of computing pairwise alignments with the star sequence takes only $O(Nn^2)$. Additional $O(N^2 n)$ is needed to compute the final score. If we start by computing all pairwise similarities (to identify the sequence that has the highest average similarity score), then the

[2]The theory of computation divides algorithmic problems into several classes, based on their computational complexity. **NP-hard** problems are problems for which there is no known polynomial time algorithm and are unlikely to have one, since the existence of such an algorithm for any one of them would entail that there is a polynomial time algorithm to many other problems that are considered difficult.

The pairwise alignments

```
Pair 1               Pair 2               Pair 3
MEDL-GENTMVLSTL      MEDLGENTM-VLSTL      MEDLGENT-MVLSTL
MEDLKGENTMVLSTL      MEDLGENTTIVLSSL      MED--ENTFIVLSTL
```

Constructing the MSA

```
MSA - step 1         MSA - step 2         MSA - step 3
MEDL-GENTMVLSTL      MEDL-GENTM-VLSTL     MEDL-GENT-M-VLSTL
MEDLKGENTMVLSTL      MEDLKGENTM-VLSTL     MEDLKGENT-M-VLSTL
                     MEDL-GENTTIVLSSL     MEDL-GENT-TIVLSSL
                                          MED---ENTFI-VLSTL
```

FIGURE 4.2: The star alignment algorithm. Three pairwise alignments with respect to the same center sequence and the construction of an MSA from these pairs.

time complexity would increase to $O(N^2n^2)$, which is still significantly faster than DP in N dimensions. Moreover, it has been shown that the score of the MSA generated by this procedure is no less than half the optimal alignment with the SP-score and a scoring function that satisfies the triangle inequality [103, p. 349]. However the algorithm explores a very small fraction of the search space, and therefore the resulting MSA might be a poor approximation of the correct MSA. It also suffers from a major drawback referred to as "once a gap always a gap", meaning, once a pairwise alignment that was added to the MSA introduced a gap, all other sequences will have a gap in that same position, precluding the possibility to fix possible errors that were introduced early on (see Figure 4.2). Therefore, this procedure might generate an inaccurate alignment.

4.2.2 Tree alignment

Tree alignment is another heuristic approach based on phylogeny, assuming we have an evolutionary tree for the input sequences (Figure 4.3). The tree alignment method computes the overall similarity based only on the pairwise alignments along tree edges. In other words, we do not need to consider all pairs the way we did with the dynamic programming algorithm and the SP scoring function. Given a tree $T = (\mathbf{V}, \mathbf{E})$, where $\mathbf{V}$ is the set of nodes and $\mathbf{E}$ is the set of edges, the algorithm works as follows:

1. Assign sequences to internal nodes that will minimize distances along graph edges (Figure 4.3).

2. Compute the weight of all edges in the tree. The weight of an edge $e = (i, j)$ that connects sequences $\mathbf{a}_i$ and $\mathbf{a}_j$ is the similarity score $S(\mathbf{a}_i, \mathbf{a}_j)$.

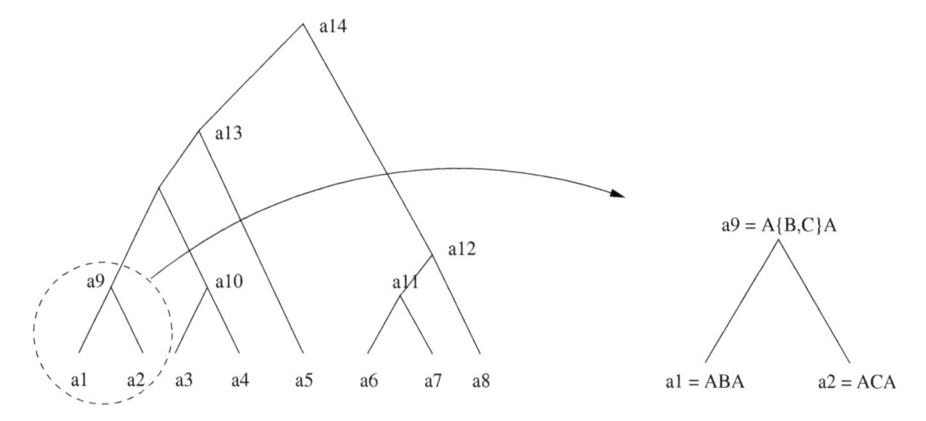

FIGURE 4.3: **Left:** A phylogenetic tree over $N = 8$ sequences. **Right:** Assignment of sequences to internal nodes.

Note that $\mathbf{a}_i$ and $\mathbf{a}_j$ can be either input sequences or ancestor (internal) sequences.

3. The sum of all these weights is the score of the MSA under the assignment of sequences to the interior nodes

$$S(\mathbf{A}) = \sum_{(i,j)\in\mathbf{E}} S(\mathbf{a}_i, \mathbf{a}_j)$$

This approach for MSA requires that the tree is given; however, to determine the evolutionary tree over a set of organisms an MSA is often needed, which is what we are trying to build in the first place. Even if a tree was established using a different dataset (typically based on highly conserved proteins, such as ribosomal proteins), we still need to determine the ancestor sequences. Finding the optimal assignment of sequences to internal nodes is an NP-hard problem [122], which deems this approach impractical for real datasets that contain more than a few proteins. A popular heuristic approach that is inspired by the idea of phylogenetic trees is to apply hierarchical

> **Pairwise Clustering** (nearest neighbor)
>
> Input: N samples with pairwise similarities $S(i,j)$ $(1 \leq i,j \leq N)$. Denote by k the number of clusters
>
> - Initialize $k = N$ and assign each sample i to its own cluster $C_i = \{i\}$
> - Loop:
>   - Find the nearest (most similar) pair of clusters, C_i and C_j
>
> $$S(C_i, C_j) = \max_{x,y} S(C_x, C_y)$$
>
>   - Merge C_i and C_j and decrement k by one.
> - Terminate if $k = 1$.
>
> The similarity of clusters is defined based on the closest pair of sequences
>
> $$S(C_i, C_j) = \max_{a\in C_i, b\in C_j} S(a,b)$$
>
> (for other alternatives see Chapter 10).

clustering to the pairwise similarity scores of the input sequences and build a **guide tree**. A basic hierarchical clustering algorithm is described on the right. If the input is a set of pairwise *distances* then the max operator is replaced with the min operator[3].

The guide tree determines the order according to which the sequences are aligned. Starting from the leaf nodes that correspond to the input sequences and proceeding according to the clustering order, the sequences are aligned to each other and aggregated as with the star alignment method. When adding a sequence to an alignment associated with an internal node, or when aggregating the alignments associated with two internal nodes, we can use the closest pair of sequences, one from each group of sequences, to decide on how to merge to sub-alignments. Note that the star alignment method can also be casted as a tree alignment algorithm, using a very simple star-shaped tree.

```
Cluster 1          Cluster 2          MSA
MEDL-GENTMVLSTL    MEDLGENTTIVLSSL    MEDL-GENTM-VLSTL
MEDLKGENTMVLSTL    MED--ENTFIVLSTL    MEDLKGENTM-VLSTL
                                      MEDL-GENTTIVLSSL
                                      MED---ENTFIVLSTL
```

FIGURE 4.4: **The tree alignment algorithm.** The example involves the same four sequences as in Figure 4.2. The sequences are clustered into two clusters, aligned and then aggregated based on the similarity of the first and third sequence. Note that the final MSA is different from the MSA that is generated with the star alignment method (Figure 4.2).

The tree alignment approach is an alternative to the star method that can produce better results (unless the tree is star shaped) since it considers the most similar pairs or groups first, leading to more accurate sub-alignments (see Figure 4.4). However, it too suffers from the same problem of "once a gap always a gap". To discuss more advanced algorithms we need first to discuss methods for MSA representation.

[3] It should be noted that this procedure does not necessarily recover the true phylogenetic tree of the sequences. This could happen, for example, when the closest leaves are not direct neighbors in the tree as a result of different molecular clocks (see Problem 4.11). An alternative is the **neighbor-joining** algorithm, which addresses the problem of different molecular clocks and can reconstruct the true evolutionary tree from pairwise distances under some additivity conditions (see [123]).

4.3 MSA representation and scoring

The methods described so far work by constructing the MSA from *pre-computed* pairwise alignments. Therefore, they are prone to errors introduced early on since they do not allow one to revisit the alignments. More advanced algorithms for MSA do exist; however, they require methods to represent and score MSAs such that we can compare them to sequences or to each other. This section discusses popular representations of MSAs and MSA-based scoring functions.

4.3.1 The consensus sequence of an MSA

The simplest representation of an MSA is the consensus sequence. There are several possible definitions for the consensus sequence. The consensus sequence can be defined as the string of the most common characters in each column of the multiple alignment, as in Figure 4.5. Since amino acids are distributed unevenly, one is advised to consider the background probability of characters $\{p_0(a)\}$, such that if $p(a) > p(b)$ but $\frac{p(a)}{p_0(a)} < \frac{p(b)}{p_0(b)}$ then b should be picked.

```
          MEDL-GENTM-VLSTL
          MEDLKGENTM-VLSTL
          MEDL-GENTTIVLSSL
          MED---ENTFIVLSTL

Consensus MEDL-GENTMIVLSTL

Position  1234567890123456
```

FIGURE 4.5: **The consensus sequence.**

Another possible definition for the consensus sequence is the string of characters that minimize the distance (or maximize similarity) to all other characters in each column

$$\mathbf{a}_{consensus}(j) = \max_{a \in \mathcal{A}} \sum_{i=1}^{N} s(a, \mathbf{a}_i(j))$$

The justification is the same evolutionary justification behind the center of gravity (CG) score (Section 4.1.1). Note that with this definition, the optimal MSA generated with the DP algorithm and the CG-score function is such that the similarity score is maximal with respect to this consensus sequence.

Depending on the pairwise scoring function used, the consensus sequence obtained with this method might differ from the one obtained based on the most common characters (Problem 4.6). Note also that the consensus sequence, based on either method, might not be in the set of input sequences $\{\mathbf{a}_1, ..., \mathbf{a}_N\}$ (as in Figure 4.5).

A similar yet different type of consensus sequence is the Steiner string [103]. This is the string $\mathbf{a}_s$ that maximizes $\sum_{i=1}^{N} S(\mathbf{a}_i, \mathbf{a}_s)$. As opposed to the previous approach, where the characters are picked individually for each column, the string $\mathbf{a}_s$ is determined based on its global similarity score with the other strings. This string does not need to be in the input set, and finding it can be computationally expensive since it might require an extensive search through the sequence space. However, it can be approximated by the center sequence, which is the sequence $\mathbf{a}_c$ in the input set that maximizes $\sum_{i=1}^{N} S(\mathbf{a}_i, \mathbf{a}_c)$. If $S(\mathbf{a}_i, \mathbf{a}_c)$ is the edit distance then this choice results in a score that is at least half of the best score obtained with $\mathbf{a}_s$ (see [103, p. 351]).

Once the consensus sequence has been determined, according to any of these definitions, it can be used when comparing a sequence to an MSA (as in the the star alignment method) or to compare two MSAs (as in the tree alignment method). This approach to MSA representation is simple and allows us to use the algorithms we have introduced so far without having to change anything. However, it is also limited since the compact representation of an MSA as a single string does not provide us with any information on the distribution of amino acids in each position, and once condensed into a consensus sequence it is not possible to distinguish between completely conserved positions of the MSA and heterogeneous positions (compare positions 1 and 10 of the MSA in Figure 4.5). The next sections will discuss other representations that preserve more of the original information stored in an MSA.

4.3.2 Regular expressions

The language of regular expressions is an efficient tool for representing groups of strings that share certain features. This is especially useful when describing relatively short and well-defined patterns. Regular expressions are basically strings over an alphabet that contains the original alphabet the strings are generated from augmented with special symbols that are used to represent classes of characters. That is, given an alphabet $\mathcal{A}$, a regular expression is a string over the alphabet $\mathcal{A}' = \mathcal{A} \cup \{\neg, *, (,), [,]\}$. A character $a \in \mathcal{A}'$ that is also in $\mathcal{A}$ represents the corresponding letter, while the character $*$ represents *any* character in $\mathcal{A}$. For example, the regular expression $A * GA$ over the DNA alphabet $\mathcal{A} = \{A, C, G, T\}$ represents the set of all strings that start with A followed by any nucleotide and then followed by GA. Brackets are used to represent classes of characters. For example, $A[A, C]GA$ represents the set of all strings that start with A followed by either A or C and then followed by GA. The expression $A[\neg A]GA$ represents the set of all strings that start with A followed by any nucleotide besides A and then followed by GA. It is possible to add additional characters to further simplify the representation of classes of strings. For example, numbers in parentheses indicate repetitions. The expression $A(4)$ represents the string $AAAA$ and the expression $A(2,5)$ represents anywhere between two and five A's.

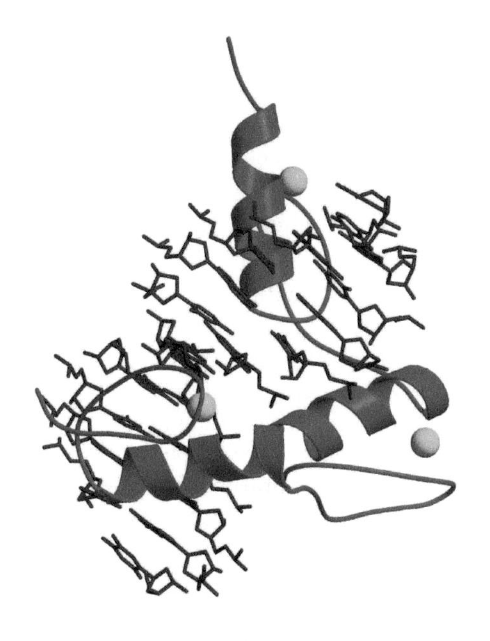

FIGURE 4.6: The zinc finger domain. A three-dimensional illustration of a zinc finger domain bound to DNA (PDB 1a1l). Zinc finger domains are between 25 and 30 residues long. By binding to the two cysteines and the two histidines at the ends of the domain, the zinc ions (light gray balls) stabilize the domain.

Regular expressions are very effective when representing short MSAs such as the one in Figure 4.5. The regular expression describing this MSA is

$$MEDL(0,1)KG(0,1)ENT[M,T,F]I(0,1)VLS[T,S]L$$

The PROSITE database [124] contains more than 1000 such patterns, many of which correspond to structural motifs or functional sites in proteins, such as catalytic or interaction sites[4]. For example, the pattern

$$C - x(2,4) - C - x(3) - [LIVMFYWC] - x(8) - H - x(3,5) - H$$

describes the zinc finger C2H2 type domain signature, where the two cysteines and the two histidines are zinc ligands (see Figure 4.6). Dashes were added

[4]Patterns in the PROSITE dictionary vary greatly in length and range from 3 to 400 amino acids. Short patterns typically represent short motifs (such as glycosylation sites, subcellular localization signals, attachment sites, etc.) which are common to many different protein families. Longer patterns are usually characteristic of specific families. Some families are characterized by more than one pattern. For many families no consensus or signature pattern can be defined since the sequences have significantly diverged. Some of these families are described in PROSITE by means of a profile, as described in the next section.

and commas were omitted, as in PROSITE, to improve clarity. The character x is used instead of $*$, to represent any character in the alphabet.

While regular expressions can be searched efficiently against a sequence database to detect occurrences of the pattern (see Problem 4.7), they are limited in the sense that sequences have to match the pattern exactly. Although it is possible to add wildcards and classes of characters, this affects the selectivity of the model as well as the search speed, and hence they are usually confined to describing short patterns and are less effective as a representation of an MSA that corresponds to a protein domain family or to a protein family. More importantly, although regular expressions are more informative than consensus sequences they still miss an important element, which is the preference of each position in the MSA to specific amino acids. For example, a column with 99 alanines and a single cysteine would be described exactly the same as a column containing 50 alanines and 50 cysteines, despite being more conserved. A more effective generalization of regular expression is discussed next.

4.3.3 Profiles and position-dependent scores

A profile is a concise, efficient and powerful representation of an MSA. It consists of the distributions of amino acids at each position along the MSA and can be used to compute the probability of other sequences belonging to the same family. Usually profiles are converted to position-specific scoring matrices and are used to search sequence databases. Since they reflect the unique properties of each position, they are capable of detecting subtle similarities between proteins that are often missed by means of pairwise comparisons. Profiles can also be compared to each other, to detect relationships between remotely related protein families. In the next sections we will discuss these applications and other issues related to profile analysis.

4.3.3.1 Generating a profile

Profiles are usually obtained directly from MSAs. Given a multiple alignment with N sequences, we start by computing the frequency $N_j(a)$ of each amino acid at each position j along the multiple alignment. These counts are normalized and transformed to probabilities

$$\hat{p}_j(a) = \frac{N_j(a)}{N}$$

so that a probability distribution over amino acids is associated with each position (see Figure 4.7). As we will see in Section 5.2.2, when formulated as a problem of finding an optimal zero-order Markov model of the MSA sequences, these probabilities correspond to the maximum likelihood estimators of the model's parameters.

Since some amino acids might not be represented in the j-th position of the MSA, we add M pseudo-counts, which are distributed as $p_j^{pseudo}(a)$ (more on this in the next section). The final profile probabilities are defined as

$$p_j(a) = \frac{N \cdot \hat{p}_j(a) + M \cdot p_j^{pseudo}(a)}{N + M} \tag{4.1}$$

Once a distribution $\mathbf{p}_j = \{p_j(a)\}_{a \in \mathcal{A}}$ is compiled for each position, a profile can be viewed as a series or a sequence of probability distributions $\mathbf{P} = \mathbf{p}_1\mathbf{p}_2 ... \mathbf{p}_L$ where L is the length of the multiple alignment. One can think of a profile as a $20 \times L$ matrix. Each probability distribution is one column in the matrix representation and hence is called a **profile column**.

MSA

A	B	C
A	A	A
A	$-$	C
$\Downarrow$	$\Downarrow$	$\Downarrow$
$\hat{p}_1(A) = 1$	$\hat{p}_2(A) = 1/2$	$\hat{p}_3(A) = 1/3$
$\hat{p}_1(B) = 0$	$\hat{p}_2(B) = 1/2$	$\hat{p}_3(B) = 0$
$\hat{p}_1(C) = 0$	$\hat{p}_2(C) = 0$	$\hat{p}_3(C) = 2/3$

FIGURE 4.7: **Generating a profile from an MSA.** A simple MSA of three sequences over the alphabet $\mathcal{A} = \{A, B, C\}$.

Some applications add an entry in the profile for gaps or extend it and include secondary structure information either from known protein structures or predicted ones. We will discuss these extensions in more detail in Section 8.4 of Chapter 8.

4.3.3.2 Pseudo-counts

When the MSA is built from a small number of sequences, we often encounter positions where only one or a few amino acids occur. If a profile is to be created (trained) from the empirical counts, then any amino acid that does not occur in a certain position will be assigned zero probability. If the profile is used for prediction, then any sequence that has in that position an amino acid that has not been observed in the MSA will be assigned a zero probability, even if it is highly similar to the sequences of the MSA in other positions. To prevent this from happening, a good practice is to add M pseudo-counts to the empirical counts, based on some prior knowledge, so as to make up for observations that are missed due to the finite datasets used for training. Adding pseudo-counts is necessary not only to avoid zero probabilities in small sample sets but also to improve the generalization and prediction power, since the dataset used to derive the profile might be biased. The addition of pseudo-counts is often referred to as **smoothing**.

The total number of pseudo-counts we add, M, determines the weight we put on the prior knowledge vs. on the empirical counts. If $M >> N$ then we believe our prior knowledge more than the training data and vice versa. There are several ad hoc approaches to setting M. The general strategy is to adjust the number of pseudo-counts to the number of empirical counts such that the weight of the prior knowledge is close to the weight of the training data for small N, while for large N the weight of the training data dominates that of the prior knowledge. For example, if M is set to 1 then the relative weight of the pseudo-counts diminishes rapidly (Figure 4.8). If M is set to $|\mathcal{A}|$ then the pseudo-counts actually dominate over the empirical counts as long as the number of empirical counts is smaller than the size of the alphabet. Alternatively, we can set $M = \min(|\mathcal{A}|, N)$. In this case the pseudo-counts weigh equally to the empirical counts as long as the number of training samples $< |\mathcal{A}|$, and their weight decreases from that point on. A popular approach is to set $M = \sqrt{N}$. In this case the decline in the relative weight of the pseudo-counts is more moderate. Yet another approach is to set the number of pseudo-counts as a function $M = f(R)$ of the diversity index R (the number of different amino acids observed in a position). This has been suggested in [125] where the linear function $M = 5R$ is used. The last two approaches seem to perform better than the others in performance evaluation tests [126].

The method used to divide the pseudo-counts among the different amino acids determines the type of prior knowledge we assume over the distribution of amino acids, and there are several common approaches.

Uniform prior. The simplest solution is to divide M equally and add $M/20$ pseudo-counts for each amino acid. That is

$$p_j^{pseudo}(a) = 1/20$$

This method of uniform prior is also called Laplace's method. In the example of Figure 4.7, adding $M = |\mathcal{A}| = 3$ pseudo-counts distributed uniformly would result in the following probabilities in the first profile column:

$$p_1(A) = \frac{3+1}{6} = \frac{4}{6} \qquad p_1(B) = \frac{1}{6} \qquad p_1(C) = \frac{1}{6}$$

Fixed prior. Another solution is to divide the M pseudo-counts based on the background probabilities $P_0(a)$ as defined from a large dataset of protein sequences

$$p_j^{pseudo}(a) = P_0(a)$$

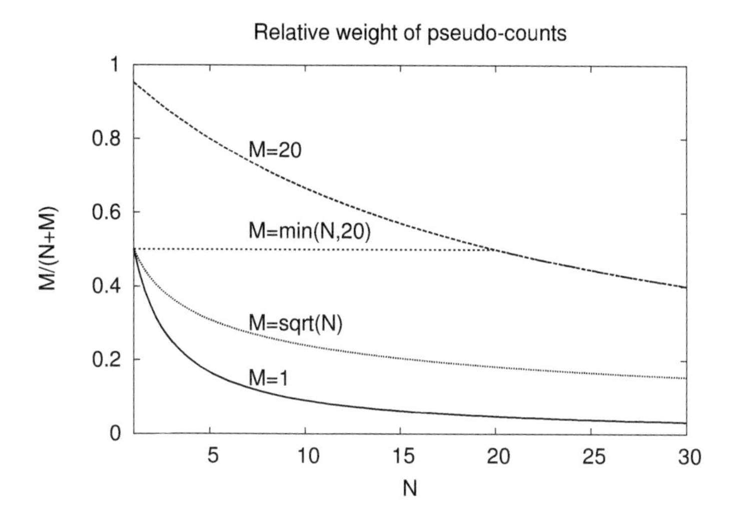

FIGURE 4.8: **The relative weight of pseudo-counts.** Here the size of the alphabet $|\mathcal{A}|$ is set to 20, the size of the amino acid alphabet.

This method, also called **Dirichlet prior**, is better than the uniform prior but is independent of the position in the MSA and assumes the same prior over all positions.

Position-dependent prior. This approach uses more of the information in the training data to derive position-specific pseudo-counts that reflect the distribution of amino acids in each position. The pseudo-counts are divided based on the empirical distribution of amino acids *and* the transition probabilities as obtained from a matrix like PAM or BLOSUM (see Section 3.5.2 in Chapter 3). Specifically, we define

$$p_j^{pseudo}(a) = \sum_{b \in \mathcal{A}} \hat{p}_j(b) \cdot p(b \to a)$$

How shall we obtain transition probabilities from a general scoring matrix? If $s(a, b)$ was derived directly from transition probabilities then the transition probabilities can be recovered from the scoring matrix

$$p(a \to b) = p_a p_b e^{s(a,b)}$$

Otherwise, the scoring matrix induces certain transition probabilities in the highest scoring matches (see Section 3.5.3) and

$$p(a \to b) = p_a p_b e^{\lambda s(a,b)}$$

where λ is a scaling factor which guarantees $\sum_a \sum_b s(a, b) = 1$.

The Dirichlet mixture approach. With a single Dirichlet prior the prior knowledge is limited to one type of pseudo-counts vector. However, there are many possible different structural environments with different priors. The position-dependent prior addresses this problem, but in order to get a good estimate for the parameters of the different environments we need many samples. This might be a problem with MSAs of a relatively small number of sequences.

A different approach that works better with small datasets is the mixture of Dirichlet priors [127]: a mixture of different priors that correspond to different structural environments (such as core positions vs. surface positions, different secondary structure elements, etc). In their paper describing this method, Sjölander and colleagues clustered a large population of profile columns, obtained from high-quality MSAs, to determine the environments and estimate their prior.

The method assumes that there are K different environments. Each environment is represented by a "typical" frequency vector $\vec{\alpha}_k$ $(k = 1..K)$, which induces a certain prior on the space of all possible frequency vectors. Frequency vectors that are similar to the typical vectors are very likely to appear in similar environments. Different vectors $\vec{\alpha}_k$ might add up to a different number of total counts M_k. As M_K increases, the distribution associated with the k-th environment narrows and the likelihood of vectors that deviate from the typical vector drops rapidly. The mixture model combines the contributions of all the environments and the total probability of observing a certain frequency vector (a profile column before normalization) is a sum over all environments, each one weighted with its own prior probability. Typical vectors that correspond to more common environments are assigned a higher weight.

Formally, each vector $\vec{\alpha}_k$ is the parameter vector of one Dirichlet distribution. The Dirichlet distribution is a popular prior distribution for multinomial distributions. It induces a prior over parameter vectors $\vec{\theta}$ that satisfy $0 < \theta_i < 1$ and $\sum_i \theta_i = 1$ (such as the parameters of multinomial distribution). This is the type of probability vector we have in profile columns. A Dirichlet distribution with a hyper-parameter vector $\vec{\alpha} = (\alpha_1, \alpha_2, ..., \alpha_m)$ is defined as

$$Dir(\vec{\theta}|\vec{\alpha}) = \frac{\prod_i \theta_i^{\alpha_i - 1}}{Z(\vec{\alpha})} \tag{4.2}$$

where $Z(\vec{\alpha})$ is a normalization factor over all possible vectors $\vec{\theta}$, given by the m-dimensional integral

$$Z(\vec{\alpha}) = \int_{\vec{\theta}} \prod_{i=1}^{m} \theta_i^{\alpha_i - 1} d\vec{\theta}$$

There is a closed-form solution for this integral, such that

$$Z(\vec{\alpha}) = \int_{\vec{\theta}} \prod_{i=1}^{m} \theta_i^{\alpha_i - 1} d\vec{\theta} = \frac{\prod_i \Gamma(\alpha_i)}{\Gamma(\sum_i \alpha_i)} \tag{4.3}$$

where the Gamma function is defined as $\Gamma(n) = \int_0^\infty e^{-x} x^{n-1} dx$ and satisfies $\Gamma(n+1) = n\Gamma(n)$. For non-negative integers $\Gamma(n+1) = n!$, however, the frequency vectors might not be integers, for example, if a weighting scheme is used (see next section). The mean of the Dirichlet distribution is the vector $\vec{\theta}' = (\theta'_1, ..., \theta'_m)$ such that $\theta'_i = \frac{\alpha_i}{\sum_i \alpha_i}$, and the distribution obtains its peak at the mean.

The Dirichlet mixture approach sums the contributions of all environments weighted with the probability to observe each environment in each position. The motivation is that if the priors can be estimated reliably, then just a few observations are enough to determine the prototype distribution that generated a specific profile column. The method uses the original frequency vectors $\vec{N_j} = (N_{j1}, ..., N_{jm})$, from which the empirical probabilities are derived[5], and the final profile probabilities in position j are defined as

$$p_j(a) = \sum_{k=1}^{K} \left(\frac{N_{ja} + \alpha_{ka}}{N_j + M_k} \right) \cdot p(k|\vec{N_j}) \tag{4.4}$$

where $N_j = \sum_{a \in \mathcal{A}} N_{ja}$ is the number of characters in the j position (excluding gaps), $M_k = \sum_{a \in \mathcal{A}} \alpha_{ka}$ and $p(k|\vec{N_j})$ is the posterior probability of environment k given the empirical counts $\vec{N_j}$. Note that the term in the parentheses is the same term we had in Equation 4.1. As we will see in Section 15.4, Equation 4.1 with pseudo-counts based on Dirichlet prior (fixed prior) is the maximum a posteriori (MAP) estimator of the profile parameters, with the Dirichlet distribution as a prior over the parameters. The Dirichlet mixture introduces one such term for each environment, weighted with its posterior probability (Equation 4.4). The posterior probability of each environment is computed using Bayes' formula

$$P(k|\vec{N_j}) = \frac{P(\vec{N_j}|k) P_0(k)}{\sum_{k'} P(\vec{N_j}|k') P_0(k')}$$

where $P_0(k)$ is the prior probability to observe the k-th environment (estimated based on its relative frequency in the training dataset). Computing the likelihood $P(\vec{N_j}|k)$ is more involved. The Dirichlet density function is a probability function on the sample space of normalized vectors $\vec{\theta}$, while $\vec{N_j}$ is

[5]Notation wise, N_{ja} is equivalent to $N_j(a)$ we used before.

a vector of counts. The likelihood $P(\vec{N}_j|k)$ can be computed by integrating over all possible normalized vectors $\vec{\theta}$. That is,

$$P(\vec{N}_j|k) = \int_{\vec{\theta}} P(\vec{N}_j|\vec{\theta})P(\vec{\theta}|k)d\vec{\theta}$$

The term $P(\vec{N}_j|\vec{\theta})$ is simply the probability of observing the actual frequencies given a multinomial distribution with a parameter vector $\vec{\theta}$. That is,

$$P(\vec{N}_j|\vec{\theta}) = \binom{N_j}{N_{j1}...N_{jm}}\theta_1^{N_{j1}}...\theta_m^{N_{jm}} = \frac{N_j!}{\prod_i N_{ji}!}\prod_{i=1}^{m}\theta_i^{N_{ji}}$$

The second term $P(\vec{\theta}|k)$ is given by the Dirichlet density of the k-th environment with hyper-parameter $\vec{\alpha}_k$

$$P(\vec{\theta}|k) = \frac{\prod_i \theta_i^{\alpha_{ki}-1}}{Z(\vec{\alpha}_k)} = \prod_i \theta_i^{\alpha_{ki}-1}\frac{\Gamma(\sum_i \alpha_{ki})}{\prod_i \Gamma(\alpha_{ki})}$$

Combined together, we get

$$\begin{aligned} P(\vec{N}_j|k) &= \int_{\vec{\theta}} \frac{N_j!}{\prod_i N_{ji}!}\prod_i \theta_i^{N_{ji}}\prod_i \theta_i^{\alpha_{ki}-1}\frac{\Gamma(\sum_i \alpha_{ki})}{\prod_i \Gamma(\alpha_{ki})}d\vec{\theta} \\ &= \frac{N_j!}{\prod_i N_{ji}!}\frac{\Gamma(\sum_i \alpha_{ki})}{\prod_i \Gamma(\alpha_{ki})}\int_{\vec{\theta}}\prod_i \theta_i^{N_{ji}+\alpha_{ki}-1}d\vec{\theta} \end{aligned}$$

and by Equation 4.3, the integral is given by

$$\frac{\prod_i \Gamma(N_{ji}+\alpha_{ki})}{\Gamma(\sum_i N_{ji}+\alpha_{ki})}$$

Therefore, all together we have a closed-form solution for the likelihood [93]

$$P(\vec{N}_j|k) = \frac{N_j!}{\prod_i N_{ji}!}\frac{\Gamma(\sum_i \alpha_{ki})}{\prod_i \Gamma(\alpha_{ki})}\frac{\prod_i \Gamma(N_{ji}+\alpha_{ki})}{\Gamma(\sum_i N_{ji}+\alpha_{ki})}$$

and we can compute the profile probabilities, including the pseudo-counts, per Equation 4.4. In Chapter 15 we will see that the Dirichlet distribution plays an important role in other contexts as well.

4.3.3.3 Weighting sequences

When training models such as profiles, HMMs, or VMMs (see Chapter 6) there are two issues we should keep in mind. One is that the datasets we are using for training might be biased. For example, the process of building a model for a protein family usually starts by comparing a member of that family with a sequence database to collect related sequences. The set of sequences detected clearly depends on the sequence database searched. No sequence database is complete and therefore it is likely that our dataset is skewed and does not represent the protein family accurately. For example, since sequencing projects focus on a very small subset of all organisms, our dataset might miss instances of the family in a class or even a phylum of organisms. On the other hand it might contain many close variations of a specific sequence as a result of special interest in the function of that gene and the effect of point mutations on the function. Furthermore, owing to the nature of sequence comparison algorithms, remote homologs of the seed sequence are usually underrepresented in these alignments (this can be mitigated to some extent by iterative algorithms such as PSI-BLAST, described in Section 4.4.1). Thus our model might represent a subfamily rather than the whole family. Another issue is that the sequences in the dataset are not independent of each other. Clearly, they are related through evolution, and training models while ignoring the dependency may result in a poor approximation of the source distribution.

To handle these issues, associating different weights with the individual sequences before training the model is advised. Our discussion so far assumed all sequences in the dataset are weighted equally. The introduction of weights serves to compensate for a possible bias in the training dataset and decrease the contributions of closely related sequences. In some cases, a bias cannot be easily fixed. For example, if our dataset lacks samples from certain organisms it is not clear how to extend it or define weights so as to make up for the missing observations. Most weighting schemes address the most common bias, which is due to the prevalence of close homologs. Identical sequences are usually eliminated, leaving only one representative in the dataset, since the other sequences do not provide additional information. With non-identical sequences weighting is more involved, and quite a few different weighting schemes were proposed over the years. The existing approaches are all heuristics, which work by decreasing the weight of highly similar sequences and increasing the weight of highly diverged sequences. The weights are usually normalized such that the sum of all weights is one.

Some methods start by building a tree. For example, the method of [128] uses the percentage of mutations in pairwise alignments as a measure of divergence $d(\mathbf{a}, \mathbf{b})$ between sequences. Sequences are merged into a tree using a procedure similar to the one described in Section 4.2.2 (the actual procedure is based on the neighbor-joining algorithm, see Footnote 4.3 on page 112). The divergence level of each internal node in the tree is defined based on its descendant sequences. That is, if node v merges cluster i with cluster j then its divergence level $h(v)$ is defined as half the distance between the two clusters (assuming equal divergence rate from the ancestor sequence). If the distance between clusters is defined as the minimum divergence over all possible sequence pairs, one from each cluster, then

$$h(v) = \frac{1}{2}d(C_i, C_j) = \min_{a \in C_i, b \in C_j} \frac{d(\mathbf{a}, \mathbf{b})}{2}$$

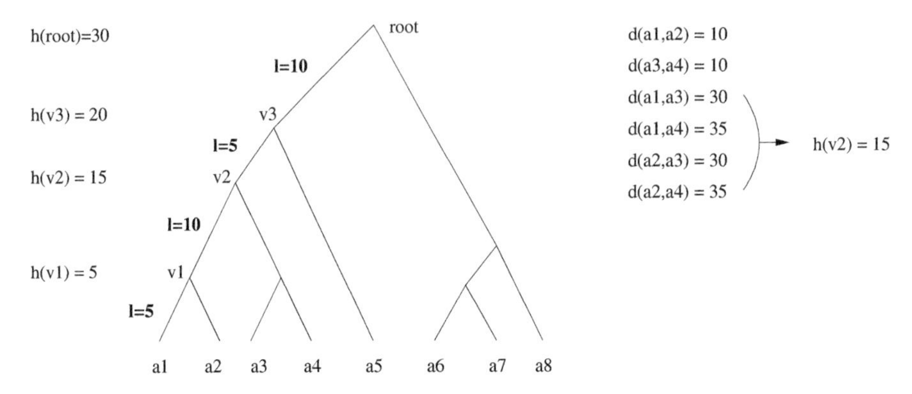

FIGURE 4.9: Assigning weights based on a tree. A subset of the pairwise distances is shown on the right. The divergence level of each internal node is defined based on these distances. For example, $h(v_2) = \min_{\mathbf{a} \in \{a_1, a_2\}, \mathbf{b} \in \{a_3, a_4\}} \frac{d(\mathbf{a},\mathbf{b})}{2} = 15$.

with leaf nodes assigned a divergence of 0 (Figure 4.9). The length of an edge connecting nodes v_1 and v_2 (where v_2 is higher up in the tree) is defined as

$$l_{(v_1,v_2)} = h(v_2) - h(v_1)$$

Note that the minimal distance is bound to increase as we move farther away from the leaf nodes; therefore branch lengths are always positive. Given the tree, the weight of each sequence is defined based on the length of the path leading from the root to the leaf node associated with that sequence and on the number of sequences sharing each edge along the path. That is, if there are k edges leading from the root to sequence $\mathbf{a}$ and edge i is shared by n_i sequences then the total weight of $\mathbf{a}$ is defined as

$$w(\mathbf{a}) = \sum_{i=1}^{k} \frac{l_i}{n_i}$$

For example, in Figure 4.9 the weight of sequence $\mathbf{a}_1$ is

$$w(\mathbf{a}_1) = l_{(a_1,v_1)} + l_{(v_1,v_2)}/2 + l_{(v_2,v_3)}/4 + l_{(v_3,root)}/5 = 5 + 10/2 + 5/4 + 10/5$$

A simple alternative method, which does not require building a tree, is to assign weights based on the average distance of each sequence from all other sequences or the profile (which can be viewed as the center of gravity of the sequence set [129]). Sequences with higher average distance are considered to be more remote and are assigned a higher weight. That is, the weight of sequence $\mathbf{a}$ is defined

$$w(\mathbf{a}) = \frac{\sum_{\mathbf{b}} d(\mathbf{a},\mathbf{b})}{\sum_{\mathbf{a}} \sum_{\mathbf{b}} d(\mathbf{a},\mathbf{b})}$$

Another interesting approach is the one by Luthy et al. [130]. This method defines the weight of each sequence in proportion to the volume of sequences in the sequence space that are similar to that sequence more than any other sequence in the dataset. Thus, the sequence space is essentially divided into cells following a Voronoi diagram with the dataset sequences being the reference set (the Voronoi sites). To approximate the volume of sequences within each cell the algorithm generates random sequences in the neighborhood of the input sequences and assigns each to the closest sequence. Specifically, let $\mathcal{A}_j$ denote the set of amino acids observed at position j of the MSA (including the gap character, if observed in that position). The algorithm initializes $w(\mathbf{a}_i) = 0$ for each sequence and applies the following procedure iteratively

1. Assemble a random sequence $\mathbf{s}$ by choosing in each position j ($j = 1..L$) one of the amino acids in $\mathcal{A}_j$.
2. Compute the distance $d(\mathbf{s}, \mathbf{a}_i)$ between $\mathbf{s}$ and each one of the input sequences $\mathbf{a}_1, ..., \mathbf{a}_N$. The distance can be defined as the percentage of mutations in the alignment.
3. Let i denote the index of the closest sequence

$$i = \arg\min_i d(\mathbf{s}, \mathbf{a}_i)$$

If the closest sequence is unique then increment the weight of $\mathbf{a}_i$ by one

$$w(\mathbf{a}_i) = w(\mathbf{a}_i) + 1$$

If there are $k > 1$ input sequences that are equally close to $\mathbf{s}$ then increment the weight of each one by $1/k$

$$w(\mathbf{a}_i) = w(\mathbf{a}_i) + 1/k$$

By repeating this procedure many times (until the changes in the weights are negligible) the algorithm converges to weights that reflect the mass of sequences in the vicinity of each input sequence. Finally, the weights are normalized such that they sum to 1.

A different approach was proposed in [125] where the weight of each sequence is defined based on the amino-acid diversity in each position of the MSA and the frequency of the amino acid the sequence has in each position. That is,

$$w(\mathbf{a}) = \sum_{1 \le j \le L} \frac{1}{\text{diversity}(j) N_j(a_j)}$$

where L is the length of the MSA and $N_j(a_j)$ is the number of times the residue a_j appears in the j-th column of the MSA. With this scoring scheme, diverse positions contribute less to the total weight of $\mathbf{a}$ as well as positions where $\mathbf{a}$ has a frequent residue. Thus, similar sequences are down-weighted while highly diverged sequences are assigned higher weights. PSI-BLAST employs a similar weighting scheme that is tuned differently in each position, as described in Section 4.4.1. Position specific sequence weighting is also closely related to pseudo-counts methods, as described in the previous section.

With sequence weights (according to any of the methods described above), the profile is defined such that the frequency of each amino acid in each position is a function of the weights of the sequences that contain that residue. That is

$$N_j(b) = \sum_{\mathbf{a}_i} w(\mathbf{a}_i)\delta(b, a_{ij})$$

where $\delta(b, a_{ij})$ is the delta function such that $\delta(b, a_{ij}) = 1$ if $b = a_{ij}$ and zero otherwise. The profile probabilities are derived by normalizing the frequencies by the sum $\sum_b N_j(b)$ as before (Section 4.3.3.1).

Not all weighting methods apply to profiles. For example, the method of Altschul et al. [118] also computes weights based on a tree but is geared toward MSA algorithms that use the sum-of-pairs (SOP) measure (see Section 4.1.1). Unlike other methods that assign weights to individual sequences, this method applies to *pairs* of sequences. As with other methods, the assumption is that the closer the pair the less information it carries on the diversity of amino acids in a certain position. On the other hand, pairs of closely related sequences help to derive more accurate alignments. Hence the method works by reducing the dependency, and not necessarily the weight, of close homologs. Each pair of sequences is assigned a weight based on the path connecting them. Overlapping paths in the tree indicate dependency, and the corresponding pairs are down-weighted accordingly. Specifically, the weight of each pair is defined based on the overlap of its path with the paths of all other pairs. The higher the total overlap, the lower the weight. More rigorous methods to deal with the dependency were proposed in [131–133], combining phylogenetic trees to model the evolutionary relationships between sequences.

4.3.4 Position-specific scoring matrices

Profiles can be used to assign probabilities to sequences that are aligned to the corresponding MSA. For example, given a profile $\mathbf{P}$ of length L and a sequence $\mathbf{a}$ of length L that was aligned to the model, then its probability is

$$\text{Prob}(\mathbf{a}) = \prod_{i=1}^{L} p_i(a_i)$$

But how should we align a sequence to a model?

So far, when constructing MSAs we used the same scoring functions that were defined for pairwise alignments (Section 3.5.2). These scoring functions are indifferent to the positions of the amino acids compared. That is, the score of aligning a with b is fixed, regardless of their positions along the sequences. However, mutations are not equally probable along the sequence and not all regions mutate at the same rate. Some regions are functionally or structurally important, and consequently the effect of mutation in these regions can be drastic. They may create a nonfunctional protein or even prevent the molecule from folding into its native structure. Such mutations are unlikely to survive, and therefore these regions tend to be more evolutionarily conserved than other, less constrained regions (e.g., loops), which can significantly diverge. In this view, it seems logical to design scoring functions with position-dependent scores. With properly designed scores it is expected that the alignment algorithm would result in alignments that are more accurate biologically than with fixed scores.

Position-specific scoring matrices (**PSSMs**) are usually built from profiles to use the information in a group of related and aligned sequences. Since the distribution of amino acids in each position already reflects the structural preferences of that position, we can derive position-dependent scores for each amino acid or the penalty for a gap from these distributions. One possible approach is to convert profiles to PSSMs using the profile's probability distributions as well as the similarities of pairs of amino acids. For example, the score for aligning the amino acid b at position i of the profile can be defined as the **weighted sum**

$$s_i(b) = \sum_{a \in \mathcal{A}} p_i(a) s(a, b)$$

where $s(a, b)$ is the similarity of a and b based on a standard scoring matrix. Another approach, inspired by log-odds based scoring matrices such as PAM and BLOSUM, is to define the scores as the log-odds ratio

$$s_i(b) \sim \log \frac{p_i(b)}{P_0(b)}$$

where $P_0(b)$ is the background probability of b. The latter is the approach used in PSI-BLAST (see Section 4.4.1).

Scores can be further modified, for instance, by incorporating structural information. If a protein's structure is known, the secondary structure should be taken into account. In the absence of such data, general structural criteria, such as the propensities of amino acids for occurring in secondary structures versus loops can be taken into account. For example, the penalty for opening a gap in a regular secondary structure (helix, strand) can be increased, while the penalty for opening/inserting a gap in loop regions can be decreased (for an extended discussion see Section 8.4).

One of the objectives when building a PSSM is that the new scores can be used with the dynamic programming (DP) algorithm of Chapter 3. Recall that one of the conditions for local alignments is that the scoring function used has both positive and negative values (see Section 3.4.3.2). The scoring functions described above satisfy this requirement and can be used with a slightly modified version of the DP algorithm, as described next. In Chapter 6 we will discuss a different approach for aligning sequences with models.

4.3.4.1 Using PSSMs with the dynamic programming algorithm

The DP algorithm we discussed in Section 3.2 uses fixed scoring matrices between characters. How should we modify the algorithm to use a PSSM, which scores matches between a profile column and a character? The fact is that the algorithm can be easily adapted to account for position-specific scores. The adaptation actually assumes an even more general scenario in which the scores are defined between positions, where the elements compared at these positions can be either individual characters or more complex elements, such as two profile columns.

Let score(i, j) be the score for aligning the i-th position of $\mathbf{a} = a_1 a_2 ... a_n$ and j-th position of $\mathbf{b} = b_1 b_2 ... b_m$. This score can be computed based on the amino acids in these positions and other information about these positions such as their structural environment or a PSSM derived from the amino-acid/nucleotide profile, if given. Let $\alpha_i^1(k)$ and $\alpha_j^2(k)$ be the position-specific gap penalty functions associated with the i-th and j-th positions in the first and the second sequences, respectively. We use recursion as before to compute the similarity of $\mathbf{a}$ and $\mathbf{b}$, and define $S_{i,j}^{local}$ as

$$\max\left\{S_{i-1,j-1}^{local} + \text{score}(i,j),\ \max_{1\leq k\leq j}\{S_{i,j-k}^{local} - \alpha_{j-k}^2(k)\},\ \max_{1\leq l\leq i}\{S_{i-l,j}^{local} - \alpha_{i-l}^1(l)\},\ 0\right\}$$

The definition of score(i, j) depends on the entities compared. When comparing two sequences, score(i, j) is reduced to a standard scoring function $s(a, b)$ such as BLOSUM or PAM. When comparing a sequence with a profile we can set score$(i, j) = s_i(b_j)$ as in Section 4.3.4. When comparing a profile column with another profile column, a simple scoring function could be the **weighted sum** of pairwise similarities

$$\text{score}(i,j) = \sum_{a\in\mathcal{A}}\sum_{b\in\mathcal{A}} p_i(a)\cdot p_j(b)\cdot s(a,b)$$

However, there are more sophisticated methods for profile-profile comparison, and we elaborate on this topic next.

4.3.5 Profile-profile comparison

When generating MSAs progressively, as will be discussed in the next sections, we often need to compare two profiles that correspond to two intermediate MSAs. We also might be interested in comparing two profiles of two protein families or subfamilies in search of a subtle sequence similarity that might imply structural and functional similarity.

Comparing position i of one profile with position j of a second profile entails comparison of two distributions, $\mathbf{p}$ and $\mathbf{q}$, which correspond to two MSA columns. Here too, there are several scoring functions we can use. We already mentioned the weighted sum, which is a generalization of pairwise scoring functions, defined as

$$\text{score}(\mathbf{p}, \mathbf{q}) = \sum_{a \in \mathcal{A}} \sum_{b \in \mathcal{A}} p(a) \cdot q(b) \cdot s(a, b)$$

However, there are many other options we should consider. The profile columns can be thought of as normalized vectors in $\mathbb{R}^m$ where $m = |\mathcal{A}|$. Given two distributions, $\mathbf{p}$ and $\mathbf{q}$, we can compare them using standard metrics, such as l_1 (the Manhattan distance) or l_2 (the Euclidean metric). Namely,

$$d_1(\mathbf{p}, \mathbf{q}) = \sum_{i=1}^{m} |p_i - q_i|$$

and

$$d_2(\mathbf{p}, \mathbf{q}) = \sqrt{\sum_{i=1}^{m} (p_i - q_i)^2}$$

Another alternative is the **Pearson distance**, based on the correlation score of the two distributions

$$d_{pear}(\mathbf{p}, \mathbf{q}) = 1 - \sum_{a \in \mathcal{A}} p(a) \cdot q(a)$$

Note that these scores are always positive and their maximal value is bounded. If a scoring function based on these functions is to be used in local alignments then it has to be shifted first such that the average score is negative, as discussed in Section 3.4.3.2.

Other scoring functions for profile-profile comparison are based on information-theoretic principles. These functions assess the statistical similarity of the two distributions. One such measure is the **Kullback-Leibler (KL)** divergence (see Section 3.12.6), which assesses the similarity of two distributions in terms of their relative entropy

$$D^{KL}[\mathbf{p}||\mathbf{q}] = \sum_{a \in \mathcal{A}} p(a) \log_2 \frac{p(a)}{q(a)}$$

Intuitively, the D^{KL} function measures how well one distribution can replace another, for example when used for compression. However, this measure is asymmetric and unbounded when $q(a) \to 0$ (Problem 4.12).

A related measure is the **Average Log Likelihood Ratio (ALLR) score** [134], defined as

$$D_{ALLR}(\mathbf{p}, \mathbf{q}) = \frac{n_p \sum_{a \in \mathcal{A}} p(a) \log \frac{q(a)}{P_0(a)} + n_q \sum_{a \in \mathcal{A}} q(a) \log \frac{p(a)}{P_0(a)}}{n_p + n_q}$$

where n_p (n_q) is the number of total counts from which $\mathbf{p}$ ($\mathbf{q}$) is derived, and $\mathbf{P}_0$ is the background distribution. This measure resembles a relative-entropy measure. Note that the only probabilities that contribute to the sum are the probabilities $p(a)$ and $q(a)$ that differ from the background probability $P_0(a)$ and that the contribution is greatest if the magnitude of the deviation is the same for $q(a)$ and $p(a)$.

Another option is the **Jensen-Shannon (JS)** divergence, which is a symmetric version of the KL-divergence. It measures the distance between the two distributions and a third, average distribution

$$\mathbf{r} = \lambda\mathbf{p} + (1-\lambda)\mathbf{q}$$

which can be considered as the most likely common source distribution of both $\mathbf{p}$ and $\mathbf{q}$, with λ ($0 \leq \lambda \leq 1$) as a prior weight (typically set to 0.5). The λ-JS **divergence** is defined as

$$D_\lambda^{JS}[\mathbf{p}||\mathbf{q}] = \lambda D^{KL}[\mathbf{p}||\mathbf{r}] + (1-\lambda)D^{KL}[\mathbf{q}||\mathbf{r}]$$

This measure is symmetric and ranges between 0 and 1, where the divergence for identical distributions is 0. Besides being symmetric and bounded, an attractive feature of the D^{JS} divergence measure is that it is proportional to the minus logarithm of the probability that the two empirical distributions represent samples drawn from the same (common) source distribution [135]. It has also been shown that $\sqrt{D_\lambda^{JS}[\mathbf{p}||\mathbf{q}]}$ is a metric [136].

The divergence score measures one aspect of the statistical similarity of $\mathbf{p}$ and $\mathbf{q}$: their relative distance. However, it does not consider the uniqueness of the two distributions. A match between two distributions that resemble the background distribution is not as significant as a match of two distributions that resemble each other, but are very different from the background distribution. The more unique the distributions are (and hence, also their common source), the more significant is a match between them. One way to measure the significance of a match is to measure the JS-divergence of the common source distribution, $\mathbf{r}$, from the base (background) distribution $\mathbf{P}_0$

$$S = D^{JS}(\mathbf{r}, \mathbf{P}_0)$$

This score reflects the probability that the source distribution, $\mathbf{r}$, could have been obtained by chance. The higher it is, the more distinctive the common source distribution, and the lower the probability that it could have been obtained by chance.
To account for both the divergence and the significance (uniqueness) of the distributions, we need to combine them both into one similarity score. A simple ad hoc combination [138] is

$$\begin{aligned} \text{score}(\mathbf{p},\mathbf{q}) &= \frac{1}{2}(1-D)(1+S) \qquad (4.5)\\ &= \frac{1}{2}(1-D^{JS}[\mathbf{p}||\mathbf{q}])(1+D^{JS}[\mathbf{r}||\mathbf{P_0}]) \end{aligned}$$

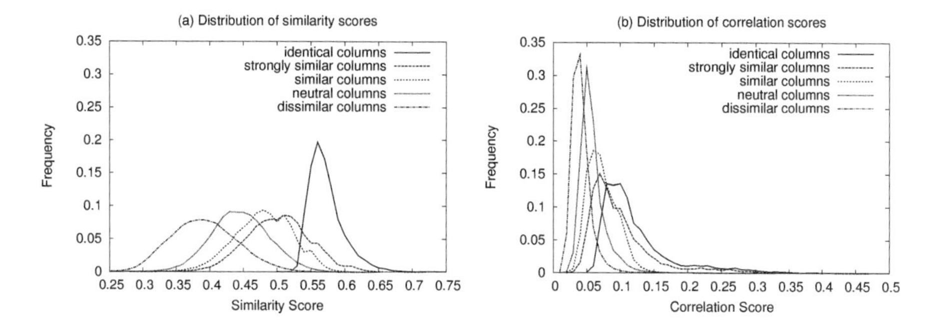

FIGURE 4.10: (a) **Distribution of information-theoretic scores (Equation 4.5) for different column types.** (b) **Distribution of correlation scores.** (Figure reproduced from [137].) The distributions are based on the largest 100 families in the SCOP 1.50 database. The pairs of profile columns are divided into five categories depending on the nature of the seed amino acids: a column with itself (**identical columns**), different columns that are associated with the same seed amino acid (**strongly similar columns**), different columns that are associated with similar seed amino acids (**similar columns**), different columns with mutually neutral seed amino acids (**neutral columns**), and different columns with dissimilar seed amino acids (**dissimilar columns**). Note that in general the distributions of correlation scores overlap more than the other scoring function. For example, the overlap between the scores of identical columns and the scores of dissimilar columns is greater (24%) than the overlap between the same types of columns, using information-theoretic similarity scores (2.1%).

With this expression, the similarity score of two similar distributions ($D \to 0$) whose common source is far from the background distribution ($S \to 1$), tends to one. On the other hand, the similarity score of two dissimilar distributions ($D \to 1$) whose most likely common source distribution resembles the background distribution ($S \to 0$) tends to zero. This scoring scheme also distinguishes two distributions that each are similar to the background distribution ($D \to 0$ and $S \to 0$ giving score $= 1/2$) from two dissimilar distributions, but whose common source is similar to the background distribution ($D \to 1$ and $S \to 0$ giving score $= 0$).

An alternative to the JS-divergence is the mutual-information based distance function, which for two random variables X, Y is defined as

$$d(X, Y) = H(X, Y) - I(X, Y)$$

where $H(X, Y)$ is the entropy of the joint probability distribution of X and Y and $I(X, Y)$ is their mutual information (see Section 3.12.6). Plugging in the probability distributions we get that:

$$\begin{aligned} d(X,Y) &= -\sum_{x\in\mathcal{X}}\sum_{y\in\mathcal{Y}} p(x,y)\log p(x,y) + \sum_{x\in\mathcal{X}}\sum_{y\in\mathcal{Y}} p(x,y)\log\frac{p(x,y)}{p(x)p(y)} \\ &= -\sum_{x\in\mathcal{X}}\sum_{y\in\mathcal{Y}} p(x,y)\log p(x)p(y) \end{aligned}$$

It is not difficult to prove that this function is a metric (Problem 4.12). When applied to profile columns, $p(x)$ and $p(y)$ correspond to the two distributions $\mathbf{p}$ and $\mathbf{q}$, respectively. However, the joint probability $P(x,y)$ has to be computed directly from the original MSAs used to derive the original profiles, so that the probability of each amino acid pair $P(a,b)$ can be computed accurately.

A couple of performance evaluation studies [137, 139] showed that scoring functions based on information-theoretic principles perform better in profile-profile comparisons of remotely related protein families. Indeed, these scores seem to distinguish better related columns from unrelated ones (see Figure 4.10). Other successful profile-profile comparison algorithms include Compass [27] and the HMM-based HMsearch [140].

4.4 Iterative and progressive alignment

Now that we discussed profiles and PSSMs we can move on to discuss more advanced algorithms for MSA. As opposed to the methods described previously, *progressive alignment* is not based just on pairwise alignments, but constructs the MSA dynamically by re-evaluating the alignments at each stage based on the current MSA. There are several variations of this approach, some of which exploit the profile representation we just discussed.

4.4.1 PSI-BLAST - iterative profile search algorithm

One of the most popular algorithms that uses the idea of a profile is PSI-BLAST [100]. Although the goal of PSI-BLAST is not to create an MSA, it inherently does so as an intermediate step. The main use of PSI-BLAST is as a database search algorithm (like BLAST). It is an iterative algorithm that is based on BLAST and uses a position-specific scoring matrix. At the end of each round the significant alignments are compiled into a PSSM, and this PSSM is used in the next round. The PSSM greatly increases the sensitivity of the algorithm and it often detects remote homologies that are missed with BLAST and the Smith-Waterman algorithm.

In more detail, given a query sequence, PSI-BLAST starts by searching the database with BLAST and collecting all the hits with evalue $< T$ (the default in the current version of PSI-BLAST is 10^{-3}). From these hits it creates a multiple alignment using the star alignment method, with the query as the central sequence. The MSA is then processed to create a profile and a position-specific scoring function. In the next iteration, PSI-BLAST searches the database again but this time with the PSSM that was generated in the previous iteration. This allows the algorithm to tune to the specific properties of the query and its homologs. Significant hits detected in this iteration are collected and are used to generate a new MSA, profile and PSSM. The algorithm repeats until convergence (no new sequences are detected) or until a pre-defined maximal number of iterations has been reached.

PSI-BLAST

Search database with the query sequence using BLAST and collect all hits with evalue $< T$.

Loop:

- Create an MSA from the hits collected using the star alignment method (with the query as the center sequence).
- Generate a profile and a PSSM from the MSA.
- Search database again with the new PSSM, and collect all hits with evalue $< T$.

Halt if the process converged (no new sequences detected). Otherwise go to loop.

Creating the profile and PSSM is a process that involves the following steps:

- **Removal of almost identical sequences.** To emphasize remote homologies and eliminate bias introduced by the particular makeup of the database being searched, PSI-BLAST ignores sequences that are very similar to another sequence in the MSA when generating a profile. There are two thresholds at which sequences are ignored. First, all sequences that are identical to the query sequence are removed from consideration. Second, for all pairs of sequences remaining under consideration in the MSA, the sequence with the higher evalue (less significant) is ignored if more than 94% of the residues of the two sequences are identical. Sequences that are ignored do not contribute to the calculation of scores; however they do remain a part of the MSA and will be re-aligned with the new profile in the next iteration.

- **Calculating sequence weights**. PSI-BLAST defines unique sequence weights for each column c. Sequences that are not used in a column (that is, their local alignment with the seed sequence either starts after c or ends before c) are given a zero weight for that column. For all other sequences (denoted as the set S_c), weights are calculated considering a subset of the columns in the alignment, which is called the **context** of

c and denoted C_c. The context is defined to be the largest contiguous set of columns containing c for which all sequences in S_c are used in each column (see Figure 4.11). Since PSI-BLAST generates gapped alignments, gaps can be observed in "used" positions and are referred to as absences.

Let $a_{c',s}$ be the residue of sequence s appearing in column c', $count_{c',S}(a)$ be the total number of times the residue a appears in column c' in the set S and $diversity_{c',S}$ be the total number of different residues that appear in the column c' in the set S. Then, the weight for each sequence $\mathbf{s}$ in column c is defined as

$$weight_c(\mathbf{s}) = \sum_{c' \in C_c} \frac{1}{count_{c',S_c}(a_{c',s}) * diversity_{c',S_c}}$$

Highly diverged sequences tend to get a higher weight since they are more likely to contain infrequent residues in more positions, especially if these residues are observed in relatively conserved positions. For subsequent computations, these sequence weights are normalized to sum to one.

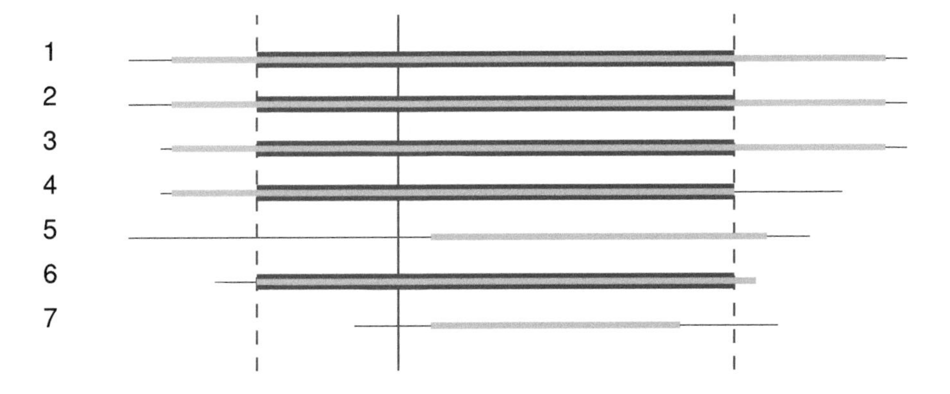

FIGURE 4.11: Defining the context of a column in an MSA. The above diagram is a schematic representation of an MSA. The black thin lines are complete sequences. The portions highlighted in light gray are those parts of the sequence included in the alignment (these positions of the sequence are considered "used"). The context of the column defined by the vertical black line is marked by the black highlights on the sequences. The leftmost and rightmost columns included in this context are further denoted with the dotted black lines. Note that the left extent of the context corresponds to the start of sequence 6 relative to the alignment. No residues from any sequence to the left of this column are considered in this context. Likewise the right extent of the context is defined by the end of sequence 4 relative to the alignment. Finally, note that sequences 5 and 7 are not included in the context because they do not exhibit residues relative to the alignment in the column in question.

- **Computing weighted observation counts.** For each column c, weighted observation counts of each residue a are computed using the sequence weights calculated above. The count $obs_c(a)$ is the sum of the weights of the sequences that exhibit the given residue in that column. Absences (gaps inside local alignments) are treated as a 21st residue with their own weighted count. Since the sequence weights sum to one, so too should these observation counts, which means they can be viewed as empirical probabilities of observing the different residues in the given column.

  At this point, absences are eliminated from consideration. The probability associated with observing an absence in the column is distributed among all other residues according to a specified background probability ($P_0(a)$) of observing each residue a.

- **Computing pseudo-counts.** For each column c, pseudo-counts are derived from the weighted observation counts using a transition probability matrix, as discussed in Section 4.3.3.2. The pseudo-count for a given residue a is

$$pseudo_c(a) = \sum_{b \in \mathcal{A}} obs_c(b) * p(b \to a)$$

  where $\mathcal{A}$ is the residue alphabet and $p(b \to a)$ is the probability of a transition from residue a to b using a matrix such as BLOSUM62 [82].

- **Generating the position-specific scoring matrix.** To compute scores, the observed and pseudo-counts for each residue a are first combined, as described in Section 4.3.3.1, to compute the final relative counts

$$p_c(a) = \frac{\alpha * obs_c(a) + \beta * pseudo_c(a)}{\alpha + \beta}$$

  where α and β are the relative weights used to combine the two types of counts. In PSI-BLAST, the constant β is set ad hoc to 9, while α is the average diversity of the columns considered when calculating the sequence weights minus one. This formula will ignore the empirical observations in columns where only one residue is observed in favor of pseudo-counts. The score for each residue a is derived from the combined count as follows:

$$score_c(a) = \frac{\log\left(\frac{p_c(a)}{P_0(a)}\right)}{\lambda}$$

  where $P_0(a)$ is the background probability of observing the residue a and λ is the known BLAST parameter (computed for a specific combination of scoring matrix and gap penalties).

- **Rescaling of the scoring matrix.** The final step is the rescaling of the scoring matrix as a whole, such that its associated lambda value is the same as that of a given scoring matrix (i.e., BLOSUM62). The rescaling is done by requiring that the PSSM satisfies

$$\sum_{a \in \mathcal{A}} \sum_{c \in C'} p(a) * \frac{1}{L} * e^{\lambda * \lambda' score_c(a)} = 1$$

  where the summation is over the alphabet $\mathcal{A}$ and all columns of the MSA, and L is the number of columns (length) in the MSA. This is accomplished using the bisection method to solve the system of non-linear equations for λ' and correcting all scores by multiplying them by the factor λ'.

  Note that this equation is a generalization of Equation 3.4 in Chapter 3. With the rescaling, the theoretical results regarding the significance of sequence matches that were obtained for pairwise comparisons (Section 3.4) apply for sequence matches scored with the PSSM.

Although PSI-BLAST is a powerful tool, it is also very sensitive to the user parameters (such as the threshold h for inclusion in the profile or the number of iterations). For example, if the evalue threshold h is too high, the program might include unrelated sequences in the profile, causing it to gradually diverge from the original query sequence. As a result it might overestimate the statistical significance of matches with unrelated sequences, and in some cases even rank the match of the query sequence with itself lower than other database sequences (see also Section 3.4.5). On the other hand, a premature halt of the program may miss many relevant and important similarities. Therefore, this tool should be used with caution. As a rule of thumb, running PSI-BLAST for three iterations with the default parameters works well in many cases although it might fall short of detecting all remote homologies, and it is advised to inspect the results manually.

If the set of homologs is known (given) and the goal is to create an MSA, then a safer and cleaner approach is to create a database of the homologs (using the `formatdb` utility, which is part of the NCBI toolkit) and search the seed sequence against this database. Since there are no unrelated sequences in the database, there is no danger that false positives will be included in the profile, although spurious matches with unrelated segments of these proteins might still occur. The main advantage of PSI-BLAST in this "cleaner" version is that it is very fast compared to other algorithms described next.

4.4.2 Progressive star alignment

While PSI-BLAST revises the scoring function in each iteration, it is not fundamentally different from the star alignment method described in Section 4.2.1. Although the PSSM is defined based on all the related sequences, the sequences are still aligned to the center sequence one at a time ignoring the other sequences in the set. Therefore the algorithm has the same problems we mentioned earlier and errors that are introduced in early stages cannot be reversed. Moreover, since each sequence is aligned while ignoring the other sequences that were already aligned, similar sequences with insertions in the same position are not aligned to each other, resulting sometimes in highly noisy and fragmented MSA. The chances of these errors happening with PSI-BLAST are quite high especially for highly diverged families. However, we can modify the star method to build the MSA *progressively*, using the following procedure:

1. Compute all pairwise similarities. (The MSA is not constructed directly from these alignments. They are used just to guide the process).

2. Pick the most similar pair and align it. Create a profile and a PSSM.

3. Pick the next sequence that is closest to a sequence that is already in the MSA (based on the pre-computed pairwise similarities).

4. Align the sequence to the MSA using the PSSM and update the MSA (and PSSM). Repeat until all input sequences were exhausted.

Step 3 can be replaced with a more refined approach where the similarity scores of all remaining sequences are recomputed with respect to the MSA, to determine the closest sequence, thus revising our decision at each stage. The MSA can be further improved by revisiting the alignment of each sequence. This is implemented by excluding one sequence at a time from the alignment, recomputing the profile from the remaining sequences and re-aligning the excluded sequence to the profile. By repeating this procedure for all sequences and iterating multiple times through the whole dataset until convergence, the alignment quality is expected to improve and reach a local maximum.

4.4.3 Progressive profile alignment

Taking the algorithm described in the previous section one step further, CLUSTALW [141] is a progressive tree alignment algorithm that uses profiles to align intermediate MSAs that are generated in the process, as opposed to sequence-to-profile alignment (as in the progressive star alignment method). The algorithm starts by computing all pairwise similarities and then builds a tree based on these scores using hierarchical pairwise clustering (Section 4.2.2). The sequences are then aligned following the nodes of the guide tree, but unlike the algorithm of Section 4.2.2, this time we employ progressive alignment. That is, if the child nodes are individual sequences then we use pairwise alignment and represent the resulting alignment as a profile. If one of the nodes is associated with a sequence and the other with a profile then we align the sequence to the profile (as in Section 4.3.4) and update the profile. If both are profiles then we use profile-profile alignment (Section 4.3.5) and update the profile.

CLUSTALW also introduces dynamic and position-specific gap penalties based on the profile content at each step, using several ad hoc rules. For example, the gap opening penalty is lowered in positions where gaps were already opened previously and is increased within a certain window from these positions to prevent gaps in proximity to each other. The penalties are also adjusted based on the residue properties at each position. For example, they are lowered in positions that are rich with residues that tend to appear in loop regions such as hydrophilic residues. The divergence level (the average number of mutations) between the groups of sequences to be aligned also affects the gap penalties. The higher the divergence, the lower the gap penalties.

CLUSTALW outperforms the classical MSA algorithms and progressive star alignment, and it has been the method of choice for many years. However it can be computationally intensive for large families, since it requires computing all pairwise similarities. Furthermore, it can still result in erroneous MSAs, as discussed in the next section.

4.5 Transitive alignment

Although progressive alignment algorithms (such as CLUSTALW) generate better MSAs than the other heuristics discussed thus far, they also suffer from the same major drawback that was mentioned before: "once a gap always a gap". That is, once introduced, a gap cannot be removed, as is illustrated in Figure 4.12.

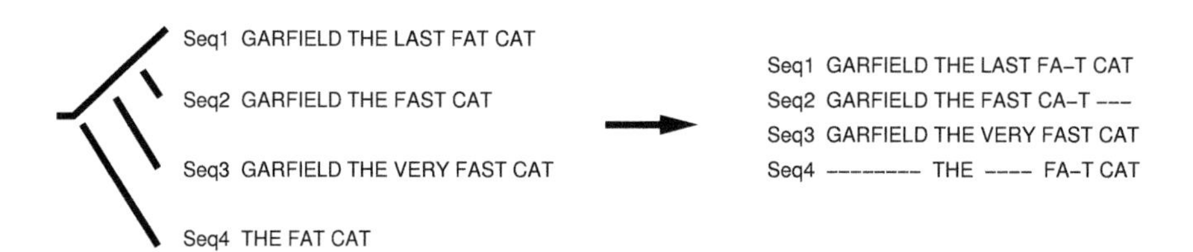

FIGURE 4.12: **Problems with progressive alignment.** (Figure adapted from [142] with permission from Elsevier.) This nice example illustrates one of the biggest problems with most MSA algorithms. Improper gaps that are introduced in early stages (for example, when aligning a pair of sequences) are carried on to subsequent iterations.

The problem does not pertain only to gaps. In general, any part that is misaligned in the beginning can affect the quality of the whole alignment.

Clearly, given the size of the search space it is very difficult to find the optimal alignment. The question is whether it is possible to reduce the chances of misalignments.

The main property that characterizes all the previous methods is that they do not use all of the information on the set of related proteins when aligning pairs of sequences in the beginning, and therefore errors are more likely to occur. In the past couple of years several algorithms were developed that attempt to address exactly that (e.g. [143–145]). In a way, these algorithms employ elaborate methods for computing position-specific sequence weights. We will discuss one such example: T-coffee [142].

4.5.1 T-coffee

The T-coffee algorithm relies initially on pairwise alignments. For each pair of residues (from two sequences) it defines a weight that reflects the collective information in the whole set of input sequences in support of this pairing (or lack thereof). These weights are then used to direct and generate new alignments.

The pairwise alignments

```
Pair 1 (50% identity)   Pair 2 (80% identity)   Pair 3 (80% identity)
MKRM                    M-KRM                   MRK-M
MRKM                    MRKRM                   MRKRM
```

FIGURE 4.13: **The T-coffee algorithm.** Using the collective information to align sequences. Consider the two sequences $\mathbf{a} = MKRM$ and $\mathbf{b} = MRKM$ and the intermediate sequence $\mathbf{x} = MRKRM$. Note that K and R are misaligned in the first pair. The residues R and K are usually assigned a positive score by standard scoring matrices. The alternative alignment that aligns K with K would introduce two gaps, and although it has higher percent identity it would have a lower score overall because of the gap penalties. However, the intermediate sequence $MRKRM$ supports aligning the second K from $\mathbf{a}$ with the third K from $\mathbf{b}$ resulting in a high pairwise score score$(a_2, b_3) =$ score$(2, 3) = 50 + 80$.

To assess the collective information in support of a specific pairing, the algorithm considers all possible intermediate sequences (in other words, all possible triplets) and assigns a weight to each. The weights are then summed over all intermediate sequences. Specifically, the algorithm works as follows:

1. Compute all pairwise similarities $\{s(\mathbf{a}, \mathbf{b})\}$. For each pair of sequences $\mathbf{a}$ and $\mathbf{b}$ define a weight $w(\mathbf{a}, \mathbf{b})$ that is the percent identity (see Section 3.4.6).
2. For each pair of sequences re-evaluate the score of aligning each pair of

residues a_i and b_j now by considering all intermediate sequences in the library

$$\text{score}(i,j) = s(i,j,\mathbf{a}) + \sum_{\mathbf{x} \neq \mathbf{a},\mathbf{b}} s(i,j,\mathbf{x}) \tag{4.6}$$

where

$$s(i,j,\mathbf{x}) = \begin{cases} \min(w(\mathbf{a},\mathbf{x}), w(\mathbf{b},\mathbf{x})) & \text{if } \mathbf{x} \text{ supports aligning } a_i \text{ with } b_j \\ 0 & \text{otherwise} \end{cases}$$

We say that $\mathbf{x}$ supports aligning a_i with b_j if there is a residue k in $\mathbf{x}$ such that x_k is aligned with a_i in the pairwise alignment of $\mathbf{a}$ and $\mathbf{x}$, and x_k is aligned with b_j in the pairwise alignment of $\mathbf{b}$ and $\mathbf{x}$ (Figure 4.13). The first term in Equation 4.6 accounts for the direct similarity of these residues and is defined as the sequence identity of $\mathbf{a}$ and $\mathbf{b}$. But note that residues can be assigned a non-zero weight even if not aligned directly, because of the second term. Each intermediate sequence that supports this pairing adds a weight that is the minimum over its sequence identity with either $\mathbf{a}$ or $\mathbf{b}$. The more intermediate sequences support the pairing of these residues, the higher their similarity score is. Residues that are never aligned are assigned a score of zero.

3. Apply the CLUSTALW algorithm with the new scores

   - Start with a guide tree (can be built using the similarity scores $s(\mathbf{a}, \mathbf{b})$ that were computed in Step 1).
   - Align the closest sequences using the algorithm of Section 4.3.4.1 with the position specific scores score(i, j) defined in Step 2.
   - Depending on the next node in the guide tree, align the next closest pair of sequences, or add a sequence to an existing alignment or align two MSAs.

The last step requires a clarification. If the node requires aligning individual sequences, then we proceed as before using the position-specific scores. Otherwise (i.e. at least one of the elements to be aligned is an MSA), we first redefine the position-specific scores as the average scores over the sets to be aligned. For example, when aligning two MSAs, over two sets of sequences, the score of aligning position i of the first MSA with position j in the second MSA is defined as the sum

$$\text{score}_{MSA}(i,j) = \frac{1}{|set1||set2|} \sum_{k \in set1} \sum_{l \in set2} \text{score}(a_{k,i_k}, a_{l,j_l})$$

where the first index (k or l) is the sequence identifier and the second index (i_k or j_l) indicates which sequence residue appears in the relevant column of the MSA (i or j). For example, when aligning the following two MSAs

$$\begin{array}{ll} \mathbf{a}_1 = A\ A\ C & \qquad\qquad \mathbf{a}_3 = B\ A\ B \\ \mathbf{a}_2 = -\ A\ C & \qquad\qquad \mathbf{a}_4 = B\ -\ B \\ \qquad\ \ i & \qquad\qquad\qquad\quad\ \ j \end{array}$$

we define $set1 = \{1, 2\}$ and $set2 = \{3, 4\}$. Position i in the first MSA contains the second residue of the first sequence ($i_1 = 2$) and the first residue of the second sequence ($i_2 = 1$). Position j of the second MSA contains the third residue of the third sequence ($j_3 = 3$) and the second residue of the fourth sequence ($j_4 = 2$). Hence, the score of aligning position i with position j is given by the average

$$\begin{aligned} \text{score}_{MSA}(i,j) &= \frac{1}{|set1||set2|} \sum_{k \in set1} \sum_{l \in set2} \text{score}(a_{k,i_k}, a_{l,j_l}) \\ &= \frac{1}{4}(\text{score}(a_{1,2}, a_{3,3}) + \text{score}(a_{2,1}, a_{3,3}) + \\ &\quad \text{score}(a_{1,2}, a_{4,2}) + \text{score}(a_{2,1}, a_{4,2})) \end{aligned}$$

Each element in this sum is defined as in Step 2 above. For example, the value of $\text{score}(a_{1,2}, a_{3,2})$ is the score (weight) of aligning the second residue (A) in the first sequence with the second residue in the third sequence (A) based on the collective information. Gaps are ignored in these calculations.

4.6 Partial order alignment

The traditional concept of alignment often fails to describe properly the relation between sequences. Consider two sequences from related organisms that have been "edited" independently after speciation, with completely unrelated fragments inserted in the middle or the beginning of the sequences. If we look back at the basic definition of alignment in Chapter 3, we conclude that the unrelated segments should not be aligned. However, the traditional MSA model cannot always accommodate this quite likely scenario, as exemplified in Figure 4.14.

As an alternative approach to the MSA problem, Chris Lee and his co-workers introduced Partial Order Alignments (**POA**) [146], following an earlier work by [147]. The POA algorithm can accommodate these kind of scenarios by using a separate DP surface along each branch where independent evolution is observed. The algorithm and the model it uses (partial order graphs) are described next.

```
...AATTGCA...AGG                    AAT...TGCA...AGG
CAC...TGCACATACG                    ...CACTGCACATACG
```

FIGURE 4.14: Ambiguous alignments with the traditional alignment model. While the gap in the middle of the alignment can be considered as a real gap (insertion/deletion), the gaps in the beginning are not true gaps and they merely serve to indicate that the first three bases of the first sequence (AAT) are not aligned to the first three bases of the second sequence (CAC). However, with the traditional row-column format of MSA there is no natural way to represent this information, and the two alternative representations are both ambiguous and misleading.

4.6.1 The partial order MSA model

Partial order alignments use Partial Order Graphs (**POG**) to represent multiple sequence alignments. Each sequence is represented as a graph where each letter corresponds to a node and directed edges are drawn between consecutive letters (see Figure 4.15). This representation is referred to as Partial Order MSA (**PO-MSA**). Upon alignment (as described below) nodes are merged into alignment rings that represent aligned residues. A single node is used to represent alignments with non-identical letters (alignment rings) and each node is associated with a probability distribution and a scoring function over amino acids that are used in subsequent alignments. The data structure maintains information on the origin of each residue, such as the sequence ID and the position within the sequence, allowing one to trace back the path of each input sequence through the POG. Branches in the graph represent insertions, which is equivalent to opening a gap in the standard MSA format. Each graph also contains start and stop nodes. The start nodes designate the nodes from which to start traversing the POG and the end nodes designate the end of the POG.

Each branch in the POG is associated with its own profile (PSSM) to which proteins can be aligned. By allowing for multiple branches at each node, the POG eliminates the need to reduce an MSA to a linear profile. This preserves information that is needed for an accurate alignment and is otherwise lost. Different branches of the graph represent segments that are unaligned to each other. Since residues in different segments are unordered with respect to each other, the model is called a partial order graph alignment.

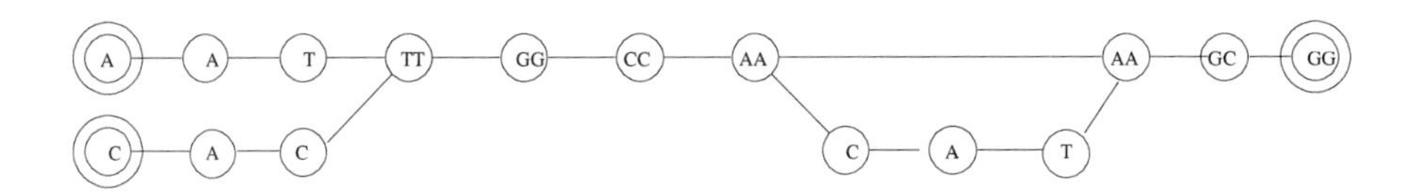

FIGURE 4.15: PO-MSA of the MSAs of Figure 4.14. Each node lists the aligned residues. Start and end nodes are double circled.

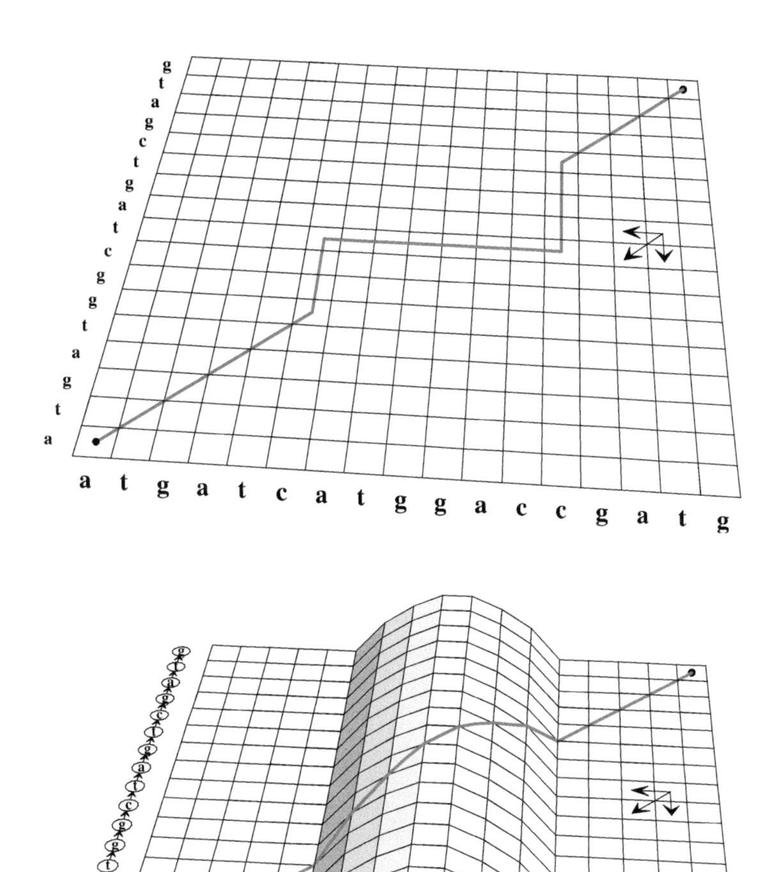

FIGURE 4.16: The POA algorithm. In the classical DP algorithm, a two-dimensional matrix is used to compute the similarity score of two sequences, and the score in each cell depends on three adjacent cells (assuming fixed gap penalty), as marked (top). When comparing partial order graphs, the DP matrix has multiple surfaces that correspond to different branches (bottom). In this example, a linear graph that corresponds to one sequence is compared with a PO-MSA. On each surface we proceed as in the classical algorithm, but when surfaces merge we need to check all possible paths that come from either surface, as depicted. (Figure is courtesy of Chris Lee. Reproduced from [146] with permission.)

As there is no loss of information in converting an MSA in row-column format to a PO-MSA, an MSA maps uniquely to PO-MSA. However, the reverse is not true, since the more expressive POA model can be mapped to multiple MSA (contrast Figure 4.14 and Figure 4.15). Once we converted an MSA to a PO-MSA we can align a sequence to the graph or compare two graphs while considering all branches, using the POA algorithm.

4.6.2 The partial order alignment algorithm

Aligning two MSAs (or two sequences) in PO-MSA format is similar to the standard Smith Waterman dynamic programming algorithm. However, each node might be connected to several incoming (and outgoing) edges, and therefore each one of the possible paths leading to a node has to be traversed. Each branch introduces another two-dimensional DP matrix, and all matrices are joined at the point of bifurcation (Figure 4.16). On each surface the algorithm operates similar to the classical algorithm, by checking the three adjacent cells as depicted. When different surfaces merge, additional cells are considered along each surface.

In the simpler case of aligning a sequence with a PO-MSA (see Figure 4.16) we define the score at node m of the graph and position j of the sequence as

$$S(m,j) = \max_{p:p\rightarrow m} \{S(p,j-1) + s(m,j), S(m,j-1) - \delta(j), S(p,j) - \delta(m), 0\}$$

where $p : p \rightarrow m$ denotes the set of all predecessor nodes p that have edges to node m, $s(m,j)$ is the score of aligning the residue at position j of the sequence with node m of the POG (using scoring functions such as those in Section 4.3.4) and $\delta(m)$ is the penalty for aligning m with a gap.

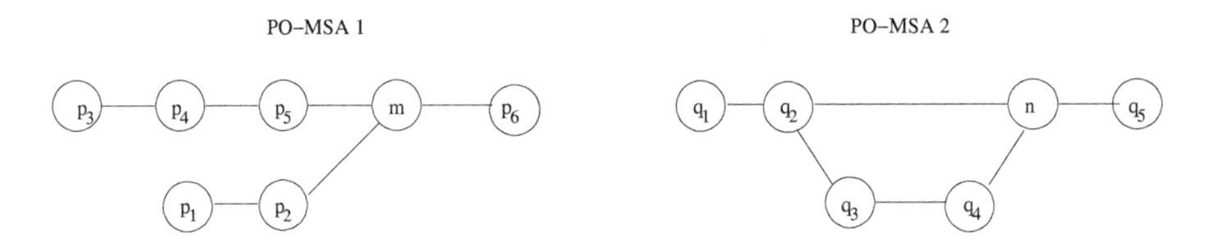

FIGURE 4.17: **Aligning two PO-MSAs.**

When aligning two PO-MSAs, the equation above has to be generalized since nodes in both PO-MSAs can have multiple predecessor nodes (depending on the number of incoming edges), and the score of a local alignment that ends in the node pair (n,m) is given by the following equation:

$$S(m, n) = \max_{p:p \to m, q:q \to n} \{S(p, q) + s(m, n), S(m, q) - \delta(n), S(p, n) - \delta(m), 0\}$$

where $s(m, n)$ is the score of aligning node m with node n using scoring functions as in Section 4.3.5, and $\delta(m)$ is the penalty for aligning m with a gap. For example, to align node m and node n of the two PO-MSAs in Figure 4.17, we need to consider the following moves:

$$\begin{aligned} S(m, n) = \max \{ & S(p_2, q_2) + s(m, n), S(p_2, q_4) + s(m, n), \\ & S(p_5, q_2) + s(m, n), S(p_5, q_4) + s(m, n), \\ & S(p_2, n) - \delta(m), S(p_5, n) - \delta(m), \\ & S(m, q_2) - \delta(n), S(m, q_4) - \delta(n), 0\} \end{aligned}$$

The algorithm is easily extended to handle gaps of length > 1 the same way the standard DP algorithm does so.

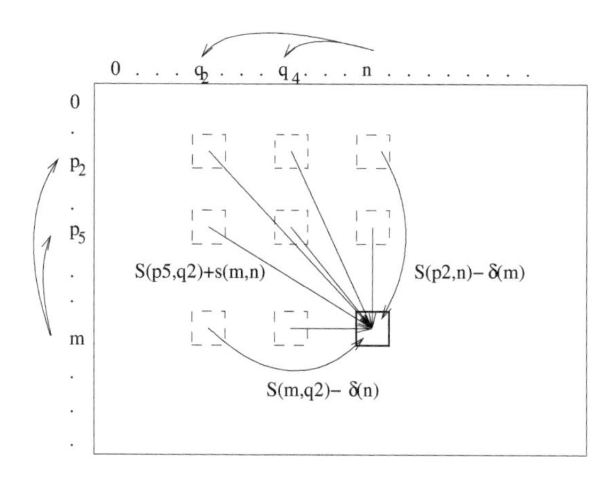

FIGURE 4.18: The POA matrix. Each PO-MSA is associated with a one-dimensional array that stores information about the edges coming in and out of each node. When computing the score of a cell in the POA matrix, we need to consider all pairs of nodes that are connected to it. For example, when aligning node m with node n of the two PO-MSAs of Figure 4.17 we need to consider all cells as depicted.

To implement the alignment of two PO-MSAs, the partial order graphs are re-represented as one-dimensional arrays. Each cell of each array corresponds to one node in the partial order graph and contains information on the directed edges coming in and out of the node, which is what differentiates it from the classical DP matrix. The 2D dynamic programming matrix is then computed with the PO-MSA representing each axis (Figure 4.18).

Finally, once the dynamic programming matrix has been scored and an alignment has been computed, the PO-MSAs are merged into a new partial order alignment graph. This graph is reconstructed by fusing aligned nodes and removing duplicate edges. If one of the nodes is aligned with a gap then a new node is added to the merged PO-MSA with edges connecting it to the nodes that appear directly before and after it. For example, if the best alignment of the two PO-MSAs in Figure 4.17 aligns node p_3 with node q_1, p_4 with q_2, p_5 with q_3, m with n and p_6 with q_5 (thus aligning q_4 with a gap), then the merged PO-MSA will result in the PO-MSA of Figure 4.19.

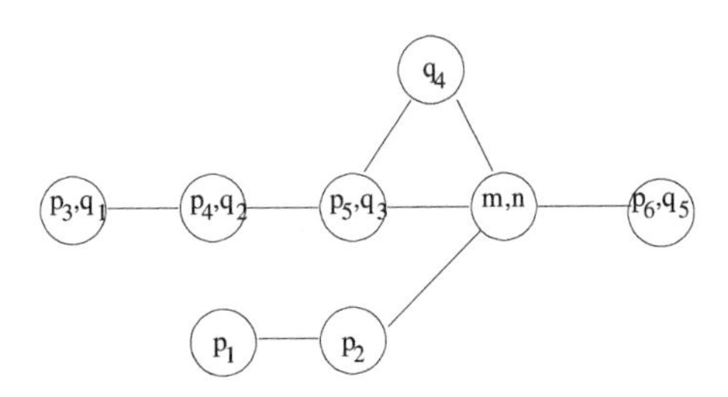

FIGURE 4.19: **Merging two PO-MSAs.**

One of the most appealing applications of POA is to the alignment of multiple identical genes from different individuals from the same species. The DNA sequences of these genes often exhibit **Single Nucleotide Polymorphisms (SNPs)**. A SNP is a change of a single base in the DNA sequence and reflects genetic variation between individuals[6]. The POA model is quite successful in displaying and highlighting these SNPs in large groups of sequences, as shown in Figure 4.20. This is useful for association studies that attempt to detect the common SNPs in subpopulations of interest (e.g., patients with a certain disease). POA is also useful for analysis of Expressed Sequence Tags

[6]SNPs are the subject of intensive research. Each SNP results in a different **allele** of the same gene. A combination of two or more SNPs within the same gene or within a group of closely related genes (i.e., genes on the same chromosome that are inherited together) is called a **haplotype**. Most of these SNPs are harmless, since they do not change the amino acid encoded in the gene sequence. Such SNPs are referred to as **synonymous**. SNPs that do result in change of amino acid are called **nonsynonymous**. Some of them affect the ability of the encoded protein to interact and bind to other molecules, others destabilize the protein or affect its ability to fold. SNPs can also fall in non-coding regions along the DNA sequence and yet have an impact on the host cell, if they alter regions that contain RNA genes or are involved in the regulation of transcription. Many of these SNPs were linked to various diseases [148–150]. Therefore, identifying SNPs is crucial to understanding the genetic factors behind many diseases.

(EST) clusters[7], since it can handle sequencing errors and alternative splicing forms of the same gene. Another effective application of POA is the alignment and display of relationships between multi-domain proteins. We will get back to this topic in Chapter 9.

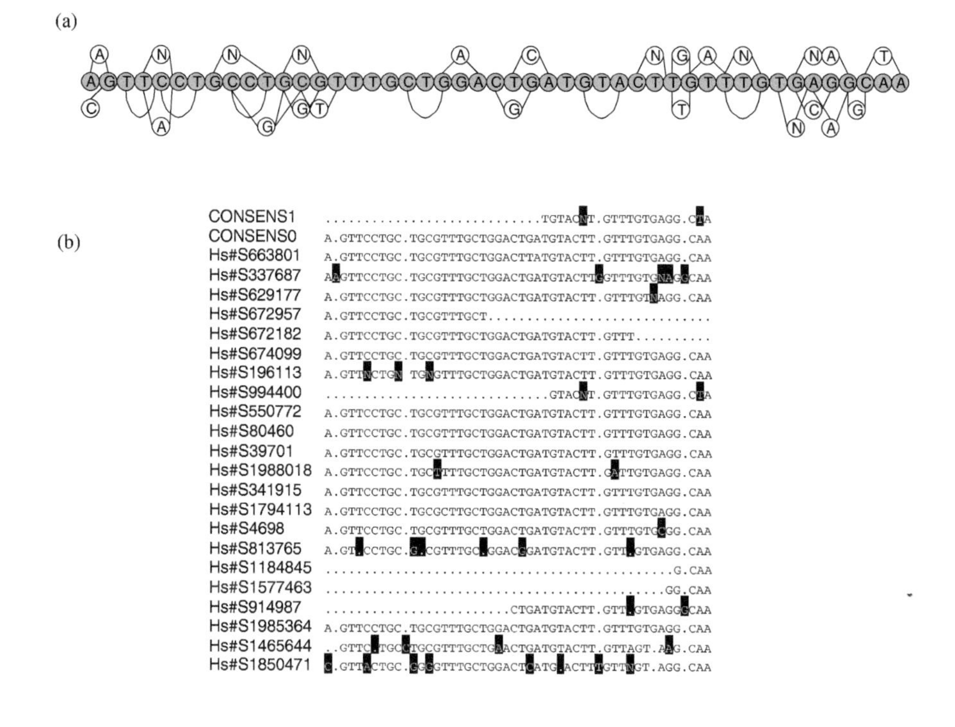

FIGURE 4.20: PO-MSA vs. MSA representation of an alignment of multiple DNA sequences with SNPs and sequencing errors. With the POA algorithm it is easier to detect the consensus sequence of the cluster. (Figure reproduced from [146] with permission.)

[7] **Expressed Sequence Tags (ESTs)** are partial DNA sequences that represent expressed gene sequences. These short fragments are usually generated by sequencing a few hundred nucleotides of a gene DNA sequence. Libraries of ESTs can be generated relatively fast and are inexpensive. Therefore, they often serve as a gene discovery tool or are used to detect genes that are linked with certain diseases or genes specifically expressed in certain tissues [151,152]. While EST libraries are potentially very informative, they are difficult to process and analyze. Since ESTs are sequenced by scanning the DNA only once, they have relatively high error rates due to either sequencing errors (about one sequencing error per 100 residues) or frameshift errors.

4.7 Further reading

The MSA problem has been addressed by many over the past three decades. Classical methods are covered in detail in [30,87,93,103]. A revived interest in the problem led to the introduction of several new heuristics in recent years, in addition to the algorithms discussed in this chapter, including [145, 153–156]. For reviews of recent methods, see [157–159]. A good review of sequence weighting methods appears in [93] and [160]. Evaluation of MSA algorithms has been done on datasets of structurally derived or manually verified MSAs [161–163], or synthetically generated MSAs [164].

In addition to their application for detection of conserved regions, mutations and remote homologs, MSA algorithms have been playing an important role in identifying and characterizing protein families in motif and domain databases. To model a motif, a domain or a protein family, many studies and databases start off by building a multiple alignment (e.g. PROSITE, Pfam, SMART and others). We will get back to these databases later in Chapter 9.

Another important application of MSA algorithms is to the construction of EST clusters. MSA algorithms can help to detect the consensus sequence of an EST cluster associated with a specific gene, as well as SNPs. SNPs that were linked with known diseases are documented in databases such as OMIM [165, 166] and dbSNP [167]. For updates and additional references check the book's website at `biozon.org/proteomics/`.

4.8 Conclusions

- Multiple sequence alignment (MSA) is an extension of the pairwise sequence alignment problem to multiple homologous sequences.
- MSAs can expose mutations and conserved positions with functional importance and regions of structural flexibility that can play an important role in interaction sites.
- Constructing an MSA from a set of homologous sequences is an NP-hard problem. Existing algorithms are heuristics that work by inspecting a reduced search space.
- PSI-BLAST is a fast iterative MSA algorithm that uses BLAST with a position specific scoring matrix (PSSM).
- The most effective MSA algorithms use progressive alignment (building the MSA progressively from smaller alignments) or transitive alignment (using collective information about residue pairings in the dataset to direct the alignment).
- MSAs can be converted to profiles and searched against sequence databases to detect remote homologies that are missed with pairwise comparison algorithms.
- Pseudo-counts and sequence weights can greatly enhance the generalization power of a profile.
- Profiles can be compared to each other to detect subtle similarities between protein families.
- The graph-based partial order alignment is an alternative to the traditional row-column alignment model. The model allows for non-linear alignments where some positions are left unaligned to each other.

4.9 Problems

For updates, additional problems, files and datasets check the book's website at `biozon.org/proteomics/`.

1. **MSA with dynamic programming**

   In Section 4.1 we discussed the dynamic programming algorithm for the multiple sequence alignment problem. Specify formally how exactly the scores should be initialized in global mode, in semi-global mode and in local mode.

2. **Scoring MSAs**

   Show that when using a log-odds pairwise scoring matrix, and assuming a direct common ancestor, then the most likely residue the common ancestor had in a position is the residue that maximizes the similarity with all other residues in that position:

$$a_{ancestor} = \max_{a \in \mathcal{A}} \sum_{i=1}^{N} s(\sigma_i, a)$$

3. **The center star method**

   (a) Use the center star method to compute a multiple alignment for the following words: BLATHER, BARTER, HEATER and THERE. Show also the intermediate steps. Use the following simple scoring matrix.

```
d(x,y)  = 0 if x and y are identical
        = 1 if x and y are nonidentical vowels
        = 2 if x and y are nonidentical consonants
        = 2 if one of x and y is a space
        = 3 if one of x and y is a consonant and the
          other is a vowel
```

   Note that Y is both a vowel and a consonant, so its score is 1 when compared to another vowel and 2 when compared to another consonant.

   (b) What is the consensus string for the resulting alignment?

4. **The computational complexity of the center star method**

   Show that the computational complexity of the method is $O(N^2 n)$.

5. **Suboptimality of the center star method**

   Give a simple example to show that the center star method does not necessarily produce the optimal sum-of-pairs alignment. Use the scoring matrix from the previous problem.

6. **The consensus sequence**

   Give an example of an MSA column where the most common character (corrected for the background frequencies) differs from the character that maximizes similarity to all other characters in the column (use the BLOSUM62 scoring matrix).

7. **Pattern matching**

   You are asked to match a string over the alphabet $\mathcal{A}$ to a regular expression over the alphabet $\mathcal{A}' = \mathcal{A} \cup \{\neg, *, (,), \{, \}\}$.

   (a) Describe a procedure to build a finite automaton that represents the pattern, and an algorithm to detect occurrences of the pattern in a sequence using the automaton.

   (b) Describe a dynamic programming algorithm that searches for occurrences of the pattern in a sequence, now assuming gaps are allowed.

8. **Partial order alignment (POA)**

   With partial order alignments, we define recursively the score for aligning (locally) a sequence $a[1..j]$ with nodes $1..m$ of a given POA

   $$S(m,j) = \max_{p:p \to m} \{S(p,j-1)+s(m,j), S(m,j-1)-\delta(j), S(p,j)-\delta(m), 0\}$$

   where the maximum is over all nodes p that have directed edge $p \to m$ in the POA graph (see Section 4.6.2).

   (a) Let m be a node with two incoming edges from p_1 and p_2. How many "paths" are there in this case? Explain the meaning of each path in the equation above and give a small diagram in each case.

   (b) In Section 4.3.4 we discussed several functions for scoring a residue against a profile column. With POA we have nodes or alignment rings.

   - How would you define $s(m,j)$ if m is a single node (which can be the result of a fusion of identical residues from several sequences)? Assume you know exactly how many and which sequences were aligned at that position.
   - How would you define $s(m,j)$ if m represents an alignment ring?

(c) How would you generalize the scoring function when aligning one POA with another POA, to score a match between two nodes (where now both can be alignment rings)?

9. **PSI-BLAST**

   (a) You are given an unknown query sequence. Use the NCBI BLAST webserver to search this sequence against the NR database. Report the top match with evalue < 0.1 that is not an hypothetical protein or unnamed protein. What can you say about the function of your query protein?

   (b) Repeat the analysis using PSI-BLAST with the threshold h set to the default value of 0.005 (the threshold for inclusion of sequences in the profile). Report the top "meaningful" match after three iterations. Are you in a better position now to infer the function of the protein?

   (c) Repeat the analysis using PSI-BLAST with the threshold h set to 0.1. Are the results better? Explain what happened.

   Query sequence and links to the NCBI webserver are available at the book's website.

10. **Profile analysis - aligning a sequence to a profile**

    Modify your program from Problem 3.9 to align a sequence with an MSA, using the profile representation of an MSA. You are given as input an MSA and a sequence. Both input and output formats are described in the book's website, as well as an example MSA, the seed sequence of that MSA, and another remotely related sequence.

    (a) Generate a profile from this MSA. Remember to add pseudo-counts (you can use any of the methods suggested in Section 4.3.3.2, and it is recommended to implement more than one method). Assign weights to sequences using one of the methods of Section 4.3.3.3.

    (b) Convert the probability profile to a position-specific scoring matrix using the weighted sum approach of Section 4.3.4.

    (c) Modify your DP algorithm to use position-specific scores.

    (d) Align the seed sequence of the MSA with the remotely related protein using your original program.

    (e) Align the remotely related protein with the MSA using your new program. Are the results different?

11. **Molecular clocks**

    You are given three MSAs of histones, ribosomal proteins and immunoglobulins over the same set of organisms. Compare the molecular clocks of these protein families by computing the average entropy along each MSA and the average SOP score.

12. **Profile-profile alignment**

    (a) Give an example of two profile columns for which the KL-divergence measure is unbounded.

    (b) Give an example showing that the KL-divergence measure is asymmetric.

    (c) Show that the mutual information based function $d(X, Y)$ of Section 4.3.5 is a metric.

Chapter 5

Motif Discovery

5.1 Introduction

The complexity of living organisms can be attributed to the complex control and regulation mechanisms they employ. One fine example is transcription regulation. As mentioned in Section 2.4, the process of transcription is carried out by RNA polymerase. However, this protein is part of a more complicated machinery that "executes" transcription, involving multiple components. Transcription of genes is regulated most predominantly by **promoters** and **Transcription Factors (TF)**. Promoters are short DNA sequences preceding genes. Transcription factors are proteins that initiate the process of transcription . Different genes are regulated by different TFs, and transcription is initiated only after a gene's TF binds to its promoter region. Thus, promoters are regulatory elements ensuring that a gene is transcribed only when its TF is active. Transcription can be "turned off" if another molecule binds to the promoter or to the TF, blocking access to the promoter or to the RNA polymerase (such proteins are called **repressors**). Other regulatory sequences, called **enhancers**, can further alter the rate of transcription by recruiting **activator proteins**. Such interventions enable the cell to control and adjust transcription under different conditions, based on the cell's needs (we will get back to this topic in Chapter 12).

In the course of evolution increasingly more complex regulation systems emerged, as a result of random duplication events. For example, some duplication events copied control elements such as promoters to other locations along the organism's genome, sometimes along with the preceding gene. Some of these events resulted in multiple genes that are controlled by the same transcription factor, giving rise to a co-regulated and coordinated set of genes (see Figure 5.1).

Duplication events could also have copied segments of functional proteins to other locations, of either coding or non-coding sequences. Some of these copies ended up being transcribed and translated (especially if the copy was embedded in a coding sequence, as in Figure 5.1). If the translated sequence was able to form a stable structure and the new protein was beneficial to the host cell, the duplication event could be considered successful and the resulting protein was likely to survive selection and pass on to the next generations.

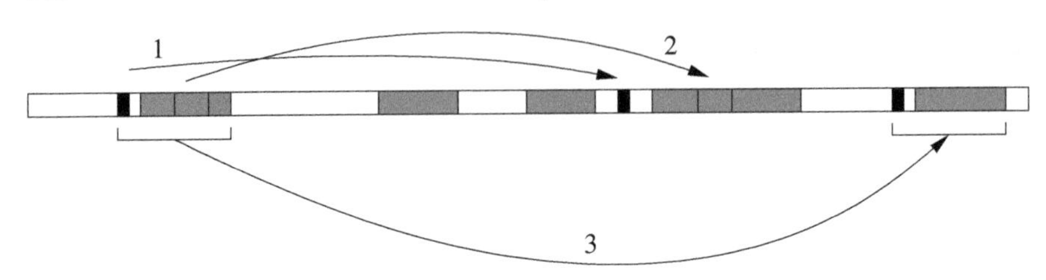

FIGURE 5.1: **Duplication events.** Three duplication events, the first of a promoter region, the second of a protein segment that was copied and inserted into another coding sequence and the third of a promoter followed by the gene sequence.

Every new copy was to evolve independently from that point on, blurring its relation with the original copy of that segment. Usually these can be detected with MSA algorithms. However, often the duplicated segments are short and are missed by sequence comparison algorithms.

Motif discovery aims to identify short sequence signatures hidden in a set of long sequences. Motif detection algorithms are most predominantly used for detection of regulatory patterns in DNA sequences, such as promoter signatures. Detection of a common regulatory element in a set of DNA sequences is an important step toward recovering the regulatory network of the cell (Chapter 15). Genes with common promoters are likely to be co-regulated and are often involved in the same molecular complex or biological process. Once detected, discovering additional occurrences of the element can expose other genes that are regulated by the same transcription factor. Other significant sequence signatures along DNA sequences include target sites of restriction enzymes and splicing sites. Motif analysis is also useful for detection of signature motifs in protein sequences and can help in characterizing their function, since many of these signatures are important functionally, such as active sites or binding sites.

Motif discovery is a challenging problem. Motifs are usually short and are hidden in long unrelated sequences, rendering algorithms for MSA ineffective. Therefore, a different type of algorithm is necessary to detect these signature patterns. In this chapter we discuss several popular approaches for motif detection.

5.2 Model-based algorithms

A typical motif finding problem starts with a set of N sequences $\mathbf{a}_1, \mathbf{a}_2, ..., \mathbf{a}_N$ of different lengths (for example, long non-coding DNA sequences, as in Figure 5.2). The sequences are assumed to contain a common short pattern.

Common does not mean identical, and the patterns can differ in some positions. The task is to find the true set of common patterns in the input sequences. Model-based approaches assume that there is a certain source that generates instances of the pattern according to a certain probability distribution and that the source can be faithfully described with a statistical model. The model induces a probability distribution over the sequence space, which approximates the true source distribution. To identify instances of the pattern, model-based approaches learn a model from the data and use it to classify substrings of the input sequences. Since the samples are unlabeled and the pattern is unknown, we have to search the space of all possible models, considering all possible combinations of short substrings, and pick the most significant one. This is an unsupervised learning problem that entails the following steps:

- **Model type:** we need to define the type of the model that will be used to represent the patterns.
- **Evaluation:** we need to specify how models are evaluated.
- **Search:** we need to outline a search strategy to find the best model (and the pattern it implies).

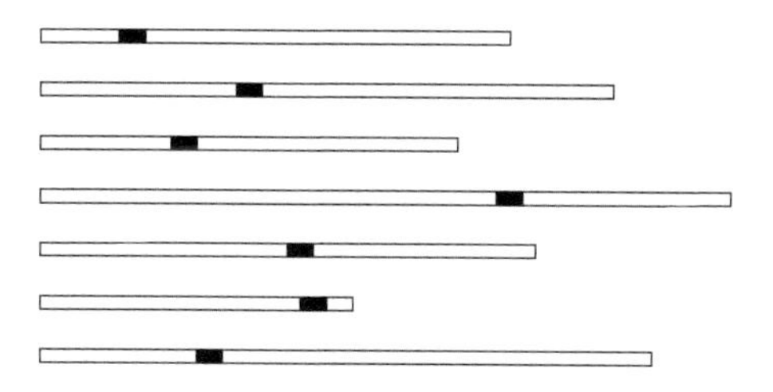

FIGURE 5.2: **A typical input with a hidden set of common patterns.**

5.2.1 The basic model

In describing short patterns, the most common approach is to assume independence between the different positions of the pattern. We assume the patterns are of fixed length l and that they are generated by a statistical source ("pattern generator") of zero Markov order, i.e the pattern generator generates one letter at a time (amino acid or nucleic acid) in each position, according to a position-specific probability distribution, which is independent of the letter chosen in the previous position. Hence, the model is essentially a sequence of distributions, or a profile, as in Chapter 4. We denote the model by **Q**. For example, a possible model **Q** of length 3 can be the following model:

	position		
letter	1	2	3
A	0.6	0.2	0.1
C	0.2	0.2	0.4
G	0.1	0.2	0.3
T	0.1	0.4	0.2
	$\mathbf{q}_1$	$\mathbf{q}_2$	$\mathbf{q}_3$

We denote the probability distribution in position j as $\mathbf{q}_j$. One possible pattern that this model can generate is ACC, and its probability is

$$\begin{aligned}\text{Prob}(ATC|\mathbf{Q}) &= \text{Prob(A at position 1)Prob(T at position 2)Prob(C at position 3)}\\ &= q_1(A) \cdot q_2(T) \cdot q_3(C) = 0.6 \cdot 0.4 \cdot 0.4 = 0.096\end{aligned}$$

A less likely pattern is GCA whose probability is

$$\text{Prob}(GCA) = 0.1 \cdot 0.2 \cdot 0.1 = 0.002$$

To distinguish the patterns from other positions along the sequences we also assume a fixed background model and denote it $\mathbf{P}_0$. In other words, the model is assumed to be position independent and the same probability distribution is used to emit characters in different positions. For example, a possible background model for nucleotides can be the uniform distribution $\mathbf{P}_0 = (0.25, 0.25, 0.25, 0.25)$. The probability of ACC according to this model is simply $P_0(ACC) = P_0(A)P_0(C)P_0(C) = 0.015625$. For amino acids, the background model can be estimated from the background distribution of amino acids in a large database of protein sequences.

More complex models can be used to describe the pattern and/or the background characters. For example, we could have used a first order Markov model in which the probability of observing a specific character at position j depends on the character we observed in position $j-1$. In that case, the probability of ACC would be $P_1(A) \cdot P_2(C|A) \cdot P_3(C|C)$. However, this model has $4 + 4 \cdot 4 + 4 \cdot 4 = 36$ parameters (as opposed to 12) and determining these probabilities reliably can be a challenge when data is sparse. In some cases there is a strong dependency between remote positions along the pattern (e.g., RNA motifs). Models that account for such dependencies were used in [168, 169].

5.2.2 Model quality

To choose between the many possible models that could generate patterns of length l (each one with a different set of parameters), we need a quality measure that would assess the fit of each model. Since the sequences are unlabeled, external quality measures that are based on a priori information are not applicable. The alternative is to use internal measures, such as the likelihood of the data given the model or the posterior probability of the model given the data (see Section 3.12.7), and pick the model that results in the maximal value. We will discuss two possible scenarios.

5.2.2.1 Case 1: model unknown, patterns are given

If we are given the set of common patterns $\mathbf{D} = \{\mathbf{s}_1, \mathbf{s}_2, ..., \mathbf{s}_N\}$ (where $\mathbf{s}_i$ is a substring of $\mathbf{a}_i$), we can estimate the best model $\mathbf{Q}$ using the Bayesian approach. This would be the model that maximizes the posterior probability of the model given the data $P(\mathbf{Q}|\mathbf{D})$. Using Bayes' formula

$$P(\mathbf{Q}|\mathbf{D}) = \frac{P(\mathbf{D}|\mathbf{Q}) \cdot P(\mathbf{Q})}{P(\mathbf{D})}$$

Assuming $P(\mathbf{Q})$ is uniform over all $\mathbf{Q}$ (this is not necessarily the case, however, estimating the prior over the hypothesis space is not trivial) and because $P(\mathbf{D})$ does not depend on $\mathbf{Q}$ then

$$\max_Q P(\mathbf{Q}|\mathbf{D}) = \max_Q P(\mathbf{D}|\mathbf{Q}) = \max_Q \log P(\mathbf{D}|\mathbf{Q})$$

The last step exploits the fact that the logarithm transformation is monotonic.

Let us look at $P(\mathbf{D}|\mathbf{Q}) = \prod_{i=1}^{N} P(\mathbf{s}_i|\mathbf{Q})$ more carefully. The model $\mathbf{Q}$ is composed of l probability distributions $\mathbf{Q} = \mathbf{q}_1 ... \mathbf{q}_j ... \mathbf{q}_l$ where each distribution $\mathbf{q}_j$ corresponds to one position and has c parameters $q_{j1}, q_{j2}, ..., q_{jc}$ (where $c = 4$ for nucleic acids and 20 for amino acids). Hence

$$P(\mathbf{s}_i|\mathbf{Q}) = \prod_{j=1}^{l} P(s_{ij}|\mathbf{q}_j) = \prod_{j=1}^{l} q_j(s_{ij})$$

and

$$\log P(\mathbf{D}|\mathbf{Q}) = \sum_{i,j}^{N,l} \log q_j(s_{ij}) = \sum_{j=1}^{l} \left[\sum_{i=1}^{N} \log q_j(s_{ij}) \right]$$
$$= \sum_{j=1}^{l} \left[\sum_{k=1}^{c} n_{jk} \cdot \log(q_{jk}) \right] \quad (5.1)$$

where n_{jk} is the number of times we observe the k-th character in the alphabet in position j. Since we assumed zero-order Markov model, positions are independent of each other. Therefore to maximize $\log P(\mathbf{D}|\mathbf{Q})$ over all $\mathbf{Q}$, we can optimize the parameters of each position separately and maximize $I = \sum_{k=1}^{c} n_{jk} \cdot \log(q_{jk})$ for each j. This is done by solving the equation $\frac{\delta I}{\delta q_{jk}} = 0$. Consider for example the two category case (i.e., the alphabet contains only two characters). Denote $q_{j1} = p_1$ and $q_{j2} = p_2 = 1 - p_1$. Solving for p_1 we get

$$\frac{\delta I}{\delta p_1} = 0 = \frac{n_{j1}}{p_1} - \frac{N - n_{j1}}{1 - p_1}$$

Hence,

$$\begin{aligned} p_1(N - n_{j1}) - (1 - p_1)n_{j1} &= 0 \\ \Rightarrow p_1 &= \frac{n_{j1}}{N} \\ \Rightarrow p_2 = 1 - p_1 &= \frac{n_{j2}}{N} \end{aligned}$$

In the general case, solving the derivative equation $\frac{\delta I}{\delta q_{jk}} = 0$ will result in the parameters $q_{jk} = \frac{n_{jk}}{N}$

Thus, the most likely model is defined simply based on the relative frequencies of letters in each position n_{jk}. To account for zero frequencies due to small sample sets it is customary to add pseudo-counts m_{jk} for a total of M pseudo-counts (see Section 4.3.3.2). With pseudo-counts, the model parameters are set to $q_{jk} = \frac{n_{jk}+m_{jk}}{N+M}$. As we will see in Section 15.4, this is the MAP estimator of the parameters, given Dirichlet prior with hyper-parameter $m_j = (m_{j1}, ..., m_{jc})$.

5.2.2.2 Case 2: model is given, patterns are unknown

The second scenario assumes we are given a set of sequences that share a common pattern and the specific model that emitted the patterns. To detect the exact occurrence of the pattern in each genomic sequence we can search for the most likely patterns that could have been emitted by the model. Using a sliding window of length l, we can compute the likelihood $Q(\mathbf{s}) = P(\mathbf{s}|\mathbf{Q})$ of each subsequence of length l in the input sequence and pick the one that maximizes the likelihood. If the pattern does not necessarily occur in all sequences, then we can revise the procedure and pick only subsequences with significant likelihood values. A simple way to determine significance is to use the zscore approach, with $zscore(\mathbf{s}) = (Q(\mathbf{s}) - \mu)/\sigma$, where the average likelihood value μ and its standard deviation σ are computed based on all subsequences in the input sequence. Another method is to compare the likelihood value to the one obtained with the background model. If $Q(\mathbf{s}) >> P_0(\mathbf{s})$ then $\mathbf{s}$ is likely to be one of the common patterns.

5.3 Searching for good models

The problem we are facing when searching for a set of common patterns in unlabeled sequences is that we neither know the model nor the patterns. In principle, we could search for the set that is the most conserved or the one that maximizes the likelihood. However, searching the space of all possible patterns is practically impossible. For example, if the dataset contains 20 sequences of length 100 each, and we are searching for a common pattern of length 10, then there are $91^{20} \sim 10^{40}$ possible combinations we need to check. In the next sections we describe two approaches that attempt to search the hypothesis (model) space without exhaustively enumerating all possible ones.

5.3.1 The Gibbs sampling algorithm

The Gibbs sampling algorithm is one of the many Monte Carlo sampling techniques. As the name implies, the Gibbs sampling algorithm [170] works by sampling the hypothesis space in search for good models. Specifically, the algorithm belongs to the family of **Markov Chain Monte Carlo (MCMC)** algorithms, which work by changing the initial model gradually, introducing small random changes at each step. In the context of motif discovery, the algorithm assumes that all sequences contain an instance of the pattern model. Given N input sequences and the pattern length l, it iterates through Case 1 and Case 2 described above, until it converges to a "good model". The exact procedure is described next.

Initialize pattern set D

- Pick a start position j in each sequence at random and mark the subsequence of length l that starts at that position as $\mathbf{s}_i = \mathbf{a}_i(j...j+l-1)$
- Define the pattern set $\mathbf{D} = \{\mathbf{s}_1, ..., \mathbf{s}_N\}$
- Set iteration number $r = 1$

Iterate until r = max-iteration

- **Predictive update step:** redefine pattern model $\mathbf{Q}$ and background model $\mathbf{P}_0$
  - Select at random one of the sequences and denote it by z
  - Exclude the segment $\mathbf{s}_z$ from the pattern set $\mathbf{D}$
  - Compile a pattern model $\mathbf{Q}$ from the set of all other subsequences as in Case 1 described in Section 5.2.2.1
  - Compute $\mathbf{P}_0$ from all positions not in the pattern set, in all sequences but z. Add pseudo-counts.
- **Sampling step:** redefine the set of patterns
  - Compute the probability of each segment $\mathbf{x}$ of length l in the sequence $\mathbf{a}_z$ according to the pattern model $\mathbf{Q}$ and the background model $\mathbf{P}_0$.
  - Assign $\mathbf{x}$ a weight $w_x = \frac{Q(x)}{P_0(x)}$. This weight is the likelihood ratio between the pattern model and the background model. If the ratio is > 1 then it is more likely that the segment was generated by the pattern model than by the background model.

– Assign a probability $p_x = \frac{w_x}{\sum w_x}$ to each segment $\mathbf{x}$ and select a segment $\mathbf{x}$ at random according to this probability distribution[1].
– Add $\mathbf{x}$ to the pattern set $\mathbf{D}$.

The algorithm works by the method of "recruitment". Once occurrences of the pattern were selected at random they influence the model that is compiled in the predictive update phase, and in the next iteration this model will tend to recruit other instances of the same pattern. These will refine the model, helping it to recruit other instances of the same pattern. The probabilistic selection of segments in the sampling step ensures that the model does not get stuck early on in a non-optimal model and allows it to jump randomly to other parts of the search space.

The algorithm can be used to detect multiple patterns in the dataset by applying it repeatedly after masking all occurrences of the motif that were detected in the last run.

5.3.1.1 Improvements

The Gibbs sampling algorithm suffers from problems of convergence and sub-optimality typical to many other sampling algorithms. The initial set, selected randomly, might divert the algorithm away from the right pattern model. By running the algorithm multiple times, starting from a different random set every time, we can increase the chances that the algorithm will find the true pattern. Another difficulty arises from the inherent randomness in the sampling step. Since the segment $\mathbf{x}$ is selected probabilistically there is no guarantee that the right segment will be picked even in advanced iterations. This can be taken care of by picking in the last iteration the segment that maximizes the likelihood ratio in each sequence. Finally, even if the algorithm detected the true pattern, the model might lock on shifted instances of the pattern (see Figure 5.3). To address this issue we can shift the final pattern model l characters at most to the left or to the right, recompute the model and the data likelihood and pick the model that maximizes the likelihood.

5.3.2 The MEME algorithm

The MEME (Multiple Expectation Maximization for Motif Elicitation) algorithm [171] uses a different approach to the same problem. It uses a mixture model with two components to explain the observations. One component describes the motif (of a fixed length l) and the second is the background model. To estimate the parameters of the two components MEME uses an **Expectation-Maximization (EM) algorithm** (see Section 5.7).

[1]Practically, this is done by associating with every segment $\mathbf{x}$ the interval $[\sum_{x'<x} p_{x'}, \sum_{x'<x} p_{x'} + p_x]$ and selecting a number in the range $[0, 1]$ at random. The range it falls in determines the segment.

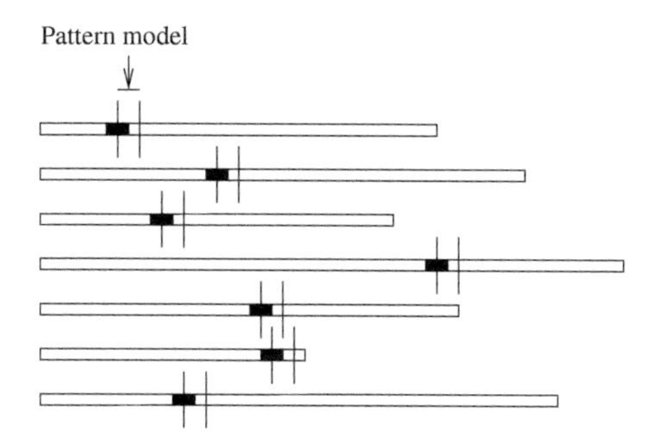

FIGURE 5.3: **Gibbs algorithm can converge to a shifted pattern.**

As a pre-processing step the algorithm takes the set of input sequences and creates a new set $\mathbf{X} = \{\mathbf{x}_1, \mathbf{x}_2, ..., \mathbf{x}_n\}$ of all subsequences (with overlaps) of length l. This is a slight variation of the original problem, and the overlapping segments might present a problem later on, but this can be taken care of by constraining the predictions.

The motif model and the background model are as defined in Section 5.2.1. The combined model assumes that one of the two is chosen first (with probability λ and $1 - \lambda$) and then a sequence of length l is generated according to the chosen model (in the case of the background model, the one-dimensional model is repeatedly applied l times to generate the sequence). The parameter λ is called the mixing parameter. Thus, the probability of a segment $\mathbf{x}$ is given by

$$P(\mathbf{x}) = \lambda_1 P(\mathbf{x}|\theta_1) + \lambda_2 P(\mathbf{x}|\theta_2)$$

where θ_1 represents the parameters of the motif model and θ_2 represents the parameters of the background model. We use θ to represent the combined set of parameters θ_1 and θ_2 (vector notation is omitted for clarity).

But how exactly can we find the parameters of the model if we do not know which class each instance belongs to? The trick is to use a set of indicator variables called class variables. These variables are later averaged out, thus making it possible to estimate the parameters that maximize $P(\mathbf{X}|\theta)$.

Notation wise, we associate the random variable z_{ik} ($i = 1..n$, $k = 1, 2$) with each sample, such that $z_{ik} = 1$ if $\mathbf{x}_i$ is in class k and 0 otherwise. These class variables are the hidden variables and are handled by the EM algorithm through averaging. We denote $\mathbf{z}_i = (z_{i1}, z_{i2})$ and let $\mathbf{Z}$ be the set of all class variables.

To understand how the algorithm works, consider the probability of a specific set of segments and a specific assignment of indicator variables that determines their class $P(\mathbf{X}, \mathbf{Z}|\theta, \lambda)$

$$P(\mathbf{X}, \mathbf{Z}|\theta, \lambda) = \prod_{i=1}^{n} P(\mathbf{x}_i, \mathbf{z}_i|\theta, \lambda) = \prod_{i=1}^{n} P(\mathbf{z}_i|\mathbf{x}_i, \theta, \lambda)P(\mathbf{x}_i|\theta, \lambda) \tag{5.2}$$

Consider the assignment $\mathbf{z}_i = (1, 0)$ then

$$P((1,0)|\mathbf{x}_i, \theta, \lambda) = P(\text{class } 1|\mathbf{x}_i, \theta, \lambda) = \frac{P(\mathbf{x}_i|\text{class } 1, \theta, \lambda)P(\text{class } 1|\theta, \lambda)}{P(\mathbf{x}_i|\theta, \lambda)}$$
$$= \frac{P(\mathbf{x}_i|\theta_1)\lambda_1}{P(\mathbf{x}_i|\theta, \lambda)}$$

Similarly,

$$P((0,1)|\mathbf{x}_i, \theta, \lambda) = \frac{P(\mathbf{x}_i|\theta_2)\lambda_2}{P(\mathbf{x}_i|\theta, \lambda)}$$

Note that this can be written in an abbreviated form as

$$P(\mathbf{z}_i|\mathbf{x}_i, \theta, \lambda) = \frac{(P(\mathbf{x}_i|\theta_1)\lambda_1)^{z_{i1}}(P(\mathbf{x}_i|\theta_2)\lambda_2)^{z_{i2}}}{P(\mathbf{x}_i|\theta, \lambda)}$$

Therefore, from Equation 5.2

$$P(\mathbf{X}, \mathbf{Z}|\theta, \lambda) = \prod_{i=1}^{n}(P(\mathbf{x}_i|\theta_1)\lambda_1)^{z_{i1}}(P(\mathbf{x}_i|\theta_2)\lambda_2)^{z_{i2}} = \prod_{i=1}^{n}\prod_{k=1}^{2}(\lambda_k P(\mathbf{x}_i|\theta_k))^{z_{ik}}$$

Taking the logarithm, we get

$$\log P(\mathbf{X}, \mathbf{Z}|Q, \lambda) = \sum_{i=1}^{n}\sum_{k=1}^{2} z_{ik} \log(\lambda_k P(\mathbf{x}_i|\theta_k))$$
$$= \sum_{i=1}^{n}\sum_{k=1}^{2} z_{ik} \log \lambda_k + \sum_{i=1}^{n}\sum_{k=1}^{2} z_{ik} \log P(\mathbf{x}_i|\theta_k)$$

All this manipulation was done so we can take the expectation value over all possible assignments of $\mathbf{Z}$, thus eliminating the hidden variables

$$E_Z(\log P(\mathbf{X}, \mathbf{Z}|\theta, \lambda)) = \log P(\mathbf{X}|\theta, \lambda)$$

Next we turn to describe the E-step and the M-step in detail.

5.3.2.1 E-step

When computing the expectation value of $\log P(\mathbf{X}, \mathbf{Z}|\theta, \lambda)$ we assume the parameters λ and θ are given from the previous iteration

$$\log P(\mathbf{X}|\theta, \lambda) = E_Z\left(\log P(\mathbf{X}, \mathbf{Z}|\theta, \lambda)\right) = E_Z\left(\sum_{i=1}^{n}\sum_{k=1}^{2} z_{ik} \log(\lambda_k P(\mathbf{x}_i|\theta_k))\right)$$
$$= E_Z\left(\sum_{i=1}^{n} z_{i1} \log(\lambda_1 P(\mathbf{x}_i|\theta_1)) + z_{i2} \log(\lambda_2 P(\mathbf{x}_i|\theta_2))\right)$$

There are only two possible assignments for $\mathbf{z}_i = (z_{i1}, z_{i2})$, which are $(0,1)$ or $(1,0)$. Therefore,

$$\log P(\mathbf{X}|\theta,\lambda) = \sum_{i=1}^{n}[1 \cdot P(z_{i1}=1)\log(\lambda_1 P(\mathbf{x}_i|\theta_1)) + 0 \cdot P(z_{i2}=0)\log(\lambda_2 P(\mathbf{x}_i|\theta_2))]$$
$$+[0 \cdot P(z_{i1}=0)\log(\lambda_1 P(\mathbf{x}_i|\theta_1)) + 1 \cdot P(z_{i2}=1)\log(\lambda_2 P(\mathbf{x}_i|\theta_2))]$$
$$= \sum_{i=1}^{n} P(z_{i1}=1)\log(\lambda_1 P(\mathbf{x}_i|\theta_1)) + P(z_{i2}=1)\log(\lambda_2 P(\mathbf{x}_i|\theta_2))$$

Note that

$$P(z_{i1}=1) = P(\text{class } 1|\mathbf{x}_i) = \frac{P(\mathbf{x}_i|\text{class } 1)P(\text{class } 1)}{P(\mathbf{x}_i)} = \frac{P(\mathbf{x}_i|\theta_1)\lambda_1}{P(\mathbf{x}_i|\theta_1)\lambda_1 + P(\mathbf{x}_i|\theta_2)\lambda_2}$$

Similarly,

$$P(z_{i2}=1) = \frac{P(\mathbf{x}_i|\theta_2)\lambda_2}{P(\mathbf{x}_i|\theta_1)\lambda_1 + P(\mathbf{x}_i|\theta_2)\lambda_2}$$

where θ and λ are given, based on the M-step estimates of the previous iteration. Using the notation $z'_{ik} \triangleq P(z_{ik}=1)$ we get

$$\log P(\mathbf{X}|\theta,\lambda) = \sum_{i=1}^{n} z'_{i1}\log(\lambda_1 P(\mathbf{x}_i|\theta_1)) + z'_{i2}\log(\lambda_2 P(\mathbf{x}_i|\theta_2)) \tag{5.3}$$

5.3.2.2 M-step

We would like to maximize $\log P(\mathbf{X}|\theta,\lambda)$ over θ and λ. To maximize over λ, we need to compute

$$\frac{\delta \log P(\mathbf{X}|\theta,\lambda)}{\delta\lambda} = 0$$

Since

$$\log P(\mathbf{X}|\theta,\lambda) = \sum_{i=1}^{n} z'_{i1}\log(\lambda_1 P(\mathbf{x}_i|\theta_1)) + z'_{i2}\log(\lambda_2 P(\mathbf{x}_i|\theta_2))$$
$$= \sum_{i=1}^{n} z'_{i1}(\log\lambda_1 + \log P(\mathbf{x}_i|\theta_1)) + z'_{i2}(\log\lambda_2 + \log P(\mathbf{x}_i|\theta_2))$$

and because z'_{ij} are constants and $\lambda_2 = 1-\lambda_1$ we get

$$\sum_{i=1}^{n}\frac{1}{\lambda_1}z'_{i1} - \frac{1}{1-\lambda_1}z'_{i2} = 0$$
$$\Rightarrow \lambda_1\sum_{i=1}^{n}(z'_{i1}+z'_{i2}) = \sum_{i=1}^{n} z'_{i1}$$

Since $z'_{i1} + z'_{i2} = 1$ we get

$$\Rightarrow \lambda_1 = \frac{1}{n}\sum_{i=1}^{n} z'_{i1} \qquad \text{and} \qquad \lambda_2 = 1 - \lambda_1 = \frac{1}{n}\sum_{i=1}^{n} z'_{i2}$$

which is an intuitive result - the probability of each class is the normalized sum of the individual posterior probabilities $z'_{ik} = P(\text{class k}|\mathbf{x}_i)$ of that class.

Maximizing $\log P(\mathbf{X}|\theta, \lambda)$ over θ is slightly more involved, but it is actually very similar to the procedure we described in Section 5.2.2.1. The only part of Equation 5.3 that is relevant when maximizing $\log P(\mathbf{X}|\theta, \lambda)$ over θ_1 is

$$\sum_{i=1}^{n} z'_{i1} \log P(\mathbf{x}_i|\theta_1)$$

The set of parameters θ_1 characterizes the pattern model **Q**. The model is identical to the model we used in Section 5.2.2.1, and therefore we proceed as in the derivation of Equation 5.1

$$I = \sum_{i=1}^{n} z'_{i1} \log P(\mathbf{x}_i|\theta_1) = \sum_{i=1}^{n} z'_{i1} \sum_{j=1}^{l} \log P(x_{ij}|\mathbf{q}_j) = \sum_{i=1}^{n}\sum_{j=1}^{l} z'_{i1} \log q_j(x_{ij})$$

$$= \sum_{j=1}^{l}\sum_{i=1}^{n} z'_{i1} \log q_j(x_{ij}) = \sum_{j=1}^{l}\sum_{k=1}^{c} w_{jk} n_{jk} \cdot \log(q_{jk})$$

The only difference is that we now have added "weights" w_{jk} such that $w_{jk} = \sum_{i=1}^{n} z'_{i1}\delta(x_{ij} = k)$. The solution to the equation $\frac{\delta I}{\delta q_{jk}} = 0$ is similar to the one of Equation 5.1, now including the weights

$$q_{jk} = \frac{w_{jk} n_{jk}}{\sum_k w_{jk} n_{jk}}$$

and as before, it is recommended to add pseudo-counts. The parameters of the background model are estimated similarly.

5.3.2.3 The iterative procedure

The MEME algorithm initializes the parameters randomly or based on prior information. Then it iterates through the E-step (to average the hidden variables) and the M-step (to re-estimate the parameters) until convergence. As with the Gibbs algorithm, MEME is also susceptible to converge to a local minimum. Moreover, it can be significantly slower than the Gibbs algorithm. However, it also has certain advantages. For example, it can accommodate zero, one or multiple occurrences of the motif in the same sequence. The algorithm can also be used to discover multiple motifs in a dataset either by increasing the number of components in the mixture or by applying the two-component mixture repeatedly and masking all occurrences of the motif modeled in each iteration.

5.4 Combinatorial approaches

Combinatorial approaches to motif discovery are non-parametric methods where the statistical formalism of the previous methods is discarded (for the most part) in favor of combinatorial analysis of the search space.

In their 2000 paper, Pevzner and Sze [172] formulated the following problem (referred to as the **planted motif problem**): Given a set of sequences, 600 nucleotides long each, find a pattern of length 15 with four mismatches at most, which is hidden in these sequences. This problem is referred to as the $(15, 4)$-signal problem. These numbers are typical of transcription factor binding sites. Therefore, given a set of co-regulated genes, the task of finding a common pattern in their downstream regions (which are usually several hundred nucleotides long) can be phrased as an (l, d)-signal problem where l is the length of the pattern and d is the number of mismatches. In general, we are given a dataset $\mathbf{D}$ of N input sequences, each of length L, and we assume each contains an instance of the (l, d) pattern.

While a $(15, 4)$-pattern is considered to be a strong signal, it is apparently missed by algorithms such as Gibbs and MEME, since the initial random model often results in a locally optimal solution. Finding a pattern of the form $(l, 0)$ is simple and can be solved through a straightforward enumeration of all possible substrings of length l using a sliding window, and can be sped up by indexing (see Problem 5.2). However, most sequence signals are variable and are often affected by mutations, insertions and deletions. In the following discussion we will consider only mutations, but the algorithms can be generalized to include deletions and insertions as well.

For short patterns, we can just enumerate all possible k-mers over the alphabet and search for occurrences of similar k-mers in the dataset. Using measures such as likelihood ratio (as in Section 5.3.1) or information content [173] we can score each k-mer and its approximate occurrences. The highest scoring one is likely to be the hidden motif. This approach is referred to as the **pattern-driven approach**, and if the pattern itself appears in the dataset, then we are almost bound to discover it. This method however does not scale well to long patterns, and exhaustive enumeration in search of the highest scoring motifs becomes unfeasible for long motifs with increasing numbers of mutations. The search space can be reduced by exploring only the substrings that are observed in the dataset (the **sample-driven approach**). However, this also reduces the likelihood that we will find the pattern. If the only occurrences of the pattern in the dataset are approximate then we are not guaranteed to reveal the pattern.

5.4.1 Clique elimination

One possible approach, as implemented in the Winnower algorithm [172], is to use a graph representation where nodes correspond to substrings of length

l in the input sequences, and edges are introduced between similar substrings. If a pattern can change by up to d mutations, then two instances of the pattern can differ in up to $2d$ positions. By comparing all pairs of sequences we can detect all pairs of substrings $\mathbf{s}_1, \mathbf{s}_2$ of length l whose hamming distance $d(\mathbf{s}_1, \mathbf{s}_2) \leq 2d$. Naturally, many similarities in this graph have nothing to do with the signal we are looking for but rather are due to chance similarities. These similarities are referred to as **spurious edges**.

Since we assumed each input sequence contains an instance of the pattern, the subgraph that corresponds to the instances of the real pattern forms a clique, where every two nodes are connected. The idea of the Winnower algorithm is to look for cliques (possibly due to multiple signals) and eliminate all other edges that are not part of large cliques.

Specifically, the algorithm starts by building a graph $G(\mathbf{D}, l, d)$ such that a node is introduced for each possible substring in each of the input sequences (that is, one for each possible start position j where $1 \leq j \leq L - l + 1$). Denote the substring that starts at position j of input sequence i by $\mathbf{s}_{i,j}$ and the character at position x within this substring by $\mathbf{s}_{i,j}(x)$. An edge is drawn between every two nodes $\mathbf{s}_{i,j}$ and $\mathbf{s}_{p,q}$ if the hamming distance between the two corresponding strings is less than d (considering only substitutions) and if $i \neq p$. That is, no edges are formed between substrings from the same input sequence. In other words, the graph is a **multipartite** graph where each part (subgraph) consists of a disconnected set of nodes (substrings) corresponding to one input sequence. Edges exist only between nodes from different parts of the graph. Given N sequences, each of length L, then the multipartite graph consists of N subgraphs of size $L - l + 1$. Note that this graph can be computed efficiently, since the distance between consecutive strings is related and can be expressed as

$$d(\mathbf{s}_{i,j+1}, \mathbf{s}_{p,q+1}) = d(\mathbf{s}_{i,j}, \mathbf{s}_{p,q}) - \delta(\mathbf{s}_{i,j}(1), \mathbf{s}_{p,q}(1)) + \delta(\mathbf{s}_{i,j+1}(l), \mathbf{s}_{p,q+1}(l))$$

where δ is Kronecker's delta function, and therefore the whole graph can be constructed in $O((NL)^2)$.

Since a signal (l, d) corresponds to substrings that are not more than $2d$ apart, then a pattern corresponds to a clique of size N in the graph $G(\mathbf{D}, l, 2d)$ where every two occurrences of the pattern are connected. Finding patterns then translates to the problem of finding large cliques in the graph. This is considered an NP-hard problem in general [174, 175]; however, assuming that there is one or a few hidden patterns the problem is actually easier since the graph consists of mostly spurious edges. Nodes that do not represent instances of the pattern can be eliminated by inspecting their connectivity. If the degree of a node is less than N then the node (and the corresponding string) cannot be part of a pattern clique and can be deleted. Note however that this still leaves many nodes in the graph that are connected to N or more other nodes but are not part of an N-clique, since the other nodes are not necessarily connected to each other. Therefore, the remaining graph is still large and spurious.

To eliminate these nodes, the Winnower algorithm tests each edge to see if it is part of an **extendable clique**. A clique over a set of k nodes is called extendable if there is at least one node in the graph that can be added to the set, such that the $k+1$ nodes also form a clique (in other words, there is a node in the graph that is connected to all k nodes of the clique). An edge is considered spurious if it is not part of any extendable clique of size k.

The key to efficient elimination is the observation that every edge in a N-clique belongs to at least $N_k = \binom{N-2}{k-2}$ extendable cliques of size k. It is easy to see that this is the case. Once we selected an edge, we selected the two nodes connecting it. To expand the edge to a k-clique we need to pick $k-2$ more nodes out of the remaining $N-2$ nodes in the N-clique. The number of possible choices is $N_k = \binom{N-2}{k-2}$. Therefore, each edge that does not belong to N_k extendable cliques can be removed.

To see if this is a feasible test, let's compute the expected running time of this algorithm. Denote the probability of a random edge to be p. This is the probability that two random strings of length l differ by no more than $2d$ characters. We can estimate p from the background distribution P_0 of nucleotides. Let $P_1 = \sum_{a \in \mathcal{A}} P_0(a)^2$ denote the probability of a match between two randomly drawn nucleotides. Let P_2 denote the probability of a mismatch

$$P_2 = \sum_{a \in \mathcal{A}} \sum_{b \in \mathcal{A}, b \neq a} P_0(a) P_0(b) = 1 - P_1$$

Then the probability of a random edge is the total probability to observe $i \leq 2d$ mismatches in l randomly drawn pairs of nucleotides, given by the sum

$$p = \sum_{i=0}^{2d} \binom{l}{i} P_1^{l-i} P_2^i$$

The probability that k nodes in different parts of the graph form a random clique is the probability that edges exist between all pairs and is given by $p^{k(k-1)/2}$. There are $\binom{N}{k} L^k$ possible sets of k nodes in the graph since there are $\binom{N}{k}$ possible subsets of size k of the input sequences, and from each we can choose any of the L nodes. Therefore the expected number of k-cliques in the graph is the number of k-sets multiplied by the probability that each one is a clique

$$\binom{N}{k} L^k p^{k(k-1)/2} \tag{5.4}$$

Note that this is an approximation, since the events are not independent.

To check whether a certain clique is extendable we need to check the graph edges from each of its nodes and see whether they lead to a node that can expand the clique (meaning, a node that is connected to all the nodes of the clique). Because of the way the graph is constructed, the set of k nodes comes from k different parts of the graph and the additional node must be in the remaining $N - k$ parts of the multipartite graph. This test takes $O(k(N - k)E) < O(kNE)$ per clique, where $E = Lp$ is the expected number of edges between a node and one part of the graph (with L nodes). All together, this algorithm takes

$$O\left(\binom{N}{k} L^k p^{k(k-1)/2} kNE\right)$$

given the list of cliques. Eliminating all edges that are not part of extendable k-cliques results in a sparse graph with one or more large cliques that correspond to hidden patterns. For $k = 3$ this takes approximately $O(N^3 L^3 p^3 3NLp) = O((NLp)^4)$, which results in reasonable running times.

Note that the whole setup assumes that l and d are given. Naturally, these are unknown in real life. A possible practical approach to determining these is to run the algorithm repeatedly, starting from $d = 0$ and fixed l and increasing d gradually until the resulting graph (after the edge elimination process) is not empty. Repeating this process for various l can help to determine the significance of the pattern since graphs tend to be either fairly empty, sparse with large cliques or dense with many spurious edges. Obviously, the success of the algorithm depends on the probability that the pattern appears by chance within the set of N sequences, which is $L^N p^{N(N-1)/2}$ by Equation 5.4. The higher it is, the larger the cliques we need to check for extendability. However, checking larger cliques can become too computationally intensive.

5.4.2 Random projections

The random projections algorithm of Buhler and Tompa [176] attacks the problem without computing the similarities between all substrings first. This algorithm can handle longer patterns with a higher number of mismatches. The algorithm works by projecting each l-mer in the input sequences onto a smaller subset of positions. This projection maps l-mers into "buckets", such that certain buckets are likely to contain only occurrences of the signal motif. The algorithm starts by choosing k out of the l positions at random and then hashes every l-mer x into a bucket $f(x)$, which is a function of its bases in these k positions. A bucket with an unusually high number of l-mers is likely to be enriched with occurrences of the motif. Any such bucket is further explored as a potential motif and refined using a MEME-like local search algorithm. By repeatedly picking a different subset of k positions and hashing the l-mers, the algorithm explores many possible seeds and increases the likelihood to discover the true motif.

Specifically, given a set of k positions that were chosen uniformly at random, without repetitions, the hash value of an l-mer x is simply the substring generated by concatenating the nucleotides in the selected k positions. Let $\mathbf{Q}$ denote the consensus string of the planted motif and $f(\mathbf{Q})$ the bucket of the planted motif (referred to as the **planted bucket**). The goal of the algorithm is to hash at least n occurrences of the planted motif to the same bucket. If $k < l - d$ then there is a good chance that this will happen. If k is not too small then the probability that other random l-mers will fall in the same bucket is small.

The dimension of the substring k is chosen such that the number of random l-mers that fall in the same bucket is minimized. There are 4^k possible substrings of length k and there are $N(L - l + 1)$ possible strings of length l in the input dataset. Assuming the nucleotides are distributed evenly in random sequences, then the expected number of l-mers in each bucket is

$$E = \frac{N(L - l + 1)}{4^k}$$

By requiring that $E \leq 1$ (that is, no more than one random l-mer per bucket) we obtain that $k \geq \log_4 N(L - l + 1)$.

The probability that an occurrence of the motif will be hashed to the planted bucket $f(\mathbf{Q})$ is the probability that the k selected positions happen to be the positions that have not changed in that specific occurrence. In the worst case, the occurrence has d mutations and hence $\binom{l-d}{k}$ of the $\binom{l}{k}$ possible ways to pick k positions fall in the non-mutated positions. Therefore, the probability that the occurrence will be mapped to the planted bucket is

$$p = \frac{\binom{l-d}{k}}{\binom{l}{k}}$$

Since we assume the occurrences are independent of each other (the mutation in each occurred after duplication) the probability that n out of N occurrences will be mapped to the planted bucket is simply given by the binomial distribution with parameter p. That is

$$P(n) = \binom{N}{n} p^n (1 - p)^{N-n}$$

and the probability that fewer than n will be mapped to the planted bucket (considered to be a "failed experiment") is given by the sum

$$f = P(i < n) = \sum_{i=0}^{n} \binom{N}{i} p^i (1 - p)^{N-i}$$

We would like to ensure with probability q that the planted bucket has at least n occurrences. If we repeat the projection again and again, every time with a different subset of positions we are bound to eventually observe n or more occurrences in the planted bucket ("success"). Specifically, each projection is considered as an independent trial, and the probability that m independent trials fail (that is, none results in n occurrences of the pattern in the planted bucket) is simply f^m. The probability that at least one of the

m trials succeeds is the complementary probability $1 - f^m$. To succeed in at least one of the experiments with confidence q (e.g., $q = 0.95$) we require that

$$1 - f^m \geq q$$

Solving for m we get that we need to repeat the projection

$$m = \frac{\log(1-q)}{\log f}$$

times. The only parameter that has not been determined is n, the minimal number of occurrences necessary to find the planted motif, but it has been suggested to set $s = 3$ or $s = 4$ for patterns occurring 4-20 times in sequences of 600-1000 long.

In a final refinement step, the projection algorithm explores each bucket with at least n l-mers and invokes an EM algorithm, similar to MEME, to locally optimize the motif. The initial model is created using the l-mers that were hashed to that bucket and estimating the parameters as in Section 5.2.2.1 with pseudo-counts following the prior of nucleotides in the background sequences. Several iterations of EM are usually enough to improve the initial model. If starting from the planted bucket, then a few iterations are likely to identify the right pattern, since k positions are already in agreement with the consensus pattern. The refined model $\mathbf{Q}$ is then used to scan each input sequence for a potential occurrence, and the l-mer with the highest likelihood ratio is picked. The total score of the set of occurrences $\mathbf{X}$ is defined as the likelihood ratio

$$\text{score}(\mathbf{X}|\mathbf{Q}) = \prod_{\mathbf{x} \in \mathbf{X}} \frac{P(\mathbf{x}|\mathbf{Q})}{P_0(\mathbf{x})}$$

The bucket that results in a set $\mathbf{X}$ with the highest score over all buckets and all experiments is reported as the planted motif

It should be noted that not all patterns can be detected with the standard pattern-driven approach, since some become indistinguishable from noise. Consider a randomly chosen l-mer pattern. The probability that another random sequence of l characters will differ from the pattern in exactly i positions is

$$\binom{l}{i}\left(\frac{3}{4}\right)^i\left(\frac{3}{4}\right)^{l-i}$$

since the probability that a randomly chosen character in one position differs from the pattern's character in the same position is $3/4$ and there are $\binom{l}{i}$ ways to choose i "mutated" positions. Therefore, the probability that a random l-mer differs from the pattern in no more than d positions is the sum

$$p = \sum_{i=0}^{d} \binom{l}{i}\left(\frac{3}{4}\right)^i\left(\frac{3}{4}\right)^{l-i}$$

Such an l-mer can be considered as a potential occurrence of the pattern. The probability that the random l-mer differs in more than d positions is the complementary probability $1 - p$. The probability that none of the l-mers in a random sequence of length L can be considered a potential occurrence is approximately (ignoring overlaps) $(1-p)^{L-l+1}$, and the probability that at least one l-mer is a potential occurrence is the complementary probability $1-(1-p)^{L-l+1}$. Therefore, the probability that there is at least one potential occurrence in each of the input sequences is

$$\left(1-(1-p)^{L-l+1}\right)^N$$

There are 4^l possible patterns of length l, and hence the expected number of such patterns that occur, just by chance, in each of the input sequences is

$$4^l\left(1-(1-p)^{L-l+1}\right)^N$$

For example, 20 sequences of length 600 can contain at least one spurious $(9, 2)$ pattern, just by chance, and therefore it is harder to detect a true $(9, 2)$ pattern that is planted in these sequences. To reveal such a hidden pattern in long unaligned sequences it is necessary to consider also higher order dependencies and correlations between remote positions, as discussed in Further reading.

5.5 Further reading

Early algorithms for motif discovery include the algorithms by [177–180]. Because of the size of the search space, these are usually limited to short patterns with limited flexibility (i.e., some constraints are imposed on the pattern, such as exact or high conservation at pattern positions and fixed spacing between the main pattern elements). Nevertheless, these algorithms can be useful for the construction of motif and domain databases. For example, the algorithm of Smith et al. [177] was used to construct the Blocks database [88] by identifying short "motif blocks" in groups of related proteins in the PROSITE database. These motifs were then extended in both directions till the score fell below a certain threshold score. Jonassen et al. [181] proposed an improved method that allows greater ambiguity at partially conserved pattern positions and limited variable spacing between pattern elements.

In addition to the pattern-driven and sample-driven approaches we discussed, there are *extended sample-driven* approaches that explore also neighborhoods of the observed substrings. A more recent work in this field further developed the work of Pevzner and Sze by combining elements from statistical models in the form of multi-dimensional profiles. These generalized profiles summarize the joint statistics of nucleotides at different positions along the pattern and can find subtler motifs in DNA sequences as they better distinguish the pattern from noise [182]. Other algorithms were suggested [183–185]. It also has been shown that combining multiple methods improves detection of motifs [186].

Motif detection algorithms provide an alternative to MSA algorithms for detection of protein motifs, since they do not require pre-aligned sequences. Among the other applications of motif discovery algorithms are the design of PCR primers [187, 188] and detection of functional sites in DNA sequences [173, 189]. Another important application is identification of RNA motifs. These motifs play an important role in post-transcriptional regulation of gene expression. They work by binding to the mRNA genes (affecting their translation) or by binding to the translated proteins and affecting their stability. RNA motifs are harder to detect since they form unique 3D structures that introduce dependencies between remote sequence positions [169]. For updates and additional references check the book's website at `biozon.org/proteomics/`.

5.6 Conclusions

- Motif discovery algorithms search for hidden patterns in sets of unaligned sequences, such as common regulatory elements or signature patterns of functional sites in protein and DNA sequences. Patterns are usually short and relatively conserved.
- Model-based algorithms use a statistical model (a profile) to describe the pattern and search for models that result in maximum likelihood of the observed data.
- Locally optimal models can be identified using sampling algorithms or the expectation-maximization algorithm.
- Combinatorial approaches search for planted patterns by exploring the similarity graph of all k-mers.
- Short patterns can be detected by enumerating all possible k-mers and looking for patterns with high information content or maximum likelihood ratio.
- Longer patterns can be detected by searching for cliques in the similarity graph or by means of random projections on a smaller pattern space.

5.7 Appendix - the Expectation-Maximization algorithm

The Expectation-Maximization (EM) method is a popular algorithmic approach for learning the parameter vector $\vec{\theta}$ of a distribution from training data with missing features.

Consider a **full** sample set $\mathbf{Y} = \{y_1, y_2, ..., y_n\}$ drawn from a single distribution that is characterized by a parameter vector θ. For simplicity we will drop the vectorial notation, though instances and parameters are assumed to be multi-dimensional. Each single example y consists of **visible features** x that can be measured and **missing (hidden) features** h

$$y = \{x, h\}$$

Notation wise, we will sometimes use z instead of h to denote the hidden features.

If there are no missing features then we can approximate the parameter vector from the training data by using the maximum likelihood (ML) estimator or the maximum a posterior (MAP) estimator (in the case of nonuniform prior, see Section 3.12.7). However, in the presence of hidden features the problem is more complicated. Consider the likelihood of a single example (the total likelihood of the dataset is the product over all instances). Our goal is to maximize the likelihood of the data given the parameter

$$p(y|\theta) = p(x, h|\theta)$$

However, only some of the data is visible, while the parameter vector characterizes the distribution of the *full* data. Therefore, we need to somehow handle the missing features h.

The approach of the EM algorithm is to maximize only the likelihood of the *observed data*, after marginalizing over the unobserved data. In other words, by averaging over all possible values for the missing features h the resulting likelihood function will reflect only the visible features x. However, to average over h we need to know the underlying distribution, but the parameter vector θ is unknown. Furthermore, h is part of y and the parameter vector characterizes the distribution of y. Therefore, averaging is not trivial.

To compute the parameter vector that maximizes the marginalized likelihood, the algorithm starts from an initial estimate for the parameters and then iteratively alternates between two modes: marginalizing over the hidden variables using the current estimate for the parameters (**E-step**), and computing θ that maximizes the marginalized likelihood (**M-step**). The iterative procedure converges to a locally optimal estimator of the function $p(x|\theta)$.

Formally, we would like to find the maximum likelihood estimator for θ, given the observed data

$$\theta = \arg\max_{\theta} p(x|\theta) \tag{5.5}$$

Given a joint probability distribution over variables X and H, marginalizing over H means averaging the joint probability distribution over all possible values of H, to obtain a marginal probability distribution over X

$$p(x) = \int_h p(x|h)p(h)dh = \int_h p(x,h)dh$$

or

$$P(x) = \sum_h P(x|h)P(h) = \sum_h P(x,h)$$

for discrete variables. Therefore,

$$p(x|\theta) = \int_h p(x,h|\theta)dh = \int_h p(y|\theta)dh \tag{5.6}$$

Maximizing Equation 5.5 is equivalent to maximizing the logarithm

$$\theta = \arg\max_\theta p(x|\theta) = \arg\max_\theta \left[\log \int_h p(x,h|\theta)dh\right] \tag{5.7}$$

which is often easier to work with since it converts products of probabilities into sums of log-probabilities, where different parameters can be maximized separately.

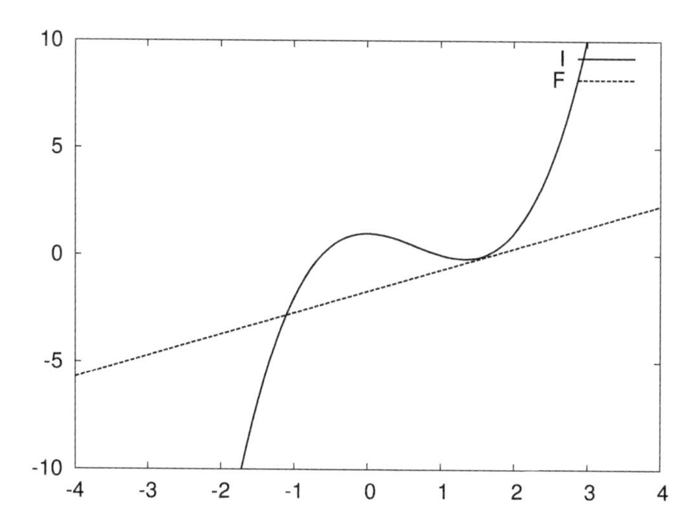

FIGURE 5.4: The EM algorithm. In each iteration the algorithm computes a lower bound $F(\theta|\theta^t)$ on the marginalized likelihood function I that we want to maximize. The lower bound is maximized locally ensuring that the new estimate for θ also results in a higher likelihood value I.

Say the parameter vector θ^t is our current best estimate (note it characterizes the **full** distribution), and θ is a candidate vector for an improved estimate. Let $p(h|x, \theta^t)$ be the likelihood of the hidden variables, given the observed data and θ^t from iteration t, then we can trivially rewrite Equation 5.7

$$I = \log \int_h p(x, h|\theta) dh = \log \int_h p(h|x, \theta^t) \frac{p(x, h|\theta)}{p(h|x, \theta^t)} dh \tag{5.8}$$

and note that $p(h|x, \theta^t) = p(h, x|x, \theta^t) = p(y|x, \theta^t)$. The logarithm of the integral (or sum, for discrete variables) in Equation 5.8 complicates the maximization, since the derivatives would involve non-linear equations. Therefore, we introduce a lower bound, which is easier to analyze than the original function, and maximize the lower bound (Figure 5.4 gives an intuitive illustration of the process). Recall that the definition of a concave function $f(x)$ is that

$$f(\lambda_1 x_1 + \lambda_2 x_2) \geq \lambda_1 f(x_1) + \lambda_2 f(x_2)$$

for any linear combination $\lambda_1 x_1 + \lambda_2 x_2$ such that $\lambda_1 + \lambda_2 = 1$ and $\lambda_1, \lambda_2 > 0$. Generalizing this basic property of concave functions to a sum of n elements, we obtain what is referred to as **Jensen's inequality**

$$f\left(\sum_i \lambda_i x_i\right) \geq \sum_i \lambda_i f(x_i)$$

which can be proved by induction. Similarly, for continuous functions

$$f\left(\int_x p(x) \cdot x \cdot dx\right) \geq \int_x p(x) f(x) dx \tag{5.9}$$

where $p(x)$ is a probability density function such that $\int_x p(x)dx = 1$. The logarithm function is a concave function, and since $\int_h p(h|x, \theta)dh = 1$ then the probabilities can serve as the coefficients of Jensen's inequality in Equation 5.9. Therefore,

$$I = \log \int_h p(h|x, \theta^t) \frac{p(x, h|\theta)}{p(h|x, \theta^t)} dh \geq \int_h p(h|x, \theta^t) \log \frac{p(x, h|\theta)}{p(h|x, \theta^t)} dh$$

Define

$$F(\theta|\theta^t) = \int_h p(h|x, \theta^t) \log \frac{p(x, h|\theta)}{p(h|x, \theta^t)} dh$$

then $F(\theta|\theta^t)$ is a lower bound on I of Equation 5.8. Note that we can write

$$\begin{aligned} F(\theta|\theta^t) &= \int_h p(h|x, \theta^t) \log \frac{p(x, h|\theta)}{p(h|x, \theta^t)} dh \\ &= \int_h p(h|x, \theta^t) \log p(x, h|\theta) dh - \int_h p(h|x, \theta^t) \log p(h|x, \theta^t) dh \qquad (5.10) \\ &= E_h[\log p(x, h|\theta)] + H(h) \end{aligned}$$

where $H(h)$ is the entropy of the distribution over the hidden variables and $E_h[f()]$ denotes the expectation value of $f()$ over h. To maximize this function we need to compute the derivative with respect to θ. That is

$$\frac{\partial F(\theta|\theta^t)}{\partial\theta} = 0$$

and since $H(h)$ does not depend on θ, this is reduced to

$$\frac{\partial E_h[\log p(x, h|\theta)]}{\partial\theta} = 0$$

The solution of this equation is the ML estimator

$$\theta_{ML} = \arg\max_{\theta} E_h[\log p(x, h|\theta)]$$

If we have a nonuniform prior over the parameter space, then we can compute the MAP estimator

$$\theta = \arg\max_{\theta} p(\theta|x) = \arg\max_{\theta} \frac{p(x|\theta)p(\theta)}{p(x)}$$

and since $p(x)$ does not depend on θ

$$\theta_{MAP} = \arg\max_{\theta} p(x|\theta)p(\theta) = \arg\max_{\theta} [\log p(x|\theta) + \log p(\theta)]$$

and the function we need to optimize in this case is

$$F(\theta|\theta^t) = E_h[\log p(x, h|\theta)] + \log p(\theta) + H(h)$$

To summarize, the basic schema of the EM algorithm consists of two steps. The E-step, during which the algorithm obtains a lower bound on the function we would like to optimize using the estimator from the previous iteration and by averaging over all possible values for the hidden variables. In the M-step the bound is maximized. The algorithm guarantees that the likelihood of the visible data (with the missing data marginalized) will increase monotonically.

The EM algorithm

Let t be an iteration counter. Let ϵ be a preset convergence criterion.

1. **Initialize:**
   - $t = 0$, set ϵ and set θ^0
2. **Do**
   - $t \leftarrow t + 1$
   - **Expectation step:** compute $F(\theta|\theta^t)$
   - **Maximization step:** $\theta^{t+1} \leftarrow \arg\max_\theta F(\theta|\theta^t)$
3. **Until** $F(\theta^{t+1}|\theta^t) - F(\theta^t|\theta^{t-1}) \leq \epsilon$
4. **Return** $\theta \leftarrow \theta^{t+1}$

5.8 Problems

For updates, additional problems, files and datasets check the book's website at biozon.org/proteomics/.

1. **The Gibbs sampling algorithm**

   You are given a set of sequences over the English alphabet (sequences). There is a hidden pattern in this set of sequences (not necessarily identical). Your goal is to learn it.

   Use the Gibbs sampling algorithm to solve this problem. Given a datafile and the pattern length, the program should output the pattern found, i.e. all the instances of the pattern in the input sequences, one per input sequence (we will refer to all instances of the pattern simply as "the patterns").

   (a) After each iteration compute the likelihood ratio of the patterns given the model (with pseudo-counts) and the background distribution of characters. Store the best model (with the highest likelihood ratio) you have found so far, and the corresponding set of patterns. Upon convergence, or when reaching the maximal number of iterations, your program should output the best set of patterns and not the last one. Explain: (1) the computation of the likelihood ratio, (2) why the likelihood ratio is better than just the likelihood, and (3) why the last model is not necessarily the best model.

   (b) To improve performance, implement the two refinements as described in Section 5.3.1.1: 1) Multiple initializations, each time starting from a different random seed. Output the best model overall. 2) Model refinement to avoid getting trapped in a shifted model. After convergence scan the sequences around the patterns of the best model, considering all possible model shifts (up to L characters apart in both sides), and output the best model.

   (c) Compare and report the best model using the vanilla version vs. the enhanced versions. How many runs on average are required to find the pattern when using the basic version vs. the enhanced versions?

2. **Indexed search**

   (a) Describe an efficient algorithm for finding exact matches of a pattern of length l with 0 mismatches in a long input sequence (the $(l, 0)$ problem), by using indexing (hashing).

(b) Extend the algorithm to handle mismatches as well. Create an extended query list by means of a Cartesian product between all positions, and represent as a finite automaton.

Chapter 6

Markov Models of Protein Families

6.1 Introduction

Statistical models are relevant to many problems in computational biology, from gene and motif prediction to analysis of pathways and gene networks. Such models can describe observations associated with certain processes or phenomena. Learning these models from data can reveal interesting patterns and explain the observed properties of the system of interest. This is especially important for classification and prediction. Once a model architecture is chosen and a model is learned, it is typically used to assess whether a new instance is consistent with the source that generated the training data and is likely to be related to the same process.

While some methods defer from defining or modeling the process directly (as will be discussed in the next chapter), statistical modeling explicitly assumes that at the core of the process we are trying to model there is a certain source that generates instances (observations) from the sample space with different probabilities. In other words, we assume the source induces a probability distribution $P(x)$ over the sample space $\mathcal{X}$. A statistical model attempts to mimic the underlying process and learn the source distribution. Such models are called **generative models**. Since we usually do not know $P(x)$, the model might induce a different probability distribution $Q(x)$ over the sample space. Our goal is to approximate the source distribution $P(x)$ as close as possible, to be able to classify instances accurately and better understand its properties.

We already discussed several statistical models in the previous chapters, in the context of various problems. Our main focus in this chapter will be on statistical modeling of protein families. Protein families are abstract entities, composed of multiple instances of homologous genes from different species. The interest in these models originates from the need to represent the information reflecting their common evolutionary origin in a concise and powerful way that would enable prediction and classification of newly sequenced proteins.

In previous chapters we already discussed several different representations of protein families, such as consensus patterns, regular expressions, position-specific scoring matrices (PSSMs), and profiles. These representations differ in their mathematical complexity, as well as in their sensitivity and selectivity. This chapter will focus on a specific type of model, called Hidden Markov Model (HMM). HMMs were initially developed and applied to tasks such as speech, handwriting and optical character recognition. They were later adopted to biological sequence analysis, with great success. Such models are at the core of several protein and domain classification systems, as will be discussed in Chapter 9.

As a motivating example we will start by discussing briefly another important problem in computational biology: the problem of gene prediction. Next we will discuss algorithms for learning and prediction with HMMs. In the second part of this chapter will discuss a related class of models, called variable order Markov models, from the perspective of codes and compression.

6.2 Markov models

In general, a Markov model describes a sequence of states that a given system might go through. Each transition from one state to another may depend on the current state and the past states. Soon we will delve into the formal description and the relevant algorithms. But to better understand Markov models and their application in computational biology, it is insightful to consider first the problem of detecting coding vs. non-coding regions along DNA sequences and see how Markov models can be naturally applied to this problem.

6.2.1 Gene prediction

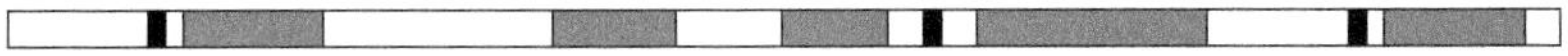

FIGURE 6.1: **Coding and non-coding sequences along a DNA sequence.** The gray regions indicate protein coding sequences, while the black regions indicate regulatory elements.

The DNA of even the smallest organism is a long sequence of nucleotides. Along this DNA sequence there are certain regions, called **coding sequences**, that code for proteins (Figure 6.1). Although proteins are considered to be the main "work-force" of the cell, the vast majority of the DNA actually does not code for proteins. Traditionally, the non-coding part has been referred to as **junk DNA**; however this name is misleading since much information is actually encoded in these regions, in the form of regulatory elements that control transcription, patterns that control DNA folding and many RNA molecules with various functions (including DNA replication and transcription, and protein translation). However, for the most part we do not know where these elements are and what their function is [190].

The protein coding regions are obviously of great interest. Detecting these regions is the key to understanding which functions exist in a given genome and the discovery of novel genes. Protein coding sequences (**genes**) are scattered along the DNA. However, their location is unknown a priori. We do know that every protein coding segment must start with a START codon ATG and end in a STOP codon (TAA, TAG or TGA). However, this in itself is not enough to determine the existence of a gene. In the course of evolution many random duplication events copied partial segments of protein coding sequences to other locations along the genome. These copies, called **pseudogenes**, can be perceived as the "playground of evolution". Many of them could not be transcribed or failed to fold into a functioning protein and were gradually degraded through random mutations. Therefore, there are many start codons that are not followed by coding regions. However, there is one major difference between coding and non-coding sequences: coding sequences are compiled of codons, and therefore there are certain structural constrains that do not exist or are weaker in non-coding sequences. Hence, we would expect the statistics of nucleotides to be different in these regions.

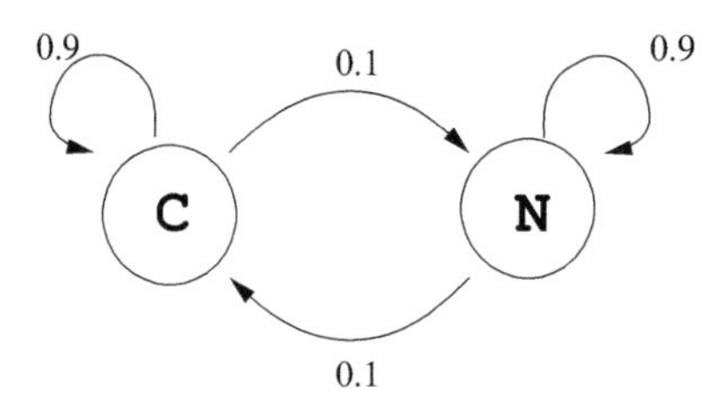

Sequence of states: **NNNCCCCCCCCCCCCNNNNNN . . .**

DNA sequence: **CGGATGCTCGGATACTAACCG . . .**

FIGURE 6.2: A two-state Markov model. This simple model has only two states, coding (C) and non-coding (N). The machine can move from one state to another or from a state to itself. Each transition occurs with a certain probability. The sum of the transition probabilities from each state is one. A sequence of moves through the model generates a sequence of states, each one corresponds to one position along the DNA sequence.

Our goal is to build and train a model that would capture these differences and would enable us to identify and predict protein-coding regions along a DNA sequence. We can think of the problem in statistical terms and imagine a state machine where the DNA moves probabilistically between states corresponding to positions in coding and non-coding regions, as in the simple Markov model depicted in Figure 6.2. This model, however, ignores the statistics of nucleotides within coding and non-coding sequences. Furthermore, we have not specified which nucleotide is observed in each position.

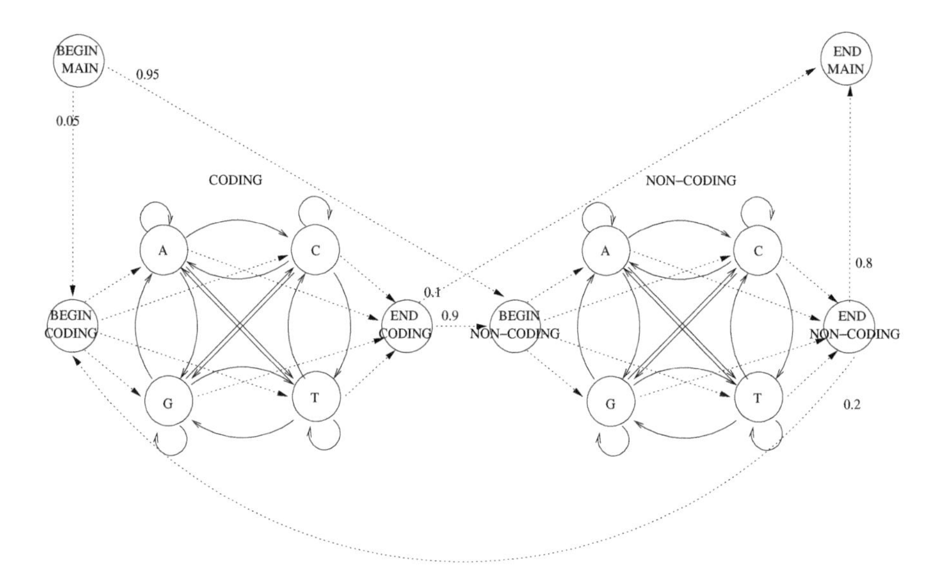

FIGURE 6.3: Markov model for gene prediction. A Markov model with two main parts, coding and non-coding and the transitions between them. Each part contains four states corresponding to the individual nucleotides, as well as begin and end states. These two states help to simplify model representation (otherwise, each of the four nucleotide states in the coding part has to be connected directly to each of the nucleotide states in the non-coding part and vice versa). The model starts from the main begin state, and then transitions to one of the begin states in either the coding or the non-coding part with probabilities 0.05 and 0.95, respectively. Assume the model transitions to the begin non-coding state. The model then continues with a sequence of moves within the non-coding part, with each state corresponding to one nucleotide, until it transitions to the end non-coding state (transition probabilities not shown). At this stage, the model can either transition to the main end state (probability 0.8) or to the begin coding state (0.2). The former ends the sequence of moves, while the latter is followed by a sequence of moves within the coding part. Once the model moves to the end coding state, it can transition to the main end state (probability 0.1) or start another non-coding sequence (0.9), and repeat.

Adding more states that correspond to nucleotides within coding and non-coding sequences we obtain the model of Figure 6.3. A specific sequence of moves through that model corresponds to a specific DNA sequence. However, this model still falls short of capturing the complexity of the problem since it ignores the codon structure and the higher-order statistical dependencies between nucleotides, which characterize coding regions. A revised model that takes the codon structure into account is depicted in Figure 6.4. The transition probabilities can be estimated from a training dataset, where the locations of the coding regions are marked (see Problem 6.1). Once these parameters were learned we can apply the model to a new DNA sequence and determine the most likely sequence of moves that generated that sequence (as will be explained later in this chapter). From that sequence of moves we can determine the most probable locations of the coding sequences.

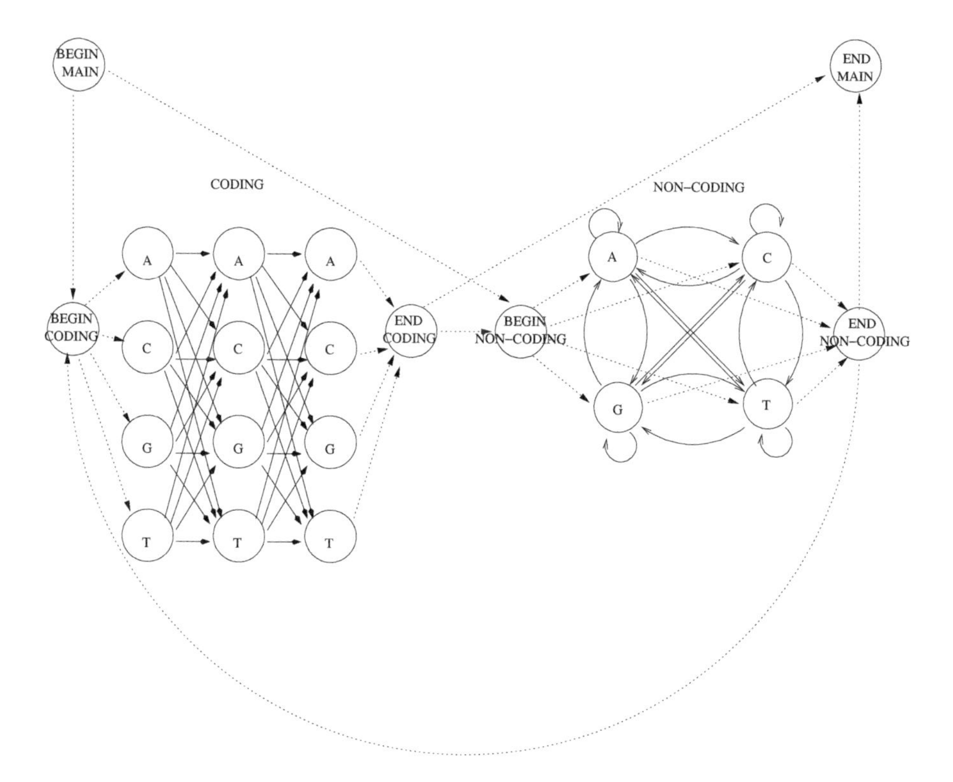

FIGURE 6.4: Markov model for gene prediction - revisited. This model is a variation on the model of Figure 6.3, where the states that make up the coding part are linked so as to reflect the codon structure.

An alternative approach is to design a model where the states correspond directly to the different processes or classes (coding vs. non-coding) as opposed to nucleotides within these classes, and introduce **visible symbols**, which are the nucleotides emitted by these states, with different probabilities. A very simple example of this type of model is given in Figure 6.5. Once we introduce multiple states that can generate the same symbols then we can no longer tell which state emitted each nucleotide in a given sequence, and this is why this kind of model is called a **Hidden** Markov Model (HMM). In HMMs, the state usually represents an internal characteristic of the source, and the visible symbol is an external feature that can be measured (such as nucleotides or amino acids, activity levels etc).

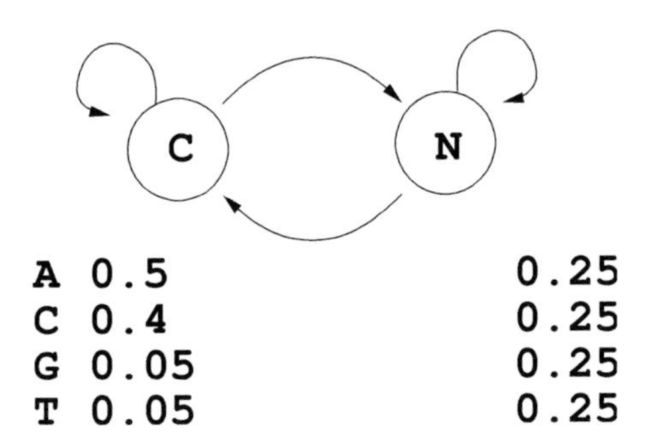

FIGURE 6.5: **A hidden Markov model for gene prediction.** The model has only two states, coding (C) and non-coding (N). Each state can emit any of the four nucleotides with different probabilities. For example, the probability to emit A in the coding state is 0.5.

The model of Figure 6.5, however, is too simple to actually be useful for gene prediction. A better model would account for the codon structure as well as unique signals (i.e., promoters) preceding the coding regions or splice sites inside the cording regions (see Problem 2.2). In fact, the problem of gene prediction requires much more complex models than the ones depicted above, and several references are given in Further reading.

Now that the intuition behind these models is hopefully clearer we will proceed to the formal definition of HMMs and their application for protein classification.

6.2.2 Formal definition

Markov models describe a sequence of "decisions" or states, where the state at a given time depends on the state(s) at previous time steps. The formalism we adopt is similar to the one used in [191]. The model has a finite number

(c) of different states, denoted by $\{w_1, w_2, ..., w_c\}$. A simple model with four states is depicted in Figure 6.6. At any given time the model may move from one state to another, or stay at the same state. We use the random variable $w(t)$ to mark the specific state at time t.

The **order** of the model is the number of previous states the current state depends on. In an n-th order model the probability of observing a certain state at time t depends on the previous n states the system visited

$$P(w(t)|w(t-1), w(t-2), ..., w(t-n))$$

Here we will focus on first-order Markov models, but the discussion can be generalized to higher order models (see Section 6.4). In a first-order Markov model the state at time t is influenced directly only by the state at time $t-1$.

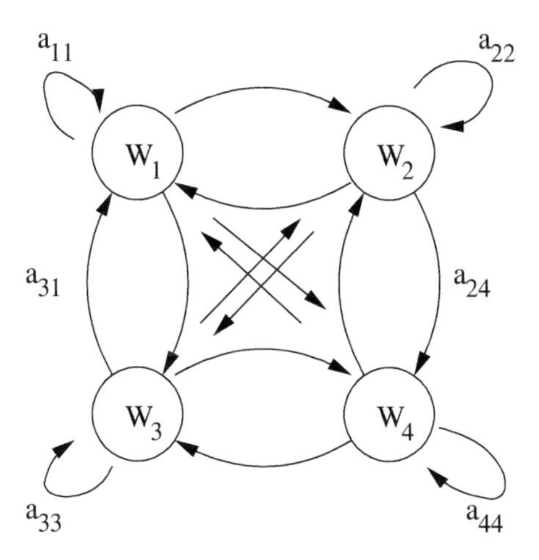

FIGURE 6.6: **An example of a first-order HMM with four states.** The model is fully connected (ergodic).

We denote a sequence of T states the model visited by $\mathbf{w} = w(1)w(2)...w(T)$. The sequence of states may include certain states more than once, depending on the model structure and parameters. We assume the sequences of states are generated according to a set of **transition probabilities**

$$P(w_j(t+1)|w_i(t)) = a_{ij}(t)$$

That is, $a_{ij}(t)$ is the time-dependent probability of moving from state w_i at time t to state w_j at time $t+1$. However, we will assume that these parameters are *time independent* and discard the time index, such that $a_{ij}(t) = a_{ij} \ \forall t$.

We denote the set of transition probabilities with the parameter vector $\vec{\theta}$. This set of parameters (together with the model structure) determines the model and is learned from the data as described in Section 6.3.3. Given a model $\vec{\theta}$ and a particular sequence $\mathbf{w}$, the probability that the model generated this sequence is given by multiplying the transition probabilities

$$P(\mathbf{w}|\vec{\theta}) = \sum_{i=2}^{T} P(w(i)|w(i-1))$$

For example, the probability of the sequence $\mathbf{w} = w_4w_3w_3w_1w_4w_3$ is

$$P(\mathbf{w}|\vec{\theta}) = a_{43}a_{33}a_{31}a_{14}a_{43}$$

We may also assume that there is a certain prior probability to observe w_i as the first state $P(w(1) = w_i) \equiv P_0(w_i)$. With priors, the probability of a sequence $\mathbf{w}$ is redefined such that

$$P(\mathbf{w}|\vec{\theta}) = P_0(w(1)) \sum_{i=2}^{T} P(w(i)|w(i-1))$$

6.2.2.1 Visible symbols and hidden Markov models

Usually we do not have knowledge of the states $w(t)$. Rather we are given a sequence of observable (i.e., measurable) symbols, where each symbol is assumed to be generated by one of the states. However, we do not know which state is associated with each symbol, and therefore we refer to the states as **hidden states** and to the model as a **hidden Markov model**. The measurable symbols are referred to as **visible symbols**.

For now we will assume that each hidden state visited must emit a visible symbol and denote the symbol emitted by the state $w(t)$ at time t as $s(t)$. Thus, the sequence of states $\mathbf{w} = w(1)w(2)...w(T)$ generates a sequence of visible symbols $\mathbf{s} = s(1)s(2)...s(T)$. We assume that the symbols are discrete, drawn from an alphabet $\mathcal{A}$; however, the model can be generalized to symbols drawn from a continuous range [192].

The emission is not deterministic - a hidden state may emit any of finite set $\mathcal{A}$ of possible symbols with different probabilities that we call **emission probabilities**

$$P(\sigma_k|w_j(t)) = b_{jk} \qquad \sigma_k \in \mathcal{A}$$

That is, b_{jk} is the probability to emit $s(t) = \sigma_k$ when we are in state $w(t) = w_j$, and as with the transition probabilities we assume the emission probabilities are time independent. These probabilities together with the emission probabilities and the prior probabilities make up the parameter vector $\vec{\theta}$ of the hidden Markov model.

6.2.2.2 The model's components

All together, a formal description of an HMM should specify the following elements:

- A set of c hidden states
- A set of d visible symbols that can be emitted from every state (the alphabet $\mathcal{A}$)
- The transition probabilities between hidden states a_{ij}

$$a_{ij} = P(w_j(t+1)|w_i(t))$$

- The emission probabilities

$$b_{jk} = P(\sigma_k|w_j(t))$$

- The prior probabilities

$$P_0(w_i)$$

In addition we require that the following three normalization constraints hold

- A probabilistic transition must take place between time t and time $t+1$, i.e.,

$$\sum_j a_{ij} = 1 \qquad \text{for all } i$$

- A symbol must be probabilistically emitted by every state visited

$$\sum_k b_{jk} = 1 \qquad \text{for all } j$$

- We must start from one of the hidden states

$$\sum_{j=1}^{c} P_0(w_j) = \sum_{j=1}^{c} P_0(j) = 1$$

It should be noted that the prior probabilities can be discarded if we introduce a Begin or Start state (see Section 6.3.5.2).

6.3 Main applications of hidden Markov models

Real life applications that use HMMs usually concern classification and characterization of samples (e.g., sequences, written characters, spoken words). Typically there are multiple processes or sources (e.g., protein families) and we would like to classify instances (sequences) to the sources that most likely generated them. If we have a model for each source (in speech recognition, a model for every spoken word, in molecular biology, a model for every protein family), then we can compute the probability that each source generated the instance and pick the model that assigns the highest probability. Computing the probability that a sequence was generated by a specific model is referred to as the **evaluation problem**.

Another type of information we are often interested in is the sequence of hidden states that most likely generated the sequence of visible symbols. For example, when analyzing DNA sequences using a model like in Figure 6.5, the sequence of hidden states determines the coding and non-coding regions. In the case of protein families, it helps to determine which amino acids correspond to core positions and which are mutations (insertions/deletions). Given the model of the correct class and the sequence of observations, determining the most probable sequence of hidden states that generated it is referred to as **the decoding problem**.

Finally, in order to answer the previous two problems we must address the **learning problem**. That is, how to train the parameters of a model from a set of observations (training data), such as protein sequences that belong to a specific protein family.

6.3.1 The evaluation problem

Given a complete HMM model with parameter vector $\vec{\theta}$ and a particular sequence of T visible symbols $\mathbf{s}$, what is the probability that the sequence was generated by this model? Since we do not know which sequence of hidden states generated the observations, we have to consider all possible paths of length T through the model. Therefore, the total probability of $\mathbf{s}$ is given by the sum

$$P(\mathbf{s}|\vec{\theta}) = \sum_{n=1}^{N} P(\mathbf{s}|\mathbf{w}_n, \vec{\theta}) P(\mathbf{w}_n|\vec{\theta}) \tag{6.1}$$

where $\mathbf{w}_n$ is a particular sequence of T hidden states $w(1)w(2)...w(T)$ and N is the total number of possible sequences of hidden states that can generate $\mathbf{s}$. With c hidden states in the model, then $N = c^T$. To simplify notation, we will ignore for now the dependency on $\vec{\theta}$ such that

$$P(\mathbf{s}) = \sum_{n=1}^{N} P(\mathbf{s}|\mathbf{w}_n) P(\mathbf{w}_n) \tag{6.2}$$

and will keep in mind that the probability is computed for a specific model and $P(\mathbf{s}) = P(\mathbf{s}|\vec{\theta})$.

To compute $P(\mathbf{s})$ we need to enumerate all possible sequences of hidden states, and for each sequence $\mathbf{w}_n$ to calculate $P(\mathbf{s}|\mathbf{w}_n)$ and $P(\mathbf{w}_n)$. The first is the probability of the visible symbols given the sequence of hidden states and is given by

$$P(\mathbf{s}|\mathbf{w}_n) = \prod_{t=1}^{T} P(s(t)|w_n(t))$$

where $P(s(t)|w_n(t))$ is determined by the emission probabilities, such that if $s(t) = \sigma_k$ and $w_n(t) = w_j$ then $P(s(t)|w_n(t)) = b_{jk}$. Similarly, we compute the probability of the sequence $\mathbf{w}_n$

$$P(\mathbf{w}_n) = P_0(w_n(1)) \prod_{t=2}^{T} P(w_n(t)|w_n(t-1))$$

using the transition probabilities such that if $w_n(t-1) = w_i$ and $w_n(t) = w_j$ then $P(w_n(t)|w_n(t-1)) = a_{ij}$. The probability $P_0(w_n(1))$ is the prior probability to start from hidden state $w_n(1)$. Combined altogether we get

$$P(\mathbf{s}) = \sum_{n=1}^{N} \prod_{t=1}^{T} P(s(t)|w_n(t))P(w_n(t)|w_n(t-1))$$

where to simplify notation we initialize $P(w_n(1)|w_n(0))$ to be the prior probability $P_0(w_n(1))$.

This is a straightforward calculation; however, it is prohibitive even for small models since it requires $O(Tc^T)$ calculations. For example, to compute the probability of a sequence of length $T = 10$ given a model with $c = 10$ states would take $\sim 10^{11}$ calculations with this algorithm. Models for protein families can contain up to several thousand states, and protein sequences can be several thousand residues long, which renders this approach impractical.

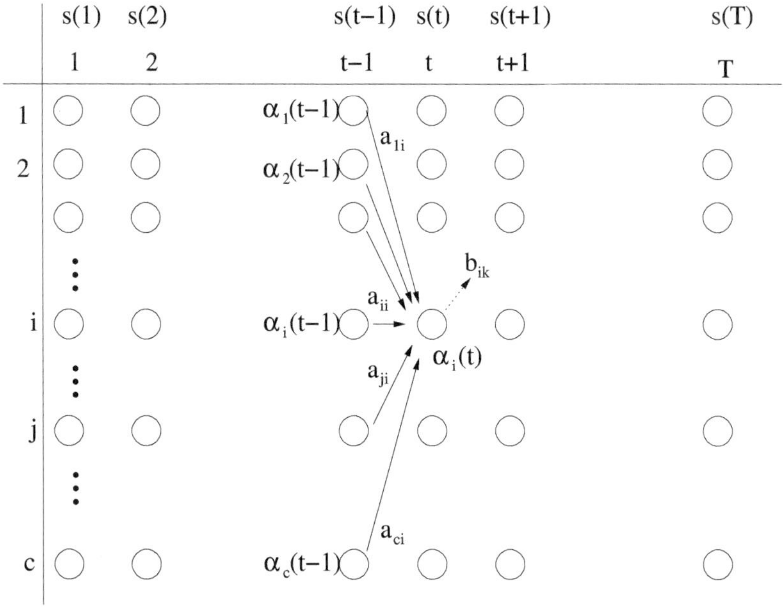

FIGURE 6.7: **The HMM forward variable.** (Figure adapted from Duda et al. [191, p. 133] with permission of John Wiley & Sons, Inc.)

6.3.1.1 The HMM forward algorithm

To tackle the computational issues with the evaluation problem, we can take advantage of the fact that many of the paths we sum over in Equation 6.1 have subpaths in common. Instead of recomputing the probability of these subpaths again and again we can store it and compute the total probability $P(\mathbf{s})$ dynamically. We first define the **forward variable** $\alpha_j(t)$ to be the probability to observe the sequence $\mathbf{s}^t = s(1)s(2)...s(t)$ *and* to reach hidden state j at time t. Note that $P(\mathbf{s}) = \sum_{j=1}^{c} \alpha_j(T)$ and therefore if we can compute $\alpha_j(T)$ efficiently then we can also determine $P(\mathbf{s})$ efficiently. The key is to use the recursive definition of the forward variable, as illustrated in Figure 6.7. Initialized to

$$\alpha_j(1) = P_0(j)b_{jk} \quad \text{where } \sigma_k = s(1) \quad (1 \le j \le c)$$

the forward variable is then defined recursively for $1 \le j \le c$ and $1 \le t \le T-1$

$$\alpha_j(t+1) = \sum_{i=1}^{c} \alpha_i(t)a_{ij}b_{jk} \quad \text{where } \sigma_k = s(t+1) \tag{6.3}$$

While the definition is recursive, the implementation is based on a dynamic programming algorithm, referred to as the **HMM forward algorithm**:

The HMM forward algorithm

Initialization: for $1 \le j \le c$
$\alpha_j(1) = P_0(j)b_{jk}$ where $\sigma_k = s(1)$
Loop: for $1 \le t \le T-1$
for $1 \le j \le c$
$\alpha_j(t+1) = \sum_{i=1}^{c} \alpha_i(t)a_{ij}b_{jk}$ where $\sigma_k = s(t+1)$
Output: $P(\mathbf{s}) = \sum_{j=1}^{c} \alpha_j(T)$

The computational complexity of this simple dynamic programming algorithm is $O(Tc^2)$, and for $c = 10$ and $T = 10$ it takes only ~ 1000 calculations to compute $P(\mathbf{s})$ and solve the evaluation problem.

6.3.1.2 The HMM backward algorithm

The backward algorithm is a time-reversed version of the forward algorithm. We define the **backward variable** $\beta_j(t)$ to be the probability to observe the remaining sequence $s(t+1)s(t+2)...s(T)$, starting from hidden state j at time t. Note that $P(\mathbf{s}) = \sum_{j=1}^{c} \beta_j(0)$, and therefore we can compute the total probability using the backward variables. While only one of either the forward or backward variables is necessary for the computation of the total probability $P(\mathbf{s})$, we will see later that both are needed for the learning problem.

As with the forward variable, the computation of the backward variable (as illustrated in Figure 6.8) relies on its recursive definition

$$\beta_j(t) = \sum_{i=1}^{c} a_{ji}b_{ik}\beta_i(t+1) \quad \text{where } \sigma_k = s(t+1) \quad (1 \le j \le c \text{ and } 1 \le t \le T-1)$$

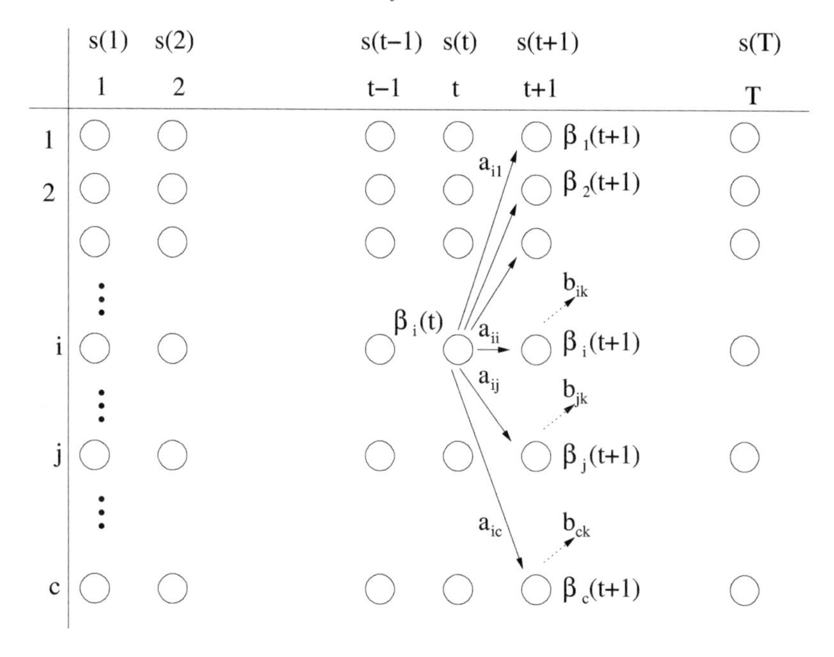

FIGURE 6.8: **The HMM backward variable.**

Some attention to details is necessary when initializing and finalizing the **HMM backward algorithm** that computes the backward variable based on this definition:

The HMM backward algorithm

Initialization: for $1 \leq j \leq c$

$$\beta_j(T) = 1$$

Loop: for $T-1 \leq t \leq 1$

for $1 \leq j \leq c$

$$\beta_j(t) = \sum_{i=1}^{c} a_{ji} b_{ik} \beta_i(t+1) \qquad \text{where } \sigma_k = s(t+1)$$

Output: $P(\mathbf{s}) = \sum_{i=1}^{c} \beta_i(0) = \sum_{i=1}^{c} P_0(i) b_{ik} \beta_i(1)$ where $\sigma_k = s(1)$

Note that the definitions of the forward and backward variables are complementary, and hence the likelihood of the visible sequence can be expressed as a combination of both:

$$P(\mathbf{s}) = \sum_{j=1}^{c} \alpha_j(t)\beta_j(t) \quad \forall t$$

where t can be picked arbitrarily anywhere from $t = 0$ to $t = T$ (for $t = 0$ initialize $\alpha_j(0) = 1$).

6.3.1.3 Using HMMs for classification

The discussion so far centered on a single model with parameter vector $\vec{\theta}$ and the probability distribution $P(\mathbf{s}|\vec{\theta})$ it induces over the sequence space, as in Equation 6.1. In a typical classification problem we have multiple models, one for each class (e.g., a protein family, a word in the dictionary). Each model has a different parameter vector and induces a different distribution over the sequence space. These distributions are called the **class-conditional** probability distributions.

To classify an instance to one of the classes we need to compare the probabilities they assign to the instance. Given a set of K models $\{\vec{\theta}_1, \vec{\theta}_2, ..., \vec{\theta}_K\}$ and a new measurement (a protein sequence $\mathbf{s}$), we can classify the sequence to the model that assigns the highest likelihood

$$class(\mathbf{s}) = \arg \max_{1 \leq i \leq K} P(\mathbf{s}|\vec{\theta}_i)$$

If the prior probability of different classes is not uniform, then the Bayesian approach is to classify the instance to the class with the highest posterior probability (see Section 3.12.8)

$$class(\mathbf{s}) = \arg \max_{1 \leq i \leq K} P(\vec{\theta}_i|\mathbf{s})$$

By Bayes' rule $P(\vec{\theta}_i|\mathbf{s}) = P(\mathbf{s}|\vec{\theta}_i)P(\vec{\theta}_i)/P(\mathbf{s})$ and we can ignore the marginal likelihood $P(\mathbf{s}) = \sum_i P(\mathbf{s}|\vec{\theta}_i)P(\vec{\theta}_i)$ since it is constant $\forall \vec{\theta}_i$. Therefore,

$$class(\mathbf{s}) = \arg \max_{1 \leq i \leq K} P(\mathbf{s}|\vec{\theta}_i)P(\vec{\theta}_i)$$

The priors $P(\vec{\theta}_i)$ are clearly problem specific. For example, different words have different probabilities to appear in a spoken language. In the case of protein families, the relative size of the family can be used to estimate its prior. However, this might lead to the "rich gets richer" phenomenon, since protein databases are often biased and do not necessarily represent the protein space accurately.

6.3.2 The decoding problem

Given an HMM and a sequence of visible symbols $\mathbf{s}$ that was emitted by the model, we sometimes want to know the most probable sequence of hidden states $\mathbf{w}$ that generated it. This is especially useful when trying to detect coding segments in a long sequence of DNA that was just sequenced, or when aligning a sequence to an HMM of a protein family. The straightforward way is to enumerate every possible path of hidden states of length T, calculate the probability of the visible sequence for each path and output the path that assigns the highest probability to the visible sequence. However, this would

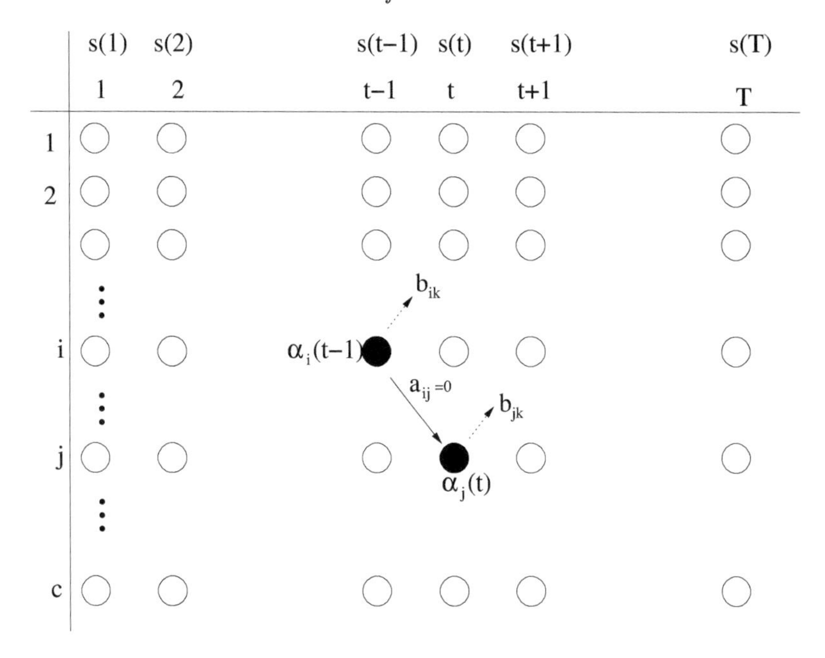

FIGURE 6.9: **Searching for the most probable sequence of hidden states.** Local decisions may result in a path with zero probability. For example, state i might be the optimal one at time $t-1$, while state j might be the optimal one at time t (based on all possible paths that end in that state at time t), but if the transition probability $a_{ij} = 0$ then a path that contains these two states in succession has a total probability of zero.

be computationally prohibitive, since it requires $O(Tc^T)$ calculations. An alternative solution is to find the path so that the state at each time step t has maximal probability of coming from the previous step and generating the observed visible symbol $s(t)$. However, this algorithm optimizes the probability only locally. That is, it ignores the maximization in the previous step and makes a local decision that we cannot revisit. Consequently, this algorithm does not guarantee that the path is allowable (for example, the chosen path might have zero probability transitions between maximally probable states in consecutive time steps, see Figure 6.9).

A better and more common algorithm is the **Viterbi algorithm.** This algorithm finds the path with the maximal *total* probability. The definition is recursive as with the forward and backward variables. At each step we maintain only the most probable path that ends in each state and keep track of the most probable path so far. The algorithm leaves all trails open until the end, and only then makes a global decision.

We define the variable $\delta_j(t)$ to be the probability to observe $\mathbf{s}^t = s(1)...s(t)$ and to reach hidden state j at time t, *maximized* over all possible sequences of hidden states of length $t-1$. We also define the pointer variable $path_j(t)$

that contains the pointer to the most probable hidden state at time $t-1$ from which we can reach state j at time t. The Viterbi algorithm works as follows:

The Viterbi algorithm

Initialization: for $1 \leq j \leq c$
$\delta_j(1) = P_0(j)b_{jk}$ where $\sigma_k = s(1)$
$path_j(1) = 0$
Loop: for $1 \leq t \leq T-1$
for $1 \leq j \leq c$
$\delta_j(t+1) = \max_{1 \leq i \leq c}[\delta_i(t)a_{ij}]b_{jk}$ where $\sigma_k = s(t+1)$
$path_j(t+1) = \arg\max_{1 \leq i \leq c}[\delta_i(t)a_{ij}]$
Output: The probability $P = \max_{1 \leq i \leq c}[\delta_i(T)]$
The final state $w(T) = \arg\max_{1 \leq i \leq c}[\delta_i(T)]$

From the final state and the path variables we can then recover the sequence of hidden states. That is $w(t) = path_{w_{t+1}}(t+1)$ for $t = T-1, T-2, ..., 1$

6.3.3 The learning problem

The learning problem is concerned with determining the model parameters (transition and emission probabilities) from the training data (set of visible sequences). Finding the most probable set of parameters is a difficult problem given the size of the search space, and there is no known algorithm that is guaranteed to solve this problem efficiently. However, it is possible to find a local optimum solution using the forward-backward algorithm (also known as the Baum-Welch algorithm [193]). The algorithm is based on an expectation-maximization technique (see Section 5.7); it starts from an initial model and iteratively improves it by updating the parameters, so as to maximize the likelihood of the observed data. There are several software packages that implement variations on the Baum-Welch algorithm and can learn an HMM from sequence data (e.g. [194, 195]).

6.3.3.1 The forward-backward algorithm

Given the coarse structure of the model (the number of hidden states and the number of visible symbols but not the probabilities a_{ij} and b_{jk}) and a set of training observations of visible symbols, the forward-backward algorithm determines the parameters of the model $\vec{\theta}$ iteratively. The process is initialized with a certain set of parameters that can be chosen randomly or estimated from the data as described in Section 6.3.5.5. The parameters are then re-estimated by considering the observations.

An intuitive explanation of the re-estimation process relies on counting arguments. To re-estimate the parameters we need to know how many times we visited a certain hidden state and moved to another hidden state and how many times we emitted a symbol from a hidden state. However, not only we do not know the parameters of the model, we also do not know which sequence of hidden states generated the observations, and therefore we cannot simply count how many times we used a transition or emitted a symbol. To circumvent the first problem, the algorithm estimates how many times we used a transition (or emitted a symbol) based on the *current model.* The second problem is addressed through averaging over all possible sequences of hidden paths.

Recall that given a model $\vec{\theta}$, we can express the likelihood of a visible sequence using the forward and the backward variables at any time point $t <= T$

$$P(\mathbf{s}|\vec{\theta}) = \sum_{i=1}^{c} \alpha_i(t)\beta_i(t)$$

Exploiting this property, we can assess how many times a certain transition was used and estimate the variable $\xi_{ij}(t)$, defined as the probability of transition between state i at time t and state j at time $t+1$

$$\xi_{ij}(t) = P(w(t) = w_i, w(t+1) = w_j|\mathbf{s}, \vec{\theta})$$

By laws of conditional probability $P(A|B,C) = P(A,B|C)/P(B|C)$ (see Section 3.12.3), and therefore

$$\xi_{ij}(t) = \frac{P(\mathbf{s}, w(t) = w_i, w(t+1) = w_j|\vec{\theta})}{P(\mathbf{s}|\vec{\theta})}$$

We estimate these probabilities from the current model (current values for $\{a_{ij}\}$ and $\{b_{jk}\}$) and the training data (the visible sequence $\mathbf{s}$) using the forward and backward variables, such that

$$\xi_{ij}(t) = \frac{\alpha_i(t)a_{ij}b_{jk}\beta_j(t+1)}{P(\mathbf{s}|\vec{\theta})} \quad \text{where} \quad \sigma_k = s(t+1)$$

Note that the numerator is the total probability that the sequence was generated by paths that go through states i and j at time steps t and $t+1$, while the denominator is the total probability that the sequence was generated by any path (summed over all paths). Hence, the fraction is the desired probability. We also define the variable $\gamma_i(t)$ as the probability of visiting state i at time t

$$\gamma_i(t) = P(w(t) = w_i|\mathbf{s}, \vec{\theta}) = \frac{P(\mathbf{s}, w(t) = w_i|\vec{\theta})}{P(\mathbf{s}|\vec{\theta})} = \frac{\alpha_i(t)\beta_i(t)}{P(\mathbf{s}|\vec{\theta})}$$

By the recursive definition of the backward variable, $\beta_i(t) = \sum_{j=1}^{c} a_{ij}b_{jk}\beta_j(t+1)$ and therefore

$$\gamma_i(t) = \sum_{j=1}^{c} \xi_{ij}(t)$$

Hence, $\gamma_i(t)$ can also be interpreted as the probability of transition from state i to *any other state* at time t.

Based on these probabilities we can now re-estimate the model parameters. Intuitively, to calculate the new transition probabilities $\{a_{ij}\}$ we need to divide the number of times we moved from state i to state j by the number of times we visited state i. Since we are interested in the time-independent parameters we first have to sum $\xi_{ij}(t)$and $\gamma_i(t)$ over the time index t. The sum $\sum_{t=1}^{T-1} \xi_{ij}(t)$ can be interpreted as the expected number of transitions between states i and j at any time, while $\sum_{t=1}^{T-1} \gamma_i(t)$ is the expected number of times we visited state i at any time. Therefore, the new probability of transition from i to j, $\hat{a}_{ij}$ is given by

$$\hat{a}_{ij} = \frac{\sum_{t=1}^{T-1} \xi_{ij}(t)}{\sum_{t=1}^{T-1} \gamma_i(t)}$$

Similarly, when calculating the new value for the emission probability $\{b_{jk}\}$, note that the total expected number of transitions from state i at any time *after* omitting the specific symbol σ_k is given by

$$\sum_{t=1,s(t)=\sigma_k}^{T-1} \gamma_i(t) \tag{6.4}$$

i.e., summing only over sequences where $s(t) = \sigma_k$. Therefore,

$$\hat{b}_{ik} = \frac{\sum_{t=1,s(t)=\sigma_k}^{T-1} \gamma_i(t)}{\sum_{t=1}^{T-1} \gamma_i(t)} \tag{6.5}$$

where the denominator is as before, the total number of times the state i was visited (each time emitting a symbol). Finally, the prior probabilities can be re-estimated as

$$\hat{P}_0(i) = \gamma_i(1)$$

which follows from the definition of the γ variable.

Once we estimated the new model probabilities $\hat{a}_{ij}$ and $\hat{b}_{jk}$ we can re-estimate the data driven transition probabilities $\xi_{ij}(t)$ and $\gamma_i(t)$ and from these estimate new transition and emission probabilities. The process is repeated until convergence (until the improvement is below a predefined threshold, or the changes in parameters are negligible).

It should be noted that the equations for $\hat{a}_{ij}$ and $\hat{b}_{jk}$ are the maximum likelihood solutions of the marginalized likelihood values if we were to write them explicitly and compute the derivative. Therefore, although the algorithm might not converge to the best model possible, the algorithm is guaranteed to improve the data likelihood with every iteration and converge to a local maximum [193].

6.3.3.2 Learning from multiple training sequences

The forward-backward algorithm was formulated assuming a single sequence of observations. This is typically not the case, as the input is usually composed of n training sequences of visible symbols $\mathbf{s}_1, \mathbf{s}_2, ..., \mathbf{s}_n$ that can be of different lengths $T_1, T_2, ..., T_n$. The modifications that need to be introduced for multiple sequences are straightforward and only require adding a summation over all sequences, such that

$$\hat{a}_{ij} = \frac{\sum_{l=1}^{n} \sum_{t=1}^{T_l-1} \xi_{ij}^l(t)}{\sum_{l=1}^{n} \sum_{t=1}^{T_l-1} \gamma_i^l(t)}$$

and

$$\hat{b}_{ik} = \frac{\sum_{l=1}^{n} \sum_{t=1, s_l(t)=\sigma_k}^{T_l-1} \gamma_i^l(t)}{\sum_{l=1}^{n} \sum_{t=1}^{T_l-1} \gamma_i^l(t)}$$

6.3.4 Handling machine precision limitations

In some applications, such as gene prediction and protein classification, the sequences of visible symbols can be very long (hundreds or thousands of nucleotides or amino acids). This might pose a problem when solving the evaluation or the decoding problem. Since we have to calculate the probability of the visible sequences by multiplying probabilities we might run into problems of underflow due to machine precision limitations. For example, if a typical transition has a probability of 0.1 then for a sequence of 1000 symbols we will need to compute the product 0.1^{1000}, which is usually below the smallest number we can store in a double precision variable.

Solving this issue for the decoding problem is easier, and when implementing the Viterbi algorithm the simplest solution is to work with the logarithm of the probabilities. That is, after initialization $\log \delta_j(1) = \log P_0(j) + \log b_{jk}$ we proceed by converting the product of probabilities to a sum of logarithms

$$\log \delta_j(t+1) = \max_{1 \leq i \leq c} [\log \delta_i(t) \log a_{ij}] + \log b_{jk}$$

However, this solution is good only for the decoding problem, where we have only products. In the evaluation problem we have also sums when computing the forward or the backward variables. One possible solution in that case is to normalize the forward variables $\alpha_i(t)$ at each time step by the normalization factor

$$factor(t) = \sum_{i=1}^{c} \alpha_i(t)$$

With this normalization, the total probability can be expressed as

$$P(\mathbf{s}) = \prod_{t=1}^{T} factor(t)$$

and we can apply the logarithm transformation $\log P(\mathbf{s}) = \sum_{t=1}^{T} \log factor(t)$. The equation above can be proved by induction (see Problem 6.2).

6.3.5 Constructing a model

So far we discussed how to use an HMM to classify instances and find the most likely sequence of hidden states. We also discussed how to train the model's parameters from training examples. However, we assumed the model's structure is given. Deciding on the model structure is the first and perhaps the most crucial step. If the architecture does not describe accurately the process we would like to simulate, then the model could perform poorly even after extensive training.

When designing the model architecture we need to have a good understanding of the underlying process. But first we need to make certain decisions about the general topology of the model.

6.3.5.1 General model topology

HMMs can be categorized into two main types. **Ergodic models** are models where all transitions are allowed (i.e., we can reach any state from any state). On the other hand, in **left-to-right models** transitions are allowed between hidden state i and hidden state j only if $j \geq i$ (Figure 6.10). These models preclude transitions "back in time". Left-to-right models are usually more appropriate in typical applications of HMMs in computational biology, where states often correspond to ordered positions along a sequence. However, specifying in advance which transitions are allowed and which are not might limit the generalization power of the model. On the other hand, if we choose an ergodic model and then rely on the Baum-Welch algorithm to learn the non-zero transitions from the training examples, we are at a higher risk of converging to local optimum since the search space is significantly larger. The better approach is to use directional models when possible, while allowing all transitions that are plausible based on the underlying process. With more constrained models, this approach eliminates many of the parameters and increases the chances that we will converge to a global maximum.

FIGURE 6.10: **Topologies of HMMs.** Ergodic (left) vs. left-to-right (right)

6.3.5.2 Model architecture

When designing the architecture of the model, now with the specific process we would like to model in mind, there are several general guidelines we can follow to construct the model from smaller substructures. For example, to model a sequence of observations that are sampled from the same distribution we can use one state with a self-transition (Figure 6.11, left). This simple model can emit sequences of any length; however, if there are other states such that the self-transition probability is $p < 1$ then the probability of emitting a long sequence from that state decreases exponentially with the sequence length since $P(l) = p^{l-1} \cdot (1-p)$.

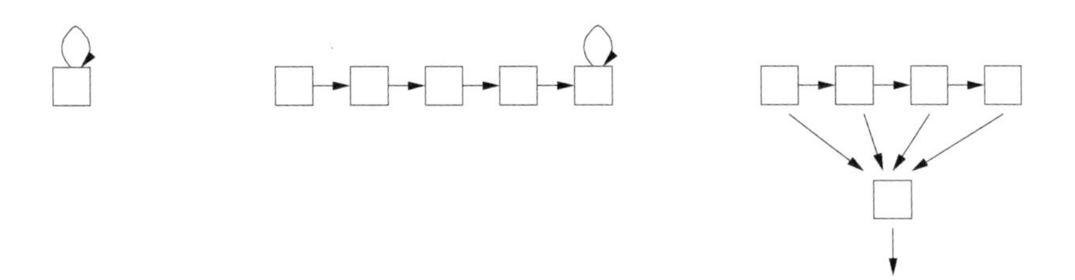

FIGURE 6.11: Basic structures of HMM. (Figure adapted from Durbin et al. 2007 [93] with permission.)

If we want to make sure the sequence is at least of length l_0 then we can introduce l_0 identical states, each with the same emission probabilities, and add a self-transition in the last one (Figure 6.11, middle). It is also simple to model other constraints such as "any length between l_0 and l_1" (Figure 6.11, right) or other types of distributions as discussed in [93].

FIGURE 6.12: Using silent states in HMM. The structure on the left can be replaced with the structure on the right that includes silent states (denoted by circles). (Figure adapted from Durbin et al. 2007 [93] with permission.)

There are several types of special hidden states that can be useful when designing the model's architecture. For example, a **start (source) state** is a single state that has zero self-transition probability and is connected to a subset of the other hidden states. Any sequence of hidden states must start with the source state. Start states simplify the treatment and the formalization and practically replace the prior probabilities over states. A **final (absorbing) state** is a single state that has self-transition probability of one (i.e., once you are there you cannot move to other states). It can be reached from a subset of the hidden states, and any sequence of hidden states must end with it. By having an end state we practically induce a certain probability distribution over the lengths of sequences that can be generated by the model. **Silent states** are states that do not emit symbols and can be used to skip and connect other states and eliminate direct transitions from all states to all states. This approach helps to build "cleaner" models, as illustrated in Figure 6.12, but it also comes with a certain cost, since we cannot always fine tune the transitions, as illustrated in Figure 6.13. Handling silent states in the algorithms we described above also requires certain adjustments, as will be described next.

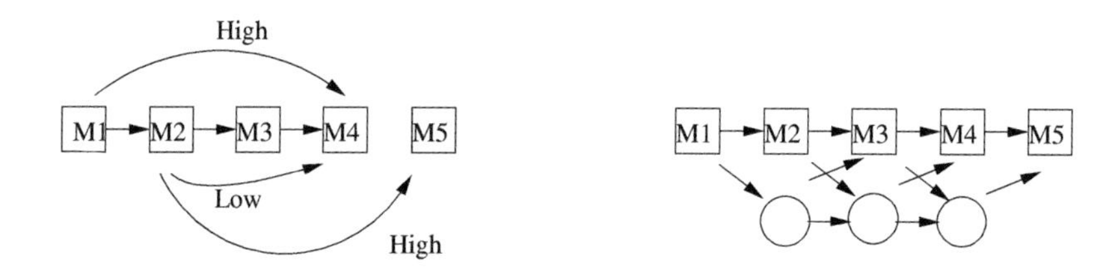

FIGURE 6.13: **Silent states in HMM.** In this example, the transition probability $a_{M2\to M4}$ is low while $a_{M1\to M4}$ and $a_{M2\to M5}$ are high. It is not possible to duplicate this scenario with the structure on the right (see Problem 6.5).

To accommodate other common situations there are several other useful strategies. For example, if we would like our model to explain multiple repeats of the same pattern then we could introduce a backward transition, as illustrated in Figure 6.16 later. If the pattern we are modeling is embedded in longer sequences (which is equivalent to the case of local alignments) then we can add states before and after the pattern model to accommodate the tails. We can also combine repeats and local matches (Figure 6.16).

6.3.5.3 Hidden Markov models for protein families

The HMMs that are commonly used to describe protein families have a specific architecture that was designed to reflect the basic evolutionary processes of deletion, insertion and substitution. A good example is the **profile-HMM** model[1] that was originally introduced in [196] (the architecture was slightly revised later, as described in [197]). Let L denote the length of a typical protein in the family being modeled, then the model has $L+2$ nodes: Begin state (B), End state (E) and L super nodes, as in Figure 6.14. Each super node

[1]The name profile-HMM reflects the resemblance of the model to the profile model we discussed in Chapter 4.

consists of three hidden states: a **match** state (M), an **insertion** state (I) and a **deletion** state (D). The exception is the L-th super node (the fourth super node in Figure 6.14) that does not have an insertion state.

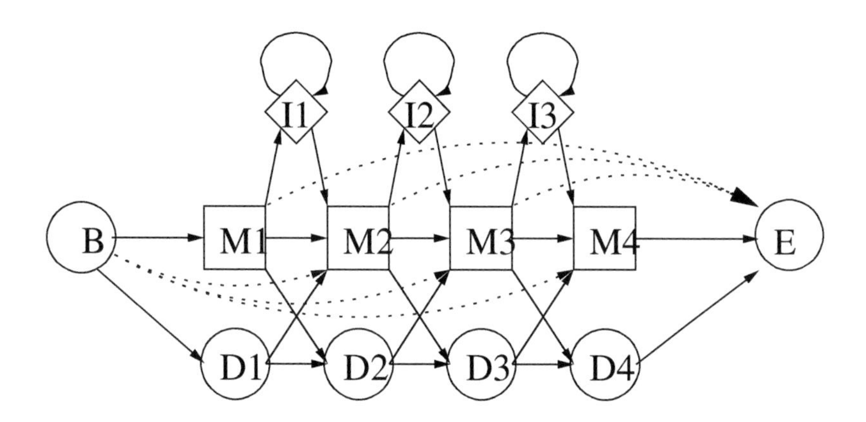

FIGURE 6.14: **Profile-HMM of a protein family.** Match states are denoted by squares, insertion states are denoted by diamonds, and deletion states are marked by circles. Note that the B and E states are silent states as well.

While match states and insertion states emit amino acids, deletion states are silent states that do not emit anything. The match states emit a "typical" protein sequence of the protein family being modeled by the HMM. Deletion states may be used to describe the event of deleting certain segments of the protein, and insertion states may be used to describe the event of an insertion. In addition, the model has a few more states, the main two are an insertion state before the begin state and an insertion state after the end state (as in Figure 6.16). These two additional states take care of parts of the sequence that cannot be explained by the model before the first match and after the last match.

The model topology is left-to-right, where allowed transitions are limited to the following set

- From begin state B to any of the match states or to D_1
- From match state M_i to match state M_{i+1} or to insertion state I_i or to deletion state D_{i+1} or to the end state E.
- From insertion state I_i to match state M_{i+1} or self-transition (to insertion state I_i)
- From deletion state D_i to match state M_{i+1} or to deletion state D_{i+1}

The transition from the last deletion state to the end state has a probability of 1. Similarly, the transition from the last match state to the end state has a probability of 1. Note that the model can explain deletions of protein segments as a series of separate events in evolution (deleting one amino acid at a time, implemented through transitions between deletion states) or as a single event where the whole segment is deleted (implemented through direct transitions from the begin state to one of the match states or a direct transition from a match state to the end state). For example to delete the first two amino acids we can move directly from B to $M3$ (a single event in evolution) or we can move through $D1$, then $D2$ and then to $M3$ (two different events, in each one only one amino acid is deleted). Both paths are plausible in evolution, and the model can accommodate both.

6.3.5.4 Handling silent states

The introduction of the deletion states deviates from the original model we described in Sections 6.3.1-6.3.3, where every state emits a symbol. Since deletion states are silent states that do not emit symbols, it might seem as if the definition of $\alpha_j(t)$ is ambiguous, which would be an issue with the forward, backward, Viterbi and Baum-Welch algorithms. This is partially true, and in the presence of silent states the forward variable has to be treated differently considering the architecture of the model. On the other hand, the forward variable has to be consistent with the original definition, according to which $\alpha_j(t)$ is the probability to end at state j and observe $s(1)s(2)...s(t)$. Since the sequence of hidden states that generated these observations might have gone through silent states, this might take more than t time steps.

Say we want to calculate $\alpha(t)$ for deletion state D_j, assuming the model of Figure 6.14. Being a silent state, D_j cannot emit a symbol at time t, and therefore we must have emitted t symbols before we moved there. Transitions to D_j are allowed from match state M_{j-1} or from the previous deletion state D_{j-1}. Therefore, Equation 6.3 of the forward algorithm is replaced with

$$\alpha_{D_j}(t) = \alpha_{D_{j-1}}(t) \cdot a_{D_{j-1} \to D_j} + \alpha_{M_{j-1}}(t) \cdot a_{M_{j-1} \to D_j}$$

Note that the time index on the right side of the equation has not changed. For match states and insertion states the definition is more straightforward, since they do emit a symbol. For example, $\alpha(t)$ for match state M_j is given by

$$\alpha_{M_j}(t) = \alpha_{D_{j-1}}(t-1) \cdot a_{D_{j-1} \to M_j} + \alpha_{M_{j-1}}(t-1) \cdot a_{M_{j-1} \to M_j} + \alpha_{I_{j-1}}(t-1) \cdot a_{I_{j-1} \to I_j}$$

With these definitions we can now compute the forward variable $\alpha_j(t)$ for every state j and for every time t.

6.3.5.5 Building a model from an MSA

When learning a model from protein sequence data we can apply the iterative Baum-Welch EM algorithm directly to the unaligned input sequences to determine the model parameters. However, depending on the initial choice of parameters and the size of the model (the number of parameters), the algorithm might converge to a mediocre model. This is especially true if the model architecture is not tuned to the input data. A better approach is to initialize the HMM based on prior information from a manually constructed MSA. This latter approach is preferred if the MSA is of high quality and was manually verified. But even if this is not true, it is advised to start with a model that is derived from an MSA and then use the EM algorithm and the unaligned sequences to refine the model.

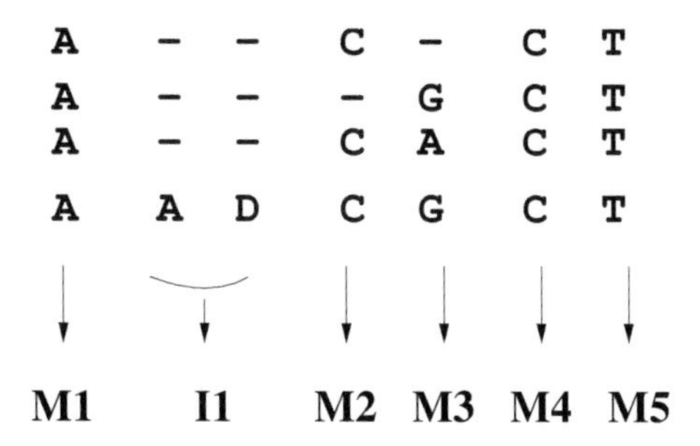

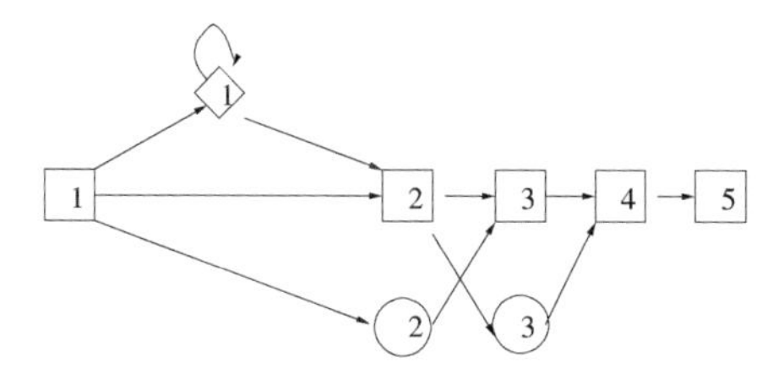

FIGURE 6.15: Constructing HMM from MSA. Assuming the architecture of Figure 6.14, the second and third columns, which have more than 50% gap characters, are mapped to an insertion state. The remaining columns are mapped to match states. Given this basic structure, the transition probabilities are initialized to $a_{M1\to M2} = 2/4 \quad a_{M1\to I1} = 1/4 \quad a_{M1\to D2} = 1/4$. Pseudo-counts are then added using methods such as those discussed in Section 4.3.3.2.

A simple procedure for deriving a model from an MSA, which was suggested in [93], is to start by assigning a match state for each column that has more than 50% non-gap characters and an insertion state for each column that has more than 50% gap characters. Then, the emission and transition parameters are estimated from the MSA, as in the example above (Figure 6.15). This procedure will result in a model that might reject sequences that differ only slightly from the input sequences, and therefore we have to introduce pseudo-counts as in Section 4.3.3.2.

Next we add the remaining insertion and deletion states as in Figure 6.14. Transition probabilities should be initialized such that transitions from match states to either deletion or insertion states have a small probability, while transitions from insertion and deletion states back to match states have relatively high probabilities. With this initialization, family members will be more likely to pass through the core sequence of match states, thus improving the parameter estimation in subsequent iterations of the EM algorithm. Self-transitions (insertion states) and transitions between deletion states should also be initialized to small probabilities and will gradually adjust if insertions or deletions are observed. Since the protein sequences that are given as input might contain more than just the domain or family signature we are trying to model, additional states should be introduced to explain unrelated sequences that appear after or before the homologous domain (see Figure 6.16).

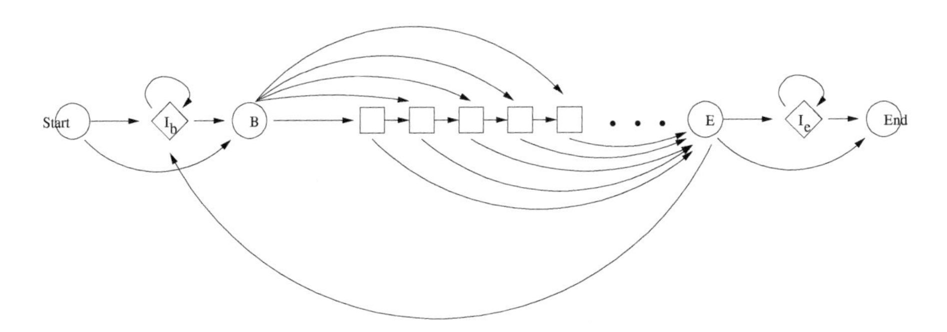

FIGURE 6.16: **Constructing HMM.** The insertion states I_b, I_e before and after the main model are used to explain segments that precede or follow the "typical" sequence and are unrelated to the protein family modeled. The unrelated segments can be potentially very long, and therefore it is recommended to set the self-transition probabilities $a_{I_b \to I_b}$ and $a_{I_e \to I_e}$ close to one, otherwise long sequences that contain an occurrence of the family modeled will be assigned a very low probability. Long sequences will be assigned low probabilities regardless, and to interpret these numbers it helps to normalize them by subtracting $\log P_0(\mathbf{s}) = \log \prod_i P_0(s(i))$ from $\log P(\mathbf{s})$, where P_0 is the background distribution of amino acids in the database (the normalization is basically a transformation of the log-scores to log-odds scores). The model starts with the first state on the left. The state marked as B is a silent state and only serves to simplify the presentation and the handling of the model. The transition probabilities from B to the match states can be initialized uniformly or decrease with the distance from B, or initialized based on data on partial matches of family members with the main part of the model. If we want the model to work only in "global mode" and reject partial matches (that is, a sequence must pass through the first and last match states) we can set all transitions but the transition to the first match state to zero. The transition from E to I_b allows the model to explain repeats.

6.3.5.6 Single model vs. mixtures of multiple models

Many protein families are very diverged and consist of two or more subfamilies with different sequence signatures. A single HMM model cannot always capture the variations observed within such protein families and might miss remote homologies when used to search a sequence database[2]. Therefore, it is recommended to use multiple models to represent highly diverged families. In a **mixture model** of HMMs $H = \{h_1, h_2, ..., h_m\}$ every model is centered around a different subfamily, and the probability of a sequence $\mathbf{s}$ is given by the sum

$$P(\mathbf{s}) = \sum_{h \in H} P(\mathbf{s}|h)P(h)$$

where $P(h)$ is the prior probability of the model h (also referred to as the **weight** of h in the mixture). The prior probability can be estimated based on size of the subfamily or the model's complexity (see Section 10.5.2.1). To simplify, a uniform prior is often assumed. However, a mixture model with uninformative prior may not necessarily improve performance over individual models.

An alternative approach is to assign each sequence a probability based on the marginal likelihood, considering also the training data. That is,

$$P(\mathbf{s}|D) = \sum_{h \in H} P(\mathbf{s}|h, D)P(h|D) = \sum_{h \in H} P(\mathbf{s}|h)P(h|D) \tag{6.6}$$

where $P(h|D)$ is the posterior probability of the model h given the training data. Using Bayes' formula, the posterior probability can be rewritten as

$$P(h|D) = \frac{P(D|h)P(h)}{P(D)}$$

Since $P(D)$ is fixed for all models and assuming uniform prior, we can define the weight of the different models in the mixture simply based on the likelihood of the training data. The weights can be defined also by means of other functions that are correlated with the data likelihood. For example, they can be approximated by measuring the performance of each model over the test set (see Section 7.4.3). Let $Q(h)$ denote the weight of h based on its performance (normalized, such that $\sum_h Q(h) = 1$), then the combined prediction of the mixture model in this case is defined as a weighted sum

$$P(\mathbf{s}|D) \simeq \sum_{h \in H} Q(h)P(\mathbf{s}|h)$$

[2]Indeed, it has been noted that some families in Pfam B have similar models to families in Pfam A (by comparing their profiles, as in Section 4.3.5), although the individual sequences of the Pfam B families were not detected when evaluated with the models of Pfam A families [198].

As with many other applications in machine learning, mixture models often improve performance. Therefore, using multiple models to represent a protein family is likely to produce better results and detect more remote homologies. This result can be explained intuitively (we will later provide a more rigorous explanation, Section 11.4.2.5). Imagine we could represent the space of all protein sequences in some real space of a certain dimension such that each protein is mapped to one point (referred to as the image point of the protein), and different families form distinct "clouds" of points (clusters) in the real space (we will discuss this mapping problem extensively in Chapter 11 and Chapter 7). A new protein can be classified by measuring the distance between its image point and the different clusters and classifying it to the closest family[3]. Imagine that each family forms a hyper-sphere in real space (circle in 2D, sphere in 3D and so on). An HMM model of a family can be perceived as the centroid (center of mass) of the sphere, or an approximation of it[4]. Measuring the likelihood of a sequence with respect to a single model is equivalent to measuring the distance between the image point and the centroid of the sphere. If the clusters form hyper-spheres then classifying instances based on their distance from the centroid of each family will minimize the error rate. However, if the family is very diverged then the cluster does not necessarily form a sphere and measuring the distance with respect to the centroid does not always produce correct results.

On the other hand, measuring the likelihood of a protein with respect to multiple models (each one representing a different subset of the family) is equivalent to measuring the distance between the image point of the query and a set of points scattered inside or on the outskirts of the cluster and classifying it based on its nearest neighbor or the k nearest neighbors. With multiple models representing each class, it is more likely that one of the models of the correct class is nearby. When the source cannot be modeled with a single model, the nearest neighbor approach often proves to be superior [191].

6.4 Higher order models, codes and compression

Throughout this chapter we focused on first-order Markov models. Such models assume that the state at time or position t depends only on the state at time $t-1$, regardless of the previous states. However, in some cases the current state depends on even earlier states. A simple example is a codon, where the

[3]Depending on the shape of the clusters, the distance might need to be adjusted based on the distribution of points in each cluster, as in the σ-distance of Section 10.9.

[4]The better the approximation, the more accurate the classification. For example, using a sequence-based model vs. a structure-based model is equivalent to different estimates of the centroid of the sphere. Structure-based models usually produce better results.

probability of observing a specific nucleotide in the third position depends on the first two nucleotides. Therefore, to model codons it is recommended to use second-order Markov models or hard code the dependency in the model architecture as in Figure 6.4. An even higher order dependency is observed in promoter regions. Protein sequences also exhibit long-range dependencies, owing to the structural arrangement of the sequence that places residues that are far apart in sequence close to each other in space. First-order models might poorly approximate the source distribution in such cases.

The HMMs that are designed to model protein families make up to some extent for their lower order by utilizing plenty of domain-specific information about the source in the model structure. The core of the model outlines the type of residues that are expected to occur in each position along the sequence of family members, and as such it essentially encodes higher order dependency between residues. HMMs are capable of modeling quite complex sources, but they require a good understanding of the properties of the source to eliminate incorrect architectures and design effective architectures, as we have seen in Section 6.3.5. They may require a high-quality multiple alignment of the input sequences to obtain a reliable model, especially for highly diverged protein families. Because of their complex structure and the many parameters they introduce, they also require large training sets. But there are other alternatives to HMMs, and we will discuss a few of them next.

6.4.1 Fixed order models

Fixed-order Markov models are models where the state or variable at any given moment depends on a fixed number of previous states. The number of past states is referred to as the **memory length** of the model. With n-order Markov chain, the value of the random variable $w(i)$ depends on the last n states, such that

$$P(w(i)|w(i-1), w(i-2), ..., w(1)) = P(w(i)|w(i-1), w(i-2), ..., w(i-n))$$

and in general, for a sequence of random variables

$$P(X_i|X_{i-1}, X_{i-2}, ..., X_1) = P(X_i|X_{i-1}, X_{i-2}, ..., X_{i-n})$$

If the n-order model is a generative model that emits observations (e.g., sequences), then each random variable corresponds to one character along the sequence and the probability of the each character depends on the past n characters. Notation wise, when dealing with strings $\mathbf{s}$ we will denote this dependency while maintaining the order of characters in the string

$$P(s_i|s_1, s_2, ..., s_{i-1}) = P(s_i|s_{i-n}, s_{i-n+1}, ..., s_{i-1})$$

or abbreviated as

$$P(\sigma|\mathbf{s})$$

where $\sigma \in \mathcal{A}$ and $\mathbf{s} \in \mathcal{A}^n$. The string $\mathbf{s}$ is called the **context** of σ. When training a model, our goal is to learn these probabilities, and they are usually estimated directly from the data.

Note that an n-order Markov process over $\mathcal{A}$ can be re-written as

$$P(X_i|X_{i-1}, ..., X_{i-n}) = P(X_i, X_{i-1}, ..., X_{i-n+1}|X_{i-1}, ..., X_{i-n})$$

using the basic property of conditional probability $P(A, B|B) = P(A|B)$. Hence, using n-tuple notation where X_i^n represents the sequence of n variables ending in X_i, an n-order Markov process over $\mathcal{A}$ can be formulated as a first-order Markov model over $\mathcal{A}^n$

$$P(X_i^n|X_{i-1}^n)$$

For example, if $\mathcal{A} = \{A, B\}$, then sequences over $\mathcal{A}$ that are generated by a second-order Markov process can be rewritten as sequences over the alphabet $\mathcal{A}^2 = \{AA, AB, BA, BB\}$ and then described using a first-order Markov model over $\mathcal{A}^2$. In that case, the string $ABBA$ would be translated to the string $AB - BB - BA$, described by the model of Figure 6.17. A similar transformation would convert a third-order model of DNA sequences to a first-order model over the alphabet of 64 trinucleotides (codons). This is especially useful for modeling protein-coding DNA sequences, as discussed in Section 6.2.1. Once translated, the mathematical formulation is the same as before, and we can use the formalism and all the algorithms that were developed for first-order models.

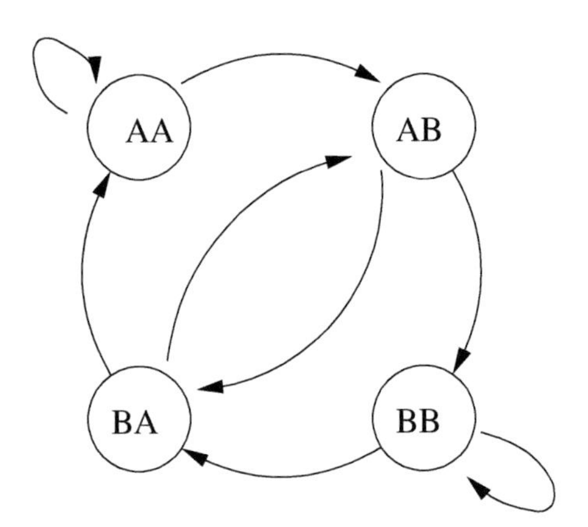

FIGURE 6.17: A first-order model describing sequences with second-order dependency. (Figure adapted from Durbin et al. 2007 [93] with permission.) Note that not all transitions are possible since the tuples are overlapping.

Higher order models can capture intricate dependencies between states, and fixed memory Markov models might be a natural choice for some problems. However, a fixed memory is not necessarily appropriate for protein families. In protein sequences, the probability of observing a certain amino acid in a certain position depends on the neighboring residues, but the context might vary in length, depending on the location in the structure. In Chapter 8 we will see how this dependency is fundamental to the prediction of secondary structures. Moreover, fixed-order models suffer from exponential growth in the number of states or the number of contexts as a function of the memory length (which is equivalent to computing high dimensional multivariate distributions) and require massive training datasets. Therefore, they are not practical for non-trivial memory lengths and are usually limited to short contexts. An appealing alternative is the family of variable-order Markov models.

6.4.2 Variable-order Markov models

Although probably no simple underlying statistical model exists for protein families, they do share a feature, termed **short memory**, which is common to many "natural" sources. That is, given a sequence of observations $\mathbf{s}$ the probability distribution of the next symbol can be quite accurately approximated by observing no more than the last L symbols in that subsequence. Often, less than L symbols are enough to predict the next symbol reliably.

Variable-order Markov Models (VMMs), as the name implies, are models that allow for a variable memory lengths. VMMs learn the conditional probability distribution of the next character $P(\sigma|\mathbf{s})$ where the length of the context $\mathbf{s}$ can vary at different positions along the string. Hence, the application of VMMs seems very appropriate for protein sequences. While less expressive than the family of HMMs, they can be surprisingly successful for classification and prediction and they are more efficient. On the other hand, they are more flexible than fixed-order models and can encode much longer memory lengths (if most dependencies are short).

All VMMs rely, in one way or another, on identifying significant segments that occur frequently in the data. If the model is used to describe a protein family, then these segments reflect some statistical properties of the corresponding protein family; they induce a probability distribution over the next character and thus can be effective in prediction. Learning the next character in a sequence of observations is also referred to as **learning sequential data**.

When designing VMMs there are several issues we need to address. For example, we need to decide how to represent context, what is the maximal memory length of the model, how to compile the contexts into a structure that can be searched efficiently, which contexts of all possible contexts of a given character should be used in prediction and how to search for them. We will discuss a few models next.

6.4.2.1 Codes and compression

Learning sequential data is closely related to problems of *lossless compression* [199, 200]. Before we continue our discussion on VMMs we will deviate briefly to discuss codes. We will soon see how this is related to our discussion.

When designing a statistical model to describe a certain process we usually do not know the true source distribution $P(x)$ or even the type of the underlying probability distribution. Therefore our model is likely to induce a different probability distribution over the sample set. However, even if our model differs from the actual process that generates the samples we might still have a good model. What really matters is whether the probability distributions over the sample space are similar or not. If the distributions are close enough, then for all practical purposes, such as prediction, classification and compression, our model can be considered a reliable model. This should be taken with a grain of salt, because if the structure or the parametric form of the model differ from those of the real process, they might lead us to wrong conclusions and false insight into the nature of the process.

How can we measure the quality of a model? It is interesting to look at this problem from a different point of view, where the goal is to obtain the best compression over instances generated by the source, using a code. A binary **source code** assigns to each possible source sequence $\mathbf{s}$ a binary code word $C(\mathbf{s})$ of length $l(\mathbf{s})$. The compression rate of the code is a function of the average number of bits needed to code observations generated by the source. Clearly, we want the length of the code words to be as short as possible, but still we would like to be able to reconstruct the source sequence from its sequence of codewords. Hence, we focus on **prefix codes**, that is, codes where no code word is a prefix of another code word. With prefix codes there is no ambiguity, and a sequence of code words can be easily decoded to recover the original source sequence. These codes are also called prefix-free codes.

A simple prefix-free code is the **Huffman code** [201]. The code is generated by repeatedly merging the least frequent instances, following a tree-like structure, until reaching the root node. Then by traversing the tree back and associating one bit with each bifurcation, the code is built. For example, assume $\mathcal{A} = \{A, B, C, D\}$ and $P(A) = 0.5, P(B) = 0.125, P(C) = 0.125, P(D) = 0.25$. The prefix-free code tree for this set is depicted in Figure 6.18, with $C(A) = 0, C(B) = 100, C(C) = 101, C(D) = 11$. In this tree no inner nodes are used as codes. That is, $C(i) \neq prefix(C(j)) \;\forall i, j \in \mathcal{A}$.

By Shanon source coding theorem [202, 203], given a source with probability distribution $P = \{p(x)\}$ over $\mathcal{A}$, then the optimal source code is the one that uses $-\log p(x)$ bits to code instance $x \in \mathcal{A}$. The average code length, when each instance is weighted with its probability, is then

$$-\sum_{x \in \mathcal{A}} p(x) \log p(x)$$

Note that this is actually the entropy of P; hence, on average we need $H(P)$

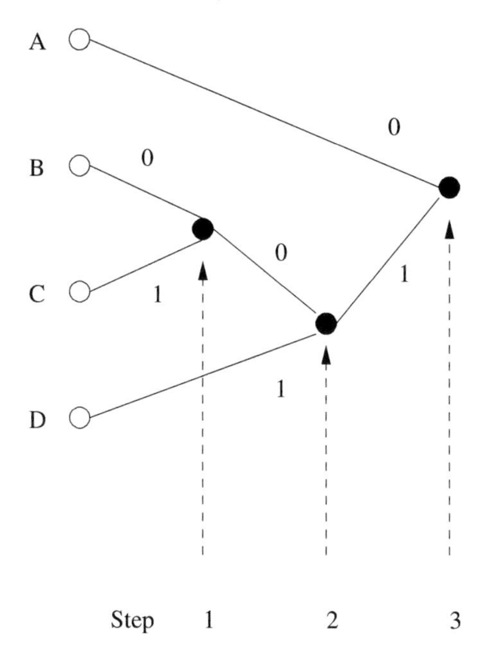

FIGURE 6.18: **Huffman code for the source:** $P(A) = 0.5, P(B) = 0.125, P(C) = 0.125, P(D) = 0.25$. The code tree is built by first merging B and C, which are the least frequent instances (step 1). The inner node has a probability of 0.25. Next this node is merged with D (step 2), as these are the least frequent events at this point. The final step merges the top inner node with A. The code is then generated by assigning zeros to all right edges and ones to all left edges (or vice versa). Each instance is associated with a code word, which is the string that is generated by traversing the path from the root node to that instance. For example, the code word for C is 101.

bits per character to describe observations that are generated by the source in the most efficient way, and any other code is less efficient. For continuous sample spaces, this equation translates to an integral over the sample space (see Section 3.12.5).

Note that the average number of bits per character in the code of Figure 6.18 is $\bar{l}(x) = 0.5 \cdot 1 + 0.125 \cdot 3 \cdot 2 + 0.25 \cdot 2 = 1.75$ and is equal to the entropy of the distribution

$$H(P) = -0.5 \log 0.5 - 0.125 \log 0.125 - 0.125 \log 0.125 - 0.25 \log 0.125 = 1.75$$

However, in general the average code length of Huffman code does not equal the entropy and is not necessarily the most efficient one[5].

[5] A method called **arithmetic coding**, which represents the complete string of characters as a single number between 0 and 1, can sometimes achieve better compression rates [204,205].

Moreover, the optimal code is not necessarily attainable. Usually we do not know the probability distribution P, but rather approximate it with a distribution $Q = \{q(x)\}_{x \in \mathcal{A}}$. What if we design a code that uses $\log q(x)$ bits to code the instance x. To see that any code other than the code derived from P is less efficient, we need to prove that the average number of bits used when assuming Q is higher than with P. That is, $AveCode(Q) - AveCode(P) > 0$. This can be proved as follows

$$\begin{aligned} AveCode(Q) - AveCode(P) &= -\sum_{x \in \mathcal{A}} p(x) \log q(x) - \left(-\sum_{x \in \mathcal{A}} p(x) \log p(x) \right) \\ &= \sum_{x \in \mathcal{A}} p(x) \log \frac{p(x)}{q(x)} = -\sum_{x \in \mathcal{A}} p(x) \log \frac{q(x)}{p(x)} \\ &\geq \sum_{x \in \mathcal{A}} p(x) \cdot \left(1 - \frac{q(x)}{p(x)}\right) \\ &= \sum_x p(x) - \sum_x q(x) = 1 - 1 = 0 \qquad (6.7) \end{aligned}$$

where we used the inequality $\log x \leq x - 1$ or $-\log x \geq 1 - x$. The term $\sum_{x \in \mathcal{A}} p(x) \log \frac{p(x)}{q(x)}$ is referred to as the relative entropy of P and Q, also called the Kullback-Leibler (KL) distance (see Section 3.12.6). What we have shown above is that the KL distance between two probability distributions is always non-negative[6].

Not all codes can be associated with a probability distribution, in the sense that the length of the code word for instance x does not necessarily equal $\log q(x)$. However, it can be shown that the inequality in Equation 6.7 holds for any code as long as it is a prefix code. If the code uses $l(x)$ bits to code x then we can define $q(x)$ such that $l(x) = -\log q(x)$ or $q(x) = 2^{-l(x)}$. It turns out that lengths of the code words of a binary prefix code satisfy Kraft's inequality [206]

$$\sum_x q(x) = \sum_x 2^{-l(x)} \leq 1$$

and hence the inequality of Equation 6.7 still holds (see last step). To see that $\sum_x 2^{-l(x)} \leq 1$, denote L as the length of the maximal code word $L = \max_x l(x)$. The number of leaf nodes in a complete binary tree of depth L is 2^L. Note that $2^{L-l(x)}$ is the number of leaf nodes in the subtree that is mounted at x (see Figure 6.19). The total number of leaf nodes in all the subtrees combined is bounded by the total number of leaf nodes in the

[6]This measure, however, is not a metric (Problem 4.12) and it does not satisfy the triangle inequality. Therefore it should not be confused with a distance measure (for a discussion see Section 3.6).

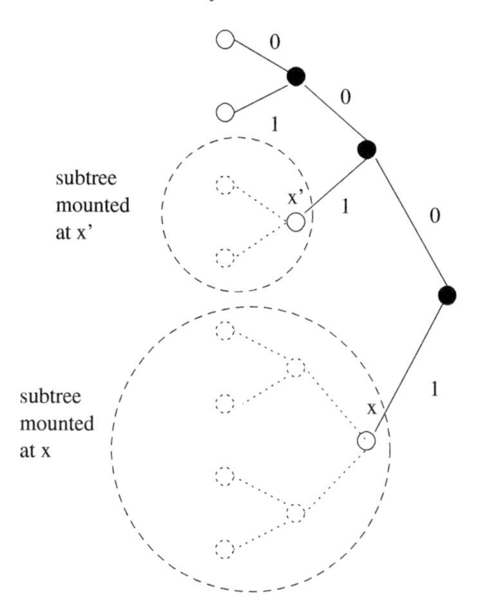

FIGURE 6.19: **Prefix codes satisfy Kraft's inequality.** The code words correspond to the empty circles. For example, $x = 1$ while $x' = 01$. Dashed edges and nodes were added to complete the code tree. The subtrees that are mounted at x and x' are circled.

complete tree and therefore when summing over x

$$\sum_x 2^{L-l(x)} \leq 2^L$$

and therefore

$$\sum_x 2^{-l(x)} \leq 1$$

Since the code is a prefix code, no two nodes x,x' are along the same branch, and therefore the subtrees mounted at x and x' are distinct and no leaf nodes are counted twice.

6.4.2.2 Compression and prediction

As mentioned above, the problem of learning sequential data is closely related to the problem of compression, which received considerable attention in the electrical engineering and information theory community, resulting in many compression algorithms that can be used for prediction as well.

In prediction, the goal is to maximize the likelihood of the string characters, given their context. That is

$$P(\mathbf{s}) = \prod_{i=1}^{n} P(s_i|s_1...s_{i-1})$$

This is equivalent to minimizing the *average log-loss function*, which is defined as

$$logloss_P(\mathbf{s}) = -\frac{1}{n}\sum_{i=1}^{n} \log P(s_i|s_1...s_{i-1})$$

In compression the log-loss function has another meaning, which is the average number of bits per character, since the term $-\log P(s_i|s_1...s_{i-1})$ is the shortest code for s_i, given the conditional distribution $P(s_i|s_1...s_{i-1})$ [204]. As discussed above, the best compression (and hence the best predictor) is obtained with the true distribution (that is, the source distribution). However, this distribution is unknown. Our goal is to learn the closest possible approximation in terms of the maximum likelihood or minimum average loss based on the empirical counts and the probability distributions that frequent strings induce over the next characters.

Over the years many compression algorithms were developed, based on different VMM structures. We will discuss a few in more detail next. Of special note are Prediction by Partial Match (PPM, Section 6.4.2.6) and the Context Tree Weighting (CTW) algorithm [207,208]. For performance evaluation over various types of classification and prediction tasks, see [209].

6.4.2.3 Lempel-Ziv compression and extensions

The Lempel-Ziv (LZ) compression algorithm is one of the first compression algorithms that were developed. It inspired many others and is used by many applications, including the gzip linux utility, which is based on the LZ77 version [210], and the compress utility, which is based on the LZ78 version [211]. The LZ compression algorithm is intuitive and works by scanning and recording unique substrings in the input and replacing each string with a pointer to its longest prefix that has already been observed.

Given a training sequence **s** the algorithm parses it into short non-overlapping substrings that are compiled into a dictionary. The algorithm parses the string from left to right, scanning the string for the next substring that has not been stored yet in the dictionary. Thus, the next substring must be a single character extension of a substring that is already in the dictionary. Starting from the empty dictionary, the first substring to be stored in the dictionary will be a single character. Subsequent substrings will gradually increase in length. For example, the string *abababababababababababababa* will be parsed into $a|b|ab|aba|ba|bab|abab|ababa|baba$. The compression is achieved by replacing the substring that has already been observed with an encoding of the pointer to its first location (or its index in the dictionary). For example, *ababa* will be represented as pointer to *abab* and the extra character *a*. For long sequences this representation becomes effective and a considerable reduction in size is obtained.

Prediction algorithms based on the LZ compression algorithm were later proposed in [212, 213]. To use it in prediction, the dictionary is complied into a tree that stores also the frequency of each substring. To compute the probability $P(\sigma|\mathbf{s})$, where $\mathbf{s} = s_1...s_{i-1}$, the tree is scanned for $\mathbf{s}$ starting from the root node. The probability is computed as the ratio between the frequencies $f(\mathbf{s}\sigma)/f(\mathbf{s})$, after smoothing (addition of pseudo-counts). If $\mathbf{s}$ is not in the dictionary tree, then the scan continues from the top with the remaining characters. Note that this algorithms does not use a preset bound on the memory length, and the maximal memory length can be as long as the longest training string.

Several variations over this algorithm were proposed over the years to address some of its weaknesses. One of them is the loss of context for strings that are not recorded in the tree, since the algorithm reset the scan from the top of the tree with the remaining characters. This might result in suboptimal estimation of the conditional probability, since there might be longer and more informative contexts that can predict the next character with a higher probability. The algorithm of [214] adds a minimal context requirement to the dictionary, so that for every character that follows strings longer than M, at least M characters are used as context.

6.4.2.4 Probabilistic suffix trees

The Probabilistic Suffix Tree (PST) model [215] draws on identifying frequent short segments in the input sequences and compiling them into a compact tree model. The key elements of the PST model are the method that determines which strings will be stored in the tree and used for prediction, and the method used to define the context of each character. While not the most powerful model for compression and prediction, it is relatively simple and intuitive.

A PST over an alphabet $\mathcal{A}$ is a non-empty tree whose nodes vary in degree between zero (for leaves) and the size of the alphabet. Each edge in the tree is labeled by a single symbol of the alphabet. Nodes of the tree are labeled by a string, which is the one generated by walking *up* the tree from that node to the root. Each node, labeled with string $\mathbf{s}$, is assigned a probability distribution vector over the alphabet that is the conditional probability distribution $P(\sigma|\mathbf{s})$ associated with the context $\mathbf{s}$. An example of a PST is given in Figure 6.20.

A PST differs from, albeit related to, the classical suffix tree, which contains all the suffixes of a given string (see [103]). In a suffix tree the father of node(tio) would have been node(ti), whereas in a PST the father of a node is a node without the first (as opposed to last) symbol. Here the father of node(tio) is node(io).

6.4.2.4.1 Learning PSTs. As with other VMMs, the PST model relies on patterns that appear frequently in the dataset. Given n training sequences $\mathbf{s}_1, \mathbf{s}_2, ..., \mathbf{s}_n$ of lengths $l_1, l_2, ..., l_n$ over the alphabet $\mathcal{A}$, the probability to observe a string $\mathbf{s}$ is defined as

$$\tilde{P}(\mathbf{s}) = \frac{f(\mathbf{s})}{n(|s|)}$$

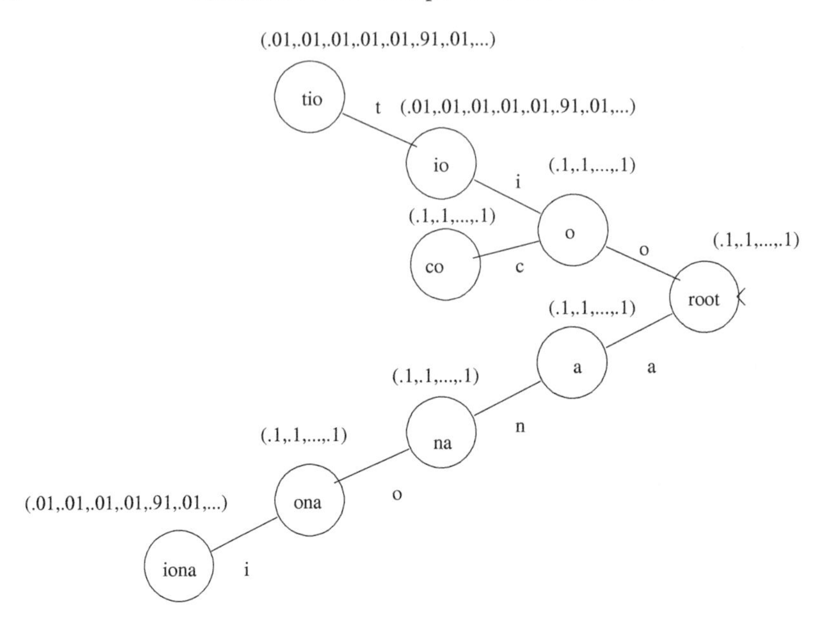

FIGURE 6.20: **An example of a PST.** This is a toy example of a tree that was trained on English text over the reduced alphabet $\mathcal{A} = \{a, c, d, i, l, n, o, r, u\}$. The tree is shown in landscape mode, which makes the prediction step easier to follow. The root is the rightmost node. The vector that appears near each node is the probability distribution over the next symbol. For example, the probability distribution associated with the subsequence *io* predicts that the next character is n with 0.91 probability and 0.01 probability to observe any other character. Similarly, after observing *iona* we can predict with high probability that the next character is l (as in "computational"). The node that corresponds to *tio* is plotted but is unlikely to be part of the tree since its predictions are identical to those of *io* (see text).

where $f(\mathbf{s})$ is the empirical frequency of $\mathbf{s}$ and $n(|s|)$ is the total number of subsequences of length $|\mathbf{s}| = l$ in the dataset, with overlaps

$$n(|s|) = \sum_{i \ s.t. \ l_i \geq l} (l_i - l + 1)$$

This definition disregards the fact that the subsequences are overlapping and therefore are not independent of each other. A more accurate estimate of the empirical probability requires knowing the maximal number of possible occurrences of the *specific* string $\mathbf{s}$ in the dataset, but computing this number is more complicated and depends on the period of that string (the minimal interval with which it overlaps itself). However, this definition seems to work well for all practical purposes [216]. It also induces a probability distribution over all strings of length l since $\sum_{s \in \mathcal{A}^l} \tilde{P}(\mathbf{s}) = 1$.

The conditional probability $\tilde{P}(\sigma|\mathbf{s})$ is also defined based on the empirical observations. If $f(\mathbf{s}*)$ is the number of non-suffix occurrences of the string

s in the data set, then the conditional empirical probability of observing the symbol σ right after the string **s** is defined as

$$\tilde{P}(\sigma|\mathbf{s}) = \frac{f(\mathbf{s}\sigma)}{f(\mathbf{s}*)}$$

With these probabilities (which are actually computed on the fly, as described next) we can now train a PST from the input data. The algorithm starts by enumerating all substrings of length 1 to L and computing their probability in the dataset. The parameter L is the maximal memory length of the model (i.e., the maximal length of a possible string in the tree), and it has to be set by the user. If the empirical probability of a string **s** falls below a certain threshold (P_{min}) then all its extensions to longer strings are ignored. The P_{min} cutoff avoids an exponentially large search space (in L).

The tree is built at the same time. The PST is initialized to a single root node, which corresponds to the empty string. Each string (context) **s** that is deemed significant, as described below, is added to the tree and its node is associated with the corresponding conditional probability distribution $P(\sigma|\mathbf{s})$. A string **s** is considered significant if it passes the following tests. First its probability must exceed P_{min}. There also must be at least one character σ in the alphabet whose conditional probability $P(\sigma|\mathbf{s})$ is surprising (as explained below) and is significantly different from the conditional probability $P(\sigma|suf(\mathbf{s}))$ of observing it after the suffix of **s**, where $suf(\mathbf{s}) = s_2 s_3 ... s_l$ and corresponds to the direct father of **s**.

The first test, for "surprise", assesses whether the probability distribution that **s** induces over the next character is informative. If it is uniform over the alphabet $\mathcal{A}$, or if it is similar to the prior probability distribution over the alphabet (e.g., the background probability of amino acids $P_0(\sigma)$ in the dataset), then there is no point in adding it to the tree, since it does not provide new information. If there is at least one character whose conditional probability $P(\sigma|\mathbf{s}) > (1+\alpha)P_0(\sigma)$, then we can say that the distribution is informative (α is a user-specified parameter). However, this condition is not sufficient to add the node to the tree. It also needs to pass the uniqueness test. That is, the conditional probabilities $P(\sigma|\mathbf{s})$ and $P(\sigma|suf(\mathbf{s}))$ must differ too. This second test ensures that the predictions induced by **s** are significantly different from the predictions induced by its father. Otherwise, there is no point in adding **s** to the tree.

Note that when testing a string **s**, its father node has *not* necessarily been added itself to the PST at this time (as in the example below). But if the two conditions hold, then the string **s** and all necessary nodes on its path are added to the PST. Therefore, it may and does happen that consecutive inner PST nodes are almost identical in their predictions. Note also that even if a string fails to satisfy the two conditions, we still need to examine its sons further while searching for significant patterns until we reached the maximal length L, or until the probability falls below P_{min}.

To better understand the two conditions, consider the following example. Say we want to train a model over an English text, so it would predict similar English texts with high probability. One of the most common suffixes in English is the suffix *tion*, and as soon as we have seen the context *tio* or even just *io* we can predict with high probability that the next character is going to be *n*. However, just observing the context *o* is not informative enough, since the distribution it induces over the next character does not differ significantly from the background distribution of letters in the English alphabet or in the input text (which is the conditional probability distribution associated with the root node). Therefore, the string *o* is unlikely to pass the first test. However it is almost certain (depending on the model's parameters) that the node *io* will be added, and its addition will require adding the node *o* as well. The string *tio* on the other hand is unlikely to be added, since the probability distribution it induces over the next character does not differ much from that of *io* and therefore it fails the uniqueness test.

In a final step, all conditional probability distributions that are associated with the nodes of the tree are smoothed by adding pseudo-counts (using any of the methods described in Section 4.3.3.2) to ensure that no single symbol is predicted with zero probability, even if the empirical counts may attest differently.

Build-PST $(P_{min},\ \alpha,\ r,\ L)$

Let $\bar{T}$ denote the tree, $\bar{S}$ denote the set of contexts to check, and $\bar{P}_s$ denote the conditional probability distribution associated with the node s.

1. **Initialization:** let $\bar{T}$ consist of a single root node (with an empty label), and let $\bar{S} \leftarrow \{\sigma \mid \sigma \in \mathcal{A} \ \text{ and } \ \tilde{P}(\sigma) \geq P_{min}\}$.
2. **Building the tree:** While $\bar{S} \neq \phi$, pick any $s \in \bar{S}$ and
   (a) Remove s from $\bar{S}$.
   (b) If there exists a symbol $\sigma \in \mathcal{A}$ such that
   $$\tilde{P}(\sigma|s) \geq (1+\alpha)P_0(\sigma)$$
   and
   $$\frac{\tilde{P}(\sigma|s)}{\tilde{P}(\sigma|suf(s))} \begin{cases} \geq r \\ \text{or} \\ \leq 1/r \end{cases}$$
   then add to $\bar{T}$ the node corresponding to s and all the nodes on the path to s from the longest suffix of s that is already in $\bar{T}$.
   (c) If $|s| < L$ then add the strings $\{\sigma' s \mid \sigma' \in \mathcal{A} \ \ and \ \ \tilde{P}(\sigma' s) \geq P_{min}\}$ to $\bar{S}$.
3. **Adding the conditional probability functions:** For each s labeling a node in $\bar{T}$, let
$$\bar{P}_s(\sigma) = \frac{N_s \tilde{P}(\sigma|s) + M_s P_{pseudo}(\sigma|s)}{N_s + M_s} \tag{6.8}$$
   where N_s is the number of observations associated with s and M_s is the number of pseudo-counts using a pseudo-counts method of choice (see Section 4.3.3.2).

The basic PST model does not make any assumption on the input data, but incorporating prior information we have about the data may improve the performance of this model, since it adjusts the general purpose model to the specific problem domain. For example, considering the amino acids background probabilities as the minimum probability $P_0(\sigma)$ in step 2b of the algorithm, or smoothing predictions with "position-specific" pseudo-counts based on amino acids substitution probabilities could bring in useful prior information. Note that position-specific in this case means node-specific, such that the number of pseudo-counts and their distribution differs for each node and depends on the number of observations $\mathbf{s}\sigma$ associated with a node and their distribution. Another possible modification is to add another test, requiring that the number of distinct sequences that include the subsequence $\mathbf{s}$ exceeds a certain threshold N_{min}, which is defined in proportion to the total number n of strings in the sample set. This would eliminate repetitive strings that are common only to a few members of the training set.

6.4.2.4.2 Prediction with PSTs. Once we trained a PST over a certain dataset, we can use it to compute the probability of new strings. Substrings that appear frequently in the dataset are likely to become part of the tree, and the distributions they induce over the next characters play a major role in identifying other members of the class.

The probability of each string is defined as the product of the probabilities of the individual characters

$$P_T(\mathbf{s}) = \prod_{i=1}^{l} P(s_i|s_1 s_2 ... s_{i-1})$$

where l is the length of the string. This is also referred to as the likelihood of the string. Each character is predicted using its longest context that is encoded in the tree. That is,

$$P_T(\mathbf{s}) = \prod_{i=1}^{l} P(s_i|\text{maxsuf}(s_1 ... s_{i-1}))$$

where maxsuf$(s_1...s_{i-1})$ is the longest suffix of $s_1...s_{i-1}$ that appears in the tree. The probabilty $P(s_i|\text{maxsuf}(s_1...s_{i-1}))$ is given by the probability distribution associated with the corresponding node in the tree. For example, if $\mathbf{s} = introduction$ then its probability using the PST of Figure 6.20 is

$$\begin{aligned} P_T(introduction) &= P(i|\text{maxsuf}()) \cdot P(n|\text{maxsuf}(i)) \cdot P(t|\text{maxsuf}(in)) \cdot \\ &\quad P(r|\text{maxsuf}(int)) \cdot P(o|\text{maxsuf}(intr)) \cdot P(d|\text{maxsuf}(intro)) \cdot \\ &\quad \cdots P(o|\text{maxsuf}(introducti)) \cdot P(n|\text{maxsuf}(introductio)) \\ &= P(i|-) \cdot P(n|-) \cdot P(t|-) \cdot P(r|-) \cdot P(o|-) \cdot P(d|o) \cdot P(u|-) \\ &\quad \cdots P(o|-) \cdot P(n|tio) \\ &= 0.1 \cdot 0.1 \cdots 0.91 = 0.1^{11} 0.91 \end{aligned}$$

where - represents the root node. Note that a random uniform distribution over the alphabet would assign a probability of 0.1^{12} to the string $\mathbf{s}$ (making $\mathbf{s}$ roughly 90 times more plausible under T than under the simple uniform model).

When computing and comparing the probability of multiple strings of different lengths, the probability of each string should be normalized by the length to avoid bias due to length differences. Another approach is to use a log-odds score, which compares the probability of an event under two alternative hypotheses and normalizes the probability of a string $\mathbf{s}$ by the chance probability to observe that string

$$\log \frac{P_T(\mathbf{s})}{P_0(\mathbf{s})}$$

where $P_0(s)$ is the probability to observe $\mathbf{s}$ as a result of independent selection $P_0(s) = \prod_{i=1..l} P_0(s_i)$. This approach is more effective for short sequences that do not trigger the long memory prediction and may be assigned a high probability simply by chance.

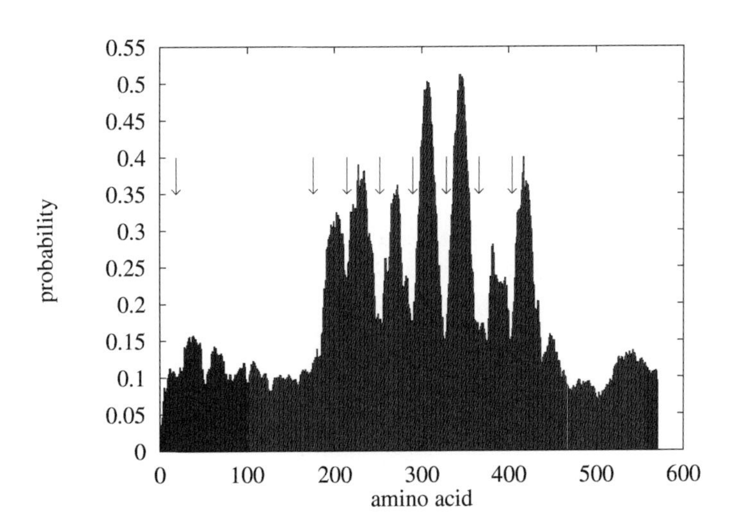

FIGURE 6.21: Prediction of EGF domains. EGF (Epidermal Growth Factor) domains are protein segments about 40 residues long that are common in secreted (extracellular) proteins or in extracellular domains of membrane-bound proteins. They are believed to play a role in cell differentiation and growth. A PST that was trained on the EGF family was used to predict the sequence of Fibropellin C (Biozon ID 570-15), letter by letter. The prediction combines both a PST that was trained on the training set and a PST that was trained on the reversed sequences (two-way prediction). The starting positions of the EGF domains (according to SwissProt records) are marked by arrows. (Figure reproduced from [216] with permission.)

It should be noted that the PST model is a prediction tool with a "global" flavor. However, in some cases what we a need is a tool that can identify partial matches. For example, the similarity of two protein sequences is often limited to a short motif or a domain. In such cases, a prediction based on the whole string could result in small likelihood values as a result of the unrelated regions. However, the model can be tweaked toward local predictions by considering only regions longer than a certain minimal length, set by the user, with a large percentage of high probability amino acids (for details see [216]). This variant can help detect significant occurrences of a domain as exemplified in Figure 6.21.

Another issue with the PST model is that patterns need to be long enough to trigger the short memory feature. Since the prediction step proceeds from left to right, letters that mark the beginning of significant patterns are predicted with non-significant probabilities. Only after a significant subsequence has been observed (i.e., once sufficient information is accumulated) the subsequent letters are predicted with high probability. Consequently, the left boundaries between significant and non-significant patterns (i.e., domain/motif boundaries) are somewhat blurred. Two-way prediction addresses this problem by training two models, T and T^R, where T is built from the sequences as they are and T^R is built from the reversed sequences. The prediction step is repeated twice. The input sequence is predicted using T, and the reverse sequence is predicted using T^R. Then, the predictions are combined by taking the maximal prediction assigned by the two models. Thus, for $\mathbf{s} = \mathbf{s}_1 \sigma \mathbf{s}_2$ where $\sigma \in \mathcal{A}$ and $\mathbf{s}_1, \mathbf{s}_2 \in \mathcal{A}^\star$,

$$P_{T,T^R}(\sigma|\mathbf{s}) = \max\left\{P_T(\sigma|\mathbf{s}_1), P_{T^R}(\sigma|\mathbf{s}_2^R)\right\}$$

6.4.2.4.3 Protein classification with PSTs. PST is a prediction model, but it can be applied to build a classifier into one of several possible classes. If we model the conditional source distribution for each class, then we can compute the likelihood of the string given each model and classify it to the model that assigns the highest likelihood, as in Section 6.3.1.3.

PSTs can be quite effective for protein classification, as is demonstrated in Figure 6.22 for the Neurotransmitter-gated ion-channels family. The graph shows the likelihood values that a PST model, which was trained on members of this family, assigns to database sequences. In the graph, the unrelated sequences show a clear linear relation in log scale while the training set and the test set samples are located far below this line, hence are well distinguished from the unrelated sequences.

Recall that the average log-likelihood of a string

$$\log P(\mathbf{s}) = \log \prod_{i=1}^{l} P(s_i|\text{maxsuf}(s_1...s_{i-1})) = \sum_{i=1}^{l} \log P(s_i|\text{maxsuf}(s_1...s_{i-1}))$$

is a measure of the number of bits per character that are necessary to encode a string (see Section 6.4.2.1). Hence, the slope of the line (in $\log_2$ base) is a mea-

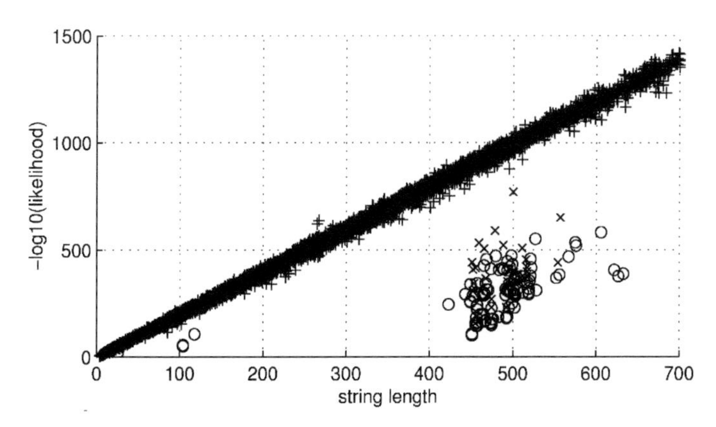

FIGURE 6.22: PST of the neurotransmitter-gated ion channels, complex transmembrane proteins that are involved in transmission of chemical signals between nerve cells. The graph plots the log-likelihood of each database sequence (including the training set and the test set) as a function of the sequence length. Training set sequences are marked with circles, test set sequences are marked with 'x' and all the other sequences in the database are marked '+'. (Figure modified from [216] with permission.)

sure of the average coding length of unrelated (essentially random) sequences using a specific PST. For example, 6.64 bits per letter are required when using the PST of the neurotransmitter-gated ion-channels family, while 8.3 bits are required when using the PST of the MHC I family (for comparison, 4.18 bits are required to code random sequences, using the background distribution). The slope is correlated with the "uniqueness" of the source distribution. The more different the amino acid distribution in the family being modeled is from the overall amino acid distribution in the database (the background distribution), the less efficient the model for coding unrelated sequences is.

On the contrary, the training set and test set can be encoded with fewer bits of information per letter. This reflects the compression obtained by the PST model and can also serve as a measure of divergence within a protein family. An upper bound on the source entropy can be derived by the effective (empirical) coding length of family members as provided by the PST model. A small average coding length reflects strong conservation within the protein family, while a large coding length reflects high divergence. The effective coding length of MHC I sequences, as predicted by the PST of that family, is 0.68 bits per letter, while the effective coding length of the neurotransmitter-gated ion channels is 2.21 bits per letter (using the PST of that family with the same set of parameters), indicating that the former is more conserved. The divergence may be related to structural diversity, suggesting that the transmembrane proteins of the neurotransmitter-gated ion channels may adopt a larger variety of shapes than the MHC I family. The average coding length

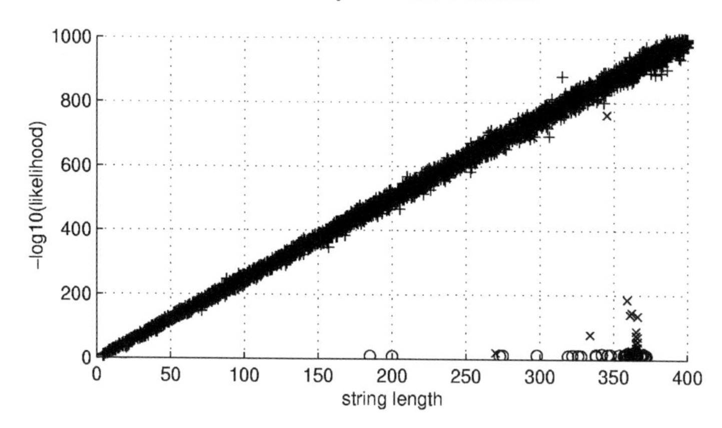

FIGURE 6.23: PST of the MHC I family. MHC proteins are part of the immune system. Their function is to "display" outside the cell fragments of proteins that are present inside the cell. If the cell is infected and contains proteins which are not part of its normal content, then their fragments will be recognized by other components of the immune system (T cells), triggering a process that eventually destroys the infected cell. The PST was trained with a lower value of P_{min} and is an example of an extreme fit of the PST to the training set. The prediction of the training set and the test set improves, while the unrelated sequences are left unaffected. When the PST of the MHC I family was trained with the same set of parameters as the PST of the neurotransmitter-gated ion channels, the results resembled those of Figure 6.22. (Figure modified from [216] with permission.)

of related sequences depends on the parameters of the model. Taking a more permissive threshold for P_{min} can improve the model with better predictions (Figure 6.23), i.e., better separation between the family members and the unrelated sequences and lower average coding length. However, it also increases the model size, and this should be taken into account when computing the average coding length.

6.4.2.5 Sparse Markov transducers

Signatures of protein sequences are hardly ever completely conserved. More typically, they contain several conserved residues spaced along a relatively diverged region (see the zinc finger signature of Section 4.3.2). Hence, PSTs that rely on repeated occurrences of perfectly conserved strings might perform poorly on highly diverged families, especially in comparison with HMMs.

Sparse Markov Transducers (SMT) or Sparse Markov Trees are a generalization of probabilistic suffix trees, which accommodates wild cards in the modeled sequences. Sequences with wild cards can be perceived as sparse sequence signatures, and hence the name. A SMT describes a sequence of observations, where ϕ is a place holder (wild card) that can be matched with any character and ϕ^n stands for n consecutive wild cards. For example, $A\phi^2A$

can be matched with $AAAA$, as well as with $ACCA$, similar to what we saw with regular expressions (Section 4.3.2). The goal is to learn the conditional probability distribution

$$P(\sigma|\mathbf{s}) = P(\sigma|\phi^{n_1}X_1\phi^{n_2}X_2...\phi^{n_k}X_k)$$

where $\mathbf{s}$ is the context of σ. The context can be the preceding characters, or it could be engineered to code for the context surrounding the position we would like to predict, both to the left and to the right. For example, to predict the middle character s_x in a sliding window of length $L+1$ (the memory length), the context can be defined as $\mathbf{s} = s_{x-1}s_{x+1}s_{x-2}s_{x+2}...s_{x-L/2}s_{x+L/2}$, starting from the closest positions to the output symbol and ending with the farthest positions.

Learning the conditional probabilities with SMTs entails learning the positions of the wild cards and their length. The algorithm suggested in [217] uses sparse Markov trees, which are more general than sparse Markov transducers since they allow a different number of wild cards along the different branches of the tree. The path from the root to the leaf node defines the context sequence, but because of the wild cards multiple strings can be mapped to a single node.

The task of automatically learning where to position the wild cards in the tree (the topology of the tree) is a difficult one and the approach that is employed in [217] is to use a mixture of trees with different topologies. Mixture models often perform well in classification tasks since they can approximate quite complex source distributions (see Section 11.4.2.5). In mixture models, the combined prediction is given by the complete probability, with component models weighted by their prior probability. That is,

$$P(x) = \frac{\sum_{i \in Trees} P(i)P_i(x)}{\sum_{Trees} P(i)}$$

where $P(i)$ is the prior probability of the i-th tree. However, estimating the prior over the space of all possible trees is difficult, as it would require an analysis of the distribution of wild cards in data of real protein families. A practical approach is to approximate the prior of trees based on their performance. Namely, each tree is assigned a weight $W(i)$ and the total probability is given by

$$P(\sigma|\mathbf{s}^t) = \frac{\sum_{i \in Trees} W(i)P_i(\sigma|\mathbf{s}^t)}{\sum_{Trees} W(i)}$$

where t designates the length of the input sequence that has been observed so far. In [217] the weights are initialized based on the complexity of the tree topologies by simulating a stochastic process that generates random trees. The weights of the trees are then updated with every new instance, based on the probabilities they assign to the instance. Specifically, $W^{t+1}(i) = W^t(i)P(\sigma(t+1)|\mathbf{s}^t)$. It is also possible to use the available information, in the form of MSA of the input sequences, to constrain the possible topologies of the trees and identify potentially successful topologies, or to train a PST first and then merge branches that share subtrees.

While conceptually appealing, the SMT model is significantly more complex than the PST model, and training this model can be quite time consuming. In addition, the inclusion of wild cards (in unknown positions) requires a significant increase in the memory usage, since there is an exponential number of possibilities to place the wild cards in the input sequences. Nevertheless, the SMT model is an interesting alternative to the hidden Markov model.

6.4.2.6 Prediction by partial matches

In the data compression community the Prediction by Partial Matches (PPM) algorithm [218] is considered one of the best, together with Context Tree Weighting (CTW), performing significantly better than the previous models we described on many prediction tasks.

The PPM algorithm is based on the same principle that characterizes the prediction algorithms we discussed earlier, namely, using the longest possible context. The novel element of the algorithm is the mixing of several fixed-order Markov models to predict the next character. The algorithm starts by analyzing the training set and collecting statistics $f(\sigma|\mathbf{s})$ on the occurrences of characters after all possible observed contexts $\mathbf{s}$. The statistics are gathered for contexts of length k, starting from the zero-order model ($k = 0$) and all the way up to $k = L$, the maximal memory length[7]. Each model induces a different probability distribution $P_k(\sigma|\mathbf{s})$ over the next character. Huffman code or arithmetic coding is implemented to actually code each character according to the distribution in the relevant context(s).

PPM combines the different models into a single predictor. Specifically, for each input character σ the algorithm searches for the longest observed context $\mathbf{s}$, like PSTs or Lempel-Ziv. However, unlike these algorithms, if the character is novel (unseen) in the context $\mathbf{s}$ then the k-th model cannot be used for prediction and the algorithm triggers an **escape** mechanism and switches to a context shorter by one. If the character has not been observed after the shorter context then the escape mechanism is triggered again and the context is shortened once again. This iterative process terminates once we reach a model where the character was observed or when we reach the zero-order model.

Triggering the escape mechanism is associated with a certain probability that can be perceived as the penalty for switching to a lower order model. Thus, the probability $P(s_i|s_{i-k}...s_{i-1})$ (denoted PPM) is defined recursively:

[7] A real-time version works by processing all the input characters observed so far to produce the probability distribution of the next character.

if the string $s_{i-k}...s_{i-1}s_i$ is observed in the training data then

$$PPM(s_i|s_{i-k}...s_{i-1}) = P_k(s_i|s_{i-k}...s_{i-1})$$

Otherwise

$$PPM(s_i|s_{i-k}...s_{i-1}) = P_k(escape|s_{i-k}...s_{i-1}) \cdot PPM(s_i|s_{i-k+1}...s_{i-1})$$

where the context on the right side is shortened by one, and the PPM function is recalled with the $k-1$ model. The escape character is necessary in compression and information transmission. Transmitting an escape code signals that the context length has changed, such that the message can be restored on the other end of the channel.

There are several variants of the PPM algorithm that differ in the way they define the escape probabilities in each context $\mathbf{s}$. The general approach is to add M counts for the escape character and normalize the counts to obtain the final probabilities. That is, the probability of the escape character is defined as

$$P_k(escape|\mathbf{s}) = \frac{M}{M + f(\mathbf{s}*)}$$

and the probability of any other character is defined as

$$P_k(\sigma|\mathbf{s}) = \frac{f(\sigma|\mathbf{s})}{M + f(\mathbf{s}*)}$$

One simple approach (called Method A) for setting M is to add one count to the escape character. This is identical to the Laplas method for introducing pseudo-counts we described in Section 4.3.3.2. Better performance was observed with another variant (called Method C), where M is set to the number of different characters σ observed in the context $\mathbf{s}$. Another popular variant (method D) increases the number of pseudo-counts by one with every unseen character.

The second element of the PPM algorithm is the exclusion mechanism. This mechanism is based on the observation that the original method for computing probabilities results in over-counting. If a prediction fails for a certain context then the unseen symbol cannot be one of those that are observed in that context, and the relevant sample space for the shorter context should be reduced accordingly. Hence, each character σ that is observed after context $\mathbf{s}$ is excluded when deriving the conditional probability distributions for all suffixes of $\mathbf{s}$. This is best explained through an example. Consider the models of Table 6.1 and the string `fortiori`. By the time we get to the second r character, the context is the strings `fortio`. The longest context that is observed in the data is `tio`; however, the character r has not been observed in that context. Therefore, the escape mode is triggered and the probability is given by

$$PPM(r|tio) = P_3(escape|tio)P_2(r|io) = \frac{2}{18+2} \cdot \frac{1}{30+7}$$

assuming the probability of the escape character is set according to method C. The exclusion mechanism adjusts the probability $P_2(r|io)$ to obtain more accurate estimates, considering only the characters that were not observed in longer contexts. That is, the characters n and u that are observed in the context tio are ignored, and the reduced sample space consists of c, d, l, r and t. This results in an adjusted probability of $P(r|io) = \frac{1}{5+7}$.

The maximal order L is a parameter of the model. It has been noted that increasing the memory length of the model does not seem to improve performance above a certain length (e.g., $L = 5$ for some standard text datasets), as the contexts become too specific and fail to predict many characters, triggering the escape mechanism many times and reducing the overall probability.

Order	Con-text	Character Frequencies										Total	Escape Counts
		a	c	d	i	l	n	o	r	t	u		
3	tio						17				1	18	2
2	io		1	1		1	21		1	1	4	30	7
1	o	3	6	4	3	10	40	7	25	7	14	119	10
0	ϵ	315	183	160	331	202	274	223	293	260	124	2365	10
-1	-	1	1	1	1	1	1	1	1	1	1	10	

TABLE 6.1: Prediction by partial matches. The table lists the frequencies of different characters following the context strings tio, io, o and the empty string ϵ, in synthetic data. The model of order -1 serves to predict alphabet characters that were not observed at all in the training set and assigns a uniform probability of $1/|\mathcal{A}|$ to all characters.

6.5 Further reading

An excellent introduction to HMMs is given in [219]. Many papers followed with applications in various fields, including speech recognition [219], optical and handwriting character recognition [220, 221], computer vision [222, 223] and networks [224]. The seminal paper by [196] introduced profile HMMs which were tailored to protein families. Variants of profile HMMs that incorporate phylogenetic information are described in [225, 226]. Perhaps the best reference for applications in computational biology is [93]. Another excellent source is [227]. Applications of HMMs to gene prediction are described in [228, 229]. Other approaches for gene prediction are discussed in [230–232].

Codes and compression are discussed extensively in books on information theory such as [203]. Variable-order Markov models are reviewed in [209]. For additional references check the book's website at `biozon.org/proteomics/`.

6.6 Conclusions

- Markov models are probabilistic state machines that describe sequences of decisions or states, with states corresponding to different internal states of the system being modeled.
- Hidden Markov models are Markov models that describe sequences of observations (visible symbols) generated by internal states we do not have access to.
- First-order profile HMMs provide a powerful representation of protein families. The model architecture and parameters reflect the core structure of the family (such as insertion and deletion events) and the distributions of amino acids along different positions.
- Given an HMM, the probability of a new sequence can be computed efficiently using a dynamic programming algorithm (the forward algorithm).
- The most likely sequence of hidden states that generated a given sequence of observations corresponds to an "alignment" of the sequence with a model.
- Learning the parameters of an HMM from training data is a difficult problem. A locally optimal model can be learned using an expectation-maximization algorithm called the forward-backward algorithm.
- Higher order Markov models can capture higher level dependencies between characters. The maximal order of dependency is referred to as the memory length.
- Variable order Markov models (VMMs) are models with a variable memory length that changes depending on the context at each position along the sequence.
- VMMs are the basis of many compression algorithms and provide an appealing alternative to HMMs.

6.7 Problems

For updates, additional problems, files and datasets check the book's website at `biozon.org/proteomics/`.

1. **Gene finding using HMMs**

   You are given a set of DNA sequences, possibly with protein coding regions. The goal is to automatically detect the coding regions. You will be using a labeled dataset for training and another dataset for testing (files are available at the book's website). A protein coding region should start with a start codon and ends in a stop codon; however, not all sequences that satisfy these conditions are really coding regions. The statistics of nucleotides that appear between the start and the stop codons will determine the potential of this region to code for a protein (inside coding regions it follows the codon structure). Note also that there might be multiple start codons before a stop codon, suggesting that the first is not necessarily the real start.

   (a) Compare the statistics of nucleotides, dinucleotides and trinucleotides in coding and non-coding regions. Do you see a difference?

   (b) Compare the 16 first-order transition probabilities (i.e., the probabilities of dinucleotides) to the probability to observe these by chance (the product of the single nucleotide probabilities), both in coding and non-coding regions. Do you see a dependency between nucleotides in coding regions?

   (c) Suggest a hidden Markov model for this problem, considering: (1) frameshifts (the reading frame can start at any one of the first three nucleotides, and it can also change between one ORF and another), (2) between a start and a stop codon there might be coding (exons) and non-coding (introns) regions, (3) the coding region(s) might be on the complement strand, (4) the existence of a promoter region (usually characterized by a relatively conserved short pattern of nucleotides) increases the likelihood that a coding region will follow not too far upstream of the pattern.

   (d) Construct an inhomogeneous Markov model, as in Figure 6.4. Estimate the transition probabilities from the data (note that not all transitions are depicted in the figure). Take special care when estimating the transition probabilities between the two models.

   (e) Test your model on the test data. To determine whether a sequence has a coding region, implement the Viterbi algorithm that finds the most likely sequence of hidden states. Run your program on each sequence in the test data and compare your results to the positions of the coding regions as given in `coding-coordinates`.

2. **Handling machine precision limitations with HMMs**

   Prove that the factor method of Section 6.3.4, which normalizes the forward variables by $factor(t) = \sum_{i=1}^{c} \alpha_i(t)$, is correct. That is, prove (by induction or other methods) that

$$P(\mathbf{s}) = \prod_{t=1}^{T} factor(t)$$

3. **HMMs of protein families**

   Consider the profile HMM of Figure 6.14. Write a program that given an HMM of a specific protein family and a protein sequence:

   (a) Calculates the probability of the protein sequence to emerge from the model (the evaluation problem).

   (b) Outputs the most likely sequence of hidden states that generated the sequence (the decoding problem).

   Input files, formats and tips are available at the book's website.

4. **Building an HMM from an MSA**

   Write a procedure that initializes a profile HMM from an input MSA, following the methodology of Section 6.3.5.5.

5. **Silent states in HMMs**

   Prove that the HMM on the left of Figure 6.13 cannot be replaced with an HMM with silent states, as depicted on the right side of Figure 6.13.

6. **Codes**

   You are given a set of sequences over the alphabet $\{A, B, C, D, E\}$.

   (a) With no other assumptions, how many bits will you need to represent each character?

   (b) Assume we know the background frequency of each character, such that $P(A) = 0.2, P(B) = 0.3, P(C) = 0.4, P(D) = 0.04, P(E) = 0.06$. Calculate the average number of bits per character used to code a long sequence of characters, with the optimal source code.

   (c) Create a codebook for this alphabet, using Huffman code. What is the average number of bits required to represent a character using this code? If it is not identical to your answer to (b) can you suggest an explanation?

7. **Prediction by Partial Matches (PPM)**

   You are given a set of sequences over the same alphabet from the previous problem. Implement the PPM algorithm to compress this dataset. Compare the compression rate you obtain with Huffman code to that obtained with PPM.

Chapter 7

Classifiers and Kernels

7.1 Generative models vs. discriminative models

Many of the problems in computational biology are classification tasks, where there is a finite set of classes $\mathbf{C} = \{c_1, c_2, ..., c_k\}$ and every data instance is assumed to have emerged from one of the classes. For example, the data instances can be protein sequences with classes corresponding to different protein families. Or the instances can be gene expression profiles measured in cells that were extracted from different tissues (e.g., different types of cancerous cells). In most cases classes are distinct and we are interested in characterizing the classes and classifying new instances correctly. That is, given an unlabeled new observation x the classification task is to determine the class $c_i \in \mathbf{C}$ the instance belongs to. The setting is that of **supervised learning** where we have labeled data that we can use to train our models.

Usually we assume that each class is associated with a statistical source that generates instances $x \in \mathcal{X}$ according to a class conditional probability distribution $P(x|c_i)$. For example, if the classes correspond to protein families then the class-conditional probability function of each class can be approximated using a hidden Markov model, as described in the previous chapter. These models are referred to as generative models.

The Bayesian approach to pattern classification is to decide that the class is c_j if the posterior probability of c_j is higher than the posterior probability of any other class (see Section 3.12.8). That is

$$P(c_j|x) > P(c_i|x) \quad \forall i \neq j$$

This is the same rule we used in Section 6.3.1.3. Using Bayes' formula, this is equivalent to the rule

$$P(x|c_j) \cdot P(c_j) > P(x|c_i) \cdot P(c_i)$$

This rule divides the sample space $\mathcal{X}$ into different regions where each region is associated with one class. Depending on the complexity of the sources, there might be multiple regions associated with the same class. The boundaries between the different classes are called **decision boundaries** or **decision surfaces** (see Figure 7.1).

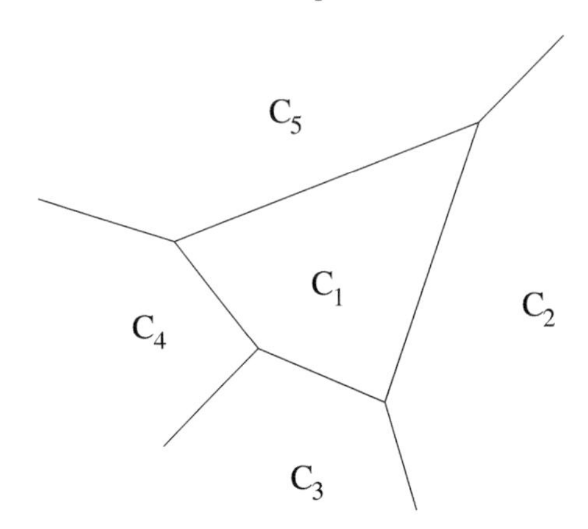

FIGURE 7.1: **Decision boundaries between different classes.**

To apply the Bayesian approach we need to know the class conditional probabilities, or accurately estimate them from the data. This entails several steps. First, we need to decide on the functional form of the model (for example Gaussian or Poisson distribution) or its architecture (if using HMMs or VMMs, for example). Given a set of labeled observations, where each instance is associated with one class, we can then train the parameters of each source using the observations in that class. For some functional forms, this can be solved by computing the parameters that maximize the likelihood of the data (ML estimator) or the posterior probability (MAP estimator). This is the case for common distributions such as the normal distribution or the binomial distribution. In other cases, such as HMMs, there are algorithms that can find a locally optimal solution (see Section 6.3.3).

However, if we do not have prior knowledge of the type of the underlying distribution we must resort to a different approach; a model that does not describe accurately the source might induce decision boundaries that deviate markedly from the true boundaries between classes, leading to erroneous predictions. The problem is especially pronounced with complex multi-modal source distributions, which cannot be approximated reliably with the common probability distributions, most of which are uni-modal.

One alternative approach is to estimate directly the posterior probability using a non-parametric method, for example based on the k-nearest neighbors. If k_j out of the k nearest neighbors of the instance x belong to class j then we can set $P(c_j|x) \leftarrow \frac{k_j}{k}$ and then apply the Bayesian decision rule as above. However, when data is sparse (which is the typically the case with high-dimensional data) this can result in poor estimates.

The approach that we will discuss in this chapter is to model the decision surface rather than the source itself. That is, instead of modeling the generative process (**generative models**), we will focus on modeling the decision boundaries they induce (**discriminative models**). These boundaries might have simpler functional forms than the sources themselves, and it may be easier and more effective to learn them. This is the underlying principle behind methods such as neural networks, linear discriminant functions (e.g., the perceptron algorithm) and Support Vector Machines (SVMs). Here we will describe briefly some of these methods with emphasis on SVMs and their application to protein classification. In the second part of this chapter we will discuss the decision tree model, which is based on a different approach to classification and classifiers and is especially appealing when the input data is not vectorial.

7.2 Classifiers and discriminant functions

In this chapter we are interested in learning classifiers. There is an extensive literature on this topic (see Further reading). Here we will use similar formalism to [191] and will tie it to our discussion on SVMs later on.

In general, a **classifier** is a function that classifies a given instance $x \in \mathcal{X}$ to one of the possible classes $c_1, c_2, ..., c_k$. Most of the attention in the literature is given to the two-category case (that is, distinguishing members of a class from non-members). However, it is insightful to start from the multi-category case. A common approach to classifier design is to utilize a set of functions $g_i(x)$, called **discriminant functions**, one for each category. The classifier compares the values assigned by each function and assigns an instance x to class c_i if

$$g_i(x) > g_j(x) \quad \forall j \neq i$$

Thus the classifier essentially divides the sample space into k regions $C_1, C_2, ..., C_k$ where $C_i = \{x \in \mathcal{X} | g_i(x) > g_j(x) \quad \forall j \neq i\}$ and all samples in C_i are assigned to c_i. In the two-category case the classifier can be reduced to a single discriminant function

$$g(x) = g_1(x) - g_2(x)$$

assigning x to c_1 if $g(x) > 0$.

If we know the generative functions $P(x|c_i)$, then the Bayesian approach as described in the previous section suggests a natural definition for the discriminant functions, setting each one to the posterior probability of the corresponding class

$$g_i(x) = P(c_i|x) = \frac{P(x|c_i) \cdot P(c_i)}{P(x)}$$

ensuring that each sample is classified to the class with the highest posterior probability. We can ignore $P(x)$ since it is fixed and independent of the class and simplify the definition of the discriminant functions such that

$$g_i(x) = P(x|c_i) \cdot P(c_i)$$

Any other monotonic transformation (such as the logarithm transformation) would give the same classification. This approach minimizes the total error rate

$$\begin{aligned} P(\text{error}) &= 1 - P(\text{correct}) \\ &= 1 - \left(\int_{C_1} P(c_1|x)p(x)dx + ... + \int_{C_k} P(c_k|x)p(x)dx \right) \end{aligned}$$

since it classifies each sample to the class i that maximizes $P(c_i|x)$ (see also Problem 3.16). However, it requires knowledge of the generative functions. In the next sections we will discuss methods for learning the discriminant functions from labeled data when we do not know the class conditional distributions.

7.2.1 Linear classifiers

In the course of this discussion we will assume initially that the samples are represented by feature vectors in $\mathbb{R}^d$ such that $\vec{x} = (x_1, ..., x_d)$. If the data samples are protein sequences then the features that make up $\vec{x}$ can be, for example, the frequencies of the 20 amino acids or other quantitative physico-chemical properties. We will elaborate on vectorial representations of proteins later on in this chapter (Section 7.3), as well as in Chapter 11.

The case we will discuss here is the one where we do not know or cannot model the generative sources. Instead, we will assume we know the functional form of the discriminant functions $g(\vec{x})$ and will review methods to learn the parameters of the functions from the input data, so we can use it for classification of new instances. We will focus on the two-category case and start by characterizing some general properties of discriminant functions.

The simplest type of a two-class discriminant function is linear. A linear discriminant function (linear classifier) is of the form

$$g(\vec{x}) = \vec{\omega}^t \cdot \vec{x} + \omega_0$$

where $\vec{\omega}$ is the **weight vector** and ω_0 is the **bias**. Thus, the value of the function is given by a linear combination of the features, with the weight vector setting the weights of the different features in the total sum, and ω_0 can be considered as an offset value. In the typical setting[1] the function is used to assign $\vec{x}$ to class c_1 if $g(\vec{x}) > 0$ and c_2 if $g(\vec{x}) < 0$. This can be simply summarized in a **classifier function** $f(\vec{x})$ using the sign function

$$f(\vec{x}) = \text{sgn}(g(\vec{x}))$$

[1]The function can also be trained to approximate the probability that $\vec{x}$ belongs to a class.

With +1 denoting class c_1 and −1 denoting c_2. If $g(\vec{x}) = 0$ then no decision is made and the instance is left unclassified or the class is chosen arbitrarily. What we are really interested in is characterizing the **decision surface**: the surface defined by the equation $g(\vec{x}) = 0$ that separates between the classes.

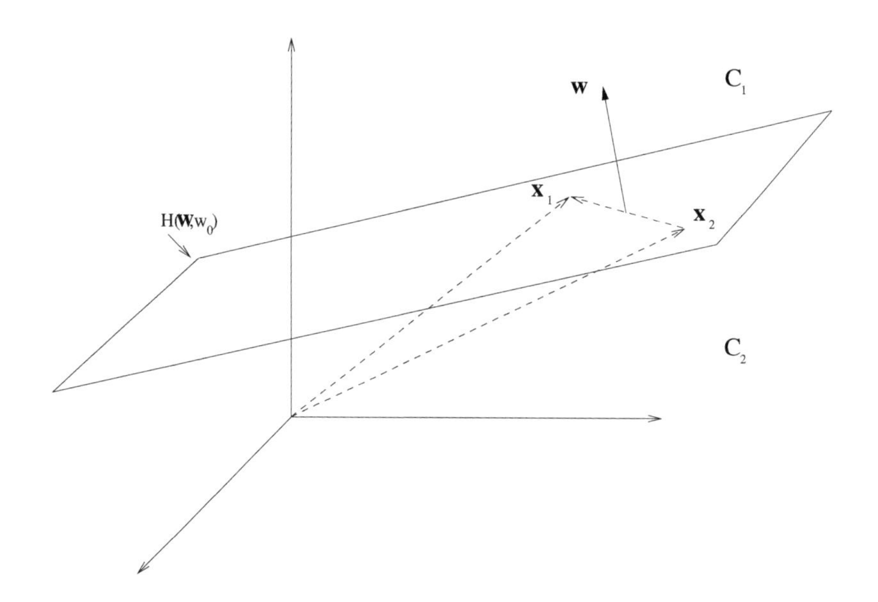

FIGURE 7.2: **The decision surface for a linear classifier.**

If $g(\vec{x})$ is linear then the decision surface is a hyperplane (line in 2D, plane in 3D and so on), as shown in Figure 7.2. We denote the separating hyperplane by

$$H(\vec{\omega}, \omega_0) = \{\vec{x} | \vec{\omega}^t \cdot \vec{x} + \omega_0 = 0\}$$

As was shown in [191], a geometrical interpretation of the parameters of a linear discriminant function is obtained by considering two points $\vec{x}_1, \vec{x}_2$ on the decision surface (Figure 7.2). All points on the decision surface satisfy $g(\vec{x}) = 0$, therefore

$$\begin{aligned} \vec{\omega}^t \cdot \vec{x}_1 + \omega_0 = \vec{\omega}^t \cdot \vec{x}_2 + \omega_0 = 0 \\ \Rightarrow \vec{\omega}^t \cdot (\vec{x}_1 - \vec{x}_2) = 0 \end{aligned}$$

A zero scalar product implies that the vectors $\vec{\omega}$ and $\vec{x}_1 - \vec{x}_2$ are orthogonal, and since the difference vector $\vec{x}_1 - \vec{x}_2$ is on the hyperplane the weight vector $\vec{\omega}$ must be **normal** to the hyperplane, pointing into C_1. The parameter ω_0 determines the relative placement of the separating hyperplane with respect to the origin, as $\omega_0 = g(\vec{0})$.

The actual value of the discriminant function for a given vector $\vec{x}$ is proportional to its distance from the hyperplane. Note that any vector $\vec{x}$ can be written as the sum of two components, one on the hyperplane and one orthogonal to the hyperplane (Figure 7.3). That is,

$$\vec{x} = \vec{x}_{\vdash} + \vec{x}_{\perp} = \vec{x}_{\vdash} + r\frac{\vec{\omega}}{||\vec{\omega}||}$$

where $\vec{x}_{\vdash}$ is the projection of $\vec{x}$ on the hyperplane, $\frac{\vec{\omega}}{||\vec{\omega}||}$ is a unit vector orthogonal to the hyperplane and r is the distance of $\vec{x}$ from the hyperplane. Therefore

$$g(\vec{x}) = \vec{\omega}^t \cdot \left(\vec{x}_{\vdash} + r\frac{\vec{\omega}}{||\vec{\omega}||}\right) + \omega_0 = (\vec{\omega}^t \cdot \vec{x}_{\vdash} + \omega_0) + r\frac{\vec{\omega}^t\vec{\omega}}{||\vec{\omega}||} = g(\vec{x}_{\vdash}) + r||\vec{\omega}||$$

Since $\vec{x}_{\vdash}$ is on the plane $g(\vec{x}_{\vdash}) = 0$ and we conclude that $g(\vec{x}) = r||\vec{\omega}||$. Hence, the function $g(\vec{x})$ is correlated with the distance r from $\vec{x}$ to the separating hyperplane. The distance is denoted by

$$r = d(\vec{x}, H(\vec{\omega}, \omega_0)) = \frac{g(\vec{x})}{||\vec{\omega}||} \tag{7.1}$$

The distance of the hyperplane from the origin is $\omega_0/||\vec{\omega}||$.

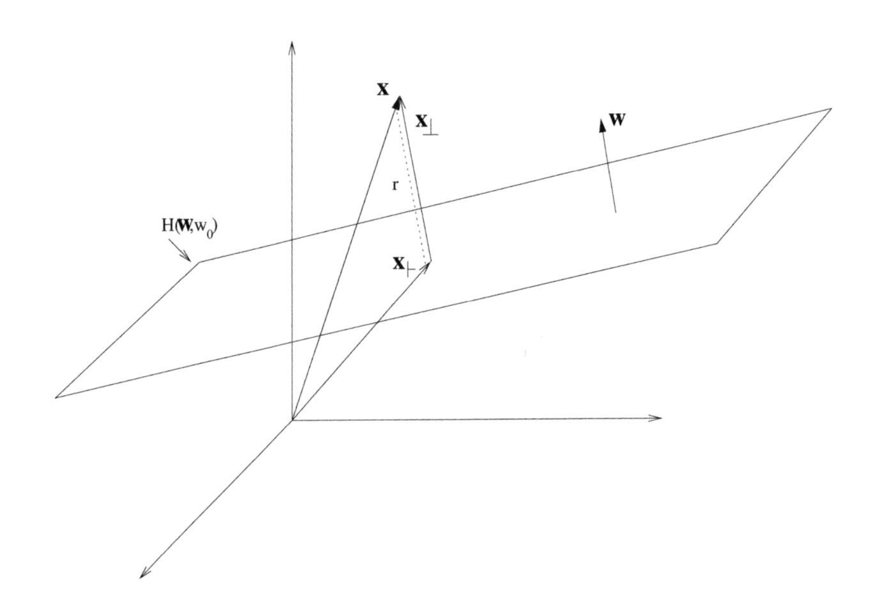

FIGURE 7.3: **Computing the distance from the separating hyperplane.**

To simplify notation, and for reasons that will become clearer later, it is sometimes convenient to represent the weight vector $\vec{\omega}$ and the bias ω_0 as a single vector $\vec{a}$ and replace $\vec{x}$ with a vector $\vec{y}$ of dimension $d+1$ such that

$$g(\vec{x}) = \vec{\omega}^t \cdot \vec{x} + \omega_0 = \vec{a}^t \vec{y}$$

where

$$\vec{a} = \begin{pmatrix} \omega_0 \\ \omega_1 \\ \vdots \\ \omega_d \end{pmatrix} \quad \text{and} \quad \vec{y} = \begin{pmatrix} 1 \\ x_1 \\ \vdots \\ x_d \end{pmatrix}$$

These vectors are referred to as **augmented vectors**. When using the augmented vectors we refer to the separating hyperplane simply as $H(\vec{a})$.

The separating hyperplane can be learned from the data, as described next. The success of this function in separating instances depends on the **separability** of the classes, and a distinction is made between the linearly separable case and the non-separable case.

7.2.2 Linearly separable case

If the data we would like to classify consists of classes that are linearly separable (see Figure 7.4), then it is possible to find a separating hyperplane using a classical algorithm called the **perceptron algorithm** [233]. Geometrically, we can easily outline where the solution should be. Given n labeled sample points $\mathbf{D} = \{\vec{y}_1, ..., \vec{y}_n\}$, the trick is to first transform all samples that are labeled c_2 to their negatives. That is, if $\vec{y}_i$ is labeled c_2 then we set $\vec{y}_i = -1 \cdot \vec{y}_i$ (using the augmented vectors). Consequently, we look for the weight vector $\vec{a}$ such that $\vec{a}^t \vec{y}_i > 0$ for all samples. Any vector that satisfies this condition is called a **separating vector**. Note that we can bound the region where the separating vector can be. Since $\vec{a}^t \vec{y}_i = |\vec{a}||\vec{y}_i| \cos \alpha$ where α is the angle between the two vectors, then the scalar product will be positive only if $|\alpha| < 90$. In other words, to satisfy the inequality $\vec{a}^t \vec{y}_i > 0$ the vector $\vec{a}$ must lie in the half space above the hyperplane normal to $\vec{y}_i$. Vectors that satisfy the inequality for all sample points must lie in the intersection of all the half spaces. The region obtained by the intersection of these spaces is called the **solution region**. How should we pick a separating vector within the solution region?

To find an algorithmic solution to the set of linear inequalities $\vec{a}^t \vec{y}_i > 0$ we can define a criterion or error function $E(\vec{a})$ that is minimized if $\vec{a}$ is a separating vector. The ultimate goal is to minimize the number of misclassified samples until we obtain zero errors (assuming the data is linearly separable). However, we cannot directly minimize the number of errors since it is not a continuous function and therefore is not suitable for optimization using derivative-based procedures, as an infinitesimal change in $\vec{a}$ may lead to a big change in the function and the derivative may be infinite. An alternative and closely related criterion function is the **perceptron function**. Using the fact that $g(\vec{y}) = \vec{a}^t \vec{y}$ is proportional to the distance of $\vec{y}$ from the separating

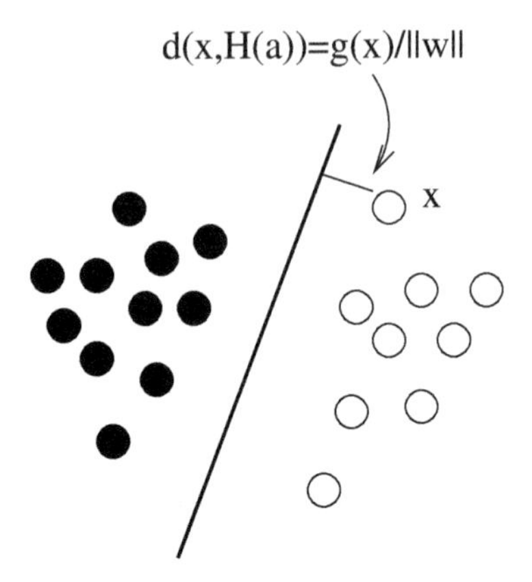

FIGURE 7.4: **The linearly separable case.**

hyperplane defined by $\vec{a}$ (Equation 7.1), the perceptron function sums $-g(\vec{y})$ over the misclassified samples, those for which $g(\vec{y}) < 0$. That is,

$$E(\vec{a}) = \sum_{\vec{y} \in \mathbf{M}} (-\vec{a}^t \vec{y})$$

where $\mathbf{M} = \{\vec{y} \mid g(\vec{y}) < 0\}$ is the set of misclassified samples and is dependent on the current value of $\vec{a}$. Note that $E(\vec{a})$ cannot be negative and is minimized (i.e., $E(\vec{a}) = 0$) for any separating vector in the solution region.

Solving the equation $E(\vec{a}) = 0$ directly is not possible since there are too many parameters and one equation. However, we can apply **gradient descent procedures** to minimize the criterion function $E(\vec{a})$. The basic gradient descent procedure works as follows:

- Start with a random weight vector $\vec{a}(1)$.
- Compute the derivative of the function E with respect to every component of the vector $\vec{a}$. The result is the **gradient vector**

  $$\nabla E(\vec{a}(1)) = \left(\frac{\partial E}{\partial a_0}, ..., \frac{\partial E}{\partial a_d}\right)$$

  The notation $\nabla E(\vec{a}(1))$ implies that the value of the criterion function depends on $\vec{a}(1)$, since the set of misclassified samples is defined by the current weight vector and the gradient (the derivative) is computed at $\vec{a}(1)$. The gradient of a function at a point is a vector that points in the direction in which the function is increasing the fastest (steepest ascent). Therefore, to minimize E we should move from $\vec{a}(1)$ some distance in the opposite direction.
- Change the weight vector in the direction of the steepest descent (i.e., along the negative of the gradient)

  $$\vec{a}(2) = \vec{a}(1) - \eta \cdot \nabla E(\vec{a}(1))$$

where η is a positive scaling factor that sets the rate the function changes (called the **learning rate**).

- Repeat

$$\vec{a}(k+1) = \vec{a}(k) - \eta \cdot \nabla E(\vec{a}(k))$$

until $\vec{a}$ converges or until the change in the weight vector is negligible.

It is easy to compute the gradient directly for the perceptron criterion function. Since

$$E(\vec{a}) = \sum_{\vec{y} \in \mathbf{M}} (-\vec{a}^t \vec{y}) = \sum_{\vec{y} \in \mathbf{M}} -(a_0 y_0 + a_1 y_1 + ... + a_d y_d)$$

then

$$\frac{\partial E}{\partial a_i} = \sum_{\vec{y} \in \mathbf{M}} (-y_i)$$

Therefore, the gradient of E is

$$\nabla E = \sum_{\vec{y} \in \mathbf{M}} (-\vec{y})$$

and the update rule is

$$\vec{a}(k+1) = \vec{a}(k) + \eta \sum_{\vec{y} \in \mathbf{M}} \vec{y}$$

where the set of misclassified samples is updated at every step. This algorithm (called the **perceptron algorithm**) terminates when the set $\mathbf{M}$ is empty.

The perceptron algorithm

Initialize: weight vector $\vec{a}$, threshold ϵ, learning rate η, index $k = 0$
do $k \leftarrow k + 1$
$\vec{a} \leftarrow \vec{a} + \eta(k) \sum_{\vec{y} \in \mathbf{M}} \vec{y}$
until

$$\left| \eta(k) \sum_{\vec{y} \in \mathbf{M}} \vec{y} \right| < \epsilon \qquad (7.2)$$

Return $\vec{a}$

The algorithm described above is the *batch perceptron algorithm*, where all samples are considered when updating the weight vector. *The single-sample perceptron* is similar, but it considers one sample at a time (selected randomly) and updates the weight vector if the sample selected is misclassified by the current vector. This process is repeated until convergence.

Note that in Equation 7.2 we wrote $\eta(k)$, meaning that the learning rate is not necessarily fixed and may change with every iteration. Indeed, the value of η can greatly affect the convergence properties of the perceptron algorithm.

If it is too small then convergence will be slow. If it is too large it will lead to divergence. An adaptive approach is to modify the learning rate gradually. For example, we can decrease the learning rate every iteration, assuming we are getting closer to a solution vector and small changes would be enough to converge to a final solution. A common practice is to set $\eta(k) = 1/k$. If the criterion function is linear or quadratic in $\vec{a}$ (such as the perceptron function) it can be shown that even a constant learning rate will guarantee convergence [191,234].

7.2.3 Maximizing the margin

In the linearly separable case, the perceptron algorithm is guaranteed to converge to a vector in the solution region. However, this separating vector would not necessarily perform well on future examples. If the vector is very close to one of the sample points then it might misclassify future samples in the vicinity of that point that belong to the same class. For future classifications it would be better if we select a solution vector (hyperplane) that maximizes the minimum distance from the samples to the plane, on both sides of the hyperplane (see Figure 7.5). This maximum-minimum distance is called the **margin** [235]. That is, the optimal hyperplane according to this principle is given by

$$\max_{\vec{a}} \min_{1 \le i \le n} d(\vec{x}_i, H(\vec{a}))$$

where $d(\vec{x}, H(\vec{a})) = \frac{g(\vec{x})}{\|\vec{\omega}\|}$ as in Equation 7.1 and n is the number of points in the training data. Geometrically, the solution is obtained by outlining the convex hull of the points in each class, and then taking the shortest line connecting them. The middle normal is the solution vector we are seeking (Figure 7.5).

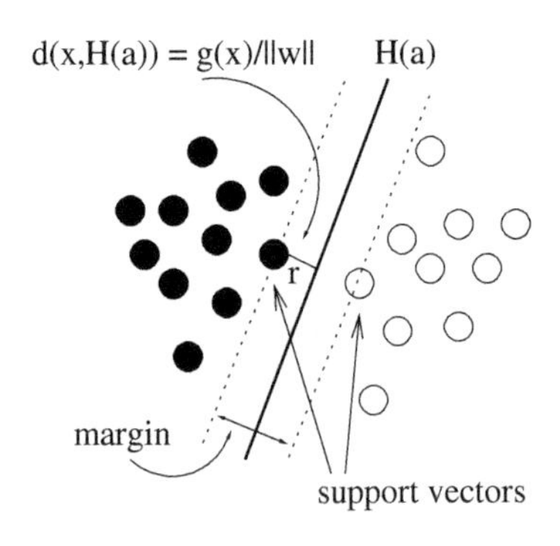

FIGURE 7.5: The margin.

To compute this vector, we can rephrase this optimization problem as a constrained minimization problem. Let $z_i = \pm 1$ be the class label of sample $\vec{x}_i$, then our goal is to satisfy

$$z_i \frac{g(\vec{x}_i)}{||\vec{\omega}||} > 0 \tag{7.3}$$

while maximizing the margin. Note that $||\vec{\omega}||$ can be arbitrarily large, since what we really care about is the sign. Therefore, if we multiply Equation 7.3 by $||\vec{\omega}||$ then we can set the threshold on the right side of the inequality as large as we want. For convenience we set the threshold to 1 and rewrite the inequality above as

$$z_i \cdot g(\vec{x}_i) \geq 1 \qquad \forall i$$

If the data is linearly separable then we can find a solution to this set of equations as outlined next. The sample points closest to the separating hyperplane from both sides are called **support vectors**. These are the vectors that satisfy $z_i \cdot g(\vec{x}_i) = 1$ and their distance from the hyperplane is $\frac{g(\vec{x}_i)}{||\vec{\omega}||} = \frac{1}{||\vec{\omega}||}$ for points labeled $z_i = 1$ and $\frac{-1}{||\vec{\omega}||}$ for points labeled $z_i = -1$. Therefore, the margin is

$$\text{margin} = \frac{1}{||\vec{\omega}||} - \frac{-1}{||\vec{\omega}||} = \frac{2}{||\vec{\omega}||} = \frac{2}{\sqrt{\vec{\omega}^t \cdot \vec{\omega}}}$$

and to maximize the margin we need to minimize the denominator. Hence the constrained minimization problem is

$$\text{minimize} \quad \vec{\omega}^t \vec{\omega} \quad \text{subject to} \quad z_i \cdot (\vec{\omega}^t \cdot \vec{x}_i + \omega_0) \geq 1 \qquad i = 1..n$$

This problem involves $d + 1$ parameters (the weight vector and the bias) and n constraints, and it can be written as a convex optimization problem with Lagrange multipliers over $n + d + 1$ parameters

$$\mathcal{L}(\vec{\omega}, \omega_0, \vec{\alpha}) = \frac{1}{2}||\vec{\omega}||^2 - \sum_{i=1}^{n} \alpha_i [z_i \cdot (\vec{\omega}^t \cdot \vec{x}_i + \omega_0) - 1]$$

where $\vec{\alpha}$ is the vector of multipliers $(\alpha_1, ..., \alpha_n)$. It is not difficult to see intuitively that solving this equation will produce the desired result. As stated, our goal is to minimize $\vec{\omega}^t \vec{\omega} = ||\vec{\omega}||^2$, which is the first part of the Lagrangian $\mathcal{L}(\vec{\omega}, \omega_0, \vec{\alpha})$. However, if we minimize $||\vec{\omega}||$ too much then some of the constraints $z_i \cdot g(\vec{x}_i) \geq 1$ will be violated. The introduction of the multipliers α_i is necessary to monitor the contribution of these constraints to the equation. Any constraint that is violated will increase the second part of $\mathcal{L}$. The balance between the constraints and the weight vector is delicate because of their mutual dependency, and it can be shown [235] that the solution to this minimization problem satisfies

$$\vec{\omega} = \sum_{i=1}^{n} \alpha_i z_i \vec{x}_i \tag{7.4}$$

such that

$$(1) \quad \alpha_i \geq 0 \qquad (2) \quad \sum_{i=1}^{n} \alpha_i z_i = 0$$

and

$$(3) \quad \alpha_i = 0 \quad \text{unless} \quad z_i(\vec{\omega}^t \vec{x}_i + \omega_0) = 1$$

The second condition above implies that the hyperplane is defined by a mix of vectors on both sides, as expected. Note, however, that only the support vectors play a role in defining the weight vector, since $\alpha_i = 0$ otherwise (condition 3). This is the origin of the name **Support Vector Machines (SVMs)**. The classifier's value for a new instance $\vec{x}$ is obtained by multiplying $\vec{x}$ by these support vectors, per Equation 7.4

$$f(\vec{x}) = \text{sgn}(g(\vec{x})) = \text{sgn}\left(\vec{\omega}^t \cdot \vec{x} + \omega_0\right) = \text{sgn}\left(\left(\sum_{i=1}^{n} \alpha_i z_i \vec{x}_i^t\right) \cdot \vec{x} + \omega_0\right)$$

Hence, when classifying a new instance the main step involves computing the scalar product of $\vec{x}$ and the support vectors. This is the key to SVM classifiers, as discussed in Section 7.2.5.

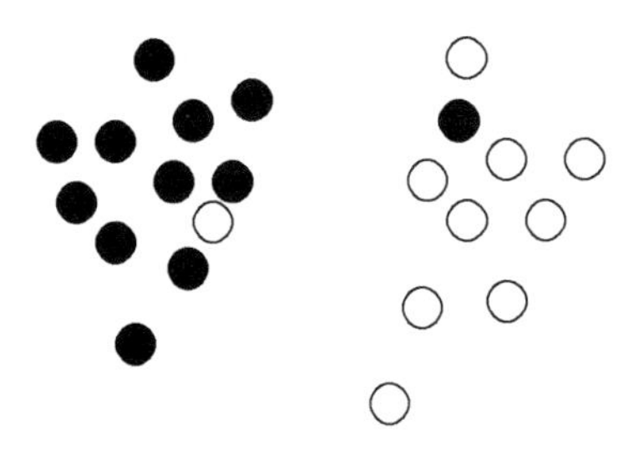

FIGURE 7.6: **The non-separable case.**

7.2.4 The non-separable case - soft margin

Often it is the case that the classes are not linearly separable (Figure 7.6), such that we cannot satisfy $z_i(\vec{\omega}^t \cdot \vec{x}_i + \omega_0) \geq 1$ for all i. In this case, we can try to find a "good" separating hyperplane that minimizes the number of errors (misclassified examples). A breakthrough solution for this problem was proposed in [236]. The idea is to introduce **slack variables** $\xi_i \geq 0$ and require that

$$\vec{\omega}^t \cdot \vec{x}_i + \omega_0 \geq 1 - \xi_i \qquad \text{for} \quad z_i = +1$$

$$\vec{\omega}^t \cdot \vec{x}_i + \omega_0 \geq -1 + \xi_i \qquad \text{for} \quad z_i = -1$$

An error occurs if $\xi_i > 1$, that is, if $\vec{x}_i$ is on the wrong side of the hyperplane. Since the value of the discriminant function is proportional to the distance of $\vec{x}_i$ from the separating hyperplane, then we can say that ξ_i measures the degree of error regarding the classification of $\vec{x}_i$. Therefore, our new goal is to maximize the margin while minimizing the total error, estimated by the sum of distances from the separating hyperplane for misclassified examples and examples within the margin. The new optimization problem becomes

$$\frac{1}{2}\|\vec{\omega}\|^2 + C\sum_{i=1}^{n}\xi_i \tag{7.5}$$

subject to the constraints $z_i(\vec{\omega}^t \cdot \vec{x}_i + \omega_0) \geq 1 - \xi_i$ and $\xi_i \geq 0$. The parameter C balances the trade-off between complexity and accuracy. Since the norm of $\vec{\omega}$ depends on the number of support vectors, small C penalizes complexity (in favor of fewer support vectors) as the first term involving $\|\vec{\omega}\|$ becomes the dominant one. On the other hand, large C penalizes errors (the second term). These properties are closely related to the generalization power of the model, and an equivalent way of interpreting Equation 7.5 is to say that C regulates overfitting. A small C promotes smaller norm $\|\vec{\omega}\|$, resulting in a larger margin that may introduce a larger number of errors over the training set but potentially improves its performance over new instances (lower generalization error). Being dependent on a smaller number of support vectors, the classifier tends to be more stable and less affected by changes in the training data. But the larger C is, the smaller number of violations that is tolerated, tailoring the separating hyperplane to the specific makeup of the training data and possibly resulting in a smaller margin and reduced generalization power.

The solution to the optimization problem is obtained again using the Lagrange multipliers α_i and is similar to the one we outlined before, but with additional constraints on α_i and ξ_i

$$\vec{\omega} = \sum_{i=1}^{n} \alpha_i z_i \vec{x}_i$$

such that

$$(1) \quad \alpha_i \geq 0 \qquad (2) \quad \sum_{i=1}^{n} \alpha_i z_i = 0$$

$$(3) \quad \alpha_i\left(z_i(\vec{\omega}^t \vec{x}_i + \omega_0) - 1 + \xi_i\right) = 0$$

and

$$(4) \quad (C - \alpha_i)\xi_i = 0$$

As before, the classifier is given by

$$f(\vec{x}) = \text{sgn}(\vec{\omega}^t \cdot \vec{x} + \omega_0) = \text{sgn}\left(\left(\sum_{i=1}^{n} \alpha_i z_i \vec{x}_i^t\right) \cdot \vec{x} + \omega_0\right)$$

and the support vectors are those for which $\alpha_i > 0$. Unlike the linearly separable case, we now have three types of support vectors

- Margin vectors are vectors for which $0 < \alpha_i < C$. In this case ξ_i must satisfy $\xi_i = 0$ (condition 4 above) and therefore $d(\vec{x}_i, H(\vec{a})) = \frac{1}{\|\vec{\omega}\|}$, which means that the vector lies on the margin (see Figure 7.7).

- Non-margin vectors ($\alpha_i = C$) for which $0 \leq \xi_i < 1$. These vectors are correctly classified but within the margin.

- Non-margin vectors ($\alpha_i = C$) for which $\xi_i > 1$. These vectors are misclassified.

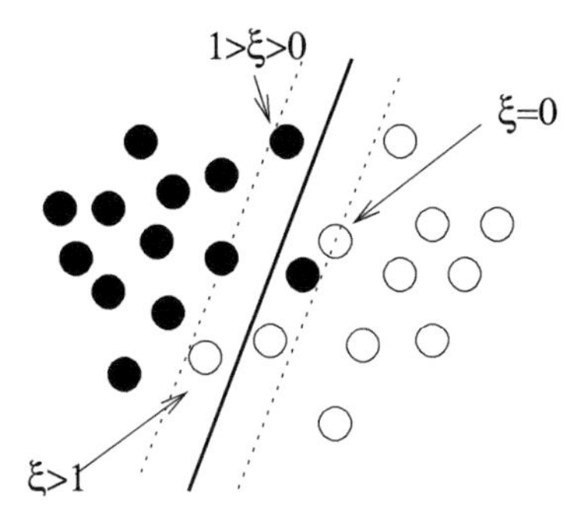

FIGURE 7.7: **Support vectors in the non-linearly separable case.** There are three types of support vectors, depending on the value of the slack variable.

Solving this optimization problem is not trivial and requires a technique called quadratic programming, but there are several software packages that given an input dataset can find the optimal hyperplane, per Equation 7.5, and provide the classification of new instances. These include SVMlight [237] and LibSVM [238], to name a few.

Note that the discussion so far assumed that there are only two classes in the data. There are two approaches to generalize a two-category classifier to a **multi-category** classifier. One is to train a classifier for each class (separating members from non-members). A new instance is then classified by applying each discriminant function and assigning the instance to the class with the highest positive value. The second approach is to train a classifier for each *pair* of classes. A new instance is processed by applying each classifier and tallying how many times each class "won". The instance is classified to the class with the highest number of wins.

7.2.5 Non-linear discriminant functions

The discussion so far assumed that the boundaries between classes can be approximated by hyperplanes. However, in most real life problems the decision boundaries are much more complicated. The idea to generalize the linear function of the previous sections, so as to accommodate such scenarios, was proposed as early as 1951 [239]. Recall that a discriminant function $g(\vec{x}) = \vec{\omega}^t\vec{x} + \omega_0$ can be re-written as $g(\vec{x}) = \vec{a}^t\vec{y}$, where each component y_i of $\vec{y}$ is a function of $\vec{x}$. Back in Section 7.2.1 the transformation was trivial with $\vec{y} = (1, x_1, x_2, ..., x_d)$, such that $y_1 = 1$ and $y_i(\vec{x}) = x_{i-1}$ for $2 \leq i \leq d$. But this framework can be generalized to arbitrary functions. That is, instead of $g(\vec{x}) = \omega_0 + \sum_{i=1}^{d} \omega_i \cdot x_i$, a **generalized discriminant function** is defined as

$$g(\vec{x}) = \omega_0 + \sum_{i=1}^{D} \omega_i \cdot \phi_i(\vec{x})$$

where each $\phi_i(\vec{x})$ is a function of $\vec{x}$ and $\phi_i \neq \phi_j$. The number of components D in our new generalized function can be arbitrarily high.

Note that the generalized function $g(\vec{x})$ is linear in the variables ϕ_i. If the ϕ_i are linear functions of $\vec{x}$ then the outcome is still a linear function in $\vec{x}$. However, if some or all of the functions ϕ_i are not linear in $\vec{x}$ then the outcome $g(\vec{x})$ is no longer linear in $\vec{x}$, and the more components we use, the more complex the function $g(\vec{x})$ is. Hence, by choosing proper functions ϕ_i we can generate a complex non-linear function that is capable of separating almost any two classes. And since the function g is linear in the variables ϕ_i we can still apply the techniques we described in the previous sections to find the parameters of the discriminant function. These two principles are what makes SVMs a very powerful supervised learning technique. By mapping *non-linearly* our input space into a higher dimensional space we will more likely have a *linearly* separable problem, as is illustrated in Figure 7.8 below[2].

The process starts by mapping the input space $\mathcal{X} \subset \mathbb{R}^d$ non-linearly into a **feature space** $\Phi \subset \mathbb{R}^D$ where $D \gg d$

$$\Phi : \mathbb{R}^d \rightarrow \mathbb{R}^D$$

$$\vec{x} \rightarrow \vec{\phi}(\vec{x}) = (\phi_1(\vec{x}), \phi_2(\vec{x}), ..., \phi_D(\vec{x}))$$

The vector $\vec{\phi}(\vec{x})$ is called the **image vector** or **feature vector** of $\vec{x}$. As before we are looking for a separating hyperplane

$$g(\vec{x}) = \vec{\omega}^t\vec{\phi}(\vec{x}) + \omega_0$$

[2]The idea is actually not that different from non-linear multi-layer **Neural Networks** (NNs). However, with neural networks the non-linear mapping and the linear discriminant function in the feature space are learned at the same time. With SVMs in the spotlight, NNs have lost some of their appeal. Although their hayday is over, neural networks are still a very powerful model that can learn quite complex decision boundaries, and they should be considered as well when tackling difficult classification tasks.

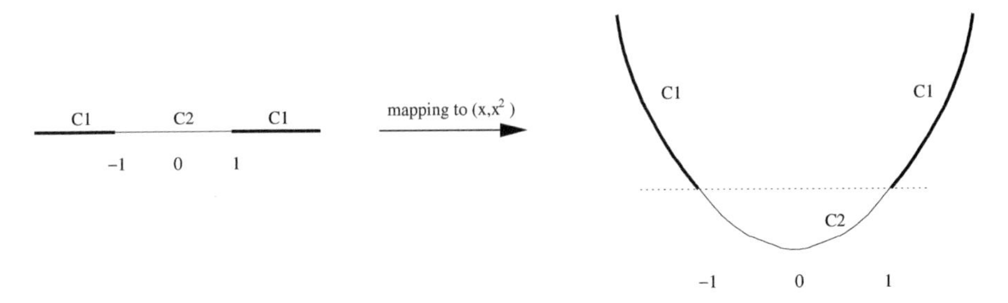

FIGURE 7.8: A non-linear mapping to a higher dimensional space. The two classes in the dataset on the left ($\mathcal{X} = \mathbf{R}^1$) are not linearly separable. The non-linear mapping of the one-dimensional vectors to a two-dimensional space with $\phi_1(x) = x$ and $\phi_2(x) = x^2$ results in the dataset on the right, which is linearly separable. The separating hyperplane (line in this case) is obtained by solving $g(x) = \sum_{i=1}^{2} a_i\phi_i(x) + a_0 = 0$. Plugging in $g(1) = g(-1) = 0$ we get that $a_1 = 0$ and $a_0 = -a_2$, or $g(x) = x^2 - 1$.

However, now we work in the feature space and the constrained minimization problem of Section 7.2.4 is formulated in terms of the feature vectors $\vec{\phi}(\vec{x})$. The solution is obtained using the same optimization technique and is also given in terms of the feature vectors. That is,

$$f(\vec{x}) = \text{sgn}\left(\sum_{i=1}^{n} \alpha_i z_i \vec{\phi}^t(\vec{x}_i)\vec{\phi}(\vec{x}) + \omega_0\right) \tag{7.6}$$

Its success in separating the classes clearly depends on the choice of Φ.

Correspondence between linear discriminant functions in the input space and their counterparts in the feature space.

	input space	feature space
input vector	$\vec{x} \in \mathbb{R}^d$	$\vec{\phi}(\vec{x}) \in \mathbb{R}^D$
$g(\vec{x})$	$\vec{\omega}^t\vec{x} + \omega_0$	$\vec{\omega}^t\vec{\phi}(\vec{x}) + \omega_0$
weight vector	$\vec{\omega} = \sum_{i=1}^{n} \alpha_i z_i \vec{x}_i$	$\vec{\omega} = \sum_{i=1}^{n} \alpha_i z_i \vec{\phi}(\vec{x}_i)$
classifier	$f(\vec{x}) = \text{sgn}(\sum_{i=1}^{n} \alpha_i z_i \underline{\vec{x}_i^t\vec{x}} + \omega_0)$	$f(\vec{x}) = \text{sgn}(\sum_{i=1}^{n} \alpha_i z_i \underline{\vec{\phi}(\vec{x}_i)^t\vec{\phi}(\vec{x})} + \omega_0)$

As before, the main effort when classifying a new instance goes into computing the scalar product between the instance and the support vectors (the underlined part in the equations above), since the factors α_i are zero for all other vectors. However, the product is computed between vectors in the feature space. If the dimension of the feature space is high then this computation can be very expensive.

The seminal paper of [240] (based on earlier work by [241]), circumvented this problem by focusing on mappings Φ that can be associated with a **kernel** function in the *lower* dimensional input space, avoiding the computation of the scalar product in the higher dimensional feature space. The definition of a kernel in general depends on the context in which it is being used. In the context of discriminant functions, a kernel is a general name for a symmetric function over pairs of vectors in the input space $K : \mathbb{R}^d \times \mathbb{R}^d \rightarrow \mathbb{R}$. The scalar product of $\vec{x}$ and $\vec{y}$ is a kernel, which is referred to as the **linear kernel**. That is,

$$K(\vec{x}, \vec{y}) = \langle \vec{x}, \vec{y} \rangle = \vec{x}^t \cdot \vec{y}$$

A specific type of kernel function that is of special interest to discriminant functions is the **inner-product kernels**. A function $K(\vec{x}, \vec{y})$ is called an inner-product kernel for a mapping Φ if it satisfies

$$K(\vec{x}, \vec{y}) = \vec{\phi}^t(\vec{x}) \cdot \vec{\phi}(\vec{y})$$

for every $\vec{x}, \vec{y} \in \mathbb{R}^d$. If there is a certain kernel function that satisfies this condition for the mapping Φ, then instead of computing the mapping to $\mathbb{R}^D$ and the inner product between the feature vectors $\vec{\phi}(\vec{x}), \vec{\phi}(\vec{y})$ (which can be very time consuming), we can compute the kernel function over the original low-dimensional vectors $\vec{x}, \vec{y}$. This is referred to as the **kernel trick**.

To illustrate the power of the kernel trick, consider for example the simple linear mapping $\vec{\phi}(\vec{x}) = \mathbf{A}\vec{x}$, where $\mathbf{A}$ is a $D \times d$ matrix and $D \gg d$. The inner product of the feature vectors is given by

$$\vec{\phi}^t(\vec{x})\vec{\phi}(\vec{y}) = (\mathbf{A}\vec{x})^t(\mathbf{A}\vec{y}) = \vec{x}^t\mathbf{A}^t\mathbf{A}\vec{y} = \vec{x}^t\mathbf{B}\vec{y}$$

where $\mathbf{B} = \mathbf{A}^t\mathbf{A}$ is a $d \times d$ matrix. Hence, if we define

$$K(\vec{x}, \vec{y}) = \vec{x}^t\mathbf{B}\vec{y}$$

we get that $K(\vec{x}, \vec{y})$ is an inner-product kernel for the transformation $\vec{\phi}(\vec{x}) = \mathbf{A}\vec{x}$. Since $K(\vec{x}, \vec{y})$ is computed in the original input space it involves only $d^2 + d$ arithmetic computations, while computing the scalar product in the feature space would require $2Dd + D$ computations.

Therefore, when using support vector machines we would prefer inner-product kernels that satisfy $K(\vec{x}, \vec{y}) = \vec{\phi}^t(\vec{x})\vec{\phi}(\vec{y})$, such that we can compute directly the value of the kernel while avoiding the actual mapping into a high-dimensional space. If we find a mapping that corresponds to an inner-product kernel, then potentially we have significantly reduced the computations involved with the classification of a new instance $\vec{x}$ since

$$f(\vec{x}) = \text{sgn}\left(\sum_{i=1}^{n} \alpha_i z_i \vec{\phi}^t(\vec{x}_i)\vec{\phi}(\vec{x}) + \omega_0\right) = \text{sgn}\left(\sum_{i=1}^{n} \alpha_i z_i K(\vec{x}_i, \vec{x}) + \omega_0\right)$$

and while the first part of the equation is computed in $\mathbb{R}^D$ the second is computed in the lower dimensional space $\mathbb{R}^d$. Two popular examples of inner-product kernels are the polynomial kernel and the Gaussian kernel.

The polynomial kernel. Given the sample space $\mathcal{X} = \mathbb{R}^d$, let $\mathbf{d}$ be the set $\{1, 2, ..., d\}$ and $2^{\mathbf{d}}$ be the power set of $\mathbf{d}$. For each subset $\mathbf{s} \in 2^{\mathbf{d}}$ we define the indicator vector $\vec{s} \in \{0,1\}^d$ such that $s_i = 1$ if $i \in \mathbf{s}$ and 0 otherwise. For example, in the set $\mathbf{d} = \{1, 2\}$ there are four possible subsets and the corresponding indicator vectors are $(0, 0), (1, 0), (0, 1)$ and $(1, 1)$. The so-called polynomial transformation maps an input vector $\vec{x} \in \mathcal{X}$ to a feature vector $\vec{\phi}(\vec{x}) \in \mathbb{R}^D$ where $D = 2^d$ such that each component corresponds to a product over a subset of $\vec{x}$. That is,

$$\vec{\phi}(\vec{x}) = \left(\prod_{i=1}^{d} x_i^{s_i}\right)_{\vec{s}\in\{0,1\}^d}$$

using an abbreviated vector notation, where every component is defined by one of the possible indicator vectors $\vec{s}$. For example, if $d = 2$ then $\vec{\phi}(\vec{x}) = (1, x_1, x_2, x_1 x_2)$ corresponding the the four indicator vectors we listed above.

The polynomial kernel is defined as

$$K_{poly}(\vec{x}, \vec{y}) = \prod_{i=1}^{d}(x_i y_i + 1)$$

Note that this kernel is an inner-product kernel for the polynomial transformation function defined above. For example, if $d = 2$ then

$$K_{poly}(\vec{x}, \vec{y}) = (x_1 y_1 + 1)(x_2 y_2 + 1) = \vec{\phi}^t(\vec{x})\vec{\phi}(\vec{y})$$

Therefore, instead of transforming the input vectors to $\mathbb{R}^D$ and computing the inner-product in the high-dimensional space we can compute the value of the kernel function in $\mathbb{R}^d$.

The Gaussian kernel. Consider a mapping that transforms each input point $\vec{x}$ to a Gaussian $N_{\vec{x},\sigma}$ that is centered at $\vec{x}$ and has a standard deviation σ. In this case the feature space is of infinite dimension and the scalar product between the feature vectors is replaced with an integral over the continuous functions, such that

$$\vec{\phi}^t(\vec{x})\vec{\phi}(\vec{y}) \to \int N_{\vec{x},\sigma}(t) N_{\vec{y},\sigma}(t) dt$$

It can be shown that this convolution results in another Gaussian and that the Gaussian kernel

$$K_{RBF}(\vec{x}, \vec{y}) = \exp\left(\frac{-\|\vec{x} - \vec{y}\|^2}{2\sigma^2}\right) = \exp\left(\frac{-(\vec{x} - \vec{y})^t(\vec{x} - \vec{y})}{2\sigma^2}\right)$$

is an inner-product kernel for the Gaussian mapping described above. This kernel is also referred to as the **Radial Basis Function (RBF)** kernel.

Note that when using RBF, only the support vectors that are localized within the vicinity of the point in question affect its classification, since the kernel value depends on the distance between the vectors and diminishes for support vectors that are farther away. In that sense, RBF kernels are local. This kind of kernel is appropriate if the data contains different classes that have unique properties and are mapped to different neighborhoods in the feature space.

7.2.5.1 Mercer kernels

Not every mapping Φ corresponds to an inner-product kernel $K(\vec{x},\vec{y})$, and not every kernel qualifies as an inner-product kernel for some mapping Φ. While certain mappings can be verified to correspond to inner-product kernels over the original vectors $\vec{x}$ and $\vec{y}$ (as we have shown above for the polynomial and Gaussian mappings), proving this property for a general mapping Φ can be difficult. However, the other way is easier and under certain conditions a function $K(\vec{x},\vec{y})$ can be proved to equal the product of the *images* of $\vec{x}$ and $\vec{y}$ under a certain transformation Φ.

Consider a function $K(\vec{x},\vec{y})$ that *is* an inner-product kernel for a certain transformation Φ. The **kernel matrix** (also called the **Gram matrix**) is the matrix of kernel values $\mathbf{K}$ over the dataset $\mathbf{D}$ where the i,j entry is

$$K_{ij} = K(\vec{x}_i,\vec{x}_j) = \langle \vec{\phi}(\vec{x}_i), \vec{\phi}(\vec{x}_j) \rangle$$

By definition, the kernel matrix is a symmetric matrix, since $K(\vec{x},\vec{y}) = K(\vec{y},\vec{x})$. It can also be shown that an inner-product kernel matrix is a positive definite matrix [242]. An $n \times n$ symmetric matrix $\mathbf{K}$ is positive definite if $\vec{v}^t\mathbf{K}\vec{v} > 0$ for all n-dimensional nonzero real vectors $\vec{v}$. This is equivalent to writing $\sum_{i=1}^{n}\sum_{j=1}^{n} v_i v_j K_{ij} \geq 0$ for all choices of real numbers $v_1, ..., v_n$.

We are interested in the reverse direction; characterizing the conditions under which $\mathbf{K}$ is an inner-product kernel matrix (or $K(\vec{x},\vec{y})$ is an inner-product kernel function, in the non-finite case). According to Mercer's theorem [242], any Symmetric Positive Definite (SPD) matrix $\mathbf{K}$ can be rewritten as an inner-product matrix over the input dataset under a certain transformation Φ. That is, there exists a mapping Φ such that

$$K_{ij} = \langle \vec{\phi}(\vec{x}_i), \vec{\phi}(\vec{x}_j) \rangle$$

The key observation is that if $\mathbf{K}$ is SPD then it can be decomposed into

$$\mathbf{K} = \mathbf{U}\mathbf{W}\mathbf{U}^T$$

where $\mathbf{U}$ is the matrix whose columns are the eigenvectors of $\mathbf{K}$, and $\mathbf{W}$ is a diagonal matrix with the eigenvalues $\lambda_1, ..., \lambda_n$ of $\mathbf{K}$ in the diagonal in decreasing order of value. Let $\vec{u}_i$ denote the i-th eigenvector of $\mathbf{U}$ (that is, the i-th column) then

$$\mathbf{K} = \begin{pmatrix} u_{11} & ... & u_{n1} \\ u_{12} & ... & u_{n2} \\ . & ... & . \\ u_{1n} & ... & u_{nn} \end{pmatrix} \begin{pmatrix} \lambda_1 & ... & 0 \\ 0 & \lambda_2 & 0 \\ 0 & ... & 0 \\ 0 & ... & \lambda_n \end{pmatrix} \begin{pmatrix} u_{11} & u_{12} & ... & u_{1n} \\ . & . & ... & . \\ . & . & ... & . \\ u_{n1} & u_{n2} & ... & u_{nn} \end{pmatrix}$$

and

$$K_{ij} = \lambda_1 u_{1i} u_{1j} + \lambda_2 u_{2i} u_{2j} + ... + \lambda_n u_{ni} u_{nj}$$

Therefore, if we define the mapping Φ:

$$\vec{\phi}(\vec{x}_i) = \left(\sqrt{\lambda_1} u_{1i}, ..., \sqrt{\lambda_n} u_{ni}\right)^t$$

then $\mathbf{K}$ is an inner-product kernel matrix for Φ and

$$K_{ij} = K(\vec{x}_i, \vec{x}_j) = \langle \vec{\phi}(\vec{x}_i), \vec{\phi}(\vec{x}_j) \rangle$$

Such kernels are called **Mercer kernels**. Obtaining the mapping itself is possible through spectral decomposition of the kernel matrix as outlined above but is of less interest since we are only interested in the value of the kernel function.

Mercer's theorem is actually stated for continuous spaces and generalizes the above result to continuous functions. That is, given a continuous and symmetric function $K(x, y)$ that is positive definite[3] then K can be decomposed into the infinite sum

$$K(x, y) = \sum_{i=1}^{\infty} \lambda_i \phi_i(x) \phi_i(y)$$

where ϕ_i are the **eigenfunctions** of $K(x, y)$ defined in analogy to eigenvectors as

$$\int_y K(x, y) \phi_i(y) dy = \lambda_i \phi_i(x)$$

where λ_i is the eigenvalue[4]. In other words, the function K can be expressed as the inner-product

$$K(x, y) = \langle \vec{\phi}(x), \vec{\phi}(y) \rangle$$

[3]A function is $K(x, y)$ is defined as positive definite if

$$\int_{x,y} K(x, y) v(x) v(y) dx dy > 0$$

for any non-zero continuous function $v(x)$ such that $\int_x v(x) dx > 0$. This is a generalization of the condition stated above for matrices, to infinite instance spaces.

[4]Vector notation was omitted for clarity, but all equations are trivially generalized to multi-dimensional input spaces.

between the infinite dimension image vectors

$$\vec{\phi}(x) = (\sqrt{\lambda_1}\phi_1(x), \sqrt{\lambda_2}\phi_2(x), ...)$$

The infinite dimensional host space is called **Hilbert space**. However, the product between the infinite image vectors need not be computed as the values of $K(x, y)$ are given. All this serves to show is that every SPD function K is an inner-product kernel for some mapping Φ.

One of the most useful properties of Mercer kernels is that we can define and build more complex inner-product kernels by combining other kernels that have been shown to be inner-product already. That is, if $k_1(x, y)$ and $k_2(x, y)$ are mercer kernels, then so are

$$\begin{aligned}
k(x,y) &= k_1(x,y) + c && \text{for } c \geq 0 \\
k(x,y) &= ck_1(x,y) && \text{for } c \geq 0 \\
k(x,y) &= k_1(x,y) + k_2(x,y) \\
k(x,y) &= k_1(x,y) \cdot k_2(x,y) \\
k(x,y) &= k_1(x,y)^n && \text{for } n \in N \\
k(x,y) &= \exp(k_1(x,y))
\end{aligned}$$

It should be noted that good kernel functions are those for which the kernel matrix does not resemble a diagonal matrix. That is, the higher number of non-zero off-diagonal elements the more relevant information we have for separating the classes. This requires an effective feature representation, where different features are not unique to certain instances.

7.3 Applying SVMs to protein classification

The discussion thus far assumed that sample points to be classified are vectors in $\mathbb{R}^d$. However, when working with biological data, such as protein sequences over the alphabet of amino acids $\mathcal{A}$, the input space is typically not vectorial and there is no obvious mapping to vector space. While it is possible to represent a protein sequence using its amino acid composition or its dipeptide composition, these representations are not very powerful since they ignore the order of amino acids along the sequence, and classes of protein sequences that are represented using these methods tend to highly overlap. Our goal is to map the sequences in $\mathcal{A}^*$ to a high-dimensional feature space where different classes are well separated.

How shall we design an effective transformation Φ from $\mathcal{A}^*$ into $\mathbb{R}^D$? Intuitively, the scalar product between vectors in Equation 7.6 measures their correlation or similarity, and therefore when designing kernels for protein sequences we would like the scalar product to reflect the similarity of the strings.

Note that in this case we cannot use the kernel trick and we have to compute the transformation and the scalar product in $\mathbb{R}^D$ because the input space is not vectorial. In the next subsections we discuss several possible kernels. In Chapter 11 we will discuss general methods for mapping non-vectorial spaces into vector space, which in conjunction with the linear kernel can be used with SVMs as well.

7.3.1 String kernels

String kernels use a simple vectorial representation of strings, based on frequencies of short sequences, similar to the **bag-of-words** approach and the related **vector space model** [243] used in many computational linguistics studies.

7.3.1.1 Simple string kernel - the spectrum kernel

One of the first kernels that was introduced specifically to accommodate protein sequences is the spectrum kernel [244]. Given sequences over the alphabet $\mathcal{A}$, we define the k-spectrum of that sequence data as the set of all k-mers ($k \geq 1$) over $\mathcal{A}$. The kernel is based on a mapping Φ that converts each sequence to the k-spectrum feature space

$$\Phi : \mathcal{A}^* \rightarrow \mathbb{R}^D$$

where $D = |\mathcal{A}|^k$. Each component in the feature vector corresponds to one k-mer, and its value for a specific string $\mathbf{x}$ is the number of occurrences of that k-mer in the string $\mathbf{x}$. Denote $n_\mathbf{s}(\mathbf{x})$ as the number of occurrences of $\mathbf{s}$ in $\mathbf{x}$ where $\mathbf{s} \in \mathcal{A}^k$, then

$$\vec{\phi}_k(\mathbf{x}) = (n_\mathbf{s}(\mathbf{x}))_{\mathbf{s} \in \mathcal{A}^k}$$

using, as before, an abbreviated vector notation. To capture significant patterns in protein sequences, the k-mers must be long enough. For large k the feature space is high-dimensional, with 20^k components. However, the maximal number of different k-mers in a sequence of length n is $n - k + 1$, and since the typical length of a protein sequence is about 350 amino acids the k-spectrum feature vectors of protein sequences are very sparse.

The k-spectrum kernel of two strings is simply defined as the inner product between their feature vectors

$$K_k(\mathbf{x}, \mathbf{y}) = \langle \vec{\phi}_k(\mathbf{x}), \vec{\phi}_k(\mathbf{y}) \rangle = \vec{\phi}_k^t(\mathbf{x}) \cdot \vec{\phi}_k(\mathbf{y})$$

If two sequences have large inner product then they share many k-mers and can be considered similar. Once this kernel has been computed for the input data we can train a classifier by solving Equation 7.5 in the feature space, as discussed in in Section 7.2.5.

7.3.1.2 The mismatch spectrum kernel

Homologous sequences diverge in the course of evolution. Therefore, we should consider mismatches when comparing feature vectors of related sequences, otherwise we will miss the relation between highly diverged sequences. The mismatch spectrum kernel [245] uses the same k-spectrum feature space described above but accounts for substitutions already when mapping a sequence to its feature vector.

Given a k-mer $\mathbf{s} = s_1 s_2 ... s_k$ where $s_i \in \mathcal{A}$ we define the m-neighborhood of $\mathbf{s}$ as the set of k-mers $\mathbf{s}' \in \mathcal{A}^k$ that differ from $\mathbf{s}$ by m mutations at most. Denote that set by $N_{(k,m)}(\mathbf{s})$. The substitutions are accounted for by analyzing every k-mer $\mathbf{s}'$ in the input sequence $\mathbf{x}$ and contributing to every coordinate that corresponds to a k-mer $\mathbf{s}$ that is in the neighborhood of $\mathbf{s}'$. The contribution is quantified based on prior knowledge about the mutability of amino acids. Let $P(b|a)$ be the probability that a mutates into b (see Section 3.5.2). If $\mathbf{s}'$ is in the neighborhood $N_{(k,m)}(\mathbf{s})$, then its contribution to the coordinate that corresponds to $\mathbf{s}$ is defined as

$$\phi_{\mathbf{s}}(\mathbf{s}') = P(s'_1|s_1) \cdots P(s'_k|s_k)$$

Denote by $\vec{\phi}_{(k,m)}(\mathbf{s}')$ the vector of contributions that are associated with $\mathbf{s}'$ (a substring of $\mathbf{x}$), where every coordinate corresponds to a different substring $\mathbf{s} \in \mathcal{A}^k$

$$\vec{\phi}_{(k,m)}(\mathbf{s}') = (\phi_{\mathbf{s}}(\mathbf{s}'))_{\mathbf{s} \in \mathcal{A}^k}$$

then the mismatch spectrum approach maps a sequence $\mathbf{x}$ to a feature vector, which is defined as the sum of vectors

$$\vec{\phi}_{(k,m)}(\mathbf{x}) = \sum_{\mathbf{s}' \in \mathbf{x} \ \ |\mathbf{s}'|=k} \vec{\phi}_{(k,m)}(\mathbf{s}')$$

Similar sequences are likely to have similar feature vectors, and as before the kernel function (the (k, m)-mismatch spectrum kernel) is defined as the scalar product between the feature vectors

$$K_{(k,m)}(\mathbf{x}, \mathbf{y}) = \langle \vec{\phi}_{(k,m)}(\mathbf{x}), \vec{\phi}_{(k,m)}(\mathbf{y}) \rangle$$

7.3.2 The pairwise kernel

The pairwise kernel [246] converts a sequence of amino acids into a real valued vector by comparing the sequence with a set of pre-selected sequences $\{\mathbf{c}_i\}$. The sequences can be, for example, representatives of different protein families. If D is the number of representative sequences then

$$\Phi : \mathcal{A}^* \rightarrow \mathbb{R}^D$$

The feature vector is defined as

$$\vec{\phi}_{pair}(\mathbf{x}) = (s(\mathbf{x}, \mathbf{c}_1), s(\mathbf{x}, \mathbf{c}_2), ..., s(\mathbf{x}, \mathbf{c}_D))$$

where $s(\mathbf{x}, \mathbf{c}_i)$ is the similarity score of $\mathbf{x}$ and $\mathbf{c}_i$ computed with BLAST or the Smith-Waterman algorithm, for example. In practice, it is recommended to use the $-\log pvalue$ instead of the raw similarity score. To eliminate biases due to different sequence lengths and improve performance the vectors are normalized to unit vectors.

The pairwise kernel is computed using a RBF kernel over the feature vectors

$$K_{pair}(\mathbf{x}, \mathbf{y}) = \exp \frac{-(\vec{\phi}_{pair}(\mathbf{x}) - \vec{\phi}_{pair}(\mathbf{y}))^t(\vec{\phi}_{pair}(\mathbf{x}) - \vec{\phi}_{pair}(\mathbf{y}))}{2\sigma^2}$$

where σ is defined as the median distance between the feature vector of any positive example to its nearest negative example. Note that the kernel can be computed slightly more efficiently by first computing the scalar product between the feature vectors

$$K(\mathbf{x}, \mathbf{y}) = \vec{\phi}^t_{pair}(\mathbf{x}) \cdot \vec{\phi}_{pair}(\mathbf{y})$$

and then transforming the linear kernel to the RBF kernel

$$K_{pair}(\mathbf{x}, \mathbf{y}) = \exp \frac{-(K(\mathbf{x}, \mathbf{x}) + K(\mathbf{y}, \mathbf{y}) - 2K(\mathbf{x}, \mathbf{y}))}{2\sigma^2} \quad (7.7)$$

RBF kernels convert kernel values into a normal distribution and improve performance by normalizing the inner product with respect to the norm (self-similarity) of individual vectors. Consider for example the following feature vectors

$$\vec{x} = (1, 1, 1, ..., 1, 1, 1) \qquad \vec{y} = (0, 0, 0, ..., 0, 1, 1, 1) \qquad \vec{z} = (0, 0, 0, ..., 0, 1, 1, 1)$$

With the linear kernel $K(\vec{x}, \vec{y}) = K(\vec{y}, \vec{z})$, which is misleading since y and z are more similar to each other. However, after the RBF transformation we get $K_{pair}(x, y) < K_{pair}(y, z)$ as desired.

7.3.3 The Fischer kernel

The SVM-Fisher kernel is an interesting approach that combines a discriminative model with a generative model based on HMMs [247]. For each sequence, the Fischer kernel generates a vector in which each component corresponds to one parameter of the HMM (such as a transition probability or an emission probability), and its value is related to the frequency with which the corresponding transition was used when generating the sequence or the corresponding symbol was emitted from a given state, depending on the parameter.

The vector of frequencies reflects the process of generating the sequence by an HMM, which was learned from the training sequences. More informative than the frequencies are the derivatives of the process. This is known as the Fischer score, defined as

$$\phi_\theta(\mathbf{x}) = \nabla_\theta \log P(\mathbf{x}|\vec{\theta})$$

where $\vec{\theta}$ is the parameter vector of the model. Each component of $\vec{\phi}(\mathbf{x})$ is the derivative of the likelihood of $\mathbf{x}$ with respect to a particular parameter θ. The magnitude of that component specifies the extent to which the parameter (and the corresponding model element) contributes to generating the sequence.

For example, it is possible to show [248] that the derivative of $\log P(\mathbf{x}|\vec{\theta})$ with respect to the emission probability b_{ik} (the probability of emitting amino acid σ_k at state i) is given by

$$\phi_{b_{ik}}(\mathbf{x}) = \nabla_{b_{ik}} \log P(\mathbf{x}|\vec{\theta}) = \frac{E_i(k)}{b_{ik}} - \sum_{k'} E_i(k') \tag{7.8}$$

where

$$E_i(k) = \sum_{t=1,x(t)=\sigma_k}^{T-1} \gamma_i(t)$$

is the expected number of times we visited state i at any position[5] t along the sequence and emitted the specific symbol σ_k (see Equation 6.4), hence the condition that $x(t) = \sigma_k$ for every value of the index t. T is the length of the sequence $\mathbf{x}$. The sum $\sum_{k'} E_i(k')$ is the total number of times we visited state i. Indeed, if we equate the derivative to zero, i.e., $\phi_{b_{ik}}(\mathbf{x}) = 0$, then the optimal solution per Equation 7.8 is

$$b_{ik} = \frac{E_i(k)}{\sum_{k'} E_i(k')}$$

consistent with Equation 6.5. For more details see Section 6.3.3.1.

Once the vectors of the Fischer scores were computed for each sequence, the Fischer kernel is defined using the same transformation into a RBF kernel as with pairwise kernels

$$K_{fischer}(\mathbf{x}, \mathbf{y}) = \exp \frac{-(\vec{\phi}_\theta(\mathbf{x}) - \vec{\phi}_\theta(\mathbf{y}))^t (\vec{\phi}_\theta(\mathbf{x}) - \vec{\phi}_\theta(\mathbf{y}))}{2\sigma^2}$$

where σ is defined as before to be the median distance between the feature vector of any positive example to its nearest negative example.

7.3.4 Mutual information kernels

Mutual information kernels [249] combine elements from both pairwise kernels and Fisher kernels. Instead of using a set of representative sequences, the feature vector is computed using an infinite set of generative models. Mutual information kernels assume that there is a certain underlying generative process (that is, a specific model such as a normal distribution). However, they

[5] In Chapter 6 we used t to denote the time index, which is identical to the position along the sequence.

do not attempt to estimate the parameters of the model. They measure the similarity between data samples based on the correlation of their probabilities according to the generative process, averaged over all possible values of the parameters.

Formally, let $\vec{\theta}_h$ denote the parameter vector of the generative model h and $p(\vec{x}|\vec{\theta}_h)$ the class conditional probability distribution. To simplify notation we will omit the vector notation as well as the dependency on h, keeping in mind that θ and $p(x|\theta)$ are dependent on a specific model h. The observation x needs not be vectorial, it can also be a sequence. In that case the model can be an HMM or VMM and θ is the set of model parameters (such as the emission and transition probabilities).

SVM kernels usually measure the similarity of data samples. We are interested in assessing the mutual similarity of x and y, assuming a generative process with a parameter set θ. Let $p(x, y|\theta)$ denote the joint probability distribution of pairs of samples x, y. Each sample is assumed to be independent of the others, hence

$$p(x, y|\theta) = p(x|\theta)p(y|x, \theta) = p(x|\theta)p(y|\theta)$$

(see Section 3.12.3). Not knowing the exact value of θ, we average over all possible values to derive the **average joint probability distribution** under the prior $p(\theta)$

$$p(x, y) = p(x, y|h) = \int_\theta p(x, y|\theta)p(\theta)d\theta = \int_\theta p(x|\theta)p(y|\theta)p(\theta)d\theta \qquad (7.9)$$

This score reflects the similarity of samples x, y with respect to the generative process with prior $p(\theta)$. When the exact prior is unknown or the integral cannot be computed easily, the score can be reduced to a sum over a discretized parameter space. Note that $p(x, y)$ is a kernel, which resembles an inner product between the infinite dimensional feature vectors

$$\vec{\phi}(x) = (\sqrt{p(\theta_1)}p(x|\theta_1), \sqrt{p(\theta_2)}p(x|\theta_2), ...)$$

Under this representation, each feature corresponds to a different generative model with a different parameter set and its value is correlated with the likelihood of the instance under that model. Similarly, we can define the **average marginal probability distribution**

$$p(x) = \int_y p(x, y)dy$$

Recall that the Mutual Information (MI) of two random variables is defined as

$$I(X, Y) = \int_{x,y} p(x, y) \log \frac{p(x, y)}{p(x)p(y)} dxdy$$

(a generalization of Equation 3.22 in Section 3.12.6 to continuous spaces). This integral considers the complete instance space. For a pair of individual samples x, y, the inner term in the integral is the **sample mutual information**

$$I(x, y) = \log \frac{p(x, y)}{p(x)p(y)}$$

This kernel reflects the information gain when the two observations compared (e.g., sequences) are compressed together, rather than independently from each other.

While similar to the Mercer kernels we discussed in Section 7.2.5.1, this kernel is not positive definite. However, it can be converted to such by applying the RBF kernel, as in Section 7.3.2, per Equation 7.7. Using RBF with $\sigma = 1$ we obtain that

$$\begin{aligned} K_{mi}(x, y) &= \exp \frac{-\left(I(x, x) + I(y, y) - 2I(x, y)\right)}{2} \\ &= \frac{\exp(I(x, y))}{\sqrt{\exp(I(x, x)) \exp(I(y, y))}} \\ &= \frac{p(x, y)}{p(x)p(y)} \Big/ \sqrt{\frac{p(x, x)}{p(x)p(x)} \frac{p(y, y)}{p(y)p(y)}} \\ &= \frac{p(x, y)}{\sqrt{p(x, x)p(y, y)}} \end{aligned}$$

So far the discussion assumed a specific generative model h (e.g., a HMM with a specific architecture and number of states). However, Equation 7.9 can be generalized to sum over many different models such that

$$f(x, y) = \sum_h P(h)p(x, y|h) \tag{7.10}$$

where each model can be, for example, an HMM or VMM with different architectures. For example, in [250] the generative model is a probabilistic suffix tree (see Section 6.4.2.4). In the more general case, the models can be trained even over different datasets (different protein families), as in pairwise kernels.

We will get back to this representation in Section 11.4.2 of Chapter 11. As we will see, the discussion on kernels is closely related to the discussion on feature spaces and effective mappings of non-vectorial data to vector space. Based on these mappings we can derive other kernels for biological data.

7.4 Decision trees

In previous chapters we discussed different methods to detect sequence similarity, from pairwise methods to statistical models based on Markov models. The function of uncharacterized proteins is usually predicted based on significant sequence similarity with proteins of known functions, or with statistical models of well-studied protein families. However, in many cases sequences have diverged to the extent that their sequence similarity cannot be distinguished from chance similarities. Furthermore, there is a growing number of protein families that cannot be defined based on sequence similarity, and are grouped based on other features. In such cases, proteins possess the same function but can have very different sequences. For example, enzyme families are defined based on the functional role of the enzymes in the cell and their catalytic activity rather than on common evolutionary ancestry (we will get back to enzymes and their functional roles in cells in Chapter 14). Hence, enzymes that perform similar functions are not necessarily homologs and might not exhibit any sequence or structural similarity. Clearly, detecting functional similarity in such cases is difficult since it involves inferring the functionality of the proteins in silico[6].

Our success in detecting functional relationships between proteins strongly depends on the representation we employ. If sequence comparison fails, other aspects of protein similarity should be considered, from basic biochemical properties to attributes extracted from biological databases. Features such as the domain content, subcellular location, tissue specificity, and cellular pathways provide a broader biological context and may suggest functional similarity when shared in common between proteins, even in the absence of clear sequence or structural similarity. These protein attributes and others are usually ignored, either because data is not available or because it is not clear how to use them to quantify functional similarity. The task of learning the mapping from features to function is further complicated by the fact that some features are numeric while others are nominal (nouns), ruling out the use of popular machine learning models such as support vector machines.

In this section we will discuss a different classifier model, called **decision tree**, which is well suited for this learning problem [251]. Decision trees can be applied to learn the mapping between basic protein properties and their function and are capable of learning subtle regularities that distinguish family members from non-members. As such, they can complement and support other techniques for function prediction based on sequence or structure information. The decision tree model is one of the very few machine learn-

[6]The term **in silico** refers to "experiments" done through computer analysis or simulation, as opposed to **in vivo** (measurements or experiments done inside living cells) and **in vitro** (experiments done in a controlled environment, such as a test tube, in the laboratory).

ing techniques that can accommodate the special nature of protein data and handle a mix of features of different types, including binary and numerical. Decision trees are especially useful when some or all of the data features are nominal, i.e., when there is no natural notion of similarity between objects. This is the case for some of the attributes we can associate with proteins, such as tissue specificity and subcellular location. Another appealing property of decision trees is that they can tolerate missing data (which is often the case with database attributes), as well as noise and errors in the training data, such as corrupted attribute values or incorrect category labels. Furthermore, they are relatively easy to construct and take no time to classify new samples.

Finally, unlike SVMs, decision trees can provide us with insight about the data. The tree structure that is learned from the data incorporates features that maximize (locally) the accuracy when predicting the class membership. Hence, it highlights properties that are strongly correlated with functional aspects of the protein family being modeled. Clearly, these features can change from one protein family to another.

We will start by discussing the basic decision tree learning model and its extensions. In Section 7.4.5 we will discuss protein representation under the decision tree model and feature extraction and processing. As we will see later in Section 14.4, decision trees will become useful also when learning regulatory modules from gene expression data.

7.4.1 The basic decision tree model

We are given a dataset of n samples $D = \{x_1, x_2, ..., x_n\}$, which belong to k different classes. Every sample is represented by m attributes, $\mathbf{A} = \{A_1, ..., A_m\}$. Each attribute can be either binary, numeric (integer or real) or nominal. For example, a protein can be represented by features A_1, A_2, A_3, A_4 where A_1 is the average hydrophobicity[7] (a continues variable), A_2 is a binary attribute indicating whether the protein has a cofactor[8], A_3 is a nominal attribute whose value is the subcellular location where the protein is present (e.g., membrane, nucleus, cytoplasm etc.) and A_4 is a nominal attribute indicating the tissues where the protein is abundant (e.g., liver, heart, skin etc). Our goal is to train a model that would classify each sample to its correct class, based on its attribute values.

A decision tree is a classifier that is designed in a tree-like structure (Figure 7.9) where every node is associated with a test over one or more of the

[7]Each amino acid has a certain hydrophobicity value, depending on its interaction with water molecules. High hydrophobicity indicates that the amino acid repels water molecules. The average hydrophobicity of a protein sequence is simply the average hydrophobicity value over all its amino acids.

[8]Cofactors are compounds that are usually required by enzymes to carry out their catalytic activity. Cofactors can be, for example, metal ions or small organic molecules including some vitamins.

data attributes and each edge (branch) corresponds to one of the possible outcomes of the test. The outcome can be a specific value of the attribute, a range of values or a subset of values, depending on the type of the attribute. Leaf nodes are terminal, and each leaf node is either associated with a class label or probabilities over different classes. A sample is classified by testing the attributes of the sample, one at a time, starting with the one defined by the root node, and following the edge that corresponds to the specific value of the attribute in that sample, until we reach a leaf node. Note that this kind of scenario does not require a metric over samples.

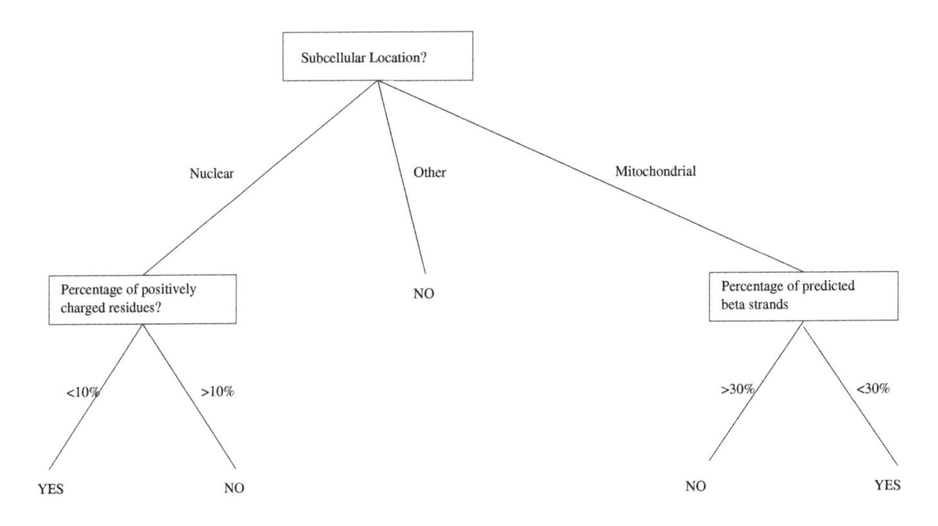

FIGURE 7.9: A decision tree for a specific class of proteins. A sample is classified by testing its attributes, starting from the test at the root node. If the final answer is yes then the sample is classified to the class. (Figure reproduced from [252] with permission.)

7.4.2 Training decision trees

When training decision trees from sample data D, the goal is to create a concise model for each class that is consistent with the training data and can distinguish members of the class from non-members. That is, we are searching for a model that within a short sequence of tests can decide with high confidence on the right category of a sample, given its attributes. Although a tree can be trained as a multi-category classifier, we will assume two classes, marked as + (class members, positives) and – (non-members, negatives).

The basic strategy is to grow the tree from the root to the leaves, by introducing additional nodes to the bottom of the growing tree with tests over new attributes. Each test splits the data based on the value of the attribute in each instance. Thus, the tree progressively splits the set of training samples into smaller and purer subsets. The process terminates when we obtain pure subsets, that is, sets that consist of samples that belong only to one class.

Usually, the selection of the next attribute to test at a node is based on how much it can reduce the **impurity** of the data after the split. The impurity of a dataset D is a measure of the class mixup in D and is denoted by $i(D)$. One possible measure of impurity is the class **entropy** (Section 3.12.5). The entropy impurity is also called the **information impurity** and is defined as

$$i_e(D) = Entropy(D) \equiv -\sum_{j=1}^{k} p_j \log_2 p_j$$

where p_j is the fraction of samples in the set D that are in class j. The class entropy measures the uncertainty in the class label for a given set D and a given categorization (labeling). The lower it is, the more certain we are about the class (in other words, the more information we have about the class). There are other possible measures of impurity [251]. For example, the **Gini impurity** is defined as

$$i_g(D) = \sum_{j=1}^{k} p_j(1 - p_j) = 1 - \sum_{j=1}^{k} p_j^2$$

measuring the error rate if the class is selected randomly based on the distribution of samples in the node.

In each node the test can involve more than one attribute, but most algorithms consider one attribute at a time. With m attributes there are $m!$ possible combinations of attributes (tests) along each branch, and finding the optimal tree is an NP complete problem [253]. The common heuristic for attribute selection is to pick at each node the attribute A that decreases the impurity the most, after the split. Let $values(A)$ denote the set of all possible values for attribute A, then the drop in impurity is defined as the difference between the impurity before the split and the weighted sum of the impurities at the leaf nodes after the split using attribute A. Using the notation of [254], that is

$$\Delta i(D) = i(D) - \sum_{v \in values(A)} \frac{| D_v |}{| D |} i(D_v)$$

where $D_v = \{x \in D | A(x) = v\}$ is the subset of samples whose value for the attribute A is v. When the impurity measure used is the entropy, the drop in impurity is called **information gain** and is denoted by $InfoGain(D, A)$

$$InfoGain(D, A) = Entropy(D) - \sum_{v \in values(A)} \frac{| D_v |}{| D |} Entropy(D_v)$$

Note that the maximal gain is obtained when the entropy of each subset D_v is zero. That is, when each subset contains samples that belong only to one class.

Over the years many algorithms for training decision trees have been proposed. The outline of one of the early algorithms (ID3) [255] is given below. This algorithm employs a depth-first approach where the last node to be added to the tree is also the first one to be tested and split, as long as the impurity can be reduced. If there are no more attributes to test on or if none of the attributes reduces the impurity then the node is declared a leaf and the recursive procedure stops.

The ID3 algorithm

Initialize: Add a root node v with the set of all samples D.
Let $\mathbf{V} = \{v\}$ denote the set of nodes to test.
Let $\mathbf{A}_v = \mathbf{A}$ denote the set of available attributes.

Recursion: Pick a node $v \in \mathbf{V}$.
For each attribute $A \in \mathbf{A}_v$ compute $InfoGain(D_v, A)$
If $InfoGain(D_v, A) \leq 0 \quad \forall A \in \mathbf{A}_v$
Node v is a leaf
Remove v from $\mathbf{V}$
Else
Let $A = \arg\max_{A \in \mathbf{A}_v} InfoGain(D_v, A)$
Split D_v into $v_1, v_2, ..., v_r$ new nodes ($v_i \in values(A)$)
Add $v_1, v_2, ..., v_r$ to $\mathbf{V}$
Set $\mathbf{A}_{v_i} = \mathbf{A}_v \setminus A$ for all $v_i \in values(A)$.

Halt: When $\mathbf{V}$ is empty.

The set $\mathbf{A}_v$ associated with each node keeps track of which attributes were already used, to make sure that each attribute is used only once along every path from a root node to a leaf node. While this guarantees optimality with multi-valued splits, it does not necessarily produce optimal results if all splits are forced to be binary splits (see Section 7.4.5.3). In the latter case, attributes can be considered multiple times, although with increased computation time.

After the learning procedure has terminated, the tree can be used to classify new samples, as depicted in Figure 7.9. During classification, a sample is repeatedly subjected to tests at decision nodes until it reaches a leaf. Note that the leaf nodes are not necessarily pure, since the data may contain multiple instances with identical attributes that belong to different classes, or as a result of pruning (see Section 7.4.2.3). While we can classify the instance to the majority class in the leaf node, it is more informative to assign a probability equal to the percentage of training samples at the leaf that are class members. That is, if the sample reaches node v then

$$P_T(+|x) = \frac{|D_v(+)|}{|D_v|}$$

where $D_v(+)$ is the subset of positive samples associated with node v. This probability reflects our confidence in the classification.

Beyond classification, decision trees can also provide a deeper understanding of the characteristics of a class. Note that every path from the tree root to a leaf corresponds to a set of conjunctions of constraints on the attribute values of samples. For example, in the tree of Figure 7.9 the rightmost path corresponds to the **decision rule**

If `(Cellular location = Mitochondrial)` *and* `(% predicted beta` $<$ `30%)`
Then `Yes.`

The whole tree can be represented as a set of if-then rules, which improves interpretability. For example, the tree of Figure 7.9 can be summarized as

If `(Cellular location = Mitochondrial)` *and* `(% predicted beta` $<$ `30%)`
OR
If `(Cellular location = Nuclear)` *and* `(% positively charged residues` $<$ `10%)`
Then `Yes.`

Thus, a decision tree can be converted to an equivalent rule set that can be considered a logical description of a category. Some of these rules can be especially interesting, since they might suggest a concise description of a family in terms of a few elements or principles. However, it should be emphasized that only on rare occasions is a single rule sufficient to produce a highly accurate classifier for a protein family.

There are many variations over the basic learning algorithm described above, and in the next sections we will discuss some of the improvements to the ID3 algorithm.

7.4.2.1 Impurity measures for multi-valued attributes

The information gain is a useful measure for attribute selection; however it is suboptimal when some of the attributes have multiple possible values, since it is biased toward selecting multi-valued attributes. Splitting a dataset based on such attributes may result in many small subsets, each one of relatively high purity, and therefore high information gain. Even a random split into many subsets will tend to reduce the impurity (Problem 7.4a). To overcome this bias one can normalize the information gain by the **attribute entropy** (also called **split information**), which measures the uncertainty in the attribute value for a given set D.

$$SplitInfo(D,A) = -\sum_{v \in values(A)} \frac{|D_v|}{|D|} \log_2 \frac{|D_v|}{|D|}$$

Split information has a high value if the attribute has many possible values, distributed uniformly (Problem 7.4b); thus normalizing the information gain by the split information should decrease the preference for multi-valued attributes. The normalized measure is called **gain ratio**.

$$GainRatio(D,A) = \frac{InfoGain(D,A)}{SplitInfo(D,A)}$$

By maximizing $GainRatio(D, A)$ we maximize the information used for classification out of the total information that the attribute contains.

While gain ratio mitigates the aforementioned bias, it suffers from another drawback: it may not be always defined (if the denominator is zero, i.e., all samples have the same value for the attribute), and it may choose attributes with very low split information, even if they do not provide much information gain (e.g., if the distribution of values is very skewed). One possible solution is to restrict the selection to attributes with information gain at least as high as the average and only then compute the gain ratio. Another solution is to group attribute values so that the new attributes are binary (see Section 7.4.5.3). Binar(ized) trees are harder to interpret; however they may eliminate the bias altogether. Other alternative impurity measures were suggested in [256, 257]. Note that these measures are of interest only when considering multi-branch splits, since their purpose is to combat high arity splitting. The basic $Gain(D, A)$ is sufficient when dealing with binary splits.

7.4.2.2 Missing attributes

Sometimes, an individual sample lacks the attribute being tested by a decision node. A simple solution to this problem is to assign the attribute the most frequent value among all samples that are associated with that node, or the most frequent one among the samples that are associated with that node *and* belong to the same class as the sample (this solution applies only during training). The more common approach to handle these samples (as in the C4.5 algorithm) is to partition them across the child nodes maintaining the same proportions as the rest of the training samples (i.e., the proportions are calculated using samples that have reached the splitting node and that have known values for the test attribute). The fractioned samples can then be used in calculations of the information gain. These fractions can be partitioned again and again with every unknown attribute. Consequently, ambiguous samples can reach several leaf nodes during classification, denoted $leaves(x)$. In these cases, the final probability is defined as the average of all leaf probabilities, weighted by the fraction of the sample that reached each leaf

$$P_T(+|x) = \sum_{v \in leaves(x)} f_v(x) \frac{|D_v(+)|}{|D_v|}$$

where $f_v(x)$ is the fraction of the sample that reached leaf v.

7.4.2.3 Tree pruning

The basic tree learning algorithm described in the previous sections continues to split nodes until they are pure. In the worst case scenario, it might generate one node per sample in the training data. While this procedure minimizes the error rate over the training data, it may perform poorly on new unseen samples. Without monitoring the learning process, the trees may

overfit the training data, capturing coincidental regularities that are not necessarily characteristic of the class being modeled. In other words, there might be another tree model that performs worse on the training data but better on average over the whole sample space. This problem is especially pronounced when the available dataset is small or when the data is noisy and contains errors.

A common strategy for improving performance and reducing overfitting is **tree pruning**. That is, eliminating nodes that degrade the generalization power of the model. There are several possible methods for pruning, which can be generally divided into **pre-pruning** and **post-pruning**.

Pre-pruning methods decide whether to further split a node or not *during* the learning process. Decisions are usually made based on measures that are correlated with the significance of the split. One simple approach is to set a threshold on the impurity reduction and reject a split if none of the available attributes reduce the impurity by more than the threshold or stop splitting when the number of points in a node is below a certain threshold. However, setting the threshold a priori is difficult. One could also estimate the probability that the reduction in impurity could have arisen by chance and accept the split only if the probability is small. While the probability can be estimated based on the empirical distribution of $\Delta(i)$ across the tree, a more rigorous approach is to assess the significance based on the null hypothesis that the split is random (see Section 7.7). An alternative approach is **validation-based pre-pruning**, which divides the samples available for learning into a **training set** and a **validation set**. The tree is built using the training set and tested on the validation set. If adding a node does not improve the impurity over the validation set, the split is rejected.

While pre-pruning methods are efficient, they may suffer from what is called the **horizon effect**. This refers to the phenomenon where splits that initially appear poor in fact lead to better performance later on during the training process. Any pre-pruning technique will be susceptible to this kind of short-sightedness [191].

Post-pruning methods work by building a complete tree first and then revisiting each node and re-evaluating its performance. For example, in **validation-based post-pruning** a complete tree is built first using the training set and then pruned using the validation set. Pruning is done by running an exhaustive search over all tree nodes. Each node is temporarily pruned (i.e., converted to a leaf), and the performance of the new tree is evaluated over the validation set (for performance assessment, see Section 7.4.4). The node that produces the largest increase in performance is selected and the corresponding subtree is pruned. The process continues as long as it does not cause a decrease in performance over the validation set.

Multi-set **cross-validation** can further enhance performance. In M-fold cross-validation a dataset of n samples is divided into M equal parts and M different trees are generated for each class, where each one uses $\frac{M-1}{M}n$ samples as its training set and $\frac{1}{M}n$ samples as its validation set. The predictions are then combined to obtain a single prediction. That is,

$$P_{Trees}(+|x) = \frac{1}{M} \sum_{T \in Trees} P_T(+|x) \tag{7.11}$$

where $P_T(+|x)$ is the class probability of the sample x according to the tree T. This method is called **model averaging**. Pruning based on multi-set cross-validation improves consistency and provides a more balanced model that learns the regularities that are common to different subsets of the data rather than those that are coincidental and occur only in a specific dataset.

One drawback of validation-based post-pruning methods is that the validation set should have a sufficiently large number of both class members and non-members to ensure high generalization power. Otherwise, many true regularities learned from the positives in the training set may be pruned, since they are not supported by the small validation set. The effect of post-pruning on decision trees in such cases can be drastic. M-fold cross-validation techniques work better with small datasets, and in the extreme can be used with the leave-one-out protocol where every training set contains $n-1$ samples and the validation set contains a single sample. However, they are more computationally intensive as they require learning trees from as many as $n-1$ datasets.

One possible solution for small datasets, where every single sample is important, is to use the **bootstrapping sampling** method [258]. Specifically, multiple training sets of size n can be generated by selecting samples from the dataset at random, with repetitions. A model is trained from each set and the predictions are then averaged, as in Equation 7.11. This method is also referred to as **bagging** (Bootstrap AGGregatING) [259]. The bootstrap method was proved to be robust in providing reliable estimates of performance in the presence of small datasets [260], and theoretical results indicate that models estimated from bootstrap datasets will converge to the optimal model at the limit of infinite bootstrap datasets [261].

Other pruning methods that use all the given data are rule post-pruning (used in C4.5) or post-pruning by the Minimum Description Length (MDL) principle [262–264], which assesses trees based on their complexity as well as their ability to explain the observations (see also Section 10.5.2.1).

7.4.3 Stochastic trees and mixture models

The basic procedure for building decision trees as described in the previous sections is deterministic. It employs a greedy approach that always selects the attribute that maximizes the information gain. However, this local maximization is not guaranteed to produce the "best" decision tree that describes the data. It may happen that several attributes have very similar information gain values. The choice of the one that marginally outperforms the others at some point during the training process may prove to be less advantageous later on. Even if the chosen attribute has a significantly better information gain, it still does not guarantee that in the optimal tree this attribute is indeed used at this point.

Several algorithms address this limitation of the deterministic approach by using ensembles of decision trees [264–266]. In particular, Breiman [267] introduced the **random forests** model, in which a large mixture of decision trees is trained from multiple bootstrap datasets. Each tree is grown such that the best split at each node is chosen from among a randomly selected subset of all the attributes. The trees are trained without pruning, and then averaged as in Equation 7.11. A related model is the mixture model of **Stochastic Decision Trees (SDT)** described in [264], which biases the selection towards higher quality attributes. The algorithm uses a probabilistic framework in which the attribute at each node is selected with probability that depends on its information gain

$$Prob(A) = \frac{InfoGain(D,A)}{\sum_{A_i \in \mathbf{A}} InfoGain(D,A_i)}$$

Thus, attributes with higher information gain are preferred, but even attributes with small information gain have a non-zero probability to get selected. To sample the space of decision trees the stochastic learning procedure is repeated multiple times to generate N different probabilistic trees. In addition, one deterministic tree is trained with the classical tree-learning algorithm. When combined with the M-fold cross-validation procedure of the previous section, then $N+1$ trees are learned from every training set for a total of $M(N+1)$ trees.

The predictions of the different trees can be combined using uniform weights, as in the model averaging method of the previous section

$$P_{Mixture}(+|x) = \frac{1}{M}\sum_j \frac{1}{N+1} \sum_{T \in Trees(j)} P_T(+|x) \tag{7.12}$$

where $Trees(j)$ is the set of trees learned from the j-th training set. However, this is not necessarily the most effective way of combining models. A simple averaging of models is equivalent to assuming a uniform (uninformative) prior over models. This may result in a suboptimal classifier, the same way an arbitrary combination of polynomials in Taylor expansion is unlikely to approximate well a target function (more on the connection between the two in Section 11.4.2.5).

The definition of the weights of the different components is a key element in a mixture model. Using informative prior (for example, based on the trees' complexity) we can rewrite Equation 7.12 as

$$P_{Mixture}(+|x) = \frac{1}{M}\sum_j \sum_{T \in Trees(j)} P_T(+|x)P(T) \tag{7.13}$$

where $P(T)$ is the prior probability of tree T. An alternative approach is to determine the weights based on the posterior probability of each model as in Equation 6.6 of Section 6.3.5.6. That is,

$$P_{Mixture}(+|x) = \frac{1}{M}\sum_{j}\sum_{T\in Trees(j)} P_T(+|x)P(T|D) \tag{7.14}$$

While it may be hard to estimate the probability $P(T|D)$ directly[9], we can approximate it based on the performance of the individual component models over the validation set, measured in terms of an evaluation or quality function Q (as discussed next). The total probability assigned by the hybrid mixture of trees is given by the performance-weighted average of the probabilities returned by each of the trees

$$P_{Mixture}(+|x) = \frac{\sum_j \sum_{T\in Trees(j)} Q(T)P_T(+|x)}{\sum_j \sum_{T\in Trees(j)} Q(T)} \tag{7.15}$$

where $Q(T)$ is the quality of tree T over the validation set. The higher it is, the better the tree separates the samples.

7.4.4 Evaluation of decision trees

The evaluation function Q is a key element of the decision tree model and plays an important role at different stages, from the learning phase to the final model definition and performance assessment. For example, during pruning a node is eliminated if the probabilities induced by the pruned tree are better at separating the positives from the negatives in terms of the evaluation function selected. In Equation 7.15 the evaluation function determines the coefficients of the different trees in the mixture model and thus affects the final probability it assigns to samples. Therefore, much thought should be given to selecting a sensible evaluation function.

A common measure of performance that is used in many machine learning applications (including decision trees learning algorithms such as C4.5) is the accuracy, i.e., the percentage of correctly classified samples

$$\text{accuracy} = \frac{tp+tn}{tp+tn+fp+fn} = \frac{tp+tn}{\text{total}}$$

where tp is the number of true positives, tn number of true negatives, fp is the number of false positives and fn is the number of false negatives (see Section 3.11). However, if the majority of the samples are negative samples, the accuracy may not be a good indicator of the discriminating power of the

[9]Decision trees are discriminative models that attempt to learn the decision boundary between classes and predict the class probability. Unlike generative models, they do not model the source directly, and therefore the likelihood of data instances $P(x|T)$ is undefined.

model. When the task is to discern the members of a specific class (family) from a large collection of negative samples, better measures of performance are the sensitivity and selectivity, defined as

$$\text{sensitivity} = tp/(tp + fn) \qquad \text{selectivity} = tp/(tp + fp)$$

The categorization of samples as true (false) positives/negatives depends on the model and on the output it assigns to the samples. Since decision trees output probabilities, we need a way to interpret these real numbers as either "positive" or "negative". Usually one sets a threshold T, a probability above which samples are predicted to be positives. Thus, if a sample reaches a leaf node j with membership probability $P_j(+)$, then it will be classified as positive if $P_j(+) > T$.

How should one define the threshold T? The naive approach would be to take a majority vote, so that all samples with probability higher than 0.5 are defined as positives. However, this may be misleading, since the number of positives and negatives may differ significantly to begin with. A more sensible approach would be to take into account the prior probabilities of positives $P_0(+)$ and negatives $P_0(-)$ and set the threshold to $P_0(-)$. A more refined approach is to calculate a different significance threshold T_j for each node, accounting for random fluctuations due to the varying sample sizes at the nodes. Specifically, if the total number of samples in a leaf node j is n_j then the number of positives that will reach this node by chance (the null hypothesis) is expected to follow a binomial distribution with mean $n_j \cdot P_0(+)$ and variance $n_j \cdot P_0(+) \cdot P_0(-)$. Using the normal approximation for the binomial distribution (see Section 3.12.4), we can set the threshold at the 95th percentile of the distribution and define a sample to be positive if the node probability $P_j(+)$ exceeds this threshold. If the sample reaches a single leaf node, then a label is assigned based on the confidence interval for this node. If it reaches more than one node, a "combined" node should be created, where the datasets are collected from the different leaf nodes, weighted by the fractions of the sample that reach each leaf node. The sample probabilities, the threshold and the label are calculated accordingly.

An alternative approach is to use the equivalence number criterion. The equivalence point is the point where the number of false positives equals the number of false negatives in the sorted list of results (see Section 3.11.1). In other words, the equivalence point is the point that balances sensitivity and selectivity. With the equivalence number criterion, all proteins that are predicted with higher probability than the probability at the equivalence point are labeled as positives.

There are alternative performance measures that do not require defining a threshold first, such as the ROC score. To compute this score, one first has to sort all samples according to their probabilities (as assigned by the model). Given the sorted list, plot the number of positives as a function of the number of negatives while scanning the list from the top. The ROC score is the area under the curve. This measure will be maximized when all positives are assigned higher scores than the negative samples (for more details see Section 3.11.1).

Another alternative is to use measures of statistical distance between the distributions of positives and negatives, such as the KL- or JS-divergence measures (see Section 3.12.6). Divergence measures take into account the complete data distribution and as such are less sensitive to outliers. In this case, the distributions compared are the distributions of class probabilities assigned by the decision tree model to positive and negative samples.

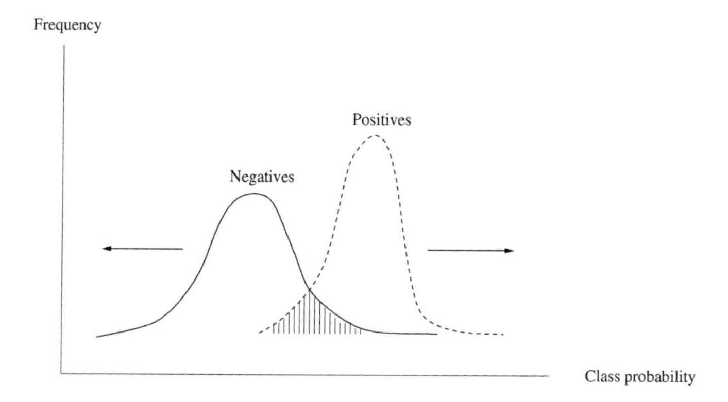

FIGURE 7.10: **Optimizing performance with statistical divergence measures.** Ideally, the model should assign high probability to class members and low probability to non-class members. The goal is to maximize the statistical distance or separation between the distribution of positives and the distribution of negatives and minimize the overlap. (Figure reproduced from [252] with permission.)

With measures like KL or JS, improved performance means maximizing the statistical distance between the distribution of positives and the distribution of negatives, as exemplified in Figure 7.10. In the SDT model (Equation 7.15), trees that maximize the separation are assigned a higher weight (a higher quality score Q) and play a more dominant role in prediction. Thus the learning process essentially maximizes the buffer between the positive and negative samples. One might notice the similarity with kernel-based methods such as support vector machines that try to maximize the margin around the separating hyperplane.

7.4.4.1 Handling skewed distributions

In typical protein classification problems the negative samples far outnumber positive samples. This poses a problem when trying to learn trees from small families. Because tests are introduced based on the reduction in impurity, most of the learning phase will concentrate on learning the regularities in the negative samples that result in significant information gain. Not only will this process cause extended learning times, it may also totally miss subtle regularities observed in the positive samples. The simple solution of presenting only a small set of negative samples equal in size to the set of positive samples is insufficient because some regularities observed in the positive set may be mistakenly considered as discriminating regularities, while if considered along with the whole set of negative samples they may prove to be uninformative.

A better approach is to present all the negative samples but assign their sum the same weight as the sum of all the positives samples (as suggested in [217]). Thus, every positive sample contributes $1/\#positives$ and every negative sample contributes $1/\#negatives$ to the total counts, and they are weighted accordingly in the calculations of the information gain.

It should be noted, however, that by doing so we introduce another problem. It may happen that we end with a tree of high purity (when using the weighted samples) but of high impurity when considering the original unweighted samples, thus increasing significantly the rate of false positives. One possible solution is to start by training the trees using the weighted samples and delay the usage of unweighted-sample entropy until the maximal information gain over all possible attributes at a given node is below a certain threshold. At that point each attribute is reconsidered using unweighted entropy and the training proceeds for as long as the information gain is below the threshold.

Note that the weighted samples are used only during the calculations of the information gain. For all other purposes, including pruning and prediction, the probabilities based on the unweighted samples should be used.

7.4.5 Representation and feature extraction

When studying the mapping between a protein sequence and its function, there is a broad range of features and properties that are related to structural and functional aspects of proteins that should be considered. These include basic compositional properties and physico-chemical features, predicted features that hint at the topology and the shape of the protein structure, and database attributes that provide additional information, such as its subcellular location or tissue preferences.

In general, we can make a distinction between internal features that can be computed directly from the sequence and external features that are based on database annotations. Internal features that can be extracted from the sequence include the composition of the 20 individual amino acids, dipeptides or longer k-mers, as well as the compositions of amino acid groups such as charged, aromatic, or hydrophobic (see Section 2.3). Each group has a unique characteristic that may play a role in defining and distinguishing family members from non-members. Other useful sequence features include the average hydrophobicity, the average molecular weight and the overall sequence length.

Certain features can be predicted from the sequence, such as the percentage of residues that are predicted to be in coil, helix, or strand, as computed by secondary structure prediction programs (Section 8.2.1.2).

External features are features that are extracted from records of sequence or structure databases such as SwissProt [16] and PDB [268], or domain databases such as Pfam and SCOP (see Section 9.8). Database attributes are often based on experimentally verified information. These could be binary features indicating whether a protein has cofactors or catalytic activity. Most database attributes are nominal features specifying the subcellular location of a protein, the tissues it is abundant in, the source organism and species. Other attributes can specify which motifs and domains are known to exist in a protein. Caution should be exercised when it comes to keywords, which many databases use to tag records. As opposed to other database attributes, keywords explicitly describe protein functionality and are based on human annotation and interpretation. If we are solely interested in obtaining the best classifier then the performance can clearly improve by integrating keywords. However, if we are interested in *learning* the mapping between proteins' properties and their function, these keywords should be excluded.

7.4.5.1 Feature processing

The transformation of features to attributes that can be used in decision trees is straightforward for binary attributes. However, numerical features must be discretized and divided into subranges before they can be used to split a node. A simple solution is to convert each continuous attribute A into a binary attribute A_t that is true if $A < t$. A natural choice for the threshold t would be one of the transition points where the class label changes (see Figure 7.11). It can be shown that the maximal information gain is obtained at one of the decision boundaries between different classes. A generalization of this approach to multiple partitions is described in [269], based on the MDL principle.

Procedures for discretizing numerical attributes can be used to generate a single global partition [270]. However, a better approach is to use them at every decision node to induce local partitions based on the given subset of sample points at that node, thus adjusting the decision to the subset of data available.

Handling of nominal features is more involved and first requires mapping every possible value of the attribute to an integer. However, the nominal database attributes that are associated with proteins are somewhat different than those typically found in the literature on decision trees. Each protein can take on several values for each nominal attribute, instead of just one, corresponding to a subset of $values(A)$. We refer to these samples as multi-value samples. For example, a protein can be abundant in the liver as well as in heart tissues. Clearly, the number of possible combinations of values is exponential in $|values(A)|$.

Instance:	x1	x2	x3	x4	x5	x6	x7	x8	x9
Attribute value:	−4.2	−2.1	−1	−0.5	1.2	2.3	2.7	3.2	4.5
Class:	c1	c1	c2	c2	c1	c1	c1	c2	c2

FIGURE 7.11: Handling continuous attributes. Discretizing a continuous attribute is done by first sorting instances by the attribute value and computing the transition points between classes (the decision boundaries). These are the mid-points between adjacent instances that belong to different classes (marked by the vertical lines). The optimal binary split, which maximizes the information gain, occurs at one of the transition points.

During training, when a node is split using a nominal attribute A, the set $values(A)$ is partitioned across the node's children. With traditional nominal attributes each sample would be assigned to the unique child that corresponds to its value. With multi-value samples, each sample may correspond to several children. To handle this scenario the sample can be divided across all children, weighting each child's portion by the number of values the sample has in common with that child.

Attributes can also be assigned probabilities, reflecting the likelihood or the confidence in their value. For example, in the tissue specificity attribute in SwissProt records, adjectives such as "abundant" or "weak" are used to indicate the abundance of the protein in the different tissues. These adjectives can be converted to probabilities over $values(A)$ for each protein, where "abundant" would be interpreted as high probability and "weak" would be interpreted as low probability.

7.4.5.2 Dynamic attribute filtering

The success of the decision tree model clearly depends on the features it explores. The more features we consider, the more expressive the model we can learn is. However, as the number of features increases so does the search space, and the learning task becomes unmanageable. For example, among the most informative sequence features are the frequencies of different k-mers (see also Section 7.3.1). However, the number of possible k-mers increases exponentially with k. Similarly, when representing the information on domains or motifs one could use an attribute for each possible motif, or a single motif attribute whose value indicates the set of motifs that is present in the protein. Either way, with more than $10,000$ known motifs and domains (see Section 9.8), the set of possible values (or combinations of values) is prohibitive. Not only does it result in extended learning times, but we are also more likely to get trapped in a locally optimal model.

When the number of attributes is too large it is recommended to narrow the scope of data that will be input to the decision tree algorithm. Not all attributes are relevant, and to reduce the search space we can filter out attributes that seem to be weak classifiers and keep only the most informative ones during learning. A simple filtering criterion is to eliminate attributes that appear with the same frequency in members and non-members, retaining only those that are more than one standard deviation from the mean (based on the distribution of frequencies among all attributes). These attributes can be selected dynamically during learning, on a per-family basis. While this method overlooks higher order correlations between attributes, it can greatly speed up the learning phase.

7.4.5.3 Binary splitting

Although many attributes are naturally suited to having multiple values, there is a case to be made for binary splitting. Binary splits preserve training data for later splits, and consequently may improve accuracy, particularly when a very small fraction of the data set represents class (family) members. Moreover, binary splits are not susceptible to the biased entropy of many-way splits (see Section 7.4.2.1). Since every decision can be expressed in terms of binary decisions, converting a tree to a binary tree does not limit in any way the expressiveness of the model.

With numerical attributes, the most straightforward procedure is also the most efficient and exhaustive: simply restrict discretization to exactly two subranges (i.e., a single cut point). For a particular attribute, the optimal split can found in $O(N)$ time, where N is the number of possible cut points [269]. However, for nominal attributes, since there is no ordering over the values, the exhaustive procedure of testing every possible binary partition of the values leads to a time complexity of $O(2^d)$ where d is the number of possible values for the attribute $d = |values(A)|$. There are at least two ways to deal with this problem. The most naive way is to examine only those binary partitions of $values(A)$ in which one value is assigned to one branch and all other values are assigned to the other branch. This is functionally equivalent to converting each value of A into a separate, purely binary attribute, and it reduces the time complexity to $O(d)$.

A more sophisticated partitioning algorithm is found in CART [271]. The problem can be expressed as identifying the subset of $values(A) = \{v_1, ..., v_d\}$ to assign to one child, such that $InfoGain(D, A)$ is maximized. Denote by D_i the subset of samples $\{x \in D | A(x) = v_i\}$. If the subsets D_i are numbered based on their class purity (with D_1 corresponding to the purest set), then the maximal information gain is obtained with the union of $D_1, ..., D_k$ for some $1 \leq k \leq d$. Therefore, we can find the optimal binary split in $O(d)$, excluding sorting time.

7.5 Further reading

An excellent introduction to classifiers is given in [191], including perceptrons, discriminant functions, neural networks and decision trees. Over the years many variations over these models were introduced. Even the simple perceptron algorithm was revived recently [272].

The subject of SVMs has been covered extensively in books on statistical machine learning such as [242, 273–275]. Excellent introductions are also given in [276, 277]. Applications of SVMs span classification problems from all domains of science, including handwritten character recognition [278], text categorization [279], face recognition [280] and regression [281], and utilize even unsupervised learning techniques such as clustering [282] and embedding [283]. In addition to protein classification discussed in this chapter, applications in computational biology include gene expression analysis and classification [284–286], subcellular localization [287], domain prediction [288], splice site recognition [289], prediction of protein-protein interactions [290] and remote homology detection [291–294] with kernels tailored to the specific problem at hand. Kernel combination has proved to be an effective way for data fusion and integration, often producing better results than individual kernels [295].

The decision tree model has also received considerable attention in the artificial intelligence and machine learning community. One of the most comprehensive resources on the model is [251]. Good summary chapters appear in [191] and [254]. For updates and additional references see the book's website at `biozon.org/proteomics/`.

7.6 Conclusions

- Classifiers are discriminative models which are designed to learn the decision boundary between different classes in the data.
- They are a powerful alternative to generative models, and are especially useful when we do not know the type of the source distributions.
- If two classes are linearly separable then a simple linear classifier can be learned with the perceptron algorithm.
- Support vector machines are a generalization of the perceptron algorithm, employing non-linear mappings into higher dimensional host spaces. Classes which are not linearly separable in the original space may be separable in the host space.
- The mapping depends on the kernel used to assess similarity between pairs of instances.
- Inner-product kernels are kernels that can be computed efficiently without actually executing the mapping to the higher dimensional host space.
- Other classifier models include neural networks and decision trees.
- Decision trees are especially useful when data is non-metric or is represented as a combination of features of different types, such as binary, numeric and nominal.

7.7 Appendix - estimating the significance of a split

In Section 7.4.2.3, when discussing pruning methods of decision trees, we mentioned the problem of estimating the significance of a split (Figure 7.12), in order to determine whether it should be accepted or not. The null hypothesis is that a split divides the samples completely at random, and our goal is to assess the probability that an observed split could have occurred by chance and determine whether we can reject the null hypothesis or not.

The methodology we will discuss here is general and applies to other similar problems we will address in sections 9.5.3 and 10.5.1.2. In more general terms, we can use this methodology whenever we have **categorical data**: a process with k possible outcomes or a random variable X that can take on a value $\mathtt{x_i}$ out of k possible values. We assume that the outcomes are mutually exclusive such that

$$\sum_{i=1}^{k} P(X{=}\mathtt{x_i}) = 1$$

When repeated n times, the number of times we observe each possible outcome follows a multinomial distribution. We will start with the case of two possible outcomes (a binary split).

FIGURE 7.12: Assessing the significance of a split. The split on the left maintains the proportions of positives (white) and negatives (black) in each child node, similar to those in the parent node. On the other hand, the split on the right separates the positives from the negatives and is unlikely to occur by chance.

Assume we have a node with N samples, of which n are class members (positives) and $N - n$ are non-members (negatives). Consider a binary split that divides the samples into N_1 and $N_2 = N - N_1$ samples in the left and right children nodes, respectively. Denote by $p_1 = \frac{N_1}{N}$ the fraction of samples that are channeled to the left node. The fraction of samples channeled to the right node is given by the complementary probability $p_2 = \frac{N_2}{N} = \frac{N-N_1}{N} = 1 - p_1$. We will denote p_1 and p_2 simply as p and $1 - p$ and refer to them as the **null probabilities**.

Let n_1 denote the number of positives that end up in the left node. The remaining $n_2 = n - n_1$ positive examples are channeled to the right node. We would like to determine whether the split divides the positives at random between the two nodes in proportion to p and $1 - p$. (We can ask the same question regarding the negatives and repeat the analysis, but as we shall see, there is no need for that.) Assuming each positive example is assigned at random to the left node with probability p and to the right node with probability $1-p$, then the number of class members in the left node is expected to follow a binomial distribution with parameter p. That is, the probability to observe n_1 out of n class members in the left node is

$$P(n_1) = \binom{n}{n_1} p^{n_1} (1-p)^{n-n_1}$$

As a binomial distribution, the expected number (the mean) of class members in the left node is $e_1 = np$ and the standard deviation is $\sqrt{np(1-p)}$. Similarly, the expected number of positives in the right node is $e_2 = n(1-p)$. However, if n_1 differs substantially from the expected number e_1 (and n_2 from e_2) then it is less likely to be the result of a random split. The more extreme the difference is, the more significant the split is[10].

Traditionally, the significance of such an experiment has been estimated by applying the Pearson **chi-square test**. The method introduces a statistic, denoted by X^2 and defined as

$$X^2 = \frac{(n_1 - e_1)^2}{e_1} + \frac{(n_2 - e_2)^2}{e_2} \tag{7.16}$$

This statistic measures the deviation of the observed numbers from the expected numbers. Since $n_2 = n - n_1$, and plugging in e_1 and e_2 we get that

$$\begin{aligned} X^2 &= \frac{(n_1 - np)^2}{np} + \frac{(n - n_1 - n(1-p))^2}{n(1-p)} \\ &= \frac{(1-p)(n_1 - np)^2 + p(np - n_1)^2}{np(1-p)} \\ &= \left(\frac{(n_1 - np)}{\sqrt{np(1-p)}} \right)^2 = \left(\frac{n_1 - \mu}{\sigma} \right)^2 \end{aligned} \tag{7.17}$$

where μ and σ are the mean and standard deviation of the binomial distribution. For large n the binomial distribution can be approximated with a normal distribution with parameters μ and σ, and therefore X^2 is a random variable whose value is the square of a standardized normal variable (that is, a

[10]The negative samples are expected to follow the binomial distribution as well; however, given N_1 and n_1, the number of negatives is completely determined and therefore it is not an independent experiment.

random variable distributed like the normal distribution with zero mean and standard deviation of one).

Recall that the chi-square distribution characterizes the sum of squares of k independent standardized normal variables, denoted by χ^2 (see Section 3.12.4). In the case discussed above the sum is reduced to one variable; therefore, the variable X^2 of Equation 7.16 is distributed like the chi-square distribution with $k = 1$ and

$$p(X^2 = x) \simeq \frac{1}{\sqrt{2}\Gamma(1/2)} \frac{1}{\sqrt{xe^x}}$$

The use of X^2 (instead of χ^2) comes to indicate that this is an approximation of the chi-square statistic. The cumulative distribution of the chi-square distribution has been characterized as well and is given by

$$p(x' \leq x) = \frac{\gamma(k/2, x/2)}{\Gamma(k/2)} \tag{7.18}$$

where γ is called the lower incomplete gamma function, defined as $\gamma(a, x) = \int_0^x t^{a-1}e^{-t}dt$. The cumulative function is available as part of many software packages. The pvalue of X^2 is the probability to obtain by chance a value of X^2 that is equal or higher than the observed value

$$pvalue(X^2 = x) = 1 - p(x' < x)$$

Hence, given the cumulative distribution we can estimate the probability that the split could have been obtained by chance. We can say that X^2 is significant and reject the null hypothesis if the pvalue of X^2 is very small. For example, the split is considered significant with confidence $T = 0.95$ if its pvalue according to the null hypothesis is less than $1 - T = 0.05$. Note that with binary splits we could have computed the pvalue directly, by summing over all the possible splits that are more extreme than the observed split. That is,

$$pvalue(n_1) = \sum_{i=n_1}^{n} P(i) = \sum_{i=n_1}^{n} \binom{n}{i} p^i (1-p)^{n-i}$$

While this becomes computationally intensive for large n, we can use the normal approximation to the binomial distribution as before. However, the discussion above on the chi-square test sets the grounds for the more general case of non-binary splits.

In the case of a binary split there are two possible outcomes but only one degree of freedom, since the outcomes are related. Therefore we end with only one standard normal variable (Equation 7.17). In the more general case a node splits into multiple nodes, with N_i examples in node i. Assume we observe n_i positives in node i. If the samples are split randomly, then the number of examples in each node is expected to follow the multinomial distribution (see

Section 3.12.4). To estimate the significance we can generalize the statistic we defined in Equation 7.16 such that

$$X^2 = \sum_{i=1}^{k} \frac{(n_i - e_i)^2}{e_i}$$

where k is the number of nodes (categories), $e_i = p_i n$ and $p_i = N_i/N$. This sum is distributed like the chi-square distribution with $k-1$ degrees of freedom (one less than the number of categories because of the constraint that $n = \sum_i n_i$). The significance can be estimated based on the cumulative distribution of Equation 7.7.

However, if the size N_i of some of the categories is small, the underlying statistical assumptions are violated and the chi-square distribution does not apply. Specifically, the normal approximation does not hold for small datasets (the rule of thumb is that the approximation is reasonable or better when the number of observations in each category is at least 10).

An alternative approach to the chi-square test is to compute the pvalue directly as the sum of probabilities of all possible outcomes that are at least as extreme as the observed set of counts. In order to discuss this approach, we will use the following terminology. A **type** is defined as a specific set of counts divided between the different possible outcomes of the random variable. We refer to the set $\mathbf{n} = (n_1, n_2, ..., n_k)$ as the **observed type** or **sample type** (the actual set of counts we observe in the different categories). The null probabilities are denoted by $\mathbf{P}_0 = (p_1, p_2, ..., p_k)$. These are the background probabilities of observing each of the possible outcomes of an experiment. We refer to the type $\mathbf{r} = (np_1, np_2, ..., np_k)$ as the **random type**. This is the most likely set of counts we would expect to obtain by chance. We would like to assess the significance of the sample type $\mathbf{n}$.

While it is easy to say whether an outcome of a two-way split is more extreme than the observed type (being a unimodal distribution, we just take the tail in the right direction), it is not so straightforward for the multinomial case, since there are many ways a type can deviate from the observed type. One approach is to consider every type whose probability is smaller than the observed type as more extreme. But since the multinomial distribution is a multi-modal distribution, this does not necessarily mean that the lower probability types are indeed more extreme. For example, a certain type might have a lower probability than the observed type, yet be more similar to the mean. Should it be considered more extreme in that case?

A common approach is to use the **likelihood ratio** statistic to compute the likelihood of types under two hypotheses (random vs. sample-based). This approach considers all types that have a higher likelihood ratio than the sample type as more extreme. We can think of it as having two sources that are characterized by two probability distributions: $\mathbf{P}_\mathbf{n} = (q_1, q_2, ..., q_k)$, which is defined based on the empirical sample type with $q_i = n_i/n$, and $\mathbf{P}_0$, which is associated with the random type. Given the two probability distributions

$\mathbf{P}_0, \mathbf{P_n}$ and a type $\mathbf{m} = (m_1, ..., m_k)$, the likelihood ratio of the type is defined as

$$L(\mathbf{m}) = \log \frac{P_n(\mathbf{m})}{P_0(\mathbf{m})}$$

where

$$P_0(\mathbf{m}) = \binom{n}{m_1, ..., m_k} p_1^{m_1} p_2^{m_2} \cdots p_k^{m_k}$$

and $P_n(\mathbf{m})$ is defined similarly. A high positive ratio for $L(\mathbf{m})$ indicates that $\mathbf{m}$ is less likely to emerge from the random type and is therefore considered more similar to the observed type. Note that

$$L(\mathbf{n}) = \sum_{i=1}^{k} \log \left(\frac{q_i}{p_i}\right)^{n_i} = n \sum_{i=1}^{k} \frac{n_i}{n} \log \frac{q_i}{p_i} = n \sum_{i=1}^{k} q_i \log \frac{q_i}{p_i} = n D^{KL}[\mathbf{P_n} || \mathbf{P}_0]$$

and hence the likelihood ratio for the sample type is proportional to the KL-divergence between the two probability distributions (see Section 3.12.6).

To assess the significance of a type we need to compute the probability that a type sampled according to the *random type* will obtain a higher likelihood ratio than that of the observed type. That is,

$$pvalue(\mathbf{n}) = \sum_{\mathbf{m} \text{ s.t. } L(\mathbf{m}) \geq L(\mathbf{n})} P_0(\mathbf{m})$$

For large k or large n exhaustive enumeration quickly becomes prohibitive. While it is possible to employ sampling methods and compute the pvalue based on the relative number of simulated types that pass the test, a pvalue computed through simulation is limited in accuracy to $1/M$ where M is the number of types simulated. However, to compute the exact pvalue efficiently there is no need to enumerate all possible types and compute their likelihood ratio. As was shown in [296], a major speedup can be obtained by using a recursive procedure and a branch and bound approach. Specifically, the method recursively explore the tree of all possible types of n counts over k categories. By bounding the likelihood ratio of certain groups of types, complete subtrees can be skipped or added to the pvalue without having to compute the likelihood ratio of each individual type. This is best explained by going over the algorithm scheme, as shown below.

Branch-and-bound pvalue computation

Input: Sample type $\mathbf{n} = (n_1, ..., n_k)$, null probabilities $\mathbf{P}_0 = (p_1, ..., p_k)$.

Initialize recursion: $totalpvalue = descend(1, (*, ..., *))$.

Procedure: $descend(i, (m_1, ..., m_k))$
$pvalue = 0$

For $m_i = 0$ to $n - \sum_{j=1}^{i-1} m_j$
    $\mathcal{M} = (m_1, ..., m_i, *, ..., *)$
    If $L_{max}(\mathcal{M}) < L(\mathbf{n})$
        Next
    If $L_{min}(\mathcal{M}) \geq L(\mathbf{n})$
        $pvalue+ = P_0(\mathcal{M})$
    If $L_{max}(\mathcal{M}) \geq L(\mathbf{n})$
        $pvalue+ = descend(i + 1, \mathcal{M})$
Return *pvalue*.

Specifically, the algorithm starts from the top level of the tree ($i = 0$), where none of the counts have been assigned yet, and calls the procedure *descend*. The first argument of the procedure denotes the level that is explored next. With $i = 1$, that means that the procedure *descend* tests each of the possible partial assignments (subtypes) of the form $\mathcal{M} = (m_1, *, ..., *)$ where $0 \leq m_1 \leq n$. The stars denote unassigned counts. Each partial assignment can evolve into many possible assignments, and at each level of the tree another category is assigned. The i-th level corresponds to partial assignments of the form $\mathcal{M} = (m_1, ..., m_i, *, ..., *)$. Each specific subtype is associated with a whole set of assignments (organized in a subtree), where the remaining categories are assigned. The idea is to take advantage of the fact that the likelihood ratio of partial assignments can be bounded. That is, it can be shown that the lower bound for the set $\mathcal{M}$ is

$$L_{min}(\mathcal{M}) = \min_{\mathbf{m} \in \mathcal{M}} L(\mathbf{m}) \geq \sum_{j=1}^{i} m_i \log \frac{m_i}{np_i} + \bar{m} \log \frac{\bar{m}}{n\bar{p}}$$

where $\bar{m} = n - \sum_{j=1}^{i} m_i$ is the number of unassigned counts and $\bar{p} = 1 - \sum_{j=1}^{i} p_j$ is the total probability of all the unassigned categories, considered as one category in the partial assignment. Similarly, it can be shown that the upper-bound is given by

$$L_{max}(\mathcal{M}) = \max_{\mathbf{m} \in \mathcal{M}} L(\mathbf{m}) = \sum_{j=1}^{i} m_i \log \frac{m_i}{np_i} + \bar{m} \log \frac{\bar{m}}{n\bar{p}_{min}}$$

where $\bar{p}_{min} = \min_{i<j\leq k}\{p_j\}$ is the minimal probability over the unassigned categories. Using the lower bound and the upper bound we can determine if there is a point in exploring each partial assignment. Specifically, if the upper-bound of all the types in $\mathcal{M}$ is lower than the likelihood ratio of the sample

type $\mathbf{n}$ then there is no need to further explore the subtree that corresponds to that partial assignment because none of the types will qualify as more extreme than $\mathbf{n}$. Similarly, if the lower-bound $L_{min}(\mathcal{M})$ is higher than $L(\mathbf{n})$ then all types in the corresponding subtree should be added, since all of them qualify. This can be done efficiently, since the combined pvalue of a partial assignment of the form $\mathcal{M} = (m_1, ..., m_i, *, ..., *)$ is given by

$$P_0(\mathcal{M}) = \binom{n}{m_1, ..., m_i, \bar{m}} p_1^{m_1} p_2^{m_2} \cdots p_i^{m_i} \bar{p}^{\bar{m}}$$

where all the unassigned categories are considered as one "other" category. If the sample likelihood ratio is between the lower bound and the upper-bound, then we need to explore the subtree further and test each of its subtrees.

This method is not the only alternative to the chi-square test, and for other methods for computing pvalues see [297, 298].

7.8 Problems

For updates, additional problems, files and datasets check the book's website at `biozon.org/proteomics/`.

1. **Protein classification with SVMs.**

   You are given a dataset (`scop.seq`) consisting of sequences from several different protein families.

   (a) Train a classifier with the **mismatch-spectrum kernel** to separate members of each family in `family-list` from all other sequences.

   - For each sequence in the dataset compute its representation using the (k, m) mismatch-spectrum model, with $k = 4$ and $m = 1$.
   - For each family in the list, divide the dataset into a training set and a test set and train a classifier using the (k, m) mismatch-spectrum kernel with the software that is linked from the book's website. Note that you have to provide a different labeling of the training set for each family.
   - Use the classifier to classify the test set. Measure the performance in terms of the ROC curve (Section 3.11.2). Use two different setups: one where a true positive is a member of the same family, and the other one where a true positive is a member of the same superfamily.

   (b) Train a classifier with the **pairwise kernel** to separate members of each family in `family-list` from all other sequences.

   - For each sequence in the dataset compute the local similarity with each sequence in the set `scop.rep` (using the program that you wrote for Problem 3.9 or the program provided in the website).
   - For each family in the list, divide the dataset into a training set and a test set and train a classifier using the pairwise kernel.
   - Use the classifier to classify the test set. Measure the performance in terms of the ROC curve, using the two different setups as in the previous problem.

   (c) Compare the performance of the two kernels by comparing the area under the ROC curves, and the ROC1, ROC10 and ROC50 scores.

2. **Mutual information kernel**

   You are given a set of HMMs representing a set of protein families. Train a classifier over the same dataset of the previous problem using the mutual information kernel, with $f(x, y)$ defined as in Equation 7.10. To compute the kernel matrix you will need to evaluate the probability to emit each pair of sequences with any of the HMMs, using the program of Problem 6.3. Assess the performance of the kernel as before using ROC curves.

3. **The decision tree model vs. naive Bayes**

   You are given a dataset (`enzyme.data`) consisting of features representing proteins from several different enzyme families. The dataset is split into a training set and a test set.

   (a) Compute the error probability when classifying instances based on a single feature. That is, given a feature f, compute the class posterior probability

   $$P(c|f = x) = \frac{n_x^c}{n_x}$$

   (where n_x is the number of instances for which $f = x$), and classify each instance to the class with the highest posterior probability.

   $$c_{MAP} = argmax_{c_j \in C}\ P(c_j|f = x)$$

   Repeat the calculation for each feature and each value. Are there certain features which can act as good classifiers for some families?

   (b) Classify the instances based on *all* features using naive Bayes classifier, where the posterior probability is approximated by

   $$P(c|f_1, \ldots, f_n) = \frac{P(f_1, \ldots, f_n|c)P(c)}{P(f_1, \ldots, f_n)} \sim \frac{P(c)}{P(f_1, \ldots, f_n)} \prod_i P(f_i|c)$$

   In other words, the likelihood is computed assuming independence between features. Compute the error probability over the test set for each family in the set.

   (c) Train a decision tree over the same dataset, using the software linked from the book's website. Compute the error probability for each family and compare to the error probabilities you obtained with naive Bayes.

4. **Attribute selection in decision trees**

   Consider attributes with uniform distributions over values.

   (a) Show that a random split of a multi-valued attribute tends to have higher information gain than a random split of a binary attribute.

   (b) Compare the split information of a multi-valued attribute to a binary attribute.

Chapter 8

Protein Structure Analysis

8.1 Introduction

So far we mainly discussed the analysis of sequences of protein or DNA molecules. However, although sequence analysis can reveal much about the functional properties of these molecules, it does not provide complete understanding. These molecules form complex structures that carry important functional information.

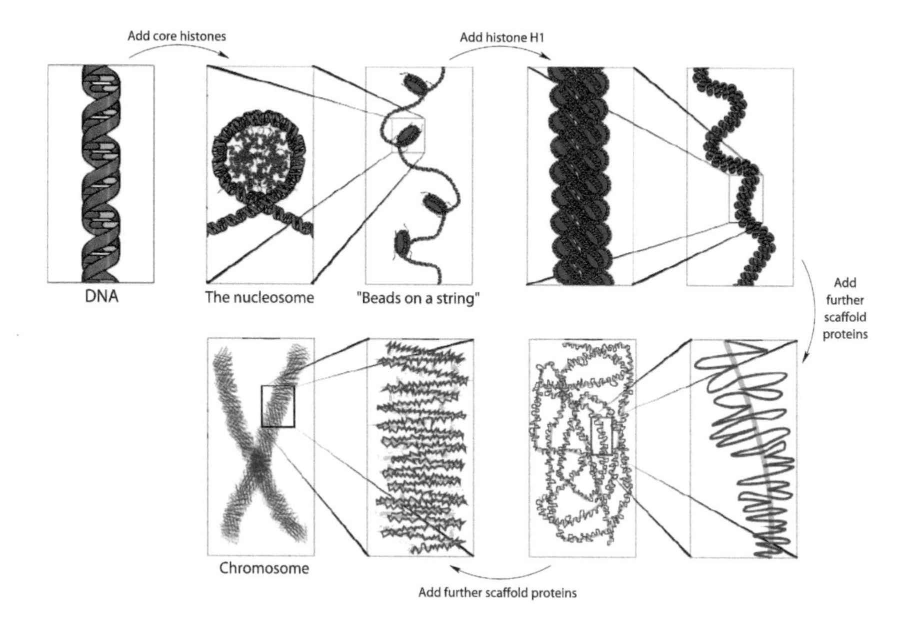

FIGURE 8.1: The chromatin structure of the DNA. Double-stranded DNA winds itself around proteins called core histones, which results in the **nucleosome** (the fundamental repeating unit of a chromatin) and the "beads on a string"-like structure. These structures are further compacted on interaction with protein histone H1. This compact structure is compressed even more by scaffolding proteins in a multi-step process resulting in a chromosome. (Figure modified from Wikipedia [299]. Original figure created by Richard Wheeler.)

For example, the DNA molecule that makes up a genome is organized in a compact space-saving structure called **chromatin**, by wrapping itself around proteins called histones, and is then packaged into chromosomes (Figure 8.1). The three-dimensional (3D) arrangement of the DNA affects transcription, influences the expression of genes (see Chapter 12) and enables complex regulation patterns. RNA molecules also form 3D structures, using a similar base pairing that underlies the double helix structure of the DNA, and perform various functions that are facilitated by these structures (Figure 2.6). Proteins fold into many different tertiary structures, dictated by their linear sequence and containing regular and irregular secondary structures. It is this great diversity of structures that enables proteins to carry out the numerous functions they perform (Figure 8.2).

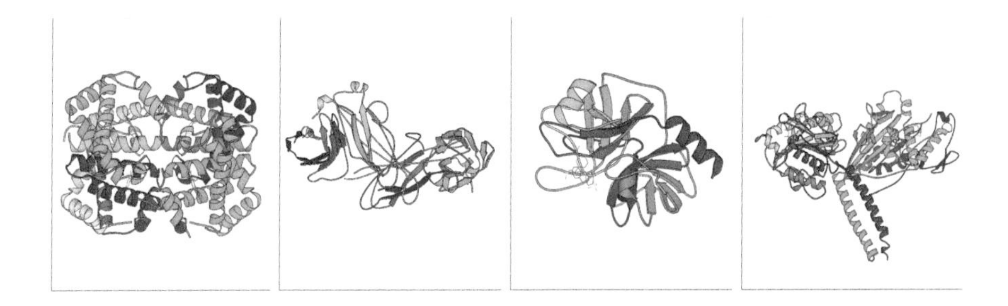

FIGURE 8.2: **Tertiary structures of proteins.** From left to right: **(a) Human hemoglobin.** This molecule is a tetramer in its active form and is responsible for oxygen transport from lungs to various tissues. The heme is responsible for the red color of blood. **(b) E. coli chaperone PAPD protein.** This chaperone helps in assembly of pili on the surface of the bacterial cell wall, which enable the bacterium to connect to another bacterium, allowing exchange of information (e.g., DNA) between the cytoplasms of the two bacteria. **(c) Human trypsin.** Trypsin is a protease secreted in the digestive system. It breaks down protein molecules found in food into smaller fragments, sometimes individual amino acids, enabling absorption. **(d) Rat kinesin.** Kinesins are eukaryotic proteins that move along microtubules, using the energy from hydrolysis of ATP. This activity is needed for important cell functions such as mitosis, meiosis and transport of cellular cargo.

To better understand the function of these molecules, we must also consider their structural properties. The chromatin is one of the more elegant, yet complex constructs in the cell that has been gaining increased attention in recent years [300], and the field of RNA structure prediction has been a very active one (see [301, 302] for a review). However, since the focus of this book is on proteins we will limit the discussion to structural analysis of proteins. There are multiple aspects to protein structure analysis, including secondary structure prediction, fold recognition (threading), protein folding, structure comparison, docking and more. Here we will discuss briefly the problems of protein folding and secondary structure prediction and will focus mostly on protein structure comparison.

8.2 Structure prediction - the protein folding problem

We were already introduced to the protein folding process in Chapter 2. After their formation as linear sequences of amino acids, proteins form local secondary structures and fold into unique 3D structures. The contacts between atoms of different molecules is what determines how proteins function, and these depend on their 3D structures. Therefore, we can say that the structure of each protein "prescribes" its function and the way it interacts with other molecules.

The protein folding process starts even before the nascent chain leaves the ribosome and usually converges into a stable conformation (referred to as the **native structure**) within a fraction of a second. Given the enormous number of possible conformations a protein can adopt, it is quite amazing that it folds into its native structure so fast (in the literature this is referred to as the Levinthal's paradox [303]). While in some cases RNA molecules or proteins called chaperones help and "guide" a protein into its native structure, the general perception is that most proteins fold without interaction with other molecules. Hence, in principle, all the information necessary to determine the structure of a protein is coded in the protein sequence. This hypothesis spawned many studies over the past four decades that attempted to predict protein structures from their sequence.

The 3D structure of a protein can be determined experimentally using **X-ray diffraction** of protein crystals or **Nuclear Magnetic Resonance** (NMR) of proteins enriched with NMR active atoms. However, obtaining diffraction quality protein crystals is difficult and not always feasible, and NMR experiments are limited to small proteins. A recent technology, based on Cryo EM[1] has improved significantly over the past several years and allows faster determination of structures, but current resolutions are still too

[1]Electron microscopes can magnify samples up to 2 million fold, thanks to the small wavelength of the electron beam they use. **Cryo Electron Microscopy** (Cryo EM) can obtain similar magnifications on samples that are cooled to cryogenic temperatures (−200 degrees Celsius or lower). This technique allows for observation of samples "rapidly frozen" in their native environments, giving a relatively realistic picture, as opposed to methods such as X-ray crystallography and NMR, in which biological molecules are in non-native environments. An advantage of the latter methods is that they can provide structural insights at much higher resolution.

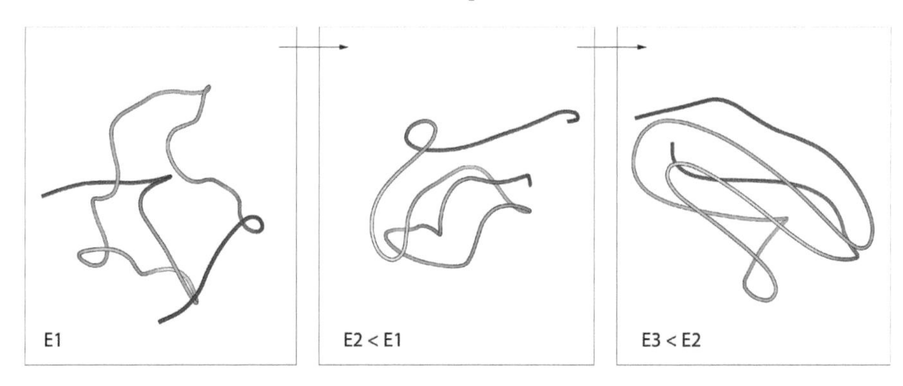

FIGURE 8.3: **Protein structure prediction.** Starting from a random configuration or one that is based on an alignment with the protein sequence of a known structure, the structure is refined until reaching a local minimum of the energy function.

coarse for detailed analysis of protein structures. Therefore, it is quite clear why there is such a strong interest in protein structure prediction. In fact, as of 2008, there are only about 50,000 structures that were determined experimentally, but many more that were predicted. Prediction methods are generally divided into **ab initio** methods, which start from a random configuration (from "scratch"), and methods based on **comparative modeling** (also called **homology modeling**), where the initial configuration is modeled based on sequence similarity with a homologous protein whose structure has been determined.

In both cases, the structure is predicted by looking for a conformation that minimizes the total free energy (Figure 8.3). This can be done by exploring conformations close to the initial conformation (especially if there is a good seed, based on sequence similarity for example) or by sampling randomly a larger search space. Since the search space is enormous (Problem 8.1), searching for the right conformation through simulation is impractical even for a single molecule. This has triggered many studies that attempt to reduce the search space through intelligent sampling of the configuration space, as well as projects such as "Folding at Home" [304], which allows owners of personal computers throughout the world to download a software that runs folding simulations when the computer is idle and sends the results to a central server that collects the results. The more successful methods thus far have been those that reduce the conformation space by combining fragments that can be associated with a local structure, using fragment libraries [305–307]. The final orientation of side-chains is usually refined and determined based on energy minimization considerations.

Structure prediction is a hard problem and requires extensive computational resources. In view of the difficulty of predicting the structure of a protein directly from its sequence, a different approach for structure prediction that

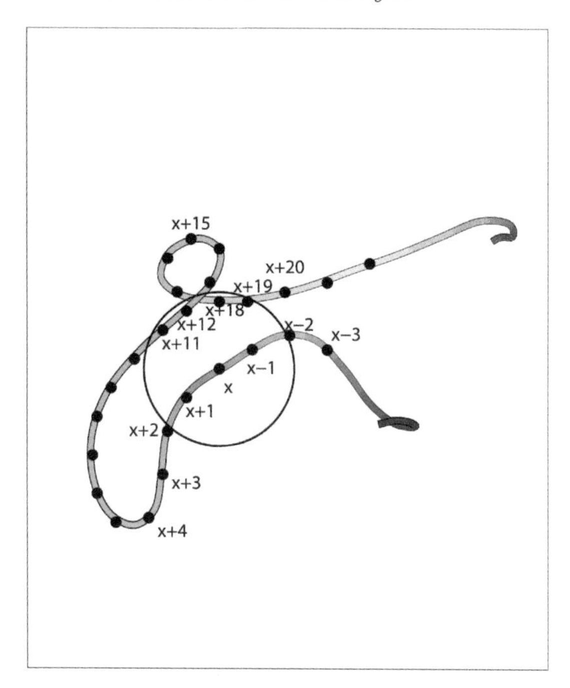

FIGURE 8.4: **Threading a sequence through a structural template.** The score of a specific sequence-to-structure alignment is the sum of all pairwise interactions between close neighbors (within a certain perimeter r) as determined by the template. For example, the score of x depends on the pairwise interactions with $x-1, x+1, x+11, x+12, x+18$ and $x+19$.

is based on **sequence-to-structure** alignment was introduced in the 1990s [308,309]. This approach, called sequence threading, or simply **threading**, is motivated by the fact that a protein might have a structure that is similar to one of the known structures even if there is no visible sequence similarity or even functional similarity. Indeed, this is often the case with newly sequenced proteins. The goal of threading is to find a known structure that best fits the native structure of the protein or identify the fold(s) that a protein sequence can adopt. Threading is often referred to as **fold recognition**.

An alignment between a sequence and a structure is computed by threading the sequence through the structural template. A score is computed for each possible alignment (placement of sequence residues in positions along the structure) based on its compatibility with the structure. The compatibility is assessed by computing the energy associated with each sequence residue and its neighboring residues, as determined by the structural template and the alignment (Figure 8.4). Many different energy functions were proposed over the years, and for a review of common approaches see [310,311]. Threading algorithms usually consider only the core elements of the structure (helices or strands) while ignoring the loops, under the assumption that the core is more likely to be conserved across distantly related proteins [312,313]. The output of the algorithms is the highest scoring alignment.

Given a library of structures representing different folds, each fold can be scored based on the highest scoring alignment. If the score of the highest scoring fold is statistically significant (for example, if it is associated with a high zscore, as in Section 3.4.2.3), then there is a high probability that the sequence adopts that fold.

Threading can be beneficial for sequences that are only weakly similar to sequences of known structures, yet adopt the same structure. However, in practice, threading algorithms do not perform as well as hoped. Performance evaluations in the Critical Assessment of Structure Prediction (CASP) meetings indicated that the correct fold is often not the top scoring fold, and even when it is, the alignment is often incorrect. Examining more than the top scoring fold and using functional considerations can help in identifying the correct fold (that is, functional similarity should be taken into account as well). Using multiple threading methods and averaging the predictions (or creating a consensus) of the query and other homologous sequences can also improve performance. For further discussion see Further reading.

8.2.1 Protein secondary structure prediction

The term **secondary structure** refers to *local* structural arrangements of short segments along the protein polypeptide chain, usually between three and ten residue long, which are formed during the protein folding process. There is a finite number of known conformations, which are categorized into three main classes: helices, strands and coils/loops (see Figure 2.8). Helices and strands are considered to be more regular shapes than coils and can be more easily characterized. The core of a protein is usually composed of regular secondary structures. Homologous proteins typically share the same core structure (called **scaffold**), and the structural differences lie mostly in the loop regions, which tend to be closer to the surface of the protein and more exposed. Loop regions typically contain coils of different shapes. The loops can vary in their length and shape; however, they are not random. This is especially true for short loops. Although there are many possible loop conformations in theory, the number of known loop conformations is much smaller than expected, and they can be organized into several hundred or a couple of thousands classes [314]. More than differences in sequence, the differences in loop conformations are mostly attributed to differences in length and the secondary structures they connect.

Knowledge of the secondary structure of a protein is essential for accurate prediction of its tertiary structure. It can help to reduce the viable conformation space and speed up the search for the optimal configuration. Therefore, prediction of secondary structures is a natural first step toward structure prediction. Unfortunately, secondary and tertiary structures are strongly dependent on each other in the sense that the secondary structures determine the global fold (given the secondary structures, there is finite number of possible arrangements into folds) and the 3D structure constrains the secondary structures that each segment of a sequence can form. Thus, the ability of a certain short sequence to form a secondary structure depends not only on its amino acid sequence but also on the 3D environment of that sequence. In fact, there are known cases of short sequences (up to five residues long) that form a helix in one structure and form a strand or a coil in another structure [315]. This mutual dependency clearly complicates prediction, but secondary structure prediction algorithms can be quite useful nevertheless.

Characterizing secondary structures can also help when comparing protein structures and can improve homology detection (as we will see in Section 8.4), even if the secondary structures are predicted.

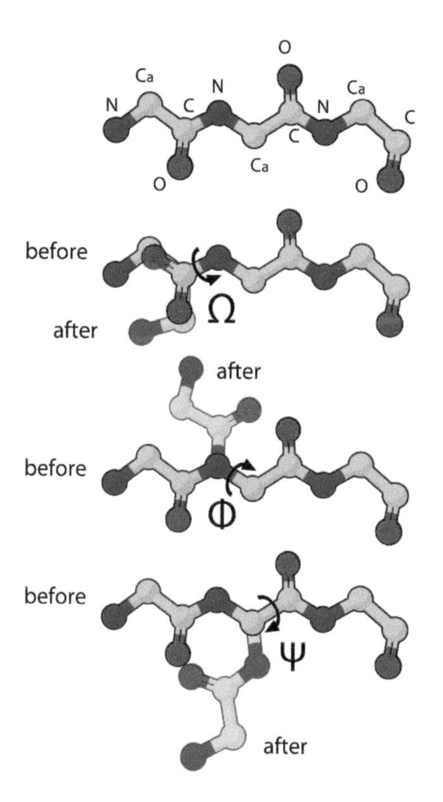

FIGURE 8.5: **The dihedral angles.** (Figure modified from Wikipedia [316].) The three dihedral angles (Ω, Φ, Ψ) that define the structure of the protein backbone, with rotation about each angle resulting in a different orientation of the peptide backbone. Omega (Ω) is the dihedral angle between atoms C_α-C-N-C_α, phi (Φ) is the dihedral angle between atoms C-N-C_α-C and psi (Ψ) is the angle between atoms N-C_α-C-N. Each angle specifies the relative orientation of two planes, where one is determined by the coordinates of the first three atoms and the other is determined by the last three atoms (for example, psi is the angle between the plane spanned by N-C_α-C and the plane spanned by C_α-C-N). Omega is usually 180 degrees because of planarity of the peptide bond, and rarely it is 0 degrees. There is a limited number of allowed combinations of Φ and Ψ, depending on the secondary structure the residue is a part of, and those allowed combinations are defined in the Ramachandran plot (Figure 8.6). Dihedral angles are also called **torsion angles.**

8.2.1.1 Secondary structure assignment

If the 3D structure of a protein is known, then it is possible to determine the secondary structure of each residue based on the Ψ and Φ **dihedral angles** between consecutive residues along the amino acid chain and hydrogen bonds between residues. These angles characterize the rotation of the bond between the N-C_α atoms and the C_α-C atoms (Figure 8.5). The seminal work by Ramachandran et al. [317] characterized the range of angles that are typical of residues in alpha helices and beta strands (see Figure 8.6). Based on this plot, programs such as STRIDE [318] or DSSP [319] assign each residue in *solved* 3D structures to one of the three main classes of secondary structures: helices, strands (extended conformation) and coils. These programs also distinguish between different types of helices (such as α-helix vs. 3_{10}-helix or π-helix, which have different periodicities) and categorize certain coils as turns (such as beta turns, characterized by a different number of residues between the hydrogen-bonding residues).

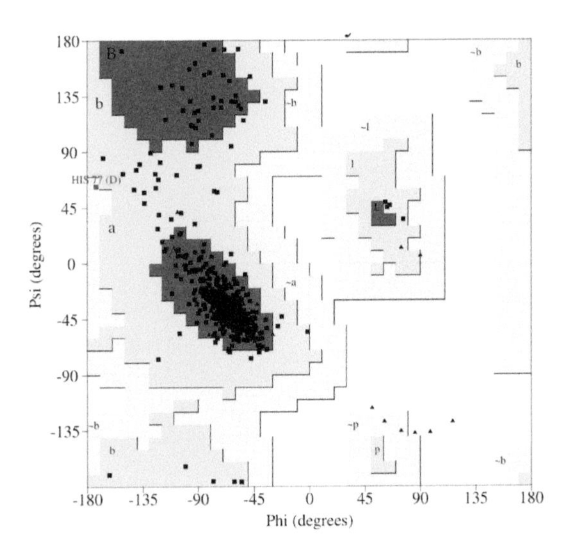

FIGURE 8.6: **Ramachandran plot.** The Ramachandran plot for all residues in the structure of hemoglobin shown in Figure 8.2. The top left corner and bottom left corner is the Phi-Psi region for residues in beta-sheet conformation. The middle left region is the Phi-Psi region for residues in right-handed α-helix conformation. The small region in the top right square is for residues in left-handed helix conformation. Dark red indicates most favored regions, yellow indicates additional allowed regions and light yellow indicates generously allowed regions. White background shows the disallowed regions. Each non glycine residue is represented by a tiny square. Glycine residues, which can lie anywhere on the Phi-Psi plot are represented by triangles. This plot represents a protein structure with a good geometry since only one of the residues is in the disallowed regions (a histidine residue in sequence position 77).

8.2.1.2 Secondary structure prediction

The process of forming secondary structures starts already while the protein is being translated from the mRNA template by the ribosome. Since it happens mostly without the intervention of other molecules, it is natural to assume that the information needed to determine these elements exists in the sequence of amino acids. The same assumption is the basis for the field of protein structure prediction, but with secondary structures the correlation with the sequence is easier to detect. Indeed, the relation between the amino acid sequence of a protein and its secondary structure was observed as early as the 1970s.

There are many algorithms for secondary structure prediction [84,320–324], most of which are based on analysis of sequences of proteins of solved structures. These methods assume that the local sequence determines the secondary structure (which is not always true because of long-range interactions). One of the first methods was developed by Chou and Fasman [320]. Through examination of known structures they created tables that list the preferences of the different amino acids to be in one of the three main states. These were estimated directly from the relative frequencies of the amino acids in each secondary structure and their average frequency. That is, if f_a is the relative frequency of amino acid a and $f_{a,s}$ is the relative frequency of amino acid a in secondary structure s then the preference (or propensity) of a to be in structure s is defined as $P_s(a) = \frac{f_{a,s}}{f_a}$.

For each secondary structure Chou & Fasman divided the amino acids into three groups, of favorable amino acids ($P_s(a) > 1$), neutral ($P_s(a) \sim 1$) and unfavorables amino acids ($P_s(a) < 1$). This is basically a likelihood test, and the propensity $P_s(a)$ is the likelihood ratio $\frac{P(a|s)}{P(a)}$, where $P(a)$ is the background probability of amino acid a. Several interesting facts emerged from the study by Chou and Fasman. For example, charged amino acids tend to appear less frequently in beta strands, and proline, glycine and asparagine tend to break helices and form beta turns. They also observed that in beta turns the propensities of different amino acids differ depending on the position within the turn. Hence, they associated different parameters with different positions in turns.

To predict secondary structures, these individual preferences are insufficient since they would lead to noisy predictions. Therefore, the propensities are first averaged over neighboring residues (three or four residues, depending on the type of secondary structure predicted), and seeds of secondary structures are formed by searching for continuous segments of residues with high propensity for helices or strands. These seeds are then extended in both directions until encountering residues that have strong preference for beta turns, or residues that have low probability for both alpha and beta. The latter are classified as coil residues. The exact procedure that locates seeds and expands them uses several ad hoc rules about the length of the segments, the densities of high propensity residues and the average propensity values.

A different approach was suggested by Garnier-Osguthorpe-Robson (GOR) [325], based on information theory principles. Instead of using the likelihood ratio, it uses the posterior probability $P(s|a)$ of s given amino acid a (estimated directly from the data as described next). The preference of each residue to be in each secondary structure is assessed based on the ratio $\log \frac{P(s|a)}{P(s)}$, which is the information residue a carries about the secondary structure s. In general, the information a random variable Y carries regarding the occurrence of the event X is defined as

$$I(X|Y) = \log \frac{P(X|Y)}{P(X)}$$

High information content indicates that the occurrence of Y increases the chances of X happening (see also Section 3.12.6). Of all possible secondary structures at position j, the GOR algorithm picks the one that results in the highest information content. The algorithm considers also the information content of neighboring residues. If R_i is the amino acid type in position i and S_j is a random variable denoting the secondary structure of residue j, then the information the sequence carries regarding the secondary structure is expressed as

$$I(S_j = X|R_1, R_2, ..., R_n)$$

That is, every residue potentially carries some information on the secondary structure at position j. However, due to insufficient statistics, it is impractical to consider the whole sequence. Assuming that the secondary structure is influenced mostly by neighboring residues, only a local region in the vicinity of the residue j is considered when predicting its secondary structure. In the original paper by GOR, eight residues in each direction are considered, such that

$$I(S_j = X|R_1, R_2, ..., R_n) \simeq I(S_j = X|R_{j-8}, R_{j-7}, ..., R_{j+8}) \qquad (8.1)$$

These information values can be computed based on solved structures. This is done by building conditional probability tables and computing for each k-mer the probability of each secondary structure. However, because of an insufficient amount of data, Equation 8.1 is reduced to first-order dependencies between the secondary structure at a certain position and residue types in neighboring positions

$$I(S_j = X|R_1, R_2, ..., R_n) \simeq \sum_{m=-8}^{8} I(S_j = X|R_{j+m})$$

where $I(S_j = X|R_{j+m})$ is referred to as *directional information* and reflects the interaction of the side chain of the amino acid in position $j+m$ with the backbone of the amino acid in position j. Side chain to side chain interactions are considered only if the equation includes second-order dependencies or higher.

The information values are computed from the observations. For example,

$$I(S_j = helix|R_{j+8} = Gly) = \log \frac{P(S_j = helix|R_{j+8} = Gly)}{P(helix)}$$

and to compute $P(S_j = helix|R_{j+8} = Gly)$ we consider all positions that are eight residues downstream (toward the N-terminus) from glycines and compute the relative frequency of helix, beta and coil. All together there are $3 \times 17 \times 20$ parameters for 3 states, 17 neighboring positions and 20 amino acids.

Interestingly, for beta strands and alpha helices, some residues exhibit a symmetric behavior around $m = 0$ with either a minimum at $m = 0$ (meaning the residue interferes with the specific secondary structure) or a maximum at $m = 0$ (the residue prefers the secondary structure). Some residues exhibit an asymmetric behavior with maximum at one side and minimum on the other side. These residues tend to either start or terminate the corresponding secondary structure. Lastly, to adjust the values to the 3D structure being analyzed, it is possible to add or subtract a constant that reflects the structural environment. For example, proteins that are rich with helices contain additional information that increases the chances of creating additional helices, and so on.

Another approach that uses information from neighboring residues is PSIPRED [324]. However, unlike previous methods, it uses information from homologous sequences to characterize the biochemical properties of each position and predict its secondary structure. It takes as input a sequence profile of the query sequence and its homologs, generated by a program such as PSI-BLAST (see Section 4.4.1). The assumption is that a sequence profile reflects the structural environment of each position better than individual residues. To learn the mapping from profiles to secondary structures PSIPRED uses neural networks. The profile matrix is fed into a standard feed-forward neural network with a single hidden layer using a window of 15 residues. This net has three output units corresponding to each of the three states of secondary structures. Another window of 15 positions of these three outputs (per amino acid) are then used as input into a second neural network to filter and smooth outputs from the main network. The final output is the probability that a certain position in the seed sequence of a profile is in a coil, helix or strand.

8.2.1.3 Accuracy of secondary structure prediction

To measure the accuracy of secondary structure prediction algorithms we can use the percentage of residues that were predicted correctly. The accuracy of these algorithms has been steadily increasing, and the most successful algorithm to date is PSIPRED [324], with an average accuracy of about 80%.

When evaluating the accuracy for different secondary structures it appears that alpha helices are usually predicted with greater success than beta strands. This is not completely surprising, since beta strands are shorter and preferences of individual residues to be in beta strands are usually weaker than of the strong alpha-favoring residues. Beta strands are part of bigger substructures (parallel or anti-parallel beta sheets), and it is often the longer-range interactions that determine the type of secondary structure within the sheet. Another problem that is common to many algorithms is that most errors occur at the ends of strands and helices at the transition to another secondary structure. Other errors tend to occur in regions where there is no clear preference to either structure. Errors of the latter type can be reduced if one considers first the total content of secondary structures. The content can be determined using an experimental method like Circular Dichroism (CD)[2]. For example, if CD analysis indicates that the protein is rich with beta strands, then short regions predicted initially as helices connecting strands are more likely to be strands, or coil regions. Indeed, the GOR method takes this information into account by adding a constant that depends on the general content of secondary structures, and their analysis showed that incorporating this information improves the accuracy of prediction. Clearly, incorporating information from homologous proteins with solved structure can also improve predictions.

Despite the many efforts to further improve predictions (including combining several methods [322, 323]), the performance seems to saturate at 80% accuracy. Hence, it seems there is an upper bound on the success of local secondary structure prediction algorithms. Interactions between residues that are far apart in sequence but close in the 3D structure often dominate the intrinsic tendency of certain residues to be in one secondary structure or another, driven by the pressure to obtain a more globular and compact structure overall. Therefore, the assumption that interactions between distant residues is negligible (as in PSIPRED and GOR) is inaccurate. Methods that do not incorporate long-range interactions at all perform even worse [322]. However, in a given structure only a few residues will be spatially close to each residue, and hence incorporating all long-range interactions might not necessarily improve prediction if done in a non-selective manner. Another issue to point out is that the success of the model depends on the type of structures used for training. Most models are trained over globular proteins, but if used to predict the secondary structures of fibrous proteins they are likely to give less accurate results, since these proteins prefer different types of secondary structures with a different arrangement (more elongated) and hence tend to form different interactions.

Nevertheless, assignments based on secondary structure prediction algorithms provide a good starting point for tertiary structure prediction algorithms and reduce the time needed to search for the true 3D structure.

---

[2]**Circular dichroism** measures the differences in absorption of light polarized under different electric fields (a right circular vs. left circular field). Upon interaction with protein structures, the different secondary structures result in different spectra of absorptions, and based on the spectra one can determine the fraction of the molecule that is in helix conformation vs. beta and coil conformations.

8.3 Structure comparison

Structure comparison is essential for functional analysis of proteins. Since the 3D structure of a protein is strongly correlated with its function, identifying structures of known function that are similar to the 3D structure of a protein with an unknown function may help to characterize the biological function of that protein. This is especially important in view of the increasing number of solved structures whose function has not been determined yet, as a result of structural genomics efforts to map the protein structure space.

Structure comparison can be particularly useful for remote homologs. Algorithms for sequence comparison often miss weak similarities between sequences that have diverged greatly [23, 25, 138]. However, structure is more conserved than sequence [44, 64, 65, 326] and homologous proteins without significant sequence similarity are still likely to have the same fold and close biological function[3]. Hence, detecting structural similarity can greatly increase the accuracy of function prediction and help infer function in cases where sequence comparison fails.

The importance of protein structure comparison pertains to other problems beyond homology detection. Methods for structure prediction often rely on structure comparison [329, 330]. Fold prediction systems might require hundreds of thousands of structure comparisons per prediction [331], and both the speed and accuracy of the algorithm are important factors in the success of these systems. Structure comparison is important for studies of structure-function relationships, since it can help to detect common and important structural and functional sites, such as active sites or binding sites. Structure comparison algorithms are also essential for automatic domain detection and classification [332, 333] and comparison of protein families [23] and are instrumental in studies of global organization of the protein space [45, 334]. We will discuss these applications later on in Chapters 9 and 10.

As with sequence comparison, structure comparison results in an alignment that assigns each residue i in the first structure a residue j in the second structure or a gap. However, as opposed to sequence comparison, there is no single natural definition of structural similarity as we will see next. Structural similarities are harder to detect and are even more elusive than sequence similarities. Simple and intuitive methods, such as comparing the Ψ and Φ angles, usually produce spurious matches (see Problem 8.7).

---

[3]This general rule does not always apply. There is a non-negligible number of sequence-similar, structure-dissimilar protein pairs in the PDB [327], usually observed in proteins whose function requires conformational flexibility (e.g. due to interactions with other molecules). Furthermore, in some cases even a single point mutation in the sequence can result in a completely different fold and function [328]. However, these cases are considered a rare exception.

FIGURE 8.7: Structural alignment of ubiquitin (PDB 1ubq) and ferredoxin (PDB 1rkf). Ubiquitin is a regulatory protein that is widely present in higher organisms. Attachment of one of more ubiquitin molecules to specific proteins is thought of as a signal for degradation of that specific protein. Ubiquitin is also known to be involved in determining localization of certain proteins in different cellular compartments. Ferredoxins are proteins that can contain different kinds of iron-sulfur clusters. These proteins are usually recruited for electron transfer in biological reactions, for example in some of the steps of photosynthesis. The two proteins have similar folds, although there is no obvious functional similarity.

Many different algorithms for structure comparison were developed over the years, based on different heuristics and definitions. Some algorithms are based on minimizing inter-atomic distances, while others match intra-atomic distance matrices, fragments or secondary structures. Their results can differ quite markedly. We will discuss several algorithms in more detail next.

All algorithms start by first reducing a structure to the set of α carbon atoms, one for each residue. The coordinate of each α carbon atom represents the position of the corresponding amino acid (Figure 8.8). That is, a structure with n residues is represented as $\mathbf{A} = \{A_1, A_2, ..., A_n\}$ where $A_i = (Ax_i, Ay_i, Az_i)$ is a vector in $\mathbb{R}^3$ of the x, y and z coordinates of residue i (for clarity, vector notation is omitted). This approach, which focuses on the **backbone**, reduces the computational complexity significantly while producing accurate enough results, although the orientations of the side chains in the two structures compared might be different. However, these can be refined later by molecular simulation [335].

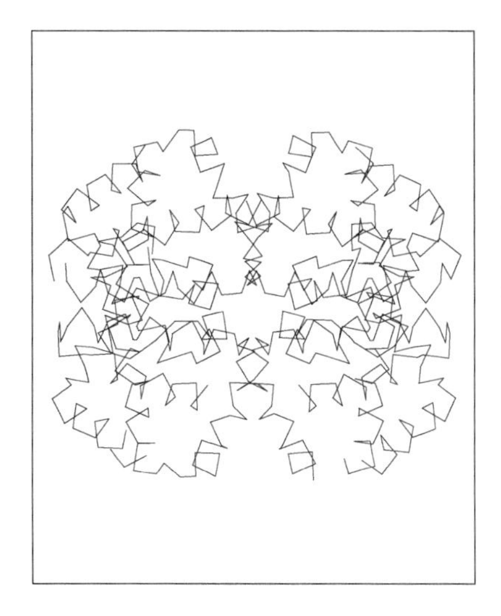

FIGURE 8.8: The backbone structure of a protein. A representation of a protein structure (human hemoglobin of Figure 8.2) as a trace of the C_α atoms only. The helical secondary structure elements are evident in this representation.

8.3.1 Algorithms based on inter-atomic distances

One of the main approaches for structure comparison is to match the two structures under a certain transformation of their coordinates so as to minimize the inter-atomic distances. **Inter-atomic** distances are the Euclidean distances between the atoms of two structures (as opposed to **intra-atomic** distances, which are internal distances within the same structure). Algorithms that belong to this category usually employ dynamic programming, similar to the algorithms for sequence comparison, where the scoring function is defined based on the distance between atoms or between secondary structures. There is more than one way to define this distance, as we will see next.

8.3.1.1 The RMSd measure

At the core of any structure comparison algorithm there is a metric that assesses the quality of the match. A popular measure that is based on inter-atomic distances is the Root Mean Square distance (RMSd) between structures **A** and **B**. Given an alignment, i.e., a set of equivalent residue pairs $(A_1, B_1), (A_2, B_2), ..., (A_n, B_n)$, it measures the average Euclidean distance between the positions of matching residues. That is,

$$RMSd(\mathbf{A}, \mathbf{B}) = \sqrt{\frac{1}{n}\sum_{i=1}^{n} \|A_i - B_i\|^2} \qquad (8.2)$$

where $\|A_i - B_i\|^2 = (A_{ix} - B_{ix})^2 + (A_{iy} - B_{iy})^2 + (A_{iz} - B_{iz})^2$. Since the coordinates of a protein structure can be in an arbitrary coordinate system and have an arbitrary orientation in that system it does not make sense to compute the distances between the arbitrary coordinates. Before we can compare the structures we need to transform both to the same coordinate system.

A **rigid transformation** of a structure consists of a linear translation T and rotation $\mathbf{R}$. These operations do not change the relative positions of atoms and their internal distances, and hence the overall shape remains intact. Therefore they are referred to as rigid transformations. A translation is a simple operation that shifts the coordinate system by a certain constant vector T (Figure 8.9).

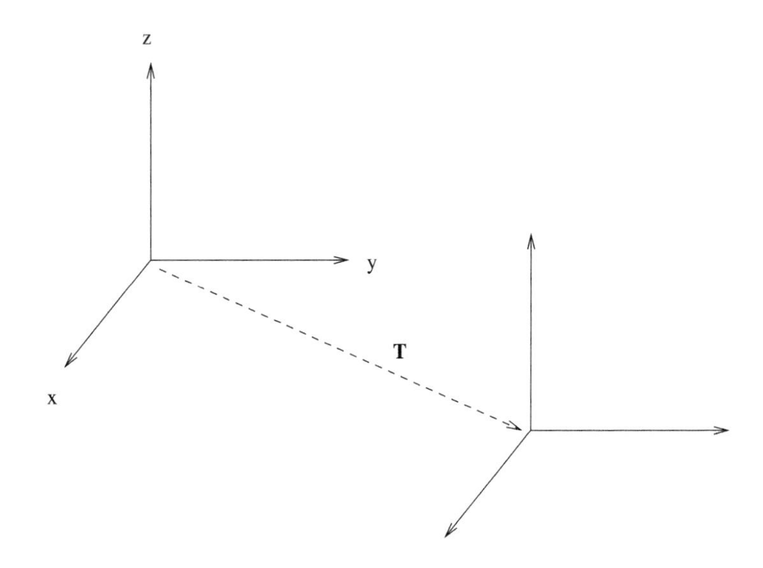

FIGURE 8.9: **Translation of coordinate systems.**

Rotation matrices in Cartesian coordinate systems are 3×3 matrices that determine the rotations α, β, γ around the x,y and z axes. It is easier to start with rotation matrices in 2D (x and y axes only), where a counterclockwise rotation of angle θ is given by the matrix

$$\mathbf{R}_\theta = \begin{pmatrix} \cos\theta & \sin\theta \\ -\sin\theta & \cos\theta \end{pmatrix}$$

Note that $\mathbf{R}_\theta$ is orthonormal, meaning that the rows are orthogonal (and so are the columns), and their norm is one[4]. This is important, to ensure that $\mathbf{R}_\theta$ does not skew the original object that is being rotated. It is easy to see that the product $\mathbf{R}_\theta \cdot \vec{v}$ of the rotation matrix and any vector $\vec{v}$ gives the coordinates of the vector in the new coordinate system (Figure 8.10).

[4]A set of vectors $\vec{v}_1, \vec{v}_2, ...$ is called orthogonal if the scalar product satisfies $\langle \vec{v}_i \cdot \vec{v}_j \rangle = 0$ for $i \neq j$. The set is called orthonormal if $\langle \vec{v}_i \cdot \vec{v}_i \rangle = 1 \quad \forall i$.

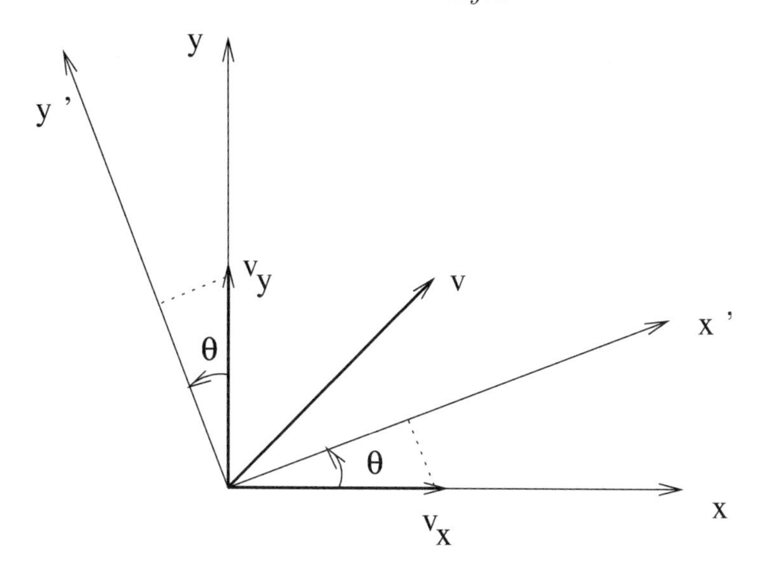

FIGURE 8.10: **A counterclockwise rotation of a coordinate system by angle θ.** The new coordinates of $\vec{v}$ are given by splitting the vector into its v_x and v_y components, applying the transformation to each and summing their contributions. For example, the new x' coordinate of $\vec{v}$ is given by $v_x \cos\theta + v_y \sin\theta$.

Rotation matrices in 3D can be written as the product of three counterclockwise rotation matrices, around the x, y and z axes:

$$\mathbf{R}_{\alpha,x} = \begin{pmatrix} 1 & 0 & 0 \\ 0 & \cos\alpha & \sin\alpha \\ 0 & -\sin\alpha & \cos\alpha \end{pmatrix} \mathbf{R}_{\beta,y} = \begin{pmatrix} \cos\beta & 0 & -\sin\beta \\ 0 & 1 & 0 \\ \sin\beta & 0 & \cos\beta \end{pmatrix} \mathbf{R}_{\gamma,z} = \begin{pmatrix} \cos\gamma & \sin\gamma & 0 \\ -\sin\gamma & \cos\gamma & 0 \\ 0 & 0 & 1 \end{pmatrix}$$

A rotation around a specific axis is a rotation in a 2D plane spanned by the other two axes. For example, $\mathbf{R}_{\gamma,z}$ is a rotation in the x, y plane which moves x toward y (as in Figure 8.10). Since a rotation around y moves z toward x (clockwise) the signs are reversed in $\mathbf{R}_{\beta,y}$.

The rotation matrix $\mathbf{R}_{-\theta}$ denotes the reverse rotation matrix of $\mathbf{R}_{\theta}$, which rotates an object in angles $-\alpha, -\beta, -\gamma$. It is not difficult to see that $\mathbf{R}_{\theta}\mathbf{R}_{-\theta} = \mathbf{I}$ where $\mathbf{I}$ is the identity matrix (Problem 8.2b), meaning that the application of a rotation and a reverse rotation brings us back to the same position. It also follows that $\mathbf{R}_{-\theta} = \mathbf{R}_{\theta}^{-1}$. (From this point on we will omit θ and will refer to rotation matrices simply as $\mathbf{R}$ and $\mathbf{R}^{-1}$.) It is also possible to show that for a rotation matrix $\mathbf{R}^{-1} = \mathbf{R}^t$, where $\mathbf{R}^t$ is the transposed matrix, and hence $\mathbf{R}\mathbf{R}^t = \mathbf{I}$ (Problem 8.2c). Another important property of rotation matrices is that $det(\mathbf{R}) = 1$ meaning that the rotation $\mathbf{R}$ is a proper rotation. If $det(\mathbf{R}) = -1$ then $\mathbf{R}$ is an improper rotation that involves also inversion, where the structure is mirrored around one of the axes (Problem 8.2d).

If the two structures compared are essentially the same structure, just translated and rotated, then determining the rotation and translation would be relatively simple (Problem 8.3). But in the general case, when we have two different structures, the best we can do is to look for the transformation that minimizes the distance between the two structures, assuming the structures are of the same length and all residues are matched. Therefore, the RMSd between two structures **A** and **B** is defined as

$$RMSd(\mathbf{A},\mathbf{B}) = \min_{R,T} \sqrt{\frac{1}{n}\sum_{i=1}^{n} \|A_i - \mathbf{R}(B_i - T)\|^2} \tag{8.3}$$

For a moment, assume we are only interested in the translation $T \in \mathbb{R}^3$ that minimizes the RMSd, then we have to minimize

$$RMSd(\mathbf{A},\mathbf{B}) = \min_{T} \sqrt{\frac{1}{n}\sum_{i=1}^{n} \|A_i - (B_i - T)\|^2}$$

By solving the equation $\frac{\partial RMSd}{\partial T} = 0$ or $\frac{\partial (RMSd)^2}{\partial T} = 0$ we get that

$$T = \frac{1}{n}\sum_{i=1}^{n}(B_i - A_i) = \frac{1}{n}\sum_{i=1}^{n} B_i - \frac{1}{n}\sum_{i=1}^{n} A_i$$

That is, the translation vector that minimizes the RMS distance between the two structures is given by the difference vector between the center-of-mass vectors of the two structures.

Finding the optimal rotation-based transformation is more complicated, but to simplify things we will assume that the center of mass vector of both structures is the zero vector. That is, the structures were already shifted such that their center of mass is at the origin of the Cartesian coordinate system. Our goal now is to find a rotation matrix **R** that would minimize the RMSd of the two structures

$$RMSd(\mathbf{A},\mathbf{B}) = \min_{R} \sqrt{\frac{1}{n}\sum_{i=1}^{n} \|A_i - \mathbf{R}B_i\|^2} \tag{8.4}$$

subject to the constraint $\mathbf{R}\mathbf{R}^t = \mathbf{I}$ or $\sum_k R_{mk}R_{nk} = \delta_{mn}$ $(1 \le m, n \le 3)$. δ_{mn} is the Kronecker delta such that $\delta_{mn} = 1$ if $m = n$ and 0 otherwise. This can be formulated as a minimization problem with Lagrange multipliers λ_{mn}, such that

$$\mathbf{R} = \arg\min \frac{1}{n}\sum_{i=1}^{n} \|A_i - \mathbf{R}B_i\|^2 + \sum_{m,n} \lambda_{mn}\Big(\sum_{k=1}^{3} R_{mk}R_{nk} - \delta_{mn}\Big) \tag{8.5}$$

This formulation is known as the Kabsch algorithm [336].

An elegant method for finding the optimal rotation, which uses matrix algebra and circumvents the Lagrange multipliers, is described in [337]. The method starts by computing the covariance matrix $\mathbf{C} = \mathbf{B}\mathbf{A}^t$ of $\mathbf{B}$ and $\mathbf{A}$. As shown in Appendix 8.7, the optimal rotation matrix is given by the product

$$\mathbf{R} = \mathbf{V}\mathbf{U}^t$$

where $\mathbf{U}$ and $\mathbf{V}$ are matrices whose rows are the eigenvectors of $\mathbf{C}$.

Given the optimal transformation (T and $\mathbf{R}$), we can compute the minimal RMSd between two structures, as in Equation 8.3. However note that in Equation 8.3 we assume a specific alignment. That is, the set of matching residues (the equivalence set) is given. Clearly, under an arbitrary equivalence set, the RMSd is meaningless. Our goal is to find the least RMSd under all possible alignments. But since the search space is exponential in the size of the molecules compared, exhaustive search is practically impossible.

The problem is further complicated by the local vs. global dilemma. Many similarities between protein structures are actually localized to a single domain or structural motif. Detecting these motifs is a hard problem and has been a major hurdle for structure comparison algorithms. The main reason is that unlike sequence comparison, where the notion of the most similar subsequences is well defined (see Section 3.2.3), no such notion has been agreed upon for protein structures, partly because there is no natural measure of similarity for individual residues as is the case for protein sequences. Nevertheless, there are many heuristics that can find "good" alignments efficiently. We will describe a few of them next.

8.3.1.2 The structal algorithm

The structal algorithm [24] is an example of transformation-based algorithms, which aligns protein structures by minimizing inter-molecule distances between aligned residues. The algorithm works by iteratively re-defining the set of equivalent residues. It starts by choosing an initial set, for example, by matching the beginnings or the ends of the two structures or by using the sequence alignment as a seed. Based on this set the two structures are superimposed so as to minimize the RMSd, as in Section 8.3.1.1, and the translation vector and rotation matrix are computed. Given the superimposed structures, the algorithm computes a position specific scoring matrix (PSSM, as in Section 4.3.4) that is then used to re-align the structures and generate a new set of equivalent residue pairs. Based on this new set, the translation vector and the rotation matrix are recomputed, again by minimizing the RMSd. The revised superposition results in a revised PSSM, which results in a revised superposition and the process continues until convergence (see Figure 8.11).

The PSSM that is used in the alignment is derived from the inter-atomic distances. If d_{ij} is the Euclidean distance between residues i in the first protein and j in the second protein, then their similarity is defined as:

$$s_{ij} = score(i,j) = \frac{M}{1 + (d_{ij}/d_0)^2}$$

where M is the maximal similarity score that can be chosen at will (set by default to 20). The parameter d_0 is the distance where the similarity score falls to half of the maximal possible score and is set to $\sqrt{5} = 2.24$ Å, which is an intermediate choice between the average distance of adjacent α carbons along the amino acid chain (3.8 Å) and the length of a carbon-carbon (C-C) bond (1.54 Å). This transformation into similarity scores assigns higher similarity values to smaller distances. Gaps are assigned a fixed penalty that is equal to $M/2$.

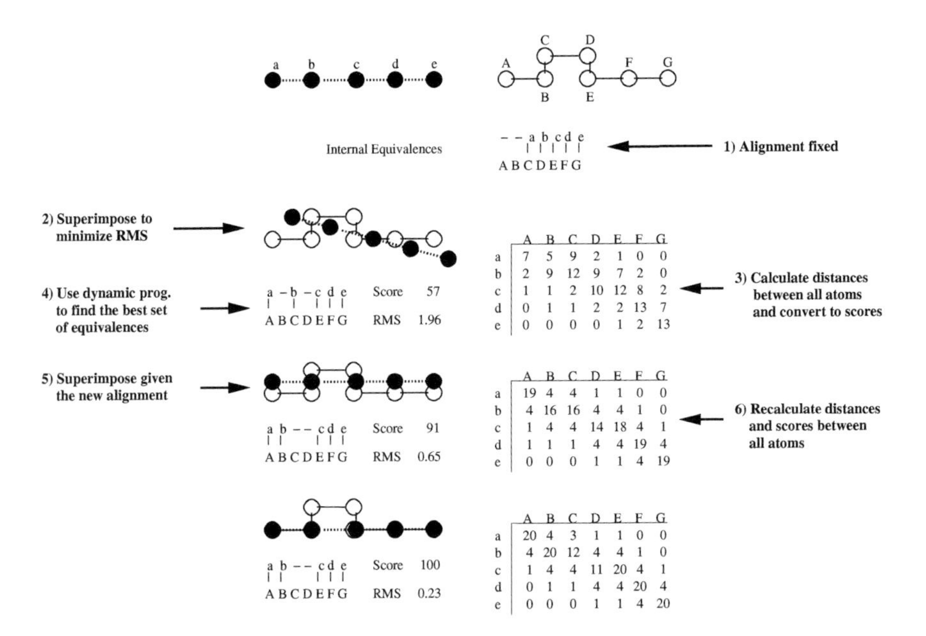

FIGURE 8.11: **The structal algorithm.**

Structal applies dynamic programming with these similarity scores to determine a new set of equivalences. The algorithm repeats the three steps (superposition, PSSM computation and alignment) until convergence, that is, when the set of equivalent residues has not changed. This whole procedure is repeated several times, each time starting with a different initial alignment, including sequence-based alignment and alignment based on torsion angles (see Problem 8.7). The highest scoring alignment at the end of the iterative process is selected as the final alignment. The significance of the alignment is estimated the same way sequence matches are estimated in BLAST, based on the extreme-value distribution (see Section 3.4.3.2).

To improve the quality of the alignment, the scoring function and the gap penalties can be modified considering secondary structures in the positions compared. For example, if the residues compared are in regular secondary structures such as helices and especially strands, then the scoring function and the gap penalties can be multiplied by a factor, to decrease the likelihood of gaps in these positions and minimize the possibility of mismatches. Similarly, internal residues can be assigned a higher weight than exposed residues, under the assumption that these residues have less freedom to move and change their position than residues on the surface. Given the position-dependent weights $w(i)$, the score of matching residues i in protein **A** and j in protein **B** is defined as

$$s'_{ij} = s_{ij} w_A(i) w_B(j)$$

Note that the similarity function of structal assigns only positive values to matches. As we already mentioned before (Section 3.2.3.1), this kind of function is not suitable if we want to detect local matches and can only be used to generate global alignments. Structal addresses this issue to some extent by post-processing the alignment and eliminating loosely aligned residues on both ends, to reduce the RMSd. Trimming is halted when one of several ad hoc conditions is met. For example, a pair can be eliminated only if it is at least 3.8 Å apart and as long as the number of pairs eliminated is not more than half of the initial number of pairs.

8.3.1.3 The URMS distance

The RMSd is a popular measure for assessing structural matches, but it has some weaknesses. The RMSd can be overly sensitive to outliers and is highly length dependent, thus promoting shorter alignments since they will naturally tend to minimize the RMSd. An alternative to the RMSd is the the **Unit Vector RMS distance** (URMSd) [338]. Rather than comparing residue positions the URMSd compares the global orientation vectors at corresponding α carbons of two proteins. Each protein is first represented by the set of direction vectors connecting adjacent α carbons along the protein backbone (Figure 8.12a). Since the direction vectors are all approximately of the same length (about 3.8 Å) they are treated as unit vectors. By placing all of the unit vectors at the origin the protein backbone is mapped into the unit sphere (Figure 8.12b). For a given a protein structure **A**, its **unit vector model** is the set

$$\mathbf{U_A} = \left\{u_i^A\right\}_{i=1}^{n-1}$$

where u_i^A are vectors in $\mathbb{R}^3$ such that $u_i^A = (A_{i+1} - A_i)/||A_{i+1} - A_i||$ and A_i is, as before, the x, y, z coordinates of the α carbon or residue i.

The URMS distance between two protein structures **A** and **B** is defined as the minimal RMS distance between their unit vector models, under rotation.

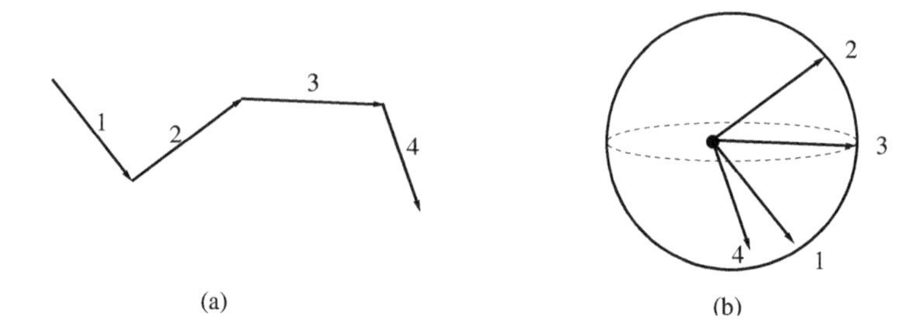

FIGURE 8.12: **The unit vector model.** (a) A sketch of the protein backbone; the unit vector i connects amino acid i to amino acid $i+1$ ($i = 1, 2, 3, 4$). (b) The unit vectors translated to the origin (into the unit sphere). (Figure reproduced from [25] with permission.)

That is,

$$URMSd(\mathbf{A}, \mathbf{B}) = \min_{\mathbf{R}}\{RMSd(\mathbf{U_A}, \mathbf{RU_B})\}$$

Note that the maximal distance between two unit vectors in the unit sphere is 2, and therefore the URMSd has an upper bound that is almost independent of the length of the proteins compared. The rotation that minimizes the URMS distance is called the **URMS rotation**.

8.3.1.4 The URMS-RMS algorithm

The URMSd measure is effective in finding global structural matches and is less sensitive to outliers than the RMSd [338]. However, as with the RMSd, local similarities are harder to detect. Furthermore, with the unit vector models of both proteins centered at the origin, the URMSd measure ignores translations and might produce incorrect alignments that are composed of several local similarities under different translations. This can be solved by considering also translations.

For example, the URMS-RMS algorithm [25] is a transformation-based algorithm that combines the RMS metric and the URMS metric to detect local structural similarities. The algorithm starts by comparing all possible ℓ-length fragment pairs $(\mathbf{a}, \mathbf{b})$ of protein structures $\mathbf{A}$ and $\mathbf{B}$, using the URMS metric. If there is a single rigid transformation under which a substructure of $\mathbf{A}$ is very similar to a substructure of $\mathbf{B}$, then one would expect to find multiple fragment pairs $(\mathbf{a}, \mathbf{b})$ that are in **structural agreement** under this transformation (Figure 8.13). If a fragment pair is in structural agreement under rotation $\mathbf{R}$, then $\mathbf{R}$ is considered a **candidate rotation** for alignment.

During the initial pass over all fragment pairs the optimal URMS rotation for each pair is computed, and the set of candidate rotations is identified. This set is then clustered into groups of similar rotations. The distance between rotation matrices used in the clustering algorithm is defined as the Euclidean distance between their representations as nine dimensional vectors (also called the **Frobenius distance**).

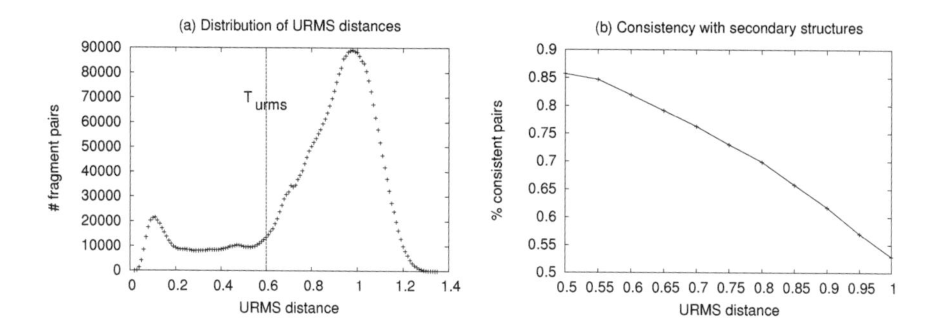

FIGURE 8.13: **(a) Distribution of URMS distances between fragment pairs.** The frequencies are computed over both all-alpha and all-beta proteins for fragments of length eight, the average length of a typical secondary structure element. The observed distribution is a mix of two distributions, one of highly similar fragments (centered at URMSd of 0.1) and the other of dissimilar fragments (centered at URMSd of 1). In the URMS-RMS algorithm, a pair of fragments $\mathbf{a}, \mathbf{b}$ is said to be in **structural agreement** if $URMSd(\mathbf{a}, \mathbf{b}) < T_{urms}$ where T_{urms} is set to 0.6 based on the distribution above. **(b) Correlation of URMS distances with secondary structure consistency.** Fragment pairs that are in structural agreement are strongly correlated with secondary structures. At a threshold of 0.6, about 82% of the fragment pairs are secondary structure consistent (93% if considering only alpha proteins and 70% for beta proteins). (Figure reproduced from [25] with permission.)

Each cluster of similar rotations corresponds to a potential alignment under a specific rigid transformation. A representative rotation $\mathbf{R}_c$ is selected for the cluster[5]. Selecting the representative translation vector is more involved. Each rotation in the cluster is associated with a different fragment pair, and the alignment of each fragment pair results in a different translation vector. Since the RMS metric depends on the absolute coordinates, finding the right translation for each cluster of rotations is crucial for proper alignment. Denote the set of translation vectors associated with the fragment pairs in the cluster by $\mathbf{T}$. Assuming that the optimal translation is in the vicinity of one of the translation vectors in $\mathbf{T}$, the algorithm searches this reduced translation space for the vector that results in an alignment with the highest number

[5]The centroid of the cluster cannot be used, since averaging rotation matrices is unlikely to generate a valid rotation matrix. Instead, the rotation that minimizes the total distance from all other rotation matrices in the cluster is selected. An alternative approach is to compute the centroid under the angles' representation, as discussed later in this section.

of consistent fragment pairs. That is, for each candidate translation vector $T \in \mathbf{T}$, the two complete protein structures are superimposed under the rigid transformation that consists of $\mathbf{R}_c$ and T. All fragment pairs in the cluster that are in structural agreement under this transformation are counted, and the vector T with the highest number of counts is selected as the cluster's translation[6].

For each cluster, the representative rotation and translation $(\mathbf{R}_c, T)$ make up a candidate transformation. The two structures are oriented and superimposed under this transformation, and all inter-atomic distances d_{ij} between the two structures are computed. From these distances the algorithm derives a PSSM, as described below. With the new PSSM, a dynamic programming algorithm is invoked (as in Section 4.3.4.1) in search for an optimal *local* structural alignment. From that point on the algorithm continues iteratively to refine the alignment (as in structal), re-defining the scoring function based on the current alignment and re-aligning the structures based on the scoring function, until convergence. This is repeated for each cluster and its candidate transformation, and all significant matches are reported. Significance is assessed as described in Section 8.3.4.

The URMS-RMS algorithm

Input: a pair of 3D protein structures.

1. **Searching the rotation space:** Identify all pairs of similar fragments using the URMS metric. For each pair of fragments determine the optimal rotation.
2. **Vector quantization:** Cluster rotations into clusters of similar rotations.
3. **Searching the reduced translation space:** For each cluster of similar rotations, identify the most consistent translation among its members. The cluster centroid's rotation and the most consistent translation define a candidate transformation.
4. **Alignment:** For each candidate transformation (rotation and translation), find the best structural alignment using dynamic programming with an RMS-based scoring matrix.
5. **Iterative refinement:** Re-define the scoring function based on the current alignment, and re-align the structures based on the scoring function. Repeat until convergence.
6. **Output:** Report the highest scoring transformations and alignments.

The URMS-RMS algorithm uses a PSSM designed to accommodate local alignments. This is done by converting the distances to scores using a simple transformation of the form

$$s_{ij} = C - d_{ij}$$

where C is called the shift parameter. Gap penalties are modeled using an affine function and are set to a range similar to that of the shift parameter C, with the gap extension being roughly one order of magnitude less than the

[6]An alternative criterion function is the total distance between the atoms in all fragment pairs, under the rigid transformation $(\mathbf{R}_c, T)$. The vector that results in the minimal total distance is selected as the representative translation.

gap opening penalty. The shift parameter and gap penalties are determined through parameter optimization, using similar indices to those described in 8.3.5. Scores can also be modulated using secondary structure information or averaged with sequence-based scoring matrices.

Since the algorithm tests multiple rotations it can detect multiple structural matches, for example when the two structures share more than one domain in common but not necessarily with the same relative orientation between domains, or when the structures are composed of several internally similar or duplicated subunits (see Figure 8.14).

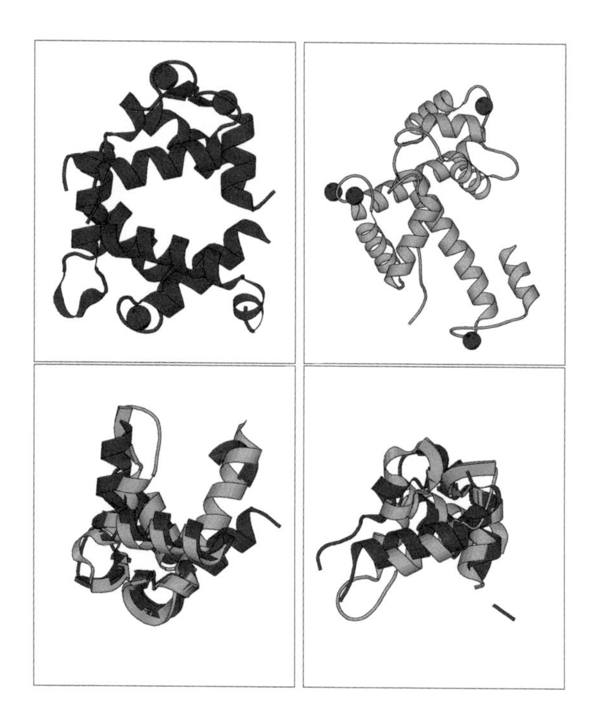

FIGURE 8.14: **Top left:** PDB structure of cardiac troponin C (PDB 1dtl, chain A), a protein that regulates calcium-dependent contraction of the heart muscle. **Top right:** structure of ALG-2 (PDB 1hqv, chain A), a protein involved in apoptosis (programmed cell death) pathways, such as those induced by T-cell receptors as an immune system response to infection. It is dependent on calcium for binding its target proteins. Both proteins are composed of two very similar domains that are glued to each other in different rotations. **Bottom:** Two of the four structural matches found between the above proteins.

Symmetry and determinism: One of the major issues with structure comparison algorithms is their non-deterministic nature. Since the search space is too large to search exhaustively, all algorithms employ some kind of a heuristic to find good alignments, through sampling or directed search that can be highly influenced by the initial configuration (e.g., the initial orientation that is used to align the two structures, as in structal or dali). One of the consequences of this strategy is that results are not symmetric. That is, when comparing structures **A** and **B** the results might differ from those one would get by comparing **B** with **A**. Obviously, one can simply run the algorithm twice, once for **A**,**B** and the other for **B**,**A**, and pick the better of the two results, but at the cost of doubling the computation time. Moreover, if there is an element of randomness in the algorithm then the symmetry is not guaranteed.

However, under some conditions symmetry can be guaranteed. For example, if rotations are clustered using the connected components clustering algorithm (see Section 10.4.1.1) with the Frobenius distance between rotation matrices, then the URMS-RMS algorithm results in exactly the same clusters with the same representative rotations, inverted, when comparing **B** with **A**. This simply follows from the fact that

$$DIST_{frob}(\mathbf{R}_1, \mathbf{R}_2) = DIST_{frob}(\mathbf{R}_1^t, \mathbf{R}_2^t)$$

and that $\mathbf{R}^{-1} = \mathbf{R}^t$ for rotation matrices.

Angles' representation: To cluster the rotations one needs a distance function between rotations. The Frobenius distance between rotation matrices is less than optimal since the high-dimensional space of rotation matrices seems to be sparsely populated even for closely related structures (Figure 8.15). Therefore, it is hard to determine typical within-cluster distances.

An appealing alternative to the matrix representation and the Frobenius distance is the compact **angles' representation** of the rotation space. In the angles' representation, a rotation is given by a three-dimensional vector of its x, y and z rotation angles. Under that representation, the distance between two rotations is defined as the angle-adjusted l_1 distance between their three-dimensional vectors. Specifically, given two angle vectors $\vec{\alpha} = (\alpha_1, \alpha_2, \alpha_3)$ and $\vec{\beta} = (\beta_1, \beta_2, \beta_3)$ their distance is defined as

$$DIST_{angles}(\vec{\alpha}, \vec{\beta}) = \sum_{i=1}^{3} f(|\alpha_i - \beta_i|)$$

where $f(x) = \min(x, 360 - x)$. Clustering rotations using that representation is more efficient when computing distances between rotations as well as when selecting a representative rotation for a cluster, since angles can be averaged (and the cluster centroid can be chosen as the representative).

The angles' representation is not only more compact. It is also insightful when characterizing the geometrical properties of the rotation space, as illustrated in Figure 8.15. The only disadvantage of the angles' representation is that the symmetry that is guaranteed when clustering rotation matrices (see previous section) is not guaranteed with the angles' representation, since for some rotations $DIST_{angles}(R_1, R_2) \neq DIST_{angles}(R_1^{-1}, R_2^{-1})$

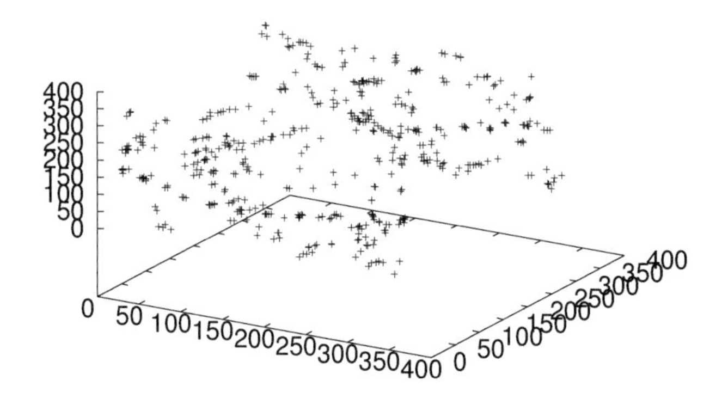

FIGURE 8.15: **The rotation space (angles representation).** All candidate rotations for domain families a.3.1.1 (SCOP:d2ycc mitochondrial cytochrome c) and a.3.1.2 (SCOP:d1qksa1 N-terminal domain of cytochrome cd1-nitrite reductase) are plotted. Although the two structures are closely related, candidate rotations (rotation matrices between fragment pairs that are in structural agreement) tend to sparsely populate the rotation space, making it sometimes difficult to determine the main clusters. Figure reproduced from [25] with permission.

Deformations: The space of all candidate rotations can be quite large. For two proteins of length 100 the number of fragment pairs that are in structural agreement can be on the order of hundreds or even thousands, each pair with its own optimal rotation. If two protein structures are similar then many of the candidate rotations are expected to aggregate into one or a few clusters of similar rotations. In practice the situation is more complex and the set of candidate rotations often forms sparse and overlapping clusters. The problem is especially pronounced when the exact transformation between two structures cannot be formulated by means of translation and rotation alone but requires **deformation** as well. Namely, when the structures are similar under a non-rigid transformation that involves a hinge, characterized by substantial changes in the relative orientation of substructures (e.g., domains) between one structure and the other (see Figure 8.14). Deformations give rise to elongated and relatively large clusters. Together with outliers, these lead to a fairly well connected rotation space.

Clustering such spaces is a challenging task, especially when computational speed is an issue. In [25] several different clustering algorithms were tested (including connected components, pairwise clustering and a grid-like approach); however, even with more advanced clustering algorithms (see Chapter 10) it is possible that the optimal rotation is overlooked. Structure comparison algorithms that attempt to account for deformations as well are described in [339–341].

8.3.2 Distance matrix based algorithms

A different approach to structural alignment is the distance-matrix based approach, which compares internal distances (intra-atomic distances) within structures. Algorithms in that category use 2D distance matrices on α carbons to identify pairs of fragments that have the same distance profiles in the two 3D structures compared [342–345]. Each structure is first represented by a distance matrix **D** where the entry (i, j) stores the C_α-C_α distance between residues i and j. That is, for structure **A**

$$d_{ij}^A = \|A_i - A_j\| = \sqrt{(A_{ix} - A_{jx})^2 + (A_{iy} - A_{jy})^2 + (A_{iz} - A_{jz})^2}$$

and d_{ij}^B is defined similarly. Given the matrices $\mathbf{D}_A$ and $\mathbf{D}_B$, these algorithms look for similar submatrices with similar distance profiles.

The idea to represent an entity through its similarity or distance to other entities (also called the **distance profile**) is actually very intuitive. In social contexts some say “Tell me who your friends are and I'll tell you who you are”. In analogy, when analyzing data the distance profile representation can help to characterize, compare and determine the whereabouts of data instances. The concept of distance profiles has proved to be quite effective in different contexts, for example in clustering [346] or homology detection [112], and we will get back to this topic in Chapter 11.

In the context of structure comparison, structurally similar segments **a**,**b** tend to have similar distance profiles with respect to other residues, and especially residues in the vicinity of the segments (see Figure 8.16). The opposite is also true. That is, similar $l \times l$ sub-matrices $\mathbf{d}_A \subset \mathbf{D}_A$ and $\mathbf{d}_B \subset \mathbf{D}_B$ correspond to segments that have the same geometry and are therefore considered structurally similar segments Hence, by searching for similar sub-matrices we can identify the seeds of a structural alignment.

Comparing distance matrices is computationally intensive and slower than comparison of inter-atomic distances. In fact, the problem of comparing distance matrices has been shown to be NP hard [347]. On the other hand, distance matrices are invariant to rotations and translations, and therefore we do not need to solve complicated equations as in Equation 8.5 when comparing structures. Two concrete examples are described next.

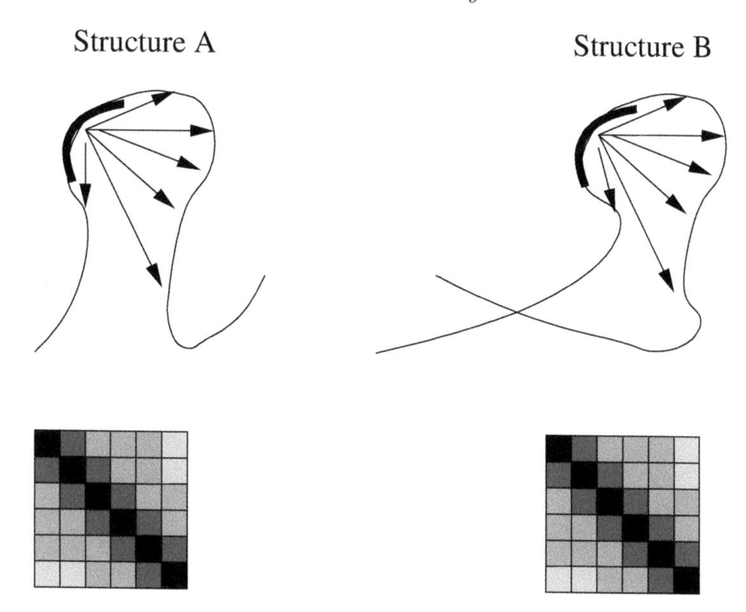

FIGURE 8.16: **Distance profiles and distance matrices**. Representing residues through their distances to other residues. Darker squares in the matrix correspond to smaller distances.

8.3.2.1 Dali

Dali [345] is a distance-matrix based algorithm (Figure 8.17). In dali, the similarity of two $l \times l$ submatrices, which correspond to the segment pair **a** and **b** of length l, is defined as

$$S(\mathbf{a}, \mathbf{b}) = \sum_{i=1}^{l} \sum_{j=1}^{l} s(i, j) \tag{8.6}$$

where the function $s(i, j)$ is the similarity score of two residues based on their distance. Dali uses two types of similarity scores. A rigid similarity score that is defined as

$$s(i, j) = s^R(i, j) = s_0^R - |d_{ij}^A - d_{ij}^B|$$

where s_0^R is a shift parameter that determines the base similarity level. This parameter is set to 1.5 Å, meaning that a pair of residues whose difference in distance is less than 1.5 Åwill be assigned a positive similarity score. Dali also uses an elastic similarity score that is defined as

$$s(i, j) = s^E(i, j) = s_0^E - \frac{|d_{ij}^A - d_{ij}^B|}{d_{ij}^*} w(d_{ij}^*)$$

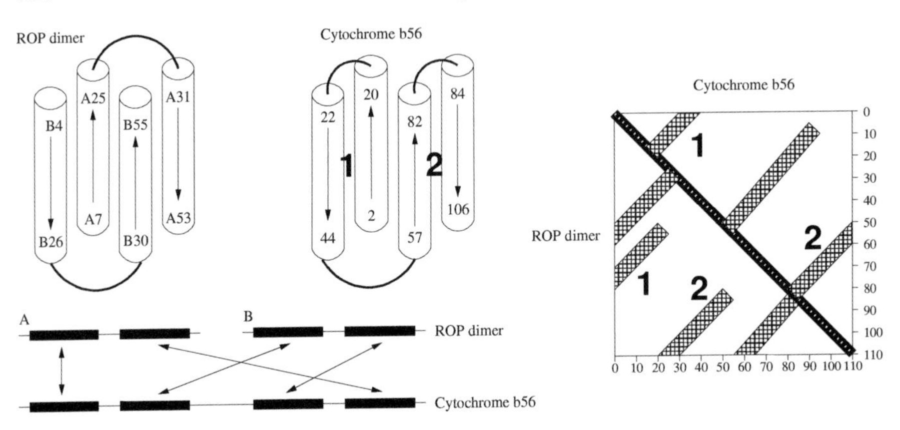

FIGURE 8.17: Dali - comparing distance matrices. (Figure adapted from [345] with permission from Elsevier.) Comparison of structures cytochrome b56 (PDB 256b), a transmembrane protein involved in electron transport across the cell's membrane, and a transcription regulation protein ROP (PDB 1rop), a protein dimer consisting of two chains, A and B, which acts as inhibitor of DNA replication. Although there is no clear functional similarity between the two proteins, their structures contain similar substructures composed of four helices. Note that the order of the helices that make up the similar substructures is different in the two structures, as illustrated schematically at the top (numbers indicate positions along the sequence). The linear representation at the bottom indicates which helices are aligned to each other. The alignment is generated by analyzing the distance matrix on the right. The residues in chain B of ROP are placed after chain A and numbered 56-110. The matrix is divided along the main diagonal so that the top part shows the pairwise distances for residues of 256b, while the bottom half shows the pairwise distances for residues of 1rop. The similar arrangement of helices results in similar distance profiles. For example, the distance profile of the residues that constitute the leftmost two helices in 256b (marked 1) corresponds to the diagonal labeled 1 in the distance matrix. Pairs of residues on this diagonal are structurally close (black dots), e.g., residues 2 and 44. The corresponding diagonal in 1rop is also marked 1.

where $d^*_{ij} = (d^A_{ij} + d^B_{ij})/2$ is the average distance between the two residues and s^E_0 is a shift parameter as before, which allows for some tolerance in distance deviation (set to 20% by default). The function $w(d) = \exp(-d^2/\alpha^2)$ is an envelope function with decay parameter α that decreases the weight of residues that are located far apart, since they play a less significant role in defining structural similarity. The advantage of the elastic score is that it adjusts the score based on the relative difference between distances rather than the absolute value. This seems to work better since variations in the distance between pairs of residues that are far apart tend to be larger even for matching pairs in homologous structures.

Dali defines the similarity of an alignment that is composed of several segment pairs simply as the sum of the scores of the individual segment pairs. To find the optimal alignment, the algorithm searches for a set of segments that would maximize the total similarity score S. This is done by first comparing all segments of fixed length l and computing their similarity. All segment pairs whose similarity score exceeds a certain threshold are considered in the next step, during which the algorithm searches for the best subset of consistent pairs (i.e., non-overlapping) using a Monte Carlo (MC) simulation (Figure 8.18). This algorithm employs a stochastic search that starts with a random combination of pairs and attempts to improve the score by replacing or adding new segment pairs. Recent variations of this algorithm use more efficient heuristics [348].

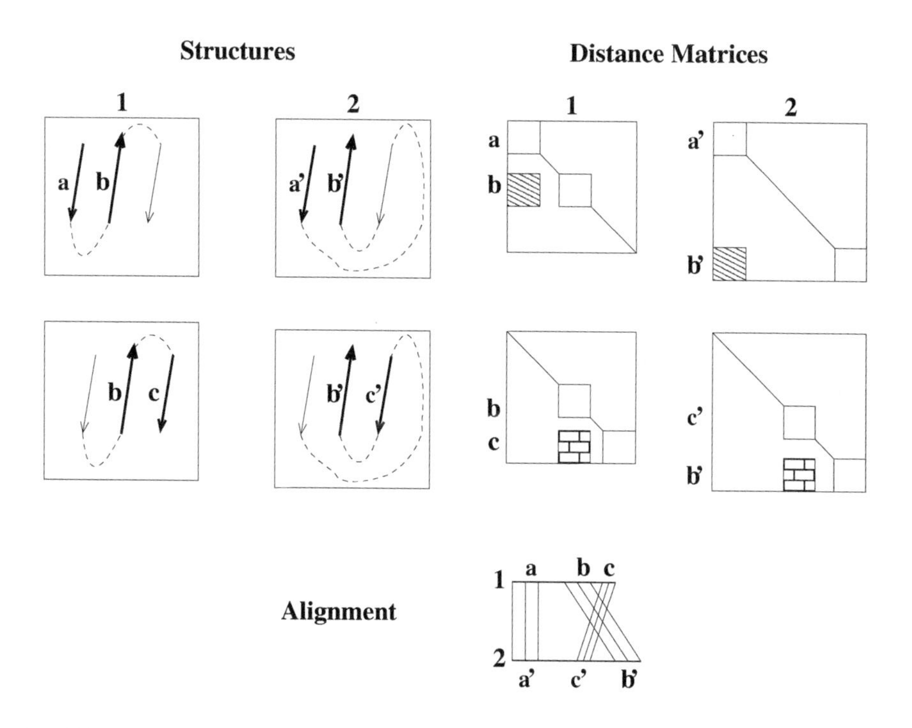

FIGURE 8.18: **Dali - combining similar segment pairs.** (Figure adapted from [345] with permission from Elsevier.) Schematic drawings of two β-sheet structures, their similarity matrices and their alignment. The segment pair a, b has a similar distance profile to that of a', b', represented schematically as submatrices with similar patterns. For clarity, only the bottom half of the distance matrix is shown for each structure, with all entries other than the similar submatrices left blank. Note that the submatrices are at different locations in the distance matrix, because of the different sequence positions of the segments a, b and a', b' (a' and b' are farther apart in sequence than a and b). Likewise, the pair b, c is structurally similar to the pair b', c' as reflected in their distance matrices. The two pairs are consistent (non-overlapping) and are combined into a single alignment as shown at the bottom.

One of the main advantages of this algorithm is that it can detect structural similarities that do not necessarily correspond to consecutive sequence residues, as is demonstrated in Figure 8.17 and Figure 8.18. Note, however, that the search space of similar sub-matrices is actually much larger than the original search space ($O(n^4)$ as opposed to $O(n^2)$) and the algorithm, although quite efficient, is not as fast as some of the other alternatives.

8.3.2.2 CE

The Combinatorial Extension (CE) algorithm [349] is another algorithm based on the distance-matrix approach, with a reduced search space. The algorithm aligns protein structures by incremental extension of alignment paths. It starts by aligning short fragment pairs that are then extended based on a set of rules that determine whether the extension results in a better alignment.

Similar to dali, an Alignment Fragment Pair (AFP) is defined as a continuous segment of protein **A** aligned against a continuous segment of protein **B**, without gaps. An alignment is defined as a path of AFPs. For every two consecutive AFPs, gaps may be inserted into either **A** or **B**, but not into both. That is, if p_i^A is the starting position of AFP i of length l in protein **A**, then for every two consecutive AFPs i and $i+1$ one of the following must hold:

$$p_{i+1}^A = p_i^A + l \qquad \text{and} \qquad p_{i+1}^B = p_i^B + l$$

meaning that there is no gap between the two fragments, or

$$p_{i+1}^A = p_i^A + l \qquad \text{and} \qquad p_{i+1}^B > p_i^B + l$$

meaning there is a gap between the two fragments in structure **B**, or

$$p_{i+1}^A > p_i^A + l \qquad \text{and} \qquad p_{i+1}^B = p_i^B + l$$

meaning there is a gap between the two fragments in structure **A**.

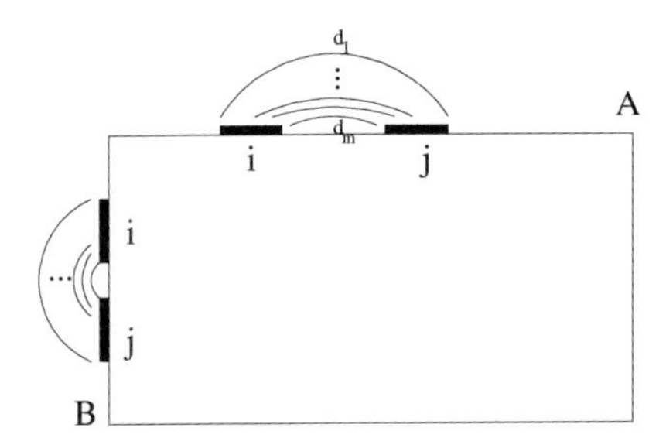

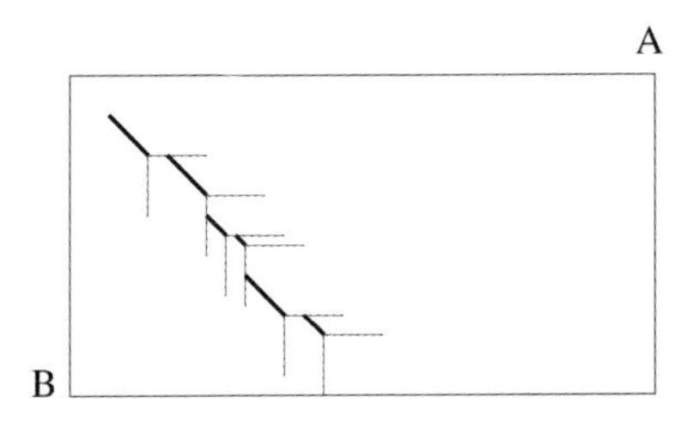

FIGURE 8.19: The combinatorial extension (CE) algorithm. Left: structural agreement between two fragment pairs is defined based on differences in pairwise distances. Right: extensions are examined along the dotted lines.

After selecting an initial "good" AFP at random the algorithm proceeds by incrementally adding good AFPs (as defined below) that satisfy the conditions on paths listed above, until the proteins are completely matched, or until no good AFPs remain. The search space is defined by the parameter M that determines the maximal distance (gap) between consecutive segments, such that an incremental extension is searched along the dotted lines as in the right side of Figure 8.19.

In CE good AFPs are defined as fragments that are in structural agreement. To determine whether the two structures **A** and **B** "agree" on a pair of AFPs i and j of length m, the algorithm associates a distance measure with the two segments, defined as

$$D_{ij} = \frac{1}{m} \sum_{k=1}^{m} \left| d^A_{p^A_i+k-1, p^A_j+m-k} - d^B_{p^B_i+k-1, p^B_j+m-k} \right|$$

where $d^A_{p,q}$ represents, as before, the distance between the α carbon atoms at positions p and q in protein **A**. That is, the distance is defined as the sum of differences between corresponding pairwise distances between the fragments, starting from the pair of residues that are farthest apart in sequence and ending with the nearest pair (Figure 8.19, left). This reduces the computation time, since it does not consider all pairs as in Equation 8.6. If the difference in distances is small, then the segment i can be considered as a good extension to segment j, or vice versa[7].

Before adding a segment to an existing alignment, the algorithm requires that it must satisfy three conditions. That is, if the alignment already contains $n-1$ AFPs and we consider adding the n-th AFP, we do so only if

$$(1)\ D_{nn} < T_0 \qquad (2)\ \frac{1}{n-1} \sum_{i=1}^{n-1} D_{in} < T_1 \qquad (3)\ \frac{1}{n^2} \sum_{i=1}^{n} \sum_{j=1}^{n} D_{ij} < T_1$$

The first condition guarantees that the self-distances within the fragment n are consistent (within the threshold T_0) between the two structures. The second condition checks whether the two structures agree on the distances between the n-th fragment and the first $n-1$ segments and requires that the average difference is below a certain threshold T_1. The last condition verifies that the average difference of all pairs of segments is still below the threshold T_1.

To assess the significance of the alignment, it is compared to the alignments of random pairs of structures, and a zscore is computed based on the RMSd and the number of gaps in the final alignment. This algorithm is faster than dali. However, note that it limits the search space as well as the type of alignments that can be constructed to sequential alignments.

[7] The criterion for the initial AFP is slightly different and is based on the difference in all pairwise distances *within* the fragments.

8.3.3 Geometric hashing

Hashing is a popular technique for fast indexing and retrieval of instances or patterns. It is usually applied to datasets sampled from large sample spaces, which cannot be realized or are too expensive (space-wise) to realize comprehensively, and inefficient if stored without indexing.

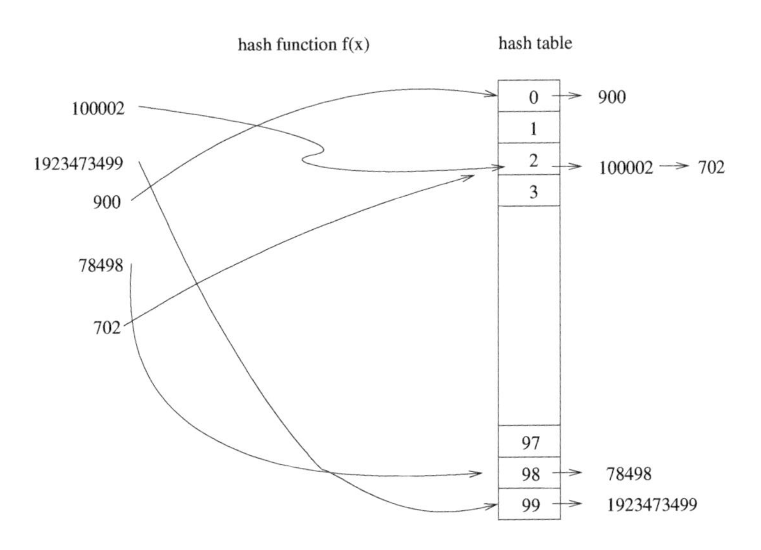

FIGURE 8.20: A simple hash function. Using the hash function $f(x) = \text{int}(x \bmod 100)$ to index data instances. Instances 702 and 10002 are mapped to the same array entry.

A hash function maps every instance in the sample space to a relatively small data structure (usually represented as an array), by computing a hash value for the instance (Figure 8.20). For example, a trivial hash function on $\mathbb{R}$is

$$f(x) = \text{int}(x \bmod 100)$$

This function maps every instance $x \in \mathbb{R}$ to an integer in the range [0..99]. Depending on the size of the dataset and the distribution of data instances, some instances might be mapped to the same hash value. To access the individual instances they are usually stored as linked lists with pointers from the array entry that corresponds to their hash value. By storing the instances in an indexed array we allow fast retrieval and comparison of similar entries. The efficiency of this representation depends on the distribution of instances in the sample space and the hash function. A good hash function will minimize the average number of instance clashes per entry. The function is usually defined based on some properties of the data instances, and can be multi-dimensional.

We have already seen an example of hashing when we discussed BLAST. BLAST pre-processes a database to be searched by using a hash table that stores the locations of all possible words of length k. Given a query sequence, BLAST compiles the set of all words of length k that score at least T with a substring of length k in the query (see Section 3.10.2). Based on this list and the hash table, database sequences that hit the list of words in the query sequence are pulled and searched.

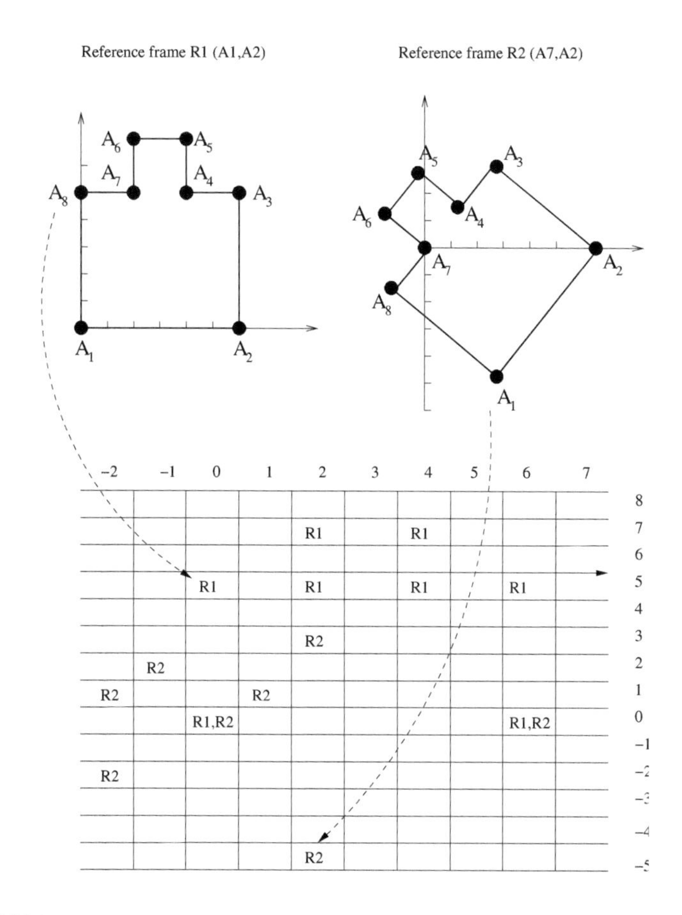

FIGURE 8.21: Geometric hashing: reference frames and the 2D hash table. Two different reference frames of the same structure. The first ($R1$) is defined by the pair (A_1, A_2) while the second ($R2$) is defined by the pair (A_7, A_2). The structure is indexed in a 2D hash table, considering all possible reference frames. In the example above, only two frames are indexed. If a certain (x, y) coordinate is associated with one of the nodes of the structure in reference frame R, then R is indexed in the hash table in the corresponding entry.

The idea behind geometric hashing is similar; by compiling small 3D patterns in the query structure into a hash table and searching other structures for occurrences of the same patterns we can identify similar structures. The key questions are how to define the hash function and how to detect similar patterns.

It is easier to start with geometric hashing in 2D. Assume we have a query structure **A** with n residues and a library structure **B**. The goal is to identify common substructures. This is done by looking for common reference frames in which the positions of many atoms are in agreement. Note that in 2D, every two points (a, b) can define a reference frame, where the origin of the coordinate system is set at point a and the x axis is going through the line (a, b). Given the reference frame (a, b) we can map all other points to its coordinate system (Figure 8.21).

An efficient representation of a point set in 2D is through a hash table, where the 2D plane is discretized and every bin stores the number of times a point in that bin was observed. Hashing helps to summarize efficiently the information on all reference frames of the query. If the structure has a residue at position (x, y) in the reference frame (a, b), then we add (a, b) to the hash table in the bin (x, y). With n residues, there are $n(n-1)$ reference frames and a total of $nn(n-1)$ points recorded in the hash table. For example, given the structure of Figure 8.21, there are 56 reference frames, two of them and their corresponding hash table are plotted.

This hash table is useful for comparing two structures. Before comparison, the coordinates of the target structure are processed in a similar way to the query but using a single reference frame (chosen at will from one of the target coordinates). Given the representation of the target in the chosen reference frame we now check the hash table of the query and determine how many matches are observed between the two structures, for any of the query reference frames. By tallying the number of times each reference frame of the query hits the coordinates of the target we can identify the best reference frame. Obviously, the hashing requires discretization, and the results depend on the resolution of the discretization. But note that there is no need to repeat that for every reference frame in the target.

The geometric hashing method that was introduced in [350, 351] for comparison of protein structures is based on 3D generalization of the technique described above for 2D. To determine a reference frame in 3D we need three points. Therefore the query structure is pre-processed by using every triplet of points as a reference frame and mapping all other points to its coordinate system, and the mappings are summarized in a 3D hash table. To save space and computation time, the structures are represented using only their α carbon atoms, as with other methods.

Given a target molecule and a reference frame, each point that coincides with one of the query points in the hash table votes for all reference frames associated with that bin. A large number of common votes indicates a possible large match between the two structures, and a rigid transformation with respect to the query structure can be computed based on the relative positions of the most common query reference frame and the target reference frame (Problem 8.6). Assuming that the points are indexed in a well-distributed hash table, the running time is $O(n^3)$. It has been suggested that it may be possible to reduce the running time for protein matching applications to a roughly quadratic bound [351].

8.3.4 Statistical significance of structural matches

As we already know, the raw similarity score of an alignment does not necessarily indicate true relationship, even if it "seems" high, since it tends to increase just by chance with the length of the proteins compared (recall our discussion in Section 3.4.3.2). To assess the biological relevance of a match, one needs to know the expected score for a match between two random proteins of the same length. Note that this is different from a random match, where the relative position of the fragments and their orientation is selected at random. Since we seek the best match between the input structures, the randomness should be introduced in the input.

There are two different measures we can use to score a match. The first and obvious one is the RMSd between aligned residues. Most algorithms also report a similarity score that is based on inter or intra-atomic distances and sometimes accounts also for gaps. To assess the significance of a match of length l with similarity score S and RMSd R between two proteins of length m and n, we need to know the probability to get by chance a match with the same score and RMSd for two random proteins of the same lengths. That is, we are interested in computing the probability P

$$P(S, R|m, n) = P(S|m, n)P(R|S, m, n)$$

Characterizing these probabilities theoretically, as was done for sequence matches, is difficult. Protein structures feature different properties than sequences and cannot be viewed as simple strings drawn from a background distribution. Unlike sequences, random structures with protein-like properties are not easy to generate, since it is necessary to consider stereo-chemical constraints to prevent residues from clashing and account for pairwise interactions that stabilize structures. Long-range interactions and physical constraints are hard to model in a simple statistical model, and a random structure is not likely be stable. However, many algorithms for structure comparison reduce the problem to comparison of generalized strings with PSSMs (e.g., structal, URMS-RMS, dali), and the probabilities can be estimated empirically based on the distributions of matches between unrelated proteins. These could be, for example, structure pairs that belong to different SCOP folds and classes [23, 352]. Note, however, that even in a population of supposedly unrelated random pairs one might observe structural similarities due to recurrence of small motifs of secondary structures, and additional pre-processing might be necessary to eliminate such local matches.

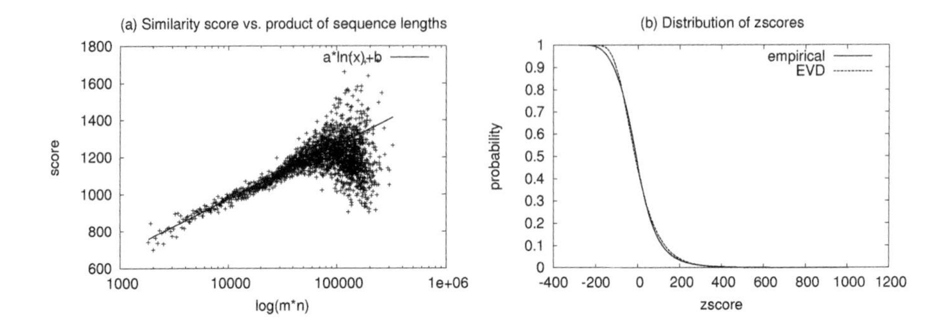

FIGURE 8.22: Statistical properties of URMS-RMS scores between unrelated proteins. (Figure reproduced from [25] with permission.)

The distributions clearly depend on the scoring functions used, but their qualitative properties are often the same. Figure 8.22 shows the distributions that were obtained with the URMS-RMS algorithm. Recall that local sequence similarity scores follow the extreme value distribution. If structures are represented as generalized sequences with PSSM, then their local similarity scores follow the extreme value distribution as well, given that certain conditions on the scoring function are met. As discussed in Section 3.4.3.2, local scoring functions should satisfy two conditions: there should be at least one positive score so that a local match is possible, but the average score should be negative to discourage random extensions of local matches. Structure-based PSSMs, as in Section 8.3.1.4, can be designed to satisfy these conditions. Clearly there are at least a few positive matches (some residues are close to each other); however, when considering all pairs of atoms the average score is negative, since only $O(n)$ positions can be aligned at the most (positive scores), while there are n^2 entries in the matrix. Under these conditions, the score of a random match between a sequence of length m and a sequence of length n has been shown to be centered around $a \cdot \ln(m \cdot n)$ where a is a function of the scoring matrix (see Section 3.4.3.2). Indeed, the distribution of similarity scores as a function of the product of the lengths of the structures compared follows a logarithmic curve, as is shown in Figure 8.22a. The transformation of these scores to zscores, by subtracting the expected average and dividing by the standard deviation, results in a distribution that follows the extreme value distribution (Section 3.4.3.2) as is shown in Figure 8.22b (this is similar to the procedure described in Section 3.4.4). Based on this distribution we can compute $P(S|m,n)$.

Estimating $P(R|S,m,n)$ is more difficult. If the score is linearly dependent on the length l of the match (see Figure 8.23) then we can use the approximation

$$P(R|S,m,n) \approx P(R|l,m,n)$$

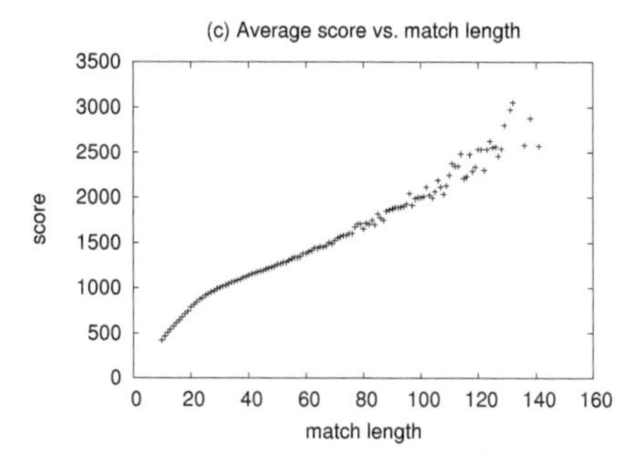

FIGURE 8.23: URMS-RMS score as a function of the match length. (Figure reproduced from [25] with permission.)

and since the RMSd depends only on the length of the match

$$P(R|l, m, n) = P(R|l)$$

The distribution of $P(R/l)$ is characterized in Figure 8.24a. Interestingly, it seems that the RMSd reaches saturation for match lengths around 100 residues, and the functional forms of the average as well as of the standard deviation (Figure 8.24b) are of tangent hyperbolic functions. The transformation to zscores results in a normal distribution, whose cumulative distribution is plotted in Figure 8.25.

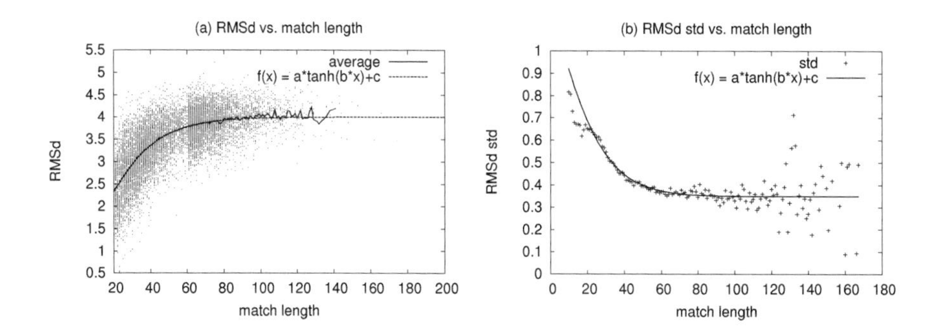

FIGURE 8.24: Statistical properties of the RMSd of matches between unrelated proteins. (Figure reproduced from [25] with permission.)

Using the cumulative distributions for the zscores (Figure 8.24b, Figure 8.25) we can estimate the probability to obtain an alignment with score $> S$ and RMSd $< R$. The joint significance can be approximated by the product of these two probabilities

$$pvalue(S, R|m, n) \approx P(S' > S|m, n)P(R' < R|l)$$

In practice, however, the events are not independent, and these estimates tend to be overly optimistic. Additional research is required to establish better statistical estimates. For further discussion see [353].

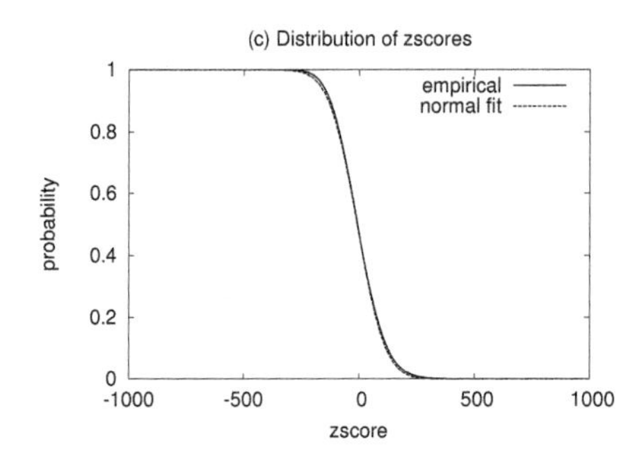

FIGURE 8.25: **Distribution of RMSd zscores.** (Figure reproduced from [25] with permission.)

8.3.5 Evaluation of structure comparison

Structure comparison is a difficult problem. Although there are many algorithms for structure comparison, none of them is guaranteed to detect all structural similarities. How should we compare the performance of different algorithms?

One of the best sources for evaluation of structure comparison algorithms is the SCOP structural classification of proteins (see Section 9.8). Being able to recover this hierarchy with an automatic structure comparison algorithm is an indication of the algorithm's sensitivity and accuracy. Families that belong to the same superfamily in SCOP have similar structures but exhibit only weak sequence similarity. A "good" structure comparison algorithm would detect many family pairs that belong to the same superfamily, with a minimal number of false positives.

To measure the *alignment accuracy* with a certain algorithm we can compute the average RMSd over all pairs of families that belong to the same superfamily. Since the RMSd depends on the length, the average RMSd per residue or a measure of statistical significance (as in Section 8.3.4) would be more informative.

To evaluate the *selectivity* and *sensitivity* of an algorithm we can use indices based on ROC curves, as discussed in Section 3.11. To apply these indices we first need to compare all pairs of structures with the algorithm we are testing. The pairs should then be sorted based on the score output by the algorithm. A ROC curve plots the number of positives vs. the number of negatives in the sorted list of results. The area under the curve can serve as a measure of performance. In the ideal scenario, where all positives are reported before the first negative, the area under the curve is maximized (see Figure 3.14). Usually we are interested in the number of true positives we detected before the first negative. This measure is called ROC-1. To compute this index, we need to compare each structure with all other structures, sort the results based on the score and count the number of positives detected before the first negative is reported. The ratio of positives detected to the total number of positives measures the relative success for that specific query, and this ratio has to be averaged over all queries.

The definition of a negative depends on what we consider to be a true relationship. Considering all structures outside of the superfamily of the query structure to be negatives might be too strict. Structures within the same fold may be significantly similar as well. A more permissive definition then would be to consider all similarities within the same fold to be positives, while considering similarities outside the fold to be negatives.

If the algorithm has parameters with unknown values (for example, the shift parameter in the scoring function or the gap penalties for the dynamic programming algorithm), one can use the same indices to find the optimal set. However, it is important that the dataset used to train the parameters is different from the dataset used for testing, otherwise performance might seem higher than it actually is, since the parameters might have been optimized for the particular makeup of the training set. The best is to divide the original dataset randomly between a training set and a test set. It is important that each subset maintains the properties of the original dataset. For example, dividing SCOP families such that alpha proteins are in the training set while beta proteins are in the test set is not a wise thing to do, because the algorithm would be trained and tuned to be optimal for one type of protein but might perform poorly on the other types. If the datasets available for training and testing are too small one can consider re-sampling techniques (see Section 7.4.2.3).

While SCOP is an excellent resource for such evaluation, one should keep in mind that it is not perfect. SCOP is a domain classification system, and domain definitions can be subjective (see Chapter 9). In some cases SCOP's classes are based on a subtle structural feature rather than on an overall similarity in conformation, and they may contain multi-domain proteins that introduce similarities between families that belong to different classes (see [354]).

8.4 Generalized sequence profiles - integrating secondary structure with sequence information

In the previous sections we discussed algorithms for comparison of protein structures. Detecting structural similarity can greatly enhance homology detection, especially for remote homologs. However, most proteins have an unknown structure, which deems these algorithms inapplicable. While we can try and predict their 3D structure first, this route is not only computationally intensive it is also a questionable one, since it is hard to assess the quality and accuracy of the predictions. Secondary structures, on the other hand, can be predicted efficiently and with relatively high accuracy. Since they hint at the topology of the protein structure, they provide us with structural information that can be utilized when comparing protein sequences to improve function prediction.

Earlier we mentioned that integrating secondary structure into models based on sequence can help detect relationships between remotely related protein families. But how should we integrate secondary structure with primary structure (i.e., sequence) in a single model? In this section we focus on extensions to the profile representation of protein families we discussed in Section 4.3.3, following the method of [137].

One possible approach to inclusion of secondary structures in profiles is a representation over a generalized alphabet that considers all possible pair combinations of amino acids and secondary structure elements. Assuming independence between positions (which is the underlying assumption of PSSMs and profiles, as well as of HMMs, which are discussed in Chapter 6), then this representation implies that for each position i we have a statistical source that "emits" amino acid a and secondary structure s with probability $P_i(a, s)$ such that

$$\sum_a \sum_s P_i(a, s) = 1$$

and assuming a three-state secondary structure model, every position can be represented by a vector of 60 probabilities (parameters) over this pair alphabet.

This representation implies that the amino acid emitted and the secondary structure are considered to be two different features of the objects generated by the source, while in reality the secondary structure is not a "character" or an independent property of the emitted objects, but rather a characteristic of the source itself that is usually unknown (very much like the hidden states of an HMM). Furthermore, the secondary structure introduces constraints on the distribution of amino acids that are emitted by the source. In other words, the secondary structure and the amino acid distribution in a position are strongly dependent on each other. Noting that

$$P_i(a, s) = P_i(a|s)P_i(s) \tag{8.7}$$

we can decompose the parameter space into the parameters of the secondary structure distribution $P_i(s)$, and the parameters of the conditional probability distributions over amino acids $P_i(a|s)$. However, the typical amino acid distributions that are available from multiple alignments of protein families differ from these conditional probability distributions by definition.

More precisely, assume we have a protein family where each protein adopts a certain structural conformation of length n (to simplify, we will ignore for now gaps and insertions). This conformation can be described in terms of the set of 3D coordinates of the n positions, or in terms of the set of distances between coordinates $\mathbf{S} = (\vec{d_1}, \vec{d_2}, ..., \vec{d_n})$ where $\vec{d_i}$ is the set of distances from the i-th residue to all other residues - the latter being more amenable for a representation as a statistical source, since it is invariant to translations and rotations. Although there is structural variation across the different instances of the protein family, it is significantly smaller than the sequence variation, and we will assume that a single **consensus conformation** $\mathbf{S}_c$ reliably describes the protein family[8] The consensus conformation determines the statistical properties of the source distribution in each position and places certain constraints on the sequence space that can be mapped to that conformation. In other words, it induces a probability distribution over the relevant sequence space of $O(20^n)$ sequences that can possibly be members of the protein family

$$P(a_1, a_2, ..., a_n|\mathbf{S}_c) \tag{8.8}$$

Note that due to convergent evolution it is possible that two disconnected regions in the sequence space (two different protein families) will be mapped to the same conformation (although experimental evidence and simulation results suggest that this is not very likely [355], and for most protein families it is reasonable to assume that the sequence space that is mapped to a structural conformation is connected).

The n-dimensional distribution of Equation 8.8 clearly introduces dependencies between remote positions, and the exact probability distribution in a position depends on the amino acids observed in all other positions

$$P(a_i|a_1, a_2, ..., a_{i-1}, a_{i+1}, ..., a_n, \mathbf{S}_c)$$

Accurate knowledge of the all-position probability distribution $P(a_1, a_2, ..., a_n|\mathbf{S}_c)$ would allow one to compare two sources of protein families theoretically by comparing these high-dimensional distributions (using methods such as those discussed in Section 4.3.5). However, because of limited data availability and for mathematical simplicity, the *marginal probabilities*

$$P_i(a|\mathbf{S}_c) = \sum_{a_1,a_2,...,a_{i-1},a_{i+1},...,a_n} P(a_i|a_1, a_2, ..., a_{i-1}, a_{i+1}, ..., a_n, \mathbf{S}_c)$$

[8]Determining the consensus structure of a protein family is closely related to the problem of multiple structural alignment (see 'Further reading').

are usually used in practice to describe the source. Given a multiple alignment of a specific protein family and the corresponding profile, the empirical distribution of amino acids at position i, denoted by $\hat{P}_i(a)$, is essentially the marginal probability of amino acids at that position, as determined by the global conformation, i.e.,

$$\hat{P}_i(a) \simeq P_i(a|\mathbf{S}_c) \tag{8.9}$$

The complete model can be represented as a set of marginal probability distributions, one per position, as we discussed in Section 4.3.3.

So far we have not considered the secondary structures explicitly. Revisiting Equation 8.7 and adding the explicit dependency on the consensus structure, we obtain

$$P_i(a, s|\mathbf{S}_c) = P_i(a|s, \mathbf{S}_c)P_i(s|\mathbf{S}_c) \tag{8.10}$$

The secondary structure sequence **s** is a reduced representation of **S** that, while incomplete, describes quite accurately the topology of the protein. However, given $\mathbf{S}_c$, the knowledge of the secondary structure s at a position does not affect the distribution of amino acids at that position, i.e.,

$$P_i(a|s, \mathbf{S}_c) = P_i(a|\mathbf{S}_c)$$

Nevertheless, the secondary structure information can still be useful when comparing protein families. This is because some information is lost if one is to use just the *marginal* amino acid distributions. For example, the same marginal amino acid distribution can be observed in different secondary structure conformations, and on the other hand, even highly similar fragments of secondary structures can be associated with different amino acid distributions. Furthermore, as mentioned earlier, convergent evolution might result in two *different* sequence sources with structurally similar conformations. These relations are usually perceived weaker than families that are similar both in sequence and structure [23]. Therefore, a proper comparison of protein families should account for both the primary and tertiary structure, or secondary structure if $\mathbf{S}_c$ is unknown.

Plugging in the empirical distribution $\hat{P}_i(a)$ for $P_i(a|\mathbf{S}_c)$ as in Equation 8.9 we get

$$P_i(a, s|\mathbf{S}_c) \sim \hat{P}_i(a)P_i(s|\mathbf{S}_c) \tag{8.11}$$

When the structure $\mathbf{S}_c$ is known then secondary structures can be assigned deterministically (Section 8.2.1.1), and $P_i(s|\mathbf{S}_c)$ assigns probability 1 to one of the structures and zero otherwise. The setup we are concerned with here is the one where we do not know $\mathbf{S}_c$. With predicted secondary structure information, each state is usually assigned a non-zero probability based on the amino acids in that position and neighboring positions. Therefore, we can approximate Equation 8.11 with

$$P_i(a, s|\mathbf{S}_c) \sim \hat{P}_i(a)P_i(s) \tag{8.12}$$

where $P_i(s)$ is the probability of observing a secondary structure s at the i-th position. As explained above, the empirical distribution of amino acids at a position, $\hat{P}_i(a)$, is conditionally independent of the distribution $P_i(s)$. Therefore, to completely describe the source one needs to provide the parameters of the marginal distribution of amino acids, *and* the parameters of the secondary structure distribution. Hence the two distributions are amenable to a representation in which their parameters are appended together in a single parameter vector.

Generalizing the profile or profile-HMM representation accordingly is straightforward. For example, the method of [137] generates a profile for a protein family using PSI-BLAST and then applies PSIPRED (Section 8.2.1.2) to predict the secondary structures using the PSI-BLAST profile as an input. The PSI-BLAST profile is then augmented with the three PSIPRED probabilities to obtain a probability distribution over 23 values (the 20 amino acids plus 3 secondary structures). Note that by doing so, each profile column is now dependent upon and contains information about its neighbors, since PSIPRED uses the profile columns surrounding each column to deduce the probability that the position in question is in a specific secondary structural conformation. This is a key element that enhances the accuracy of this tool in protein family comparisons.

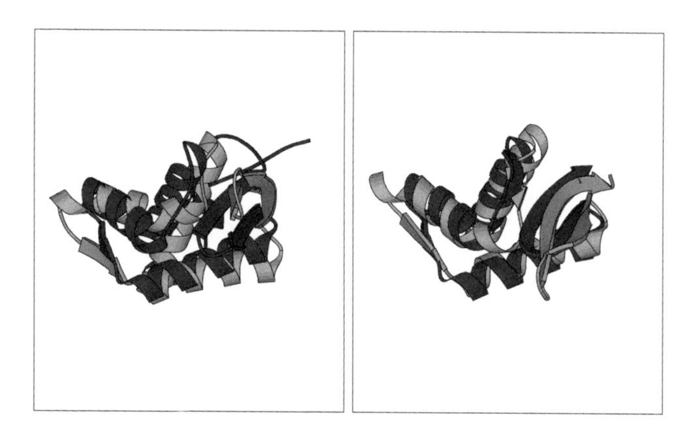

FIGURE 8.26: **Profile-profile alignment with and without secondary structure information.** Comparison of two families from SCOP superfamily a.4.5, which is part of the DNA/RNA-binding 3-helical bundle fold. Although designated as all-alpha, proteins in this superfamily contain a small beta-sheet at the core. The similar substructures have three alpha helices and a couple of beta strands. Profile-profile alignment based on just primary sequence (*prof_sim*) is able to roughly match up the helices but not the beta strands, with RMSd of 11.96 (left). When the true secondary structure is considered (*prof_ss*), the alignment includes the helices as well as most of the strands with a much better RMSd of 4.45 (right).

To use the profile-profile metrics described in Section 4.3.5, the 23-dimensional profile columns have to be normalized to conform with probability distributions. However, a priori it is not clear whether the primary information and the secondary structure information should be weighted equally. To control the impact of the secondary structure information on the representation one can introduce a mixing parameter γ that ranges from 0 to 1 and normalize the secondary structure probabilities such that they sum to γ and the amino acid probabilities such that they sum to $1 - \gamma$. The higher γ is, the more dependent the profile column is upon secondary structure information. This parameter can be optimized using indices similar to those described in Section 8.3.5. Based on such empirical evaluations, it has been suggested that the optimal γ can be as little as 0.05 [137].

Nevertheless, the effect of the additional information can be substantial. For example, in [137] it is demonstrated that profile-profile comparison of profiles augmented with secondary structures is not only more sensitive to remotely related families, but also produces more accurate alignments (see Figure 8.26).

The above discussion touches upon some of the main issues with integrated models of protein families; however, it is by no means complete and serves only as a motivating example. We already know that there is a dependency between positions that is especially pronounced with secondary structures. For example, if a certain residue is in an alpha helix, it is unlikely that the residue on its right side (or its left) is in a beta strand. Rather it is likely to be in an alpha helix as well, or in a coil. While secondary structure prediction algorithms take that into account, considering this kind of dependency directly requires more complicated models. Other ways to represent secondary structure information are discussed in [356–359].

8.5 Further reading

The problem of structure prediction has fascinated many since the late 1960s. This is a grand problem and complete books have been written about this topic including [360–362]. Beyond the application to function prediction, understanding protein folding has also medical applications, as it can help to identify the specific mutations that cause certain proteins to misfold (as happens in Alzheimer and mad-cow disease). It is also beneficial for drug design, as it can suggest ways to interfere with the function of certain proteins and the pathways they are involved in, based on their structure (for example, by designing complementary ligands).

Our improved understanding of the rules of folding resulted in steady progress in the accuracy of ab initio structure prediction. Thanks to rapid advances in computing power, molecular dynamics simulations these days are capable of performing extensive sampling of the energy landscape to find the structure with the lowest energy. Two notable examples of special-purpose supercomputers designed specifically for molecular dynamics simulations are Blue Gene [363] and Anton [364]. However, despite the great strides in protein structure prediction and progress in fold recognition, the success of existing methods is still limited [365]. This is explained, among other reasons, by the existence of other factors that can affect folding, such as the presence of other molecules in the cellular environment of the protein and the involvement of chaperones.

Homology modeling tends to be more successful than ab initio structure prediction, especially if there is a close homolog whose structure is known. Use of low-resolution cryo-EM or NMR data (which is more easily attainable than X-ray diffraction data) can also help to obtain improved models [366].

Threading algorithms were hailed in the 1990s as an appealing alternative to ab initio structure prediction (for reviews see [367,368]). Although they were met with only limited success [369], they are still an active field of research [370–374], with recent applications geared toward detection of novel molecular interactions [375–378].

The problem of structure comparison has also received considerable attention. The early method by [379] is based on a local representation of protein chains by means of discrete analogues of curvature and torsion (a similar approach is described in [380]). Later algorithms extended classical sequence alignment ideas to handle structural alignment. We discussed a few in Section 8.3.1. Other methods employ double dynamic programming (DP) techniques [381–383]. For example, in [383] the first level of DP is applied to create a residue-residue scoring matrix using chemical information on residues, and the secondary structure information is used in the second level DP. In [382], linear representations of secondary structures are derived and their features are compared to identify equivalent elements in the two input structures. The secondary structure alignment then constrains the residue alignment, which compares only residues within aligned secondary structures.

The second category of structure comparison algorithms is based on matching distance matrices. In addition to the algorithms we discussed in Section 8.3.2, this category includes [384, 385]. Other methods that do not fall under either of these categories include the geometric hashing method of [350, 351] which we discussed in Section 8.3.3. Several algorithms perform fast structure comparison by means of reduced representations and transformations [386–391]. Extensive evaluations of structure comparison algorithms are reported in [392, 393].

The problem of multiple structure alignment has been addressed in [394–401]. For example, in [395] the average or median structure is obtained, after rotating all the structures to a single coordinate system based on their alignment with one of the structures.

While all the methods we discussed in this chapter use rigid transformations, in some cases conformational changes might deform the overall backbone structure such that no single rigid transformation can map one structure to the other [402]. Flexible structure comparison algorithms introduce hinges that connect regions that can be matched under rigid transformations [401, 403–406]. For example, the algorithm of [407] uses partial order graphs (see Section 4.6) to align multiple structures with hinges. Among other things, such deformations have been linked with changes in rates and specificity of biochemical activity [408]. For updates and additional references see the book's website at `biozon.org/proteomics/`.

8.6 Conclusions

- Protein sequences fold into three-dimensional structures. The structure determines the intricate way a protein functions.
- The information necessary to form the 3D structure is encoded, for the most part, in the protein sequence.
- Secondary structures can be predicted quite reliably from the sequence.
- Tertiary structure prediction is more complicated and existing methods are only moderately successful. There are other factors at play, such as chaperons, which are involved in the protein folding process and affect the final structure.
- The best ab initio structure prediction methods to date are those that utilize fragment libraries. Homology modeling and threading algorithms are usually more successful when averaging predictions from multiple models.
- Structure comparison of solved structures can help in function prediction, since proteins with similar structures usually have similar functions.
- Methods for structure comparison are generally divided into atom-based and distance-based methods.
- Atom-based methods seek rigid transformations that minimize the RMSd between the atoms of the two superimposed structures.
- Distance-based methods look for similar matrices of intra-atom distances in the two structures compared.
- In some cases structural similarity is evident only under a flexible transformation, involving deformation of the backbone of one of the structures.
- Even in the absence of tertiary structure, integrating predicted secondary structure with primary structure (sequence) can improve homology detection between remote homologies.

8.7 Appendix - minimizing RMSd

To solve the optimization problem of Equation 8.4, the algorithm of Coutsias et al. [337] uses matrix algebra. If we think of the structure $\mathbf{A}$ as a $3 \times n$ matrix $\mathbf{A}$, where column i corresponds to the coordinates A_i of the i-th residue, then we can write

$$\sum_{i=1}^{n} \|A_i - \mathbf{R}B_i\|^2 = Tr((\mathbf{A} - \mathbf{RB})^t(\mathbf{A} - \mathbf{RB})) \tag{8.13}$$

where the trace of a square $n \times n$ matrix $Tr(\mathbf{X})$ is the sum of its diagonal elements $Tr(\mathbf{X}) = \sum_k X_{kk}$ (Problem 8.4). The trace function is defined only for square matrices and is additive $Tr(\mathbf{X} + \mathbf{Y}) = Tr(\mathbf{X}) + Tr(\mathbf{Y})$ and symmetric $Tr(\mathbf{X}^t) = Tr(\mathbf{X})$. Note that the trace of the product of an $m \times n$ matrix $\mathbf{X}$ and $n \times m$ matrix $\mathbf{Y}$ is commutative such that $Tr(\mathbf{XY}) = Tr(\mathbf{YX})$.

Using properties of matrix multiplication and the trace function, we can rewrite Equation 8.13 as

$$\sum_{i=1}^{n} \|A_i - \mathbf{R}B_i\|^2 = Tr(\mathbf{A}^t\mathbf{A}) + Tr((\mathbf{RB})^t(\mathbf{RB})) - 2Tr(\mathbf{A}^t\mathbf{RB}) \tag{8.14}$$

Note that

$$Tr((\mathbf{RB})^t(\mathbf{RB})) = \sum_{i=1}^{n} \|\mathbf{R}B_i\|^2 = \sum_{i=1}^{n} \|B_i\|^2$$

since $\mathbf{R}$ is a rotation matrix and as such it does not skew the structure $\mathbf{B}$. Therefore, the first two terms in Equation 8.14 do not depend on $\mathbf{R}$, and to minimize $\frac{1}{n}\sum_{i=1}^{n} \|A_i - \mathbf{R}B_i\|^2$ we only need to maximize $Tr(\mathbf{A}^t\mathbf{RB})$, or

$$\mathbf{R} = \arg\max Tr(\mathbf{BA}^t\mathbf{R}) = \arg\max Tr(\mathbf{CR}) \tag{8.15}$$

utilizing the commutative property of the trace function. The matrix $\mathbf{C} = \mathbf{BA}^t$ is the 3×3 covariance matrix of $\mathbf{B}$ and $\mathbf{A}$, where

$$C_{mn} = \sum_{i=1}^{n} B_{im}A_{in}$$

This matrix can be decomposed using Singular Value Decomposition (SVD) into the product of three matrices (see Section 11.2.1.2) $\mathbf{C} = \mathbf{UWV}^t$ where $\mathbf{U}$ and $\mathbf{V}$ are matrices whose rows are the eigenvectors of $\mathbf{C}$ and $\mathbf{W}$ is a diagonal matrix with the eigenvalues of $\mathbf{C}$ along the diagonal, in decreasing value. That is

$$\mathbf{W} = \begin{pmatrix} \lambda_1 & 0 & 0 \\ 0 & \lambda_2 & 0 \\ 0 & 0 & \lambda_3 \end{pmatrix}$$

where $\lambda_1 \geq \lambda_2 \geq \lambda_3$. Plugging in $\mathbf{C} = \mathbf{UWV}^t$ in Equation 8.15 we get

$$\mathbf{R} = \arg\max Tr(\mathbf{UWV}^t\mathbf{R}) = \arg\max Tr(\mathbf{WV}^t\mathbf{RU})$$

using again the commutative property of the trace. Denoting $\mathbf{D} = \mathbf{V}^t\mathbf{RU}$ we get

$$\mathbf{R} = \arg\max Tr(\mathbf{WD}) = \arg\max \sum_{k=1}^{3} \lambda_k D_{kk} \tag{8.16}$$

Note that $\mathbf{D}$ (which depends on $\mathbf{R}$) is the product of three 3×3 orthonormal matrices, and therefore $\mathbf{D}$ is also a 3×3 orthonormal matrix and its determinant is $det(\mathbf{D}) = 1$ or $det(\mathbf{D}) = -1$. Hence, $-1 \leq D_{mn} \leq 1 \quad \forall m, n$, and it follows that the maximal value of $\sum_{k=1}^{3} \lambda_k D_{kk}$ is obtained when $D_{kk} = 1 \ \forall k$. Since $\mathbf{D}$ is orthonormal we conclude that in this case all other components of $\mathbf{D}$ are zero and hence $\mathbf{D} = \mathbf{I}$, the identity matrix. That is,

$$\mathbf{D} = \mathbf{V}^t\mathbf{RU} = \mathbf{I}$$

and multiplying both sides of the equation by the orthonormal matrix $\mathbf{V}$ on the left and the orthonormal matrix $\mathbf{U}^t$ on the right we get

$$\mathbf{VV}^t\mathbf{RUU}^t = \mathbf{VU}^t$$

$$\mathbf{IRI} = \mathbf{VU}^t$$

and we have a solution for the rotation matrix

$$\mathbf{R} = \mathbf{VU}^t$$

that is orthonormal since it is the product of two orthonormal matrices (the eigenvectors of the covariance matrix $\mathbf{C}$).

We showed how to compute $\mathbf{R}$, which minimizes the RMSd. However, we need to make sure that the matrix represents a proper rotation. What if it turns out that $det(\mathbf{R}) = -1$? In this case the matrix $\mathbf{R}$ is not a proper rotation matrix, as it also inverts the structure around one of the axes to create a mirror reflection around that axis (Problem 8.2d). To resolve this problem we can go with the second best solution to Equation 8.16. This solution is obtained when

$$\mathbf{D} = \begin{pmatrix} 1 & 0 & 0 \\ 0 & 1 & 0 \\ 0 & 0 & -1 \end{pmatrix}$$

and repeating the last few steps with this matrix we get that the solution for the rotation matrix is

$$\mathbf{R} = \mathbf{V} \begin{pmatrix} 1 & 0 & 0 \\ 0 & 1 & 0 \\ 0 & 0 & -1 \end{pmatrix} \mathbf{U}^t$$

8.8 Problems

For updates, additional problems, files and datasets check the book's website at biozon.org/proteomics/.

1. **Protein folding**

   Assume a simple 3D lattice model, where each alpha carbon atom of a protein chain can occupy a discrete position in the lattice. Show that the number of possible configurations for a protein of length n increases as 5^n (use induction).

   The actual number of possible conformations is smaller and increases as $\sim 3.99^n$. Explain why is it smaller. How many conformations are there for a chain of length 300?

   Now assume an HP lattice model, where the chain is composed only of hydrophobic and polar residues. How many possible sequences in theory can occupy each configuration? Explain why the actual number is much smaller.

2. **Rotation matrices**

   (a) Show that a rotation matrix in two dimensions that rotates a vector $\vec{x}$ counterclockwise in an angle θ is given by

   $$\begin{pmatrix} \cos\theta & -\sin\theta \\ \sin\theta & \cos\theta \end{pmatrix}$$

   Note that this is the opposite to rotating a coordinate system counterclockwise by an angle θ.

   (b) Show that the product of a rotation matrix and its reverse rotation results in the identity matrix:

   $$\mathbf{R}_\theta \mathbf{R}_{-\theta} = \mathbf{I}$$

   (c) Show that $\mathbf{R}_\theta^t = \mathbf{R}_\theta^{-1}$ (hence $\mathbf{R}_\theta \mathbf{R}_\theta^t = \mathbf{I}$).

   (d) Show that the determinant of a rotation matrix $det(\mathbf{R})$ equals 1 for proper rotation and equals -1 for rotations that involve inversions (reflections).

3. **Rigid transformation under exact match**

   You are given two *identical* structures whose coordinates differ as a result of a linear translation T and a rotation $\mathbf{R}$. Outline a procedure to determine the translation vector and the rotation matrix from the coordinates of the two structures.

4. **Minimizing RMSd**

   Verify the equality in Equation 8.13

$$\sum_{i=1}^{n} \|A_i - \mathbf{R}B_i\|^2 = Tr((\mathbf{A} - \mathbf{RB})^t(\mathbf{A} - \mathbf{RB}))$$

5. **The RMS distance**

   Write a program to compute the RMSd between two structures under a fixed (given) alignment. Input files and formats are available at the book's website.

6. **Geometric hashing**

   You are given a query structure and a library structure (see book's website for files).

   (a) Create a 3D hash table and index all possible reference frames in the query structure.

   (b) Select one of the reference frames in the library structure and assess its match with each of the reference frames in the query structure.

   (c) Compute the rigid transformation from the coordinates of the matched reference frames, using the procedure of Problem 8.3.

   (d) Calculate the RMSd between the two structures per Equation 8.2, after applying the transformation you computed.

7. **Comparing torsion angles**

   One way to compare structures would be by comparing the ξ and ϕ angles between consecutive residues (see Section 8.2.1.1). Since the ξ and ϕ angles do not depend on the coordinate system, this method is invariant to rotations and translations.

   (a) Assume each protein is given as a sequence of torsion angle pairs $(\xi_1, \phi_1), (\xi_2, \phi_2), ..., (\xi_n, \phi_n)$. Design a scoring matrix for comparing angles. Consider carefully the cyclic nature of angles.

   (b) Design an algorithm for local comparison of two protein structures given that representation. Do you need to introduce a threshold on the maximal deviation in torsion angles? If so, why?

   (c) What is the time complexity of this algorithm?

   (d) Write a program that compares the torsion angle distance to the RMS distance.

   (e) Apply the program to compare the two pairs of structures given in the book's webpage. Why is the algorithm effective only on small and highly similar structures?

Chapter 9

Protein Domains

9.1 Introduction

Proteins can vary in length greatly. They can be as short as several amino acids (short protein sequences are usually called **peptides**) or as long as several thousands amino acids. The longest protein known as of today is a heart muscle protein called Titin (Biozon[1] ID 26926-1), which is more than 30,000 amino acids long.

Theoretically, if all possible combinations of amino acids into strings could form functional proteins then the protein space would consist of more than $20^{30,000}$ proteins. But not every sequence of amino acids can fold into a stable three-dimensional (3D) structure, and even if a sequence can fold in vivo, it might not form a protein with a useful function. Therefore, not all possible sequences are bound to exist, and the potential number of viable proteins is much smaller. In practice, the number of known proteins and the number of proteins that are expected to exist (proteins that are encoded in the DNA of all organisms on earth) are many magnitudes of order smaller, and among these proteins the redundancy is very high since many proteins are similar to each other.

It is unreasonable to assume that all the other possible strings were tested and failed to create functional proteins. More reasonable is to assume that evolution has explored only a small fraction of the space of all possible proteins, starting from a set of elemental building blocks that were proved to work "beneficially". These building blocks, referred to as **domains**, are the subject of this chapter.

The existence of domains in proteins was observed already in the 1970s. As data on protein sequences and structures started to accumulate, researchers observed that certain patterns tend to appear in multiple proteins, either as sequence motifs or as structural substructures [409,410]. Clearly, domains did not appear at once, and their formation was the result of a long evolutionary process starting from shorter elements.

[1] To view the profile page of a protein with Biozon ID x follow the link `www.biozon.org/db/id/`x

The common perception is that amino acids evolved gradually into the vocabulary that exists today, with valine being the first one in some primitive form (indeed, valine serves as the start codon in some lower organisms [411]). Once the mechanisms to form bonds evolved (i.e., the peptide bond) the amino acids started to bind to each other, gradually forming peptides and longer molecules. Molecules that were stable structurally and acquired a function that proved beneficial to the host organism were more likely to survive natural selection and eventually were passed on to successor organisms. These molecules are generally regarded as domains[2]. Current studies estimate the number of these ancient domains to be around several thousands or a few tens of thousands. They are of different lengths, and during evolution they have changed much by insertions, deletions and mutations.

With time, duplication and transposition events resulted in combinations of these building blocks that were capable of folding and forming new **multi-domain** proteins, with more complex functions. One such protein, a cell-surface protein that is part of the immune system, is depicted in Figure 9.1. We will see more examples throughout this chapter. While many of the proteins known today are single-domain proteins, it has been estimated that the majority of proteins are multi-domain proteins [413, 414]. The number of domains in multi-domain proteins ranges from two to several hundred. However, most proteins contain only a few domains (see Section 9.4.1.1). Consistent with the evolutionary aspect of the domain hypothesis, more complex organisms have a higher number of multi-domain proteins [414–416].

A protein can contain a single domain or multiple domains, some of which may appear more than once. It is the domain combination that determines the function of a protein, its subcellular localization, the cellular pathways in which it is involved and the molecules it interacts with. Therefore, domain characterization is crucial for functional and structural analysis of proteins, and one of the first steps in analyzing proteins is to detect their constituent domains or their **domain structure**. Specifically, understanding the domain structure of proteins is important for several reasons:

- **Functional analysis of proteins.** Each domain typically has a specific function, and to decipher the function of a protein it is necessary first to determine its domains and characterize their functions. Since domains are recurring patterns, assigning a function to a domain family can shed light on the function of the many proteins that contain this domain, which makes the task of automated function prediction feasible. This is especially important in light of the massive sequence data that is generated these days.
- **Analysis of protein interactions.** Some domains have the ability to interact with other large molecules. Certain ligands bind between

[2]It is hard to pinpoint when exactly domains were formed. More precise definitions will be provided in the next section.

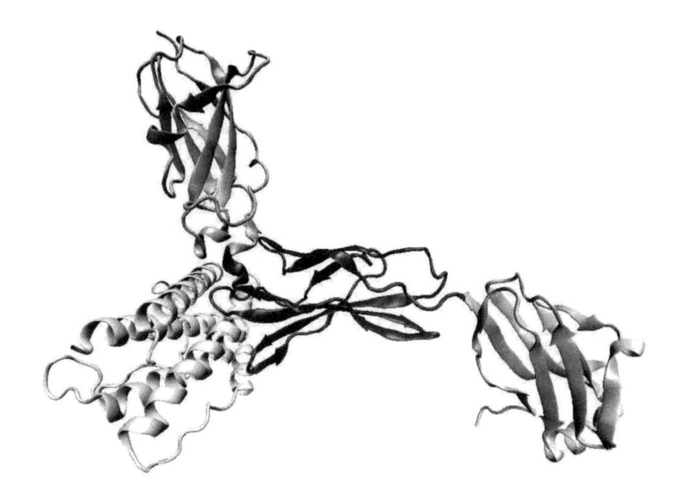

FIGURE 9.1: **Domain structure of a cytokine/receptor complex.** (Figure reproduced from [412] with permission.) 3D rendering of a cytokine/receptor complex solved by X-ray diffraction (PDB 1i1r). This complex is composed of two protein chains. Chain B is an all alpha helical bundle of an interleukin-6 cytokine protein (light gray, on the lower left). It is secreted by certain T-cells of the immune system, in response to tissue damage (e.g., a burn). Chain A is a receptor consisting of three domains (per SCOP definitions) colored dark gray (positions 1-101, lower right), black (positions 102-196, middle) and dark gray (positions 197-302, upper left). The middle domain binds to the extracellular cytokine. Upon binding the receptor is "activated", triggering a chain reaction to transfer the signal inside other cells of the immune system (such a cascade is called **signal transduction**, see Chapter 14).

domains. Determining the domain structure of proteins can help to identify potential interacting partners and ligands (see Chapter 13).

- **Structural analysis of proteins.** Determining the 3D structure of large proteins using NMR or X-ray crystallography is a difficult task due to problems with expression, solubility, stability and more. If a protein can be chopped into relatively independent units that retain their original shape (domains) then structure determination is likely to be more successful. Indeed, protein domain prediction is central to the structural genomics initiative, whose long-term goal is to solve the structures of all proteins. Domain detection is also important for structure prediction. Structure prediction algorithms (Section 8.2) are more successful for smaller amino acid chains that correspond to individual domains [330, 417, 418].
- **Protein design.** Knowledge of domains and domain structure can greatly aid protein engineering (the design of new proteins and chimeras).

Much of the knowledge on domains has been organized in domain databases (Section 9.8). These databases have become an important tool in the analysis of newly discovered protein sequences. By representing domains as consensus patterns, profiles and hidden Markov models (HMMs) and searching them against sequence databases, patterns of biological significance can be detected in new protein sequences. In many cases, such discoveries can lead to the assignment of the new sequences to known protein families. These attempts are extremely important in view of the large number of newly sequenced proteins that await analysis.

In the following sections we will discuss methods for domain prediction and analysis of multi-domain proteins. When discussing the modular structure of proteins it is important to make a distinction between motifs and domains. Motifs are typically short sequence signatures. As with domains, they recur in multiple proteins; however, they are usually not considered as "independent" structural units and hence are of less interest here (the problem of motif detection is discussed in Chapter 5).

9.2 Domain detection

Unlike a protein, a domain is a somewhat elusive entity and its definition is subjective. A domain usually combines several secondary structure elements and motifs, not necessarily contiguous, which are packed in a compact globular structure. Over the years several different definitions of domains were suggested, each one focusing on a different aspect of the domain hypothesis [419–421]:

- A domain is a rigid protein unit that can fold independently into a stable 3D structure.
- It forms a specific cluster in 3D space.
- It performs a specific task/function.
- It is a recurring unit that was formed early on in the course of evolution.

While the first two definitions focus on structural aspects, the other definitions do not necessarily entail a structural constraint. Some definitions are more subjective than others. For example, the definition of a cluster in 3D space is dependent on the algorithm used to define the clusters and the parameters of that algorithm, as discussed in the next section. The structural definitions can also result in a domain that is not necessarily a continuous sequence (see Figure 9.16). However, this somewhat contradicts the evolutionary viewpoint that considers domains to be elementary units that were combined to form multiple-domain proteins through duplication events.

Detecting the domain structure of a protein is a challenging problem. To some extent, it can be determined experimentally using proteolysis, the process of protein degradation with proteases [422]. Proteases are cellular enzymes that cleave bonds between amino acids, but they can only access relatively unstructured regions of proteins. By carefully manipulating experimental conditions to make sure that a protein is in native or near native state (not denatured), this method can be applied to chop a protein into fragments that can be considered potential domains. However, experimental verification of domains is time consuming and not necessarily reliable.

Despite the many studies that investigated protein domains over the years and the domain databases that were developed, deciphering the domain structure of a protein remains a non-trivial problem. Structural information can help in detecting the domain structure, and domain delineation is currently best done manually based on the 3D structure of proteins. The SCOP domain classification [23], which is based on expert knowledge, is an excellent example (see Section 9.8). However, structural information is available for a relatively small number of proteins. Therefore, there is a strong interest in detecting the domain structure of a protein directly from the sequence. Unfortunately, there are no obvious signals in the sequence that indicate when one domain ends and another begins (referred to as **transition points** or **domain boundaries**).

There are many methods for protein domain prediction, which can be divided into several categories. In the next subsections we will give a few examples of each. We will discuss in length sequence-based methods for the reasons mentioned above. But a word of caution first. Domain prediction is not only one of the important problems in computational biology, it is also one of the more complicated and difficult ones since there is no precise and consistent definition of domains that is widely accepted. Different definitions lead to different algorithmic problems and different answers. Indeed, predictions of sequence-based methods can vary a great deal from structure-based methods. The goal of this chapter is to discuss the current state of domain prediction, the existing methods and the open problems.

9.2.1 Domain prediction from 3D structure

The first methods that were developed for protein domain detection were based on structural information. All algorithms in this category are based on the same general principle that assumes domains to be compact and separate substructures, with higher density of contacts within the substructures than with their surroundings. The differences lie in the algorithms employed to search for these substructures, most of which are basically clustering algorithms. In Chapter 10 we will discuss clustering algorithms extensively, but for now recall the clustering procedure we described in Section 4.2.2. Early methods by [420, 423] used a similar, bottom-up agglomerative procedure to cluster residues into domains based on their α-carbon distances. Given a 3D structure with n residues, the algorithm works as follows

- **Initialize:** Compute pairwise distances between all α-carbons. Set the number of clusters to k. Initialize each residue in one cluster $\mathbf{C}_i = \{i\} \quad i = 1..n$
- **Loop:**
  - Find the closest pair of distinct clusters, say $\mathbf{C}_i$ and $\mathbf{C}_j$.
  - Merge $\mathbf{C}_i$ and $\mathbf{C}_j$ and decrement k by one.
  - If $k > 1$ go to loop.

The distance between clusters can be defined as the minimal or average distance over all pairs of residues. This simple procedure will result in all residues being eventually in one cluster. However, by traversing the clustering tree and computing different statistics for clusters at each level of the tree (that is, before and after every merger) we can detect compact substructures and report them as potential domains (see Problem 9.1). These statistics can be, for example, the maximal distance, the contact density or the ratio of contacts between and within clusters, as explained next.

The number of **contacts** between two clusters $\mathbf{A}$ and $\mathbf{B}$ is defined as the number of residue pairs that are closer than a certain threshold distance (typically set to 8 Å, which is the average distance between the C_α atoms of two neighboring amino acids considering their side chains). That is,

$$contacts(\mathbf{A}, \mathbf{B}) = \sum_{i \in \mathbf{A} \ j \in \mathbf{B}} \delta_{i,j}$$

where $\delta_{i,j} = 1$ if residues i and j are in contact. When considering a merger between two clusters we can compare the number of contacts within and between clusters and accept or reject the merger based on the density of contacts

$$\frac{contacts(\mathbf{A}, \mathbf{B})}{|\mathbf{A}||\mathbf{B}|}$$

The merger is accepted as long as the density is higher than a certain preset threshold. Alternatively, we can use the following disassociation measure between clusters, as in the normalized cut clustering algorithm (Section 10.4.3)

$$dis(\mathbf{A}, \mathbf{B}) = \frac{contacts(\mathbf{A}, \mathbf{A})}{contacts(\mathbf{A}, \mathbf{V})} + \frac{contacts(\mathbf{B}, \mathbf{B})}{contacts(\mathbf{B}, \mathbf{V})}$$

where $\mathbf{V}$ is the set of all residues in the structure. The higher the value of $dis(\mathbf{A}, \mathbf{B})$ is, the more distinct the clusters are. Note that $dis(\mathbf{A}, \mathbf{B})$ obtains its maximal value of two when there are no contacts between the two clusters. As with the contact density, the strategy is to merge clusters as long as their disassociation measure is below a certain preset threshold. The thresholds can be set by studying the distribution of contact density or disassociation values in domain databases such as SCOP.

Other algorithms use a top-down divisive approach to split a protein into its domains. For example, the method of [424] represents a protein 3D structure as a flow network, with residues as nodes and residue contacts as edges of capacity related to the distance between them. Their algorithm iteratively locates bottlenecks in the flow network using the classical Ford-Fulkerson algorithm [425]. These bottlenecks represent regions of minimum contact in the protein structure, and the protein is partitioned into domains accordingly as long as the domains are longer than a certain minimal sequence length. In Section 10.4.2 of Chapter 10 we will see that minimizing contacts is closely related to maximizing disassociation as defined above.

There are many variations on these two general approaches. Since there is no single objective definition of a cluster in 3D, each algorithm in this category uses a slightly different definition of compact substructures, resulting in different partitions into domains. Methods can also be sensitive to the parameters of the algorithm used to determine the clusters/domains (such as the maximal distance, the threshold contact-density value, etc.). However, to this day the most effective methods are still those that utilize the 3D information to define domain boundaries, as manifested in perhaps the two most important resources on protein domains: SCOP [23] and CATH [332] (see Section 9.8). This is not surprising, since the definition of domains relies most naturally on the 3D structure of a protein.

9.2.2 Domain analysis based on predicted measures of structural stability

The methods of the previous section are of limited applicability, since structural data is available for only a relatively small number of proteins. For proteins of unknown structure we can approach the problem of domain prediction by using structure prediction methods first. However, structure prediction is a hard problem (see Chapter 8). Existing algorithms are computationally intensive, and more often than not they give wrong predictions.

There are other types of predicted 3D information that can be useful for domain prediction. For example, we can search for regions that are potentially structurally flexible. Such regions often correspond to domain boundaries, where the structure is usually exposed and less constrained. We can identify such regions by inspecting multiple sequence alignments (MSAs) of sequences that are related to the query sequence, whose domain we would like to predict. Positions along the MSA with many indels indicate regions where there is a certain level of structural flexibility. The larger the number of insertions and the more prominent the variability in the indel length at a position, the more flexible we would expect the structure to be in that region. To quantify this variability we can define measures such as **indel entropy** based on the distribution of indel lengths as

$$E_{gap} = -\sum_i p_i \log_2 p_i$$

where p_i is the empirical probability of a gap of length i. Regions with high values of E_{gap} might indicate domain boundaries (Figure 9.2).

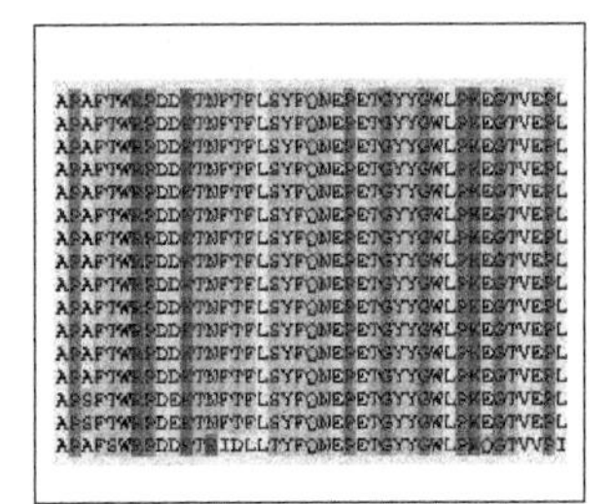
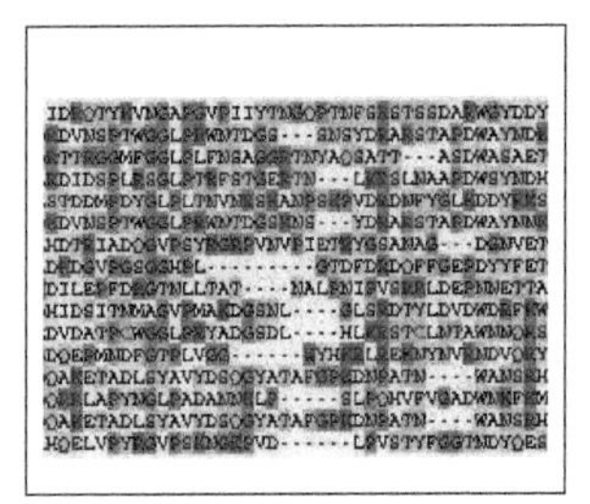

FIGURE 9.2: **Gaps and structural flexibility.** Regions with many insertions and deletions often correspond to loops and are more likely to appear near domain boundaries. (Figure reproduced from [354] with permission.)

The organization of residues into **secondary structures** can also hint at the domain structure of a protein. Most inter-domain regions are composed of flexible loops. On the other hand, beta strands (usually organized in sheets) and alpha helices are relatively rigid units that constitute the core of protein domains, and domain boundaries rarely split these secondary structure elements.

Several algorithms use the secondary structures of a protein to predict domains [426, 427]. The secondary structures can be clustered using similar algorithms to those described in Section 9.2.1 or can be analyzed and converted to scores that increase the likelihood of a domain boundary in coils while decreasing the likelihood in regular secondary structure elements [354]. Other algorithms use the secondary structure propensities as a feature in multivariate model, as discussed in Section 9.3. All these algorithms can feed on either secondary structures determined from the 3D structure of a protein or predicted from its sequence.

Another way to identify regions of structural flexibility is by examining the **contact profile** of a protein [428]. The contact profile depicts the number of contacts between residues along the protein sequence. Points of local minima along this graph correspond to regions where fewer interactions occur *across* the corresponding sequence positions, implying relatively higher structural flexibility and suggesting a domain boundary (see Problem 9.2).

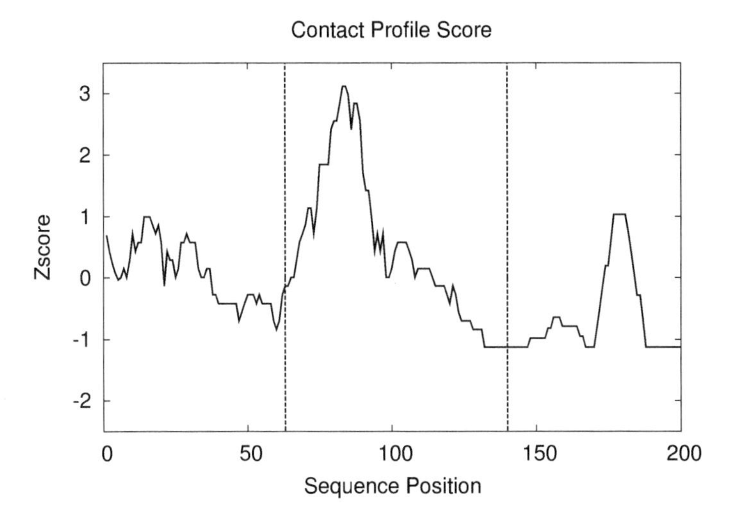

FIGURE 9.3: **Predicted contact profile.** (Figure reproduced from [354] with permission.) A significant correlated mutation score between two positions of an MSA is considered as indication of contact. Significance can be measured by means of a zscore with respect to a background distribution (see text). The contact profile is generated by counting for each sequence position the number of contacts between residues that reside on opposite sides of that position and normalizing that number by the total number of possible contacts. Alternatively, we can sum the zscores associated with all pairs of columns across each position to generate a weighted profile. The weighted profile shown above was generated for PDB 1bvs, a DNA binding protein. Vertical lines mark domain boundaries according to SCOP. Note that the minima in this graph correspond to domain boundaries.

When the 3D structure is unknown we cannot determine which residues are within contact distance from each other, but we can sometimes predict contacts if correlations are observed between the amino acids in certain positions. The underlying assumption is that if two residues are in contact, then a mutation in one is likely to result in a compensatory mutation in the other, to maintain the stability of the structure. For example, if a hydrophobic residue mutates into a positively charged residue, then the other will sometimes mutate into a negatively charged residue. The **correlated mutations** score [429] uses the columns of an MSA to assess whether pairs of positions show signs of adaptive mutations and hence are likely to be in contact. The correlation coefficient for two positions k and l is defined as

$$corr(k,l) = \frac{1}{n^2}\sum_{i=1}^{n}\sum_{j=1}^{n}\frac{(s(a_{ik},a_{jk}) - \bar{s}_k)(s(a_{il},a_{jl}) - \bar{s}_l)}{\sigma_k \cdot \sigma_l}$$

where a_{ik} is the amino acid in position k of sequence i and $s(a,b)$ is the similarity score of amino acids a and b according to a scoring matrix of choice.

The term $\bar{s}_k$ is the average similarity in position k, and σ_k is the standard deviation. Here n is the number of sequences that participate in both columns. Positions with significant signals of correlated mutations are more likely to be physically close to each other and are probably part of the same domain. By considering all pairs of columns across each sequence position and summing their correlated mutations score we can generate a contact profile for a protein as in Figure 9.3.

To predict a contact based on a correlated mutation score it is necessary to have a reliable statistical significance measure that would help to discern true correlations from random coincidental regularities. This can be obtained by computing the correlation scores for a large collection of random alignment columns. Based on the background distribution of scores we can then associate a *zscore* with each correlated mutation score. If the average correlated mutation score for random columns is μ and the standard deviation is σ then the zscore of a correlated mutation score r is defined as $zscore(r) = \frac{r-\mu}{\sigma}$.

Beyond structural stability, correlated mutations provide another piece of evidence for the domain structure of a protein from an evolutionary point of view. Positions that are strongly correlated through evolution imply that the sequence in between must have evolved in a coordinated manner as one piece. As such, the sequence qualifies as a building block and it is less likely to observe a domain boundary in between. However, while conceptually appealing, signals of co-evolution are often too weak to be detected (see discussion in Chapter 13), and the method is computationally intensive for large MSAs.

Another related method is the **correlated divergence** approach. This method tests independence between domains and appraises whether regions to the right and left of a position are part of the same domain or not, based on sequence divergence patterns in the MSA. If the two regions belong to two different and independent domains then the patterns of sequence divergence for each one might be different due to different structural constraints. The divergence pattern can be characterized based on the evolutionary distances between sequences in each region. Imagine that each domain is associated with a different statistical source that generates instances according to some probability distribution over the protein sequence space. Then the set of pairwise distances between instances is a function of the source's distribution, and by comparing these distances one can indirectly measure the statistical similarity of the two sources. In Chapter 13 we will see that a similar approach is used to predict whether two proteins interact.

The evolutionary distances between pairs of sequences in the MSA can be estimated by counting the number of point mutations (normalized, e.g., per 100 amino acids, to eliminate bias due to length differences). To assess to what extent the source distributions are unique we can compute the Pearson correlation between the two divergence patterns over the set of common sequences (that is, sequences that participate in both regions). Zero correlation indicates two unique sources and hence two regions that evolved independently or under independent sets of constraints (for more details, Section 13.3.2.2). As with correlated mutations, this measure is computationally intensive and it is better to apply it to a set of transition points suggested by other methods or to a reduced subset of positions along the sequence (e.g., every x residues) and test the independence assumption over neighboring domains.

9.2.3 Domain prediction based on sequence similarity search

Algorithms for domain prediction that are based only on sequence information are very popular since they are relatively fast (compared to structure-based and structure prediction methods) and provide an appealing solution for large-scale domain prediction of sequence databases.

Sequence information can be used in many ways, one of which is sequence similarity. If a domain indeed existed at some point as an independent unit, then it is likely that traces of the autonomous unit exist in other database sequences, possibly in lower organisms. Thus a database search can sometimes provide us with ample information on the domain structure of a protein. Furthermore, by analyzing homologous sequences we can often determine domain boundaries that apply to almost all instances of the corresponding protein family. Indeed, many algorithms use homologous sequences detected in a database search to predict domains. We will discuss a few techniques next (for additional references see Further reading). It should be noted that all of the methods discussed in this section can be applied to MSAs as well, but throughout this section we will assume that the data is given as pairwise similarities, without having to generate an MSA first.

A simple procedure to chop proteins based on sequence similarity is MKDOM [430], an automatic domain prediction algorithm based on repeated application of the PSI-BLAST algorithm (see Section 4.4.1). The procedure iteratively looks for the shortest repeat-free sequence in the database, which is assumed to represent a single domain. After removing it and all its PSI-BLAST homologs, the algorithm is applied again with the next shortest sequence, until the database is exhausted. This approach is fast, but it tends to truncate domains short of domain boundaries, since the seed sequences do not necessarily correspond to domains [354].

Another approach to domain prediction is to utilize **sequence termination** signals. Sequence termination is a strong indicator of domains in sequence alignments. Since duplication events of incomplete domains were unlikely to fold and function properly, most of them did not withstand evolution. Therefore start and end positions of sequences along alignments can be considered to be potential domain termination signals (see Figure 9.4). Sequence termination has been shown to be one of the main sources of information on domain boundaries [354]. Indeed, this is the basis of methods such as DOMO [431], Domainer [68], DOMAINATION [432], PASS [433] and others.

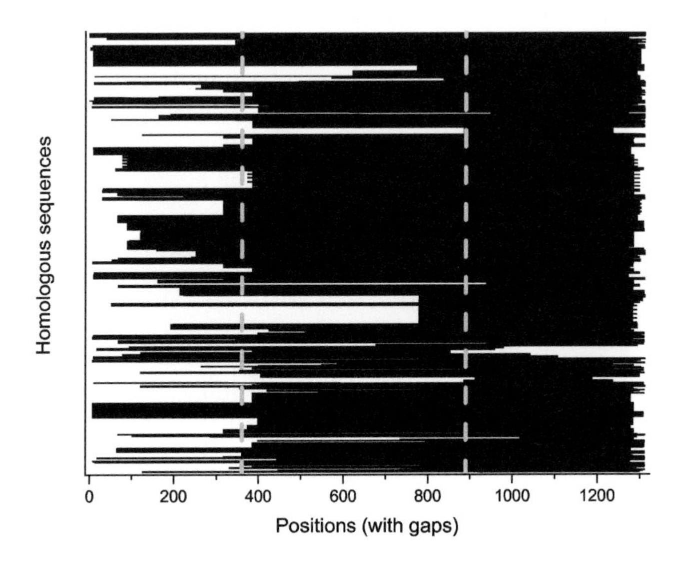

FIGURE 9.4: A schematic representation of homologous sequences to interleukin-6 receptor protein (PDB 1i1r, chain A). The structure of this protein has three domains (see Figure 9.1). Alignments were generated using PSI-BLAST (see [354] for details). Homologous sequences are in order of decreasing evalue. The dashed gray lines mark the positions where SCOP predicts domains, adjusted due to gaps (360 and 890). Note the correlation with sequence termini in the MSA. (Figure reproduced from [412] with permission.)

A simple way to extract plausible domain boundary signals from a database search is to count the number of sequences that start or end at each residue position and generate a histogram. Proteins can then be partitioned at positions with the maximal number of termination signals. A closely related approach is to partition a protein at positions that minimize the number of similar sequences that are aligned to *both* sides of the cut (see Problem 9.4). This approach for domain prediction, which focuses on minimizing contacts, was used in [434].

However, one should be cautious in analyzing database matches in search for such signals. A sequence can contain multiple domains even if there are no sequence termination signals in a database search. For example, pairs of sequence domains may appear in many related sequences, thus hindering the ability to tell the two apart. If the search results in just highly similar sequences (which have the same domain organization) then similarity-based methods are ineffective. On the other hand, sequence termination is not necessarily indicative of a true domain termination. Many sequences in protein databases are fragments that may introduce spurious termination signals that can be misleading. Even if all sequences that are documented as fragments are eliminated from the database, some sequences may still be fragments of longer sequences without being documented as such. Moreover, the termination may be premature since coils at the end of domains are often loosely constrained and tend to diverge more than core domain positions, especially in remotely related sequences. These diverged subsequences may be omitted from the alignment if they decrease the overall similarity score. Mutations, insertions and deletions further blur domain boundaries and make it hard to distinguish a signal from background noise.

For the above reasons, histograms of sequence termination signals tend to be quite noisy, especially for large alignments with many sequences (see Figure 9.5a) and may be misleading if used simple-mindedly. The histogram can be smoothed by averaging over a sliding window (see Figure 9.5b); however, even after smoothing it is still unclear which signals should be considered representative of true domain boundaries.

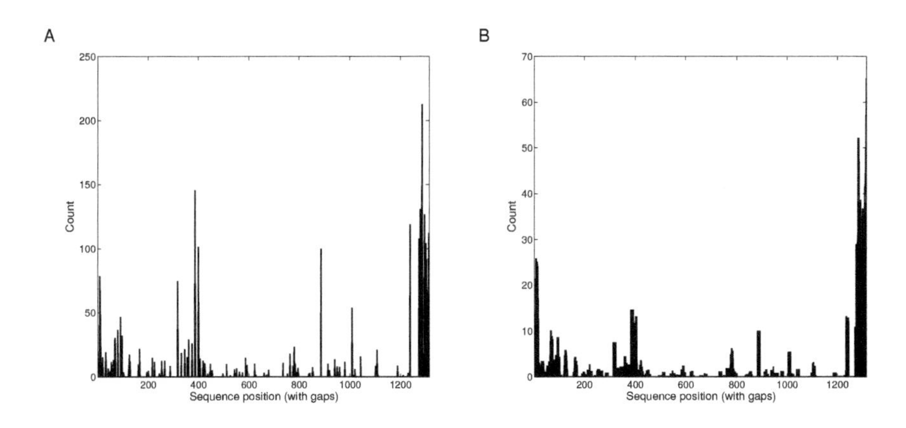

FIGURE 9.5: Analysis of sequence termination signals in sequence alignments. Sequence termination signals for interleukin-6 receptor signaling protein (PDB 1i1r, Figure 9.1) based on the alignments of Figure 9.4. **A:** Original histogram of sequence termini counts. **B:** Smoothed histogram (window size $w = 10$). (Figure reproduced from [412] with permission.)

To extract probable domain boundary signals from noisy alignments it is necessary to assess the statistical significance of each signal by estimating the probability of observing this signal by chance. For simplicity, we assume that the distribution of sequence termination signals for randomly aligned sequences is uniform over all positions (a more accurate analysis would require more complex distributions). That is, the probability of a randomly aligned sequence to start or end at any position is $p = \frac{1}{L}$ where L is the length of the alignment (which is the length of the query sequence or the MSA). To assess the significance of a position with m termination signals, we ask what the probability that m or more sequences terminate at the same position by chance is. Assuming for a moment independence between the sequences[3], we can use the binomial distribution with parameter $p = \frac{1}{L}$. The probability that m or more sequences (out of n sequences) terminate in one position is given by

$$pvalue(m) = P(i \geq m) = \sum_{i=m}^{n} \binom{n}{i} p^i (1-p)^{n-i} \qquad m = 0, 1, \ldots, n \quad (9.1)$$

Applying this procedure to the histogram of Figure 9.5b results in the graph of Figure 9.6, where peaks are suspected as signals of true domain boundaries. The two peaks on both ends of the alignment are due to the assumption of uniformity along the alignment, which does not hold at the start and end of a protein due to the presence of closely similar sequences, but they can be ignored.

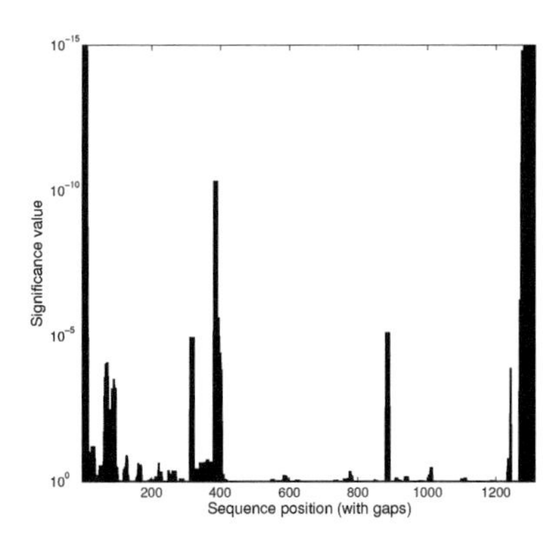

FIGURE 9.6: **Significance of sequence termination signals (PDB 1i1r).** Pvalue (in logscale) of sequence termination signals (from Figure 9.5) computed using the binomial distribution. (Figure reproduced from [412] with permission.)

If the dataset contains highly diverged sequences, then the alignment is likely to include many gaps, and some positions might be very sparsely popu-

[3]Unfortunately, the independence assumption is clearly inaccurate, since the sequences are homologous and hence related to each other. However, one can take certain measures to reduce the dependency between sequences by grouping highly similar sequences and picking one representative of each.

lated. Therefore Equation 9.1 might grossly overestimate the pvalue in positions with low density. To correct for this bias we can adjust p locally based on the number of non-gap characters in each position and its local neighborhood. Define the indicator variable $c_j(i)$ such that $c_j(i) = 1$ if $seq_j(i) \neq$ gap and zero otherwise. Let $\hat{d}$ denote the average density of the alignment

$$\hat{d} = \frac{1}{L}\sum_{i=1}^{L}\sum_{j=1}^{n} c_j(i)$$

and define $d(k)$ as the smoothed density at position k over a window of size w

$$d(k) = \frac{1}{w}\sum_{i=k-w/2}^{k+w/2}\sum_{j=1}^{n} c_j(i)$$

then the probability of a randomly aligned sequence to start or end at position k is estimated as $p_k = d(k)/\hat{d}$. The smoothing improves the statistical estimates; however, it also results in a large number of neighboring positions that could be deemed significant, since isolated peaks of low probability are now spread over their local neighborhood. In each such region the position with the minimal probability can be considered as a candidate domain boundary.

A different heuristic for assessing the significance of sequence termination signals is based on the evalue of each pairwise alignment. The evalue indicates the alignment's significance and also reflects its reliability. For example if a certain position in the query sequence is aligned in n sequences, of which c terminate or originate at that position, and the evalues of the corresponding alignments are $e_1, e_2, ..., e_c$ then we can define the termination evalue score as

$$E_{termination} = -\log(e_1 \cdot e_2 \cdot ... \cdot e_c) = -\sum_i \log e_i$$

where again, to assess the significance, we assume the sequences are independent. The higher the score (the lower the evalue), the more significant the signal is. As before, it is necessary to average these scores over a window surrounding the position, since signals of termination are noisy.

Another type of sequence termination signal comes from a different source. In Chapter 2 we mentioned briefly the mechanism of alternative splicing. This mechanism is used extensively in higher organisms to generate multiple mRNA and protein products from the same DNA strand. By sampling and sometimes shuffling the set of exons encoded in a gene DNA sequence, the cell generates different proteins that are composed of different combinations over the same set of exons (see Figure 2.7).

Several studies tested the hypothesis that **intron-exon** data at the DNA level is correlated with domain boundaries [435, 436]. As building blocks, domains are believed to have evolved independently. Therefore it is likely that each domain has a well-defined set of exons associated with it, and if a protein contains multiple domains, we expect exon boundaries to coincide with domain boundaries.

Intron-exon data is available in databases such as EID [437], although it is often a mix of experimentally determined and predicted exons. By comparing a query sequence with all sequences in such a database and recording all significant ungapped matches we can detect exon-intron boundaries. To quantify the likelihood of an exon boundary we can use similar measures to those suggested above for sequence termination. For example, if a certain position is aligned with n sequences, of which c coincide with exon boundaries and the evalues of the corresponding alignments are $e_1, e_2, ..., e_c$, then we can define the exon termination score as

$$E_{exon} = -\log(e_1 \cdot e_2 \cdot ... \cdot e_c)$$

Alternatively, the signals can be assessed using Equation 9.1.

A different way to utilize similarity relations for domain prediction is by analyzing the all-vs-all similarity matrix, using for example **transitivity tests** [438]. By identifying triplets of proteins A, B, C, where A has no direct link with C but is transitively related to C through an intermediate protein B (Figure 9.7), we can detect the existence of multi-domain proteins and partition them accordingly. This, however, should be treated with caution, as transitivity can also fail when proteins are only remotely related, and to rule out this possibility it is necessary to check whether the similarity of A and C with respect to B overlaps.

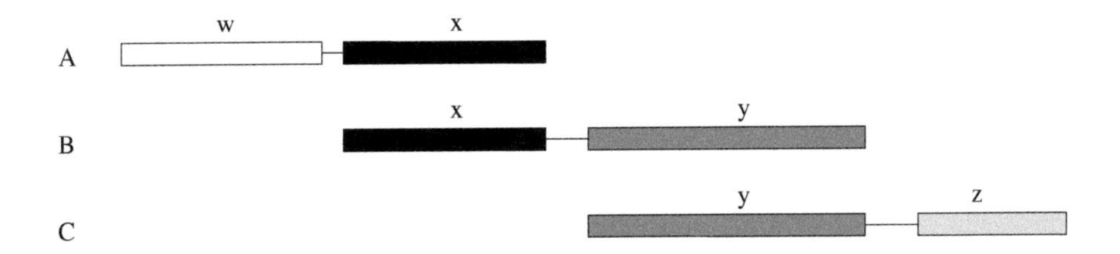

FIGURE 9.7: **Transitive chaining in multi-domain proteins.** In the example above, protein A shares domain x with protein B, and B shares domain y with protein C. Transitivity of similarity relations will link proteins A and C, although they have no domains in common.

Enumerating all possible triplets is computationally intensive. An alternative way to identify these violations on a large scale is by starting with cluster analysis. Clustering algorithms aggregate proteins into groups of similar sequences (we will discuss clustering algorithms in Chapter 10). Clusters can be split into domains if they contain non-overlapping sequences. We can analyze and split clusters based on the histogram and profile of sequence matches, and sequence termination signals as discussed above. There is plenty of additional information we can extract from clusters, and in the next sections we will discuss methods for domain prediction based on MSAs of sequence clusters.

9.2.4 Domain prediction based on multiple sequence alignments

A similarity search provides information on pairwise similarities; however, these similarities are often erroneous, especially in the "twilight zone" of sequence similarity. A natural progression toward more reliable domain detection is to use MSAs. Not only do MSAs align sequences more accurately, but they are also more sensitive and they often detect remote homologs that are missed in pairwise comparisons, when converted into a profile or an HMM and searched against sequence databases.

We already mentioned several MSA-based measures that can hint at the domain structure of a protein, such as the correlated mutations score, which assesses the potential of contacts between different positions, or the indel entropy and the secondary structure propensities that can reveal regions of higher structural flexibility (see Section 9.2.2). The methods of the previous section (such as sequence termination analysis) can be naturally applied to MSAs as well, although MSA is not required as input. But there are many more features we can extract from MSAs. For example, MSAs can expose the core positions along the backbone that are crucial to stabilize the protein structure or play an important functional role in an active site or in an interaction site. These positions tend to be more conserved than others and strongly favor amino acids with similar and very specific physico-chemical properties because of structural and functional constraints.

To detect these positions we can define various **conservation measures** of alignment columns. One such measure is the entropy of the amino acid distribution (Figure 9.8). For a given probability distribution $\mathbf{p}$ over the alphabet $\mathcal{A}$ of 20 amino acids $\mathbf{p} = \{p_1, p_2, ..., p_{20}\}$, the entropy is defined as

$$E_{amino\ acids}(\mathbf{p}) = -\sum_{i=1}^{20} p_i \log_2 p_i$$

This is a measure of the disorder or uncertainty we have about the type of amino acid in each position (see also Section 3.12.5). For a given alignment column, the probability distribution $\mathbf{p}$ can be defined from the empirical counts, but pseudo-counts should be added, as discussed in Section 4.3.3.2, to avoid zero probabilities due to undersampling.

Some positions in MSAs have a preference for a *class* of amino acids, rather than for a single amino acid, all of which have similar physico-chemical prop-

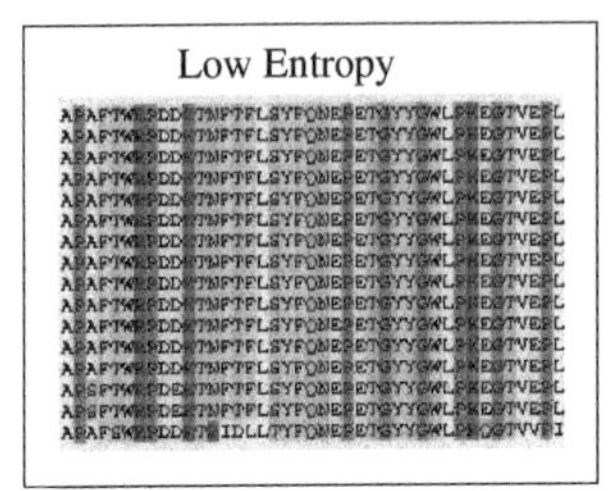

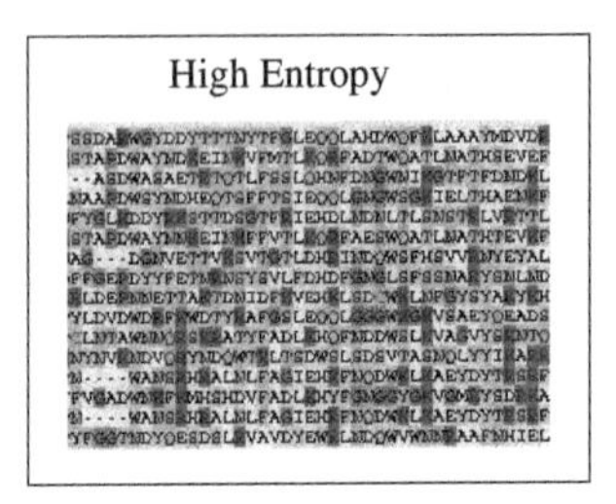

FIGURE 9.8: Conservation measures. High entropy vs. low entropy alignments. (Figure reproduced from [354] with permission.)

erties. Amino acids can be classified into several different categories. For example, the classification of [439] groups amino acids based on their similarity score and divides them into classes such as hydrophobic (MILV), hydrophobic aromatic (FWY), neutral and weakly hydrophobic (PAGST), hydrophilic acidic (NQED), hydrophilic basic (KRH) and cysteine (C). There are other ways to divide amino acids into classes (see Section 2.3), each one according to slightly different properties.

The amino acid entropy measure is not effective in such cases since it ignores amino acid similarities, but an entropy measure based on suitably defined classes may reveal positions with subtle preferences toward classes of amino acids. Given the set $\mathbf{C} = \{C_1, ..., C_k\}$ of amino acid classes and their empirical probabilities (augmented with pseudo-counts) $\mathbf{p} = \{p_1, ..., p_k\}$, the class entropy is defined in a similar way to the amino acid entropy

$$E_{class}(\mathbf{p}) = -\sum_{i \in C} p_i \log_2 p_i$$

However, even with this measure we might still underestimate the level of conservation in certain positions. Similarities between amino acids exist across classes, and it is difficult to define an exclusive set of classes. After all, amino acids are all evolutionary related. Entropy measures ignore similarities between different classes of data, and therefore they only approximate the total divergence of an alignment column. An alternative measure that is correlated with the divergence is the sum of pairwise similarities of amino acids in a column (similar to the SP scoring function used in MSA algorithms, as discussed in Section 4.1.1). If the number of sequences participating in an alignment column k is n then the SP-score of this column (also referred to as *evolutionary span*) is defined as

$$sp(k) = \frac{2}{n(n-1)} \sum_{i=1}^{n} \sum_{j<i} s(a_{ik}, a_{jk})$$

where a_{ik} is the amino acid in position k of sequence i and $s(a,b)$ is the similarity score of amino acids a and b according to a scoring matrix of choice.

The higher the SP-score, the lower the divergence and the more conserved is the position.

Physico-chemical properties of residues may also help in predicting domain boundaries since residues around domain transitions tend to have different characteristics than in the core of a domain. For example, hydrophobic (water-repelling) residues tend to cluster inside domain cores while hydrophilic residues occupy more exposed locations in protein structures and therefore are more likely to be in inter-domain regions. Similarly, certain amino acids such as glycines, cysteines and prolines, which play a significant role in stabilizing protein structures, tend to occur in different frequencies inside domains vs. inter-domain regions. Other classes of residues (aliphatic, aromatic, negatively charged, etc.) might also be instrumental in characterizing protein structures and can help to determine domain boundaries. Given an MSA of the query sequence and related proteins, we can generate a *class-profile* based on the average frequency of class residues along a sliding window. If the class residues tend to terminate domains, then maxima in this graph are likely to correspond to domain boundaries. The merit of considering residue composition in detecting domain boundaries has been demonstrated in [354,440–442].

We can extract other features from MSAs that measure **consistency and correlation.** Since protein domains are believed to be stable building blocks of protein folds, it is reasonable to assume that all appearances of a domain in database sequences will maintain the domain's integrity. By integrating the information from multiple sequences we can detect changes in sequence participation along the query sequence. For example, the **sequence correlation** measure assesses the correlation of two columns i and j based on sequence participation

$$corr(k,l) = \sum_{j=1}^{n} \delta(a_{kj}, a_{lj})$$

where $\delta(a_{kj}, a_{lj}) = 1$ if $a_{kj}, a_{lj} \in \mathcal{A}$ (the alphabet of amino acids) or if both a_{kj} and a_{lj} are gap characters, and 0 otherwise. High correlation values reflect consistent sequence participation while low correlation values signal discontinuity of some sequences and possible domain boundaries. Other, related measures are described in [354].

While MSAs provide ample information on domain boundaries, they should be carefully constructed to maximize their performance. Data preparation can make a big difference in the quality of the results and the success of the method. Advanced MSA algorithms (such as T-coffee or POA) should be used when possible. Alignments should also have "enough" sequences (a few tens at least) and should be free from fragments and loosely aligned remote homologs (as discussed in Section 9.2.3). Picking an appropriate evalue cutoff for the similarity search can help to eliminate questionable alignments and reduce the noise level. On the other hand, sequences that are too similar usually provide little information on domain boundaries and bias predictions since they mask other equally important but less represented sequences. Defining the correct balance between distantly related sequences and highly similar sequences is not trivial, and it is advised to weight the sequences based on their similarity (see Section 4.3.3.3). Finally, to smooth random local fluctuations and enhance the signal, it is recommended as before to smooth the scores by calculating the average over a window of size w around each sequence position and use the average as the score at that position. The window size w is a parameter that can be optimized as described in Section 9.3.1.

9.3 Learning domain boundaries from multiple features

In the previous sections we discussed many measures and features (based on MSAs or other sources) that reflect structural properties of proteins and could therefore be informative of the domain structure of a protein. While simple and fast, predictions produced based on individual features are noisy and are only moderately successful, at best. However, by integrating these scores and measures into a single model we can devise an effective domain prediction algorithm that can estimate the likelihood that a sequence position is part of a domain or at domain boundary.

A simple approach is to generalize Equation 9.1 to handle combined termination signals based on alignments and exon-intron signals, as well as domain boundaries predicted with other methods or domain databases. However, this approach for integration of signals may result in poor predictions, especially if the different signals are associated with different confidence values (boundaries predicted by domain databases are much more reliable than alignment-based signals). More importantly, this method is limited to analysis of possible transition points (which are a small subset of all sequence positions), while most of the features we discussed in this chapter are defined for all positions and are associated with a continuous range of possible values.

A more flexible and expressive method for integrating multiple features into a single predictor is to use a learning algorithm. For example, given a training set of structurally verified protein domains, the features can be fed into a model such as a Neural Network (NN) or an SVM to learn the optimal complex decision boundary and label single positions as domain or boundary positions. An example of such a model is shown later in Figure 9.10. Software packages for learning NNs or SVMs from input data are available in several programming languages, including C and Matlab. In the next sections we will discuss some of the issues one encounters when designing such models.

9.3.1 Feature optimization

Extracting features from datasets is central to many learning and classification tasks. The goal is to define features that well characterize the data and distinguish between different classes. Specifically, an effective feature for domain recognition should assign different scores to domain positions (positives) vs. boundary positions (negatives) such that the distributions of feature scores for positives and negatives are well separated. However, it is hardly ever the case that the two distributions are completely disjoint, and the parameters the features depend on (such as the smoothing window w) may greatly affect the separation of these distributions. While the interdependency between features makes this a complex optimization problem, a simple strategy toward improving the prediction accuracy is to optimize the parameters of each feature separately so as to maximize the separation between the classes in the data based on that feature.

To find the best set of parameters for a given feature we can measure the statistical similarity of the two probability distributions (negatives and positives) for different sets of parameters and select the one that minimizes the similarity (Figure 9.9). This is the same strategy we discussed in Section 7.4.4. Statistical similarity can be assessed using one of the measures discussed in Section 4.3.5. By repeating this procedure for each feature individually, we can optimize and also identify the best features, those with the maximal separation. In the context of domain prediction, better separation was observed with features such as sequence termination, sequence correlation, class entropy, coil frequency, intron-exon boundary, and proline and glycine frequencies [354]. However, note that even features with near-identical distributions may be informative in a multi-variate model where higher level correlations can result in an effective boundary surface.

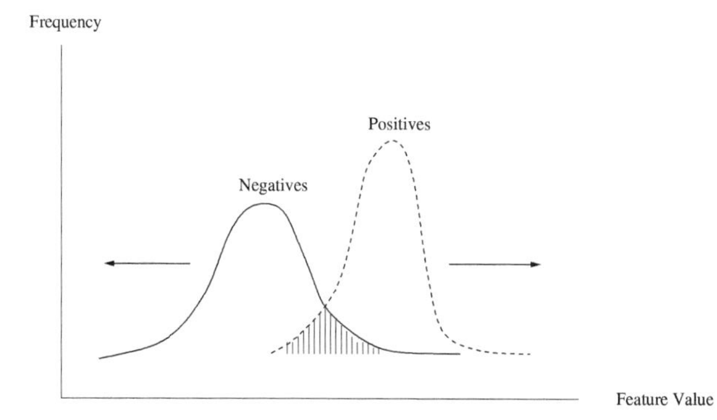

FIGURE 9.9: **Optimizing features.** To improve the model's performance, parameters are optimized by maximizing the statistical distance between the distribution of positives and the distribution of negatives.

9.3.2 Scaling features

Since different features are measured in different units, a straightforward combination of scores may introduce a strong bias toward one or a few of them. Therefore, is it necessary to normalize features and convert them to a uniform scale. A common scaling technique, called the **whitening transformation**, is to transform every score to a zscore based on the background distribution of scores (see Section 3.4.2.3). For example, if the scores are computed from MSA columns, then the background distribution can be compiled from the scores along all alignment positions or a large set of randomly selected positions. The normalization should be invoked separately for each alignment to adjust for the particular makeup of that alignment. The zscore not only serves as a universal scale where all features have the same units, it also provides a measure of statistical significance for each position in the alignment, helping to detect atypical positions. If converted to probability values, then features can be integrated into a single predictor using methods as discussed in Section 10.6.4 of Chapter 10.

The whitening transformation implicitly assumes a normal distribution, but the distribution of scores is not always normal. For example, for features such as sequence termination or exon-intron boundaries the distribution of scores is far from normal, which deems the zscore normalization inappropriate. In such cases a uniform scaling can be used where the scale factor is the maximal value observed, and scores are transformed to the $[0, 1]$ range, after dividing them by the scale factor.

9.3.3 Post-processing predictions

Domain prediction methods (such as those based on neural networks or SVMs) may produce an output for each position, reflecting its probability to be a core domain position or a boundary position. These predictions can be noisy, and it is important to consider local information from multiple positions to smooth and refine the predictions[4]. Smoothing can eliminate spurious transition points, as demonstrated in Figure 9.11 (page 368). Each minimum in the smoothed graph that is below a certain threshold can then be considered as a candidate transition point. The predictions can be further filtered by requiring that a significant fraction of the positions around each candidate transition are also predicted as transition points (this fraction can be altered as a threshold parameter to give different levels of accuracy and sensitivity as discussed in [354]).

[4]We already mentioned smoothing as an important step in feature processing (Section 9.2.3). The smoothed (averaged) features can then be analyzed by systems such as the one depicted in Figure 9.10 and given as input to the learning algorithm. Here we are concerned with the *output* of the learning algorithm.

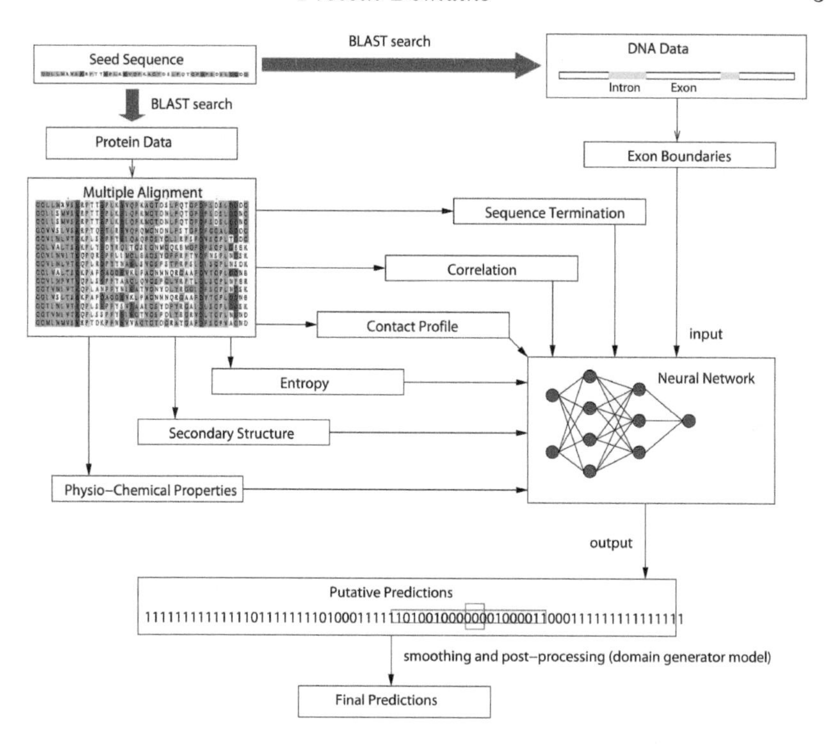

FIGURE 9.10: Overview of a domain prediction system. The method, as proposed in [354], works by analyzing MSAs that are derived from a database search. Multiple measures are defined to quantify the domain information content of each position along the sequence and are combined into a single predictor using a neural network. The output of the network is further smoothed and then post-processed using a probabilistic model (Section 9.4) to predict the most likely transition positions between domains. (Figure reproduced from [354] with permission.)

Yet this procedure does not necessarily produce accurate predictions of domain boundaries and may overly fragment proteins into domains. For example, candidate positions A, B and C of Figure 9.11 are fairly close to each other, and it is unlikely that all three are true domain boundaries. In fact, each possible combination of candidate transition points is a possible partitioning of the protein into domains, with some of them being more likely than others.

Given multiple hypotheses, i.e., alternative partitions of the query sequence into domains, we would like to determine the most likely one. A simple greedy approach that seems to work pretty well for many proteins [354] is to select the candidate transition point with the minimal probability and then eliminate all other minima that are within a window of l amino acids from the selected point (where l is the minimal domain length). This process is repeated iteratively with the remaining minima until all candidate points are processed or when the next minimum is above a certain threshold probability (e.g., 0.5). For example, applying this procedure to the candidate transition points of Figure 9.11 results in B and D as the final predictions. This however, might not be the optimal partition when considering also prior knowledge on known domains. In Section 9.4 we will discuss a more elaborate model to test multiple hypotheses and output the most likely partition, using a Bayesian approach that accounts for the likelihood of the data given the suggested partition, as well as the prior probability to observe the partition. When prior information is included, the procedure results in C and D as the final predictions.

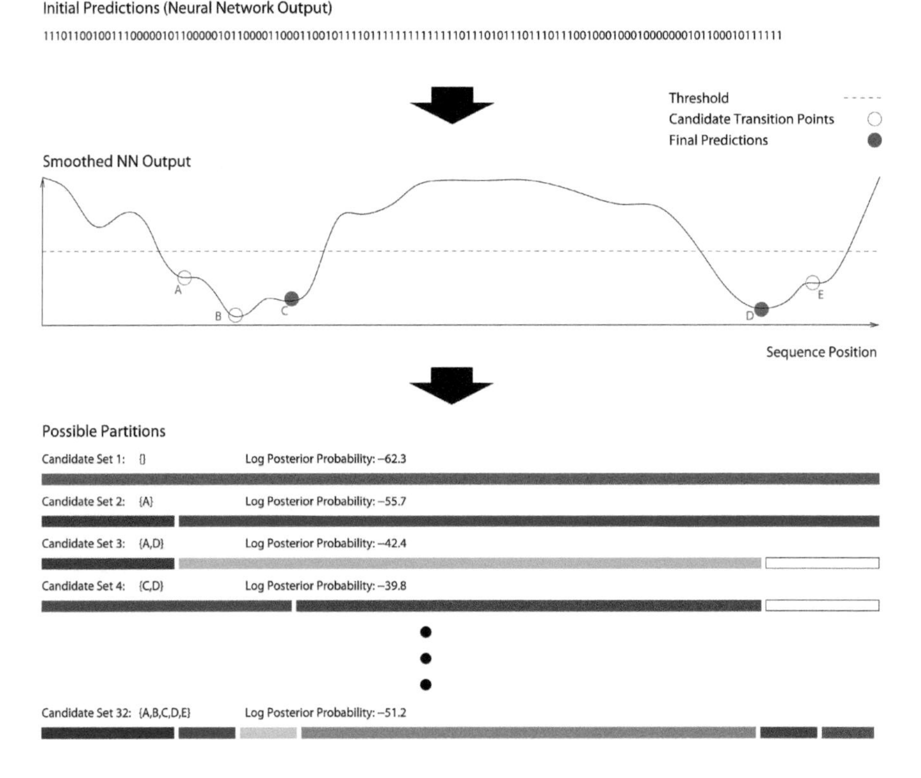

FIGURE 9.11: Selecting candidate transition points. Processing the output of the domain prediction system of Figure 9.10. The output of the neural network is noisy. The **initial** predictions are smoothed by averaging the value at each position over its local neighborhood. All minima in the smoothed curve are defined as **candidate** transition points (marked A-E). Every combination of the candidate points is a possible partitioning into domains. The posterior probability of each partitioning is assessed using the model of Section 9.4, and a **final** set of transition points is predicted. (Figure modified from [354] with permission.)

9.3.4 Training and evaluation of models

Evaluation is an important part of the design of any successful prediction system. In the absence of general rules or principles that define domain boundaries, one must rely on existing knowledge of protein domains to calibrate the model used for prediction. There are several useful resources on domains that can be used as reference benchmarks (see Section 9.8). The knowledge within, in the form of complete protein chains and their partition into individual domains, can be used to both train and test the learning models. Note that domain databases can be dominated by single domain proteins, and to avoid bias toward such proteins, the proportions of single and multi-domain proteins should be the same more or less, in both the training set and the test set. To avoid bias toward large families, sequences from the same family should be grouped and only one representative should be picked from each one, for example, the one with the maximal number of homologs.

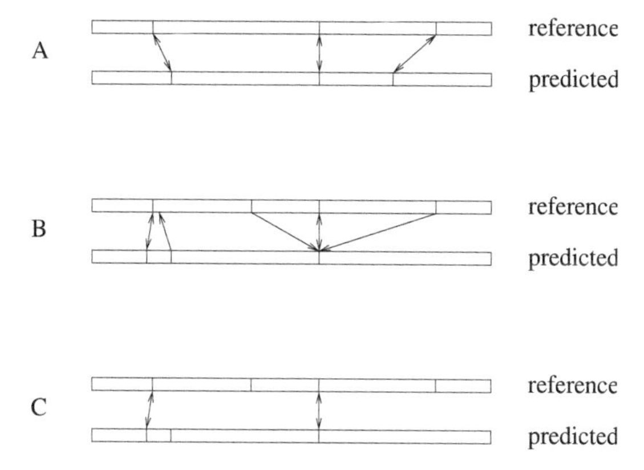

FIGURE 9.12: Correspondence between predicted and reference domain boundaries. A simple protocol is to associate each predicted point with the closest reference point and vice versa. In the first example, this results in one-to-one correspondence. However, in the second example this results in many-to-one correspondence, both ways. To avoid such mappings, one can use a more strict protocol, as demonstrated in the third example. Starting by computing the distances between all pairs of reference and predicted points, the pairs are then considered in increasing distance order starting from the closest pair. Given a pair of a predicted point and a reference point, a correspondence is established if both have not been assigned already.

Since the predictions obtained from different algorithms are often incomplete it is necessary to design an evaluation procedure that has different scores for accuracy and coverage. In addition, the predictions may disagree with the reference set on the number of domains in each protein. Therefore one needs to define a procedure for associating predicted transition points with their most probable counterparts in the reference set and vice versa. The simplest choice is to assign every transition point that is being considered to the closest reference transition point (see Figure 9.12). Based on this procedure we can define measures such as *distance accuracy*, which is the average distance of the predicted transitions from their associated transition points in the reference set, or *distance sensitivity*, which is the average distance of domain transitions in the reference set from the associated predicted transitions. Measures of selectivity and coverage can help to assess the overall performance of a prediction algorithm. Designating predictions that are within x residues (e.g., $x = 10$) of a real transition as correct predictions, the *selectivity* would be the percentage of predictions that are considered correct. Similarly, the *coverage* would be the percentage of transitions in the reference set that are correctly predicted (real transitions associated with a predicted transition point within x residues).

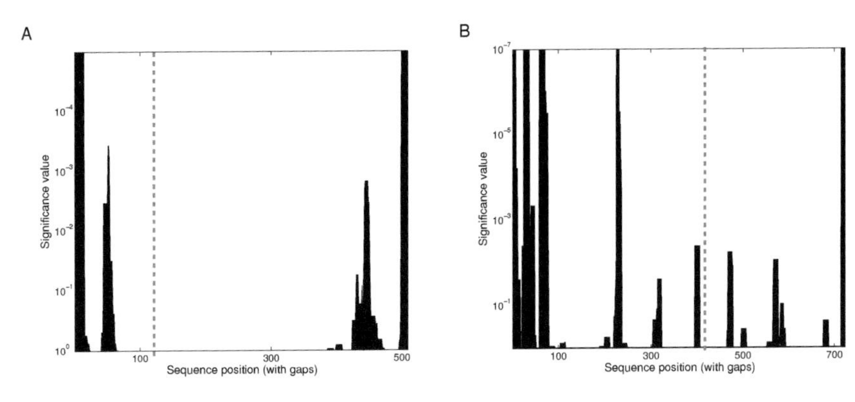

FIGURE 9.13: Noisy signals of sequence termination in MSAs. Two examples where the signal is too noisy to predict domain boundaries. **A:** Dihydroorotate dehydrogenase B (PDB 1ep3_B). **B:** Tetanus toxin Hc (PDB 1fv3_B). SCOP domains are marked with dashed vertical lines at the corresponding positions (after correcting for gaps). (Figure reproduced from [412] with permission.)

9.4 Testing domain predictions

Despite the many domain databases and the numerous domain prediction algorithms, the domain prediction problem is by no means solved. The query protein whose domain structure we would like to solve may contain new domains that have not been characterized or studied yet. Therefore the protein might be poorly represented in existing domain databases with limited information about its domain structure or no information at all. One can attempt to predict the domain structure by applying any of the algorithms described in the previous sections. However, not all methods are applicable. For most proteins (and especially newly sequenced ones) the structure is unknown, thus ruling out the application of structure-based methods. If the protein has no homologs then methods that are based on sequence similarity and MSAs are ineffective. Other methods (such as those that are based on structure prediction) might be too computationally intensive.

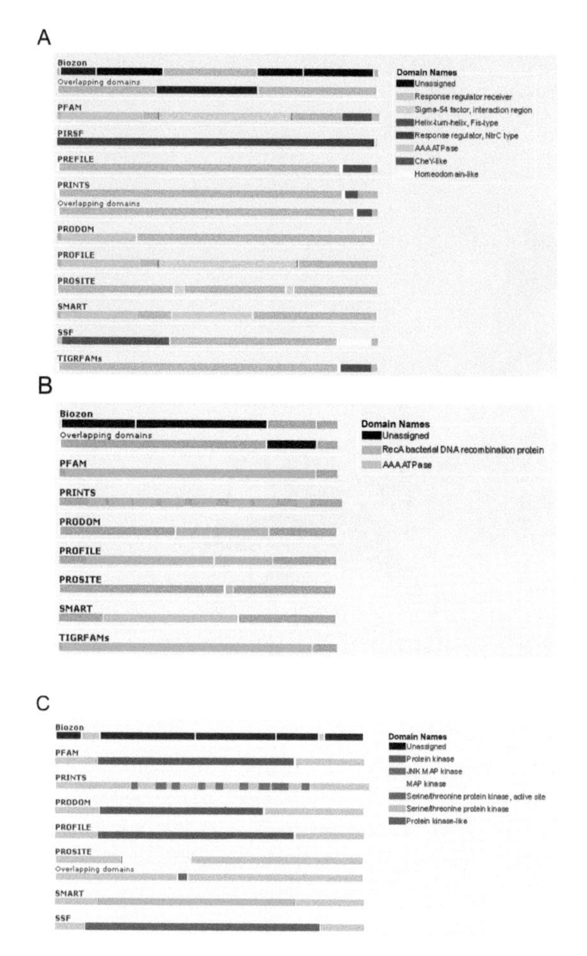

FIGURE 9.14: Examples of disagreement between different domain databases. (Figure obtained from Biozon and reproduced from [412] with permission.) **A:** Nitrogen regulation protein ntrC (Biozon ID 457-90). **B:** Recombinase A (Biozon 355-123). **C:** Mitogen-activated protein kinase 10 (Biozon 464-54). To view the profile page of a protein with Biozon ID x follow the link www.biozon.org/db/id/x. As these examples demonstrate, different domain databases have different domain predictions, and domain boundaries are often ill defined. Domain databases can also generate overlapping predictions (displayed in separate lines). This can happen, for example, when domain signatures in the source database overlap, where the longer one could be a global protein family signature and the shorter one a local motif.

Nevertheless, usually at least one of the methods we discussed can be applied. However, as with any prediction, the output of domain prediction algorithms should be considered with caution. For example, misalignments and fragments might introduce noise that can lead to many erroneous predictions by MSA-based methods, often in proximity to each other (see Figure 9.13).

Slight variations in domain lengths due to post-duplication mutations and truncations further obscure true signals. Consequently, some of the true transitions between domains might be completely missed by the specific method used. On the other hand, some of the predicted transition points might be accurate but overlooked by domain databases. These problems are typical of all domain prediction algorithms.

The problem is further complicated by the fact that domain prediction algorithms and domain databases can be very inconsistent in their domain assignments (see Figure 9.14). However, without structural information it is difficult to tell which ones (if any) are correct. Even when the structure is known, determining the domains can be difficult and domain definitions might be in dispute (Figure 9.15).

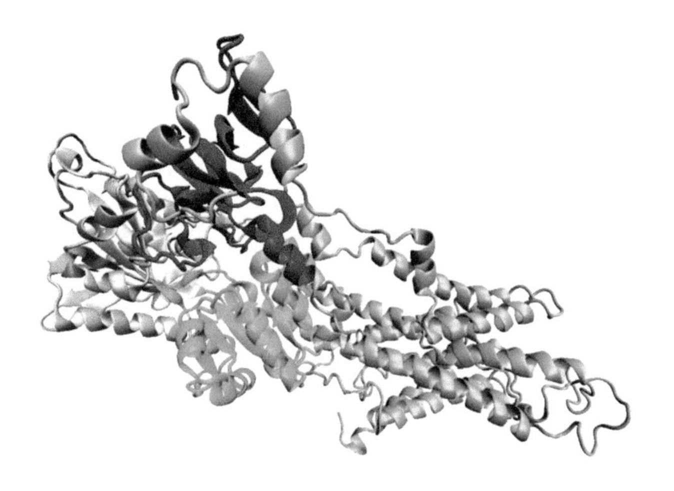

FIGURE 9.15: **Domain structure of calcium ATPase** (PDB 1iwo, chain A). According to SCOP, this structure is composed of four domains: a transmembrane domain (d1iwoa4, positions 1-124, colored cyan), a transduction domain (d1iwoa1, positions 125-239, colored red), a catalytic domain (d1iwoa2, positions 344-360 and 600-750, colored green), and an unlabeled fourth domain (d1iwoa3, positions 361-599, colored gray). Note that the domains are not well separated, structurally. (Figure reproduced from [412] with permission.)

Although structure-based databases of protein domains such as SCOP and CATH are considered to be the gold standard by many, there is no assurance that their domain definitions are accurate and correspond to "real" domains, as becomes evident by the disagreement between the two. While the two databases are in pretty good agreement overall, they do differ in many cases. For example, CATH often assigns small structural fragments from one sequence domain to another based on structural compactness considerations (Figure 9.16). However, such sequence discontinuous domains need accurate structural information to delineate them correctly, and it is not clear if it is possible to detect these domains based on sequence information alone. These issues make the identification of discontinuous domains harder.

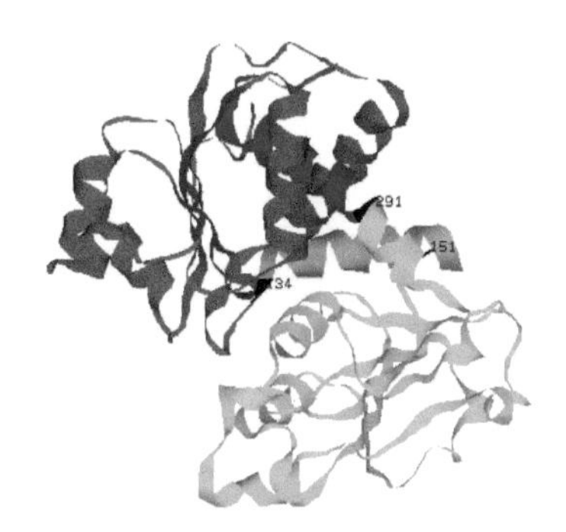

FIGURE 9.16: Inconsistent domain definitions. Domain definitions for PDB 1ekx, an enzyme involved in biosynthesis of certain nucleic acids. The differently colored segments on the top left and bottom right define the two domains of the protein. CATH assigns the fragment between positions 292 and 310 (in blue) to the domain on the top based on compactness considerations, while SCOP assigns it to the domain on the bottom. (Figure reproduced from [354] with permission.)

A potential improvement in domain prediction accuracy can be achieved by using **majority voting** between multiple methods (as in [332,443]). However, majority voting does not necessarily produce correct results either. The different domain-prediction methods are correlated and often use similar types of information (e.g., predicted secondary structures, MSAs) and therefore might provide similar but wrong predictions. In the next section we will discuss a more rigorous approach for assessing the likelihood of different partitions, which can improve the accuracy of domain prediction and provide a framework for assessing alternative partitions. When the likelihood of the top scoring hypothesis is only marginally higher than alternative partitions, then one might want to consider also other partitions as suggested by the competing hypotheses.

9.4.1 Selecting more likely partitions

Given different predictions that partition a protein into domains, we would like to determine which one is the most plausible one. We already mentioned a simple method for selecting a "good" partition given a set of candidate transition points (Section 9.3.3). Here we will describe a more elaborate approach, which uses a model that considers the likelihood of the data as well as prior information on domains. Once such a model is defined, the best partition can be selected by enumerating all possible combinations of predicted transitions and

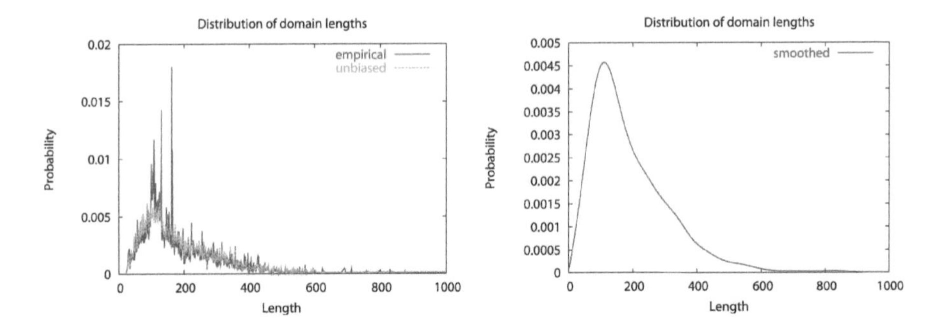

FIGURE 9.17: **Distributions of domain lengths.** (Figure reproduced from [354] with permission.) The empirical distribution of domain lengths in SCOP is very noisy, sparse for domains longer than 600 amino acids and biased because of uneven sampling of the protein space, even after eliminating redundancy. To overcome the bias, only one entry of the same length from each protein family is retained (left). Noise and sparse sampling for domains longer than 600 amino acids are handled by running a few smoothing cycles (right).

picking the subset of positions (and hence domain assignments) that results in maximum likelihood or maximum posterior probability. With k candidate transition points there are 2^k possible combinations of points. Each combination is considered an **hypothesis** regarding the domain structure of the query protein as it induces a certain partition. Note that the predicted transitions can be a combination of ones obtained by different methods. Hence, such a model can also be used to integrate the results from multiple domain prediction methods.

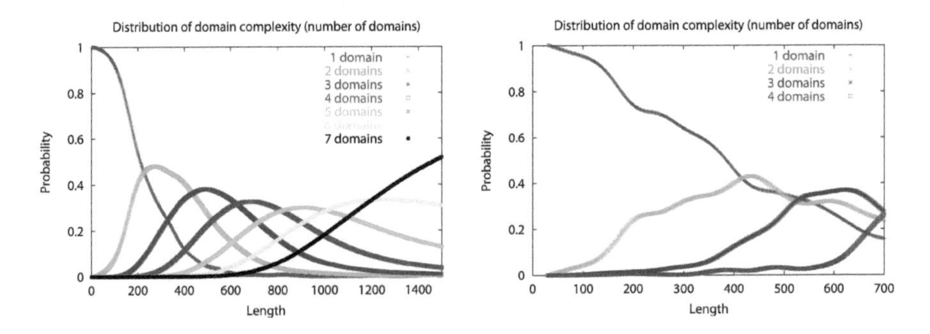

FIGURE 9.18: **Distributions of number of domains:** (a) extrapolated, (b) empirical (based on SCOP). The extrapolated distributions are normalized assuming that the maximal number of domains is seven (the maximal number of domains observed in SCOP). (Figure reproduced from [354] with permission.)

Bayesian decision theory calls for selecting the hypothesis with the maximal posterior probability (see Section 3.12.8). Assume we are given a protein sequence and a multiple alignment S of length L and a possible partition $\mathbf{D}$ of S into n domains $\mathbf{D} = D_1, D_2, ..., D_n$ of lengths $l_1, l_2, ..., l_n$. We would like to compute the posterior probability of the hypothesis $\mathbf{D}$ given the data $P(\mathbf{D}|S)$. The posterior probability can be expressed in terms of the prior probability of the partition $P(\mathbf{D})$ and the likelihood of the data (such as the MSA S) given the partition $P(S|\mathbf{D})$, using Bayes formula

$$P(\mathbf{D}|S) = \frac{P(S|\mathbf{D})P(\mathbf{D})}{P(S)}$$

Since $P(S)$ is fixed for all partitions then

$$\max_D P(\mathbf{D}|S) = \max_D P(S|\mathbf{D})P(\mathbf{D})$$

and we are looking for the partition that will maximize the product $P(S|\mathbf{D})P(\mathbf{D})$

9.4.1.1 Computing the prior $P(D)$

The following likelihood model was developed for MSA-based methods [354] but is applicable to other methods as well. Imagine a "domain generator" model that moves repeatedly between a domain state and a linker state (which corresponds to the transitions between domains) and emits each time a domain sequence or a linker sequence respectively. The sequences are generated according to different source probability distributions.

To calculate the prior $P(\mathbf{D})$ we have to estimate the probability that an arbitrary protein sequence of length L consists of n domains of the specific lengths $l_1, l_2, ..., l_n$. That is,

$$P(\mathbf{D}) = P((d_1, l_1), (d_2, l_2), ..., (d_n, l_n) \textbf{ s.t. } l_1 + l_2 + ... + l_n = L)$$

This can be estimated by considering known domain partitions of proteins of length L in domain databases. However, the amount of data available is insufficient to accurately estimate these probabilities for all possible partitions. We can approximate this probability by using a simplified model; given the length of the protein, assume that the domain generator selects the number of domains first and then selects the length of one domain at a time, considering the domains that were already generated. For a partition into n domains there are $n!$ possible orderings of the domains, and therefore the prior probability of the partition is approximated by

$$P(\mathbf{D}) \simeq \text{Prob}(n|L) \cdot \sum_{\pi(l_1,l_2,...,l_n)} P_0(l_1|L)P_0(l_2|L-l_1)...P_0(l_{n-1}|L-\sum_1^{n-2} l_i) \quad (9.2)$$

where Prob$(n|L)$ is the prior probability to observe n domains in a protein of length L and $P_0(l_i|L)$ is the prior probability to observe a domain of length l_i given a sequence of length L. The term $\pi(l_1, l_2, ..., l_n)$ denotes all possible permutations of $l_1, l_2, ..., l_n$.

Estimating $P_0(l_i|L)$ can still be difficult, given the limited amount of empirical data. A reasonable approximation is the background distribution of domain length $P_0(l_i)$, which can be estimated from the empirical distribution of domain lengths in a domain database such as SCOP and normalized to the relevant range $[0..L]$ (see Figure 9.17 on page 374). Interestingly, the obtained distribution follows closely the extreme value distribution (see Section 9.4.2 for discussion).

The first term in Equation 9.2, Prob$(n|L)$, can be estimated empirically from domain databases or it can be extrapolated from $P_0(l_i)$ since Prob$(n|L) =$ Prob$(n, L)/P(L)$ where Prob(n, L) can be computed directly considering all possible combinations of n domains over a sequence of length L

$$\text{Prob}(n, L) = \sum_{x_1=1}^{L} P_0(x_1) \sum_{x_2=1}^{L} P_0(x_2)... \sum_{x_{n-1}=1}^{L} P_0(x_{n-1}) \cdot P_0(L - x_1 - x_2... - x_{n-1}) \tag{9.3}$$

These probabilities can be precalculated using a dynamic programming algorithm (Problem 9.5). The probability $P(L)$ is simply given by the complete probability formula

$$P(L) = \sum_{i=i}^{L} \text{Prob}(i, L)$$

The extrapolated distributions Prob$(n|L)$ for $n = 1..7$ are plotted in Figure 9.18a (page 374). Interestingly, the empirical distributions (Figure 9.18b) differ quite markedly from these extrapolated distributions. However, since the empirical data is noisy, sparse and possibly biased, the extrapolated distributions could be more reliable than the empirical ones. For one, note that the empirical probability for a protein to be a single domain dominates all other scenarios up to proteins of length 400, while the curves meet much earlier (around 200) in the extrapolated distributions. Of note is that different empirical distributions are observed in the CATH database.

9.4.1.2 Computing the likelihood $P(S|D)$

Computing the likelihood of the data (e.g., a sequence, an MSA or a structure) given the hypothesis (a certain partition $\mathbf{D} = (d_1, l_1), (d_2, l_2), ..., (d_n, l_n)$) depends on the model used to represent domains and transitions. For example, the model of [354] considers domains and the transitions between domains (the linkers) as two different sources (states), generating different types of MSA columns. MSA columns can be characterized using many different features $f_1, f_2, ..., f_r$, such as entropy, secondary structure propensities, conservation scores and others. Each of the two sources induces a unique probability distribution over the features of the MSA columns it emits. That is, the joint probability distribution of the r features is state dependent and $P(f_1, f_2, ..., f_r|domain)$ differs from $P(f_1, f_2, ..., f_r|linker)$.

These probability distributions can be trained from MSAs of proteins whose domains were identified in databases such as SCOP (see Section 9.8). Domain positions are defined as the positions that are at least $x = 10$ residues apart from a domain boundary. All positions that are within x residues from domain boundaries are considered linker positions. By collecting all MSA positions that are labeled "domains" per this definition, the distribution of features for domain positions can be characterized. The distribution for linker sequences is characterized from all the remaining positions.

Based on these empirical distributions, new MSAs can be analyzed to estimate the probability of each position according to either source. Given a position j with features $f_{j1}, f_{j2}, ..., f_{jr}$ we can compute the likelihood that it was emitted by the domain source

$$P(j|domain) = P(f_{j1}, f_{j2}, ..., f_{jr}|domain)$$

and the likelihood $P(j|linker)$ that it was emitted by the linker source. The likelihood of the data (a complete MSA) given a certain partition is the product over the MSA positions, using either $P(j|domain)$ or $P(j|linker)$ depending on the position and the partition tested. That is,

$$P(S|\mathbf{D}) = \prod_{j=1}^{L} P(j|state_D(j))$$

where $state_D(j)$ is the state (linker or domain) of position j according to partition $\mathbf{D}$.

Estimating the joint probability distributions $P(f_1, f_2, ..., f_r|domain)$ and $P(f_1, f_2, ..., f_r|linker)$ from the training data might be impractical if the number of features is more than three or four, since available data on known domains is limited. On the other hand, feature scores are often correlated and therefore we cannot assume independence between features. In other words, we cannot approximate the joint probability distribution as a product of one-dimensional distributions over individual scores

$$P(f_1, f_2, ..., f_r) \neq P(f_1)P(f_2) \cdots P(f_r)$$

An intermediate approach is to use **pair statistics** (which usually can be calculated reliably from the training datasets) and use first-order dependencies in the expansion for the joint probability distribution

$$P(f_1, f_2, ..., f_r) \sim P(f_1)P(f_2|f_1) \cdots P(f_r|f_{r-1})$$

However, the order of the features in this expansion can greatly affect the quality of the approximation. One possible heuristic is to start the expansion from the two most dependent features and continue by selecting every time the variable that has the strongest dependency with variables that are already used in the expansion. The dependency between a pair of variables X, Y can be measured in terms of the relative entropy between the joint distribution of the two variables P_{XY} and the product of the two marginal probability distributions (also called the mutual information, see Section 3.12.6)

$$I(X,Y) = D^{KL}[P_{XY}||P_X P_Y]$$

This heuristic results in an expansion of the form

$$P(X_1)P(X_2|X_1) \cdot P(X_3|PILLAR(X_3))...$$

where $PILLAR(X)$ denotes the random variable that X is most dependent on (of the random variables that are already in the expansion). This heuristic attempts to minimize the errors that are introduced by relaxing the dependency assumption to a first order dependency, and maximize the support for each random variable we introduce in the expansion. Thus, highly correlated variables affect the total probability only marginally, while under the independence assumption they might introduce a substantial error (other, alternative methods for approximating the joint probability distribution from the marginal distributions are described in [444] and [445]). Note that the expansion for domain regions can be different from the expansion for linker regions, since the source distributions differ. However, once the two expansions (for domains and linkers) are defined based on the pair statistics, the same two expansions can be used for all domains and all linkers.

9.4.2 The distribution of domain lengths

One of the most interesting observations that emerges from domain analysis is that the distribution of domain lengths follows closely the extreme value distribution (EVD), as in Figure 9.19. This distribution has been studied extensively in the context of sequence similarity and has been used by software packages such as BLAST and FASTA to associate statistical significance measures (evalues) with similarity scores (see Section 3.4.3.2). However, its appearance in the context of domain lengths is surprising and suggests the presence of an underlying maximization process.

These fairly ancient domains, as the traces of sequence matches indicate, are perhaps the earliest form of fully functional proteins. Their formation is the result of a random process that put together shorter coding elements. One possible explanation for the EVD is the fact that this random process of generating a protein domain favored *longer* coding regions, as is explained next.

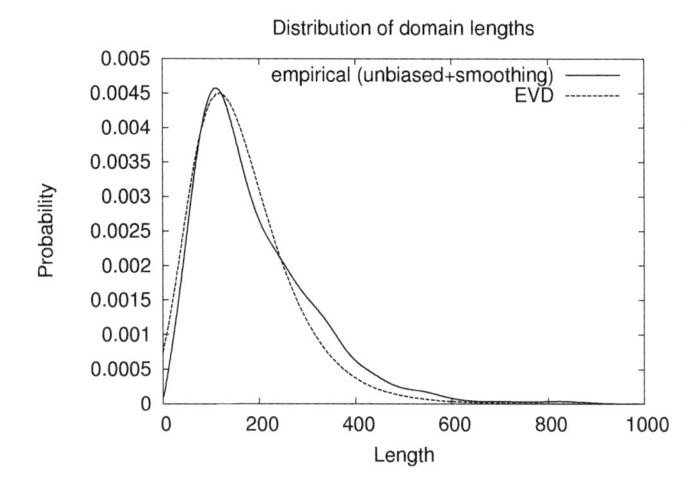

FIGURE 9.19: Distributions of domain lengths after eliminating bias and smoothing. The distribution closely resembles the extreme value distribution (EVD). See Figure 9.17 for details. (Figure reproduced from [354] with permission.)

The random process itself is carried at the DNA level and manifests itself in events such as duplications and insertions. We can hypothesize that in the beginning, most duplications and insertions introduced fairly short elements that were transformed into local secondary structures. With time, longer domains were formed by insertions and duplications that put a few of these motifs adjacent to each other. However, the outcome of such a process is likely to end up with a non-transcribable sequence of DNA. For a new gene to be created, several more conditions must be met. For one, a start and a stop codon must be present. Secondly, a promoter or a RNA-binding site should exist just before the start codon to enable transcription initiation. The likelihood of these two conditions to be met increases as the length of the elements used in the construction (the subsequences that are being duplicated or inserted) increases. Since it is more likely that a random copy will include a promoter site as the length of that copy increases, it is hypothesized that this process led to the formation of longer and longer domains. In other words, this process inherently creates a bias toward longer fragments. If we associate with each duplication event a random variable whose value is the length of the duplicated segment, then of all these random processes, the longest ones are those that are more likely to survive and eventually be transcribed and translated into new domains. Thus, this process is implicitly, by natural selection, a maximization process and the extreme value distribution follows. While this theory may explain the observed distribution of lengths, it is speculative, and it is worth re-examining this phenomenon from other possible viewpoints.

9.5 Multi-domain architectures

So far we discussed methods for domain prediction. Many proteins are single domain proteins, and successful domain prediction for these proteins translates to a single domain hypothesis that is consistent with the actual structure. However, the majority of proteins are multi-domain proteins. That is, they are composed of multiple domains, arranged together in what is called a **domain architecture**.

Domain analysis for multi-domain proteins does not end in detecting the constituent domains. It is the domain architecture of a protein that prescribes its function, as it affects its binding capabilities and thus the molecules it interacts with and consequently the cellular pathways it participates in. Therefore, to properly analyze the function of a protein we also need to study the properties of the specific architecture the constituent domains are arranged in. However, the study of multi-domain proteins poses many challenges. Domain-based classifications can provide useful information on the functional role of domains, but are not always effective for the analysis and characterization of multi-domain proteins since they focus on individual domains. On the other hand, protein-wise studies that cluster proteins based on sequence or structure comparison of the *complete* protein chains ignore the domain structure altogether and overlook the nature of proteins as modular entities (see Chapter 10). Therefore, they lack the precision of domain-based studies.

Multi-domain proteins result from duplication or recombination of several domain units [446–448], and it has been estimated that they comprise about two-thirds of the proteins in prokaryotes and 80% in eukaryotes [414, 446]. The distribution of the number of different types of neighbors for each domain family follows a power law, where the probability to observe k neighbors decreases as $k^{-\gamma}$ and $\gamma > 0$ is a constant. This means that most domain families partner up with one or two other families; however, some are versatile and can work together with many domain partners (such domains are sometimes referred to as **promiscuous domains**). The duplication process seems to occur more frequently for large domain families [416, 447–450]. In [451] it has been shown that multi-domain proteins have significantly higher functional divergence than single-domain proteins. This illustrates the difficulty in determining the functional roles of multi-domain proteins.

9.5.1 Hierarchies of multi-domain proteins

To study the function of multi-domain proteins it is necessary to organize them into classes and analyze their properties. In view of the evolutionary process that led to the formation of multi-domain proteins, it seems most natural to organize domain architectures into hierarchies that convey the relations between different architectures.

Architectures can be characterized based on **sequential** signatures of domains (consecutive in the protein sequence) or **non-sequential** signatures, as described in [452]. If a protein p is composed of k domains $p = d_1 \ldots d_k$ then each subset of n domains ($n \leq k$) is an n-gram domain architecture that is observed in p. The length $length(D)$ of a domain architecture D is the number of domains it contains. For example, the protein $p = d_1 d_2 d_3$ contains three different 1-gram domain architectures (D=d_1, D=d_2 or D=d_3), three 2-gram domain architectures (D=$d_1 d_2$, D=$d_2 d_3$ or D=$d_1 d_3$) and one 3-gram domain architecture. The architecture $D = d_1 d_3$ is a non-sequential one. We are interested in functional characterization of all domain architectures, including the non-sequential ones.

In order to establish a hierarchy over domain architectures, it is necessary to define relations between pairs of architectures. The **group-based** hierarchy is based on relations between groups of domains (*subgroups* and *supergroups*). An architecture D_1 is said to be a *subgroup* of another architecture D_2 (i.e., D_2 is a *supergroup* of D_1) if the domain sequence in D_1 can be extracted from the domain sequence D_2 while preserving the ordering. For example, $d_1 d_2 d_3$ is a supergroup of $d_1 d_3$, $d_1 d_2$ or d_1 but not of $d_3 d_1$.

The **string-based** hierarchy is more restrictive and is based on relations between strings (*substrings* and *superstrings*). D_1 is said to be a *substring* of D_2 if the sequence of domains D_1 can be found in D_2 in the exact same order without extraneous elements appearing in between adjacent elements of D_1. For example, $d_1 d_2$ is a substring of $d_1 d_2 d_3$ but not of $d_1 d_3 d_2$.

One can represent in a tree-like graph an entire database of protein sequences based on these relations. For example, Figure 9.20 shows the complete string-hierarchy for a databases of three proteins p_1, p_2 and p_3 whose domain architectures are $d_1 d_2 d_6 d_8$, $d_3 d_1 d_4 d_5$ and $d_7 d_1 d_2$, respectively.

The two different hierarchies help to study two different aspects of domain organization: signatures (group-based) vs. sequential modules or "phrases" (string-based). The main difference between the two is that if D_1 is a subgroup of D_2, then D_2 can be obtained from D_1 by inserting domains between any two adjacent domains or appending them on either end of D_1. On the other hand, if D_1 is a substring of D_2, then D_2 can only be generated through left or right extension of additional domains to D_1. Note also that if D_1 is a substring of D_2 then D_1 is also a subgroup of D_2.

9.5.2 Relationships between domain architectures

Studying relationships between different domain architectures can add insight into their evolutionary history, help to track functional divergence of architectures and characterize important functional junctions in the hierarchy of all domain architectures.

We can differentiate between two types of relationships among architectures in these hierarchies; relationships through domain deletion/insertion and relationships through domain shuffling. These can be perceived as the elemental

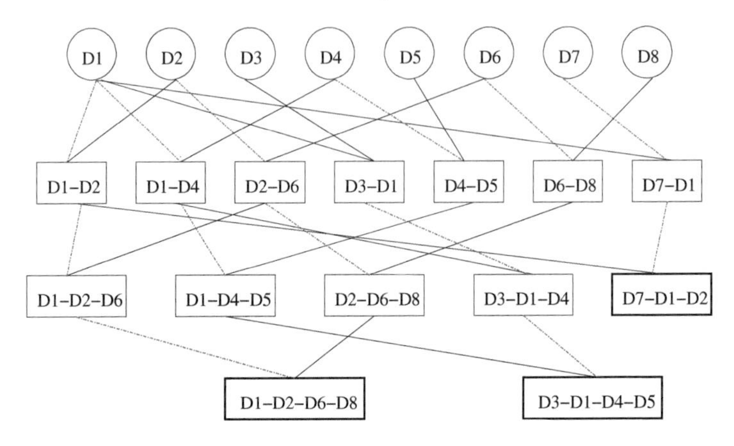

FIGURE 9.20: String-hierarchy for a database of three proteins $\{p_1, p_2, p_3\}$ with $D(p_1) = d_1 d_2 d_6 d_8$, $D(p_2) = d_3 d_1 d_4 d_5$ and $D(p_3) = d_7 d_1 d_2$. Dashed lines: left parent/child relationships. Solid lines: right parent/child relationships. The hierarchy has eight single domains $d_1, \ldots, d_8$ at the top level and the three protein domain architectures as terminal nodes. Notice that each inner level node has exactly two direct parents (left and right), but it can have zero or more direct children domain architectures depending on the domain architectures of the proteins in the database. (Figure reproduced from [452].)

edit operations at the domain level.

The most basic hierarchical relationship between domain architectures is the **parent-child relationship**. This relationship corresponds to domain deletion and insertion events (see Figure 9.20). In the string hierarchy, a domain architecture D_1 is a *parent* of another domain architecture D_2 (i.e., D_2 is a *child* of D_1) if D_1 is a substring of D_2. Similarly, in the group hierarchy, D_1 is a parent of D_2 if it is a subgroup of D_2. A pair of parent-child domain architectures can be characterized by its length difference $L = length(D_2) - length(D_1)$, i.e., the difference in the number of domains between the parent and the child. A child domain architecture possesses a larger number of domains. As we shall see next, through the addition of domains architectures usually evolve to perform more specific functions.

How shall we measure functional similarity of proteins with different architectures? In previous chapters we discussed algorithms for sequence and structure comparison. Clearly, if there is a significant sequence or structural similarity between two proteins, then the proteins are likely to have similar functions. However, it is unclear how to quantify functional similarity. Does higher sequence similarity always entail stronger functional similarity? Is it possible that a weaker similarity along a longer sequence is more important from a functional standpoint than a stronger similarity along a shorter sequence? To answer these questions we need a method that can assess, quantitatively, the functional similarity of proteins.

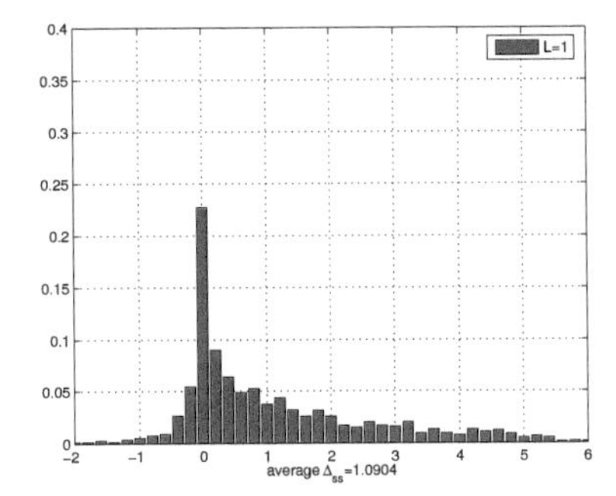

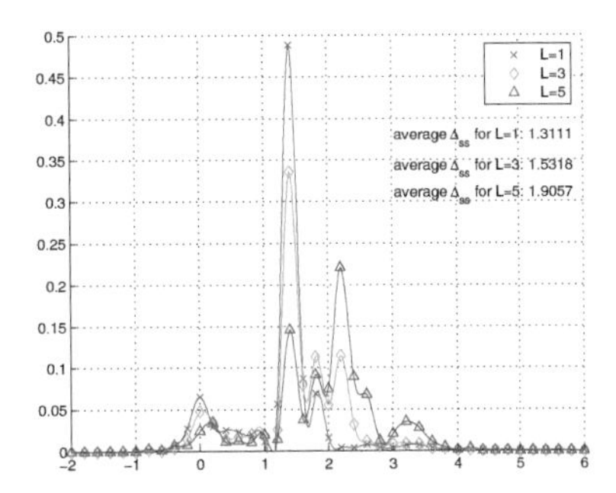

FIGURE 9.21: **Differences in functional similarity between parent and child architectures.** Left: Histogram of Δ_{ss} for parent and child architectures in the string hierarchy of the Pfam database (release 13), with length difference of $L = 1$. The average *ss* score in the child group is usually higher than in the parent group. Right: Functional similarity between parent-child architectures at varying length differences, in the group hierarchy of the Pfam database. Δ_{ss} increases as L increases. Similar results are obtained for domain architectures in the Biozon database, suggesting that the connection between parent-child relationship and functional similarity is consistent independent of the choice of the domain database. (Figure reproduced from [452].)

One such measure is the **semantic similarity** (ss) score [453], which is defined based on Gene Ontology (GO) annotations [454]. We will discuss GO and this measure in detail in Section 12.5.3.1, but for now we will just mention that this measure is defined in correlation with the probability that two proteins x, y that are annotated with functional keywords $\mathbf{k}_x$ and $\mathbf{k}_y$, respectively, share a keyword by chance. Specifically, each keyword (GO term) k is first associated with a significance score that is the empirical probability $p(k)$ that two proteins selected at random will be associated with that keyword. The semantic similarity of two proteins x and y is defined as

$$ss(x, y) = -\log\left(\min_{k \in \mathbf{k}_{x,y}} p(k)\right)$$

where $\mathbf{k}_{x,y} = \mathbf{k}_x \cap \mathbf{k}_y$ is the set of common keywords. Thus, if two proteins share more than one keyword, then their semantic similarity is defined as the minus logarithm of the probability of the least probable keyword over all common keywords. The least probable keywords are associated with smaller groups of proteins, and the intuition is that smaller groups are likely to correspond to proteins that were studied more closely and characterized in more detail, and therefore they share stronger functional similarity. While this is not always the case, as discussed in Section 12.5.3.1, it serves as a reasonable approximation. The functional similarity among a group of proteins $\mathbf{S}$ can be simply defined as the average pairwise semantic similarity score $\mu_{ss}(\mathbf{S})$. Thus a higher similarity score entails that the protein sequences share closer functional properties. The term Δ_{ss} represents the difference in functional similarity of two groups $\mu_{ss}(\mathbf{S}_1) - \mu_{ss}(\mathbf{S}_2)$.

Figure 9.21a shows the histogram of Δ_{ss} under the string hierarchy for the Pfam database. Not surprisingly, in most cases the average *ss* score of the child domain architecture is higher than that of the parent domain architecture. Figure 9.21b illustrates the evolution of Δ_{ss} as the length difference L between the parent and the child domain architectures increases. As the graph shows, most of the time Δ_{ss} is positive and its magnitude increases with L, which indicates that the functional similarity between proteins increases as additional domains are added to the parent domain architecture.

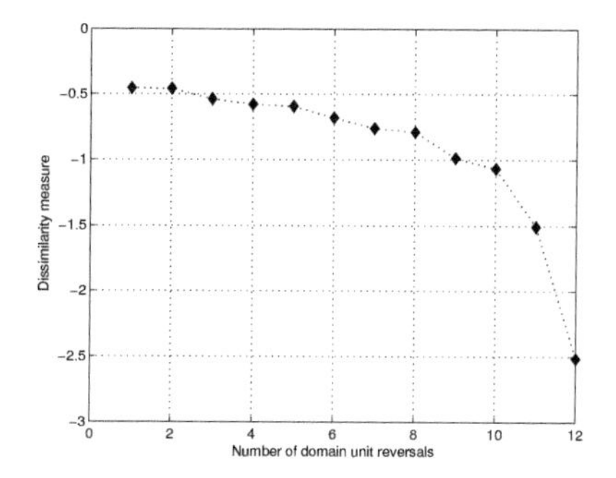

FIGURE 9.22: The sibling relationship. Average dissimilarity vs. the number of reversals for sibling domain architectures in the group hierarchy of the Pfam database. The dissimilarity of two architectures D_1 and D_2 is defined as the difference between the average inter-cluster and intra-cluster similarities of proteins associated to these architectures $\Delta(D_1, D_2) = s_{12} - \min\{s_1, s_2\}$, where $s_i = \mu_{ss}(D_i)$ and s_{12} is the average *ss* score between the two sets of proteins. (Figure reproduced from [452].)

Another interesting relationship in the domain architecture hierarchy is the **sibling** relationship, which corresponds to domain shuffling events. Two domain architectures D_1, D_2 are said to be siblings if one can be derived from the other by reshuffling the order of the constituent domains. In general, different siblings of a given domain architecture may not be similar in terms of their structures and biological functions [447, 451]. But is the functional similarity correlated with the distance between siblings? To answer this question we need a distance measure between architectures. Since a shuffling operation on a string of domains can be understood as a succession of domain reversals of the adjacent elements in the string, one may define a distance measure between two sibling architectures based on the minimum number of adjacent domain reversals separating them. For instance, three reversals are necessary to transform ABC to CBA.

Intuitively, it is expected that there is a higher degree of functional similarity between domain architectures that are separated by a fewer number of shuffling operations, and empirical results are consistent with this hypothesis (Figure 9.22). Indeed, a very different arrangement of domains is likely to result in a substantially different structure, affecting the accessible area for interaction with other molecules (from proteins and DNA to small ligands), and consequently the functional role of the protein in the cell.

9.5.3 Semantically significant domain architectures

When studying the parent-child relationship, one might raise the question whether the insertion of an arbitrary domain will always result in increased specificity and functional similarity. Clearly, some combinations are more functionally "meaningful" than others, but how can we identify those domain insertions that specify a particular group of proteins from the functional standpoint? Interestingly, the problem of finding key domain combinations associated to some particular groups of proteins has parallels in Natural Language Processing (NLP), where the goal is to identify the semantic structure of sentences and to understand their underlying grammar. This problem is also known as the *knowledge acquisition* problem, i.e., learning the linguistic structures from the dataset, which is crucial to the induction of grammar rules [455–457].

Representing protein sequences in terms of their domain architecture may be regarded as parallel to the representation of sentences as sequences of words: the proteins can be perceived as equivalent to sentences and the domains play the role of constituent words in the protein space. By studying the *semantics* of protein architectures we wish to discover the "grammar" of protein sequences and understand the principles that govern the formation of complex architectures from simple ones. This would enable inference of the biological function of a protein based on its domain architecture.

To identify "meaningful" domain combinations that can be associated with a specific functional role, we can search for architectures that impose a significant constraint on the structure of the adjacent domains in the protein architecture. As an illustrating example, consider the sentence

"mouse eats cheese"

The question is whether the word "eats" (or equivalently, the verb "to eat") places a particular constraint on the occurrence of the neighboring words. Obviously, it has many possible extensions on both sides; its left extension may be anyone or any living organism that feeds on other organisms, and the right extension could be any food or living organism as well. Once we append the word "mouse" to the left of "eats", the number of possible right extensions significantly decreases. Similarly, when the word "cheese" is appended to the right-hand side of "eat", the distribution of the possible eaters is considerably different from the background distribution, since many animals do not eat cheese. We note that the limitation on the structure of the adjacent word(s) is directional; "eat + cheese" imposes constraints on the extension to the left of "eat", but not necessarily to the right of "cheese". When studying multi-domain protein sequences we seek similar characteristics of domain architectures.

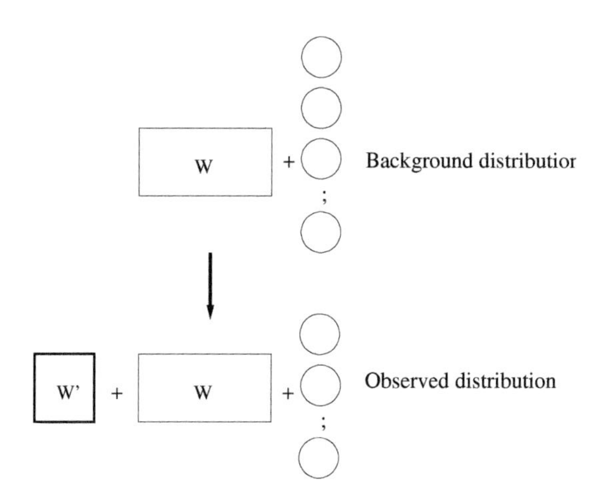

FIGURE 9.23: Testing semantic acquisition with respect to a given domain architecture. This example tests whether the addition of the domain w' to the architecture W (that results in the architecture w'-W) acquires functional semantics by comparing the empirical counts of the right-end extensions to w'-W with the background distribution of the right-end extensions to the architecture W. (Figure reproduced from [452].)

The equivalent of grammatical and semantical constraints in sentences is structural constraints in proteins. These constraints can be detected by comparing the background distribution of the right extensions to an architecture W against the observed frequencies of the right extensions to w'-W (see Figure 9.23). Should the addition of w' impose functional semantics by restricting the nature of the extension to the right of W, the two distributions will significantly deviate from each other. The significance can be measured in terms of the probability (pvalue) of observing such deviation by chance, as described in Section 7.7. Architectures w'-W with small pvalues may be considered to be semantically significant. However, additional research is necessary to determine their biological significance.

9.6 Further reading

The first studies that coined the term "domain" date back to the 1960s. One of the earliest studies that discovered protein domains is by Phillips [458], who reported the existence of distinct substructures of lysozyme, one of the first protein structures that were determined. He noted that the substructures have a hydrophobic interior and somewhat hydrophilic surface. A later study by Cunningham *et al.* [409] separated immunoglobulins into structurally distinct regions that were called domains. They hypothesized that these regions evolved by gene duplication/translocation. A different explanation was suggested by Wetlaufer [410], who was the first to examine multiple proteins and compile a list of their domains. Unlike the work of Cunningham *et al.* that suggested a separate genetic control for each region, Wetlaufer proposed that the structural independence is mostly due to rapid self-assembly of these regions. Later studies established the evolutionary aspect of domains and argued in favor of separate genetic control [459]; however, there is also evidence in support of Wetlaufer's approach [415, 460].

Domain prediction methods can be divided into several categories, as discussed in this chapter. Methods which are based on 3D structure include [421, 461–464], in addition to the algorithms we discussed in Section 9.2.1. Closely related are methods that are based on structure prediction (Section 9.2.2). For example, the algorithm of [465] uses predicted 3D models and then applies the algorithm of [463]. The algorithm of [466] is based on clustering of secondary structure units. Secondary structure information is also utilized in [467], where domain predictions are based on the propensity of amino acids to be in ordered secondary structures (helices and strands) or disordered structures (coils). The method of [468] also uses secondary structure information. It predicts the secondary structure of a protein using PSIPRED [324], aligns the protein to CATH domains based on the secondary structure sequence and then uses the most similar domains to chop the protein.

Methods that use sequence-based features include [440], which uses a neural network to learn and predict linker regions between domains, based on their composition. In a follow-up study [441] they report that low complexity regions also correlate with domain termini and combine both methods to predict domain boundaries. The method of [442] also predicts domain linkers from amino acid composition. Some methods combine several different sequence features. For example, the method of [288] combines three support vector machines (SVMs) that are trained over different feature sets. Biozon uses a neural network trained over 20 different features derived from multiple sequence alignments, intron-exon data and more [354].

Methods based on similarity search include [68, 430, 431, 434, 438, 469, 470]. MSA-based approaches are the basis of several popular domain databases, such as Pfam, SMART and TigrFam (see Section 9.8), as well as of the methods described in [354, 432, 433]. For a review and more details on these algorithms see [412].

Functional analysis of multi-domain proteins has attracted increasing attention over the past several years. For example, in [471] statistical language modeling methods were applied to improve domain recognition in protein sequences, by using a variable-order Markov model to describe the domain sequence of a protein and its context. Another NLP-inspired approach to the analysis of multi-domain structures is presented in [472]. Studies of the evolution of multi-domain proteins are discussed in [416, 450, 473, 474]. Other interesting studies are discussed in [475, 476], focusing on homology inference and function prediction of multi-domain proteins. The study by [475] utilizes a combination of multiple HMMs representing the domains that make up the domain architecture of a protein. The study by [476] analyzes the similarity network of proteins to devise a measure of "neighborhood correlation", reflecting the local structure of the network in the vicinity of each protein, and distinguish between proteins with common ancestry and proteins that share only a domain in common. For updates and additional references see the book's website at `biozon.org/proteomics/`.

9.7 Conclusions

- Protein domains are the building blocks of proteins.
- Although there is no single definition of domains which is widely accepted, domains are generally presumed to be capable of folding independently into compact and stable structures. Each domain is usually associated with a specific function.
- A protein can consist of a single domain or multiple domains, arranged in a specific domain structure (called domain architecture).
- Structure prediction is more successful if the domain structure is known, since the task can be broken into smaller problems (per domain).
- Knowledge of the domain structure of a protein is very important to functional characterization of proteins.
- The most accurate methods for domain delineation are based on the 3D structure of a protein.
- Predicting the domain structure from sequence is still a largely open problem.
- Signals of the domain structure are sometimes reflected in sequence termination signals, correlated mutations, and other structural features.
- Different domain prediction algorithms often produce very different domain structures. Hypotheses regarding the domain structure can be evaluated considering prior information on domains, such as their length distribution and the typical number of domains.
- Multi-domain proteins comprise the majority of proteins in genomes of higher organisms.
- The function of a protein is dictated by its domain architecture. Multi-domain proteins can be organized into a hierarchy of architectures.
- Similar architectures tend to have similar functions.

9.8 Appendix - domain databases

There are many databases of protein domains. Early databases of recurring patterns in proteins focused on short signature patterns or motifs. As was mentioned earlier, these patterns are typically shorter than domains and are usually not considered to be independent folding units. However, it is worth mentioning a few important resources on protein motifs, such as PROSITE [477,478], PRINTS [479,480] and Blocks [481,482].

Domain databases can be categorized into two categories: sequence-based and structure-based. One of the first sequence-based domain databases was ProDom [68]. More recent versions of this database [483] were generated using the MKDOM procedure we described in Section 9.2.3. This DB was later followed by HMM-based domain databases such as Pfam [484,485], SMART [486,487] and TigrFam [488]. In the Pfam database, the construction of new families starts from an HMM model derived from multiple alignment of related proteins, which is then improved iteratively by searching for further related sequences in the database. These sequences are iteratively incorporated into the model, until the process converges. After each iteration the alignment is checked manually to avoid misalignments (the exact procedure has changed over the years). Pfam is separated into two parts, Pfam-A and Pfam-B. Pfam-A is a manually curated version, constructed from high-quality manually verified multiple alignments, while Pfam-B is a fully automated version over the remaining sequences, derived from other domain databases. SMART is a domain database that concentrates on signaling and extracellular proteins. As in Pfam-A, each domain is represented by a profile HMM that is constructed from a manually verified multiple alignment. TigrFam is another database that uses HMMs generated with HMMER [197]. It is similar to Pfam-A, but while Pfam focuses on the sequence-structure relationship when defining domains, TigrFam focuses on the functional aspects and domain families are defined such that sequences have homogeneous functions.

Domain databases that are based on structure analysis of proteins include SCOP [23,352], CATH [332,489], FSSP [333], 3Dee [490] and DDBASE2 [491]. SCOP (Structural Classification of Proteins) is based on an expert analysis of protein structures in PDB [268]. Proteins are visually inspected and chopped into domains based on their compactness, the contact area with other parts of the protein and resemblance to existing domains. The domains are organized into a hierarchy consisting of seven classes, several hundred folds and more than one 1000 protein families. At the bottom level homologous domains are grouped into families based on sequence similarity. At the next level, structurally similar families are grouped into super-families based on functional similarity (common residues in functional sites). Next, structurally similar super-families are grouped into folds. Finally, folds are grouped into seven general classes: all-alpha, all-beta, alpha/beta, alpha+beta, multi-domain proteins, membrane and cell surface proteins and small proteins. SCOP is a popular benchmark for structure comparison algorithms, remote homology detection and domain prediction.

CATH is based on a mixture of automated and manual analysis of protein structures and sequences. Proteins that have significant sequential and structural similarity to previously processed proteins inherit their classification. For other proteins, domain boundaries are manually assigned based on the output of a variety of different structural- and sequence-based methods and the relevant literature. Domains are hierarchically classified into four major levels: Class (all domains with similar secondary structure composition), Architecture (domains with similar secondary structure orientation), Topology (domains whose secondary structure shape and connectivity are similar) and Homology (domains that share significant sequence, structural and/or functional similarity). Hence the acronym CATH.

Dali/FSSP is another structural classification of proteins; however, unlike SCOP and CATH, it classifies complete PDB structures and it is fully automatic. It uses the structure comparison program Dali to perform an all-vs-all structure comparison of all entries in PDB (see Section 8.3.2.1). The structures are then clustered hierarchically based on their similarity score into folds, second cousins, cousins and siblings.

TABLE 9.1: Domain databases

Database Names	Reference	URL
Motif databases		
PROSITE	[478]	http://ca.expasy.org/prosite/
PRINTS	[480]	http://www.bioinf.man.ac.uk/dbbrowser/PRINTS/
Blocks	[482]	http://blocks.fhcrc.org/
Sequence- and MSA-based, automatically generated		
ProDom	[483]	http://prodom.prabi.fr/prodom/current/html/home.php
Biozon	[354]	http://biozon.org/
MSA-based, manually verified		
Pfam	[485]	http://pfam.sanger.ac.uk/
SMART	[487]	http://smart.embl-heidelberg.de/
TigrFam	[488]	http://www.tigr.org/TIGRFAMs/
CDD	[492]	http://www.ncbi.nlm.nih.gov/entrez/query.fcgi?db=cdd
3Dee	[490]	http://www.compbio.dundee.ac.uk/3Dee/
Integrative databases		
InterPro	[493]	http://www.ebi.ac.uk/interpro/
Biozon	[494]	http://biozon.org/
Structure-based		
SCOP	[352]	http://scop.mrc-lmb.cam.ac.uk/scop/
CATH	[489]	http://www.cathdb.info/
Dali/FSSP	[333]	http://ekhidna.biocenter.helsinki.fi/dali
DDBASE2	[491]	http://caps.ncbs.res.in/ddbase/
SBASE	[495]	http://www.icgeb.trieste.it/sbase

A partial list of motif and domain databases is given in Table 9.1. Some of them integrate domain definitions from multiple sources, such as InterPro [493,496], CDD [492], SBASE [495], CDART [497] and Biozon [354,494]. For more information see [412].

9.9 Problems

For updates, additional problems, files and datasets check the book's website at `biozon.org/proteomics/`.

1. **Structure-based domain prediction**

   Write a program that given a protein structure identifies clusters of close residues, using the agglomerative clustering algorithm of Section 9.2.1. Use the distribution of pairwise distances to determine a threshold distance above which pairs of residues are more likely to belong to different domains. Input files, formats and tips are available at the book's website.

2. **Contact profile from structure**

   Given a protein structure, compute the contact profile of each residue (i.e., the number of neighboring residues within a certain threshold distance) and predict domain boundaries based on minima in the contact profile. Repeat the procedure for different threshold distances and compare the results to the true domain boundaries.

3. **Contact profile based on correlated mutations**

   You are given an MSA. Compute the correlated mutations score for each pair of columns and generate a predicted contact profile based on the score, as in Section 9.2.2. Analyze the MSAs of the structures given in the previous problem, and predict domain boundaries based on their predicted contact profile. Compare the predictions you obtained to those you got based on structural information.

4. **Contact profile based on sequence termination**

   Given an MSA, compute for each position the number of sequences which are aligned to both sides of that position. Normalize and generate a predicted contact profile. Predict domain boundaries based on this contact profile. Are the predictions consistent with those you get based on correlated mutations?

5. **Computing the prior probability of a domain partition**

   Describe a dynamic programming algorithm to calculate $\text{Prob}(n, L)$ (as in Equation 9.3) efficiently.

Part II

Putting All the Pieces Together

Chapter 10

Clustering and Classification

10.1 Introduction

Clustering is one of the most popular exploratory data analysis techniques and is fundamental to all domains of science. The goal of clustering is to identify clusters in the data that may correspond to classes and subclasses with specific properties. Identifying these classes and characterizing their properties can provide insight into the structure and nature of the data and help with classification of new instances.

We already discussed the problem of classification in Chapters 6 and 7, but the setting there was of supervised learning, where the samples are labeled and the goal was to design a "good" classifier (e.g., a discriminant function that assigns each sample to its correct class). In cluster analysis the samples are **unlabeled** and the goal is to explore the structure of the data without any assumptions regarding the type and the structure of the model, such as the functional form of the classifier or the type of the underlying distributions. As such, clustering is an **unsupervised learning** technique, which can identify patterns and classes in unlabeled data. With cluster analysis, we first search for the presence of groups of data instances with similar properties. Each group is considered a class, and instances are classified based on these classes.

The global organization of the protein space is inherently involved with problems of unsupervised learning. One of our main goals is to detect classes in the data that correspond to protein families (i.e., **protein classification**). The underlying assumption is that proteins that belong to the same class have similar physical properties and share a common function. If the function of some of the proteins that belong to a certain class is known, then it can be extrapolated to other, uncharacterized members of the same class (this type of inference is sometimes called **"guilt by association"**). With sequence databases expanding in almost an exponential rate, the ability to automatically infer a function from sequence is of major importance. As we will see in Section 10.6, functional analysis of proteins based on clustering can be much more effective and refined than inference based on a simple database search.

Another important goal when mapping the protein space is to study higher order organization that might link several protein families together, suggesting a possible common ancestry and similar biological functions, or hierarchical organization that would reveal the existence of subgroups with slightly different functions within a protein family. Hierarchical organization of families may also shed light on the evolution of the protein space.

Clustering is one of the tools we can apply to address these problems. More generally, clustering is useful whenever we want to find groups of similar entities and is relevant to many other problems in computational biology, such as expression data analysis (see Chapter 12) and motif discovery. Since it does not require any prior information, it can be applied quite easily to data sets of varying complexity. Some clustering algorithms can work even on massive datasets.

Clustering can have different roles, depending on the task. For example, clustering can be used to compress data. By grouping data points into clusters and representing each cluster using its centroid, data can be compactly represented and efficiently transmitted. This procedure is often called **vector quantization**. To minimize loss of information with vector quantization, data is clustered so as to minimize certain global distortion measures. The number of clusters depends on external parameters, such as the capacity of the channel through which the data is to be transmitted, the desired speed of data transfer, or the desired compression ratio. Therefore, clusters do not necessarily correspond to natural subclasses in the data and might not have actual meaning. The more common use of clustering is to obtain a reliable model of the data that can be applied for classification of new instances. In this case, our goal is to cluster data so as to maximize the success of the model in prediction and **generalization** to new instances.

Clustering is strongly affected by the representation of the data. Particularly, the shapes and the sizes of the clusters depend on the specific set of attributes that are used to represent data instances and on the function that measures the similarity or distance between instances. We touched upon the issue of representation in Chapter 7. In the next chapter we will discuss another popular unsupervised learning technique called *embedding*, which is very relevant to the issue of representation. Unsupervised learning techniques are sometimes referred to as **pattern recognition**.

In this chapter we will discuss several different approaches to clustering, their properties, advantages and disadvantages. The assessment and interpretation of the results is another important issue we have to address. The results can be highly sensitive to the choice of the algorithm or the algorithm's parameters, and often it is unclear if the results obtained represent the "correct" clustering. Even the concept of "correct" clustering is not well defined. In the second part of this chapter we will discuss how to evaluate clustering solutions. Finally, in the third part we will discuss the application of clustering algorithms to the protein space.

10.2 Clustering methods

All clustering algorithms attempt to group data points into clusters of similar patterns. A cluster can be defined as a set of entities that resemble one another more than they resemble entities in other clusters. It can also be defined as a set of sample points such that the maximum **intra-cluster** distance (i.e., the distance between points within the cluster) is smaller than the minimum **inter-cluster** distance (i.e., the distance between points from different clusters). In the literature on clustering one can find other definitions as well, and there is no single definition that is universally accepted.

Many clustering algorithms were proposed over the years. Different algorithms have different properties, and since not all algorithms are applicable to all problems, it is important to consider these differences when choosing the algorithm to study a specific dataset of interest.

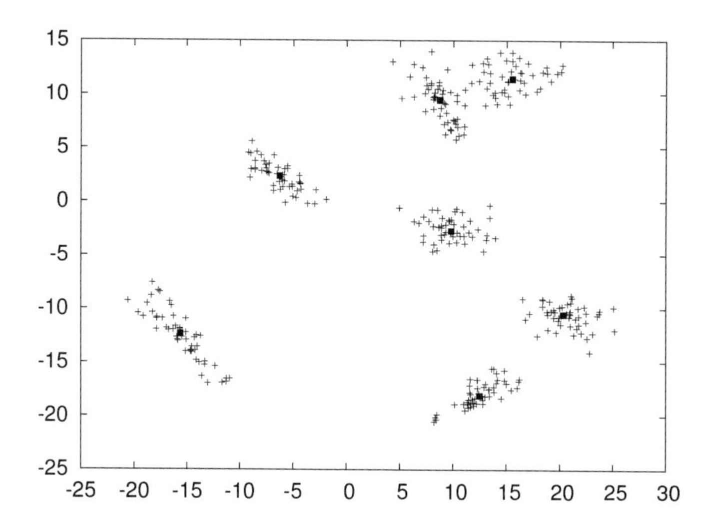

FIGURE 10.1: **Clusters in 2D.** A mixture of seven Gaussians with different shapes and orientations, some of them are partly overlapping. The means of the Gaussians are denoted by black boxes.

For example, algorithms can be distinguished by the type of the input data they can process. Usually the dataset is composed of objects, where each object is represented as a point in a d-dimensional metric space $\mathbb{R}^d$ (measurement space) and clusters can be viewed as dense sets of points in this space, with different shapes and orientations (Figure 10.1). To measure the similarity or dissimilarity between objects we can use the Euclidean distance between the points, or other norms over $\mathbb{R}^d$. In some cases the objects are represented only by their pairwise distances/similarities. The first case is referred to as **pattern matrix** or vector data, while the second case is referred to as **proximity matrix** or proximity data. Note that every pattern matrix can be converted into a proximity matrix by computing the distances or similarities between all pairs first. Hence, all clustering algorithms can be applied to the first type of data. Algorithms for clustering data of the first type are referred to as **vector-space clustering**, while algorithms of the second type are often called **pairwise clustering** or **graph-based algorithms**. When clustering proteins, the second type is generally of greater interest, although vector-space clustering algorithms can be used as well if the data is represented as feature vectors or embedded in real-normed space as discussed in Chapter 11.

Clustering algorithms also differ by the availability of the data, i.e., **on-line** (the samples are introduced one by one) vs. **batch** (all samples are stored in the memory and are accessible at any step of the algorithm). On-line algorithms are important when the size of the input data becomes unmanageable by batch algorithms, in terms of memory and CPU resources.

Some algorithms are based on parametric modeling of clusters and work by minimizing or maximizing a certain function that depends on the model. Some algorithms do not assume a model but optimize a certain cost function that depends on the pairwise distances. Other algorithms are completely heuristics and do not attempt to optimize any specific function.

Another distinction is made between exclusive (hard) vs. non-exclusive (fuzzy) clustering. In **hard clustering** each point belongs to exactly one cluster, while in **fuzzy clustering** the assignment is probabilistic as a result of uncertainty. That is, each point belongs to all clusters with finite probabilities. Most algorithms assume a hard clustering model but usually can be generalized to the fuzzy clustering model. Probabilistic assignments can also be generated in a post-processing step. An example of a fuzzy clustering model would be a mixture of k distributions, one for each cluster, so that the probability of a sample $\vec{x}$ is

$$p(\vec{x}) = \sum_{j=1}^{k} p(\vec{x}|C_j)p(C_j)$$

where $p(C_j)$ is the prior probability of cluster j and $p(\vec{x}|C_j)$ is the class conditional probability distribution. In Section 10.3.2 we will discuss such a model where the class conditional probability distributions are normal distributions.

Clustering algorithms can also differ in the algorithm scheme, i.e., hierarchical vs. partitional clustering. **Partitional clustering** algorithms partition a finite set of objects into clusters so as to optimize a certain criterion function. The term **hierarchical clustering** is used to describe techniques for obtaining a sequence of nested clusters. Most pairwise clustering algorithms (see Section 10.4.1) follow this hierarchical scheme. However, many clustering algorithms employ hierarchical schemes that do not necessarily result in nested groupings (Section 10.3.3).

It is important to note that there is no single clustering algorithm or scheme that outperforms all other clustering algorithms on all possible datasets (see also the "no free lunch" theorem of Section 10.5.2.1). Therefore, the decision on which algorithm to use is data dependent. In this chapter we will divide clustering algorithms into two main categories based on the type of data we have to analyze. If the data is represented as a proximity matrix we can apply **graph-based algorithms**. If data is vectorial and represented as a pattern matrix, then we can *also* apply **vector-space clustering algorithms**. We will start the discussion with the second category. In Section 10.5 we will discuss methods for assessing the "success" of different clustering algorithms.

10.3 Vector-space clustering algorithms

Vector-space clustering algorithms assume that the data reside in a real-normed space, such that each sample point is represented as a vector in $\mathbb{R}^d$. Almost all these algorithms also assume that the input space is Euclidean, and the metric used to assess the similarity of data points is the Euclidean metric (see Section 3.13). Other metrics can be plugged in, but unlike the Euclidean metric, for which some of these algorithms have closed-form solutions or proven properties, they are not guaranteed to converge to solutions of established properties. We will start with partitional algorithms.

Given a dataset $\mathbf{D} = \{\vec{x}_1, \vec{x}_2, ..., \vec{x}_n\}$ the goal of partitional clustering algorithms is to partition "optimally" the dataset into k mutually disjoint sets $\mathbf{D} = C_1 \cup C_2 \cup ... \cup C_k$, such that each sample is assigned to one set, and $C_i \neq \emptyset \quad \forall i$. This definition is very general; it requires optimality without specifying when clustering is considered optimal and without addressing the means by which a clustering should be evaluated.

In the classical clustering schemes one usually selects a clustering criterion or an objective function and sets the desired number of clusters k. The best **configuration** (the best partition of sample points into clusters) is the one that optimizes the objective function. For example, the **squared error** objective function is defined as follows. Given n samples $\mathbf{D} = \{\vec{x}_1, \vec{x}_2, ..., \vec{x}_n\}$, where $\vec{x}_i \in \mathbb{R}^d$, and an initial partition of $\mathbf{D}$ into sets $C_1, C_2, ..., C_k$, define the **center** of the j-th cluster as the average of the samples that are classified to the j-th set, i.e.,

$$\vec{\mu}_j = \frac{1}{n_j} \sum_{\vec{x}_i \in C_j} \vec{x}_i$$

where n_j is the number of samples in C_j. The center (also called **centroid**) is a representative of the cluster. The squared error for cluster j is defined as

$$e_j^2 = \sum_{\vec{x}_i \in C_j} \|\vec{x}_i - \vec{\mu}_j\|^2$$

and the **total squared error** (also called **global distortion**) of a specific clustering configuration is given by the sum over all clusters

$$E_k^2 = \sum_{j=1}^{k} e_j^2 \tag{10.1}$$

Note that of all possible choices of cluster representatives, the centroids is the choice that minimizes the total squared error (see Problem 10.2).

Once an objective function has been chosen, the optimal partition into clusters is defined as the one that optimizes that function. Theoretically this optimization problem can be solved by evaluating the objective function for all possible partitions. However, the size of the search space (number of possible configurations) for n sample points and k clusters is at least exponentially large (see Problem 10.1). Therefore, an exhaustive enumeration is impossible in practice, and the search techniques that are usually employed are either iterative, "hill-climbing" techniques or are based on sampling the search space.

10.3.1 The k-means algorithm

The **k-means** algorithm is a simple and popular clustering algorithm that partitions the data so as to minimize the squared error function. It is a heuristic algorithm that is based on an iterative hill-climbing scheme. The algorithm has one parameter that needs to be set by the user: the number of clusters k. Later we will discuss methods to determine k (Section 10.5). The outline of the algorithm is as follows:

The k-means algorithm

Initialize: Select a centroid for each cluster $\vec{\mu}_1, \vec{\mu}_2, ..., \vec{\mu}_k$.
Iterate:

(1) For each sample $\vec{x}_i$
  For each cluster $j = 1..k$
    Compute the distance between the sample and the cluster:
    $d(\vec{x}_i, C_j) = \sqrt{(x_{i1} - \mu_{j1})^2 + ... + (x_{id} - \mu_{jd})^2}$
  Classify $\vec{x}_i$ to the closest cluster C_j:
    $class(\vec{x}_i) = \arg\min_j d(\vec{x}_i, C_j)$
(2) For each cluster $j = 1..k$
  Recompute the centroid $\vec{\mu}_j = \frac{1}{n_j} \sum_{\vec{x}_i \in C_j} \vec{x}_i$

Terminate: if centroids have not changed

During initialization, the centroids can be selected arbitrarily in $\mathbb{R}^d$ or by selecting randomly k of the n data points as the centroids, thus ensuring that the initial centroids are confined within the input dataset. Note that a selection of centroids is equivalent to selecting an initial partition into k clusters. Once initialized, the algorithm iterates between classifying samples and recomputing the centroids. If cluster memberships haven't changed then the centroids haven't changed either and we can terminate the loop.

For massive datasets, a stepwise procedure that is similar to the k-means algorithm can be devised by considering one example at a time and updating the means every time. This procedure is useful for online clustering but it is more susceptible to noise.

The time complexity of the k-means algorithm is $O(c \cdot n \cdot k \cdot d)$ where c is the number of iterations until convergence. Since c is usually much smaller than the number of data points n (as are also the number of clusters k and the dimension d), the algorithm is quite fast with time complexity $< O(n^2)$. However, for large datasets with overlapping clusters the algorithm might take many iterations to converge with a few points changing their assignment every iteration. These small changes are usually insignificant, and we can relax the termination criterion and stop when the differences between the current centroids and the centroids of the previous iteration are smaller than a preset small threshold ϵ:

$$d(\vec{\mu}_j(t), \vec{\mu}_j(t-1)) < \epsilon \qquad \forall j$$

For the squared error function, the optimal set of representatives is the cluster centroids, and the k-means algorithm ensures that the global distortion decreases or remains unchanged with every iteration (see Problem 10.3). Since the error depends only on the distance from the center, this function favors spherical distribution of points around the centers where samples are equally distributed in all directions. This objective function, however, is not appropriate for every dataset. For example, it might perform poorly if the dataset contains many elongated non-spherical clusters. We can approximate other ellipsoid distributions by assuming a model for the underling distribution (e.g., normal) and estimating the parameters of the distribution empirically, as described in the next section. In general, there is no single "best" objective function that works well on all datasets. Therefore, the optimal clustering configuration (for a specific function) may not recover the true clusters.

The success of this algorithm also depends on the initial choice of the parameters: the number of clusters and the initial centroids. While the algorithm is deterministic, the output may vary a great deal depending on the initial centroids and small changes may result in completely different clusters upon convergence. In practice, the algorithm often converges to a locally optimal solution of the criterion function rather than the global optimum. Yet, k-means remains popular because of its simplicity and tractability[1].

10.3.2 Fuzzy clustering

One of the issues with the k-means schema described above is that the assignments are "hard", meaning each sample is classified to a single cluster. If clusters overlap significantly, the k-means algorithm might create arbitrary partitions, depending on the initialization. In this case, **fuzzy clustering** is recommended. Fuzzy clustering assumes a mixture of k distributions so that the probability of a sample $\vec{x}$ is

$$p(\vec{x}) = \sum_{j=1}^{k} p(\vec{x}|C_j)p(C_j) \tag{10.2}$$

where $p(C_j)$ is the prior probability of cluster j and $p(\vec{x}|C_j)$ is the class conditional probability distribution. This function is called a **mixture density** function.

We mentioned earlier that the goal of clustering is to detect structure without making assumptions on the nature of the data and the underlying distributions. However, from practical considerations, such assumptions are often necessary to simplify the mathematical formulation. With the mixture density function we assume that we know the functional form of the class conditional distributions, and every distribution is characterized by a parameter vector $\vec{\theta}_j$ such that

$$p(\vec{x}|\vec{\theta}) = \sum_{j=1}^{k} p(\vec{x}|C_j, \vec{\theta}_j)P(C_j)$$

Finding a fuzzy clustering solution means determining the parameter vector $\vec{\theta}$, which is composed of the parameter vectors of all clusters $\vec{\theta}_1, \vec{\theta}_2, ..., \vec{\theta}_k$. For example, if $p(\vec{x}|C_j, \vec{\theta}_j)$ is a normal distribution then $\vec{\theta}_j$ consists of the mean vector $\vec{\mu}_j$ and the covariance matrix Σ_j. Once the parameters are defined

[1]Closely related to k-means is the **Self-Organizing Maps (SOMs)** algorithm which was introduced by Kohonen [498]. SOMs combine elements from clustering algorithms and embedding algorithms (see Chapter 11). This neural-network based algorithm utilizes an unsupervised learning approach inspired by the k-means algorithm to generate a topographic grid-like mapping of the input dataset in N dimensional lattice, where neighboring input points are mapped to neighboring points in the grid. Thus, this algorithm reduces infinite input spaces into a finite number of representative points, which may correspond to classes or categories of the data (similar to the vector quantization approach we mentioned in the introduction). Usually, the lattice resides in a low-dimensional space (2D or 3D), which also helps in visualization of the data. Applications of this algorithm are especially common in computational neurobiology, when modeling stimuli and response in biological systems. For example, in sensory systems, physical proximity in the input (e.g. retina, skin) translates to physical proximity in the neural system in the corresponding cortex. Mappings obtained with SOMs ensure robustness with regards to slight changes in the input, and as such capture one of the main aspects of many real neural networks.

we can calculate the probability of each sample to belong to each one of the clusters.

The Bayesian approach to solving this problem is to look for the parameters that have the maximal posterior probability given the dataset D

$$p(\vec{\theta}|D) = \frac{p(D|\vec{\theta})p(\vec{\theta})}{p(D)}$$

If the prior probability over the parameter space is uniform then this is reduced to searching for the set of parameters that maximizes the likelihood of the data given the model $p(D|\vec{\theta})$ where

$$p(D|\vec{\theta}) = \prod_{i=1}^{n} p(\vec{x}_i|\vec{\theta})$$

or the logarithm of the likelihood, which is easier to work with

$$\ln p(D|\vec{\theta}) = \sum_{i=1}^{n} \ln p(\vec{x}_i|\vec{\theta}) \tag{10.3}$$

A rigorous solution can be computed, as discussed in [191]. Denote by $\hat{\theta}$ the value of θ that maximizes $p(D|\theta)$ and $\ln p(D|\theta)$ (vector notation is ignored for clarity). To find $\hat{\theta}$ we need to compute the derivative of Equation 10.3 with respect to each component of the parameter vector

$$\nabla_{\theta_j}[\ln p(D|\theta)] = \sum_{i=1}^{n} \frac{1}{p(x_i|\theta)} \nabla_{\theta_j} \left[\sum_{j'=1}^{k} p(x_i|C_{j'}, \theta_{j'}) P(C_{j'}) \right]$$

Assuming that θ_j is independent from the parameters of all other distributions in the mixture density, we can discard the other components leaving us with

$$\nabla_{\theta_j}[\ln p(D|\theta)] = \sum_{i=1}^{n} \frac{1}{p(x_i|\theta)} \nabla_{\theta_j} [p(x_i|C_j, \theta_j) P(C_j)]$$

Using Bayes' formula

$$P(C_j|x_i, \theta) = \frac{p(x_i|C_j, \theta_j) P(C_j)}{p(x_i|\theta)} \tag{10.4}$$

then

$$\nabla_{\theta_j}[\ln p(D|\theta)] = \sum_{i=1}^{n} \frac{P(C_j|x_i, \theta)}{p(x_i|C_j, \theta_j) P(C_j)} \nabla_{\theta_j} [p(x_i|C_j, \theta_j) P(C_j)]$$

or

$$\nabla_{\theta_j}[\ln p(D|\theta)] = \sum_{i=1}^{n} P(C_j|x_i, \theta) \nabla_{\theta_j} \ln p(x_i|C_j, \theta_j)$$

and as an extremum of the likelihood function, the maximum likelihood estimator $\hat{\theta}_j$ must satisfy

$$\sum_{i=1}^{n} P(C_j|x_i, \hat{\theta}) \nabla_{\theta_j} \ln p(x_i|C_j, \hat{\theta}_j) = 0 \qquad j = 1..k$$

This solution assumes that the priors $P(C_j)$ are known or given but it can be generalized to the case of unknown priors. The Maximum Likelihood (ML) estimators for θ_j and $P(C_j)$ are obtained by maximizing $\ln p(D|\theta)$ subject to the constraints that $P(C_i) \geq 0$ and $\sum_{j=1}^{k} P(C_j) = 1$. As before, the optimal parameters must satisfy

$$\sum_{i=1}^{n} \hat{P}(C_j|x_i, \hat{\theta}) \nabla_{\theta_j} \ln p(x_i|C_j, \hat{\theta}_j) = 0 \qquad j = 1..k \tag{10.5}$$

and it can also be shown that

$$\hat{P}(C_j) = \frac{1}{n} \sum_{i=1}^{n} \hat{P}(C_j|x_i, \hat{\theta}) \tag{10.6}$$

Thus the prior for a category is the average of its posterior probability per sample over the entire dataset, where samples are weighted equally.

These solutions are general without specifying the type of class conditional distributions. Once the parametric form of these distributions is chosen, we need to plug them in and solve Equation 10.5, which depending on the type of the distributions might be difficult. A case of special interest is the mixture of normal distributions where each cluster is associated with a multivariate normal distribution

$$p(\vec{x}|C_j) \sim N(\vec{\mu}_j, \Sigma_j) = \frac{1}{(2\pi)^{d/2}|\Sigma_j|^{1/2}} \exp\left[-\frac{1}{2}(\vec{x} - \vec{\mu}_j)^t \Sigma_j^{-1} (\vec{x} - \vec{\mu}_j)\right]$$

where $\vec{\mu}_j$ is the mean vector and Σ_j is the covariance matrix. A priori there is no reason to assume that the underlying distributions are normal. However, by assuming a mixture of normal distributions it is possible to approximate quite complicated sample distributions (see Section 11.4.2.5).

Plugging in the normal distributions in Equation 10.5 we can get closed-form solutions for the priors and the parameters $\vec{\mu}_j$ and Σ_j (see [191])

$$\hat{P}(C_j) = \frac{1}{n} \sum_{i=1}^{n} \hat{P}(C_j|x_i, \hat{\theta})$$

$$\hat{\mu_j} = \frac{\sum_{i=1}^{n} \hat{P}(C_j|x_i, \hat{\theta}) x_i}{\sum_{i=1}^{n} \hat{P}(C_j|x_i, \hat{\theta})}$$

$$\hat{\Sigma}_j = \frac{\sum_{i=1}^{n} \hat{P}(C_j|x_i, \hat{\theta})(x_i - \hat{\mu}_j)(x_i - \hat{\mu}_j)^t}{\sum_{i=1}^{n} \hat{P}(C_j|x_i, \hat{\theta})}$$

where the parameters are defined as weighted averages of the samples ($\hat{\mu_j}$) or their covariance ($\hat{\Sigma}_j$). Note that these solutions depend on the posterior probabilities $\hat{P}(C_j|x_i, \hat{\theta})$ (which act as the weights in the equations above) where

$$\hat{P}(C_j|x_i, \hat{\theta}) = \frac{p(x_i|C_j, \hat{\theta}_j)\hat{P}(C_j)}{\sum_{j'=1}^{k} p(x_i|C_{j'}, \hat{\theta}_{j'})\hat{P}(C_{j'})}$$

and $p(x_i|C_j, \hat{\theta}_j)$ is determined by $\hat{\mu_j}$ and $\hat{\Sigma}_j$. Therefore the solution to equations 10.5 and 10.6 depends on the solutions for the posterior probabilities, which in turn depend on the solutions for the parameters. Hence, these exact solutions involve complex non-linear equations and are far from being trivial. A practical solution is to apply an iterative Expectation-Maximization (EM) procedure to find locally optimal solutions (see Section 5.7). One such algorithm is the **fuzzy k-means** algorithm, which generalizes the k-means algorithm and results in a fuzzy assignment of samples to clusters. The algorithm works iteratively, first defining the set of parameters that maximize the likelihood according to the current configuration, and then updating the configuration based on these parameters.

The fuzzy k-means algorithm

Initialize: $t = 0$, k, ϵ, $\mu_1(0)$,...,$\mu_k(0)$, $\Sigma_1(0)$,...,$\Sigma_k(0)$, $P(C_1)$,..., $P(C_k)$.

Do $t = t + 1$
  For $j = 1..k$
    For $i = 1..n$
      Compute $P(C_j|x_i, \theta)$
    Recompute μ_j, Σ_j and $P(C_j)$
Until change in parameters is below threshold ϵ

Return: μ_1, μ_2,..., μ_k, Σ_1, Σ_2,..., Σ_k, $P(C_1)$,..., $P(C_k)$

Why is this an expectation-maximization procedure? The hidden variables are the assignments of samples to clusters, the expectation step is the average over all possible assignments as manifested in the computation of $P(C_j|x_i, \theta)$, and the maximization step is the computation of the solutions to Equations 10.5 and 10.6, which maximize the likelihood of the data.

This algorithm can be computationally intensive and might fail if one or more of the covariance matrices are singular (especially if the number of samples is small), since the inverse matrices are required to compute $p(x_i|C_j)$. A simpler version of this algorithm is obtained by assuming diagonal covariance matrices and ignoring off-diagonal elements when computing the probabilities $p(x_i|C_j)$. In other words, we assume there is no correlation between the different coordinates or features. Computing the inverse matrices is much simpler in this case, since $\Sigma^{-1}(l, l) = 1/\Sigma(l, l)$. (The complete covariance matrices can still be computed and used later to assess the validity of this assumption.) Since the covariance matrices are not fully utilized in the iterative process, the termination test checks for convergence only for the means.

Note that the k-means algorithm is an approximation of the fuzzy k-means, as pointed out in [191]. To simplify the calculations the k-means algorithm implicitly assumes scalar covariance matrices, with a fixed scalar. Therefore, during the iterative calculation of the means all we need to know is the previous means. Exploiting the fact that the posterior probability $\hat{P}(C_j|x_i,\hat{\theta})$ is large when the **Mahalanobis distance** $(x_i-\hat{\mu}_j)^t\hat{\Sigma}_j^{-1}(x_i-\hat{\mu}_j)$ is small, the algorithm approximates the Mahalanobis distance with the Euclidean distance $||x_k - \hat{\mu}_i||^2$, and sets $\hat{P}(C_j|x_i,\hat{\theta}) = 1$ for the closest cluster and 0 otherwise. However, when the procedure converges, we can estimate the actual distribution around each mean, and usually we will observe different standard deviations for each cluster. The assumption of scalar covariance matrices is a strong assumption, yet it works fairly well in many cases. When the overlap between clusters is small the EM approach and k-means clustering give similar results. Otherwise, k-means can be used as a starting point for fuzzy k-means.

Since fuzzy k-means can be computationally intensive, it is worth mentioning that we can also produce fuzzy assignments even with hard clustering algorithms by post-processing the results. For example, we can apply an algorithm like k-means and after convergence model each cluster with a Gaussian, estimating its parameters from the samples that were assigned to that cluster. These distributions can then be used to compute class-conditional probabilities for all sample points. However, these solutions do not have any established property and are not guaranteed to maximize the likelihood of the data.

10.3.3 Hierarchical algorithms

As opposed to partitional clustering algorithms, hierarchical clustering algorithms work by *gradually* building a clustering solution for the input dataset, by either dividing or merging clusters. **Agglomerative** procedures follow a bottom-up approach where the initial configuration consists of singleton clusters, which are gradually merged to form bigger clusters. On the other hand, **divisive** procedures follow a top-down approach, starting from one big cluster containing all samples that is gradually split into smaller clusters.

One of the appeals of hierarchical clustering algorithms is that they reduce the dependency of the results on the initial configuration or the initial parameters. In some applications the hierarchical structure itself is correlated with features of the data and is therefore of interest as well. Two representative examples of divisive procedures are discussed next. In Section 10.4 we will discuss other, graph-based hierarchical algorithms.

10.3.3.1 Hierarchical k-means

There are several variations to the k-means algorithm that operate hierarchically. All work more or less the same, starting with a single cluster and then iteratively splitting one or more of the clusters in each round. At each level of the hierarchy we can apply a basic clustering procedure, such as k-means, to partition the data. The initial points for the centroids are defined based on the centroids at the previous level, with the addition of duplicate copies of the clusters to be split. To make sure that a cluster splits, its duplicated centroids should differ slightly.

Splitting clusters arbitrarily might result in suboptimal solutions and will force partitions even if they are not necessary. Splitting one cluster at a time is safer, as the change to the configuration would be mostly local and involve only a fraction of the sample points, although the split is likely to affect neighboring clusters as well, especially if the clusters overlap. To improve the chances we converge to a "good" solution we can analyze the data before splitting, to decide whether and which cluster to split. For example, assuming that clusters with large scatter consist of samples from more than one class, we can compute the covariance matrix of each cluster and pick the cluster with the largest variance along any of the axes (in Section 11.2.1.1 we will see that the largest variance equals the principal eigenvalue of the covariance matrix). However, this is not always the case, and clusters might form arbitrary shapes. We can stop splitting if clusters fail certain tests, for example if clusters are spherical or are too small. We will touch on these issues again in Section 10.5.2.2.

10.3.3.2 The statistical mechanics approach

It is interesting to consider a different perspective of clustering based on analogy with physical systems. The **deterministic annealing** algorithm is one such example [499]. It is a hierarchical clustering algorithm that applies concepts from statistical mechanics, such as free energy and entropy, to the clustering problem. The dataset is viewed as a physical system where every possible partition of data points into clusters is considered as one possible configuration of the system. A specific fuzzy clustering solution, with a given set of representatives and a total distortion E, is equivalent to a state of the system whose total free energy $F = E - TS$ is defined in terms of the temperature T, the distortion E (energy) and the entropy S of the distribution of partitions (configurations) with average distortion E. The higher the number of possible configurations with distortion E, the greater the uncertainty we have about the solution.

It is well known that physical systems are likely to adopt the state with the minimum free energy, since this solution satisfies the fundamental principles of minimal energy and maximum entropy. The optimal state changes when the temperature of the system changes, since the degree of order within the system increases as the temperature decreases (the entropy decreases). At zero temperature the unique configuration with the minimum energy E is also the configuration with the minimum free energy F. This state is equivalent to a hard clustering solution. At temperature $T > 0$ there is no single dominant configuration. Rather, the state of the system is described in terms of a distribution over all possible configurations. Consequently, the samples have different probabilities of being assigned to the different clusters and hence the fuzziness. This distribution changes with the temperature.

To find the optimal state at each temperature, the algorithm uses an annealing process starting from the solution at a very high temperature, where all samples form what is essentially a single large cluster. The temperature is then lowered gradually, and a solution is found at each temperature. The process traces the system through a series of **phase transitions**, each of which corresponds to a solution at a different temperature (meaning, by analogy, a different scale or resolution of the analysis). For more details see [499].

The deterministic annealing approach suggests that clusters split at a temperature that depends on the distribution of points within the cluster, and that the split temperature is proportionate to the reciprocal of the largest eigenvalue of the covariance matrix. Although this relation was established only for a single cluster, the qualitative characteristics remain when multiple clusters exist [499]. During the cooling schedule, the first cluster to split is the one that has the largest eigenvalue. By replicating its centroid and allowing the two copies to be positioned along the principal direction, we increase the chances that the split happens along this axis. The split criterion for the hierarchical k-means algorithm described in the previous section is based on this intuition. By splitting the cluster with the longest aspect ratio in each iteration the hierarchical k-means algorithm essentially jumps from one phase transition to another without repeating the process for temperatures at which the clustering solution does not change.

10.4 Graph-based clustering algorithms

Graph clustering algorithms are of great interest when the data is given as a proximity matrix containing pairwise similarities or dissimilarities. However, they are not confined to proximity data and can be applied to vector data as well, simply by first calculating the distances between the vectors.

In terms of graph theory we can think of the sample set as a weighted directed graph $\mathbf{G} = (\mathbf{V}, \mathbf{E})$, where the nodes of the graph (the set $\mathbf{V}$) are the sample points and an edge is formed between pairs of nodes. The set of all edges is denoted by $\mathbf{E}$. Graph-based algorithms consider clustering as a graph partitioning problem over the graph $G(\mathbf{V}, \mathbf{E})$. A clustering $\mathbf{C} = \{C_1, C_2, ..., C_k\}$ is a partitioning of $\mathbf{V}$ into nonempty mutually disjoint subsets $C_1, C_2, ..., C_k$ where $k \leq n$.

The clusters are determined based on the edges in the graph. Each edge is associated with a weight $w(i,j)$, which is a function of the similarity/dissimilarity of the two sample points it connects, i and j. To simplify the analysis, we usually assume that the proximity matrix $\mathbf{W}$ is symmetric, i.e., $w(i,j) = w(j,i)$. If this is not the case, then the matrix should be symmetrized first, for example by averaging or setting $w(i,j) = \min\{d(i,j), d(j,i)\}$ or $w(i,j) = \max\{s(i,j), s(j,i)\}$, depending on the measure used to assess the proximity of sample points.

Graph clustering algorithms are very popular since they are relatively fast. Moreover, unlike k-means or fuzzy k-means, which assume normal distributions and introduce many parameters, graph algorithms do not make assumptions on the nature of the data and therefore are less likely to impose a structure on the data. In that sense, they are considered to be **non-parametric** methods. Reinterpreting the definition of clustering as stated in Section 10.2, graph clustering algorithms work by seeking groups of data points that possess strong mutual similarities. In the next sections we will discuss several types of graph algorithms.

10.4.1 Pairwise clustering algorithms

Pairwise clustering algorithms are among the most popular clustering algorithms. The steps of a pairwise clustering algorithms are equivalent to operations on the graph (e.g., deleting/adding edges), and the resulting clusters are equivalent to sets of nodes in the graph.

Almost all pairwise clustering algorithms share the same basic hierarchical scheme, which partitions the data into nested groupings. Traditionally these algorithms were called hierarchical clustering, but as was mentioned earlier, the term hierarchical clustering is also used to describe hierarchical schemes producing non-nested groupings (Section 10.3.3). Hierarchical algorithms can be divided into two classes: agglomerative and divisive. A few divisive algorithms were already described above in Section 10.3.3. The agglomerative scheme requires fewer computations to proceed from one level to another, and therefore it is used more frequently. Most pairwise clustering algorithms follow the agglomerative scheme.

Given n samples with pairwise dissimilarities $d(i,j)$ $(1 \leq i,j \leq n)$, the basic agglomerative clustering is very simple. Denote by k the number of clusters, and let k_0 be the desired number of clusters. Then the algorithm works as follows:

Pairwise clustering

Initialize: Let $k = n$ and initiate the clusters $C_i = \{i\}$ $\; i = 1..n$
Loop: If $k = k_0$, stop.

- Find the nearest (most similar) pair of distinct clusters, say C_i and C_j.
- Merge C_i and C_j and decrement k by one.
- Go to Loop.

Letting $k_0 = 1$ terminates the process when all samples are classified to one cluster. Usually the algorithm is terminated when the desired number of clusters has been reached or when the distance between the nearest clusters exceeds a certain preset threshold.

There are different ways to measure the distance between clusters. Each method can result in an algorithm with different properties. Three common distance functions and their corresponding algorithms are described next. An excellent discussion of these algorithms appears in [191].

10.4.1.1 The single linkage algorithm

This algorithm (also called the **nearest neighbor algorithm**) uses the following distance measure between clusters:

$$d_{min}(C_i, C_j) = \min_{x \in C_i \ y \in C_j} \{d(x, y)\}$$

i.e., the distance is defined as the least of all possible pairwise distances between the clusters. If the input is given as a similarity matrix then we replace $d_{min}(C_i, C_j)$ with

$$s_{max}(C_i, C_j) = \max_{x \in C_i \ y \in C_j} \{s(x, y)\}$$

We can view pairwise clustering as a graph building process, starting from a set of isolated nodes with no edges. Merging clusters according to the nearest neighbor criterion is equivalent to introducing an edge between the nearest pair of nodes in the nearest pair of clusters C_i and C_j. By adding one edge at a time, and always between two distinct clusters, the process eventually results in a **spanning tree**: a cycle-free graph where every node is connected to every other node (directly or indirectly). The tree is actually a minimal spanning tree, meaning the sum of edge lengths is the smallest of all possible spanning trees over the same set of nodes (see Problem 10.4). If a threshold is set, and the clustering stops when the minimum distance between two clusters exceeds this threshold, the resulting graph splits into subgraphs called **connected components**.

If we are interested in just determining the connected components, then we can use a more efficient algorithm than the pairwise algorithm described above. Given the proximity matrix and the distance threshold we can scan the matrix, starting with one instance (one row in the matrix) and adding all its neighbors that are closer than the threshold to the same cluster. The neighbors of these instances are added as well, and the process continues until there are no more neighbors to check. A new cluster is then initialized with an instance that has not been marked yet, and its neighbors are added in the same way. The whole procedure terminates when all sample points are marked. This procedure visits every cell in the matrix only once; therefore the complexity of this algorithm is $O(n^2)$, while that of the pairwise clustering is $O(n^3)$. The full procedure is described next.

Connected components

Input: proximity matrix $\mathbf{D}$, threshold d_{max}
Initialize: test set $\mathbf{T} = \{1..n\}$, cluster number $k = 0$, queue $\mathbf{Q}$

Loop: While $\mathbf{T}$ not empty
  Increment cluster number $k = k + 1$
  Pick (and remove) sample i from $\mathbf{T}$
  Initialize $C_k = \{i\}$ and $\mathbf{Q} = \{i\}$
  While $\mathbf{Q}$ not empty
    Pick (and remove) sample i from $\mathbf{Q}$
    For $j = 1..n$
      If $j \in \mathbf{T}$ and $d(i,j) \leq d_{max}$
        Add j to cluster C_k and to the queue $\mathbf{Q}$
        Remove j from $\mathbf{T}$

Note that it is impossible that $j \notin \mathbf{T}$ and $d(i,j) \leq d_{max}$ because otherwise i must have been added to the same cluster j is in, and it would have been eliminated from the set $\mathbf{T}$ already.

The single linkage and connected components procedures are effective when the clusters are well separated. They work well on compact sets of points, but the sets do not necessarily have to resemble spheres or ellipsoids. These algorithms can detect sets that form arbitrary shapes (see Figure 10.3), as long as they do not overlap with other clusters and the largest inter-cluster distance is less than the minimal intra-cluster distance. However, these algorithms are very sensitive to noise or to slight changes in the distances between points. For example, in the presence of a few outliers these algorithms may connect unrelated clusters (Figure 10.2). Thus elongated unnatural clusters can be formed (the **chaining effect**).

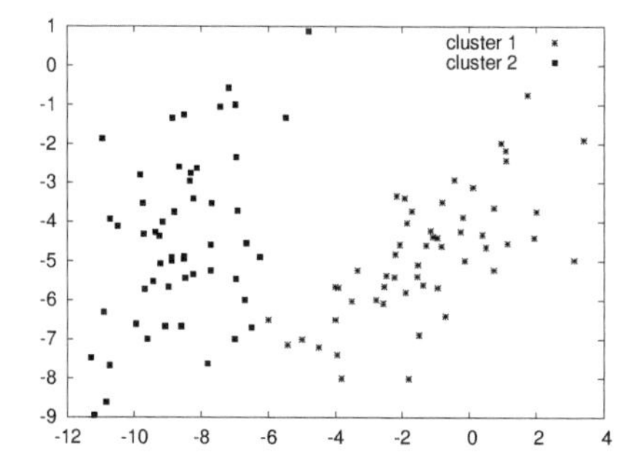

FIGURE 10.2: **The chaining effect.** Outliers can link adjacent clusters in single linkage clustering.

10.4.1.2 The complete linkage algorithm

This algorithm (also called the **farthest neighbor algorithm**) uses the following distance measure between clusters:

$$d_{max}(C_i, C_j) = \max_{x \in C_i \; y \in C_j} \{d(x, y)\}$$

In this case the distance between two clusters is determined by the farthest pair of nodes. Merging two clusters means introducing an edge between the farthest pair and implies that all other pairs are connected as well. As such, this procedure essentially builds a graph where every cluster is a complete subgraph (and hence the name). Therefore, the complete linkage algorithm strongly favors compact clusters, and compared to the single linkage algorithm it is less accommodating to other data distributions. On the other hand, being more conservative, the algorithm may prove more successful when clustering noisy data, such as gene expression data (see Chapter 12).

10.4.1.3 The average linkage algorithm

The average linkage algorithm uses the average distance between points from the two clusters to define their distance,

$$d_{ave}(C_i, C_j) = \frac{1}{n_i n_j} \sum_{x \in C_i} \sum_{y \in C_j} d(x, y)$$

which is a compromise between the other two measures d_{min} and d_{max}. Consequently, this algorithm tends to generate more stable clusterings. If data is noisy (as is often the case with biological data), it might be useful to consider also other, more strict criteria for merging. For example, a simple modification of the average linkage algorithm is to merge clusters only if the length of *all* edges within the merged cluster do not exceed a certain threshold. Note that this is not the same as complete linkage.

When sample points are vectors in a real normed space then another measure of distance, which produces the same results with fewer computations (see Problem 10.5), is $d_{mean}(C_i, C_j) = \|\vec{m}_i - \vec{m}_j\|$ where $\vec{m}_i$ ($\vec{m}_j$) is the center of cluster C_i (C_j).

When should we use pairwise clustering? Pairwise clustering algorithms are very simple to implement and are quite fast, although they can get computationally intensive for large datasets, since their running time is $O(n^3)$. They make no assumption regarding the type of the source distributions and can handle even complex data distributions, as in Figure 10.3. Moreover, they are not confined to vector data and can be applied directly to proximity data (pairwise distances or similarities). As such, pairwise algorithms are

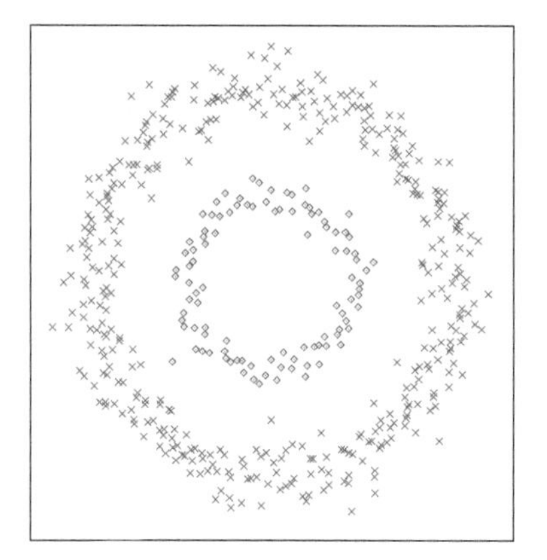

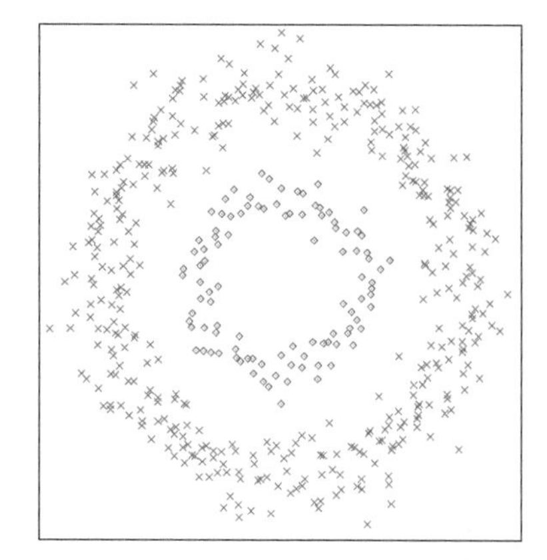

FIGURE 10.3: **Two concentric clusters.** (Figure is courtesy of Yoram Gdalyahu.) The single linkage algorithm is non-parametric and can handle complex data distributions, such as the two concentric clusters on the left. However, it is likely to fail in the presence of noise (right). Concentric clusters, however, are atypical distributions for biological data.

very appealing for clustering of biological data that is naturally represented as proximity data, and in Section 10.6 we will discuss a few concrete examples.

On the other hand, classical pairwise algorithms can be sensitive to noise and outliers as illustrated above. It is also harder to assess the validity of their results. The partitions do not necessarily optimize any particular criterion function, and hence solutions have no established properties. The performance is usually assessed using external indices (see Section 10.5.1). It is also not clear what is the best way to generalize the solutions to new instances, and new instances are usually classified simply based on their average or minimal distance from existing clusters.

10.4.2 Collaborative clustering

Pairwise similarities between biological entities are often unreliable and noisy. This might greatly affect clustering algorithms that are based purely on pairwise distances or similarities, as discussed above. Collaborative clustering algorithms make use of collective properties of the data graph to eliminate suspicious connections, reduce noise and produce better clusterings. They work by pre-processing and manipulating the proximity matrix to obtain a new representation of the data that reflects the global geometry of the dataset (in Section 11.2.3.1 we will further discuss similar approaches in a different context, where the goal is to map proximity data into vector space). The graph nodes are then clustered using one of the algorithms we already discussed, or related algorithms.

A simple example of this approach is the mutual neighborhood clustering algorithm [500], which utilizes near neighbor information. The algorithm defines the weight of the edge between sample i and sample j as the sum of their mutual neighborhood ranks, i.e., if j is the n-th neighbor of i and i is the m-th neighbor of j, then the weight is defined as $m + n$. This transformation can increase or decrease the significance of a pairwise similarity, based on the local neighborhood of each sample.

A straightforward transformation of the proximity matrix is to replace it with its power to a certain degree. Consider for example the **adjacency matrix A** of a graph (Figure 10.4). In this matrix $A_{ij} = 1$ if there is an edge between node i and node j and $A_{ij} = 0$ otherwise. As with the proximity matrix, we assume that the matrix is symmetric, meaning the graph is undirected and $A_{ij} = A_{ji}$. Note that for a symmetric matrix $\mathbf{A}^t = \mathbf{A}$.

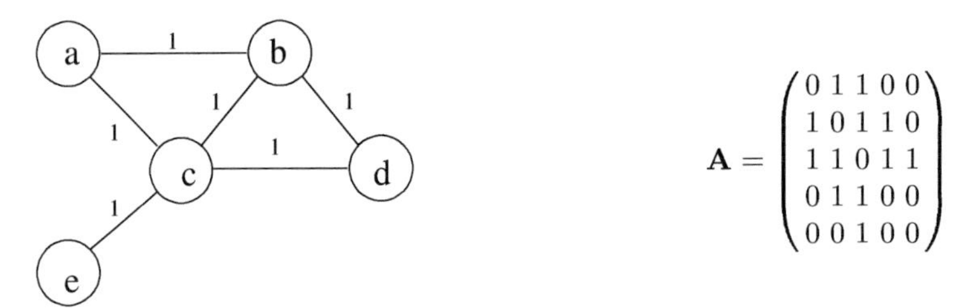

FIGURE 10.4: A sample data graph and the corresponding adjacency matrix. (Figure reproduced from [501].)

The adjacency matrix provides information on direct contacts in the data graph. However, there are other ways to characterize the connectivity in the graph. For example, note that the product of the adjacency matrix by itself results in a **connectivity matrix** where the (i, j) entry is the number of paths of length two linking nodes i and j. In other words the transformation $\mathbf{B} = \mathbf{A}^2$ creates a new graph in which i and j are considered adjacent if they are connected by a path of length two in the original graph, and the (i, j) edge in the new graph is weighted by the number of paths of length two connecting i and j in the original graph (Figure 10.5). The higher the number of paths that connect i and j through intermediate nodes, the stronger their mutual association is. Hence, the connectivity matrix $\mathbf{B}$ contains collective information on the relations between nodes. Similarly, the connectivity matrix $\mathbf{B} = \mathbf{A}^3$ uses paths of length 3, and $\mathbf{A}^k$ exercises paths of length k.

Note that a connectivity matrix $\mathbf{A}^k$ considers *only* paths of length k. Therefore, an analysis that uses a connectivity matrix of order k ($k > 1$) ignores all shorter paths (including direct relations between nodes) and possibly eliminates useful information. For example, the edge (e, c) in Figure 10.4 is eliminated in the connectivity matrix $\mathbf{B} = \mathbf{A}^2$ (Figure 10.5). A better approach is

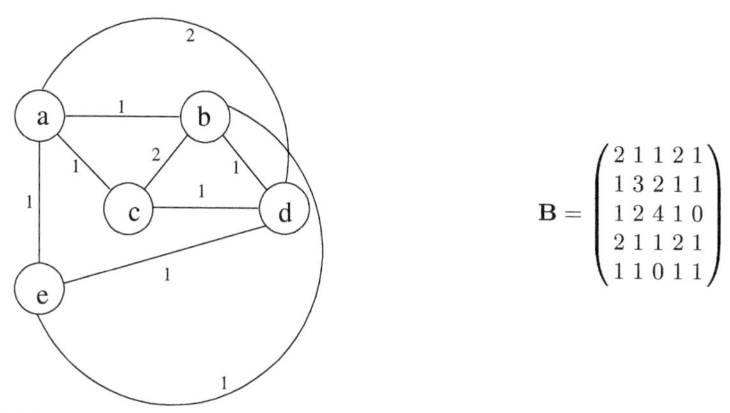

FIGURE 10.5: **The connectivity matrix $\mathbf{B} = \mathbf{A}^2$** for the data graph from Figure 10.4. The (i, j) entry corresponds to the number of length 2 paths between nodes i and j, as depicted in the graph on the left. Note that with undirected graphs the diagonal elements (i, i) of $\mathbf{B}$ correspond to the number of edges occurring at node i in the original data graph. (Figure reproduced from [501].)

to combine the contributions of paths of different lengths, essentially building a connectivity matrix $\mathbf{B}$ where B_{ij} is defined based on all paths from i to j. However, connections through longer paths are usually weaker and less significant and should be down-weighted. A possible solution is to fix the maximal path length k and define

$$\mathbf{B} = \sum_{i=1}^{k} w_i \mathbf{A}^i$$

where w_i, the weight of paths of length i, is a decaying function of the path length.

A related transformation is the **transitive closure** of the adjacency graph, $\mathbf{A}^*$, where $A^*_{ij} = 1$ if there is a path of any length between node i and node j. The transitive closure can be written as

$$\mathbf{A}^* = \delta\left(\sum_{k \geq 1} \mathbf{A}^k\right)$$

where the $\delta(\mathbf{X})$ function converts a matrix $\mathbf{X}$ to a matrix of zeros and ones such that $\delta(X_{ij}) = 1$ if $X_{ij} > 0$ and zero otherwise. In practice there is no need to compute all powers of $\mathbf{A}$, and there are efficient algorithms for computing the transitive closure of a graph. For example, the Floyd-Warshall algorithm [502] uses dynamic programming with time complexity of only $O(n^3)$.

Transitive closure

Input: Adjacency matrix $\mathbf{A}$
Initialize: $A^*_{ij} = A_{ij}$

For $k = 1..n$
    For $i = 1..n$
        For $j = 1..n$

$$A^*_{ij} = \max\{A^*_{ij} \ , \ A^*_{ik} \cdot A^*_{kj}\} \tag{10.7}$$

The transitive closure of a matrix only indicates whether there is a path between nodes i and j. But using a similar algorithm we can also compute the length of the shortest path between every two nodes of the graph. In that case we initialize $A^*_{ij} = A_{ij}$ if $A_{ij} = 1$ and $A^*_{ij} = \infty$ otherwise and replace Equation 10.7 with

$$A^*_{ij} = \min\{A^*_{ij} \ , \ A^*_{ik} + A^*_{kj}\} \tag{10.8}$$

These transformations to the transitive closure, however, ignore important collective properties of the graph. Under this transformation, for example, two nodes that have multiple paths connecting them have exactly the same weight as two nodes that are connected by a single path.

All the transformations we discussed so far started with the adjacency matrix, but they can be applied to the proximity matrix of the graph as well. For example, the transitive closure can be used to compute the shortest path, now taking into account the edge lengths, or to determine whether there is a path between two nodes where all edges are shorter than a certain threshold. Similarly, connectivity matrices of the form $\mathbf{W}^k$ provide information on the strength of intermediate connections between nodes. For example,

$$\mathbf{W}^2(i,j) = \sum_{k=1}^{n} w(i,k)w(k,j) \tag{10.9}$$

can be viewed as the cumulative weight of all paths of length two connecting nodes i and j.

A more elaborate transformation of the proximity matrix is used in the **minimal cut clustering algorithm** [503]. This algorithm uses the notion of cuts in graphs and seeks partitions into subgraphs such that the maximum flow across subgroups is minimized. A **cut** $(\mathbf{A},\mathbf{B})$ in graph $G(\mathbf{V},\mathbf{E})$ is a partition of $\mathbf{V}$ into two disjointed sets of vertices $\mathbf{A}$ and $\mathbf{B}$ such that $\mathbf{V} = \mathbf{A} \cup \mathbf{B}$. The **capacity** of a cut is the sum of the weights of all edges that cross that cut, i.e.,

$$cut(\mathbf{A},\mathbf{B}) = \sum_{i\in\mathbf{A} \ j\in\mathbf{B}} w(i,j)$$

where $w(i,j)$ here denotes the *similarity* of i and j. Note that the capacity of a cut can be viewed as a similarity measure between the subsets **A** and **B**. There are many possible cuts in a graph, and the **minimal cut** is the one that has the minimal capacity. The bi-partitioning strategy that is adopted by the minimal cut algorithm is to split the graph along minimal cuts. In other words, the minimal cut splits the graph into two subsets whose mutual similarity is minimal.

The minimal cut clustering algorithm starts with a preliminary step in which weights (the pairwise similarities) are transformed to the maximal flow between the corresponding vertices. According to one of the famous theorems in graph theory, the maximal flow between two vertices equals the capacity of the minimal cut between them. Finding the maximal flow is a well-studied problem, and there are efficient algorithms for solving it, including deterministic algorithms with time complexity of $O(n^2)$ [502] and probabilistic algorithms with complexity $O(n)$ [504].

After the transformation, the new weights reflect the overall possible intermediate connections between vertices, with high flow values indicating strong interactions. Based on these new weights the algorithm divides the graph into subgraphs of connected components by removing edges of low capacity in order of increasing capacity until an external threshold value is met. In practice, the maximal flow need not be calculated for all possible edges. Specifically, according to the Gomory-Hu theorem [505], the minimal cut of a graph splits the graph into two subgraphs, each with a maximal flow value that is higher than the maximal flow value *across* the subgraphs (the one determined by the minimal cut). Therefore, by applying this algorithm recursively to each subgraph we can identify the next minimal cuts until the subgraphs contain single nodes. With n nodes, this means solving $n-1$ minimal cut problems. The process can terminate once the capacity of the minimal cut exceeds a certain preset threshold.

A very different approach to clustering, based on a dynamic transformation of the proximity matrix, is described in [346]. This algorithm, which is quite effective for noisy data, iteratively modifies data representation. It operates in two steps. In the first step each row in the proximity matrix is normalized such that its Euclidean norm is one. Next, the distances between instances are re-calculated based on the proximity matrix. Specifically, the distance between instances i and j is defined as the distance between rows i and j of the proximity matrix. In other words, each instance is represented by its vector of distances to all other instances. The choice of the metric when normalizing and comparing the rows can vary, including l_1, the Euclidean metric and statistical measures such as KL- or JS-divergence that can assess the similarity of probability distributions (see Section 4.3.5).

The procedure is repeated until the data matrix converges. Empirical tests show that this process almost always converges to a binary matrix containing only zeros and some positive constants. In most cases this final matrix can be permuted to a block diagonal matrix with two or three blocks, each one corresponding to a cluster (Figure 10.6). This process can then be repeated with each one of the clusters from the previous step.

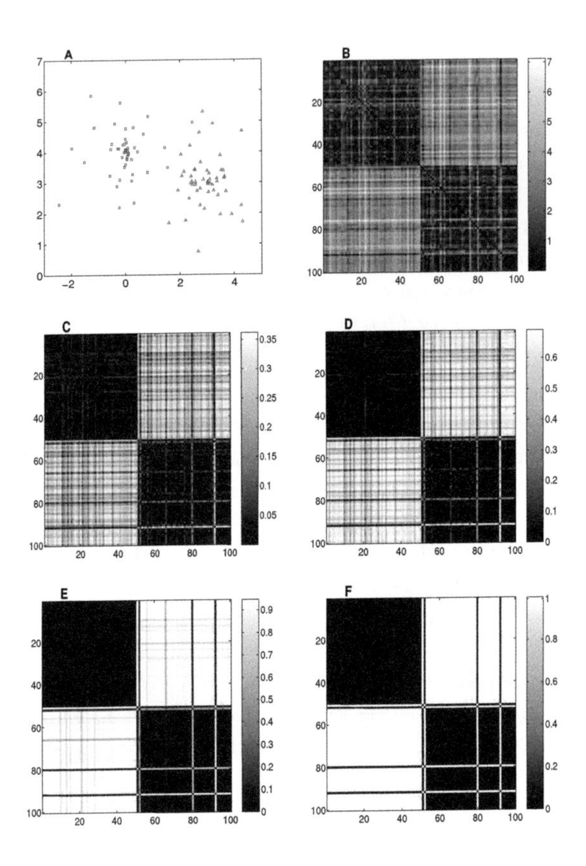

FIGURE 10.6: **Iterative modification of data representation.** A sequence of transformations applied to 100 data points generated from two $2D$ Gaussians (A), starting from the original matrix of pairwise distances (B), and after 1, 2, 5 and 27 iterations, at convergence (C-F). The original proximity matrix is based on Euclidean distances, but subsequent matrices use the JS-divergence measure (note the change in the range of values, shown on the right of each figure). The final matrix is a binary matrix, which can be permuted into a block-diagonal matrix. (Figure reproduced from [346] with permission from Springer.)

A key step of this algorithm is the representation of instances through their vector of distances to other instances. We already mentioned briefly a similar method when we discussed distance-based structure comparison algorithms in Chapter 8. In Chapter 11 we will see that a representation based on inter-distances between data points (i.e., **distance profile**) is a common and effective method for representing non-vectorial data in vector space. By modifying the representation of the data iteratively, the clustering algorithm described above effectively reassesses the position of each individual data point with respect to all other points in the dataset, and random fluctuations or sparse connections tend to average out.

Collaborative clustering utilizes more information about the relationships between entities when clustering instances. However, with protein data these transformations should be applied carefully. For example, transitivity often fails with protein sequences (see Section 3.4.7). Moreover, depending on the type of similarity measure used, taking the product $w(i,k)w(k,j)$ in Equation 10.9 or the sum $w(i,k) + w(k,j)$ in Equation 10.8 does not necessarily make sense. With some measures (such as evalue), the significance of a path is actually determined by the weakest similarity along the path, and the existence of multiple intermediates between two proteins does not necessarily increase the significance of their mutual association, since the different intermediate proteins can be close homologs that cannot be considered as independent observations. Therefore, the application of any of these transformations to protein data should be carefully scrutinized and soundly justified.

10.4.3 Spectral clustering algorithms

Spectral clustering is a term referring to graph-based clustering algorithms that explore global properties of the proximity matrix. These algorithms work by analyzing the eigenvectors of the input matrix. They can be applied directly to the proximity matrix or to connectivity matrices that are derived from the proximity matrix by means of one of the collaborative transformations described in the previous section.

To understand how solving the eigenvectors of a connectivity matrix is relevant to the graph partitioning problem, recall that an n-dimensional vector $\vec{x}$ is called an eigenvector of an $n \cdot n$ matrix $\mathbf{A}$ if $\mathbf{A}\vec{x} = \lambda\vec{x}$ where λ is a scalar. The scalar λ is the eigenvalue associated with $\vec{x}$. A matrix can have multiple eigenvectors (the number of eigenvectors equals $rank(\mathbf{A})$), and the eigenvectors are orthogonal to each other.

There are many methods for computing the eigenvectors of a matrix. Of special interest is the fast **power iteration method** [506, 507], which is insightful to understanding the role of eigenvectors in clustering. The power iteration method computes the principal eigenvalue/eigenvector pair of a matrix $\mathbf{A}$ by repeatedly multiplying $\mathbf{A}$ to some random non-zero starting vector $\vec{x}(0)$ until convergence (see Figure 10.7). Namely,

$$\vec{x}(t+1) = \mathbf{A}\vec{x}(t)$$

To avoid overflow or underflow, the vector should be normalized after every iteration, such that $\|\vec{x}(t+1)\| = 1$. It can be proved that for large t, the vector $\vec{x}(t+1)$ converges to the first eigenvector[2]. The eigenvalue is given by $\lambda = \vec{x}'\mathbf{A}\vec{x}$. While exact matrix diagonalization takes $O(n^3)$, the power method takes only $O(n^2)$ and is the only feasible method for computing eigenvectors of very large and dense matrices (like those used to represent large protein datasets).

When applied to a connectivity matrix $\mathbf{A}$, then one could think of the initial vector $\vec{x}(0)$ as a mass distributed among the graph nodes, and the entries $a(i,j)$ in the connectivity matrix as weights proportionate to the probabilities to move from one node to another. Assuming the input matrix is symmetric (i.e., undirected graph), the i-th row of $\mathbf{A}$ characterizes the connectivity of the i-th node with all the other nodes. The iterative process then will redistribute the mass, following the edges of the graph, until convergence to a steady state vector. Hence, each entry in this final vector is proportionate to the probability that a random walk through the graph edges will end in the corresponding node. With strong connections (large $w(i,j)$) dominating the iterative process, the eigenvector tends to have similar values for nodes that are connected directly or indirectly. Thus, this process essentially discovers a cluster of related instances.

Informally, we can say that the eigenvector partitions the matrix $\mathbf{A}$ into different subspaces. The instances that are associated with high x_i values make up the subspace that corresponds to $\vec{x}$. Thus, by analyzing the components of an eigenvector we can induce a partition of the graph, where all instances that are associated with values higher than a certain threshold are classified to one cluster, and all other instances are classified to another.

This intuition is the basis for the factorization method [509], which partitions graphs based on the eigenvectors of the proximity matrix. Similarly, we can apply spectral analysis to other connectivity matrices, which are derived from the adjacency or proximity matrix, as discussed in the previous section.

One of the best known spectral algorithms is the **minimal normalized cut algorithm** [510]. This algorithm improves over the minimal cut algorithm described in the previous section. Since the cut capacity increases with the number of edges crossing the cut, the minimal cut algorithm tends to create partitions with small sets of isolated nodes as subgraphs. To avoid this bias, the normalized cut algorithm normalizes the capacity of the cut $(\mathbf{A}, \mathbf{B})$ by the total association of the set $\mathbf{A}$ (which is the sum of weights incident on $\mathbf{A}$), and the total association of $\mathbf{B}$. This disassociation measure is called the

[2]To compute non-principal eigenvectors, the procedure starts with a random set of orthonormal vectors. After each vector is multiplied by $\mathbf{A}$, the vectors are orthonormalized using the Gram-Schmidt algorithm [508].

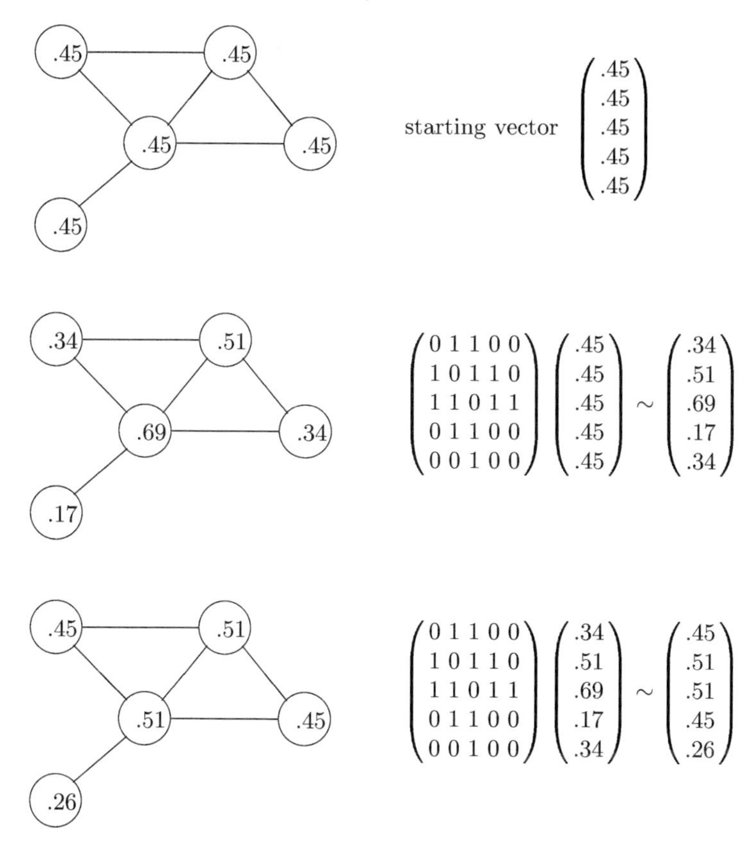

FIGURE 10.7: **Computing eigenvectors with the power iteration method.** Starting from a uniform set of values, the vector is iteratively redefined by multiplying it by the matrix, until convergence. The power iteration computes a matrix's principal eigenvector, and a more elaborate manipulation is required to compute non-principal eigenvectors (see text). (Figure reproduced from [501].)

normalized cut:

$$Ncut(\mathbf{A},\mathbf{B}) = \frac{cut(\mathbf{A},\mathbf{B})}{assoc(\mathbf{A},\mathbf{V})} + \frac{cut(\mathbf{A},\mathbf{B})}{assoc(\mathbf{B},\mathbf{V})}$$

where

$$assoc(\mathbf{A},\mathbf{V}) = \sum_{i \in \mathbf{A} \;\; k \in \mathbf{V}} w(i,k)$$

This measure has a maximum value of 2. The algorithm looks for partitions with small $Ncut$ value, and the graph is partitioned recursively so as to minimize the normalized cut criterion at each level until $Ncut$ exceeds a predefined threshold. With this normalization, small sets of isolated points are no longer preferred since the cut capacity for such sets is a large percentage of the total flow from the set to all other nodes in the graph, resulting in a large $Ncut$ value.

The $Ncut$ disassociation measure has a dual measure of association *within* groups which is defined as

$$Nassoc(\mathbf{A},\mathbf{B}) = \frac{assoc(\mathbf{A},\mathbf{A})}{assoc(\mathbf{A},\mathbf{V})} + \frac{assoc(\mathbf{B},\mathbf{B})}{assoc(\mathbf{B},\mathbf{V})}$$

Note that

$$\begin{aligned} Ncut(\mathbf{A},\mathbf{B}) &= \frac{cut(\mathbf{A},\mathbf{B})}{assoc(\mathbf{A},\mathbf{V})} + \frac{cut(\mathbf{A},\mathbf{B})}{assoc(\mathbf{B},\mathbf{V})} \\ &= \frac{assoc(\mathbf{A},\mathbf{V}) - assoc(\mathbf{A},\mathbf{A})}{assoc(\mathbf{A},\mathbf{V})} + \frac{assoc(\mathbf{B},\mathbf{V}) - assoc(\mathbf{B},\mathbf{B})}{assoc(\mathbf{B},\mathbf{V})} \\ &= 2 - Nassoc(\mathbf{A},\mathbf{B}) \end{aligned}$$

Therefore, minimizing the disassociation $Ncut$ between different groups is equivalent to maximizing the association within groups $Nassoc$.

The problem of finding the minimal normalized cut (or maximal association) is NP-hard. However, with some approximations it can be computed efficiently. In their paper, Shi and Malik show that the $Nassoc$ criterion function can be computed approximately by reformulating the problem and solving the eigenvalue problem

$$\mathbf{Z}^{-1}\mathbf{W}\vec{x} = \lambda\vec{x}$$

where $\mathbf{W}$ is the $n \times n$ symmetric similarity matrix and $\mathbf{Z}$ is an $n \times n$ diagonal matrix with $z(i) = \sum_j w(i,j)$. Since

$$\mathbf{Z}^{-1}\mathbf{W} = \begin{pmatrix} 1/z_1 & ... & 0 \\ 0 & ... & 0 \\ . & ... & . \\ 0 & ... & 1/z_n \end{pmatrix} \begin{pmatrix} w_{11} & ... & w_{1n} \\ w_{21} & ... & w_{2n} \\ . & ... & . \\ w_{n1} & ... & w_{nn} \end{pmatrix} = \begin{pmatrix} w_{11}/z_1 & ... & w_{1n}/z_1 \\ w_{21}/z_2 & ... & w_{2n}/z_2 \\ . & ... & . \\ w_{n1}/z_n & ... & w_{nn}/z_n \end{pmatrix}$$

then the sum of each row is 1, and the vector $\vec{1}$ is an eigenvector of $\mathbf{Z}^{-1}\mathbf{W}$ with the (maximal) eigenvalue of 1. However, this vector is not informative for clustering. Instead, the normalized-cut algorithm uses the eigenvector associated with the *second* largest eigenvalue of $\mathbf{Z}^{-1}\mathbf{W}$. As noted in [510], the second eigenvector approximates the optimal normalized cut solution, making the optimization problem tractable.

Once the eigenvectors are computed, the optimal partition is determined by sorting the components of the eigenvector in decreasing value and computing the $Ncut$ value for each possible bipartition. Specifically, if $\mathbf{C}_i = \{1, ..., i\}$ and $\mathbf{C}'_i = \{i+1, i+2, ..., n\}$, then the optimal partition is obtained at i_0 where $i_0 = \min_i Ncut(\mathbf{C}_i, \mathbf{C}'_i)$, and the graph is partitioned into two clusters accordingly. The algorithm then works as follows:

Spectral clustering (the normalized cut algorithm)

1. Set up the weighted graph with the scoring matrix $\mathbf{W}$ and compute $\mathbf{Z}$.
2. Solve $\mathbf{Z}^{-1}\mathbf{W}\vec{x} = \lambda\vec{x}$
3. Use the eigenvector associated with the second largest eigenvalue to partition the graph into two subsets.
4. Set up two new weighted graphs corresponding to the two subsets and repeat steps 1 to 3 until k clusters are obtained.

Note that $\mathbf{Z}^{-1}\mathbf{W}$ is basically the original similarity matrix where the i-th row (which corresponds to the vector of similarity values of the i-th sample with all other samples) is normalized by the sum of the similarity values $z(i)$. Other connectivity matrices or different normalizations result in different matrices with different spectral properties. For other variations, including k-way partitioning algorithms, see [511–514].

Spectral clustering algorithms appeal to intuition. By analyzing global properties of the graph, they increase the robustness against spurious noise. However, they are computationally intensive, and handling large datasets can be a challenge. Moreover, as with all other clustering algorithms, they are not designed to address the specific problems that are inherent to the analysis of proteins, such as the transitive chaining that is observed for multi-domain proteins (we will get back to that issue in Section 10.6.3). Nevertheless, they can be very effective for smaller datasets and other clustering problems, and in Chapter 12 we will demonstrate the application of these algorithms to expression data.

10.4.4 Markovian clustering algorithms

Markovian graph clustering algorithms are collaborative algorithms that are closely related to spectral clustering. These algorithms explicitly formulate the clustering problem in terms of a Markovian random walk, where the probability to reach node i in time $t + 1$ depends on the node visited at time t.

Denote by $\mathbf{T}$ the transition probability matrix of the graph, where each column corresponds to one node and contains the normalized probabilities of moving from that node to any other node in the graph, such that $\mathbf{T}(j, i) = \text{Prob}(i \rightarrow j)$. The transition matrix can be derived from the proximity matrix, for example by means of a simple normalization

$$\mathbf{T}(j, i) = \frac{w(j, i)}{\sum_{j'} w(j', i)}$$

If the proximity matrix consists of distances, we can apply exponential transformation of the form

$$\mathbf{T}(j,i) = \frac{\exp(-\lambda w(j,i))}{\sum_{j'} \exp(-\lambda w(j',i))}$$

where λ is a scaling factor that determines the decay rate and the effective radius of the local neighborhood where transitions occur (see Section 10.6.4 for further discussion on transformations and normalizations). The normalization may produce a matrix that has different properties than the original matrix it was derived from.

Using the transition matrix $\mathbf{T}$, a random walk traverses the similarity network following the edges of the graph. Specifically, starting from a source node x, a random walk moves to one of the neighboring nodes with probability that is proportional to the weight of the edge connecting it with x. The process then continues iteratively from the neighboring node. As a result, the probabilities to reach each node are updated. This property of the algorithm is referred to as **expansion** [515]. The basic random walk algorithm is equivalent to multiplying the transition matrix repeatedly by itself.

$$\mathbf{T}^{t+1} = \mathbf{T} \cdot \mathbf{T}^{t} \tag{10.10}$$

The product will eventually converge to a steady state matrix where the (j,i) entry corresponds to the steady state probability of reaching node j from node i. Thus, Markovian algorithms can be considered as another way of transforming a proximity matrix. Similar to the dynamic transformation of a proximity matrix illustrated in Figure 10.6, the random-walk process terminates at a matrix that essentially partitions the graph into components, potentially with some nodes connected to multiple clusters. This matrix can then be analyzed and clustered using any of the graph algorithms we described before, including spectral algorithms. Note that if we repeatedly multiply the transition matrix to a start vector $\vec{x}$, as in the power method described in the previous section, then the procedure results in the principal eigenvector of the transition matrix. Therefore, upon convergence the vector $\vec{x}$ contains the steady state probabilities of reaching any node in the graph, regardless of the source node.

The problem becomes more interesting once we consider variations over the random walk process described above, by controlling the extent to which the algorithm explores the global graph structure. One such variation is the **inflation** mechanism. Inflation operates on the columns of the transition matrix, by taking each column to the power c where c is a parameter of the algorithm. That is,

$$\mathbf{T}(j,i) = \frac{\mathbf{T}^{c}(j,i)}{\sum_{j'=1}^{n} \mathbf{T}^{c}(j',i)}$$

For $c > 1$ the effect of inflation is to redistribute the weight in each column such that higher probabilities become even more dominant while small probabilities gradually diminish. With inflation, the transition matrix changes dynamically after every multiplication in Equation 10.10. Since higher probability transitions are considered more reliable, we can assume that they correspond to nodes within the same cluster. Hence, inflation effectively emphasizes local walks within the same cluster over cross-cluster walks. The higher the parameter c the more substantial the inflation is, and as a result, the more compact the clusters are.

It is sometimes useful to consider the effect of localized random walks on individual nodes (as we will see later in Section 10.6.4). A variant with a similar effect to the inflation mechanism is the random walk with a **restart probability** r [516]. The restart probability corresponds to self-transitions (loops) in the graph. The presence of loops means that a random walk can also return directly to the source node. The restart probability determines how far from the source node the process wanders off. If r is close to 1 the process remains in the close vicinity of the source and reflects the local structure of the network, while for r closer to zero the process reflects the global structure. For $r = 0$ this procedure is identical to the basic random walk process described before. The random walk algorithm with restart mechanism is outlined below.

Random walk with restart

Input: graph proximity matrix $\mathbf{W}$, query node x, restart probability r
Initialize:

- Compute the column normalized transition probability matrix $\mathbf{T}$
- Set $\vec{s} = (0, 0, 1, 0, ...)'$ such that $s_x = 1$ and 0 otherwise
- Initialize $\vec{x}(0) = \vec{s}$

Iterate: while $\vec{x}$ has not converged

$$\vec{x}(t + 1) = (1 - r)\mathbf{T} \cdot \vec{x}(t) + r\vec{s} \tag{10.11}$$

Once the difference between consecutive vectors is less than a certain threshold we say that the vector $\vec{x}$ converged. Convergence is slower for smaller r since the process samples larger neighborhoods. Note that the start vector sets all the initial mass at node x. Therefore, upon convergence the vector $\vec{x}$ contains the steady state probabilities of reaching any node *from the source node x*. This vector can be viewed as the **affinity vector** of x with all other nodes of the graph. The process can be repeated for each node to obtain its unique affinity vector.

10.4.5 Super-paramagnetic clustering

Super-paramagnetic clustering [517] is another graph clustering algorithm, based on an analogy with physical magnetic systems (inhomogeneous ferromagnets). It has been used in several studies in computational biology [518–520]. Like other collaborative algorithms, the algorithm utilizes collective properties of the graph to cluster the data.

In this model each data point is associated with a "state" out of a finite number of possible states, but points are able to "interact" with each other and change states until they reach a steady state. The strength of the interactions decreases with the distance. Points may influence each other directly or through other mediating points, hence the mutual influence between points is a collective effect. Specifically, for a given configuration of states, the **energy** of the system is defined as $-\Sigma J(i,j)\delta(s_i, s_j)$, where the summation is over all interacting pairs of points, s_i (s_j) is the state of the sample i (j), and $J(i,j)$ is a positive interaction strength between sample i and sample j, which decays with the dissimilarity between the points. The Krönecker delta $\delta(s_i, s_j)$ equals one if $s_i = s_j$ and zero otherwise. Hence, only interacting samples in the same state are counted.

The system has minimal energy when all the samples are in the same state (a single cluster). However, the statistical properties (e.g., the total energy) of this thermodynamic system are affected by the temperature, and such a perfect configuration can exist only at zero temperature. At every finite temperature there is some level of noise, and the state variable s associated with each sample will fluctuate. As a result, the sample points are distributed in different states. The level of the noise decreases as the temperature decreases. Consequently, the fluctuations decrease as well, and clusters of samples of different states may merge (a **phase transition**). Therefore, as the temperature is decreased the system undergoes a series of phase transitions, and by tracing this process, it is possible to obtain hierarchical classification of the samples.

10.5 Cluster validation and assessment

Clustering algorithms can be very effective in discovering patterns in unlabeled data. However, they are also susceptible to various factors, and when applied simple mindedly, they can lead to false conclusions. The application of *any* clustering algorithm will result in some partitioning of the data into groups. But the choice of the clustering algorithm may greatly affect the outcome, and without further assessing the results it is unclear which one we should believe. Naturally, with such a large variety of clustering algorithms to choose from one could ask which method is the "best" and whether the results produced with one algorithm are more meaningful than the other. This is referred to as the problem of **cluster validity**.

In search for the best clustering solution, we should also consider the problem of (in)stability. Since most clustering algorithms can become trapped in a local minimum, their output may vary substantially depending on the initialization of the parameters. Therefore, it is important to repeat the procedure with different parameters and evaluate the results to identify the optimal solution.

Finally, there is the problem of determining the correct number of clusters (classes) in the dataset. This number depends on the scale or the resolution of the analysis (see Figure 10.8), and multiple scale analysis (as in Section 10.3.3) may help to reveal the cluster structure. However, in the absence of a natural criterion by which to stop the process of splitting or merging clusters, one usually uses an external parameter (often preset arbitrarily) such as the maximal distance between clusters or the maximal number of clusters. Consequently, the "recovered structure" (the clustering solution) does not necessarily fit the "true structure" of the data, and its validity may be questionable.

The problem of cluster validity is particularly important when the dataset is a sample of a larger instance space, and clusters are to serve as a reliable model for **generalization**, i.e., new samples are expected to be assigned correctly to these clusters. Without validating and assessing the cluster structure, we are at risk of **overfitting** the model to the training data. When overfitting occurs the model can be misleading. It might split a real class or type into multiple clusters or merge different classes into one cluster. This is especially pronounced if the data is noisy or contains outliers and if clusters are not well separated. Such a model will perform poorly on new instances.

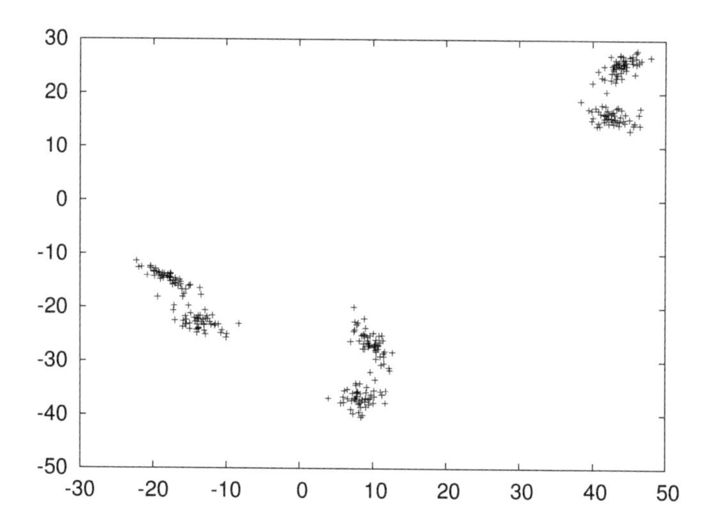

FIGURE 10.8: **Analysis at different resolutions suggests either three or six clusters.**

The problem of cluster validation has occupied the machine learning community for many years now, and several different strategies have been proposed. In general, methods to assess the quality of the cluster structure and adjust the number of clusters can be divided into external indices and internal indices. **External** validation compares the recovered structure to an a priori structure (i.e., known classification) and quantifies the match between the two. An **internal** validation test seeks to determine whether the structure is intrinsically appropriate for the data, usually by assessing the likelihood it could have occurred by chance.

Many indices of validity were proposed over the years, and in the next subsections we will give only a few examples of each type of validation. In general, any of the established methods can be effective in detecting "good" structures, but each one is fraught with its own problems, as we will see next. Consequently, there is no single approach that is guaranteed to work well under all scenarios. Therefore, it is recommended to use more than one index or test when assessing clustering results. We will demonstrate this approach in Chapter 12 in the context of a specific clustering task, based on mRNA expression data.

10.5.1 External indices of validity

External indices rely on external information, such as an established classification or prior knowledge of relationships between data instances, to evaluate the validity of clusters. We will discuss two cases next.

10.5.1.1 The case of known classification

If an external classification exists, we can assess the clustering results in terms of the mutual agreement between the two classifications. There are several quality indices one can think of when comparing a new classification $\mathbf{B}$ to a reference classification $\mathbf{A}$ over the same dataset $\mathcal{X}$. We assume the two classifications are "hard", i.e., each instance is classified to exactly one group (the discussion can be generalized to the case of "soft" classifications, where each instance can be classified to more than one group).

Before assessing the mutual agreement we need to characterize the relations between groups and devise a protocol for associating groups from classification $\mathbf{B}$ with groups from classification $\mathbf{A}$. For two groups, $A \in \mathbf{A}$ and $B \in \mathbf{B}$, we define the set of true positives (tp) as the subset of instances both groups agree upon (see Figure 3.11). The number of true positives equals $|A \cap B|$. The set of false positives (fp) is defined as $B \setminus A$. These are the instances that are in B but not in A. Similarly, the set of false negatives (fn) is defined as $A \setminus B$, and their number is $|A \setminus B|$.

The procedure suggested in [431] associates each group $A \in \mathbf{A}$ with the group $B \in \mathbf{B}$ that maximizes the quantity $tp - fp - fn$. The quality of the match is defined as the ratio of true positives to the total number of instances in the union of A and B. That is,

$$Q_1(A) = 100 \cdot \frac{|A \cap B|}{|A \cup B|}$$

Note that this index accounts for both types of error (false positives and false negatives), as $|A \cup B| = tp + fp + fn$. This procedure should be repeated for every group $A \in \mathbf{A}$, and the total percentage of true positives is given by the average over all groups in $\mathbf{A}$.

This index, however, is ineffective when one classification is a refinement of the other (e.g., each group in $\mathbf{A}$ splits perfectly into several groups in $\mathbf{B}$). In this case, Q_1 underestimates the performance since for each group in $\mathbf{A}$ only one group in $\mathbf{B}$ will be accounted for. A possible solution is to associate with $A \in \mathbf{A}$ all groups $B \in \mathbf{B}$ for which $tp > fp$. In other words, we consider $B \in \mathbf{B}$ as related to A if more than 50% of B's members are also members of A (see Figure 10.9). The union of all groups B that are related to A is denoted $\mathbf{B}_A$. Under this procedure an instance is misclassified by classification $\mathbf{B}$ if it is a member of A but not a member of $\mathbf{B}_A$ (false negative) or is a member of $\mathbf{B}_A$ but not a member of A itself (false positive). The intersection of $\mathbf{B}_A$ and A defines the group of correctly classified instances. The quality index Q_2 is defined as the percentage of the true positives in the union $A \cup \mathbf{B}_A$. Namely,

$$Q_2(A) = 100 \cdot \frac{|A \cap \mathbf{B}_A|}{|A \cup \mathbf{B}_A|}$$

The index Q_2 gives more favorable evaluations when $\mathbf{B}$ is a refinement of the partition $\mathbf{A}$. However, it ignores another undesirable situation in which each group $A \in \mathbf{A}$ splits into many small clusters in $\mathbf{B}$ (and in the extreme, to singletons clusters). The Q_3 index accounts for this scenario by penalizing for excess in the number of clusters in $\mathbf{B}$ which are related to A

$$Q_3(A) = 100 \cdot \frac{|A \cap \mathbf{B}_A| - (r-1)}{|A \cup \mathbf{B}_A|}$$

where r is the number of related clusters. With this index, a class with n elements that has a single and identical relative cluster in classification $\mathbf{B}$ has quality of 1 (or 100%). On the other hand, if the relatives are n singletons then the quality is close to zero.

All three indices Q_1-Q_3 are based on the ratio of the true positives to the total number of samples $tp + fn + fp$ in $A \cup B$. However, in some cases we might want to ignore the false positives when computing the ratio. This is the case for example when the reference dataset is partial and classifies only some of the data instances or if it detects only some of the pairwise relationships. In the context of protein classification this is a common problem. For example, reference classifications that are based on protein sequence analysis may not detect all homologies. Clearly, a proper evaluation should be based only on data instances that are classified by both the new classification and the

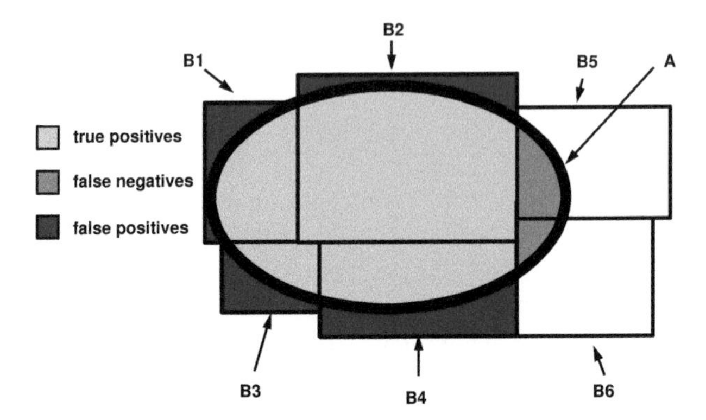

FIGURE 10.9: Association of groups in classification B with groups in the reference classification A. Groups $B1$-$B4$ are relatives of the dataset enclosed in the ellipse (which is a group of $\mathbf{A}$), while groups $B5$ and $B6$ are not, as the overlap is too small. The set $\mathbf{B}_A$ is defined as $\mathbf{B}_A = B1 \cup B2 \cup B3 \cup B4$, and the quality Q_2 is defined as $Q_2(A) = 100 \cdot \frac{|A \cap \mathbf{B}_A|}{|A \cup \mathbf{B}_A|}$. For comparison, $Q_1(A) = 100 \cdot \frac{|A \cap B2|}{|A \cup B2|}$. (Figure reproduced from [45] with permission from John Wiley & Sons, Inc.)

reference classification. However, since some instances can belong to multiple classes (e.g., a protein can contain multiple domains), the scenario in which the reference classification lists only some of the classes an instance belongs to is one we have to reckon with. The presence of short fragments and hypothetical proteins that may or may not be related to certain protein families further complicates the evaluation problem. Unfortunately, there is no automatic way to discern the relations that are biologically meaningful from those that aren't. Therefore, some of the false positives may actually be true positives [45,216]. These examples illustrate the difficulty of assessing a new protein classification by means of another, man-made, classification[3].

An alternative quality measure to consider in such cases is the ratio

$$Q_4(A) = \frac{tp}{tp + fn}$$

This index is based on the overly generous assumption that all false positives are in fact instances potentially related to group A. Obviously, this is not true in general, and we should be extra cautious in interpreting the results with this index. However, this index can provide a useful upper bound on the

[3]Choosing the right reference classification is crucial to proper evaluation. When clustering proteins this is especially important, since most known classifications are motif or domain-based classifications where the "class" is usually determined on the basis of short subsequences that represent various signals, local motifs or individual domains. These classes are not necessarily appropriate for evaluation of multi-domain proteins whose cellular function depends on their domain architecture. See also Section 10.6.3 and Chapter 9.

quality of the new classification, since typically some of the false positives are indeed related instances.

There are many other external validity indices see (for reviews, see [521, 522]), especially in the context of biological data analysis [523–526]. When assessing the validity of the clustering structure according to any of them what we really need to know is the significance of the index value. In other words, how unusual or surprising is the index value assuming the null hypothesis that the classification of samples into clusters is random (**random clustering**). To answer this question we need to characterize the **null distribution** (also called the background distribution or baseline distribution) of the index for random clustering. A clustering may be considered valid if the index is unusually high or surprising, given the null distribution (see Section 3.12.8).

For individual clusters we can establish the theoretical null distribution, assuming a multinomial process and using the methodology described in the next section. However, assessing the significance of the *overall* correspondence between clusters with any of the indices mentioned above is more involved, since not only the number of clusters may differ between the two classifications. The clusters' sizes are likely to differ as well. If the null distribution cannot be easily characterized theoretically, then it has to be approximated empirically by simulation. Depending on the size of the dataset, this might require extensive statistical sampling.

10.5.1.2 The case of known relations

An external classification of the dataset is not always available. The more common scenario is that some prior information is available on relations between entities (e.g., gene ontology, as discussed in Section 12.5.3.1). We can use this information to construct a relation graph and check whether clusters conform to the graph structure. For example we can count the number of true edges (i.e., known relations) within each cluster and estimate the probability to obtain this partition of edges among clusters by chance. The significance of the clustering is determined by how much this partition deviates from the partition one would expect to get by assigning edges to clusters and between clusters at random (the null hypothesis).

Assume we are given a cluster set $\mathbf{C} = C_1, C_2, ..., C_k$ with $n_1, n_2, ..., n_k$ samples and $e_1, e_2, ..., e_k$ edges (representing similarity relations), which partitions the total set of T true edges such that t_i edges fall in cluster i ($1 \leq i \leq k$). In addition we have e_{cross} edges across clusters, of which t_{cross} are true. If we place a true edge at random in one of the clusters, then there is a probability $p_i = e_i/E$ that it will be placed in cluster C_i where $E = \sum_i e_i + e_{cross}$ is the total number of edges. Therefore, the probability of obtaining this partition by chance is given by the multinomial distribution

$$P_{\mathbf{C}}(t_1, t_2, ..., t_k, t_{cross}) = \binom{T}{t_1, t_2, ..., t_k, t_{cross}} p_1^{t_1} p_2^{t_2} ... p_k^{t_k} p_{cross}^{t_{cross}} \quad (10.12)$$

where $T = \sum_i t_i + t_{cross}$ is the total number of true edges.

This model of edge distribution may seem odd at first, since one is used to thinking about partitioning the set of samples and not the set of edges. It is easier to think of choosing an edge at random as the same action as choosing two samples at random. This must result in either the two samples landing in the same cluster or in these samples lying in different clusters. The multinomial distribution is a model of this edge picking process.

While Equation 10.12 gives the probability of a certain partition of samples into clusters, it does not ascertain the significance of the partition. To assess the significance one needs to know how probable it is to get partitions that are less likely by chance than the one in question. In other words, we need to compute the weight of the "tail" of the null distribution (the pvalue), consisting of all partitions that are more extreme than the one observed. The clustering with the minimal pvalue (maximum significance) can be considered the best clustering.

The tail distribution of some unimodal distributions has been characterized (e.g., normal). However, the edge distribution model in this case is more complex. A common approach for assessing the significance of an experiment as described above, assuming the multinomial distribution as the null hypothesis, is by using the statistic

$$Q = \sum_i \frac{(t_i - Tp_i)^2}{Tp_i} + \frac{(t_{cross} - Tp_{cross})^2}{Tp_{cross}}$$

This statistic has an approximate chi-square distribution with k degrees of freedom, and the pvalue can be computed from the chi-square cumulative distribution. An alternative approach that is more accurate for small datasets is to compute the pvalue directly, using the likelihood ratio test. Both approaches are described in Section 7.7.

10.5.2 Internal indices of validity

As opposed to external tests, which can be applied only when we have a prior knowledge on the cluster structure of the data, internal indices of validity rely on the data itself to assess the clustering configuration and are therefore more generally applicable. Some indices measure the stability of solutions under re-sampling [527, 528]. Others measure various geometric properties of clusters (e.g., diameter, contact density within and between clusters). Such measures can be easily defined even when the dataset is represented as a graph [521, 529–533]. In the next sections we will focus on two popular approaches.

10.5.2.1 The MDL principle

The **Minimum Description Length (MDL)** [262, 534, 535] is a popular heuristic for selecting a "reasonable" model out of many possible models. The MDL approach assesses the quality of a clustering solution by taking into account how well it describes the data (the distortion) as well as the model's complexity.

Inspired by a more general principle called **Occam's razor** (named after the 14th century mathematician who suggested it, William of Ockham), the MDL principle is often used to prefer one type of model over another. The idea is that of all possible models that can explain the data, we should prefer those that are simpler and reject the complex ones. Naturally, simpler models are easier to describe. However we should also take into account how well the model describes the data. The **description length** of a model is a measure of the complexity of a model and how well it explains the observations. Measured in bits, the description length of a given model (hypothesis) *and* a dataset is defined as the description of **the model** h plus the description of the **data** D **given the model**

$$l(h, D) = l(h) + l(D \text{ using } h)$$

The *minimum* description length heuristic favors models that minimize the overall description length, thus compromising between complex, overly specific models and simple models that are too general and fail to provide a reliable and accurate description of the data. In other words, the MDL method seeks a model h^* such that $h^* = \arg\min_h l(h, D)$.

One could argue that simpler models are more likely to generalize better to new instances. For example, given the dataset of Figure 10.10 we could perfectly fit a polynomial function of order 10, *or* try to fit a line and add a normally distributed noise to explain the observations. The second model is much simpler with only four parameters (two for the line and two for the normal distribution), while the first has 10 parameters. The MDL principle suggests that the second approach might generalize better to new instances, and in practice this is often the case. The meaning of this claim in the context of the example above is that if we do not have any prior information on the type of source that emitted these instances, then it is "more reasonable" to assume that the source is linear (with some white noise) and new instances are more likely to follow the line than a polynomial function of order 10. We will get back to this argument at the end of this section.

One way to define the description length of a model is by its algorithmic complexity, also known as **Kolmogorov complexity**. The Kolmogorov complexity of a string x is defined as

$$l(x) = \min_{|y|}[U(y) = x]$$

which is the size of the shortest program y that computes the string x and halts. Both x and y are binary strings and U is the universal Turing machine, which can implement any algorithm and compute any computable function. By relying on the universal machine, this definition compares the complexity of the strings regardless of the specific details of the implementation. For

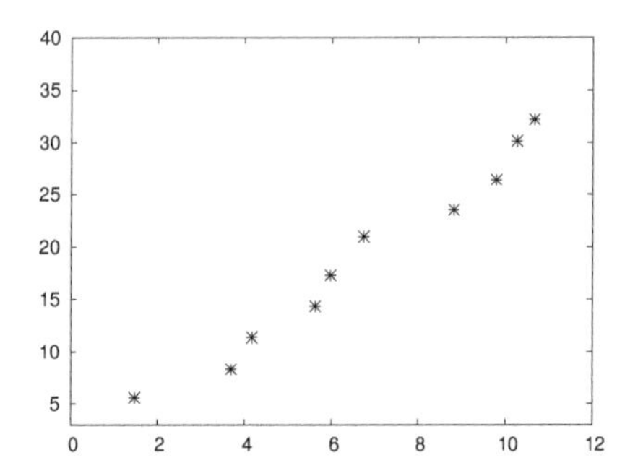

FIGURE 10.10: **Occam's razor.** If we do not have any prior information on the type of the model that generated the data, representing it as a line with normal noise might generalize better to new instances over a polynomial function of order 10.

example, the Kolmogorov complexity of a string of n 1's is $\log_2(n) + c$ where c is the number of bits needed to describe a general machine that contains a loop for printing 1's, and $\log_2(n)$ is the number of bits needed to code the condition for halting. In this case $l(x) = O(\log_2(n))$. On the other hand, the complexity of the number π with an arbitrarily large number of digits is a constant because it is computable by a fixed length program (ratio between a circle's circumference and its diameter). For a random string x the complexity is $O(|x|)$.

However, the Kolmogorov complexity approach is not very practical, since to define the algorithmic complexity of a model we have to find the shortest program that encodes for this model, something that is not computable in general. A more practical approach that formulates a similar principle in different terms is the **Bayesian approach**. According to this approach the most probable model is the one that maximizes the posterior probability of the model given the data

$$P(h|D) = \frac{P(h)P(D|h)}{P(D)}$$

That is, the hypothesis h^* is given by

$$h^* = \arg\max_h [P(h)P(D|h)] = \arg\max_h [\log_2 P(h) + \log_2 P(D|h)]$$

which is equivalent to minimizing $-P(h)P(D|h)$, thus

$$h^* = \arg\min_h [-\log_2 P(h) - \log_2 P(D|h)] \tag{10.13}$$

By Shannon's coding theorem $-\log_2 P(x)$ is related to the shortest description of a string x (see Section 6.4.2.1), and therefore the first term can be considered the model description length, while the second term $\log_2 P(D|h)$

is the likelihood of the data given the model and can be considered the description of the data given the model (also referred to as **conditional data**). Hence, we have a link between the MDL principle and the Bayesian approach.

Let's take a closer look at these two terms. The second term in Equation 10.13 is the minus log-likelihood of the data given the model, and assuming the samples are i.i.d.

$$\log_2 P(D|h) = \log_2 \prod_{i=1}^{n} p(x_i|h) = \sum_{i=1}^{n} \log_2 p(x_i|h)$$

This is a measure of the uncertainty that is left in the data, given the model. If the likelihood is 1 (no uncertainty) then $\log_2 P(D|h) = 0$.

Clearly, the likelihood depends on the underlying model. Consider for example the k-means algorithm, where every instance is classified to a single cluster C_i. If we assume that clusters can be modeled as Gaussians (see Section 10.3.2), then the probability of an individual sample to belong to cluster C_i can be approximated using the normal density function

$$p(\vec{x}|C_i) = \frac{1}{(2\pi)^{d/2}|\Sigma_i|^{1/2}} \exp^{-\frac{1}{2}(\vec{x}-\vec{\mu}_i)^t\Sigma_i^{-1}(\vec{x}-\vec{\mu}_i)}$$

where $\vec{\mu}_i$ and Σ_i are the mean vector and the covariance matrix of cluster C_i. The likelihood can be computed efficiently once the parameters for each cluster are determined.

Note that the likelihood is maximized when the number of clusters is n, that is, when all clusters are singletons such that each data point is in its own cluster. In that case, the likelihood of each sample point is one. However, this model is overly complex (with $n \cdot d$ parameters) and would generalize poorly to new instances, since it would assign zero probability to any future sample that does not coincide with one of the points in the training set. Equation 10.13 balances between maximal likelihood and maximal complexity by considering also the prior probability $P(h)$ for observing such a model.

As mentioned above, the first term in Equation 10.13 can be viewed as the approximated model description length. In some cases it is simpler to estimate the shortest description than the actual prior distribution $\log P(h)$. Recall that $-\log_2 P(x)$ is a typical term that we have in an entropy function measuring the uncertainty we have about the value of x, and as such $-\log_2 P(h)$ measures the uncertainty we have about the model. In the common hard clustering model, such as k-means, each cluster is represented by a prototype (the centroid), which is computed as the average of all samples that are classified to that cluster. If the source distribution of each cluster is a Gaussian (or can be closely approximated as such), then the prototype is an estimate of the Gaussian's mean. In the limit of infinite data samples, the centroid will converge to the true mean. However, since the prototypes are estimated from a finite dataset, there is a certain uncertainty about the exact value of the mean. We can estimate the first term in Equation 10.13 by calculating how many bits are needed to specify the coordinates of the prototypes (that is, localize the means), taking into account this uncertainty.

How can we estimate the uncertainty? Applying the simplifying assumption that the covariance matrix of each cluster is diagonal, then the multivariate normal distribution can be written as a product of d normal distributions,

$$p(\vec{x}|C_i) = p((x_1, x_2, ..., x_d)|C_i) = p(x_1|\mu_{i1}, \sigma_{i1})p(x_2|\mu_{i2}, \sigma_{22}) \cdots p(x_d|\mu_{id}, \sigma_{id})$$

where each $p(x_j|\mu_{ij}, \sigma_{ij})$ is a one-dimensional normal distribution. The means and the standard deviations characterize the source distributions that generated the samples. If the standard deviations within clusters are not explicitly used in the clustering algorithm, as is the case with most clustering algorithms (including k-means), we can assume that each cluster in the model h is parametrized by only d parameters (the mean vector). The standard deviations quantify the scatter of sample points around each mean and can serve as a measure of uncertainty regarding the exact whereabouts of each one.

Specifically, consider a one-dimensional cluster distribution, assumed to be generated by a normal distribution $N(\mu_{ij}, \sigma_{ij})$. Its estimated mean is given by

$$\hat{\mu}_{ij} = 1/n \sum_{i=1}^{n} x_{ij}.$$

According to the central limit theorem, if a parameter is estimated as a sum of i.i.d. random variables X_i drawn from a probability distribution with mean μ and variance σ^2 as in

$$\hat{\theta} = \frac{1}{n} \sum_{i=1}^{n} X_i$$

(which is exactly how $\hat{\mu}_{ij}$ is estimated), then $\hat{\theta} \to \mu$ as $n \to \infty$ with standard deviation $\sigma_\theta = \sigma/\sqrt{n}$. Hence, σ_θ is a measure for the uncertainty in the value of the parameter $\hat{\theta}$. The $-\log_2$ function measures the uncertainty in bits, and the overall uncertainty is a sum over the uncertainties of all parameters for all clusters. For example, for k clusters in d dimensions, with n_i samples in cluster i the total uncertainty is estimated as

$$-\sum_{i=1}^{k} \sum_{j=1}^{d} log_2 \left(\frac{\sigma_{ij}}{\sqrt{n_i}} \right)$$

Note that the number of samples differs for each cluster, and therefore the uncertainty estimates use different n_j for every parameter. For singleton clusters the uncertainty could be set to $\log n$ bits per parameter, to indicate which of the data instances is classified to this singleton.

The application of the MDL principle in this setting requires a parametric model, but it can be applied with some success even when no such model was used to cluster the data, if the final clusters can be well approximated by Gaussians. The model can also be generalized to the soft clustering case, where the underlying assumption is that the data comes from a mixture of Gaussians and each sample has a nonzero probability to belong to each one of the clusters (see Section 10.3.2). In this case, we have to calculate the probability of each sample according to each normal density, multiply by the prior, and sum over all components. Note that with fuzzy clustering all samples contribute to all parameters (weighted by their probability), and therefore we have to use the same n in all uncertainty estimates, which is the total number of samples. It should be noted that the MDL approach may be over-sensitive to perturbations in the data or in the parameters. Nevertheless it is a practical approach that often helps to identify better models.

Although it has been shown in [536] that a model designed with the MDL principle is guaranteed to converge to the "ideal" or true model in the limit of infinite data, there is no proof to support this argument in the case of finite data, which is always the case in reality. In fact, it has been argued (the so-called "no free lunch" theorem [537]) that this principle will not perform better on average than other methods, not even better than a random guess, when considering all possible learning tasks[4]. However, the method does work well in practice. The success of this principle could be attributed to the bias in the types of problems science has been addressing thus far, and it could be that most are relatively simple on average (when considering all possible learning tasks), or at least historically they were. The two theorems do not contradict each other, since the no free lunch theorem was formulated in the case of a finite training set, while the theorem about the MDL principle holds in the infinite case.

10.5.2.2 Cross-validation

Cross-validation is a common approach for optimization and assessment of clusters. Cross-validation compares the clusters against independent validation data and applies statistical tests to verify that the match between the two sets is statistically significant.

For example, the validity of a clustering solution can be monitored by splitting the data into two random subsets, applying the algorithm independently to each set and comparing the clusters. The underlying assumption is that patterns that are common to two large independently chosen subsets of the same original set capture true features of the original space. Therefore, this kind of test can be considered an objective assessment of the clustering struc-

[4] The "no free lunch" theorem argues that there is no single algorithm or principle that performs better on average over all learning problems.

ture. It can help to determine whether a structure is meaningful and is unlikely to have occurred by chance, or an artifact of the clustering algorithm. The quality of the match between clusters can be measured by means of different statistical tests, and a few examples are described in Section 10.9.

Cross-validation is actually not limited to evaluation of clustering, and it can be used in unsupervised as well as supervised learning problems, to test models during training and parameter optimization (see Chapter 7). It can usually reduce the risk of overfitting the model to the training data. Furthermore, it can help to estimate the generalization error of the model [258], since a statistically significant correspondence between models trained from two independent samples implies an upper bound on the probability that the models disagree on the classification of new instances [235]. However, cross-validation does not necessarily help to find the best model overall, especially if the clustering algorithm makes certain assumptions about the functional form of clusters. This technique also requires that the validation set is relatively large, since a small validation set might not display all the patterns that exist in the training set.

10.6 Clustering proteins

We mentioned earlier the motivation for clustering proteins. Clustering can help to identify groups of related proteins. It can also help to characterize proteins with unknown functions by observing other members in their cluster and extrapolating their properties. By using collective information, clustering can help to reduce noise and provide more accurate functional predictions. This is especially useful for the characterization of remote homologs.

Clustering also aims to detect higher level organization within the input dataset. Such organization can reveal relationships among protein families, such as the existence of subfamilies and superfamilies, and yield deeper insights into the nature of newly discovered sequences.

Protein classification has been an active research field since the early 1990s. In general there are two main approaches. Some studies focus on finding significant motifs, patterns and domains within protein sequences. Other studies focus on complete proteins. We already discussed some of the domain and motif-based studies in Chapters 5 and 9. In this section we will discuss algorithms of the second type.

10.6.1 Domains vs. complete proteins

Clustering complete protein sequences is instrumental to the study of the protein space. However, is it the most effective method for functional analysis of proteins? In Chapter 9 we discussed the problem of domain prediction and the significance of multi-domain architectures. The important role of motifs and domains in defining protein's function is unquestionable, and functional analysis of proteins should always consider their domain structure.

However, in many cases characterizing a new protein only by its domain content is insufficient. This happens, for example, when the protein does not contain known domains. In some instances, only a few related sequences are available, too few to define a reliable prototype signature or a profile of the common domains. Therefore, a proper analysis of a new protein sequence should compare it against domain-based databases, as well as sequence databases.

Moreover, comparing a protein against a set of protein clusters can amplify the outcome of a database search. When close hits are already grouped together based on mutual similarity, it may highlight a similarity with a group that could otherwise be missed by a simple manual scanning. Furthermore, if the query sequence is related to several different groups, this may indicate the existence of several distinct functional/structural domains. Most importantly, clustering complete proteins can reveal new protein families and domains. In fact, domains are often discovered by clustering sequence databases first (e.g., ProDom, Pfam).

For these reasons, and the reasons mentioned before, clustering of complete proteins is essential for an accurate analysis of the protein space and, together with domain analysis and multi-domain analysis (as discussed in Chapter 9), is part of a comprehensive approach for functional analysis of proteins.

Clustering proteins is a challenging problem. When clustering proteins, what we essentially search for is groups of homologous proteins. However, the similarity data we work with does not necessarily imply homology (as discussed in the introduction to Section 3.4 and in Section 3.4.7). Therefore, if proteins are clustered just based on similarity we will end with unrelated proteins classified to the same cluster. Multi-domain proteins make the deduction of homology particularly difficult, since they often cause **transitive chaining** where two unrelated proteins are connected through an intermediate sequence, as illustrated in Figure 9.7. Nevertheless, even with relatively simple clustering algorithms one can obtain biologically meaningful results, by taking proper precautions and pre-processing the data or post-processing the results. In the next two sections we will discuss some of the algorithms that were applied to cluster complete protein sequences.

10.6.2 Graph representation

Many protein-based clusterings draw directly on pairwise comparisons [13, 67, 69, 70, 538–540]. A natural way to represent pairwise similarities is as a weighted graph whose vertices are the sequences, and edge weights reflect the degree of similarity between the corresponding sequences, as in Section 10.4. Rather than raw similarity values, we can use the statistical significance values, which are more informative. For example, the weight of an edge between two proteins, p_1 and p_2 can be set to $evalue(S)$ or to $-\log evalue(S)$, where S is their similarity score. Recall that a low expectation value reflects a significant, strong connection, whereas a high expectation value reflects an insignificant, weak connection. Intuitively, with these weights the term distance can be used instead of expectation value, to indicate that two proteins are either close or far, but the evalue is not a proper distance function (see Section 3.6) and practically no metric is defined.

Not all edges should be retained in the graph, and edges of statistically insignificant similarity scores should be discarded, so that if an edge does exist between sequences p_1 and p_2 then the corresponding proteins are likely to be related. Note also that the graph is *directed*, since the evalue of a similarity score depends on the background distribution of the similarity scores of each node, and the distribution for p_1 might differ from that for p_2. Specifically, the weight of the edge (p_1, p_2) is based on the distribution associated with p_1 while the weight of (p_2, p_1) is based on the distribution associated with p_2. To simplify the analysis one is advised to transform the graph into an undirected graph by replacing the directed edges from p_1 to p_2 (with weight w_1) and from p_2 to p_1 (with weight w_2), with one undirected edge whose weight is defined as the maximum, minimum or average of w_1 and w_2, depending on the practitioner's preference.

A graph is not the only representation of the protein space we can employ. In Chapter 11 we will discuss algorithms for embedding biological data in Euclidean spaces, allowing us to apply clustering algorithms in vector space.

10.6.3 Graph-based protein clustering

Graph clustering algorithms aim to detect strongly connected sets of vertices that correspond to groups of related instances. Most studies that cluster complete sequences use the basic single linkage algorithm to cluster sequence databases. Even recent studies, such as [541], are still based on this simple procedure. Single linkage clustering is equivalent to transitive closure. If all edges of significance below a certain threshold are eliminated, the transitive closure of the similarity relations among proteins splits the graph into connected components (clusters), wherein every two members are either directly or transitively related.

The idea of transitive closure is simple and appealing; however, the protein space has unique properties that prevent this procedure from being fully successful. There are two complications that should be considered: (i) multi-domain proteins can create undesired connections among unrelated groups, as discussed previously; (ii) chance similarities become more abundant as significance levels decrease. As a result, applications of the single linkage algorithm usually lead to a few gigantic clusters that contain many different and completely unrelated protein families[5].

Over the years methods more advanced than single linkage were applied to cluster protein sequences, including spectral clustering algorithms (Section 10.4.3), supermagnetic clustering (Section 10.4.5) and Markovian graph clustering (Section 10.4.4). While computationally intensive, these algorithms tend to produce more stable solutions that reflect the collective effect of all pairwise similarities. On the other hand, all these algorithms face the same challenges mentioned above and are not fundamentally different from single linkage in that respect.

There is no simple solution to the domain chaining issue when clustering complete protein sequences. The presence of "promiscuous" domains transforms the protein space into a complex manifold that does not conform with standard norms (e.g., Euclidean). Such spaces are not easily amenable to clustering and post-processing of clusters seems inevitable. However, the problems of multi-domain proteins and chance similarities have been addressed to some extent in several studies [45, 67, 68, 71, 438, 538]. The common strategy is to apply additional tests and monitor closely the process of merging clusters. For example, seed clusters of high quality can be detected by searching for pseudo-cliques, where *all* pairwise similarities are more significant than a preset threshold [45] or are based on global similarities with similar overall domain architectures [539]. To address the transitive chaining problem it is advised to use a moderate version of transitive closure where clusters are merged gradually, lowering the significance threshold by a certain degree at every step. We can further require that clusters merge only if the sequence overlap between clusters is substantial (e.g., more than 80% with respect to each) and apply transitivity tests to all triplets within the same cluster, as in [515, 538] (see also Section 9.2.3). In a post-processing step, clusters that contain non-overlapping sequences can be split at positions that minimize the number of neighbors in the similarity graph that are aligned to both sides of the cut (as in [434]). This method, combined with other termini analysis and domain prediction methods (see Chapter 9) can help to partition clusters to smaller building blocks.

Graph clustering algorithms that follow an hierarchical scheme provide additional useful information, as the gradual process of merging clusters can be insightful in itself. It can reveal the evolutionary history of protein families and suggest the existence of subfamilies. Even rejected mergers are of interest since they might indicate a true relationship between clusters. Based on these connections or the hierarchy of clusters it is possible to draw schematic

[5] Another immediate observation when clustering proteins is the presence of many singletons. Singletons might be the result of lateral gene transfer from ancient organisms, or loss of function of a protein after speciation (resulting in mutational drift without any selective pressure).

local maps for the neighborhoods of protein families that can expose remote relationships (see [45]).

It should be noted that evaluating the validity of classifications based on complete protein sequences is difficult. When clustering domains, a comparison with the manually derived domain databases SCOP or Pfam is natural and essential for testing the biological significance of the results. However, when clustering complete protein chains that contain multiple domains, no standard benchmarks exist to assess the quality of the results, and one often resorts to comparisons against domain-based databases. Obviously, this may bias the assessment, a fact that should be kept in mind when evaluating the results. Nevertheless, protein-based benchmarks can be generated from external domain databases by compiling all domain combinations in these databases and the proteins associated to these combinations. Databases such as Gene Ontology (GO) can provide alternative benchmarks; however, GO is more suited for qualitative evaluation of clusters (see Sections 3.11.4 and 12.5.3.1). An example of a quantitative analysis based on GO is given in [58].

10.6.4 Integrating multiple similarity measures

In Section 10.6.2 we briefly described the process of building a protein similarity graph from pairwise similarity data. We assumed that the similarity data is generated by applying one of the protein comparison algorithms we discussed in previous chapters. However, assigning weights to edges in the graph can be done based on more than just one similarity measure. It can be based on sequence or structure (if the structures of some proteins are known) or other features. With more information about relationships between entities, we expect the clustering solution to be more accurate and reflect the true cluster structure of the data.

Furthermore, when comparing *clusters* of protein sequences we can use more than just the pairwise similarities. We can represent each cluster using a sequence-based or structure-based profile or an HMM, for example, and compare clusters by comparing their models[6]. Using multiple models and similarity measures to represent and compare individual proteins and protein clusters is almost guaranteed to produce better results.

However, different methods detect different subsets of similarities and might report the same similarities with different scores. In other words, each method results in a different similarity graph **G**. Some methods use raw similarity scores, while others use zscores or statistical significance measures. How should we compare scores that are assigned by different methods and integrate them into a single similarity graph?

[6]Note that this is similar to progressive alignment (see discussion on MSAs in Section 4.4), which uses profiles to compare intermediate alignments.

A good practice is to first transform the scores of each method to statistical significance values. If the scores are given as statistical significance values already, then a simple solution is to assign each edge the maximal statistical significance according to any of the methods. However, the statistical estimates might be computed based on different sample spaces, and different algorithms have different statistical properties; therefore they might result in very different distributions.

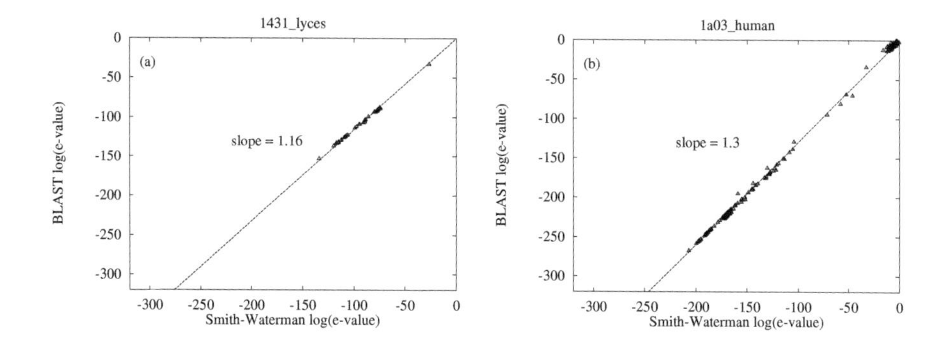

FIGURE 10.11: Correlation of BLAST evalues and Smith-Waterman (SW) evalues. (a) BLAST evalues of neighboring sequences of 14-3-3 protein (SwissProt 1431_lyces, Biozon ID 260-76) vs. the SW evalues of the same neighbors. The graph is plotted in log-log scale. Note the strong linear correlation between the scores assigned by the two methods, where the slope is 1.16, i.e., $evalue_{\mathsf{BLAST}} = (evalue_{\mathsf{SW}})^{1.16}$. (b) BLAST evalues of neighboring sequences of MHC class I antigen (SwissProt 1a03_human, Biozon ID 365-226) vs. the SW evalues of the same neighbors. The slope is 1.3 in this case. (Figure reproduced from [45] with permission from John Wiley & Sons, Inc.)

If the methods use the same statistical framework over the same dataset we can try to place them on the same numerical scale. For example, in [45] the scores are normalized based on the list of close neighbors for each sequence. When plotting these similarity scores against one another for two methods at a time, the scores show a strong linear relation on a log-log scale (Figure 10.11). The differences between methods can be due to different scoring matrices that are being used and can be corrected by multiplying the original score by the relative entropy of the two matrices (see Section 3.5.3). Differences may also be due to approximations in estimating the parameters (e.g., λ and K of BLAST). The scores are normalized to the same scale by introducing a usually small correction factor per each protein and per method.

Once all scores are normalized, the weight of an edge can be defined based on the strongest relationship. If edge weights represent expectation values (evalues), then small values indicate high similarity and the weight of an edge can be set to the minimum evalue associated to it by any of the methods.

However, in most cases different scores that are based on different features or models are likely to have very different distributions. In addition, some methods cannot be easily associated with statistical significance scores. One possible solution in this case is to use the zscore transformation (see Section 3.4.2.3), adjusting each pairwise similarity score to the specific distribution of scores in each network. Another alternative solution is to convert all the scores (or zscores) in each network to naive probability values, by means of simple normalization

$$w(x,y) \rightarrow \frac{w(x,y)}{\sum_{y'} w(x,y')}$$

With this normalization, the new weight of an edge (x,y) can be viewed as the probability $\text{Prob}(x \rightarrow y)$ to move from node x to node y. The probabilities assigned by m different methods can then be combined, assuming independence, by computing the complete probability

$$\text{Prob}(x \rightarrow y|\mathbf{G}_1, \mathbf{G}_2, ..., \mathbf{G}_m) = \sum_i \text{Prob}(x \rightarrow y|\mathbf{G}_i)P(\mathbf{G}_i) \quad (10.14)$$

where $P(\mathbf{G}_i)$ is the prior probability of the network $\mathbf{G}_i$ and can be either set uniformly to $1/m$ or defined based on our confidence in the i-th similarity measure, and $\text{Prob}(x \rightarrow y|\mathbf{G}_i)$ is the normalized weight of the edge (x,y) in the i-th network.

Recall our discussion on collaborative and Markovian clustering. By utilizing collective properties of the graph we can often obtain better estimates of the similarity between data instances. This is also the basis of the approach described in [516], which redefines the transition probabilities $\text{Prob}(x \rightarrow y)$ based on the global graph structure, by emulating a random walk process (see Section 10.4.4)[7]. Consider the affinity vector that contains the steady state probabilities of moving from x to any other data point. Each network $\mathbf{G}_i$ results in a different affinity vector $\vec{x}_i$. We can combine the predictions of the affinity vectors, assuming that the methods are independent, and compute the complete probability as in Equation 10.14. With $x_i(y)$ representing the steady state probability of reaching the node y from node x we obtain that

$$\begin{aligned}\text{Prob}(x \rightarrow y|\mathbf{G}_1, \mathbf{G}_2, ..., \mathbf{G}_m) &= \sum_i \text{Prob}(x \rightarrow y|\mathbf{G}_i)P(\mathbf{G}_i) \\ &= x_1(y)P(\mathbf{G}_1) + x_2(y)P(\mathbf{G}_2) + ... + x_m(y)P(\mathbf{G}_m) \quad (10.15)\end{aligned}$$

[7]The algorithm of [516] does not use $\mathbf{T}$ as described in Section 10.4.4, but instead normalizes the number of edges per method using a score threshold or considering a fixed number of neighbors (the top-k scoring proteins). Similarity values are then converted into transition probabilities by normalizing the weights to the range $[0,1]$ such that the smallest similarity corresponds to 0 and the highest similarity to 1.

These complete probabilities can then be processed using any of the clustering algorithms we discussed in the previous sections.

The simulation of random walks is not limited to transformations of the proximity matrix, as in Markovian graph clustering. By including a known classification, such as SCOP, as one of the networks, we can view the combined network as a *classifier* and compute the affinity of each protein with each family (category) in the external classification. This is done by adding a node for each family and connecting it to its members. The affinity vector for each family is then computed in each similarity network. The total affinity of a protein x with category c_j can be computed considering the affinity vectors from all the different networks.

Specifically, given a set of m similarity networks and m affinity vectors $\vec{x}_1, ..., \vec{x}_m$, let $c_i(x)$ denote the class with the highest affinity in the affinity vector of x using method i

$$c_i(x) = \arg\max \vec{x}_i$$

and let $a_i(x)$ be the affinity value

$$a_i(x) = \max \vec{x}_i$$

then the posterior probability that the class of x is c_j is given by

$$P(c(x) = j | c_1, a_1, ..., c_m, a_m)$$

where we removed the dependency of $a_i()$ and $c_i()$ on x for clarity. Only the maximal affinity categories and their affinities are used, instead of the complete affinity vectors, to minimize the number of variables. Using Bayes' formula we obtain that

$$P(c(x) = j | c_1, a_1, ..., c_m, a_m) = \frac{P(c_1, a_1, ..., c_m, a_m | c(x) = j) P(c(x) = j)}{P(c_1, a_1, ..., c_m, a_m)}$$

where $P(c(x) = j)$ is the prior probability of observing class j. Assuming independence between the sources, the likelihood is reduced to

$$P(c_1, a_1, ..., c_m, a_m | c(x) = j) = \prod_i P(c_i, a_i | c(x) = j)$$

and therefore

$$\begin{aligned}
P(c(x) = j | c_1, a_1, ..., c_m, a_m) &= \frac{P(c(x) = j)}{\prod_i P(c_i, a_i)} \prod_i P(c_i, a_i | c(x) = j) \\
&= \frac{P(c(x) = j)}{\prod_i P(c_i, a_i)} \prod_i \frac{P(c(x) = j | c_i, a_i) P(c_i, a_i)}{P(c(x) = j)} \\
&= P(c(x) = j) \prod_i \frac{P(c(x) = j | c_i, a_i)}{P(c(x) = j)} \qquad (10.16)
\end{aligned}$$

Therefore, the class probability depends on the prior probability and the conditional probabilities $P(c(x) = j|c_i, a_i)$.

The prior probability can be computed based on the relative size of the j-th class, while the conditional probabilities can be computed from the data. However, usually there is not enough data to produce reliable estimates of these probabilities. With n categories and a discretization of the affinity values to k bins, there are $k \cdot n^2$ possible combinations of $P(c(x) = j|c_i, a_i)$. Each protein in the dataset is associated with one combination of c_i, a_i and $c(x)$. With n on the order of 10,000, the total number of proteins in the database N is typically much smaller than the number of possible terms. The solution used in [516] is to introduce a global prior over all bins and modify the estimates based on the training data. Specifically, the probability is initialized to the prior probability $\hat{P}(c(x) = j|c_i, a_i) = P(c(x) = j)$ and is then incremented by one for each training sample that falls in that category. At the end, these counts are normalized such that

$$P(c(x) = j|c_i, a_i) = \frac{\hat{P}(c(x) = j|c_i, a_i)}{\sum_j \hat{P}(c(x) = j|c_i, a_i)}$$

to create a new set of probabilities $P(c(x) = j|c_i, a_i)$. The probabilities can be initialized such that $\hat{P}(c(x) = j|c_i, a_i) = M \cdot P(c(x) = j)$, where the number of prior counts M determines the weight of the prior information in the estimation (see the discussion on pseudo-counts in Section 4.3.3.2 of Chapter 4).

Note that Equations 10.14-10.16 are computed assuming conditional independence between the sources. However, this assumption does not necessarily hold in practice, since different similarity measures are often correlated and tend to report the same similarities. Therefore the joint and complete probabilities per these equations are only an approximation of the "true" probabilities.

10.7 Further reading

There are numerous papers describing different approaches to clustering, some of which date back to the 1940's. The book by Jain & Dubes [521] discusses extensively various algorithms as well as validation techniques. An excellent summary of data clustering appears in Chapter 10 of [191]. Another excellent survey is the review by [542]. For other, recent reviews see [543,544]. For reviews and analysis of spectral clustering and their Markovian interpretation see [513,514,545]. Another related approach is the information bottleneck algorithm of [546]. The paper by [547] has an interesting discussion on the relation between super-paramagnetic clustering and the min-cut clustering algorithm.

The MDL criterion has been discussed in many papers [262, 535, 536, 548, 549]. The MDL function is closely related to the Bayesian information criterion (BIC) function [550] and the Akaike information criterion (AIC) function [551], which are variants with different weightings of the prior probability. An evaluation of model selection algorithms was conducted in [552]. Recent clustering techniques based on kernel methods are discussed in [553].

Many studies approached the problem of protein clustering, but the problem is far from being solved. In addition to the studies we mentioned in Section 10.6, more recent ones include [554–557]. For updates and additional references see the book's website at `biozon.org/proteomics/`.

10.8 Conclusions

- Clustering is a popular unsupervised data analysis technique, which seeks to identify groups of similar instances in the input data.
- Clustering is an important tool in the analysis of biological data, and is especially fundamental to the analysis of the protein space.
- Clustering of proteins often reveals new domains and protein families.
- The many different clustering algorithms can be generally divided into vector-space algorithms and graph-based algorithms. Vector-space algorithms assume the data resides in a real-normed space of some dimension. Graph-based algorithms utilize only the pairwise distances between instances, and therefore can be applied to a broader class of datasets.
- Algorithms can be further categorized based on the scheme (hierarchical vs. partitional), mode (online vs. batch) and exclusiveness (hard vs. soft clustering).
- Vector-space algorithms partition instances so as to minimize a certain criterion function, such as the sum-of-squared error.
- Graph-based algorithms treat the dataset as a graph where nodes represent instances and edge lengths denote their similarity. They work by means of removing or adding edges between nodes of the graph.
- Collaborative clustering algorithms utilize collective properties of the dataset to cluster instances, usually by redefining the mutual similarity of instances based on transitive relations.
- Spectral and Markovian clustering algorithms utilize the eigenvectors of the proximity matrix to derive partitions of instances into clusters.
- Cluster validation is important in order to assess the significance of the results. Clusters can be validated using internal indices which measure various geometric properties of clusters, or by comparing them to external classifications when available.
- Graph-based approaches, including collaborative algorithms, are more naturally suited for clustering protein sequences.
- Integration of multiple similarity measures can greatly improve the quality of the clustering.
- The multi-domain nature of proteins complicates the task of protein classification.

10.9 Appendix - cross-validation tests

Cross-validation works by comparing the parameters of a model against the results obtained over an independent validation data. When applied to assessment of clustering results, the first step in cross-validation is to establish **one-to-one correspondence** between clusters in the training set and clusters in the validation set. Finding the correspondence in general can be a difficult problem; however, one can take certain measures to simplify it. For example, we can apply hierarchical clustering algorithms (as in Section 10.3.3) to both sets (independently of each other) and establish the correspondence at each level of the hierarchy before splitting clusters. Given the correspondence for k clusters, establishing the correspondence between $k+1$ clusters usually involves only the two clusters that result from the split. Once the correspondence is established, cross validation uses certain tests to assess the agreement between the clustering solutions over the two different datasets. A few examples are given next, following in part the algorithm of [99].

Clusters' sizes: The cluster size is one of the simplest tests we can use to check if clusters in the cross-validation set agree with the model derived from the training set. If the data is split randomly between a training set and a validation set of equal size, then we expect that every cluster would split into two clusters of nearly identical sizes. If the cluster does not overlap with other clusters then we expect the clustering algorithm to detect both subclusters when applied independently to each dataset. On the other hand, outliers and sets of samples that do not form distinct clusters will not necessarily split into two similar subclusters, and the clustering algorithm is more likely to produce different results in that case.

If corresponding clusters are of nearly identical sizes, then we can say that the clusters are in agreement. But what difference in size should we expect by chance and when can we say that the match is plausible? Given a cluster of m data points, a random split into two subsets would result in two clusters of size $n_1 = \frac{m}{2} + \delta$ and $n_2 = \frac{m}{2} - \delta$, where δ is the deviation from a perfect match. The number of points that fall in the first subset is distributed as the binomial distribution with $p = 1/2$ such that

$$\text{Prob}\left(n_1 = \frac{m}{2} + \delta\right) = \binom{m}{\frac{m}{2}+\delta} \frac{1}{2}^{\frac{m}{2}+\delta} \frac{1}{2}^{\frac{m}{2}-\delta} = \binom{m}{\frac{m}{2}+\delta} \frac{1}{2}^{m}$$

The pvalue of a certain partition with deviation δ is the probability to observe by chance a more "extreme" partition with a greater deviation $\delta' \geq \delta$ (see Section 3.12.8). That is,

$$pvalue(\delta) = \text{Prob}(\delta' \geq \delta) = \sum_{\delta'=\delta}^{m/2} \text{Prob}\left(n_1 = \frac{m}{2} + \delta'\right)$$

and the bigger the deviation, the smaller is the pvalue.

What is the maximal difference δ_{max} that we are likely to observe in our dataset? If n is the total number of data points, then there are at most n/m clusters of size m. In other words, we can repeat the "split experiment" up to n/m times. The maximal difference can be set by requiring that the expectation value is greater than one

$$\frac{n}{m} \cdot pvalue(\delta_{max}) \geq 1 \tag{10.17}$$

meaning, in a dataset of size n we expect to see at least one cluster of size m splits into two subsets where the larger one is of size $\leq \frac{m}{2} + \delta_{max}$. Such splits are deemed valid, and the clusters are said to be in agreement on their size.

For large m we can use the normal approximation for the binomial distribution, where $\mu = m \cdot p = \frac{m}{2}$ and $\sigma = \sqrt{mp(1-p)} = \sqrt{\frac{m}{4}}$. There are closed-form solutions for the cumulative distribution of the normal distribution, and therefore $pvalue(\delta)$ can be computed efficiently. By solving Equation 10.17 for δ_{max} we can derive the maximal "allowed" deviation. In other words, two corresponding clusters are considered of nearly equal sizes, and therefore assumed to come from the same prototype cluster, if the difference in their size $\leq \delta_{max}$.

For small clusters ($m < 20$) the normal approximation does not hold and the bound has to be computed directly by plugging in the binomial coefficients in Equation 10.17 and testing each δ. Note, however, that the smaller the cluster the higher the ratio n/m (the number of "experiments") and the relative accepted variation in size might be substantial. Therefore, an agreement according to that test may not necessarily indicate that the subclusters indeed represent the same cluster.

Clusters' prototypes: Another simple test we can invoke is to check whether the centroids of corresponding clusters are close to each other. If data is vectorial we can use the Euclidean distance and assert that two clusters are nearly identical if the Euclidean distance between their centroids is small. However, there is no general definition of what should be considered small Euclidean distance. Sometimes the distribution of distances between sample points may suggest a natural threshold. But the Euclidean distance may be misleading, and the bound should depend also on the clustering resolution we are aiming for and on the distribution of data points around the centroids.

Consider for example the configuration in Figure 10.12. Although the Euclidean distance between the two centroids is small (compared with the average distance within clusters), the two centroids stand for two different clusters. However, by observing the distribution of data points in these clusters we can immediately distinguish between the two. In such cases it is useful to use the **σ-distance** to decide if the two centroids are nearly identical. The σ-distance measures the distance between centroids in terms of the distributions of data points in the clusters.

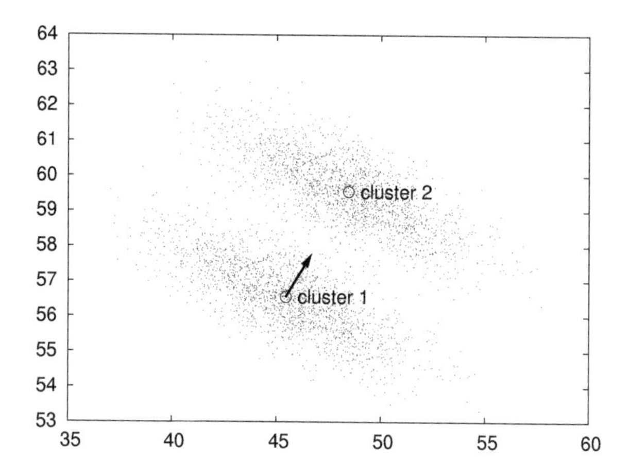

FIGURE 10.12: **The σ-distance.** The Euclidean distance between the centroids of the two clusters is small compared to the average distance within the clusters. Therefore, we cannot distinguish between the two based on just the Euclidean distance. The σ-distance, on the other hand, indicates that there are two distinct clusters. The projection of the distribution of points in cluster 1 on the axis that connects the two centroids defines σ_1 (shown with arrow). The σ-distance from cluster 1 to cluster 2 is 2.6; therefore the clusters are distinguishable.

The *σ-distance* from $\vec{\mu}_1$ to $\vec{\mu}_2$ (the notation implies the asymmetric characteristic of this measure) is defined by first taking the projection of the distribution of cluster 1 on the axis that connects the two centroids, $\vec{\mu}_2 - \vec{\mu}_1$, and calculating the variance of this projection, denoted by σ^2

$$\sigma^2 = \frac{1}{n_1} \sum_{i=1}^{n_1} \left[(\vec{x}_i - \vec{\mu}_1) \cdot \frac{(\vec{\mu}_2 - \vec{\mu}_1)}{||\vec{\mu}_2 - \vec{\mu}_1||} \right]^2$$

The *σ-distance* is defined as

$$d_\sigma(\vec{\mu}^1, \vec{\mu}^2) = \frac{||\vec{\mu}^2 - \vec{\mu}^1||}{\sigma}$$

When the σ-distance between two centroids is small (say, less than 1), the distributions highly overlap, and the centroid of the second cluster falls well within the cloud of points surrounding the first centroid. Therefore these centroids can be considered nearly identical.

Alternatively, if clusters are associated with a certain parametric model, then we can use the model to estimate the probability that the centroid $\vec{\mu}_1$ belongs to cluster C_2 and vice versa. If the probability is high we can say that the clusters agree on their centroids with confidence that depends on the class-conditional distribution.

It is harder to assess how well the clusters match if data is not vectorial, since there is no natural definition of a centroid in non-vectorial spaces. In such cases, we can instead pick as center "point" the instance whose total distance from all other instances is minimal. That is,

$$x_i = \arg\min_{x \in C_i} \sum_{x' \in C_i} d(x, x')$$

The quality of the correspondence between clusters is then assessed by considering the original dissimilarity (or similarity) of their center points. The significance can be assessed based on the distribution of dissimilarities.

Clusters' shapes: The centroid is one property (the first moment) of a cluster, averaged over its data points. What we really want to compare is the "shapes" of the clusters. If data is vectorial we can compare the distributions of data points around the centroids. For example, assuming clusters form ellipsoids in $\mathbb{R}^d$ then we can compare the covariance matrices of the clusters (second moment). But the random makeup of each subset will most definitely result in many differences between the entries of the two matrices, and as with the Euclidean distance between centroids, determining what should be considered a small variation is not trivial. Moreover, for high-dimensional data the number of entries we need to compare grows as d^2. But there is really no need to compare all entries. The distribution in each cluster can be characterized by the principal components of the samples covariance matrix, which are the principal directions of the distribution, where the largest variances are observed. These directions characterize the geometric orientation of the distribution. We will talk about Principal Component Analysis (PCA) in the next chapter. For now, all we need to know is that the principal components are the eigenvectors of the covariance matrix, and the corresponding eigenvalues denote the variance along these eigenvectors[8]. The term principal components is often used to denote just the main eigenvectors, those with the largest eigenvalues (largest variance).

If the distributions of two clusters are similar, then their principal directions are expected to be nearly identical. We can verify that by computing the scalar product between the principal directions. To assess the significance of the match we need to determine what kind of match we should expect by chance. The background distribution in this case is defined as the distribution of scalar products for random vectors in $\mathbb{R}^d$, drawn uniformly over the unit sphere in $\mathbb{R}^d$.

The scalar product of two vectors $\vec{x}$,$\vec{y}$ in $\mathbb{R}^d$ is given by $\sum_{i=1}^{d} x_i \cdot y_i$. For random vectors, each coordinate is chosen independently at random, hence x_i and y_i are random variables. The product of two random variables is a random variable, and the sum of d i.i.d. random variables is distributed

[8]If the covariance matrix is singular (which is often the case for high-dimensional data) then PCA will fail. But we can determine the principal directions using Singular Value Decomposition (SVD) as described in Section 11.2.1.2.

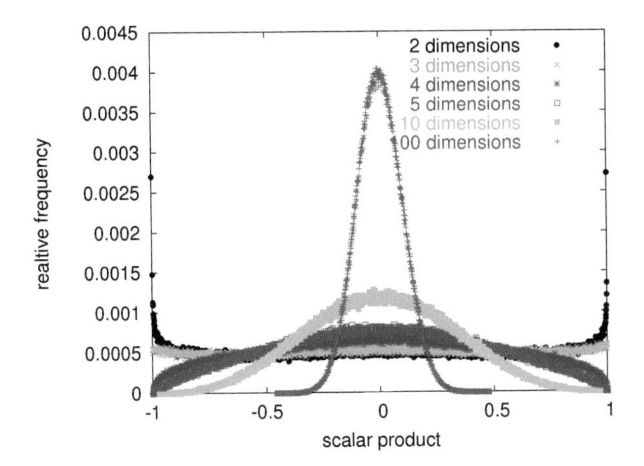

FIGURE 10.13: **Distribution of scalar product for random vectors in $\mathbb{R}^d$.** Distributions are plotted for d=2 d=3 d=4 d=5 d=10 d=100. The distribution changes drastically as the dimension increases. For random vectors in 2D the distribution is mostly flat besides two peaks at the boundaries. Indeed, for vectors in 2D the scalar product is determined by the cosine of the angel between the vectors. Therefore, the distribution simply follows from the characteristic of the cosine function (which is denser in the boundaries). For higher dimensions (4,5) the distribution is a broad unimodal concave function that narrows down as the dimension increases. For large d the distribution is clearly normal.

normally for large d. Therefore, the scalar product of random vectors in $\mathbb{R}^d$ is distributed normally. However, for **unit** vectors, the vector components are not independent since the vectors are normalized to have unit length. Nevertheless, the normal distribution still holds for large d (see Figure 10.13). The distribution of scalar products of random unit vectors in $\mathbb{R}^d$ follows a normal distribution that is centered at zero and has a standard deviation that depends on the dimension $\sigma = \frac{1}{\sqrt{d}}$ (see the Johnson-Lindenstrauss theorem in Section 11.2.2.2).

Therefore, for high-dimensional Euclidean spaces it is possible to assess the statistical significance by means of the standard deviation σ of the distribution (since the distribution is quite broad for lower dimensions, the test is not effective for lower dimensional spaces). For example, if the absolute value of the scalar product of two principal components from two different clusters exceeds 3σ then there is a probability < 0.01 of observing this match by chance and we can say that the clusters agree on their principal directions with 99% confidence. This probability should be corrected for multiple tests (the number of clusters) using, for example, the Bonferroni correction (see Section 3.12.8).

If data is not vectorial we can compare the distributions of distances *within* clusters, as shown in Figure 10.14. Although different shapes can result in

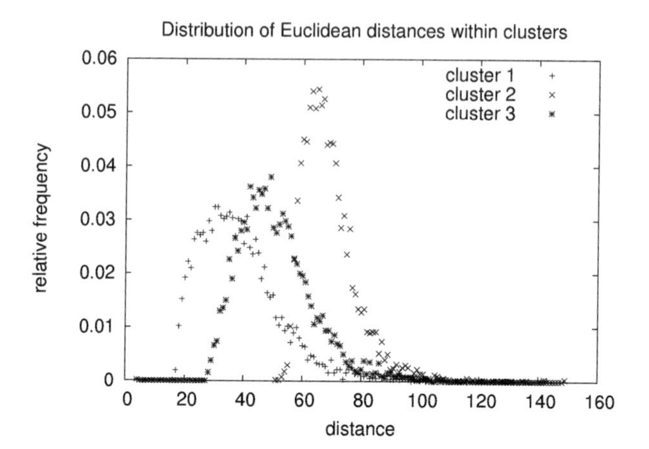

FIGURE 10.14: **Distributions of pairwise distances within clusters.** Clusters with different distributions are more likely to have different geometries.

the same distribution of pairwise distances within a cluster, this test can be applied to reject agreement when the distributions are different. We already discussed several measures for comparison of distributions, such as the KL-divergence and the JS-divergence measures (see Section 4.3.5). Other measures, such as the l_1 or l_2 distances or the earth-mover distance (Section 11.2.3.3), can be used as well.

10.10 Problems

For updates, additional problems, files and datasets check the book's website at `biozon.org/proteomics/`.

1. **The complexity of clustering.**

   Show that the number of possible clustering configurations for n sample points and k clusters is at least exponentially large.

2. **The squared error function.**

   Show that of all possible choices of cluster representatives under the hard clustering model, the centroid is the choice that minimizes the total squared error function.

3. **Hill-climbing procedures.**

   Show that the k-means algorithm decreases the global distortion (total squared error) with every iteration, until convergence.

4. **The single linkage algorithm.**

   Prove that the single linkage algorithm generates a minimal spanning tree over the data graph.

5. **The average linkage algorithm.**

   Show that if sample points are vectors in real-normed space then

   $$d_{ave}(C_i, C_j) = d_{mean}(C_i, C_j)$$

6. **Time complexity.**

   Compare the time complexity of the k-means algorithm vs. the average linkage algorithm.

7. **Induced metrics by hierarchical pairwise clustering**

   In cases where we do not have a metric but only dissimilarity values for the data points, the hierarchical pairwise clustering algorithm can also induce a distance function d_k on the sample set.

   Define cluster dissimilarity as the minimum over all possible pairwise dissimilarities between the two clusters. Initialize $d_1 = 0$ and let d_k be the minimum dissimilarity between pairs of distinct clusters at level $k-1$. Note that d_k increases as k increases, and assuming that no two samples are identical then $0 = d_1 < d_2 \leq d_3 \leq ... \leq d_n$.

   Define the function $d(i,j)$ between sample point i and sample point j as the value d_k of the lowest level k for which i and j are in the same cluster. Show that this function is a distance function that satisfies the three metric requirements.

8. **The k-means algorithm**

   Write a program that implements the k-means clustering algorithm of Section 10.3.1. You are given data points in R^d that come from k clusters. Assume that in the mixture density all clusters are equally likely, and that each cluster can be described as a Gaussian (normal distribution) with mean μ_i and a scalar covariance matrix $\Sigma_i = \sigma I$. For example, with $\sigma = 2$ and $d = 5$,

$$\sigma I = \begin{pmatrix} 2\,0\,0\,0\,0 \\ 0\,2\,0\,0\,0 \\ 0\,0\,2\,0\,0 \\ 0\,0\,0\,2\,0 \\ 0\,0\,0\,0\,2 \end{pmatrix}$$

   (a) Given the data file and the maximal number of (possible) clusters K, apply the k-means algorithm to cluster the samples into k clusters for each $1 \leq k \leq K$.

   (b) Use the MDL principle (in its Bayesian interpretation) to decide what is the most probable number of clusters.

   Input files are provided in the book's website.

9. **Clustering protein sequences**

   You are given the same set of protein sequences from Problem 3.10 and the set of pairwise distances between them.

   (a) Use your implementation of the single linkage algorithm (the agglomerative pairwise clustering of Section 10.4.1) to cluster the proteins. Can you tell at what point incorrect clusters start to form? What are the reasons the clustering fails at some point?

   (b) Modify the algorithm to use the average linkage algorithm. Are the results any better?

10. **Clustering and multi-domain proteins**

    Analyze the similarity matrix of the previous problem to identify violations of the transitivity test. Modify the matrix accordingly and cluster the data with the new matrix. Do you still get large clusters? What is the downside of this approach?

Chapter 11

Embedding Algorithms and Vectorial Representations

11.1 Introduction

In Chapter 10 we discussed clustering algorithms, which are instrumental to the analysis of the protein space. Some of the algorithms we discussed, including the popular k-means algorithm, assume that data instances reside in a real-normed space[1]. In other words, these algorithms are applicable only if the dataset is given as vectors of attributes or features. However, biological data, such as protein sequences, does not naturally lend itself to such representation. This is one of several reasons to look into embedding algorithms.

Embedding is concerned with mapping a given space or a dataset into another space, often Euclidean, in order to study the properties of the original dataset. Traditionally, embedding algorithms have been used to visualize complex data by mapping it to a Euclidean space of a lower dimension. Data visualization in 2D or 3D can provide a global view that might reveal high-order organization within the data, such as existence of classes or other constructs that are not obvious by other means of analysis, as illustrated in Table 11.1 and Figure 11.1.

Embedding is of special interest when the dataset we would like to analyze is a set of abstract objects such as strings, trees, or graphs, represented as a set of proximities. Proteins fall exactly in that category. Such objects cannot be handled directly by many useful data analysis tools and learning algorithms, including some clustering algorithms as mentioned above, SVMs and principal component analysis. Embedding algorithms can map proximity data to vectorial data, thus making it possible to apply these algorithms.

Embedding can also help to assess the proximity of remote entities, when no such measure exists in the original space. For example, similarity measures between protein sequences are effective for related sequences but are meaningless for unrelated sequences (see Chapter 3). However, if we can represent entities as vectors in real space of some dimension then we can measure their

[1]For glossary of terms see Section 3.13

	x_1	x_2	x_3	x_4	x_5	x_6	x_7	x_8
x_1	0	1.33	1.75	1.39	1.01	1.42	0.97	0.98
x_2	1.33	0	1.04	1.40	0.99	1.40	1.70	0.94
x_3	1.75	1.04	0	1.02	1.48	1.05	1.47	1.49
x_4	1.39	1.40	1.02	0	1.03	1.45	0.99	1.76
x_5	1.01	0.99	1.48	1.03	0	1.80	1.44	1.44
x_6	1.42	1.40	1.05	1.45	1.80	0	1.07	1.06
x_7	0.97	1.70	1.47	0.99	1.44	1.07	0	1.47
x_8	0.98	0.94	1.49	1.76	1.44	1.06	1.47	0

TABLE 11.1: Proximity data of a 3D construct. Datasets are often given only as a proximity matrix, as in the table above, which shows the pairwise distances between eight data instances. The dataset has a unique and meaningful geometry in 3D, which is not so obvious from the pairwise distances.

distance based on that representation and look for statistical regularities that would be otherwise hard to detect.

Embedding can also be applied when we are given a dataset that is already represented as vectorial data, but the dimensionality of the data is too high for the application of certain algorithms. For example, when sample points reside in a high-dimensional space, clustering becomes susceptible to the famous "curse of dimensionality", leading to problems of convergence and reliability[2]. In such cases, embedding can be employed to reduce the dimensionality of the original space before the application of clustering algorithms.

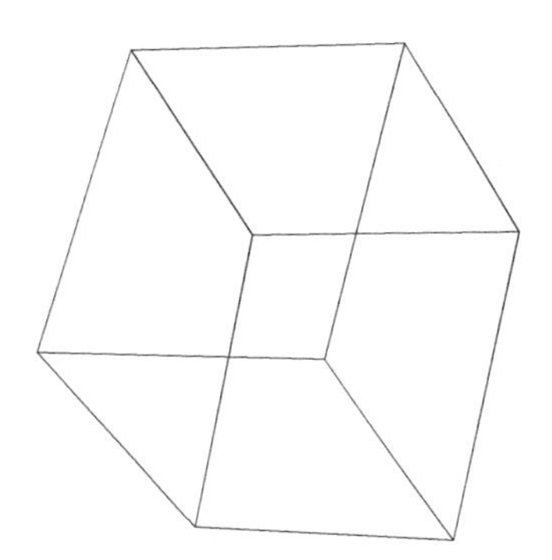

FIGURE 11.1: Recovering structure from proximity data. Embedding algorithms can recover higher-level structures from proximity data. After embedding, it is apparent that the data instances of Table 11.1 are the eight vertices of a distorted cube. (Example was suggested by Nathan Linial.)

In this chapter we discuss different approaches to embedding. We will use the term **embedding** in a very general sense to refer to any technique that

[2]The **curse of dimensionality** refers to the problem of estimating the parameters of high-dimensional models. Consider for example the problem of estimating a d-dimensional distribution from sample data. In order to obtain a coverage of the sample space at a fixed resolution c per dimension (the "bin size"), we need at least $1/c^d$ samples. Therefore, the number of samples needed to estimate the parameters "reliably" increases exponentially with the dimension d.

maps a space $\mathcal{X}$ that is not necessarily real-normed or metric to a vector space $\mathcal{Y} \subseteq \mathbb{R}^d$, called the **host space**. Embedding maps each instance $x \in \mathcal{X}$ to a point $\vec{y} \in \mathcal{Y}$, which is called the **image point** of x. We will make a distinction between two types of embedding algorithms. The first type, which we refer to as **structure-preserving embedding**, consists of algorithms that embed data in real-normed spaces, usually low-dimensional, while trying to preserve specific properties of the original space. The first part of this chapter will discuss algorithms of this type and demonstrate their application to protein data.

The second type, which we will simply call **vectorial representations**, consists of algorithms that do not attempt to preserve any specific property, at least not explicitly. We already discussed several such algorithms in Chapter 7, when we discussed SVMs and kernels. This free-form approach to embedding was motivated by our need to represent proteins as vectors so we can compute various kernels. We will further discuss this approach in the second part of this chapter, in a more general context than SVMs.

11.2 Structure preserving embedding

Structure preserving embeddings are algorithms that attempt to preserve a certain global property of the input space $\mathcal{X}$ in the host space. If the data originally resides in a real-normed space then we can characterize the geometry of the data using for example the directions with maximal variance, or other quantitative features. Since the dataset may be non-Euclidean or even non-metric, it is hard to speak in general terms about the structure of the data, and the concept of geometry might not be clearly defined for the input space. Yet we can use properties that reflect topological ordering of the dataset, such as the set of pairwise distances between data instances or cluster memberships. We refer to such properties as **structural properties** of the space $\mathcal{X}$.

In general, the embedding techniques we will discuss in this section fall into two categories: linear and non-linear. The various algorithms differ also in their (usually implied) interpretation of structure. Each embedding method is best suited for a certain type of problems, and we will present several techniques next.

11.2.1 Maximal variance embeddings

The first two techniques we will discuss are Principal Component Analysis (PCA) and Singular Value Decomposition (SVD). These are classical linear embedding techniques that strive to preserve a property we call **variance**.

The term *linear* means that they work by forming linear combinations of features, essentially rotating the coordinate system. They do not distort the original space and can be reversed.

11.2.1.1 Principal component analysis

PCA is a popular technique for visualization and dimensionality reduction of *vectorial data*. It is useful for feature detection and identification of the most meaningful features. It can also help to detect certain regularities in the data that are otherwise not so obvious.

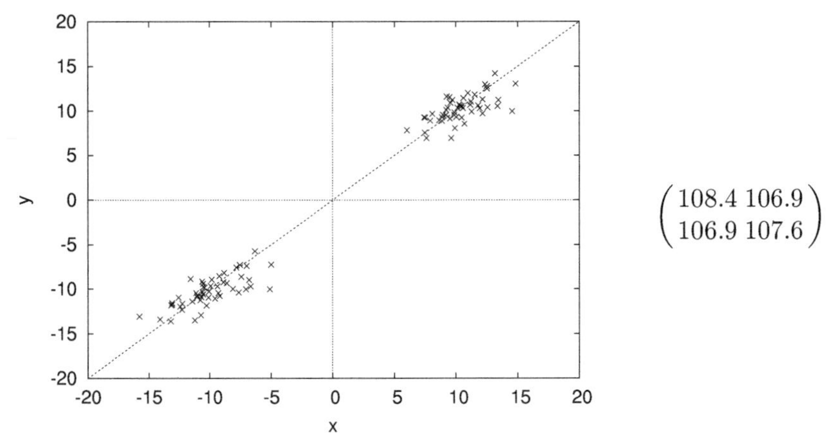

$$\begin{pmatrix} 108.4 & 106.9 \\ 106.9 & 107.6 \end{pmatrix}$$

FIGURE 11.2: **Principal component analysis.** The dataset in this figure can be projected on the $y = x$ axis, where the maximal variance is observed. The data covariance matrix is shown on the right.

PCA seeks projection in Euclidean space that best represents the data in terms of the least sum-of-squared error. It reduces dimensionality by rotating the axes to coincide with the eigenvectors of the data covariance matrix and keeping only those eigenvectors that are associated with the largest eigenvalues (Figure 11.2). As we will see, the eigenvalues measure the variance along these new axes. The variance can be viewed as a structural property of the space $\mathcal{X}$, and the goal of PCA is to identify the axes that maximize the variance of the data. Note that this is not necessarily best for classification, as illustrated in Figure 11.3.

To illustrate how PCA works, we will show (following the intuitive proof of [558]) that the maximal variance projections are obtained by projecting the data samples on the eigenvectors of their covariance matrix. We are given a sample set $\mathcal{X} = \{\vec{x}_1, \vec{x}_2, ..., \vec{x}_n\}$ where each example is a d-dimensional vector $\vec{x} \in \mathbb{R}^d$ sampled from a certain unknown distribution. We assume that the

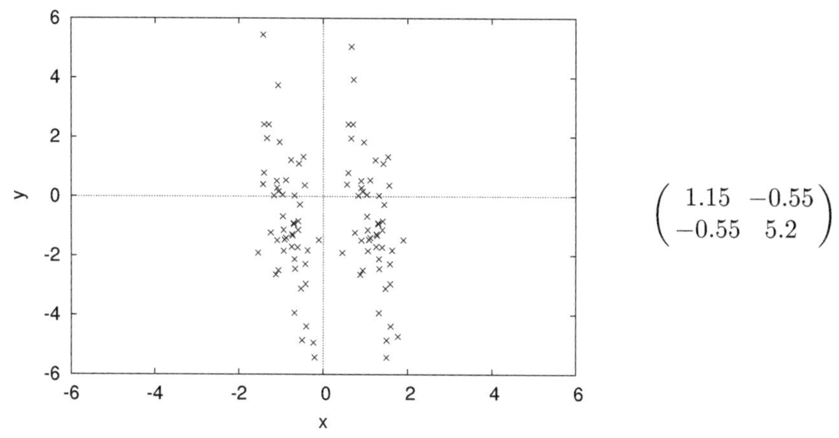

$$\begin{pmatrix} 1.15 & -0.55 \\ -0.55 & 5.2 \end{pmatrix}$$

FIGURE 11.3: PCA and clustering. Maximal variance is observed along the y axis, but reducing the dimensionality by keeping only the y coordinate will completely diminish the existence of clusters in the data.

mean (expectation value) of all data points is the zero vector

$$\vec{\mu} = E[\vec{x}] = \vec{0}$$

and if this is not the case we start by first transforming $\vec{x} \to \vec{x} - \vec{\mu}$. Let $\vec{e}$ be a unit vector in $\mathbb{R}^d$. The **projection** of $\vec{x}$ on $\vec{e}$ is denoted by

$$\vec{x}_e = a\vec{e}$$

where a is the length of the projection, given by the scalar product

$$a = (\vec{x}^t \cdot \vec{e}) = (\vec{e}^t \cdot \vec{x})$$

Since $\vec{e}$ is a unit vector

$$||\vec{e}|| = \sqrt{\vec{e}^t \cdot \vec{e}} = 1$$

We are interested in the mean and standard deviation of the projection of $\vec{x}$ on $\vec{e}$. Since $E[\vec{x}] = \vec{0}$ then $E[\vec{x}_e] = \vec{0}$ as well because $\vec{e}$ is constant

$$E[\vec{x}_e] = E[a] \cdot \vec{e} = E[\vec{x}^t \cdot \vec{e}] \cdot \vec{e} = (E[\vec{x}^t] \cdot \vec{e}) \cdot \vec{e} = 0$$

The variance is given by

$$\sigma^2 = E[a^2] = E[(\vec{e}^t \cdot \vec{x})(\vec{x}^t \cdot \vec{e})] = \vec{e}^t E[\vec{x} \cdot \vec{x}^t]\vec{e} = \vec{e}^t \mathbf{C}\vec{e}$$

where $\mathbf{C} = E[\vec{x} \cdot \vec{x}^t]$ is the $d \times d$ covariance matrix of the sample set. Note that the variance in the projection's length depends on the unit vector $\vec{e}$, namely, $\sigma^2 = f(\vec{e})$.

We would like to determine the unit vectors $\vec{e}$ for which the variance in projection $f(\vec{e})$ is maximal. If $\vec{e}$ is an extreme point of f then

$$\frac{df(\vec{e})}{d\vec{e}} = 0$$

$$\Downarrow$$

$$\frac{f(\vec{e}+\delta\vec{e}) - f(\vec{e})}{\delta\vec{e}} \backsimeq 0$$

$$\Downarrow$$

$$f(\vec{e}+\delta\vec{e}) - f(\vec{e}) \backsimeq 0$$

By the definition of $f(\vec{e})$

$$f(\vec{e}+\delta\vec{e}) = (\vec{e}+\delta\vec{e})^t\mathbf{C}(\vec{e}+\delta\vec{e}) = \vec{e}^t\mathbf{C}\vec{e} + \vec{e}^t\mathbf{C}(\delta\vec{e}) + (\delta\vec{e})^t\mathbf{C}\vec{e} + (\delta\vec{e})^t\mathbf{C}(\delta\vec{e})$$

and since $\mathbf{C}$ is symmetric

$$= \vec{e}^t\mathbf{C}\vec{e} + 2(\delta\vec{e})^t\mathbf{C}\vec{e} + (\delta\vec{e})^t\mathbf{C}(\delta\vec{e})$$

Dropping second order terms and plugging in $f(\vec{e})$ for $\vec{e}^t\mathbf{C}\vec{e}$ we get

$$f(\vec{e}+\delta\vec{e}) = f(\vec{e}) + 2(\delta\vec{e})^t\mathbf{C}\vec{e}$$

Since we require $f(\vec{e}+\delta\vec{e}) - f(\vec{e}) = 0$ we conclude

$$\textbf{(1)} \quad (\delta\vec{e})^t\mathbf{C}\vec{e} = 0$$

We are interested in unit vectors; therefore we require

$$||\vec{e}+\delta\vec{e}|| = 1 \quad \rightarrow (\vec{e}+\delta\vec{e})^t(\vec{e}+\delta\vec{e}) = 1$$

since $\vec{e}^t\vec{e} = 1$ and ignoring second-order terms we get

$$\textbf{(2)} \quad (\delta\vec{e})^t\vec{e} = 0$$

meaning that the deviation should be perpendicular to the unit vector $\vec{e}$. But from (1) and (2) it follows that $\mathbf{C}\vec{e} = \lambda\vec{e}$ and the solutions for this equation are the **eigenvectors** of the covariance matrix $\mathbf{C}$ and λ are the **eigenvalues**[3]. Note that $\sigma^2 = E[a^2] = \vec{e}^t\mathbf{C}\vec{e} = \vec{e}^t\lambda\vec{e} = \lambda$, and hence the unit vectors $\vec{e}$ for which the variance in projection $\sigma^2 = f(\vec{e})$ is maximal are the eigenvectors of the sample covariance matrix, and the variance values are the corresponding eigenvalues.

[3]This intuitive proof concentrates on the first component, but it is possible to extend the proof to the remaining principal components.

The eigenvectors and eigenvalues of the covariance matrix can be computed using methods such as the QR algorithm [559] or the Jacobi transformation [560, pp. 456-462], which are implemented as part of libraries available in all popular programing languages. A projection on m ($m < d$) eigenvectors will map $\vec{x}$ to $\vec{y} = \sum_{i=1}^{m} a_i \vec{e}_i$ with a residual squared error

$$\text{error}(\vec{x}) = \sum_{i=m+1}^{d} a_i \vec{e}_i$$

The total squared error is the sum over all examples

$$J = \sum_{j=1}^{n} ||\vec{x}_j - \vec{y}_j|| = \sum_{j=1}^{n} ||\vec{x}_j - \sum_{i=1}^{m} a_{ji}\vec{e}_i|| = \sum_{j=1}^{n} || \sum_{i=m+1}^{d} a_{ji}\vec{e}_i|| = n \sum_{i=m+1}^{d} E[a_i^2]$$

$$= n \sum_{i=m+1}^{d} \lambda_i$$

Therefore, to minimize the squared error we first have to order the eigenvectors $\vec{e}_1, \vec{e}_2, ..., \vec{e}_d$ by their eigenvalues $\lambda_1, \lambda_2, ..., \lambda_d$ and project the data on the top m ($m < d$) most prominent vectors. For example, if we want 95% accuracy, we should sum over the m most prominent eigenvalues until $\frac{\sum_{i=1}^{m} \lambda_i}{\sum_{i=1}^{d} \lambda_i} > 0.95$.

The PCA algorithm

Given a dataset $\mathcal{X} = \{\vec{x}_1, ..., \vec{x}_n\}$

1. Transform the data such that the mean is the zero vector

$$\vec{\mu} = \frac{1}{n} \sum \vec{x}_i$$

$$\Rightarrow \vec{x}'_i = \vec{x}_i - \vec{\mu}$$

2. Compute the covariance matrix $\mathbf{C}(\mathcal{X})$

$$\mathbf{C}_{jk} = \frac{1}{n-1} \sum_{i=1}^{n} x'_{ij} x'_{ik}$$

or in terms of the original vectors

$$\mathbf{C}_{jk} = \frac{1}{n-1} \sum_{i=1}^{n} (x_{ij} - \mu_j)(x_{ik} - \mu_k)$$

3. Compute the eigenvectors of the covariance matrix
4. Order the eigenvectors based on the eigenvalues $\lambda_1, \lambda_2, ...$
5. To embed in m dimensions, pick the top m eigenvectors and project on the subspace spanned by these eigenvectors

$$\vec{x}'_i \to \vec{y}_i = \sum_{j=1}^{m} (\vec{x}'_i \cdot \vec{e}_j)\vec{e}_j$$

The most common application of PCA is for dimensionality reduction. This is useful for example in data compression, where there is a channel with limited capacity and we are interested in reducing the volume of the data transmitted through the channel, while preserving the variance in the data as much as possible so that we can recover and distinguish between examples on the other end of the channel. By projecting the sample set on the eigenvectors with the largest eigenvalues and dropping the vectors with the small eigenvalues (small variance) we can reduce the dimensionality of the data while preserving maximum information. If there are a few (m) large eigenvalues and the rest are small, then this could imply that the **intrinsic dimensionality** of the dataset is m. The intrinsic dimension is the dimension of the subspace containing the signal, while the rest is likely to contain noise (for further discussion see Section 11.3).

Another important use of PCA is to create an independent "feature" set. By projecting the sample set on the new set of orthogonal unit vectors we essentially switch to a new rotated coordinate system in which every feature (axis) is a combination of the old features. In the new coordinate system the off-diagonal elements of the covariance matrix vanish and there is no correlation between the different components, since

$$\begin{aligned} E[y_1 \cdot y_2] &= E[(\vec{x}^t \cdot \vec{e}_1)(\vec{x}^t \cdot \vec{e}_2)] \\ &= E[(\vec{e}_1^t \cdot \vec{x})(\vec{x}^t \cdot \vec{e}_2)] \\ &= \vec{e}_1^t E[\vec{x}\vec{x}^t]\vec{e}_2 = \vec{e}_1^t \mathbf{C}\vec{e}_2 \\ &= \vec{e}_1^t \lambda_2 \vec{e}_2 = 0 \end{aligned}$$

Note that this is a linear combination of features and therefore can be reversed and recovered.

The main advantage of PCA is that it can easily identify significant coordinates and linear correlations in high-dimensional data. It is appropriate when searching for a global rule that characterizes the data, which can be applied to new data instances as well (by projecting them on the same eigenvectors). Since it is based on linear combinations, it is simple to compute, analytically tractable and reversible. However, PCA can be applied directly only to data that already resides in a real-normed space. For a space like the protein space, which is represented by proximities, PCA is inapplicable[4]. However, it can be applied at a later stage, after the space has been embedded in a real-normed space. PCA is unsuitable when the correlations are non-linear or when no simple rule exists.

[4]PCA can also be applied to proximity data that has been appropriately pre-processed, under certain spectral conditions on the matrix of pairwise distances [561], but as discussed in Section 10.4.3 this analysis has a different meaning.

11.2.1.2 Singular value decomposition

In principal component analysis the covariance matrix has to be nonsingular, since the inverse matrix is required to compute the eigenvectors. However, for small datasets the empirical covariance matrix may be singular, which rules out the application of PCA.

An alternative solution for singular matrices is the Singular Value Decomposition (SVD) technique. The technique is based on a theorem from linear algebra: Any $m \times n (m \geq n)$ matrix $\mathbf{A}$ can be written as the product of $m \times n$ column-orthogonal matrix $\mathbf{U}$, an $n \times n$ diagonal matrix $\mathbf{W}$ with positive or zero elements (the singular values), and the transpose of an $n \times n$ orthogonal matrix $\mathbf{V}$, i.e.,

$$\mathbf{A} = \mathbf{U} \cdot \mathbf{W} \cdot \mathbf{V}^T$$

If $\mathbf{A}$ is a square nonsingular matrix then $\mathbf{U} = \mathbf{V}$ is simply the matrix whose rows are the eigenvectors of $\mathbf{A}$ and $\mathbf{W}$ is a diagonal matrix with the eigenvalues of $\mathbf{A}$ in the diagonal. The inverse of $\mathbf{A}$ is then

$$\mathbf{A}^{-1} = \mathbf{V} \cdot [diag(1/w_j)] \cdot \mathbf{U}^T$$

For singular matrix $\mathbf{A}$, one or more of the $w_j's$ is zero and $\mathbf{A}$ does not have an inverse. In practice, similar problems arise also if the entries in $\mathbf{W}$ are smaller than the machine's floating point precision. In SVD these components are ignored and only columns of $\mathbf{U}$ that correspond to the nonzero components of $\mathbf{W}$ are retained. These columns span the range of $\mathbf{A}$ and constitute the main principal directions of the corresponding subspace (for more details see [560, pp. 60-72]).

11.2.2 Distance preserving embeddings

When mapping one metric space into another, one of the structural properties we often would like to preserve is the set of pairwise distances. If the distances in the host space are identical to the distances in the original space, then the topology of the dataset, as well as its exact geometry, is preserved[5].

The best scenario is if the distances are completely preserved. Such an embedding is called an **isometry**. That is, isometry is a mapping Φ from a metric space $(\mathcal{X}, \delta)$ to a metric space $(\mathcal{Y}, d)$ that preserves distances, i.e., $d(\phi(x), \phi(y)) = \delta(x, y)$ for all $x, y \in \mathcal{X}$.

Any metric space $(\mathcal{X}, \delta)$ with n points $x_1, x_2, ..., x_n$ can be embedded isometrically in an n-dimensional space with the max norm l_∞^n. This can be

[5]**Topology** characterizes certain structural properties of sets, spaces and shapes, such as connectedness and continuity. These are qualitative properties which are invariant under continuous deformations, such as stretching and twisting. **Geometry**, on the other hand, is concerned with more quantitative properties, such as size, distance and curvature. For example, a circle and an ellipse have the same topology, but different geometries. A line and a circle have different topologies.

done by associating with each point $x_i \in \mathcal{X}$ a vector $\vec{x}_i$ of dimension n whose components are the distances from all the n points. Let $\delta_{ik} = \delta(x_i, x_k)$ denote the distance from the k-th point, then

$$x_i \to \vec{x}_i = (\delta_{i1}, \delta_{i2}, ..., \delta_{in})$$

From the definition of the max norm (see Section 3.13) it follows that

$$\|\vec{x}_i - \vec{x}_j\|_\infty = \max_k |(\vec{x}_i)_k - (\vec{x}_j)_k| = \max_k |\delta_{ik} - \delta_{jk}| \geq |\delta_{ij} - \delta_{jj}| = \delta_{ij}$$

On the other hand, from the triangle inequality it follows that

$$\|\vec{x}_i - \vec{x}_j\|_\infty = \max_k |(\vec{x}_i)_k - (\vec{x}_j)_k| = \max_k |\delta_{ik} - \delta_{jk}| \leq \delta_{ij}$$

Therefore, $\|\vec{x}_i - \vec{x}_j\|_\infty = \delta_{ij}$ for all $x_i, x_j \in X$. We will later see that this method resembles the distance profile method of Section 11.4.2.

Aside from the fact that the max norm is not very handy with most applications, a major disadvantage of this mapping is that the host space is of very high dimension. While in some cases it is possible to isometrically embed the input space in a lower dimensional space, isometries are too restricted in general and this ideal situation cannot be attained. Allowing the embedding to distort the metric may add some flexibility and enable us to obtain approximated solutions to isometry.

Distance preserving algorithms attempt to embed the input space in real normed spaces while minimizing the dimension of the host space as well as the **distortion**, i.e., the extent to which the distances between the original points differ from the distances between their representatives in the host space. We will focus on embeddings in Euclidean spaces $\mathbb{R}_2^d$, since the Euclidean metric is geometrically intuitive and better suited to clustering[6].

11.2.2.1 Multidimensional scaling

Multidimensional scaling (MDS) is a classical *non-linear* embedding method. Unlike PCA, it works by analyzing the set of pairwise distances between sample points rather then their coordinates, and it can be applied to non-vectorial data as well. Traditionally MDS has been used to visualize high-dimensional data in two or three dimensions. It has long been employed for data analysis in the social sciences [562], and more recently it has been applied to biological and chemical data [99, 563, 564].

[6]The Euclidean metric has a few advantages when it comes to clustering. For example, when we think of clusters we often imagine ellipsoids, which are well characterized in Euclidean space. Furthermore, if we want to select a typical representative for a cluster, the centroid is a natural choice in Euclidean space since it minimizes the squared error (see Section 10.3), while with other metrics it is harder to define a representative in terms of an intuitive criterion function.

MDS represents data instances as points in some lower-dimensional Euclidean space, while attempting to preserve the pairwise distances. The quality of the mapping is measured using a cost function (called a **stress function**), which compares the distances in the original space and the host space. Formally, we are given a set of n objects $\mathcal{X} = \{x_1, \ldots, x_n\}$. Let δ_{ij} denote the dissimilarity between x_i and x_j. We also define $\vec{y}_i$ as the low-dimensional representation of x_i in $\mathbb{R}^d$ and d_{ij} as the Euclidean distance between $\vec{y}_i$ and $\vec{y}_j$. The goal is to find a configuration $\mathcal{Y} = \{\vec{y}_1, \ldots, \vec{y}_n\}$ such that the $\frac{n(n-1)}{2}$ pairwise distances $\mathbf{D} = \{d_{ij}\}$ are as close as possible to the original dissimilarities $\mathbf{\Delta} = \{\delta_{ij}\}$. This problem can be formulated as a continuous optimization problem, where the goal is to optimize a certain cost function that depends on the distortion in the pairwise distances. There are several popular **sum-of-squared-error** cost functions such as

- $S_{global}(\mathcal{Y}) = \frac{1}{\sum_{i<j} \delta_{ij}^2} \sum_{i<j} (d_{ij} - \delta_{ij})^2$: this function emphasizes the largest errors (also referred to as global mode).
- $S_{local}(\mathcal{Y}) = \frac{2}{n(n-1)} \sum_{i<j} \frac{(d_{ij} - \delta_{ij})^2}{\delta_{ij}^2}$: this function emphasizes the largest fractional errors (local mode).
- $S_{mid}(\mathcal{Y}) = \frac{1}{\sum_{i<j} \delta_{ij}} \sum_{i<j} \frac{(d_{ij} - \delta_{ij})^2}{\delta_{ij}}$: this function is a compromise between global and local modes (semi-local mode).

Since the cost functions involve only distances between points, these transformations are invariant under rotations and translations of the data configurations. Note that all cost functions are normalized such that $S(\{\vec{0}\}) = 1$.

Given proximity data[7] the embedding is initialized by a random projection or a projection by PCA. A gradient descent procedure (as in Section 7.2.2) is then applied to improve the mapping until a local optimum of the stress function is reached. Specifically, the position of each point is updated such that

$$\vec{y}_i(t+1) = \vec{y}_i(t) - \eta \nabla_{\vec{y}_i(t)} S(\mathcal{Y}(t))$$

where η is the learning rate. In order to derive the update formula, we need to compute the gradient of the cost function with respect to the image points $\vec{y}_i$. Consider for example the gradient in the case of $S_{mid}(\mathcal{Y})$. Note first that for a fixed value of i, the dependence on $\vec{y}_i$ is only through $\{d_{ij}\}$ $(i < j)$. By definition,

$$d_{ij} = \|\vec{y}_i - \vec{y}_j\| = \sqrt{\sum_{k=1}^{d} (y_{ik} - y_{jk})^2}$$

[7]Unless noted otherwise, we will assume that the proximity data is a set of dissimilarity values.

therefore

$$\frac{\partial d_{ij}}{\partial y_{ik}} = \frac{y_{ik} - y_{jk}}{d_{ij}} \quad for\ k = 1 \ldots d,$$

and it follows that

$$\nabla_{\vec{y}_i} d_{ij} = \frac{\vec{y}_i - \vec{y}_j}{d_{ij}}$$

therefore

$$\nabla_{\vec{y}_i} S_{mid}(\mathcal{Y}) = \frac{1}{\sum_{j<k} \delta_{jk}} \sum_{j \neq i} \frac{2(d_{ij} - \delta_{ij})}{\delta_{ij}} \cdot \frac{\vec{y}_i - \vec{y}_j}{d_{ij}}$$

Multidimensional scaling

Given a dataset with n instances and pairwise distances $\boldsymbol{\Delta} = \{\delta_{ij}\}$, a desired dimension d for the host space, a cost function $S(\mathcal{Y})$, a learning rate η, maximal number of iterations T and maximal distortion ϵ

Initialize the embedding by selecting at random n vectors in $\mathbb{R}^d$ such that $\mathcal{Y} = \{\vec{y}_1, ..., \vec{y}_n\}$.

Loop:

1. For each point, compute the gradient

$$\nabla_{\vec{y}_i} S(\mathcal{Y})$$

2. Update the position of the point

$$\vec{y}_i(t+1) = \vec{y}_i(t) - \eta \nabla_{\vec{y}_i(t)} S(\mathcal{Y}(t)) \tag{11.1}$$

3. Stop if the distortion (or relative distortion) dropped below the threshold ϵ

$$d_{ij} - \delta_{ij} < \epsilon \qquad \forall i, j$$

or if t exceeds the maximal number of iterations.

Multidimensional scaling techniques may be relatively slow but are simple and straightforward. Contrary to linear embedding techniques, they can be applied to proximity data as well as to high-dimensional feature vector data. They can even work on non-metric data. On the other hand, embedding with MDS is not reversible, and the dimension of the host space has to be set a priori by the user. It also does not produce a rule for embedding new data instances.

Another issue with MDS is that the technique does not guarantee optimality. The gradient-descent algorithms that are usually applied to find the configuration of points in the host space tend to get trapped in a local minimum, which can be far from optimal. Stochastic techniques like simulated annealing (see Section 15.5.2.2) or deterministic annealing [563] can reduce the probability of being trapped in a local minimum. However, these procedures are even more computationally intensive than standard MDS[8]. Note that even with the optimal solution, MDS does not provide a bound on the *expected* distortion of the embedding.

11.2.2.1.1 Examples: Figure 11.4 demonstrates the application of MDS to protein data. The input in this case is a set of about 1000 clusters of protein sequences, defined in the study of [334]. Each point in this map corresponds to one cluster, and the dissimilarity between clusters is defined based on all pairwise similarities between their members. Specifically, the dissimilarity of clusters C_i and C_j is defined as

$$\delta_{ij} = \frac{1}{n_i n_j} \sum_{x \in C_i} \sum_{x' \in C_j} \log s(x, x') \tag{11.2}$$

where n_i is the number of proteins in cluster i and the score $s(x, x')$ is defined as

$$s(x, x') = \min\{1, evalue(score(x, x'))\}$$

assuring that insignificant similarities are assigned zero score after the logarithm transformation. Note that the average in Equation 11.2 is the logarithm of the geometric mean of the evalues since

$$\log \left(\prod_i a_i \right)^{1/n} = \frac{1}{n} \sum_i \log a_i$$

which has a negative value in this case. To convert the scores to positive dissimilarity values the scores are shifted by subtracting a constant

$$\delta_{ij} = \delta_{ij} - \texttt{minscore},$$

which is the minimum score observed in the dataset

$$\texttt{minscore} = \min_{x, x' \in \mathcal{X}} \{\log s(x, x')\},$$

such that the most significant similarity is associated with zero distance.

Lower dimensional maps of complex spaces (as in Figure 11.4) often help to detect higher-level structure or topological features of the input space. However, their success depends on the metric properties of the input space and the dimension of the host space. While the topology of the data in the two-dimensional host space of Figure 11.4 is in general agreement with the

[8]One way to reduce the time complexity is to apply MDS incrementally, as in [565]. This approach starts by carefully embedding a small subset of objects (using one of the stochastic methods or by applying classical MDS multiple times with different initializations). The positions of the remaining objects are determined from the skeleton embedding. This method is especially effective for large datasets.

SCOP classification of these proteins, the overlap between different classes suggests that the host space should be of a higher dimension to faithfully represent the topology of the input space in Euclidean space.

In some cases the embedding can be assessed by means of external datasets, such as the SCOP classification in this example. However, in the general case no such information is available, and additional tests are necessary to evaluate the accuracy of the representation in the lower dimensional space (see Section 11.3).

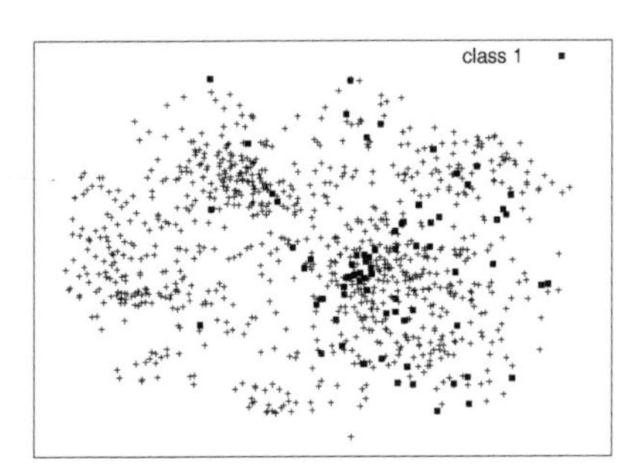

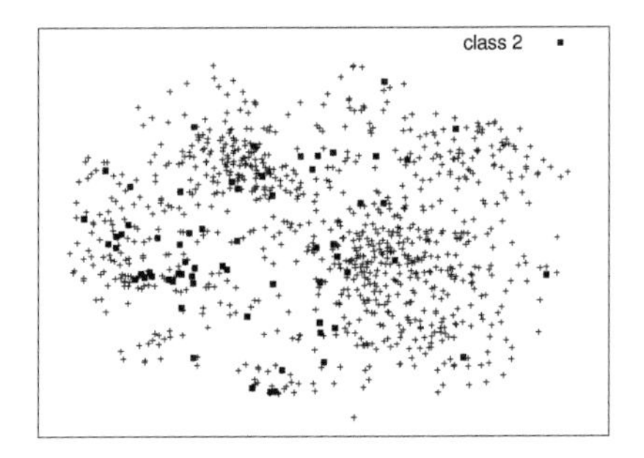

FIGURE 11.4: **A 2D projection of protein clusters.** Tests with respect to SCOP classification show that clusters that belong to the same fold are generally mapped to the same vicinity in this map. All-alpha proteins occupy one part of the space, while all-beta proteins occupy another part, and the overlap between the two parts is small. (Figure reproduced from [334] with permission from AAAI Press.)

When applying MDS to protein data we have to consider the special properties of this type of data. As we already discussed in Section 3.6, it is difficult to define effective distance measures between protein sequences. The evalues can be considered as pseudo-distances and used as dissimilarity input for MDS. However, these scores are uninformative once the similarity between two sequences falls below a certain value. In other words, they cannot distinguish the far from the very far. Therefore, there is no point in preserving all the pairwise similarities, and insignificant similarities should be either ignored or downweighted. This can be done for example by modifying the update rule of Equation 11.1 to preserve small distances while allowing larger distortion for large distances.

Initialization can also make a difference. For example, to reduce the dependency on the initial seed, the starting configuration in Figure 11.4 is created by mapping close clusters to adjacent points within a certain radius, based on the distance between the clusters. Another way to obtain a better configuration is to start with a non-random seed, based on some a priori information we may have about the space. For example, PCA can provide a reasonable initial seed; however it requires that data resides in a real-normed space while the space of protein sequences is represented by proximities. Later in this chapter we will discuss the distance profile method (Section 11.4.2), which can produce an initial mapping into vector space, where the dataset can be analyzed using PCA.

Another possible application of embedding algorithms is to visualize complex networks, such as protein interaction networks (we will discuss protein-protein interactions in Chapter 13). These networks can be viewed as large graphs, with many interconnected elements. Many of the connected components in these graphs have intricate and biologically meaningful structures, such as feedback loops, signal transduction cascades and multi-component complexes. However, without visualizing them it could be difficult to figure out the nature of these networks (recall also our example from Figure 11.1). Embedding is very useful in that respect, and protein interaction maps can provide a bird's eye view of networks and insight into the underlying biological systems, as illustrated in Figure 11.5.

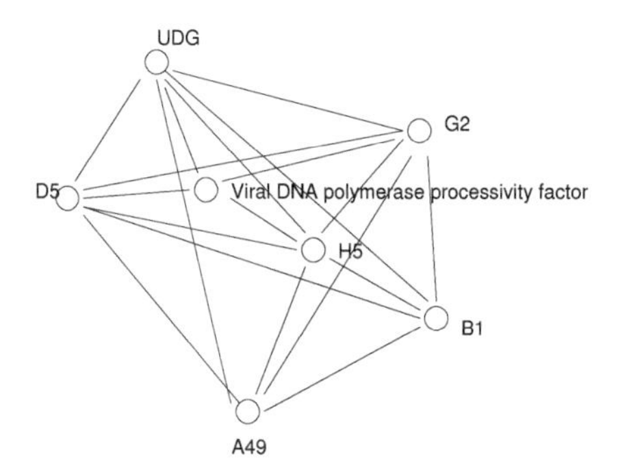

FIGURE 11.5: **An interaction map of vaccinia virus proteins.** (Figure reproduced from [566].) The map shown is a subnetwork of vaccinia virus proteins that seem to control its activity through a series of mediated interactions or by forming a complex. For example, the inactivation of protein G2 (Biozon ID 220-201) renders the virus dependent upon isatin-beta-thiosemicarbazone for growth. This protein interacts with envelope protein H5 (Biozon 203-40), which interacts with protein A20 (Biozon 426-170) whose function is unknown, as well with viral DNA polymerase processivity factor. The latter interacts with uracil-DNA glycosylase (UDG, Biozon 218-247), as well as with protein D5 (putative DNA replication factor). Proteins that directly interact are positioned closely in this map, while proteins that are connected through mediated interactions are positioned farther apart. The set of seven proteins in this connected component form an interesting subgraph that was exposed by embedding the data in 2D space, using the graph distances as input.

11.2.2.2 Embedding through random projections

Random projection algorithms work by selecting at random a reference set of vectors or data points and projecting the dataset on this representative set. While they do not try to optimize a specific cost function or preserve any property explicitly, they do have certain geometric properties that can help to characterize their performance. A common measure of performance is the maximal distortion. An embedding Φ of metric space $(\mathcal{X}, \delta)$ in a metric space $(\mathcal{Y}, d)$ is said to have a **distortion** $\leq c$ if every two points $x, y \in \mathcal{X}$ satisfy

$$\delta(x,y) \geq d(\varphi(x), \varphi(y)) \geq \frac{1}{c} \cdot \delta(x,y)$$

One of the simplest random projection algorithms is the Johnson-Lindenstrauss algorithm. Given a dataset $\mathcal{X}$ in $\mathbb{R}^n$ the algorithm projects each sample point on a lower dimensional subspace of dimension d. Let $\{\vec{e}_i\}$ be a set of d orthogonal unit vectors in $\mathbb{R}^n$ (i.e., **base vectors**), then the projection is obtained by mapping each $\vec{x} \in \mathbb{R}^n$ to $\vec{y} \in \mathbb{R}^d$ such that

$$\vec{y} = \sqrt{n/d}(\langle \vec{x}, \vec{e}_1 \rangle, \langle \vec{x}, \vec{e}_2 \rangle, ..., \langle \vec{x}, \vec{e}_d \rangle)$$

where $\langle \vec{x}, \vec{e}_i \rangle = \vec{x} \cdot \vec{e}_i$ is the scalar product between $\vec{x}$ and the unit vector $\vec{e}_i$. In their paper [567], Johnson and Lindenstrauss show that the length of the projected vector is strongly concentrated around $\sqrt{d/n}$. Intuitively, one can easily verify that this is the case. Without loss of generality assume $\vec{x}$ is a random unit vector. That is, $\|\vec{x}\| = 1$ and a projection on any set of n orthogonal unit vectors must satisfy then

$$\sum_{i=1}^{n} \langle \vec{x}, \vec{e}_i \rangle^2 = 1.$$

Since $\vec{x}$ is a random vector, $\langle \vec{x}, \vec{e}_i \rangle^2$ is a random variable. Hence, on average

$$E_{\vec{x} \in \mathbb{R}^n} \left(\langle \vec{x}, \vec{e}_i \rangle^2 \right) = \frac{1}{n}$$

and the sum over d randomly selected orthogonal unit vectors is

$$E \left(\sum_{i=1}^{d} \langle \vec{x}, \vec{e}_i \rangle^2 \right) = \frac{d}{n}.$$

Therefore, the length of a projection of $\vec{x}$ on the d-dimensional subspace is centered around $\sqrt{d/n}$. This claim can be verified rigorously by writing an explicit formula for the scalar product and computing the expectation value, which is obtained by integrating over the distribution of the angle [567, 568]. If $d = O(\frac{\log n}{\epsilon^2})$, then it is possible to prove that this projection in l_2^d has a distortion $\leq 1+\epsilon$ for every $0 < \epsilon < 1$. In practice, this method is effective only for very high-dimensional spaces, and to obtain low distortion the dimension of the host space must be high as well (see Figure 11.6).

Another algorithm that is based on an extension of the random projections idea is the algorithm of [569]. Unlike the Johnson-Lindenstrauss algorithm, it can be applied to non-vectorial data as well. Instead of base vectors, the algorithm uses random sets of samples from the input dataset as reference sets. The embedding in Euclidean space is obtained by measuring the distance of each sample from the reference sets and representing the sample through its vector of distances. Specifically, given a metric space $(\mathcal{X}, \delta)$ with n points, the algorithm starts by picking at random $\log n$ reference sets $\mathbf{A} \in \mathcal{X}$ of size k for every $k < n$ that is a power of 2. That is, $\log n$ sets of size 1, $\log n$ sets of size 2, $\log n$ sets of size 4, and so on. The largest sets are of size $2^{\lfloor \log n \rfloor - 1}$. Every sample x is then mapped to a vector of dimension $\log^2 n$, where one coordinate is associated with each reference set and its value is the distance of x from the set

$$x \rightarrow (\delta(x, \mathbf{A}_1),\ \delta(x, \mathbf{A}_2),\ ...,\ \delta(x, \mathbf{A}_{O(\log^2 n)}))$$

The distance $\delta(x, \mathbf{A})$ is defined as minimum distance from the points in the set $\mathbf{A}$, i.e., $\delta(x, \mathbf{A}) = \min\{\delta(x, y) | y \in \mathbf{A}\}$. This method resembles the distance profile approach of Section 11.4.2, but unlike other vectorial representations this algorithm provides a bound on the expected distortion under this representation. As was shown in [569], this mapping into $d = \log^2 n$ dimensional space with the l_1-norm has an $O(\log n)$ distortion for almost all choices of subsets $\mathbf{A} \subset \mathcal{X}$. By proper normalization this holds also for other norms $p > 1$.

As with the Johnson-Lindenstrauss algorithm, to obtain embeddings with the guaranteed distortion the data needs to be embedded in a high-dimensional space. And although the distortion is bounded, it is not negligible, especially if the dataset is large and the algorithm cannot be implemented without approximations, as demonstrated in the next example. However, it is a simple approach that can provide a good starting point for other methods.

It should be noted that not every space can be faithfully embedded in Euclidean space while preserving the structure of the data, and even with optimal embedding the host space is not guaranteed to have zero distortion. Actually, for some metric spaces it is possible to prove that the optimal embedding in Euclidean space has a finite non-zero distortion. For example, take the graph-metric of the graph $K_{3,1}$ that consists of four vertices y, x_1, x_2, x_3, so that $d(y, x_i) = 1$ and $d(x_i, x_j) = 2$ for $i \neq j$. It is easy to see that this metric cannot be embedded isometrically in Euclidean space (regardless of the dimension). Since $d(x_1, y) + d(x_2, y) = d(x_1, x_2)$, then in any Euclidean realization of this metric, y must be on the line between x_1 and x_2. The same argument shows that y must be on the line between x_1 and x_3; thus all four points must lie on a line. But then we have no choice but to map x_2 and x_3 to the same point; hence the distance between this pair has not been preserved. This argument is independent of the dimension, and therefore no further improvement is expected by increasing the dimension of the host space. In view of this example, we probably should expect some non-zero distortion for complex spaces (such as the protein space), even with the optimal embedding.

11.2.2.2.1 Example - embedding protein sequences: The application of the random projections algorithm of [569] to protein data has a few practical limitations. First, we need a distance measure between sample points. While there are no good distance measures between proteins, we can use one of the approximated solutions suggested in Section 3.6. However, the main difficulty stems from the need to compute coordinates associated with large reference sets. In practice, when applied to the space of all protein sequences, the large sets contain tens of thousands of sequences (as big as half of the dataset, if the algorithm is implemented as intended). Calculating the distance from these sets involves a massive number of pairwise comparisons. Therefore, a straightforward implementation of the algorithm is not feasible.

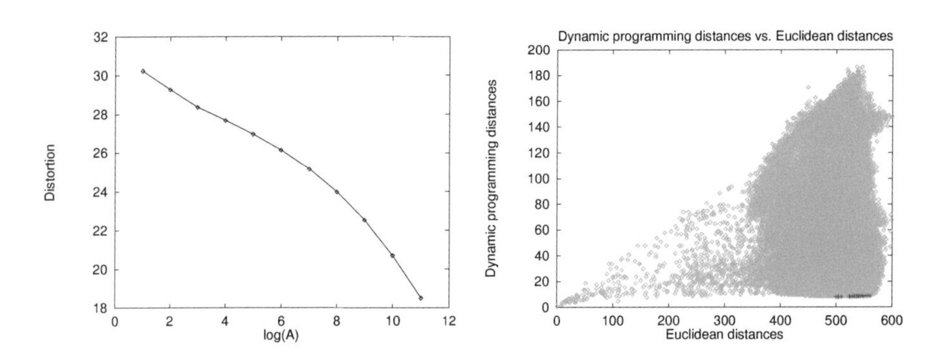

FIGURE 11.6: Left: average distortion of pairwise distances vs. the size of the largest reference set. The random projection algorithm of [569] was applied to a dataset with $n = 540,000$ protein segments of length 50 (see [99]). The dimension of the host space is defined to be $\lfloor \log n \rfloor^2 = 19^2 = 361$. That is, the embedding is obtained by selecting at random 19 sets of size $2^0 = 1$, 19 sets of size 2^1, and so on, for every power of 2, up to 19 sets of size 2^{18}. Practical considerations have limited the size of the largest reference set to 2^{10}, and therefore the dimension of the host space was set to $19 \cdot 11 = 209$. Further dimensionality reduction, by discarding even smaller sets, increases the average distortion. **Right: correlation of distances before and after embedding.** All pairwise distances between 5000 randomly chosen protein segments are calculated using dynamic-programming and plotted vs. the distances between their representatives in the Euclidean host space $\mathbf{R}^{209}$. The average distortion is 18.5.

Given the matrix of all-against-all distances, one can extract the distance of each sample x from each reference set $\mathbf{A}$ simply by scanning this matrix. However, the size of the matrix is unmanageable when all known protein sequences are considered. One possible solution is to pre-process each reference set $\mathbf{A}$ by clustering its members (using pairwise clustering, for example) and reducing the set to a smaller set of representative sequences $\mathbf{A}_r$, one from each cluster. Given the representative sequences, there is no need to compute the distance of a sample x from each member of a reference set in order to determine its distance from the set. We can just compute the distance from the representative sequences, $\mathbf{A}_r$. Depending on the coverage of the representative set, this could be a good approximation.

Another possible solution is to randomly sample the reference sets and reduce them to smaller representative sets. While computationally this is much simpler than clustering the reference sets, these representative sets would probably result in noisier approximations of the true distances from the reference sets, with the quality of the approximation depending on the size of the representative set.

The approach that was taken in [99] was to discard the coordinates associated with large reference sets, assuming this can be done without a significant distortion of the structure of the original space. In practice, the distortion can be substantial, as shown in Figure 11.6, and the distributions of distances before and after the embedding differ quite markedly. Interestingly, the distribution of Euclidean distances changes as larger sets are included in the embedding, i.e., as the dimension of the host space increases (see Figure 11.7). As discussed above, the host space is not necessarily an exact replication of the original space, and especially so for complex input spaces. However, one might ponder whether the distortion we observe in the host space is a negative result, or perhaps a reflection of a genuine high-order global structure, which was not so obvious based on the original distances.

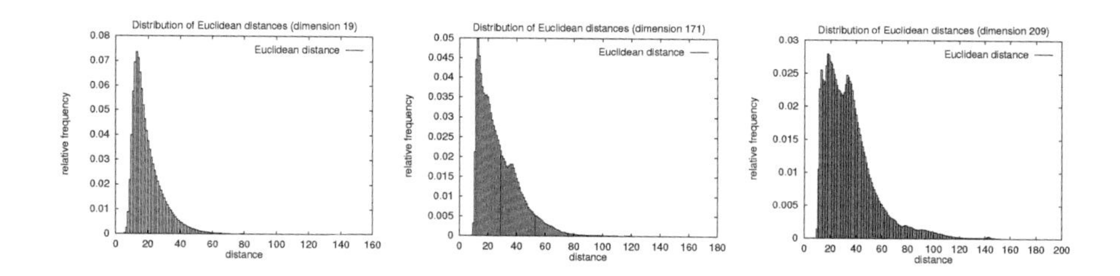

FIGURE 11.7: Distribution of Euclidean distances for different dimensions of the host space. The distribution starts as a unimodal distribution but changes into multi-modal distribution.

This could be the case for example if the data is originally represented as proximity data, and the pairwise similarities or dissimilarities are noisy. This is exactly the case with proteins. Similarity measures between proteins are not only noisy, they are also meaningless for unrelated proteins. Yet it is possible that the global structure that is encoded in these pairwise distances has been exposed during the embedding process. The intuition is simple and applies to other embedding algorithms as well: every pairwise distance with respect to a certain protein places a constraint on the possible position of that protein in the Euclidean host space. The total number of constraints grows as the square of the number of points. Hence, even if the original distance measure is not very informative and some distances are noisy, when all constraints are met together the collective may reveal hidden regularities. Therefore, in principle, it is possible that this procedure recovers hidden geometric information from the proximity data. Indeed, empirical results indicate that different protein families are mapped to different "quadrants" of the Euclidean host space [334] and suggest that the embedding procedure can detect regularities that were not so evident before the embedding. However, it also adds noise that cannot be attributed to true relationships between entities.

In conclusion, random projections can be used to quickly embed data in real-normed spaces to help discover patterns or trends [99, 570], but because of the high levels of distortion they introduce, conclusions on the properties of the input space should be drawn with caution.

11.2.3 Manifold learning - topological embeddings

Manifold learning is a new class of non-linear embedding techniques that are designed to discover the structure of high-dimensional data that lies on or near a low-dimensional manifold. A **manifold** is a space that is characterized by complex geometry but can be locally approximated using the Euclidean metric. That is, each local patch is relatively "flat" and has similar properties to Euclidean spaces.

Usually, when we visualize in our mind high-dimensional spaces, we still tend to think in terms of Euclidean geometry. What differentiates general non Euclidean manifolds from Euclidean geometries is that the shortest distance between two points need not be along the line connecting them. If the manifold has a complex topology then certain paths are not allowed. Therefore, the shortest path is often longer than the Euclidean distance since it must pass through and be confined within the manifold (a good example is Möbius strip, see Figure 11.8).

By using a collection of local neighborhoods or by exploiting the collective properties of the adjacency graph, manifold learning algorithms extract information about local manifolds from which the global geometry of the manifold can be reconstructed. Since they focus on preserving *local* manifolds, these algorithms often result in non-linear embeddings.

FIGURE 11.8: **Möbius strip.** This shape can be created by taking a paper strip, twisting it half-twist and joining the ends. Starting from any point on this single surface manifold, one has to "walk" twice the length of the strip before ending at the same point. (Figure reproduced from Wikipedia [571]. Original figure created by David Benbennick.)

Manifold-learning techniques are especially appropriate when the input data is non-vectorial and when the distances between data instances do not conform to any of the common norms (for example, when MDS results in significant distortion regardless of the dimension). Protein data, which is characterized by complicated topology and unique geometry (see discussion in Section 3.6 and Section 10.6) is exactly the kind of data that manifold learning algorithms are suited for. As with other embedding methods, they may fail if the data is intrinsically non-metric.

There are several different approaches for manifold learning, as described next.

11.2.3.1 Embedding with geodesic distances

The **isomap algorithm** [572] attempts to retain the information on the topological data structure in the original space by focusing on the *geodesic* distances between points. The **geodesic distance** between two points is defined as the length of the shortest path from one point to another while remaining on the manifold on which the data resides.

Consider for example the dataset of Figure 11.9, with 10 three-dimensional data points confined to a manifold. We are interested in an embedding $\Phi : \mathbb{R}^3 \rightarrow \mathbb{R}^2$ that can best preserve the similarity relationships among the 10 data points. Note that the 10th point is the farthest from the 1st point on this manifold (considering its topology) and therefore their geodesic distance is large compared to their Euclidean distance. As Figure 11.10a shows, an embedding algorithm based on Euclidean distances mistakenly places the images of the two extreme points close to each other and thus destroys the topology of the original data.

This example is exactly the kind of dataset the isomap embedding algorithm is well suited for. The structural property of the input space $\mathcal{X}$ that

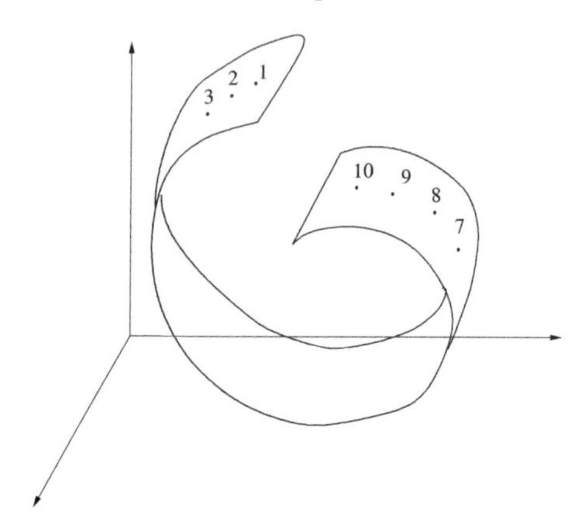

FIGURE 11.9: **Data samples confined to a manifold in 3D.**

isomap attempts to preserve is the set of geodesic distances. By utilizing the geodesic distances, the algorithm essentially "encodes" the manifold structure into the distances, and therefore produces embeddings that preserve better the geometry of the original dataset (Figure 11.10b).

To define the geodesic distances the algorithm constructs the local neighborhood graph $\mathbf{G}$ of $\mathcal{X}$, with one node per data instance. The graph is initialized such that the weight of an edge connecting nodes i and j is set to the original distance δ_{ij} between the corresponding instances if the instances are close enough, or infinity otherwise. The definition of "close" depends on a certain user-defined parameter. There are two variants: in the ϵ-isomap algorithm all instances that are within ϵ distance from each other are considered close, while in the k-isomap algorithm instances are considered close if either one is among the k nearest neighbors of the other.

Once initialized, the algorithm then approximates the geodesic distances d_G between any two nodes based on the shortest path between them, using Dijkstra's algorithm to compute the transitive closure over the graph edges:

$$d_G(i,j) = \min_k \{d_G(i,j), d_G(i,k) + d_G(k,j)\} \tag{11.3}$$

Equation 11.3 is computed iteratively until convergence of all the geodesic distances. Upon convergence, isomap invokes the MDS algorithm described in the previous section with $\mathbf{\Delta} = \{d_G(i,j)\}$ as the set of input distances. Thus the isomap algorithm is not fundamentally different from the traditional MDS algorithm. The uniqueness of this algorithm is in the way the input distances are processed, and the power of the algorithm stems from the use of geodesic distances. By focusing on local neighborhoods, the geodesic distances sample the geometry of the input manifold while avoiding invalid paths. This is the case in Figure 11.9, where computing the direct Euclidean distance between remote points entails using paths that are external to the manifold.

The success of this approach depends on the complexity of the original space and the dimensionality of the host space. In the case of the toy example of Figure 11.10 the topology of the manifold is known, but in the general case we do not know the properties of the manifold and the quality of the embedding has be assessed in terms of a stress function, as discussed in Section 11.2.2.1.

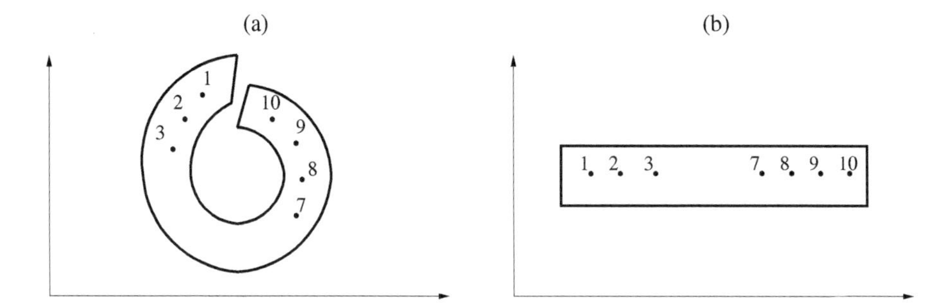

FIGURE 11.10: Embedding of the dataset of Figure 11.9 in 2D. (**a**) Embedding using the Euclidean metric. The data structure is not well preserved since the images of points 1 and 10 are close to each other. (**b**) Embedding using the geodesic metric. The original data structure is well preserved in the sense that points 1 and 10 are kept far away from each other after embedding. The lines denote the manifold boundaries (which are unknown, in the general case).

The **stochastic proximity embedding** (SPE) [573] algorithm attempts to achieve the same mapping as the isomap algorithm without actually computing the geodesic distances. Instead, an approximated geodesic distance is evaluated based on the two following assumptions: (1) The geodesic distances are always greater than or equal to the Euclidean distances in the input space, if the space is a metric space. (2) The Euclidean distances are good approximations to the geodesic distances when the objects are close to each other (within a threshold distance δ_c).

The algorithm follows the MDS schema, but unlike the MDS algorithm it does not try to preserve all pairwise distances, but only the distances between close points. The algorithm uses a stress function that is similar to the $S_{mid}(\mathcal{Y})$ semi-local function of Section 11.2.2.1 and minimizes the function by repeatedly selecting a random pair of points and adjusting their coordinates. The random selection of pairs is the stochastic element. The SPE algorithm can be summarized as follows:

1. Initialize the embedding by selecting at random vectors $\mathcal{Y} = \{\vec{y}_1, \vec{y}_2, \ldots, \vec{y}_n\} \subseteq \mathbb{R}^d$. Select the threshold distance δ_c and the learning rate η.

2. Choose two points i, j at random and compare δ_{ij} to $d_{ij} = \|\vec{y}_i - \vec{y}_j\|$. If

$\delta_{ij} \leq \delta_c$ or $\{\delta_{ij} > \delta_c$ and $d_{ij} < \delta_{ij}\}$, update the coordinates:

$$y_{ik} \leftarrow y_{ik} - \eta \cdot \frac{1}{\sum_{i<j} \delta_{ij}} \cdot \frac{(d_{ij} - \delta_{ij})}{\delta_{ij} + \epsilon} \cdot \frac{(y_{ik} - y_{jk})}{d_{ij} + \epsilon'}$$
$$y_{jk} \leftarrow y_{jk} - \eta \cdot \frac{1}{\sum_{i<j} \delta_{ij}} \cdot \frac{(d_{ij} - \delta_{ij})}{\delta_{ij} + \epsilon} \cdot \frac{(y_{jk} - y_{ik})}{d_{ij} + \epsilon'}$$

where the addition of ϵ, ϵ' in the denominators serves to avoid division by zero.

3. Repeat step (2) m times where m is recommended to be much larger than n, the number of data samples (e.g., $m \sim 100n$ but less than n^2).

4. Reduce the learning rate η.

5. Repeat steps (2) through (4) several times or until convergence.

The main step of the algorithm is step (2). If the input distance between two instances is smaller than the threshold distance then the coordinates are updated in an attempt to preserve the distance, while if the instances are farther apart than the threshold, then the coordinates are updated only if their Euclidean distance is smaller than the input distance, to ensure that the instances do not get any closer. If the Euclidean distance between two dissimilar instances ($\delta_{ij} > \delta_c$) exceeds their original distance, then we do not bother updating the coordinates. These rules help to preserve small geodesic distances, which are deemed more reliable, while relaxing the constraints for large distances.

The stochastic element helps to speed up the algorithm significantly, and although not all pairs are tested in every iteration the algorithm can still produce "good" embeddings. The assumption is that a large enough sample of the pairwise distances of an instance is sufficient to figure its relative position and map it accurately. Indeed, a similar principle underlies the methods of Section 11.2.2.2 and Section 11.4.2.

11.2.3.2 Preserving local neighborhoods

Another manifold learning algorithm that was introduced around the same time as isomap is the **Locally Linear Embedding** (LLE) algorithm [574, 575]. The underlying assumption of LLE is that each point and its close neighbors lie on a nearly linear patch on the manifold from which the data is sampled. Therefore, each point can be approximated as a linear combination of its neighbors

$$\hat{x}_i(\vec{w}) = \sum_{j \in \mathcal{N}_i} w_{ij} \vec{x}_j$$

where $\mathcal{N}_i$ denotes the set of indices of the neighbors of $\vec{x}_i$ and w_{ij} denotes the contribution of the neighbor $\vec{x}_j$ to the combination (Figure 11.11). As such, the LLE algorithm is applicable only to vectorial data. Since the patch is not necessarily linear, some reconstruction error exists and it can be expressed as

$$\epsilon(\vec{w}) = \sum_i |\vec{x}_i - \hat{x}_i(\vec{w})|^2$$

To determine the weights w_{ij} the LLE algorithm first minimizes the error $\epsilon(\vec{w})$ subject to the following constraints:

1. Each data point is constructed only from its k nearest neighbors (or all points that are within ϵ-radius from the point), setting w_{ij} to 0 if $\vec{x}_j$ is not a neighbor of $\vec{x}_i$ according to this definition.
2. $\sum_j w_{ij} = 1$ for every i.
3. $w_{ij} \geq 0$ to ensure that each point resides inside the convex hull of its neighbors[9].

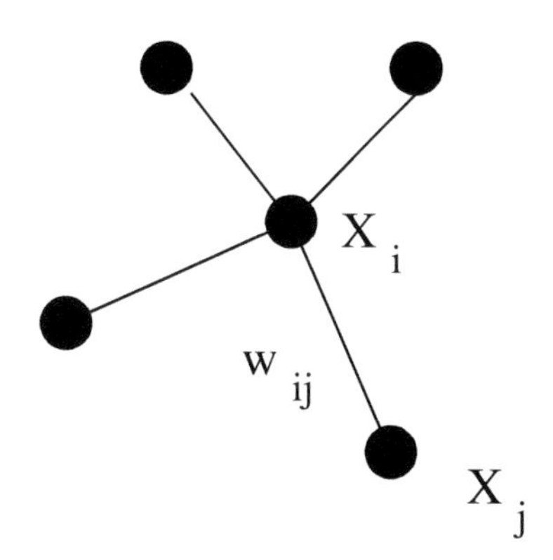

FIGURE 11.11: Approximating samples through local neighborhoods.

Solving the constrained optimization problem

$$\min_{\vec{w}} \epsilon(\vec{w}) \tag{11.4}$$

for every point provides us with the optimal parameters $\{w_{ij}\}$. These weights characterize locally linear manifolds that are invariant to translations and rotations. LLE works by preserving these coefficients in the host space. In other words, the structural property of $\mathcal{X}$ that LLE attempts to preserve is the geometry of local neighborhoods. Specifically, LLE embeds $\mathcal{X}$ in a lower dimensional space $\mathcal{Y}$ by looking for a set of $\{\vec{y}_i\}$ that minimizes the local distortion in the host space $\mathcal{Y}$

$$\min_{\{\vec{y}_i\}} \epsilon(\mathcal{Y}) = \sum_i \left| \vec{y}_i - \sum_{j \in \mathcal{N}_i} w_{ij}\vec{y}_j \right|^2$$

As before, this is a constrained minimization problem but this time with the coefficients w_{ij} fixed (as determined by solving Equation 11.4). Solving this equation for $\{\vec{y}_i\}$ will produce the desired embedding.

[9]The **convex hull** of a set of points is the tightest enclosing convex polygon. Its outline emerges when we connect every point in the set to every other point.

11.2.3.3 Distributional scaling

The **distributional scaling algorithm** [576], also referred to as distributional MDS algorithm, is an embedding algorithm that considers collective aspects of the data when mapping datasets into lower dimensional Euclidean spaces. The structural property that distributional scaling attempts to preserve is the cluster structure of the data (Figure 11.12). This is done by implicitly encoding the cluster structure into the cost function that is being optimized. The cluster structure is a strong indicator of self-organization of the data and reflects the presence of groups that usually can be mapped to specific subcategories of the data.

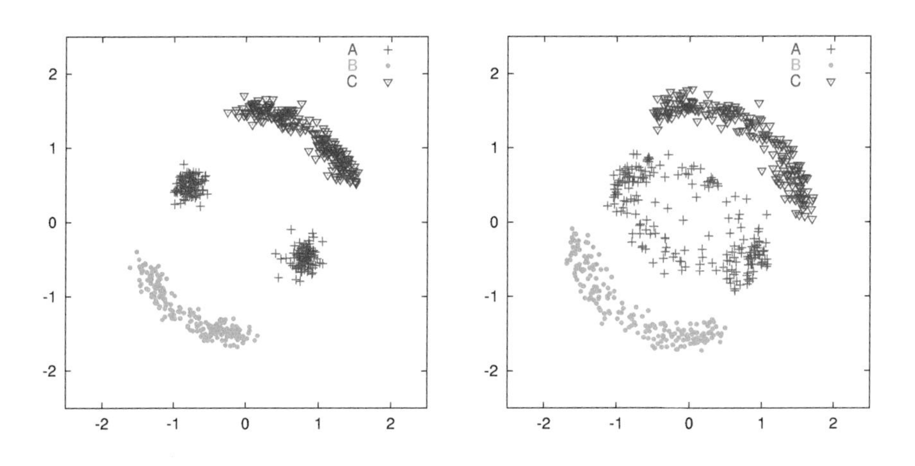

FIGURE 11.12: **Left: structural artifact generated by MDS.** Three 200-point clusters (A, B and C) were embedded in two dimensions using MDS. The central cluster (A) was split into two subclusters by the embedding process. The input was a synthetic dissimilarity matrix, where each dissimilarity value δ_{ij} was drawn from one of three χ-distributions. If i and j are in the same cluster, $\delta_{ij} \sim \chi_2(1.0)$. If $i \in A$ and $j \in B, C$ then $\delta_{ij} \sim \chi_2(1.5)$. If $i \in B$ and $j \in C$ then $\delta_{ij} \sim \chi_2(2.0)$. **Right: improved map using distributional scaling.** Starting from the embedding obtained with MDS, the objective function defined in Equation 11.5 (with $\alpha = 0.1$) was numerically optimized. The artifact seen on the left is largely corrected: cluster A now appears as a single cluster, as it should. (Figure reproduced from [576].)

The algorithm takes as input the pairwise dissimilarity data $\mathbf{\Delta} = \{\delta_{ij}\}$ and cluster assignments, which may be either known *or* estimated. The cluster structure of the data is characterized in terms of the distribution of distances between and within clusters. That is, for each pair of clusters A and B define

ρ_{AB} as the distribution of distances between the points in A and the points in B:

$$\rho_{AB}(x) = \frac{\sum_{i \in A} \sum_{j \in B} w_{ij} \delta(\delta_{ij} - x)}{\sum_{i \in A} \sum_{j \in B} w_{ij}} \tag{11.5}$$

where $\delta(x)$ is the Dirac delta function and the coefficients w_{ij} are optional weights that can be introduced to emphasize or deemphasize certain distances (for example, if large distances are less reliable they can be down-weighted such that $w_{ij} = \frac{1/\delta_{ij}}{\sum_{m,n} 1/\delta_{mn}}$). Similarly, denote $\hat{\rho}_{AB}$ as the distribution of target distances between A and B (in the original input space).

One could use other features to describe the cluster structure, such as the cluster diameters, or the mean and variance of the sample points in each cluster. Note however that cluster assignments alone are not sufficient, since they do not provide information on the geometry of clusters.

The cost function for the distributional MDS algorithm is a combination of the traditional stress function that depends on the distortion in the pairwise distances *and* a distributional component that measures the geometric distortion, as reflected in the discrepancies between the distributions of distances between and within clusters before and after embedding. Specifically,

$$S_d(\mathcal{Y}) = (1 - \alpha) S(\mathcal{Y}) + \alpha \sum_{A \leq B} W_{AB} D[\rho_{AB}, \hat{\rho}_{AB}]$$

where $S(\mathcal{Y})$ is the classical cost function (e.g., $S_{mid}(\mathcal{Y})$), and the second term is the distributional component that considers all cluster pairs. The parameter α determines the relative weight of the distributional information. The distributional component penalizes situations in which the distributions of distances before and after the embedding are inconsistent (see Figure 11.13). The function $D[p, q]$ is a measure of the difference between the distributions, and the coefficient W_{AB} is the weight of the distribution $\hat{\rho}_{AB}$ in the sum, which is defined based on the entropy of the target distribution. Specifically,

$$W_{AB} = 2^{-S(\hat{\rho}_{AB}(x))}$$

where

$$S(\hat{\rho}_{AB}) = -\int_x \hat{\rho}_{AB}(x) \log_2 \hat{\rho}_{AB}(x) dx$$

Recall that a uniform distribution over a dataset has the maximal entropy of all possible distributions over that set. Intuitively, high-entropy distributions are more likely to arise by chance, while non-uniform low-entropy distributions are unique, informative and more likely to reflect a true pattern in the data. This weighting scheme attempts to minimize the sensitivity of the model to noise by emphasizing the low-entropy target distributions.

As we have seen before, the discrepancy between two probability distributions $D[p, q]$ can be quantified by statistical measures such as the Kullback-Leibler divergence or the Jensen-Shannon divergence. However, the first

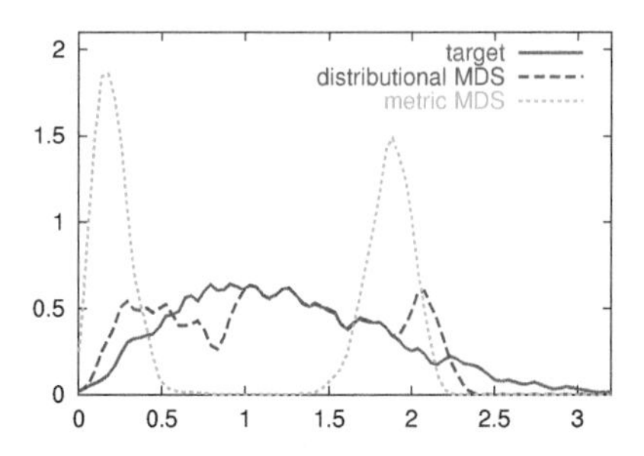

FIGURE 11.13: Distribution of intra-cluster distances within a split cluster. Distributions of distances in cluster A before and after embedding (see the caption to Figure 11.12). Target distances δ_{ij} are the original distances. (Figure reproduced from [576].)

is asymmetric and unbounded while the second tends to be trapped in local minima during minimization of $D[\rho_{AB}, \hat{\rho}_{AB}]$. Both are local measures of the difference between two distributions and are therefore less suited for shape matching. Another distance measure between probability distributions that is more amenable to optimization is the **Earth-Mover's Distance (EMD)**. Given two distributions over an interval $[0, K]$, which can be thought of as the distributions of *earth* and *holes*, the EMD measure is defined by means of a transport problem. Let $f(x, y)$ be the amount of earth that is carried from $x \in [0, K]$ to $y \in [0, K]$ such that every hole is filled and no new holes are created. In other words, the function f should satisfy

$$f(x,y) > 0 \qquad p(x) = \int_0^K f(x,y)dy \qquad q(y) = \int_0^K f(x,y)dx$$

Let $dist(x, y)$ be the ground distance between x and y (in this case it is simply $|x - y|$), then the EMD is the minimum total distance traveled by the earth:

$$EMD[p\|q] = \min_f \int\int dist(x,y)f(x,y)dxdy$$

In other words, it is the minimal amount of "work" required to match p with q. Intuitively, with one-dimensional distributions the minimization is achieved by filling the leftmost hole with the leftmost available dirt (earth). Indeed, for one-dimensional distributions, the EMD is reduced to the l_1 distance between the corresponding cumulative distributions:

$$EMD[p\|q] = \int_0^K dx \left| \int_0^x (p(y) - q(y))dy \right|$$

where the inner integral compares the cumulative distributions with dy denoting the linear distance we have to move the dirt.

Since distributional scaling strives to maintain the distribution of dissimilarities as well as the pairwise distances, it can be more successful in preserving the cluster structure of the data, as demonstrated in Figure 11.12, which shows the embeddings produced with MDS and distributional scaling on the same dataset.

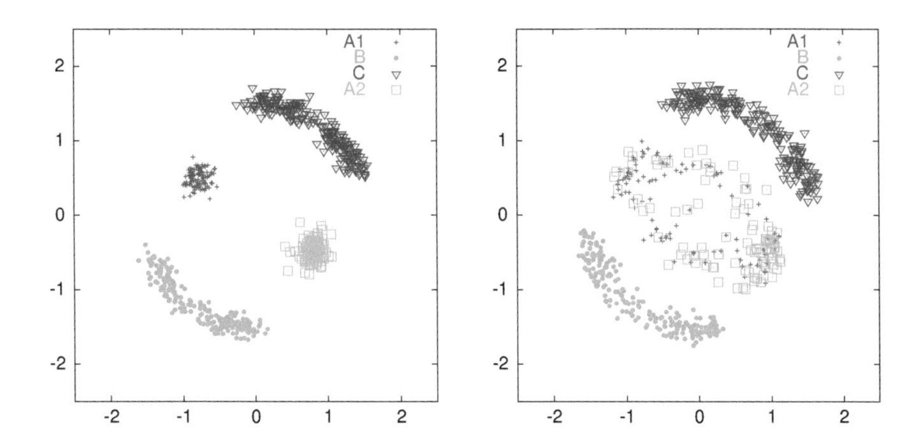

FIGURE 11.14: **Left: naive cluster assignments** generated by the application of k-means clustering ($k = 4$) to the points in Figure 11.12a. This is a case of overclassification, where a single true cluster is broken into two classes. **Right: distributional scaling with naive cluster assignments.** The figure was generated in the same way as Figure 11.12, except that the k-means cluster assignments (left) were used in place of the true ones. The process merges the two central groups of points, while keeping them separate from the remaining two groups. For more examples, see [576].

As mentioned earlier, the cluster assignments need not be known in advance. One can apply traditional MDS techniques to generate a preliminary embedding and use simple clustering algorithms (such as k-means) in the host space to generate cluster assignments, or cluster the dataset directly based on the original proximity matrix. If the data is sufficiently ordered, the estimated clusters can provide a rough snapshot of the manifold topology (the quality of which depends on the clustering algorithm and the dataset). These clusters can then be used to guide the embedding process. Since the algorithm considers the distributions between all pairs of clusters, it tends to avoid embeddings that grossly distort the cluster structure, even when the higher-order structure of the data is misrepresented (e.g., when a real cluster is split into two by a clustering algorithm), as demonstrated in Figure 11.14.

Note that model validity is not an issue here as it was when we discussed clustering algorithms in Chapter 10, since the clusters are not used for classification. Given the choice between a conservative clustering and a more permissive one (e.g., hierarchical clustering with different thresholds), one might prefer the conservative algorithm. Opting for structure-preserving embeddings, smaller and more compact clusters (i.e., overclassification) can be considered entities of high confidence and are more likely to undergo this process successfully.

11.3 Setting the dimension of the host space

One of the challenges with embedding algorithms is determining the intrinsic dimensionality of the data. In classical MDS techniques, the embedding dimension must be set by the user. When the dimension of the host space is increased, the stress will always decrease as the search space is enlarged. One would like to know when the embedding dimension is sufficiently large, i.e., when any additional improvement is insignificant. There are many methods for estimating the intrinsic dimensionality of the data (for a review see [577]) and here we will mention only a few.

Principal component analysis can sometimes suggest the appropriate embedding dimension, based on the number of "large" eigenvalues. In some cases there are a few large eigenvalues, while the rest are significantly smaller, indicating that the variance in the remaining dimensions is negligible. If only the top d principal directions are retained, then the sum over the remaining eigenvalues is a measure of the residual error

$$\text{error} = \frac{\sum_{i=d+1}^{m} \lambda_i}{\sum_{i=1}^{m} \lambda_i}$$

However, in many cases the distribution of eigenvalues is relatively flat and uninformative (Figure 11.15) and the subtlety then lies in setting the correct eigenvalue threshold. Moreover, as a linear embedding technique, PCA explores only a small subset of all possible embeddings.

The accuracy of an embedding can be assessed based on external datasets, for example by clustering the data points in the host space and comparing the clusters to an established classification as discussed in Section 10.5.1. Increasing the dimension of the host space is likely to separate clusters better and improve performance. However, we can stop once the performance plateaus. Monotonicity tests over pairwise distances can also help to determine the effective dimension of the host space. This can be done by plotting the number of distance pairs δ_{ij}, δ_{kl} that are reversed in the host space (i.e., $\delta_{ij} < \delta_{kl}$ but $d_{ij} > d_{kl}$) as a function of the dimension.

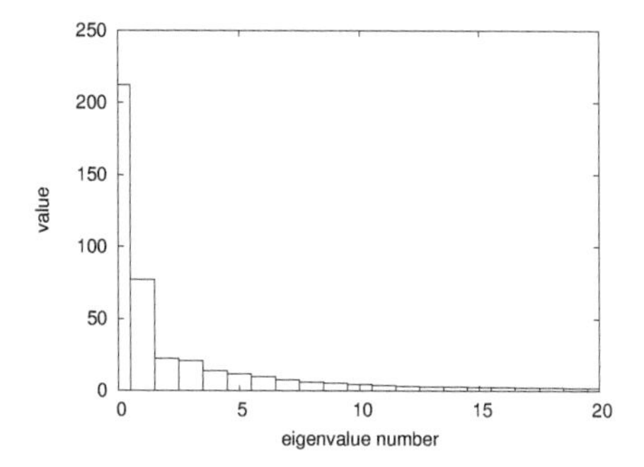

FIGURE 11.15: Determining the dimension of the host space. Eigenvector analysis of the cluster set of [334] (see also Figure 11.4). To compute the directions with the maximal variance, clusters were first mapped to vector space using the distance profile method (see Section 11.4.2). The eigenvalues were then calculated by applying the singular value decomposition method (SVD) to the data covariance matrix. Only the first 20 eigenvalues are shown. The first two eigenvalues are significant, but more than two dimensions are necessary to capture most of the variance that is observed in this set. (Figure reproduced from [334] with permission from AAAI Press.)

The statistical approach of [576] is based on the assumption that once we reached the "right" dimension, the residual error is distributed evenly in the remaining dimensions, and increasing the dimension of the host space would result only in an insignificant reduction of the error. Specifically, the statistical null model assumes that the remaining discrepancies between the target distances Δ_{ij} and the embedded distances are *independent* and *identically distributed* Gaussian random variables. By comparing the measured decrease in stress associated with an increase in embedding dimension to the stress predicted under the null model, we can assign statistical significance to the decrease in stress. When the statistical significance becomes too low, we can conclude that we may well be "fitting noise," and terminate the iterative method. This statistical approach is applicable to squared-error stress functions of the form discussed in Section 11.2.2.1.

A heuristic alternative, which is more widely applicable, is an information-theoretic approach that compares embeddings based on the principle of minimum description length (MDL). Recall that the description length of a given **model and data** is defined as the description length of the **model** plus the description length of the **data given the model** (see Section 10.5.2). In this case the model is a specific embedding $\mathcal{Y} \in \mathbb{R}^d$. The model description is a function of our uncertainty in the positions of the data points in the host space, and the description of the data given the model is a function of the error, assessed in terms of the distortion or the relative distortion

$$E_{ij} = \frac{d_{ij} - \delta_{ij}}{\delta_{ij}}$$

The original data (be it data matrix or proximity matrix) can be reconstructed from the model and the pairwise distortions.

The description of the model (the set of points $\mathcal{Y} = \{\vec{y}_1, \vec{y}_2, ..., \vec{y}_n\}$ in Euclidean space $\mathbb{R}^d$) has to account for the uncertainty in the exact position of each of the $n \cdot d$ coordinates. Since we determine $\mathcal{Y}$ implicitly, through optimization of the stress function, it is difficult to specify the uncertainty in each coordinate. One can estimate the uncertainty from the gradient of the stress function in the vicinity of the point or from the overall distortion in pairwise distances associated with it. For simplicity we can assume a constant uncertainty for all coordinates, such that the description length of the model is $\alpha \cdot n \cdot d$ where α represents the description length per coordinate of the model. For example, $\alpha = 10$ means 10 bits per coordinate, which corresponds to a relative precision of $1/2^{10} \sim 0.001$.

The description of the data given the model (conditional data) is a function of the $n(n-1)/2$ pairwise distortions in the distances (the errors). Computing the exact description (as a sum over all pairs) can be time consuming. However, asymptotically it can be approximated based on the entropy S of the error distribution. The error distribution can be estimated empirically from the observed pairwise distortions. Recall that the entropy of a distribution is a lower bound on its average coding length (see Section 6.4.2.1). Hence the combined description length is

$$\alpha n \cdot p + \frac{1}{2} n(n-1) S \tag{11.6}$$

It should be noted that the shortest encoding of the data does not necessarily coincide with the data's original dimensionality. When embedding complex non-metric data in lower dimensional spaces, distortion is often inevitable even as the dimension of the host space is increased. Therefore, the error distribution may never become very narrow, resulting in non-negligible entropy. On the other hand, each additional coordinate in the model costs the same amount. Therefore, unless α is unreasonably small (say, less than 2 bits per coordinate), the MDL heuristic is likely to select a lower dimension than it would for metric data. Nevertheless, this approach is still informative, and by playing with the parameter α it is possible to approximately identify the dimension above which the error rate drops (for more information see [576]).

11.4 Vectorial representations

Recall our discussion on SVMs in Chapter 7. In order to apply SVMs to protein classification we needed to devise an effective mapping of proteins to feature space, where hopefully they can be separated into classes with a hyperplane. We discussed several mappings, which were the basis for the definition of kernel functions such as string kernels and pairwise kernels. But as we have seen in the previous sections, mapping biological data into vector space is useful not just for the computation of kernel functions. There are other data analysis algorithms we could apply to non-vectorial datasets, once they were mapped to vector space.

The goal of the embedding algorithms we discussed in this chapter so far is to map the input data into low-dimensional spaces while preserving certain properties of the original space (e.g., low distortion in pairwise distances). The mappings developed for SVMs do not make such an attempt, and in fact often achieve their success by mapping the data into very *high* dimensional spaces and deliberately distorting its geometry to obtain maximal separation between classes and the best classifier[10]. In this section we will discuss other approaches for mapping proteins into feature space that do not have the same objectives as embedding algorithms. These mappings are not necessarily geared toward SVMs either (although they can be), and they can be used in other contexts, such as homology detection, structure and function prediction and clustering. We refer to these representations generally as **vectorial representations**.

When choosing a vectorial representation we hope to capture characteristic properties of the input space, and map similar objects to close points. For example, a "good" representation of proteins will map distinct families of proteins to separate clusters. Clearly, the choice of the representation is important and can greatly affect the results.

A vectorial representation of a set $\mathcal{X}$ associates a vector with every instance $x \in \mathcal{X}$, where the coordinates are features that are computed based on the properties of the instance

$$x \rightarrow \vec{x} = (x_1, x_2, ..., x_d)^t$$

If d is the number of features, then this vectorial representation maps $\mathcal{X}$ to $\mathbb{R}^d$. If $\mathcal{X}$ is already a vector space, then any other vectorial representation must be derived from the the coordinates of instances and depending on the nature of the mapping (linear or non-linear) might be reversed. Mappings of this type were discussed in Section 11.2. In this section our discussion focuses on vectorial representations of non-vectorial spaces $\mathcal{X}$. For example, the protein sequence space is a non-vectorial space. Another example is graphs, which are represented as sets of nodes and edges.

A vectorial representation directly embeds the input space into a real-normed space of choice. However, it does not necessarily attempt to preserve certain properties of the input space or minimize certain functions, as is usually the case with the embedding algorithms we discussed in the previous

[10]Note that not only the primary goal of SVMs is different. The setting is different too. SVMs are a supervised learning technique, where the data is labeled. Embedding, on the other hand, is an unsupervised learning technique since the samples are unlabeled.

sections, and therefore we refer to the procedures discussed in this section simply as *representations*. We make a distinction between two types of representations: internal representations and external or collective representations.

11.4.1 Internal representations

Internal representations are vectorial representations that use only the object's properties to determine the mapping. The most trivial representation is the one that maps a protein of length l to a vector of dimension $20 \cdot l$, with residue positions and residue types as the basic dimensions (based on some mapping of amino acids to integers in the interval $[1, 20]$). That is, each block of 20 coordinates corresponds to one position, where only one coordinate is set to one and the others to zero, depending on the residue type at that position. Such representation was used in [578] to identify functional residues in multiple sequence alignments of protein families, based on PCA of the data covariance matrix. This analysis suggested that the direction of the main principal component is correlated with the consensus pattern of the entire family, while other principal components point out individual residues and positions that are characteristic of the different subfamilies within the family. However, this representation is of limited practical use in general, since the vectors are not only high-dimensional and sparse, they are also of different dimensions for proteins of different lengths.

One of the simplest vectorial representations of proteins is based on the **composition** of amino acids, where each protein is represented by the vector of relative frequencies of the 20 amino acids in its sequence. Similarly, we can map proteins to vector space based on the relative frequencies of short k-mers, such as dipeptide or triplets of amino acids. This approach was employed by several studies to cluster sequences or search for close homologs [439,579,580]. We have already seen this representation when we discussed string kernels (Section 7.3.1) and their application to protein classification. String kernels are based on the same principle and can be extended to account for potential mutations in homologous sequences (e.g., the mismatch spectrum kernel).

Proteins that are represented as composition vectors over k-mers can be compared by means of the correlation of their representations (using measures such as Pearson correlation), or their distance using measures such as the Euclidean distance or the Manhattan distance (see Section 3.13). Alternatively, we can measure their similarity using statistical measures of similarity between probability distributions (see Section 4.3.5). Note that with this vectorial representation proteins can be compared even if they are unrelated. This is an advantage of this approach compared to sequence alignment, which does not produce a meaningful measure of similarity or dissimilarity in such cases. The representation can be extended to include other physical and chemical properties of the sequence, such as the molecular weight, the average hydrophobicity or the isoelectric point (e.g., [581]), but this precludes the use of statistical measures to compare vectors.

Internal representations such as the amino acid composition of a sequence are simple and easy to compute, and they encode some information that pertains to a protein's function. But they often fall short in homology detection since they do not capture the essence of proteins as ordered sequences of amino acids. Statistics of longer substrings in the sequence encodes local order within the sequence, but using k-mers longer than dipeptides or tripeptides would result in high-dimensional and sparse vectors, since already for $k = 2$ the number of possible dipeptides is 400, whereas the length of an average protein is around 350 amino acids.

11.4.2 Collective and external representations

External representations are vectorial representations that use more than just the object itself to derive its mapping to vector space. They can be based on collective properties of the input dataset or external datasets or models. As discussed in Chapter 10, encoding data through collective or transitive relations can be very effective for data representation as well as for clustering.

The **distance profile** method is a general framework for vectorial representations of datasets (either vectorial or not). This representation is based on mapping entities of the input space $\mathcal{X}$ to their vector of distances with respect to members of a certain **reference set** $\mathcal{Y}$. For example, when applying the distance profile method to proteins, the set $\mathcal{Y}$ can consist of representative proteins from different protein families, or generative models of protein families, or protein structures, and so on.

The distance profile method can be applied to arbitrary spaces $\mathcal{X}, \mathcal{Y}$ if there exists an **association measure** between instances of $\mathcal{X}$ and instances of $\mathcal{Y}$. The association measure can be a distance function, similarity function or a probability measure. Given an instance $x \in \mathcal{X}$, a reference set $\mathcal{Y} = \{y_1, y_2, ..., y_n\}$ and an association measure $S(x, y) : \mathcal{X} \times \mathcal{Y} \rightarrow \mathbb{R}$, the distance profile maps the instance x to a position in $\mathbb{R}^n$, where every coordinate is associated with one member of the reference set and its value is the association value with that particular reference object

$$x \rightarrow \vec{x} = \begin{pmatrix} S(x, y_1) \\ \vdots \\ S(x, y_n) \end{pmatrix}$$

This simple representation induces a new measure of distance or similarity among samples based on their vectorial representation. It can be used in classification to compute various kernels, and it is very effective for tasks such as homology detection, database search and clustering. One might observe the resemblance of this method with pairwise kernels of Section 7.3.2. Text categorization and web-page classification techniques often use similar representations (called **bag-of-words** [243, 279]), where the text is represented as a histogram vector over a vocabulary of words.

Under the proper transformations the distance profile representation has mathematical and statistical interpretations that have other implications. For example, when the reference set is identical to the input space ($\mathcal{X} = \mathcal{Y}$), an iterative application of this representation can boost weak signals and recover structure in noisy data (see Section 10.4.2). The relation of the distance profile method to mixture models is especially interesting. The components of distance profile vectors can be considered the coefficients of a mixture model where the component models correspond to the reference points (e.g., protein structures or protein families), each one inducing a different probability distribution over the instance space (e.g., the protein sequence space). This interpretation emphasizes the similarity with mutual information kernels, which also resort to a probabilistic representation of protein families. We will get back to the relation with mixture models in Section 11.4.2.5.

The distance profile method has several degrees of freedom, such as the reference set, the association measure and the metric used to compare vectors. In the next sections we will discuss different variations over this representation depending on these different choices.

11.4.2.1 Choosing a reference set and an association measure

The choice of the reference set is the first and most influential one on the power of the distance profile representation. The simplest reference set would be to choose $\mathcal{Y} \subset \mathcal{X}$ or $\mathcal{Y} = \mathcal{X}$. However, depending on the task, it could be wise to use a different reference set. For example, if our goal is to improve homology detection, we might benefit from enriching our reference set and using representatives of all known protein families and not just the input set. This choice would result in initial vectors that are similar to the feature vectors of pairwise kernels.

If both $\mathcal{X}$ and $\mathcal{Y}$ are sets of protein sequences, then the association measure is restricted to similarity measures between sequences such as BLAST score or evalue. Instead of using representative sequences of protein families we can represent families using generative models, such as HMMs or profiles. The association measure in that case could be the likelihood or the profile alignment score (such as the one reported by PSI-BLAST). This approach results in a considerable improvement in detection of remote homologies [112].

The input set and the reference set need not be of the same type, as long as there is an association measure between the two types. For example, if $\mathcal{X}$ is a set of protein sequences, we can use a set of protein structures as the reference set $\mathcal{Y}$ and derive the mapping using an association measure between sequences and template structures, based for example on threading (see Section 8.2). Remotely related proteins usually share little sequence similarity; however, they are expected to have similar structures. Feature vectors with respect to a reference set that represents the protein structure space could expose such similarities [112].

11.4.2.2 Transformations and normalizations

The choice of the association function $S(x, y_i)$ can have an impact on the effectiveness of the representation. The default choice is the score reported by the algorithm that compares entities of $\mathcal{X}$ with entities of $\mathcal{Y}$. However, if the association measure is distributed over a wide range, the most significant scores will inevitably shadow other numerically less important but still significant matches, thus reducing the sensitivity of the representation. This is the case with most types of similarity scores used in protein analysis.

To address this problem we can convert the scores to their underlying cumulative distribution function (cdf) value, where the amplitude of outlier scores is reduced to within a reasonable range. For example, if the association measure is based on a local alignment then the scores follow the extreme value distribution (see Section 3.4.3.2) whose cumulative function $F(s)$ takes the form

$$F(s) = \text{Prob}(s' \leq s) = e^{-\phi(s)} \qquad \text{where } \phi(s) = e^{-\lambda(s-\mu)}.$$

Based on this background distribution we can replace the original scores with a new association measure such that $S'(x, y_i) = F(S(x, y_i))$. With this transformation, all coordinates are bounded between 0 and 1, with high scores or zscores transformed to values close to 1. If the scores are already reported as evalues (e.g., BLAST and PSI-BLAST) then they can be transformed to $pvalue(s)$ as $pvalue(s) = 1 - \exp(-evalue(s))$ (see Section 3.4.3.2). Note that $F(s) = 1 - pvalue(s) = 1$ for large s because of machine precision limitations. To avoid this, the pvalues that are associated to significant scores can be approximated by their empirical distribution, thus allowing distinction between a pair of highly significant yet numerically disparate scores.

Normalizing each feature vector to form a probability distribution can be very useful as well. This transformation enables us to explore distance measures that are suited for probability vectors, as described in Section 11.4.2.4.

11.4.2.3 Noise reduction

Feature vectors can contain many entries that are essentially random and meaningless. For example, if the reference set is composed of proteins that belong to different protein families and folds then each protein x will have only a few significant similarity values in its feature vector $\vec{x}$ (those evalues that correspond to related proteins) and the rest will be chance similarities. The chance similarities result in random numbers that contribute to the differences between feature vectors, thus masking possibly significant similarities. To reduce noise due to unrelated proteins one should eliminate all entries with a score below a certain threshold T, or pvalue above a certain threshold T', to reflect the fact that the corresponding feature is considered irrelevant. The parameter T (or T') should be optimized to maximize performance. Note that entries with low scores that are filtered in this step (assigned 0 score) remain zero under the transformation to the cdf pvalue as described in the previous section. Another alternative is to weight the differences by the significance of the measurements. However, the threshold approach speeds up the processing and comparison of high-dimensional feature vectors.

Representation	Sequence	2163-th to 2170-th entries of the feature vectors							
zscore	*c.47.1.10.3*	0	5.17	63.21	6.42	15.79	4.15	0	1.59
	c.47.1.10.4	1.98	1.73	7.07	53.53	6.38	3.29	0	0
noise reduction	*c.47.1.10.3*	0	5.17	63.21	6.42	15.79	4.15	0	0
	c.47.1.10.4	0	0	7.07	53.53	6.38	0	0	0
pvalue conversion	*c.47.1.10.3*	0	0.9989	0.9999	0.9991	0.9996	0.9968	0	0
	c.47.1.10.4	0	0	0.9992	0.9998	0.9991	0	0	0
normalization	*c.47.1.10.3*	0	0.0528	0.0528	0.0528	0.0528	0.0526	0	0
	c.47.1.10.4	0	0	0.0772	0.0772	0.0772	0	0	0

TABLE 11.2: Distance profile - transformations and normalization. Distance profiles of two proteins (c.47.1.10.3 and c.47.1.10.4) that belong to the same SCOP family. Features are computed by threading a sequence in each member of the reference set of structural domains. Only eight entries of the feature vectors are displayed. The table illustrates the effect of noise reduction, pvalue conversion and normalization on the feature vectors. The zscore cutoff value T is set at 3.5. The feature vector for c.47.1.10.3 reaches its maximum at the 2165-th position, which corresponds to the self-alignment zscore. (Table reproduced from [112].)

Table 11.2 demonstrates the effect of these steps on the feature vectors of two closely related proteins: thioredoxin peroxidase 2 and peroxiredoxin 5, both classified to the same SCOP family (c.47.1.10). Sequences that belong to this family are involved in protecting an organism from oxidative stress (overabundance of free radicals, including reactive oxygen atoms), which can cause major damage to DNA, proteins and lipids (see also Figure 12.1). The sequences were threaded against the SCOP library of structural domains using the FUGUE algorithm [582], and feature vectors were compiled from the zscores reported by the threading algorithm. The table shows a subset of eight entries of the original feature vectors as well as their transformations after noise reduction, pvalue conversion and normalization. As this example demonstrates, the zscore entries are noisy and spread over a wide numerical range. These large differences will inevitably result in large distances between the feature vectors despite the fact that they have significant zscore values in the same positions. The pvalue conversion and normalization (third and fourth rows in the table) resolve this problem by rescaling scores to within a fixed interval.

11.4.2.4 Comparing distance profiles

Under the distance profile representation, the distance (similarity) between two instances p and q is defined as the distance (similarity) of their corresponding feature vectors

$$D(p,q) = d(\vec{p},\vec{q})$$

The function d can be a distance function such as the l_2 norm (the Euclidean metric) or the l_1 norm (the Manhattan distance). If the feature vectors are normalized, we can use similarity measures between probability distributions as discussed in Section 4.3.5.

All these measures are simple and straightforward. But in some cases we might want to consider alternative measures that take into account the specific makeup of the input dataset. With norms such as l_1 or l_2, the distance between features is a function only of their absolute difference. However, a finite dataset occupies a subspace in $\mathbb{R}^d$, and the distribution of feature values in each dimension is usually nonuniform. Moreover, often most of the distribution is noise, and the signal lies in the extremes[11]. In such cases, the similarity of samples depends not only on the relative nominal difference between feature values but also on the magnitude of each feature. Consider for example the distribution of Figure 11.16. Since there are fewer measurements with score between the feature values p_1 and q_1, compared to x_1 and y_1, then we can say that the two measurements p_1 and q_1 are statistically more similar to each other than x_1 and y_1, although $|p_1 - q_1| = |x_1 - y_1|$.

One possible solution is to apply the whitening transformation (see Section 3.4.2.3), which standardizes feature values and converts them to zscores based on the mean and standard deviation of the data distribution. This linear transformation helps to calibrate feature values to the background distribution, but in itself it does not affect the distance since $|zscore(p_1) - zscore(q_1)| = |zscore(x_1) - zscore(y_1)|$. A potential simple fix is to weight the distance between the scaled feature values by the maximal, minimal or average zscore of the two. However, the zscore-based approach is effective only when the distribution is Gaussian-like.

A more generally applicable approach is the **mass-distance** measure [29]. This measure assesses the distance between samples p and q based on the (empirical) data distribution. Specifically, it estimates the probability mass to observe a random point with a feature vector inside the volume delimited by their two feature vectors $\vec{p},\vec{q}$. The smaller the probability, the more significant the match is and the two sample points are considered more similar. Consider the i-th coordinate (feature). Denote by $mass(p_i, q_i)$ the total probability mass of samples whose i-th feature is bounded between the feature values p_i

[11]For example, given a reference protein that belongs to a specific protein family, a dataset of protein sequences from protein families unrelated to the reference protein, and an association measure based on sequence alignment, then the association value of most proteins in the dataset with the reference protein can be considered random, or noise, since it reflects a chance similarity.

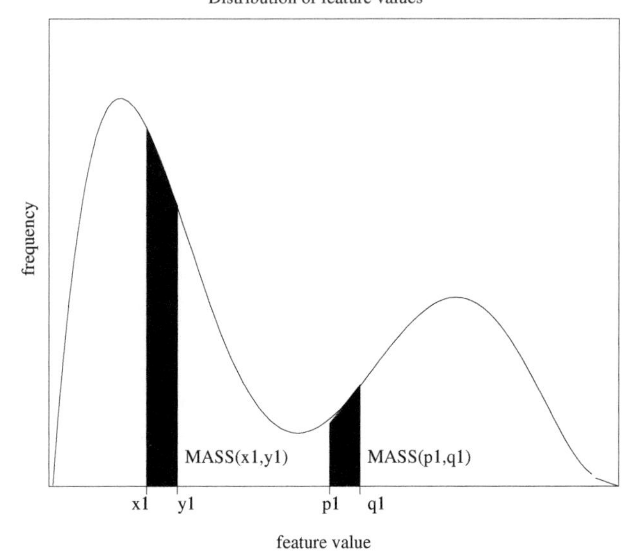

FIGURE 11.16: Assessing distance between samples. Standard distance measures depend only on the nominal difference between features. In this example, $|p_1 - q_1| = |x_1 - y_1|$. However, based on the distribution of values, the pair p_1 and q_1 may be considered "closer", since the probability of observing by chance samples with similar values is smaller. The mass-distance between two samples is defined as the probability mass of samples that are bound between the feature values of the two.

and q_i (as illustrated in Figure 11.16). Then

$$\begin{aligned} mass(p_i, q_i) &= \text{Prob}(\min\{p_i, q_i\} \le x_i \le \max\{p_i, q_i\}) \\ &= \frac{1}{|D|} \sum_{x \in D} \delta(\min\{p_i, q_i\} \le x_i \le \max\{p_i, q_i\}) \end{aligned}$$

Given two feature vectors $\vec{p}$ and $\vec{q}$, the total probability is computed assuming independence between the features[12]. That is, the total volume is computed by taking the product over all coordinates

$$MASS(\vec{p}, \vec{q}) = \prod_{i=1}^{n} mass(p_i, q_i). \tag{11.7}$$

To avoid issues with machine precision limitations, it is easier to work with the logarithm transformation of the mass-distance

$$-\log MASS(\vec{p}, \vec{q}) = -\sum_{i=1}^{n} \log \max(p_i, q_i)$$

[12]Choosing sources of independent nature as representatives in the reference set (e.g., proteins from different protein families) is more likely to produce independent feature values. As we will see in Section 11.4.2.5, the choice of uncorrelated sources is important for the success of the distance profile method.

The mass-distance measure can be applied to assess the similarity of any two vectors sampled from a finite dataset and is especially useful for comparison of expression profiles, as we will see in Section 12.4.1.

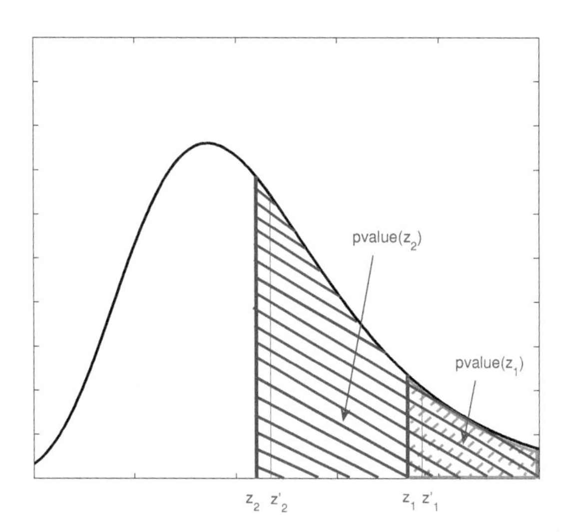

FIGURE 11.17: **The mass-distance for feature values that correspond to pvalues.** (Figure reproduced from [112].) A simulated distribution of zscores that indicate the significance of a match between a sequence and a structural template (the *source*). For a given zscore z, the pvalue measure $pvalue(z)$ is an estimate of the total probability mass in the protein sequence space of sequences that match the structural template with $zscore \geq z$. When comparing two instances, the least significant pvalue of the two associated with their zscores is an estimate of the mass of sequences with similar properties.

What if the features are measures of statistical significance? That is, if p_i and q_i represent the pvalues of observing the feature values or similarity scores (as in PSI-BLAST, for example) rather than the raw feature values. Does it make sense to compare them directly? Recall that the pvalue is the probability of obtaining by chance a score as high as the observed score (see Section 3.12.8). In other words, it characterizes the null distribution, rather than the distribution of feature values in the input dataset. Consider two insignificant but very different measurements that happen to have similar pvalues (e.g., z_2 and z_2' in Figure 11.17). Taking the difference in this case will overestimate the significance of the match between the two, since both are essentially chance similarities with respect to the source and there is nothing that suggests they are similar to each other. In general, each representative in the reference set induces a complex high-dimensional distribution over the instance space, and the one-dimensional pvalue measure p_i can only serve to approximate a certain perimeter in the instance space of instances that are as similar or more similar to the i-th source than instance p (recall also our discussion on pvalue estimation for the multinomial distribution in Section 7.7). Taking the least significant pvalue of p_i and q_i places an upper bound estimator on the volume of relevant instances, as illustrated in Figure 11.17. That is, the best we can do is to say that both features are at least as significant as the less significant one (the higher pvalue). In that case, the total mass can be estimated by the product

$$MASS(\vec{p}, \vec{q}) = \prod_{i=1}^{n} \max(p_i, q_i). \tag{11.8}$$

It has been shown [112] that this function performs well, compared to standard distance measures, when assessing the similarity of remote homologies from pvalue-based distance profiles, as described above.

11.4.2.5 Distance profiles and mixture models

We mentioned earlier that the distance profile representation bears a similarity with mixture models. The reference set $\mathcal{Y}$ can be considered to be a set of functions or statistical sources, each inducing a certain probability distribution over the instance space $\mathcal{X}$, and the association measure determines the coefficients for each instance $x \in \mathcal{X}$. Clearly, the choice of the sources can greatly affect the power of this representation. In this section we will touch upon the issue of reference set selection, in view of the relation with mixture models.

In the basic setup of a mixture model we are given a set of models, all with the same general structure over a certain hypothesis space (a set of decision trees, a set of Gaussians, etc.). We mentioned mixture models when discussing HMMs (Section 6.3.5.6), decision trees (Section 7.4.3) and clustering (Section 10.3.2). The common thread among the different mixture models is that the output is given by integration (or summation) over many possible component models. Other learning algorithms, such as Bayes' optimal classifier [254], bagging [259] and boosting [583] are based on similar principles where the output is the average of multiple predictions (for a review of mixture models or ensembles in general see [584]).

It has been noted for some types of models that given a large set of "sufficiently diverse" models, a mixture model can approximate almost any distribution [191]. This observation seems to apply for other types as well. The hypothesis space we start with assumes a particular representation or model type, possibly based on our prior set of beliefs. This choice might impose a structure on the data and constrain the expressiveness of the models. Namely, it may work well only on distributions or functions of a similar type. However, a non-linear combination of the component models enables us to model complex target functions that cannot be modeled effectively in the original hypothesis space. By combining the predictions of all possible hypotheses we overcome the restriction we imposed initially, and we are no longer confined to the original representation.

This is similar to the Bayesian approach for estimating the likelihood of a new instance x directly from the training data D. We usually assume that the instances are generated by a source with a certain parametric model (the choice of the model determines the hypothesis space), and the parameters are estimated from the training data. However, the Bayesian approach considers the parameters as random variables, and the likelihood of new instances is estimated by integrating over all possible values for the parameters, weighted with the proper priors. That is,

$$p(x|D) = \int_{\theta} p(x|\theta)p(\theta|D)d\theta$$

where θ is the model's parameter vector. Thus, we can estimate distributions far more complex than those that follow the original parametric model. For discrete models the integral is reduced to summation over a finite set of models. For example, in the case of decision trees, θ represents the set of tests in a specific tree, and the integration is replaced with summation.

Unlike mixture models, the distance profile representation does not sum the contributions of the different components (it does so only when comparing profiles). However, the analogy with mixture models is relevant to the choice of the reference set. One can consider the mixture model as an expansion, similar to the Taylor polynomial expansion. Given a set of basis functions such as polynomials we can span the complete space of continuous, "well-behaved" functions, with the right coefficients. The same principle applies here. The set of models in a mixture model defines the set of basis functions. Although these functions are not necessarily orthogonal (one of the requirements from basis functions), a "sufficiently diverged" set of functions is expected to have the desired properties. With each model inducing a certain probability distribution over the instance space $P(x|\theta)$, the orthogonality requirement between two models θ_1 and θ_2 translates (under one interpretation) to zero joint probability distribution

$$\int_{x} P(x|\theta_1)P(x|\theta_2)dx = 0$$

Less correlated hypotheses should be preferred, where completely uncorrelated hypotheses can be considered as orthogonal basis functions. Indeed it has been observed that the performance of certain mixture models improves for uncorrelated models [252, 585].

11.5 Further reading

An excellent review of classical embedding techniques, such as PCA and MDS, appears in [191]. PCA is also known as the basis of classical scaling techniques [586,587]. Many variants of the MDS approach are reported in the literature, and a broad overview of the field is given by [561]. In addition to the algorithms we discussed in Section 11.2.3, manifold learning algorithms include [588], which is an incremental approach to LLE, as well as the spectral methods of [589, 590] and the Hessian eigenmaps approach of [591], which is an alternative to LLE that is applicable to non-convex datasets.

Embedding algorithms are closely related to distance geometry [592], a method for producing isometries for finite datasets. Isometries or near isometry algorithms of manifolds are also discussed in the early works by [593–595], but in host spaces of high dimension. Identifying the right dimensionality of the data is the subject of many studies, including [596–604]. For a survey of dimensionality estimation methods see [577, 605]

Vectorial representations based on composition and other features are reported in [439, 579–581, 606]. The introduction of kernel methods (see Chapter 7) revived the interest in vectorial representations of proteins; however, SVMs usually employ high-dimensional representations. For updates and additional references see the book's website at `biozon.org/proteomics/`.

11.6 Conclusions

- Embedding algorithms are concerned with mapping an input space or dataset into a real-normed space, usually Euclidean, while preserving certain structural or geometric properties of the data.
- Unlike support vector machines, embedding algorithms seek mappings in lower dimensional spaces.
- Embedding algorithms are often employed to visualize high-dimensional data in two or three dimensions.
- Embedding algorithms are especially useful when applied to datasets whose geometry is represented only as a set of pairwise distances or similarities (such as biological sequence data), enabling the application of various data analysis tools.
- Linear embedding algorithms project a dataset on a lower dimensional space by means of a linear combination of features. The mapping obtained is reversible, but they can be applied only to data that already resides in a real-normed space.
- Multidimensional scaling algorithms are more generally applicable. They embed datasets in low dimensional Euclidean spaces while trying to minimize the distortion in the pairwise distances before and after the embedding.
- Random projection algorithms work by projecting the input dataset on a randomly selected set of instances. They are fast and can provide a good starting point for other embedding algorithms.
- Manifold learning algorithms focus on preserving the local geometry of the input dataset, attempting to maintain global properties of the underlying topology.
- Manifold learning algorithms are useful when analyzing spaces with complex topology, such as the protein space.
- Vectorial representations map datasets to feature vectors in vector space, without trying to optimize or preserve any specific property. They rely on the simple assumption that similar instances will be mapped to close feature vectors.

11.7 Problems

For updates, additional problems, files and datasets check the book's website at `biozon.org/proteomics/`.

1. **MDS**

   To illustrate that embedding can recover a topology and structure from a set of pairwise distances we will test it on a simple dataset. Implement the MDS algorithm and apply it to the dataset provided in the book's website.

2. **Embedding protein data with the SPA algorithm**

   (a) You are given a matrix of pseudo-distances between protein sequences (`scop.dist`). The data is not necessarily symmetric, so you have to symmetrize it (for each two proteins A, B define their distance to be the minimum of $dist(A, B)$ and $dist(B, A)$).

   (b) Modify the MDS algorithm of the previous problem to implement the stochastic sampling of the SPA algorithm. Apply it to the pairwise distances and embed the protein data in two dimensions.

   (c) Draw the two-dimensional space. Mark three regions of high density in this space that "seem interesting" and label them (using either the family.superfamily designation or the family names, such as "protein kinase"). The labels are given in `scop.labels`.

3. **Embedding protein data with the Isomap algorithm**

   (a) Compute the geodesic distances for the protein data of the previous problem.

   (b) Apply the SPA algorithm to the geodesic distances you computed and embed the protein data in two dimensions.

   (c) Draw the two-dimensional space as before. Compare to the map you obtained without the geodesic distances.

4. **Vectorial representations**

   Use the dataset of the previous problem and the set of representatives in `scop.rep` to generate a distance profile for each protein. Apply the PCA algorithm to this dataset and project on the first two eigenvectors. Does the map resemble the maps you obtained in the previous problem? Can you explain the results?

Chapter 12

Analysis of Gene Expression Data

12.1 Introduction

In previous chapters we focused on individual molecules. We studied their sequence and structure and discussed algorithms to compare them and models to represent them. Our main goal was and remains to better understand the function of genes[1]

While we have a fairly good understanding of the function and properties of many genes, we cannot ignore the fact that they do not work alone. They are part of a complicated cellular network that is composed of various molecules, including the chromosomes of the DNA, RNA molecules, proteins and other cellular components and smaller molecules. These molecules interact with each other to regulate activity and trigger responses to outside stimuli and changes in the molecular content of the cell. The main mechanism for regulating the activity of genes is by controlling their **expression levels** using **Transcription Factors** (TFs). Transcription factors are proteins that initiate gene transcription and thus affect the quantity of genes in the cell (see Section 5.1). There are many TFs in a cell and each one initiates the transcription of a different group of genes. By activating or deactivating TFs (as illustrated in Figure 12.1 on page 507), a cell can control which genes will be transcribed (expressed) and their expression levels[2].

The expression level of each gene can vary at different stages during cell formation and in different types of cells. It has been estimated that out of all genes in a mammalian genome, each cell expresses only between 15% and 40%

[1]Throughout this book we occasionally use the term gene, when actually referring to its protein product(s). In this chapter we will predominantly use the term gene, as the data we will be analyzing is measured over the mRNA gene sequence. However, in most cases the conclusions of this analysis pertain to the protein products as well. We will get back to this issue in Section 12.5.4.

[2]The process of transcription regulation is quite complicated. In addition to TFs, it involves promoter and enhancer sequences and proteins called repressors and activators (see Section 5.1). The chromatin structure of the DNA (see Section 8.1) can also affect transcription, as well as feedback loops (Section 13.4.3). In this chapter we will focus on the collective effect of all these components on expression. Learning the interactions and regulatory relationships that underlie this complex system is the subject of Chapter 13 and Chapter 15.

of the genes. The subset of genes that are expressed changes from one type of cell to another, and different genes are expressed in different quantities, resulting in a unique **activity profile** in each cell. About 30% of the genes are considered rare with only 1-5 copies of mRNA molecules per cell. Genes with 10-200 copies per cell are considered moderately expressed, while 200 copies per cell and up is considered abundant.

With the DNA content of different cells being identical, it is the process of modulating the expression levels of genes that translates the genetic information in the DNA to different types of cellular machineries. How a multicellular organism develops different types of cells of different activity profiles and different functions is not completely understood, but it is apparently the result of a gradual developmental process during which changes in the molecular content of cells affect neighboring cells and differences in expression levels of genes emerge.

Changes in the expression levels of genes clearly affect their ability to perform their function and can disrupt or change the function of the cell as a whole. However, tracing an effect to its cause has been a challenging goal. For decades, biologists who were trying to identify the main genes that are linked to a certain disease or cellular process had to go through a time-consuming and frustrating task, equivalent to searching for a needle in a haystack. By measuring the expression level of one gene at a time and comparing the levels across different cells, finding the "right" genes was a matter of perseverance and a lot of luck.

Microarrays changed all that. The introduction of microarray technology enabled researchers to measure the expression levels of hundreds or thousands of genes in a genome at once (referred to as **gene expression profiling**). It enabled detection not only of individual genes that differ in expression, but of groups of genes that are expressed similarly, thus gradually shifting the interest from analysis of single molecules to large complexes and gene networks[3].

Since its introduction, microarray expression data has played an essential role in functional analysis of genes and is accompanied with a myriad of possible applications. Microarray analysis has been used to monitor the expression levels of genes as a cell undergoes a normal physiological process, such as the cell cycle [607, 608], in an attempt to determine the genes involved in this

[3]The term **gene networks** is a general term which can refer to several different types of cellular networks or their combinations. Throughout the next four chapters we will discuss four types of gene networks. **Interaction networks** are graphs representing protein-protein interactions, where nodes correspond to proteins and edges represent interactions. **Metabolic networks** are graphs of substrates, with edges denoting enzymatic reactions (or enzymes) which can lead from one substrate to another. In **regulatory networks**, nodes correspond to regulators (such as transcription factors) and the genes they regulate. Edges denote regulatory relationships. **Signaling networks** consist of elements from all three networks. The cell is a complicated web consisting of all these networks, intermingled together.

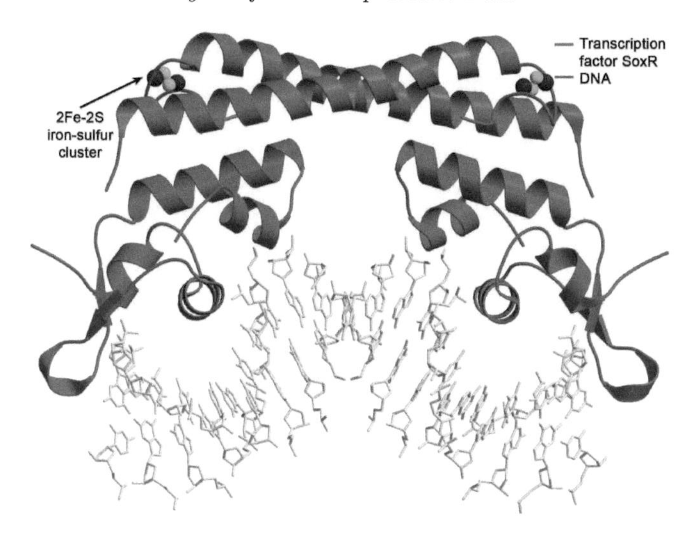

FIGURE 12.1: **The SoxR transcription factor** (PDB 2zhg). In bacteria, the TF SoxR senses oxidative stress (presence of a strong oxidizing agent like superoxide or nitricoxide), which can cause DNA damage. In humans, oxidative stress has been linked to many diseases, including Parkinson and Alzheimer. The oxidizing agents bind close to the iron-sulfur clusters and cause a change in the conformation that allows the TF to bind to the DNA. Upon sensing, SoxR gets activated and enhances expression of several antioxidant proteins, which in turn protect the cell from oxidative stress.

process. It has been used to study differential gene expression patterns under different environmental and cellular conditions, or different developmental stages [609–613]. It has been utilized in comparative genomics, to compare the activity profiles of genomes of related organisms [614]. Gene knockout experiments followed by microarray assays were conducted to determine the role of different genes in various cellular processes [615].

Microarrays also have important applications in medicine. Microarrays can be used to detect SNPs (see Section 4.6.2), often indicating predisposition to certain diseases [616]. Furthermore, gene expression profiling can help in developing assays that are designed for medical diagnosis of specific diseases and studying the effect of different treatments. For example, they have been used to detect different types of cancers based on unique expression patterns of genes [617–620]. Beyond functional analysis and tuning of medical diagnosis and treatment, microarrays can help to identify regulatory networks and cellular procedures [621,622], a topic we will elaborate on in Chapters 14 and 15. While some have criticized the usefulness of microarray technology (as will be discussed in Section 12.5.4), expression data is still considered a significant source of information on cellular activity and regulation, and the data collected from such studies is often used to suggest possible functional links between genes.

Expression data can be analyzed in several ways. Analysis of *individual* genes looks for relative differences in expression level, in search of critical genes that are involved in a specific process or related to a specific disease. *Pairwise analysis* of genes looks for correlation in the expression profiles of gene pairs, which might be attributed to an interaction or other functional links. At the next level we can look for groups of similarly expressed genes, which might correspond to complexes, functional or regulatory modules. In the next sections we will discuss each type of analysis, after a brief description of the technology itself.

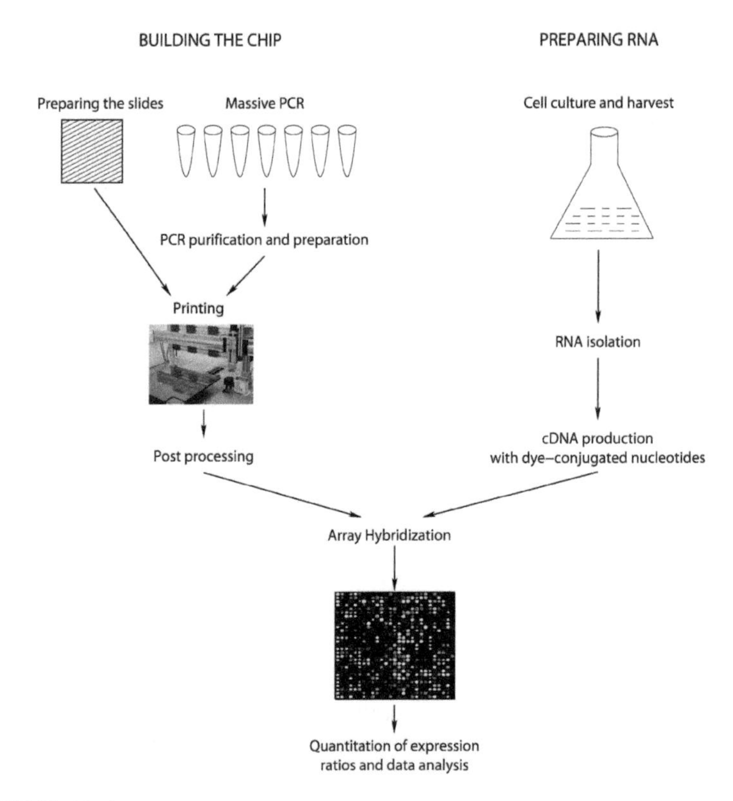

FIGURE 12.2: **cDNA arrays.** The first step of chip preparation is PCR of all known genes in the cell of interest, to amplify their quantities. These samples are then printed on a slide using an arrayer (left). Each spot on the slide corresponds to one gene. The arrayer has multiple pins and can print thousands of samples in a matter of hours on multiple slides. Separately, the mRNA content of the cell under study is purified (right) and used to produce complementary DNA (cDNA), which is more stable than the short-lived mRNA. The cDNA molecules are tagged (colored), as explained in Figure 12.3. Finally, the slide is hybridized with the colored cDNA. The intensity of each spot depends on the abundance of the corresponding RNA molecule in the sample. (Photo of a slide is modified from Wikipedia [623]. Photo of an arrayer is courtesy of David Lin.)

12.2 Microarrays

Microarray technology utilizes the hybridization principle that underlies the double strand structure of the DNA and the DNA replication process. By exposing a cell extract[4] to genome-derived probes, this technology enables measurement of the expression levels of genes by monitoring the extent of hybridization. The basic idea is to use DNA sequences complementary to the gene DNA sequences (probs), "print" them on a chip (usually referred to as an **array**, because of the layout of the probs on the chip) and then expose the chip to the cellular content of the cell of interest. Genes that are abundant in the cell will bind to the chip at the location of their prob, and by "coloring" the molecules, differences in gene expression levels become visible.

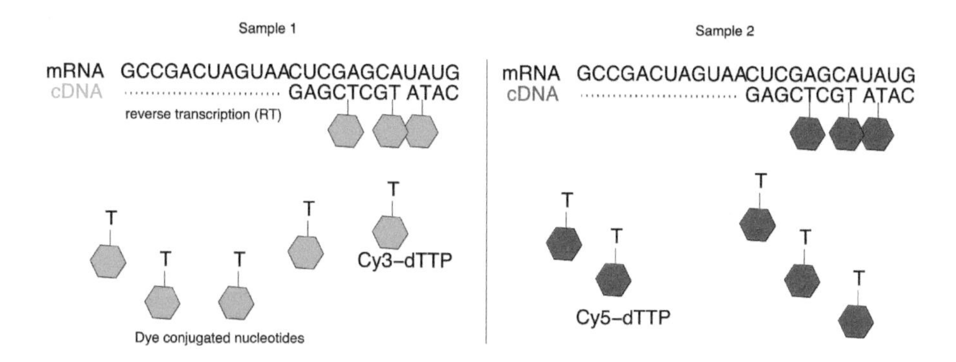

FIGURE 12.3: **RNA labeling.** Complementary DNA (cDNA) is synthesized from the RNA samples (a process called reverse transcription). Given two RNA samples (e.g., from two different tissues), the cDNA of each is labeled with fluorescent dyes Cy3 (green) or Cy5 (red) via a chemical coupling to the nucleotide thymine.

There are several types of arrays (also called **microarrays**, **gene chips** or **DNA chips**), based on variations on this general procedure [624]. Among the most common ones are cDNA arrays, which work by printing short complementary DNA fragments (usually several hundred nucleotides long) on glass slides, followed by hybridization with mRNA-derived target molecules and fluorescence-based detection of under or over-expressed genes. The process

[4]**Cell extract** is a sample of the cell's content. To obtain a cell extract, the cells of interest are grown to a certain "cell density" in a nutritious liquid growth medium. Separating the cells from the medium (harvesting) is done usually by centrifugation. To extract the content, cell walls are broken down usually by applying high pressure. Cell extract is also referred to as **cell lysate** .

starts with PCR[5] of each gene, followed by preparation of a chip that serves as a footprint for the genome under study (Figure 12.2 on page 508). Separately, the RNA content of the cell of interest is purified and prepared for hybridization with the chip. To detect differences in quantities the molecules are tagged with fluorescent dyes (Figure 12.3). Once the chip and the RNA sample are ready they are hybridized, and depending on the amount of each RNA molecule, each spot in the chip will illuminate in different intensity.

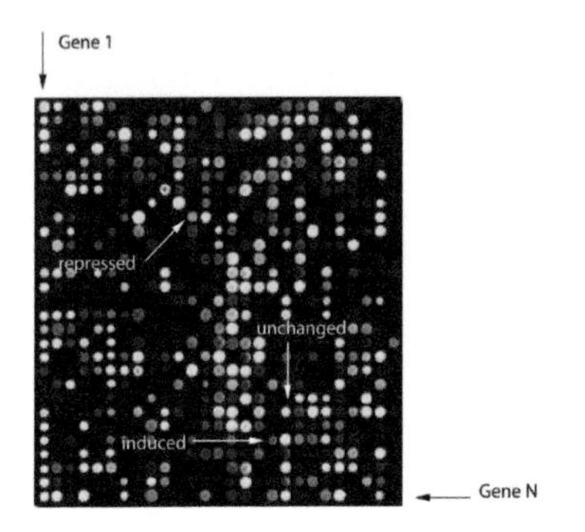

FIGURE 12.4: **A microarray slide.** (Figure modified from Wikipedia [623].) Each spot corresponds to one gene. In addition to the genes of interest, several spots are reserved for a control set. These are usually genes that are known to be unaffected by the experiment, such as ribosomal genes. The control set helps to calibrate the expression values in the rest of the slide.

Usually two RNA samples (for example, from two different tissues) are prepared and tagged using different fluorescent dyes (Figure 12.3) and are simultaneously hybridized to the chip. For each gene the color of the corresponding spot on the chip will depend on the ratio between its quantities in each of the two samples, thus resulting in color variation across the hybridized chip as seen in Figure 12.4. By measuring the emission values it is possible to quantify the ratio, and data analysis starts. If $E1$ is the expression of one gene in experiment 1 (or sample 1) and $E2$ is the expression of the same gene in experiment 2 (sample 2) then the intensity of the spot that corresponds to that gene is converted to a number representing the ratio $r = E1/E2$. Usually the logarithm is used such that each gene is associated with a number that is proportional to $\log(E1/E2)$.

If multiple experiments are carried out on the same set of genes, then each gene is associated with multiple measurements. An **expression profile** of a gene is generated by collecting the measurements for that gene from all experiments to form a vector. That is, the expression profile of a gene is a vector $\mathbf{e}= (e_1, e_2, ..., e_d)$ where d is the number of experiments. The values

[5]**PCR (Polymerase Chain Reaction)** is a technique used to amplify DNA fragments. The process starts by splitting the double strand DNA sequences into denatured single strand DNA sequences (referred to as **templates**). Next, short single-strand sequences (called **primers**), which are designed to bind to the templates, are introduced. Upon binding, DNA polymerase molecules extend the primer based on the complementary strand, to form a double stranded molecule. The process (denaturation, binding and polymerisation) is controlled by temperature and is repeated multiple times to obtain exponential amplification of DNA.

$e_1, ..., e_d$ are usually the logarithm of the ratio between the expression value of the gene in each experiment and the expression value in a control set or in a normal cell.

Expression profiles serve as functional blueprints of genes and convey information on their responses to different experimental conditions or the cellular processes they are part of. However, in practice microarray data is often noisy, as experiments are fraught with many problems. For example, genes are not equally amplified by PCR, resulting in uneven quantities of samples. The pins that print the samples on the slide might load excess sample, leading to over-exposure of certain spots. They can also get clogged or carry sample over from a previous cycle, thus tainting the slide. Some of the noise is due to physical imperfections; there might be differences in the sizes of print tips and physical differences between slides. Slides can also be damaged during the process. All these factors affect the spot size and density and add noise to the results.

Signal analysis is also not a trivial task. Images of the arrays are taken and then processed to determine spot intensities. But determining the spots might prove difficult, as the background is usually uneven and spot size varies. There is also a dye effect, since the red dye is stronger than the green, and conversion to ratios requires appropriate calibration. All these issues require careful scale normalization across slides (for extensive discussion of all these issues and others see [625]). But despite all these problems, expression data does carry useful information on genes, as will be demonstrated in the next sections.

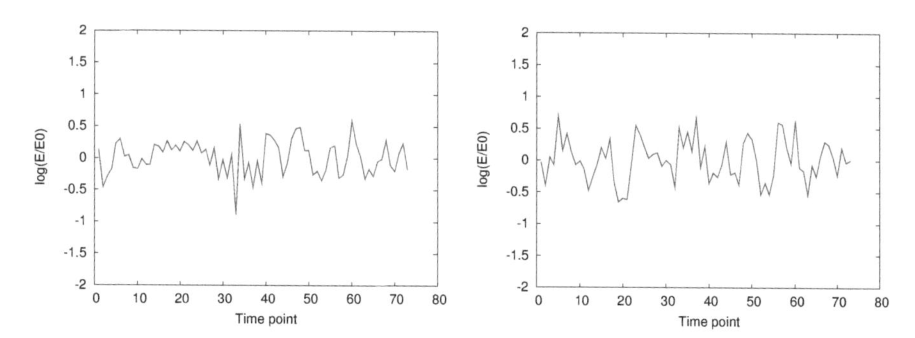

FIGURE 12.5: Time series cell cycle data. The dataset of [607] measures the expression levels of genes in four time series of synchronized *S. cerevisiae* cells, going through the cell cycle. The four datasets are concatenated together to form expression profiles over about 70 different measurements. Profiles of two genes are shown. Left: transcriptional activator HAP2 (Biozon ID 265-117). Right: hypothetical protein YLR156w (Biozon 114-581). The dataset has been normalized to adjust for experimental variation between the different microarrays.

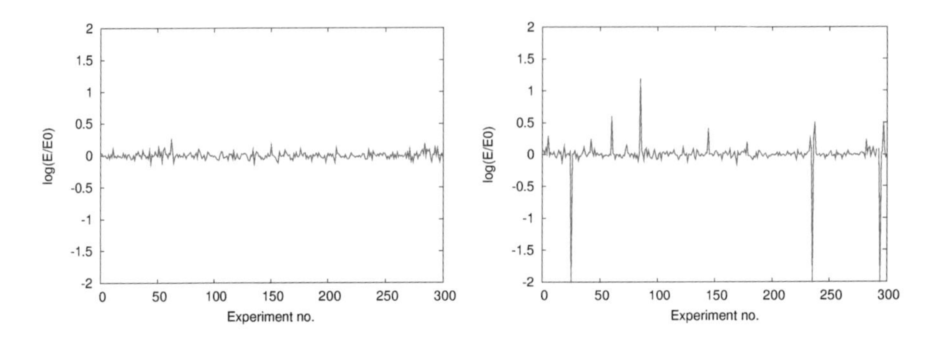

FIGURE 12.6: **The Rosetta microarray data.** Expression profiles for the same two genes shown in Figure 12.5. The Rosetta dataset was generated with oligonucleotide arrays, which use shorter probs ($\sim$25 nucleotides long) than those used in the cDNA arrays of the yeast cell cycle data. Most commercial microarrays are based on oligonucleotides, and can pack more "features" (probs) per slide. However, with shorter probs there is a higher risk of non-specific cross hybridization between sequences of high sequence identity.

12.2.1 Datasets

Microarrays are used to measure expression levels under many different scenarios. One of the first datasets that became famous in the literature as the *yeast cell cycle data* was published by [607]. This dataset contains time-series expression data, where the expression levels of genes in synchronized *Saccharomyces cerevisiae* cells (also known as baker's yeast) were measured at different times along the cell cycle. An example is shown in Figure 12.5.

Another typical setup for microarray experiments is a **knockout experiment**, where the expression profiles are measured after one or more genes were eliminated from the cell under study[6]. For example, the Rosetta Inpharmatics yeast compendium data [615] consists of 300 different conditions, most of which are deletion mutants. Two examples are shown in Figure 12.6 where each gene is represented by a 300-dimensional vector of the expression values associated with these different conditions.

There are many other types of microarray datasets, some measure the expression of genes across different tissues or cells, others compare samples taken from individuals who suffer from a certain disease to samples from healthy individuals. Cancer related datasets are especially common. As the cost of

[6]In a knockout experiment, a gene may be "eliminated" by either deleting it from the genome of the host organism, thus preventing its *transcription*. It may also be eliminated by introducing a small **interfering RNA molecule** (also known as **siRNA** or **silencing RNA**), which is an RNA molecule complementary to the mRNA of a gene. By binding to the mRNA of a gene, siRNA stops *translation*. A gene may also be interfered at the level of the *protein function* (i.e., after transcription, translation and folding), by introducing competitive ligands, repressors or activators.

the technology to produce such arrays is constantly dropping, it is not long before these little chips will become affordable enough to be part of standard medical diagnostic procedures.

12.3 Analysis of individual genes

The first step in the of analysis of microarray data is **differential analysis**. Differential analysis looks for differences in the expression levels of genes due to different cellular conditions or different cell types in an attempt to identify key genes that play an important role in determining the function of cells. For example, if there is a two-fold difference in the expression level of a specific gene between a normal cell and a cancerous cell, then the gene might be linked to that type of cancer. There are many interesting examples in the literature of genes that were characterized this way [607,610,611,615,618,619,626,627].

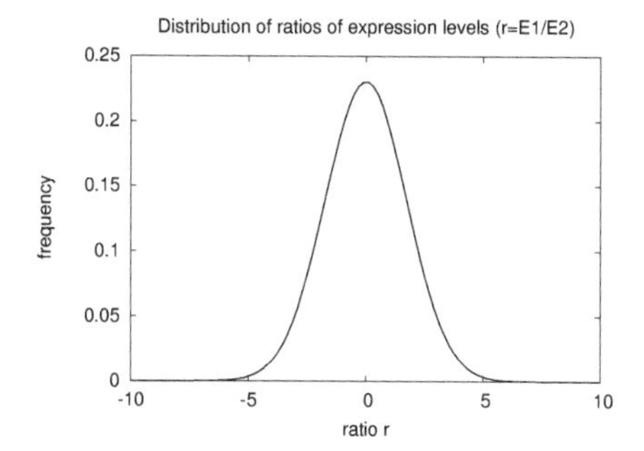

FIGURE 12.7: **Inferring differential expression.** Distribution of ratios of expression levels in two experiments. In this synthetic example, a gene with a ratio of 3 (which could have been considered as significant) is not that uncommon.

Detecting differential expression under different conditions can give insight into the function of genes. However, inferring differential expression from raw ratio values should be done with caution. At what ratio should we consider a gene to be differentially expressed? Differences in expression between experiments are unavoidable, even for genes whose expression does not change in vivo. As discussed in Section 12.2, there are many factors that can affect the results, and therefore it is important to calibrate measurements, filter out noise and establish statistical measures to assess the significance of each observation. For example, assuming that a small number of genes are actually related to the process or disease that differentiates between the cells, then the differences in the expression levels of most other genes reflect random fluctuations that are likely to be the result of the collective effect of all other factors (experimental setup, differences in slides etc.). By removing genes with extreme ratios and characterizing the background distribution of the remaining observations (or of a control group) we can usually get reasonable estimates of the probability of observing a certain ratio by chance (see Figure 12.7).

There are many variations on this general approach, depending on how outliers are defined and the statistical model used to characterize the background distribution. Gaussians are often the default choice. Based on the mean μ and the variance σ^2 of the distribution we can associate a zscore with each ratio, $zscore(r) = \frac{r-\mu}{\sigma}$, which can serve as a significance measure (see Section 3.4.2.3). A simple outlier detection method such as the **jackknife test** excludes one measurement at a time to recompute the parameters of the background distribution based on all other measurements and assess the significance of the excluded measurement [628].

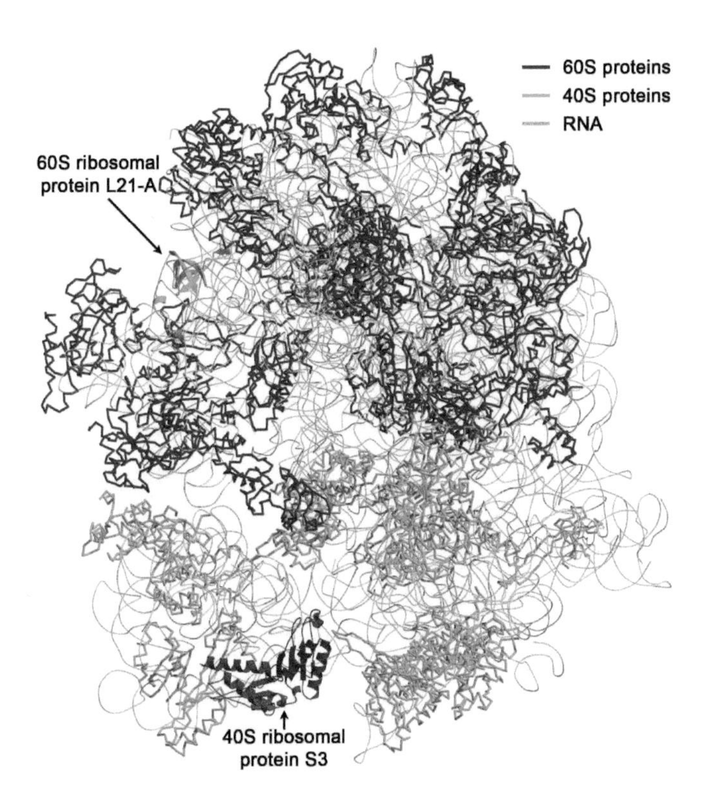

FIGURE 12.8: A model of the 80S ribosome complex from yeast. Atomic models of RNA and protein components were docked into a low-resolution cryo-EM map. The figure shows both the 60S subunit (PDB 1s1i) and the 40S subunit (PDB 1s1h). The 60S subunit protein L21-A (green) and 40S subunit protein S3 (red) have similar expression profiles (Figure 12.9). Although they do not interact directly, they are part of a single multi-component complex.

Another common procedure starts by estimating the parameters from all samples. All measurements that are more than c standard deviations from the mean are then excluded (typically $c \geq 3$), and the estimates of the parameters are revised based on the remaining samples. The process (outlier exclusion, parameter estimation) is then repeated until convergence.

Alternatively, we can assess the significance of a specific differential expression value by computing the empirical pvalue, which is the probability to obtain by chance a value as high as the one observed. One should remember to correct for multiple testing using, for example, the Bonferroni correction or the FDR method (see Section 3.12.8). In some cases there might be very few genes, if any, that stand out above the noise level after correction. In Section 12.5 we will get back to this problem and discuss methods that enhance detection of differential gene expression by considering sets of genes. For an in-depth discussion of differential analysis of genes see [625, 629, 630].

12.4 Pairwise analysis

Individual differential analysis of genes can be insightful, but it tells us only part of the story. Many genes work in concert with each other in a coordinated manner and hence are expected to be expressed similarly. The reverse is also generally accepted, and genes that are similarly expressed under different experimental setups are assumed to be functionally related. Many studies have tested and confirmed this hypothesis and functional links that manifest themselves in **co-expression** can be attributed to several possible cellular "constraints". Specifically, genes might be expressed similarly because they are part of the same complex as interacting partners (Figures 12.8 and 12.9), or they can participate in the same cellular pathway without interacting directly. In both cases it is expected that the genes are synchronized since they have to be available at the cell's disposal concurrently or under the same conditions to sustain the normal function of cells and tissues.

Genes are likely to be co-expressed also if one activates the transcription of the other or if both have similar regulatory elements (promoters) that bind to the same transcription factors. Because they are controlled by the same TFs, any change in the activity of the regulator TFs will have the same or similar effect on the expression of these genes. Interestingly, genes that are co-expressed often share significant sequence similarity [29]. The role of sequence homology in co-expression is believed to be through **fail-safe mechanisms** that evolved in the cell in the course of evolution. The simplest form of such mechanisms is redundancy, since it provides the cell with improved immunity to gene malfunction [632]. This mechanism can evolve at random through a series of duplication events at the single gene level, or in some cases by

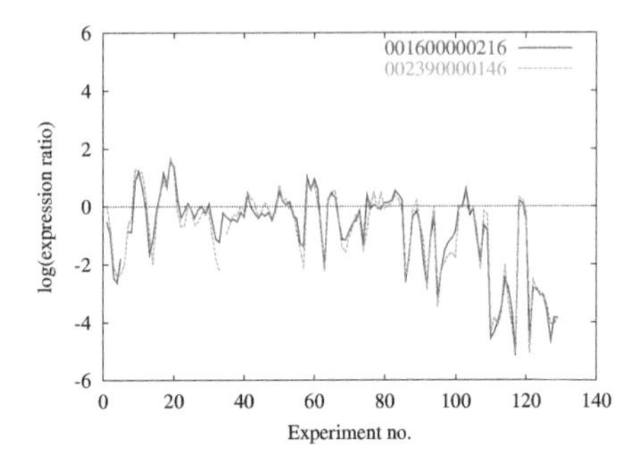

FIGURE 12.9: **Similar expression profiles.** Time series expression profiles of 60S ribosomal protein L21-B (Biozon ID 160-216) and 40S ribosomal protein S3 (Biozon 239-146). Both proteins are part of the ribosome complex (see Figure 12.8). Expression profiles are derived from the stress time series of [631].

duplicating groups of genes or even complete genomes. These duplication events may also preserve the promoter region that precedes the gene, thus generating **"backup genes"** that are concurrent with the original genes. This process might be the underlying explanation behind co-expression if a protein is used as an alternative or as a backup protein ("plan B") for another protein. Many examples are observed in known systems and are documented in the literature. For example, two-thirds of the fly genes have no observable loss of function phenotype[7] under knockout experiments [633].

It should be noted that co-expressed genes that are regulated by the same TFs might be but are not necessarily functionally related. The co-regulation could be attributed to a gene duplication event proceeded by the loss or change of function of one of these genes, or a duplication event that copied and transposed only the regulatory element. It is also possible that the co-regulation is due to the physical location of the genes, whose adjacency has been maintained throughout evolution without an explicit functional constraint.

Since the existing knowledge on regulatory networks and cellular processes is limited, it is quite clear why expression data has generated such a great

[7]**Phenotype** refers to the observable traits of an organism, for example, the shape of the leaves in plants, the color of the hair or eyes in humans, the body physique and so on. Each one of these is referred to as a **phenotypic trait** (some of the traits are not so easily visible, but are measurable, e.g., developmental stages). The **genotype** is the molecular underpinnings of an organism. It refers to the DNA and the "genetic instructions" it contains. But usually the term is used with respect to the molecular counterpart of a specific phenotypic trait. Each observable trait is the result of one or more genes in the organism's DNA. Recall that each individual specimen has a specific allele of each gene (see Section 4.6.2). The observed differences between individuals are often the combined effect of multiple variations at the genome level. The genotype of a phenotypic trait is the collective set of genes that result in that trait.

interest. By analyzing expression data we can detect genes that are similarly expressed and thus expose important and unknown functional links in gene networks. The most straightforward approach to discovery of such links is to compare the expression profiles of pairs of genes. If the profiles are similar then we might have just detected a new functional link. We will get back to the topic of gene networks in Chapter 14 and Chapter 15. Defining similarity between profiles and discerning true similarities from chance similarities is the subject of the next subsections.

12.4.1 Measures of expression similarity

Two common similarity measures for expression profiles are the Euclidean metric and the Pearson correlation. Given two expression vectors $\mathbf{v}$ and $\mathbf{u}$ of dimension d (that is, d measurements or experiments), their **normalized Euclidean distance** is defined as

$$dist_{euc}(\mathbf{v}, \mathbf{u}) = \sqrt{\frac{1}{d}\sum_{i=1}^{d}(v_i - u_i)^2}$$

The division by d serves to normalize the distance, since some of the measurements might be missing (see Section 12.4.2). The **Pearson correlation** is defined as

$$corr(\mathbf{v}, \mathbf{u}) = \frac{1}{d}\sum_{i=1}^{d}\frac{(v_i - \bar{\mathbf{v}})(u_i - \bar{\mathbf{u}})}{\sigma_{\mathbf{v}}\sigma_{\mathbf{u}}}$$

where $\bar{\mathbf{v}}$ is the mean of the values in the vector $\mathbf{v}$ and $\sigma_{\mathbf{v}}$ is the standard deviation. The Pearson correlation coefficient is bounded between -1 and 1, with identical profiles having a correlation value of 1. The Pearson correlation can be converted into a pseudo-distance function by applying the transformation $1 - corr(\mathbf{v}, \mathbf{u})$.

Another possible measure is the **Spearman rank correlation** [634]. To define this measure we first rank the experiments in each vector based on the expression values. Given the ranks, the correlation value is defined as

$$Spearman(\mathbf{v}, \mathbf{u}) = 1 - \frac{6\sum_{i=1}^{d} x_i^2}{d^3 - d}$$

where x_i is the difference in rank for the i-th experiment. This measure also ranges between -1 and 1. If the ranks of the experiments are perfectly correlated (the i-th experiment is ranked at the same place in both profiles, for every i) then the Spearman rank correlation is 1, and if they are completely reversed the rank correlation would be -1. Note that this measure is more lenient than the Pearson correlation, since it does not require that the values in each experiment deviate proportionally from the mean to obtain a correlation value of 1.

The three measures described above are **global measures** that take all measurements in the expression profiles into account. However, often genes are expressed similarly only along a subset of the experiments. To compute the **local similarity** of two expression profiles we can use a dynamic programming (DP) algorithm almost identical to the one we discussed in Chapter 3 to compare sequences. However, to detect local similarities one needs a scoring function that can assign positive and negative scores to individual pairs of measurements v_i and u_j (see discussion in Section 3.4.3.2). One way to achieve that, as suggested in [28], is by taking the product $v_i \cdot u_j$ that measures the correlation between the two values. Given this scoring function, a DP algorithm can then be invoked to find the most similar subsets of measurements. Unlike sequence comparison, gaps should be avoided. With non time-series data, a gapped match might align arbitrarily different conditions. If the data is a time-series dataset, the meaning of a gap is that a time delay is introduced. However, if two genes are co-expressed along a certain set of experiments, it is unlikely that one will re-synchronize itself to the other once they departed (a gap was introduced). Therefore, it is advised to set the gap penalty to ∞ such that gaps are not allowed.

To eliminate bias due to differences in magnitudes between measurements, the vectors can be first normalized such that each has zero mean and standard deviation of one. However, if the individual experiments were already normalized then profile normalization might diminish meaningful differences in expression. Alternatively, the individual measurements can be converted to zscores based on the mean and standard deviation in each experiment and assuming a normal-like distribution (see Section 3.4.2.3). Zscores take on more conservative values than the original raw values while providing information on the importance or uniqueness of the measurement relative to the distribution of all measurements in that experiment.

Another measure that is effective for both local and global comparison of expression profiles is the **mass-distance** (see Section 11.4.2.4). The mass-distance (MD) assesses the dissimilarity between two profiles by estimating the probability mass of observing by chance a vector inside the volume delimited by the two high-dimensional profiles. The underlying assumption is that the smaller the probability, the more similar the two profiles are.

For two expression profiles $\mathbf{u}$ and $\mathbf{v}$, the mass variable is estimated one coordinate (experiment) i at a time, based on the background distribution for that experiment, as illustrated in Figure 12.10. Its value is computed by integrating or adding the probability mass of samples (genes) whose i-th feature is bounded between the expression values u_i and v_i. Hence the measure adjusts to the specific distribution that characterizes each experiment.

Often, the background distributions can be quite reliably modeled using normal distributions. Once the parameters μ, σ of these distributions are

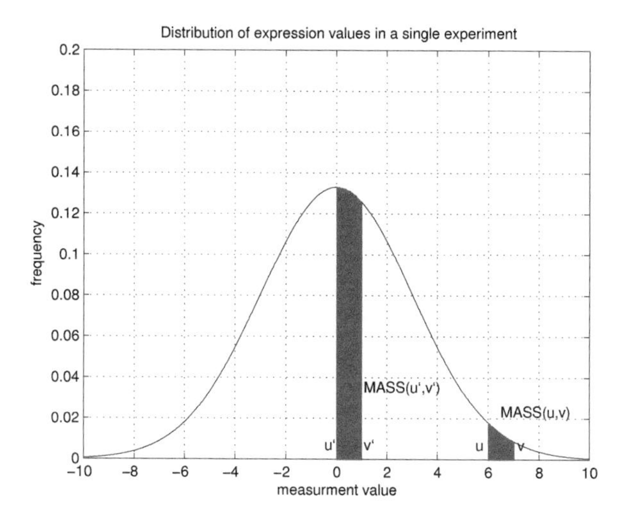

FIGURE 12.10: **The mass-distance.** (Figure reproduced from [29] with permission.) Often it is the case that the distance between two measurements depends not only on the relative nominal difference between the measurements but also on the background distribution. For example, two measurements u and v are statistically more similar to each other than the two measurements u' and v' although $u - v = u' - v'$. That is to say that there are fewer measurements between u and v and therefore fewer instances that have similar properties. The probability mass (the shaded area) in each case is given by the integral over the background distribution.

estimated, the probability mass between u_i and v_i is computed by the integral

$$mass(u_i, v_i) = \text{Prob}(\min\{u_i, v_i\} \leq x \leq \max\{u_i, v_i\}) = \int_{\min\{u_i,v_i\}}^{\max\{u_i,v_i\}} \text{Prob}_i(x)dx$$

where $\text{Prob}_i(x) = N(\mu_i, \sigma_i)$ is the normal distribution for experiment i. However, the background distribution does not always follow closely a normal distribution. Two-way experiments that measure, for example, the difference in expression between two tissues (e.g., brain vs. liver) might exhibit a bimodal or other distribution. In such cases one could use the empirical distributions when computing the *mass* variables

$$mass(u_i, v_i) = \sum_{\min\{u_i,v_i\} \leq x \leq \max\{u_i,v_i\}} freq(x)$$

where $freq(x)$ is the empirical relative frequency of the measurement x.

The mass-distance of $\mathbf{u}, \mathbf{v}$ is defined as the total volume of samples bounded between the two expression profiles. Assuming independence between the experiments, it is estimated by the product over all coordinates

$$MASS(\mathbf{u},\mathbf{v}) = \prod_{i=1}^{d} mass(u_i, v_i). \tag{12.1}$$

It is more convenient to work with the logarithm of the term above

$$MD = -\log MASS(\mathbf{u},\mathbf{v}) = -\sum_{i=1}^{d} \log mass(u_i, v_i) \tag{12.2}$$

The higher it is, the smaller the probability of observing a nearby profile by chance, and the two profiles are considered more similar. Note that the mass-distance measure can be local or global, since even one significant match will tend to result in a small probability overall (large $-\log MASS(\mathbf{u},\mathbf{v})$) and multiple ones will tend to strengthen the signal.

When comparing expression profiles, the distinction between global and local metrics is not as substantial as for sequence comparison. With time-series datasets the temporal aspect is important, and one usually expects to detect co-expression along a continuous set of experiments. All the measures described above are appropriate for this kind of analysis. Other datasets are generated over a set of experiments that are not expected to form continuous patterns of co-expression (such as the Rosetta dataset, Section 12.2.1). With the exception of the DP-based local similarity measure, the other measures can work as well on such datasets.

It is important to point out that not all expression similarities can be detected with every measure [29] and the choice of the similarity function can have a great impact on the analysis. For example, when clustering genes based on microarray data in search of groups of co-expressed genes, different representations and similarity measures can have significant effect on the quality of the clustering results, since most clustering algorithms rely directly on pairwise distances or similarities between instances. This includes k-means, pairwise clustering and spectral clustering algorithms (see Chapter 10). Therefore, better pairwise measures are likely to produce better results, i.e., clusters that better correlate with complexes or cellular processes. In Section 12.4.5 we will discuss some of the issues relevant to performance evaluation of these measures.

12.4.1.1 Shifts

In time-series data we might observe similar expression patterns that are shifted, for example if one gene activates another (directly or indirectly). Time-delayed responses can be accommodated when comparing expression profiles by considering shifts. Since the typical cellular reaction times are almost instantaneous, delays of a couple of minutes or hours (which is the typical time interval between consecutive measurements in time-series expression data) are unlikely, and therefore shifts of more than one time point are probably meaningless, although they might still result in significant matches if the process is periodic. With global measures, shifts can be handled by simply shifting one of the profiles and comparing the shifted profiles. With the local similarity measure, which is based on the DP algorithm, shifts can be limited by setting a hard bound and exploring only a small area around the main diagonal in the DP matrix or by introducing a modulation that decreases the similarity score as a function of the time delay, for example using an exponentially decaying function

$$score(i,j) = v_i u_j e^{|i-j|}$$

12.4.2 Missing data

Missing data is a common problem with microarray expression data (Figure 12.11). Slides are often damaged and scratched in certain spots, making it impossible to determine the expression levels of all genes in each experiment. Owing to missing data the dimension of each vector (expression profile) can vary, and this has to be taken into account when comparing profiles, or otherwise the profiles have to be normalized properly.

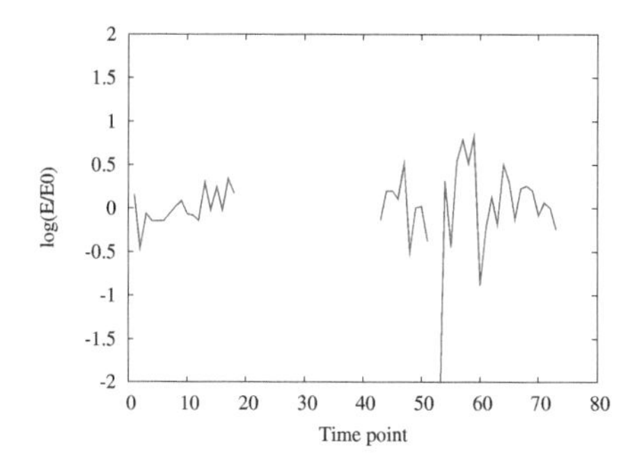

FIGURE 12.11: Missing data. An example of an incomplete expression profile of 60S ribosomal protein L35 (Biozon 120-272) over the time series dataset of [607].

A simple approach is to add a constant where entries are missing, but there are better solutions. Most studies deal with missing data using different variations of extrapolation methods (e.g., [635]). For example, a missing value can be replaced with the average value in that specific experiment over all other genes, or with the value its nearest neighbor has in that experiment. An intermediate approach it to use the local neighborhood and weigh nearby genes such that their contribution to the average decreases with the distance (using a factor proportional to the inverse of the distance or an exponentially decaying function). However, the neighboring genes might belong to multiple cellular processes and their expression values might vary, hence extrapolation might not work well in all situations. A more sophisticated approach is to use the average value in the missing experiment only over genes that belong to the same class [636], that is, genes with similar expression profiles that are likely to correspond to the same regulatory module or process. In Section 12.5 we will discuss the application of clustering algorithms to determine gene classes based on expression data. For now, we note that determining the classes naturally requires comparison of expression vectors, and if the similarity or distance measure used does not tolerate missing values then we cannot apply this method. However, this can be solved by initializing the missing values using one of the other methods (e.g., the average value in the corresponding experiments), use the augmented vectors to compare genes and determine the classes, and recompute the missing values based on the genes that belong to the same class. This EM-like procedure can be applied iteratively until convergence.

The problem with extrapolation methods is that they might introduce errors, especially when there is a large number of missing values, and result in similar expression profiles although the corresponding genes are not truly similarly expressed. With proper normalization, using just the available data would be the safer way to go in most cases. Note that the Euclidean, Pearson and Spearman measures defined in Section 12.4.1 are already normalized by the dimension d, which is defined as the number of features the pair of vectors compared has in common. With the mass-distance measure the dimension is not an issue since it is a measure of probability.

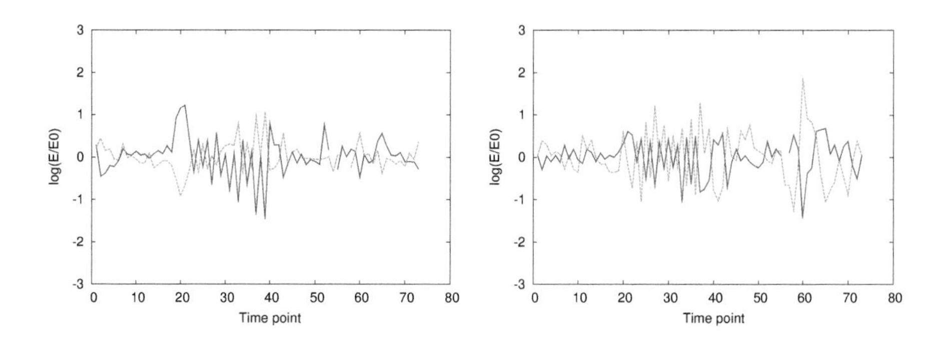

FIGURE 12.12: **Anti-correlated gene expression.** Pairs of genes that are significantly anti-correlated in the time series dataset of [607]. Left: FK506-binding protein (Biozon 392-54) and endoplasmic reticulum protein with unknown function (Biozon 321-315). Right: chorismate synthase (Biozon 376-120) and a probable glucose transporter (Biozon 592-20).

12.4.3 Correlation vs. anti-correlation

Not all functional links between genes result in similar expression profiles. While in most cases it is the genes with highly similar expression profiles that we are interested in, one might find the genes that are strongly *anti-correlated* even more interesting (Figure 12.12). That might happen for example when there is a regulatory relation between two genes (e.g., when one gene suppresses another), as seen in Figure 12.13.

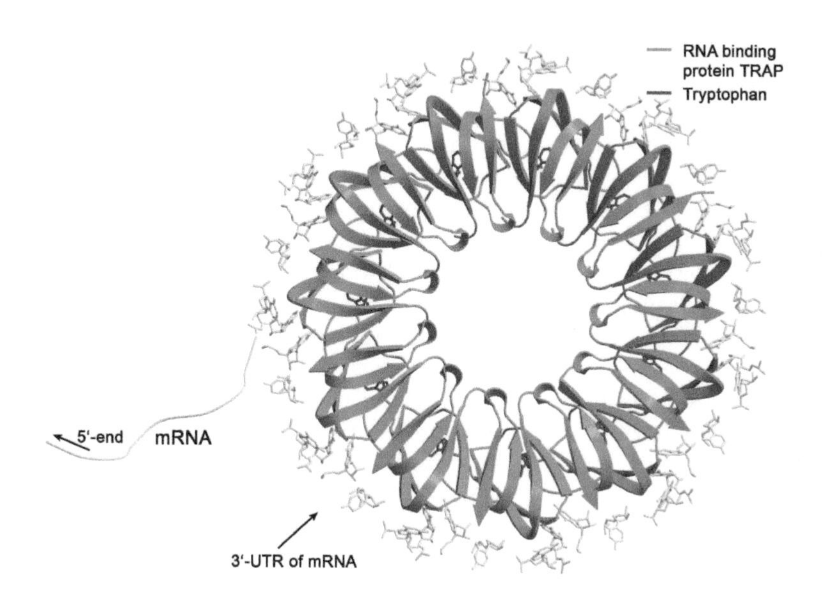

FIGURE 12.13: **Anti-correlated genes.** A complex of a 11-mer of an RNA binding protein TRAP (PDB 1gtn) bound to mRNA. The protein TRAP (Tryptophanyl RNA Attenuator Protein) binds to 11 GAG repeats in the 3'UTR (UnTranslated Region) of mRNA of tryptophan biosynthesis genes. This interaction is possible only in the presence of tryptophan (which enables the formation of the multimer) and prevents translation of the mRNA. In the absence of tryptophan, TRAP cannot bind to the mRNA, resulting in their translation and hence biosynthesis of tryptophan. This is an example of regulation of protein expression at the translation level. For an example of regulation of protein expression at the transcription level, see Figure 12.1.

Of the global similarity measures described above the Pearson and Spearman correlation measures are well suited for detection of anti-correlation, assigning negative values to anti-correlated genes. The closer the value is to -1, the stronger the anti-correlation. The DP-based local similarity algorithm can also detect locally anti-correlated genes after a small modification to the scoring function such that the score of matching measurement i with measurement j is defined as $-v_i \cdot u_j$. If the vectors are normalized first as described in Section 12.4.1, then this definition converts anti-correlated measurements to positive scores. As before, a time-shift modulation can be introduced by adding an exponentially decaying function

$$score(i,j) = -v_i u_j e^{|i-j|}$$

Note that if the expression values are given in the form of log-ratio then other measures can be adjusted to measure anti-correlation as well by mapping one of the expression profiles to its negative.

12.4.4 Statistical significance of expression similarity

As mentioned previously, analysis of gene expression data is difficult since data is noisy and in many cases unreliable. Many factors may affect the experiments and the measurements, thus obscuring signals that might indicate relations between genes. Therefore, it is often unclear whether genes are truly co-regulated or are functionally linked even if they seem to be similarly expressed. Measures such as the Pearson correlation or the Euclidean distance do not necessarily entail correlation or co-regulation even for pairs of genes that score well under these measures. Therefore, to determine which similarities are more likely to be due to an underlying biological phenomena it is important to have a statistical measure for assessing the significance of similarity based on expression profiles.

Methods to assess the significance are usually based on permutation methods [29, 637]. To assess the significance of a similarity score between two vectors, according to any one of the measures suggested in Section 12.4.1, we can permute the entries of one of the vectors and recompute the distance between the two vectors. By repeating this permutation step over and over again we can determine the background distribution of distances between random vectors. For most measures the empirical distributions can be very well approximated by the unimodal normal distribution, as expected by the central limit theorem considering the functional form of the similarity measures[8]. Given the average distance and the standard deviation of distances between the permuted vectors, the significance of the distance between the non-permuted vectors can be assessed in terms of the zscore (the distance from the mean in units of standard deviation). The distribution can differ quite markedly for different pairs of genes (see Figure 12.14).

[8]Measures such as the Euclidean distance, the Pearson correlation and the mass-distance are defined as sums over many experiments. For example, the mass-distance measure in its logarithmic form is a sum of all the *mass* values over all experiments. Since each experiment is typically distributed normally, so are the *mass* random variables, and so is the total sum. Therefore, the resulting mass-distance is also normally distributed (Figure 12.15 on page 526).

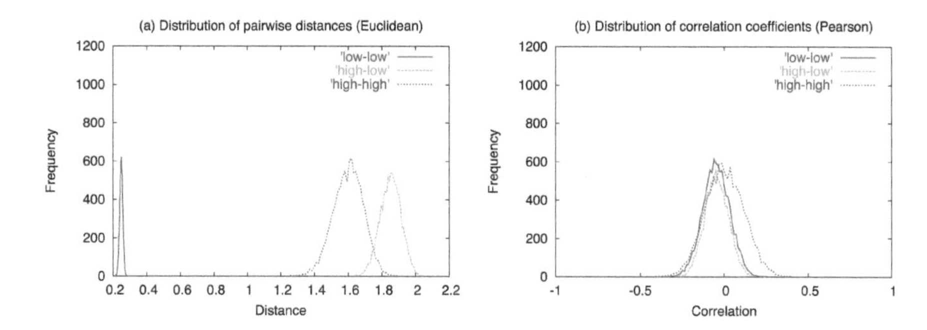

FIGURE 12.14: Background distribution of distances for three pairs of genes: (a) Euclidean metric (b) Pearson correlation. Each plot shows the distributions for three pairs of genes of different norms. The Euclidean norm of the common features in these gene pairs was calculated to determine whether the gene had high or low norm. For example, the low-high curve corresponds to a pair of expression profiles, one with low norm and one with high norm, and is based on 100 permutations of the expression vectors. Despite dimension-normalization, all measures are sensitive to the norm and can differ markedly from one pair of genes to another (especially with the popular Euclidean metric), emphasizing the inadequacy of the raw similarity or distance measures. For example, an Euclidean distance of 1.5 is significant for the pair of high-norm vectors but insignificant for the two other pairs.

This self-calibrating approach is computationally simple and provides reliable measures of significance as it adjusts to the specific "compositions" of the pair of vectors compared. Moreover, such a permutation method converts all distances to a uniform scale, independent of the norm and the dimension. This is especially useful for the analysis of incomplete expression data, with many missing values. Note that for the Euclidean distance highly negative zscores (small distances) are significant, while for Pearson correlation highly positive zscores are significant.

While the zscore is a useful and intuitive measure of significance, it is usually more meaningful to work with a probability measure such as pvalue, which is the probability of obtaining by chance a score which is higher than the observed score. Each of the similarity measures described above can be associated with a pvalue based on the distribution of pairwise similarities for randomly selected pairs. Most measures follow the normal distribution as is expected by their functional form. For scores that are distributed like the normal distribution, the probability $pvalue(S)$ of a score S can be computed rigorously given the parameters μ and σ, using a closed-form solution (see the book's website at `biozon.org/proteomics/`).

For *local* similarity scores the background distribution is different and follows the extreme value distribution, as we have already seen with local se-

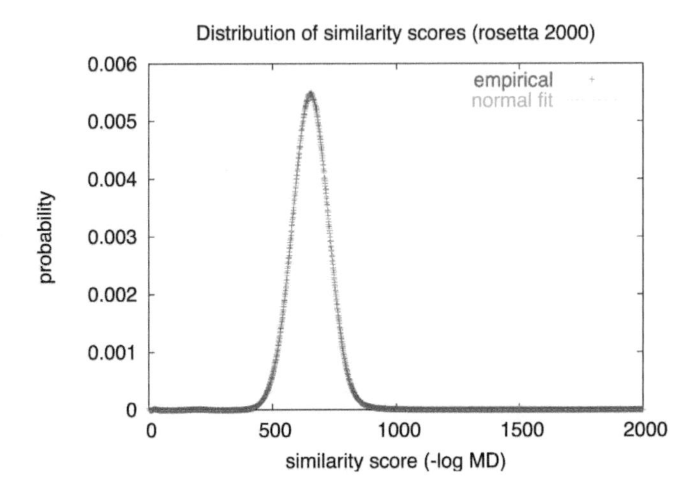

FIGURE 12.15: **Distribution of mass-distances in the Rosetta dataset.**

quence similarity scores (see Section 3.4.3.2). However, the size of the search space (the number of common features) can differ from one pair to another if missing values are ignored, and to properly assess the significance of a similarity score it is necessary to adjust for the variable search space and derive different background distributions for different dimensions. Note that while a dimension-independent significance measure (based on a single background distribution for all dimensions) maintains the monotonicity of scores, a dimension-normalized measure might change the order in which hits are reported as higher score no longer entails more significant match.

Computing the statistical significance of the Spearman correlation is more involved, as described in [638]. The statistical significance of a specific correlation value is estimated by computing the probability p of getting a value greater than or equal to $\sum_i x_i^2$ in d experiments. For identical profiles, the Spearman rank correlation is 1 and so is p.

A typical analysis of a microarray dataset involves many pairwise comparisons of expression profiles. When performing multiple comparisons, the probability of observing by a chance a certain similarity score increases with the number of tests. Therefore, the pvalue should be corrected accordingly. A simple and useful measure of significance in this case is the evalue (see Section 3.4.3.2). The evalue of a similarity score S can be obtained by applying the Bonferroni correction and multiplying the pvalue by the number of tests, or the number of pairwise comparisons (see Section 3.12.8). If we run an all-vs-all comparison of all expression profiles in a microarray dataset then

$$evalue(S) = \frac{N \cdot (N-1)}{2} pvalue(S)$$

where N is the number of genes in the dataset. The evalue is the expected number of occurrences of chance similarities with a score S or higher. An evalue of 0.1 for example means that on average one needs to perform $10 \cdot \frac{N \cdot (N-1)}{2}$ pairwise comparisons (e.g., compare all pairs of genes in 10 datasets as big as the dataset analyzed) before encountering a similarity with that similarity score or higher by chance.

12.4.5 Evaluating similarity measures

When using any of the measures we discussed in Section 12.4.1 to analyze microarray expression data we are interested only in the set of significant similarities (above a certain significance threshold). This set differs from one measure to another. Naturally, we would prefer a measure that detects the maximal number of similarities that are due to true functional links, while minimizing the number of false positives due to chance similarities. But how can we compare the performance of different measures? Unlike other types of similarity (such as sequence similarity), it is harder to verify similarities based on expression profiles. Experimentally verified functional links are available in databases that store information on cellular pathways (see Chapter 14), protein-protein interactions (Chapter 13) and promoter data, among others. However, it is very hard or even impossible to determine which pairs of genes are totally *unrelated*. In other words, no similarity based on expression data can be confidently designated as a false positive. The more likely scenario is that significant expression similarities between pairs of genes that seem unrelated correspond to functional links that have not been established yet.

One way to assess the performance under such a scenario is to compute the fraction of verified relationships out of all significant similarities. However, the order in which these relationships are reported is also important. That is, if the experimentally verified relationships are the less significant similarities in the pool, then it should raise a red flag. To monitor the fraction and the order of experimentally verified relationships, we can compute all pairwise expression similarities, sort them in decreasing order of significance and plot the number of known relationships detected as a function of the total number of pairwise similarities as we scan the sorted list from the top. This curve is similar to a ROC curve (see Section 3.11.2), where the number of positives is plotted vs. the number of negatives, but with the x-axis replaced with the total number of relationships. An example of such an evaluation is shown in Figure 12.16.

To improve sensitivity one can combine several measures. However, combining measures is not always advantageous, as the combined measure might overestimate the similarity of unrelated profiles, thus leading to a higher error

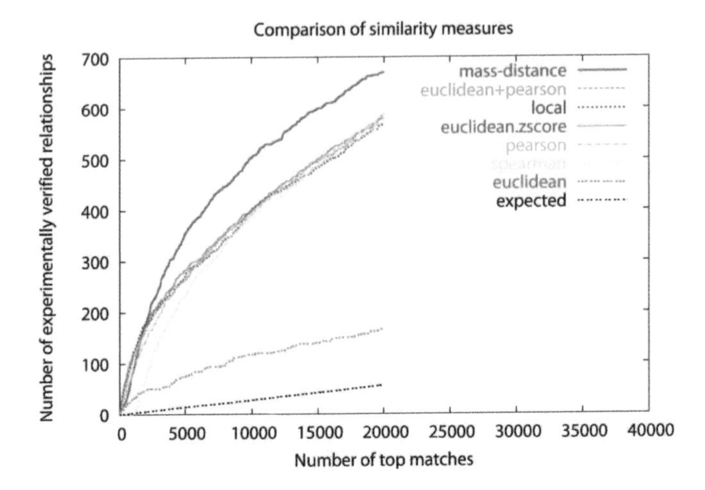

FIGURE 12.16: ROC-like performance evaluation of similarity measures. (Figure reproduced from [29] with permission.) Performance of selected measures on the 1998 yeast cell cycle data, as reported in [29]. These results give an idea of the effectiveness of different methods. For clarity, the labels are ordered according to the performance of the corresponding measures. Of all measures tested, the mass-distance measure gives the best results, followed by the Pearson correlation and the Spearman rank correlation. Surprisingly, despite its popularity in studies of expression arrays, the Euclidean measure performs poorly. Zscore-based measures always improve over the original global measures, and the most drastic improvement was observed with the Euclidean-based zscore measure. All measures contain some information about the true relationships, when compared to what is expected by a random guess (see Section 12.4.5.1). Similar results were reported with other expression datasets (for more details see [29]), and while in general no single measure can be crowned as the best for all possible datasets, the mass-distance measure and the Spearman rank correlation seem to produce better results overall. The Euclidean metric seems to be performing the worst.

rate. For example, two unrelated expression profiles that are only slightly similar might end up higher in the list if reported by multiple measures. Several different combinations were tested in [29], and although they performed quite well on average, they did not improve over the best individual measures.

12.4.5.1 Estimating baseline performance

When assessing the performance of any algorithm on a classification or prediction task it is important to know what to expect just by a random guess. This is often referred to as the **baseline performance**. When assessing the effectiveness of a certain similarity measure in detecting functional links based on expression data, the baseline performance would be the number of true relations e that are expected to occur at random. To estimate e, it is useful to think of the set of all genes and relations as a graph (a *relation graph*) in which each node represents a gene and an edge exists between the nodes if the genes are functionally linked. A **uniform random setup** assumes that all the edges in the relation graph are placed uniformly, at random, between genes. Given this setup, one can easily compute e analytically. That is, the number of true edges that are expected to occur in a set of n randomly chosen edges is $e = n\frac{T}{N(N-1)/2}$ where N is the total number of genes in our set and T is the total number of true relationships (total number of edges in the graph).

While computationally very simple, this random setup does not necessarily preserve the structure of the gene network. Gene networks are far from being random, since some genes are more connected than others, either because they are truly functionally related to multiple genes or because they were better studied than others (see Chapter 13). In the **structure preserving random setup** the graph structure is frozen (in other words, the edge structure is kept intact and the connectivity of each node remains the same) while the expression profiles associated with the different nodes are shuffled. Given the shuffled graph, we can approximate the number of expected relations by computing the number of true edges that exist between the n most similar pairs of expression profiles. To obtain a good estimate of the expected number, the process should be repeated multiple times to derive the average number of relationships and the standard deviation. If the results are similar to those obtained with the uniform random setup then we can conclude that the network connectivity is fairly low and that the theoretical estimates based on the simplified model are accurate enough.

12.5 Cluster analysis and class discovery

Individual (differential) or pairwise analysis of genes can be insightful when studying functional links and gene networks. But the complete picture is more complicated. What differentiates expression data analysis from protein sequence or structure analysis is the coordinated nature of cell regulation mechanisms. We already know that many genes work in a correlated manner, and for a reason. Groups of co-expressed genes often have related biological functions, and many of them correspond to fundamental cellular processes or functional modules (we will discuss these processes in detail in Chapter 14). The coordinated patterns of expression hint at global processes, the nature of which cannot be determined based on individual gene pairs. Clustering is instrumental in detecting these groups of similarly expressed genes and the cellular processes they participate in.

Cluster analysis is especially effective for analysis of expression data because of the high level of noise typical of this data. Clustering algorithms help

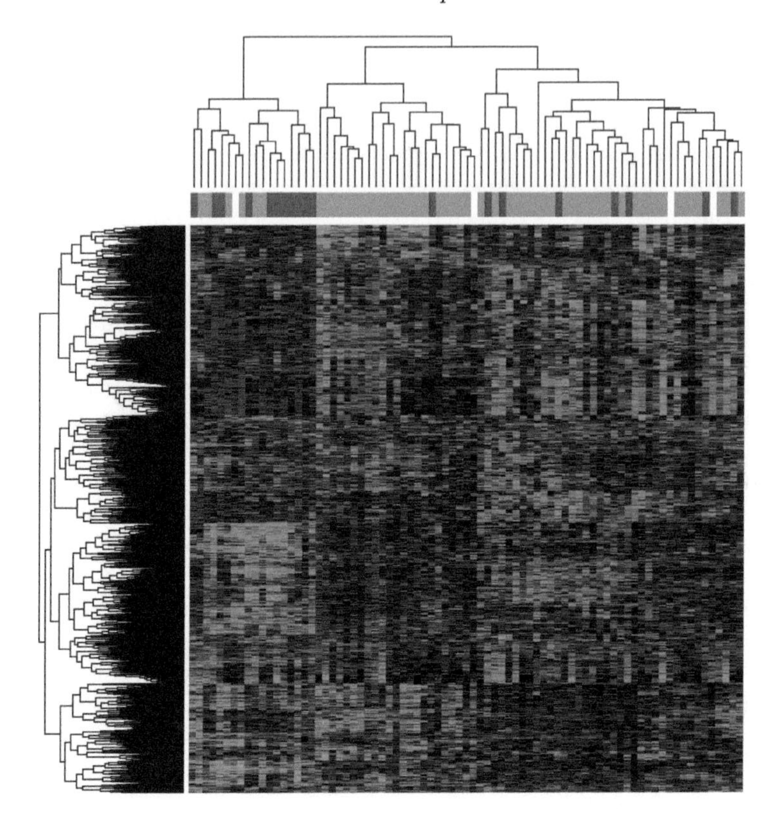

FIGURE 12.17: **Clustering genes based on mRNA expression data.** Cluster analysis of the dataset of [639], which contains samples from patients with a type of blood cancer called Acute Lymphoblastic Leukemia (ALL). Each row corresponds to one gene and displays the expression levels of that gene in different samples (patients). Each column corresponds to one patient. Such images are often referred to as **heatmaps**. Gene expression profiles were clustered using the complete linkage algorithm and organized in a dendrogram (on the left) so as to reflect the cluster structure. Similarly, samples were clustered based on gene activity profiles and were organized in a dendrogram (at the top). The horizontal bar below the top dendrogram indicates whether the sample was identified in a separate bioassay as having extra chromosomes (hyperdiploidy). Samples are flagged as hyperdiploids (red), normal (gray), and unknown (white). The top horizontal dendrogram divides the samples into three major clusters. The samples in the leftmost cluster are predominantly hyperdiploid, suggesting that the most conspicuous factor affecting expression levels is whether the sample has extra chromosomes [640]. The left vertical dendrogram divides genes into four major groups. The top group of genes is predominantly genes from the X and 21 chromosomes, which are overwhelmingly the ones with extra copies in the hyperdiploid samples of ALL. This example illustrates both the utility of visual descriptive tools such as the heatmap, and the importance of taking hyperdiploidy into account when analyzing cancer microarray datasets. (Figure was generated by Assaf Oron, using the R statistical software: `www.r-project.org`.)

Cluster Number	Pvalue	Number of Genes	Cluster Summary
201	6.51e-294	44	Large diverse cluster, cytochrome c oxidase subunits ATP synthase subunits, mitochondrial proteins
66	6.72e-166	14	Dehydrogenases
152	2.42e-147	88	DNA replication, DNA repair, mitosis, meiosis
119	1.50e-143	141	Translation related proteins, ribosomal proteins tRNA synthetases, translation initiation factors
193	2.20e-140	113	Mitochondrial proteins
192	3.43e-136	116	Large diverse cluster, transferases, hydrolases
120	3.13e-129	15	Diverse cluster, lysases, transferases, ligases
206	4.04e-123	19	Hydrolases
139	3.17e-117	8	Mitosis
129	3.36e-117	9	tRNA synthetases
153	1.99e-116	50	Ribosomal proteins, helicases
202	6.01e-111	5	Diverse cluster with oxidoreductases
56	2.40e-100	31	Diverse cluster, hydrolases, oxidoreductases Mitochondrial proteins
115	6.94e-99	21	RNA processing, growth regulation
63	1.68e-95	8	Oxidoreductases, Fe containing proteins

TABLE 12.1: Summary of gene clusters. (Table reproduced from [641] with permission.) Clusters were obtained by applying the clustering algorithm of [346] to the time-series expression data of [607]. For details see [641]. The top most significant 15 clusters are listed, with their significance, size and type of genes in each cluster. Significance is estimated as described in Section 12.5.2.

to enhance signals and expose subtle functional links between genes since the coordinated expression profiles of multiple functionally related genes are more easily detectable than a single pair of similarly expressed genes. This could be extremely useful, for example, when assessing the effect of mutations or when evaluating new drugs. Many drugs fail because of their toxicity (such as induced oxidation, see Figure 12.1), resulting in failure of cells and tissues. Often the effect of these drugs is not so visible on individual genes, but becomes evident when considering the aggregated expression profile of the affected group of genes, which may correspond to a specific pathway or a functional module. Similarly, a mutation in a specific gene might have a significant impact on a certain cellular process, but it may be hard to establish the connection without observing the collective expression profile of the genes involved in that process.

Motivated by the potential of new biological discoveries, a surge of papers followed the introduction of microarray technology, describing clustering algorithms and their application to expression data. We already discussed clustering algorithms in Chapter 10. Almost all of these algorithms and many more were applied to expression data and with much success. The clusters detected with these algorithms often reveal biologically meaningful collective patterns of co-expression. To demonstrate, Table 12.1 lists some of the clusters in the yeast time-series dataset that were detected with the clustering algorithm of [346]. Most clusters have common "themes", which sometimes can be deduced by simply inspecting the annotations of cluster members. For example, the second cluster of Table 12.1 contains mostly dehydrogenases, while the third cluster contains genes that are involved in DNA repair, mitosis and meiosis. The fourth contains many ribosomal genes as well as translation initiation factors and tRNA synthetases, and the fifth contains mitochondrial genes. The collective expression profiles for the four most significant clusters are shown in Figure 12.18.

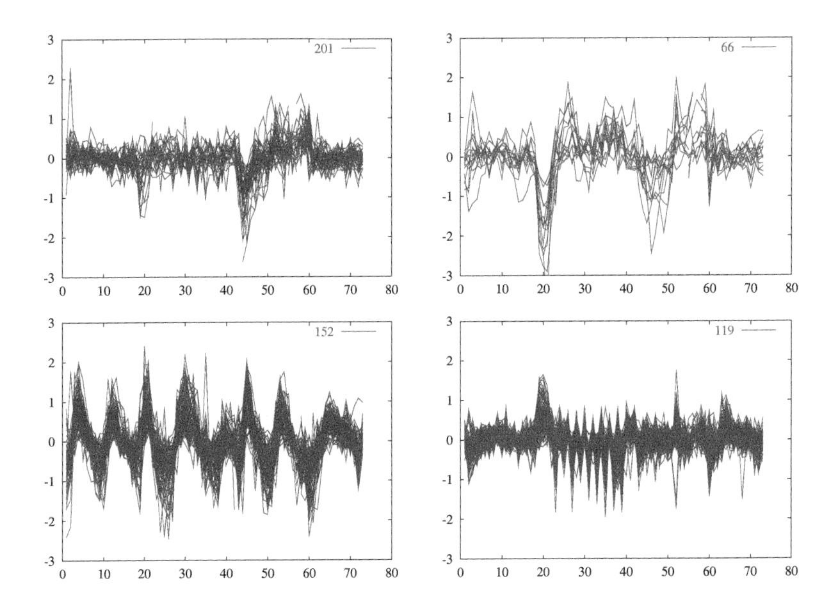

FIGURE 12.18: **Clusters of expression profiles.** The collective expression profiles of the four most significant clusters of Table 12.1 are shown. Note the very distinctive expression profile of each cluster. The **consensus profile** of a cluster can be derived by averaging the expression profiles of the cluster's genes. (Figure reproduced from [641] with permission.)

Clusters can also be visualized as a combination of a **heatmap** and a **dendrogram**, as illustrated in Figure 12.17 (page 530). This example demonstrates the application of clustering to samples taken from cancer patients. The dendrogram on the left organizes the genes in a tree-like structure, reflecting the order in which they were clustered (recall the discussion on agglomerative pairwise clustering algorithms in Section 10.4.1). Each leaf node corresponds to one gene. The measurements in the heatmap are ordered to coincide with the tree structure, such that genes corresponding to close-by leaf nodes are placed adjacent to each other. This representation makes it easier to detect groups of co-expressed genes. In this specific example, a second dendrogram is displayed over samples (at the top of the figure), clustering patients based on the similarity in their gene activity profiles. Samples are ordered accordingly. This kind of analysis is useful for class discovery, as discussed next, and can help to identify genes that are highly over-expressed or under-expressed in specific subgroups of samples.

Cluster analysis is one way of searching for patterns in large datasets, and a very successful one at that. But, as we saw in Chapter 11, it is not the only unsupervised learning technique we can employ in search of "trends" or characteristic patterns. Principal component analysis (PCA) or singular value decomposition (SVD) can also detect such patterns by deriving the covariance matrix of the data and computing its eigenvectors. The eigenvectors with the largest eigenvalues characterize the directions that span most of the variance in the data (see Section 11.2.1). Alter et al. [642] used SVD to analyze the yeast cell cycle data, referring to the base vectors as **eigengenes** or **eigenarrays**, depending on the matrix analyzed. For example, an eigengene is a "typical gene" representing a group of similarly expressed genes, while an eigenarray is a "typical experiment" representing a group of microarrays with similar expression patterns over the gene set. This type of characterization can be useful for array classification.

Another type of analysis which is closely related to clustering is **class discovery**. Class discovery is mostly concerned with classification of samples (e.g., cells), rather than genes. Unlike clustering, however, samples are labeled and the goal is to learn a classifier that can classify new samples to one of the predefined classes. As such, this is a supervised learning problem. In a typical setup, samples are extracted from two or more classes (e.g., cell extracts from cancer patients vs. healthy people, or samples extracted from patients with different types of a certain cancer). For example, the dataset of [643] contains samples of two types of leukemia patients: Acute Myeloid Leukemia (AML) and Acute Lymphoblastic Leukemia (ALL), measured over more than 7000 genes in samples from bone marrow as well as blood samples. The dataset of [635] contains samples from patients with two types of diffused large B-cell lymphoma. The goal of class discovery is to design reliable classifiers that can predict the right class given the activity profile of a new sample. Designing and training classifiers from expression data is the subject of many studies (e.g. [523, 635, 643–645]), employing methods similar to the techniques we discussed in Chapter 7 and Chapter 10.

Class discovery is extremely important for medical diagnosis and prognosis, such as characterizing whether a sample is tumorous, or when diagnosing patients with different variants of a certain disease, variants that require different treatments. Often, patients with a certain variant do not respond well to drugs and treatments that were developed for other variants. This kind of analysis can help to identify the variants of a disease, better understand the differences between them and develop more effective treatments for patient groups of each variant.

12.5.1 Validating clustering results

With the many clustering algorithms that are available at the disposal of the practitioner in computational biology, it is not clear which one should be used to cluster expression data and which one would produce the most meaningful results. This problem is not unique to microarray data analysis but rather is common to any learning task that is involved with clustering.

As we already discussed in Chapter 10, clustering algorithms can sometimes lead to false conclusions and their results should be carefully inspected. To determine whether a clustering configuration is meaningful we can use any of the validity indices discussed in Section 10.5, but as pointed out previously no single index is optimal for all tasks and therefore it is recommended to use a combination of more than one index. In this section we will demonstrate the application of two criteria to assess the quality of different clustering models. The results of the two indices can be corroborated to select the better models. For further reading on other evaluation techniques see Section 10.5.

There are two types of evaluations when assessing clustering algorithms. One could compare the performance of the same algorithm with different sets of parameters (**parameter optimization**), for example, the $Ncut$ parameter of the normalized cut spectral algorithm or the number of clusters with k-means and pairwise clustering. The second type of evaluation considers the best clustering obtained with each method and compares them to each other to find the best clustering overall (**best clustering**). Naturally, parameter optimization precedes the second type of evaluation. The left side of Figure 12.19 demonstrates the application of the Bayesian index of Section 10.5.2.1 to parameter optimization with three different clustering algorithms on the yeast time-series data. The Bayesian index is an MDL-like internal index of validity, assessing clustering solutions by measuring their complexity as well as their ability to explain the observed data. Note that in each case, the graph suggests a range for the optimal parameter value.

While the internal validity index can help in discerning better models it does not necessarily produce clusterings that best describe the experimental data. As a heuristic it can only guide the designer by pointing to more effective models. However, usually the designer is left with several models, all seem to describe the data equally well (Figure 12.19).

To reduce this set to a single model, we can re-evaluate the clustering results using an external index of validity that assesses the correlation of the results with experimentally verified data and select the most significant one. When analyzing clusterings based on expression similarity, the experimental data can consist of relationships established from data on protein-protein interactions, cellular pathways, promoters and other sources [29]. The right side of Figure 12.19 demonstrates the application of the pvalue-based index (Section 10.5.1.2) to the same clustering results that were assessed with the MDL criterion. This second criterion helps to reduce a set of "viable" models to a smaller set or a single best one.

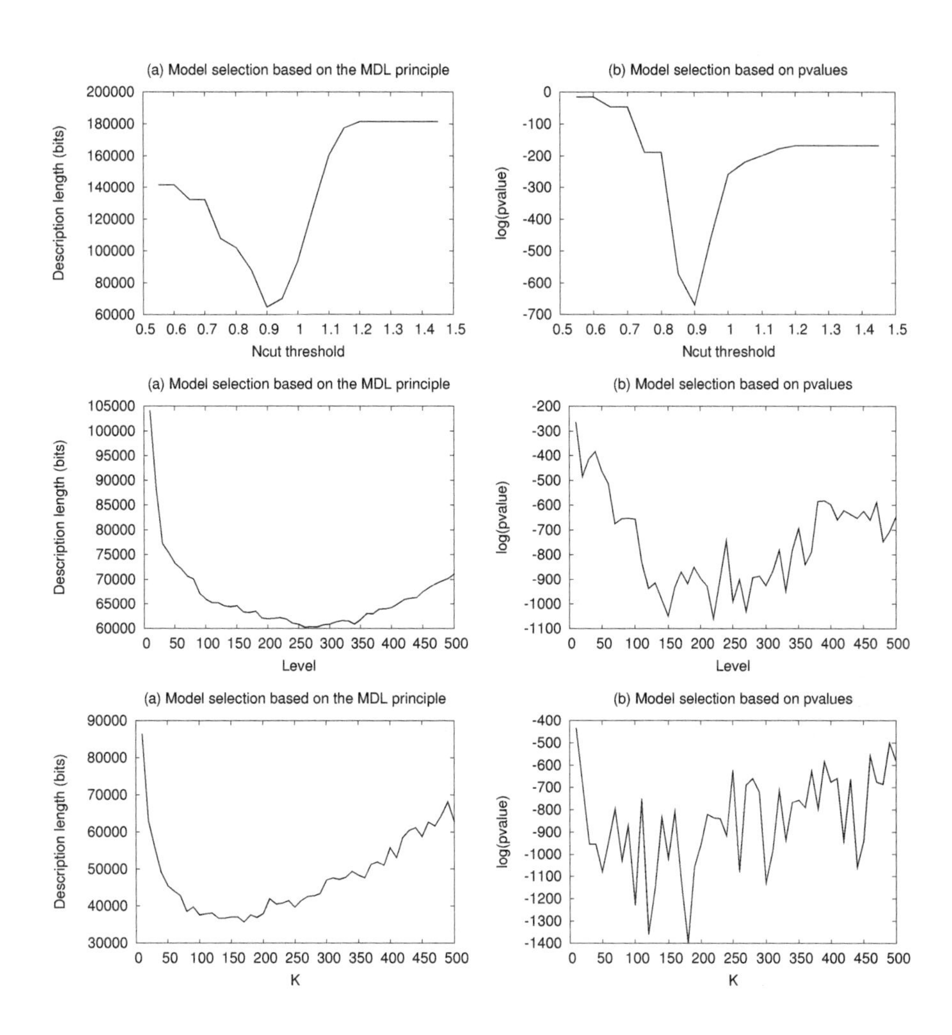

FIGURE 12.19: Model selection based on the MDL principle (left) and the pvalue measure (right). Three clustering algorithms were applied to the yeast time-series data, with various values of their main parameter. Top row: normalized cut spectral clustering (parameter: $Ncut$ value). Middle row: average linkage (parameter: level in the hierarchy). Bottom row: k-means (parameter: k). (Figure reproduced from [641] with permission.)

In the example of Figure 12.19 the good correlation between the two indices makes it easier to pick the best model in each case. The two indices assess two different aspects of the clustering results. The external measure directly assesses the correlation of the clusters with the underlying structure of functional links, while the internal measure addresses issues such as model complexity and generalization. It is the combination of the two measures that leads to a selection of a better model overall.

Note that the pvalue graph for the k-means algorithm fluctuates substantially more than with other algorithms. This is because the k-means algorithm starts from a different random configuration for every k, while other algorithms gradually refine the clustering at a given level to obtain the clustering at the next level. Hence, clusterings obtained for close values of k can be quite different. Running the algorithm repeatedly for the same k with different starting points and picking the best model or employing a hierarchical scheme as in [646] can improve the stability of the algorithm (see Section 10.3.3).

Determining the best clustering overall is much more difficult and requires extensive tests of many clustering algorithms with multiple configurations. As with parameter optimization, conclusions are dataset dependent. On the yeast time-series dataset, the normalized-cut spectral clustering algorithm produces interesting results but is very sensitive to the $Ncut$ parameter value, and to obtain the right model one has to repeatedly apply this algorithm with different threshold values. K-means is fast and is more controllable (as far as the number of clusters) but is very susceptible to the initial configuration. Furthermore, it inherently uses the Euclidean metric, which performs poorly on expression data compared to other metrics (Figure 12.16). Nevertheless, it produces good results. Hierarchical clustering algorithms such as average linkage and single linkage are deterministic and as such always produce the same results when applied to the same dataset. To determine the best model one has to keep checkpoints of all the intermediate clusterings. However, these algorithms can be sensitive to noise and outliers in the dataset itself and not all variations perform equally well. For example, single linkage generates many singletons. Since it is unclear a priori what the typical shape of the clusters in the dataset is, testing different variants is recommended (for spherical clusters average linkage will perform better, while for elongated clusters single linkage will perform better). For further reading on evaluation of clustering based on expression data, see Section 12.7.

12.5.2 Assessing individual clusters

Since expression data is noisy, not all clusters that are detected with cluster analysis are equally meaningful and important. This is true for any clustering algorithm and regardless of the specific parameters. One can compare the clusters obtained with different algorithms and focus on the clusters that are more stable (meaning clusters detected with significant overlap by multiple algorithms). However, this requires running many clustering algorithms on the same dataset. Moreover, in reality perfect agreement is hardly ever obtained, and we might end up being more confused with multiple possible solutions.

If external data is available, we can assess the significance of clusters based on their correlation with the experimental data. Clusters that contain many experimentally verified functional links between genes can be considered more reliable or meaningful. The significance of individual clusters can be computed using a pvalue measure similar to the one we described in Section 10.5.1.2. Given a cluster i with n_i genes and e_i edges, of which t_i edges are experimentally verified edges, and $e_{i,cross}$ cross edges of which $t_{i,cross}$ are experimentally verified, we can compute the probability of the observed type $(t_i, t_{i,cross}, \bar{t}_i)$ where $\bar{t}_i = T - t_i - t_{i,cross}$, assuming a multinomial distribution with parameters $p_i = e_i/E$, $p_{i,cross} = e_{i,cross}/E$ and $\bar{p}_i = 1 - p_i - p_{i,cross}$. That is,

$$P(i) = \binom{T}{t_i \; t_{i,cross} \; \bar{t}_i} p_i^{t_i} \, p_{i,cross}^{t_{i,cross}} \, \bar{p}_i^{\bar{t}_i}$$

Similarly, we can compute the likelihood of the observed type assuming a multinomial distribution with parameters $q_i = t_i/T$, $q_{i,cross} = t_{i,cross}/T$ and $\bar{q}_i = 1 - q_i - q_{i,cross}$. The likelihood ratio indicates to what extent the observed type deviates from the random type. The significance of the observed type can be estimated exactly as in Section 7.7, by computing the total probability of types that are sampled according to the random type $(p_i, p_{i,cross}, \bar{p}_i)$ and have higher likelihood ratio than the observed type.

Many of the pairwise similarities that are detected based on expression data are between proteins that participate in the same pathway [29]. It is interesting to analyze the association between pathways and clusters to see whether specific pathways can be mapped to a single or a few clusters. If the mapping is unlikely to occur by chance, then the corresponding clusters can help in refining the pathway structure. For example, if the consensus profiles of clusters (as in Figure 12.18) are correlated when a time shift is introduced, then the pathway may be composed of several modules (each one associated with one cluster) that activate or inhibit each other. Associating clusters with pathways can also help in functional annotation of genes. If multiple genes from the same cluster belong to a specific pathway then the other cluster members that have not been characterized yet can be predicted to belong to the same pathway as well, or interact with the pathway genes. This type of inference is sometimes referred to as "guilt by association".

Given a pathway with n genes, any cluster that contains a disproportional number of these genes might be linked with that pathway. That is, if n_i genes are classified to cluster C_i and $n_i/n >> |C_i|/N$. However, since the typical number of genes in a pathway is relatively small, random fluctuations could result in what might seem to be a significant link. The significance (pvalue) can be estimated more precisely based on the binomial distribution with the parameter $p = |C_i|/N$, as

$$pvalue(n_i) = \sum_{k=n_i}^{n} \binom{n}{k} p^k (1-p)^{n-k}$$

It should be noted that clusters of similarly expressed genes do not correspond to pathways in general, but possibly to functional modules. We will get back to this topic and elaborate on the relation between pathways and clusters in Chapter 14.

12.5.3 Enrichment analysis

Cluster analysis can expose groups of co-expressed genes, but determining the functional role of each cluster requires further assessment, and clustering is usually followed with an **enrichment analysis**. The straightforward approach is to inspect the annotations associated with genes in each cluster and search for common descriptors. Clusters often have "themes", as exemplified in Table 12.1. However, the vocabulary of molecular biology is so diverse and these free-text annotations are often confusing, even to an expert biologist. Different annotators might use different terms to describe similar biological functions or provide different levels of detail. Parsing and processing this information automatically is a nontrivial task. Moreover, sometimes a cluster may contain genes with different functions. In such cases it is even more difficult to characterize the theme of the cluster.

The inconsistent use of attributes and terms to describe the function of a protein makes it difficult to "connect the dots", even in obvious cases of functionally related genes. This has triggered the creation of the **Gene Ontology** (GO). The gene ontology is a community-wide effort to standardize the nomenclature of molecular biology [454]. This resource has become a standard tool for annotators and bioinformaticians alike and is especially helpful in automatic evaluation of new algorithms and automatic genome annotation.

12.5.3.1 The gene ontology

The gene ontology is a collection of terms, referred to as **GO terms**, used for functional annotation. Each GO term represents some mutual property of all proteins sharing it. This property could be a cellular location (such as membrane), a molecular function (such as receptor activity), a cellular process (e.g., neurotransmitter binding) or other. GO terms are organized in an acyclic graph, as is demonstrated in Figure 12.20. A node's parent represents a property that is more general than the node's property. For example, cytoplasm is a parent of the more specific location "golgi apparatus". At the top of the hierarchy, there is a single root node ("gene ontology").

This node branches into three main categories: "biological process", "cellular component" and "molecular function". Each one of these second-degree terms denotes a major class of GO terms. A protein may be assigned more than one GO term, on different branched of the graph[9].

All paths in this graph begin at the same root node and end in one of the graph's leaves or inner nodes. The level or the **depth** of a node is defined as its shortest distance from the root node. The amount of knowledge available on a protein is correlated with the depth of its GO terms, and proteins that were more closely researched are likely to be assigned to a lower GO term in the graph. Unlike in a tree form of a graph, in the GO graph it is possible to have more than one path leading from the root to a node; this implies that inner or leaf nodes may have more than one parent. For example, the molecular function "neurotransmitter receptor activity" is linked to the two general categories "binding" and "signal transducer activity" (see Figure 12.20).

By definition, each pair of proteins must share some mutual ancestor; if the two proteins are not biologically related in any way, they will probably share the root GO term, or one of the second-degree GO terms ("biological process", "cellular component" and "molecular function"). The **common level** of two proteins is defined as the level of their deepest common GO term. To compute the common level of two proteins it is necessary to first compile the complete list of GO terms associated with each one them, starting from the leaf nodes, and adding to the list all the parent nodes. These two lists are then compared to find the deepest common GO term.

The popular assumption is that the deeper the shared GO term, the more biologically significant the result is. However, while the common level is a good indicator of a valid relationship, it does not perfectly correlate with functional similarity. Indeed, it has been shown in [453] that different GO terms at the same level are not equally significant. For example, the GO term "binding" at level 3 is associated with hundreds of thousands of proteins, while only several are associated with "chaperone regulator activity" at the same level. In other words, the common level does not necessarily designate the level of the functional similarity.

To account for the variability in the size of GO families one can use a different approach as in [453], which attempts to quantify the **semantic similarity** based on the least frequent common GO term. A significance measure is associated with each GO term, which is the probability that two proteins selected at random will be associated with that GO term. Given a GO term g we first count how many proteins are associated with it (or any of its children), denoted by N_{cg}. The probability is then given by

$$p_g = \frac{N_{cg}(N_{cg} - 1)}{N(N - 1)},$$

[9]Similar principles were applied to construct ontologies of structured vocabulary over other types of biological entities or processes, such as cells, phenotypic traits, diseases and others.

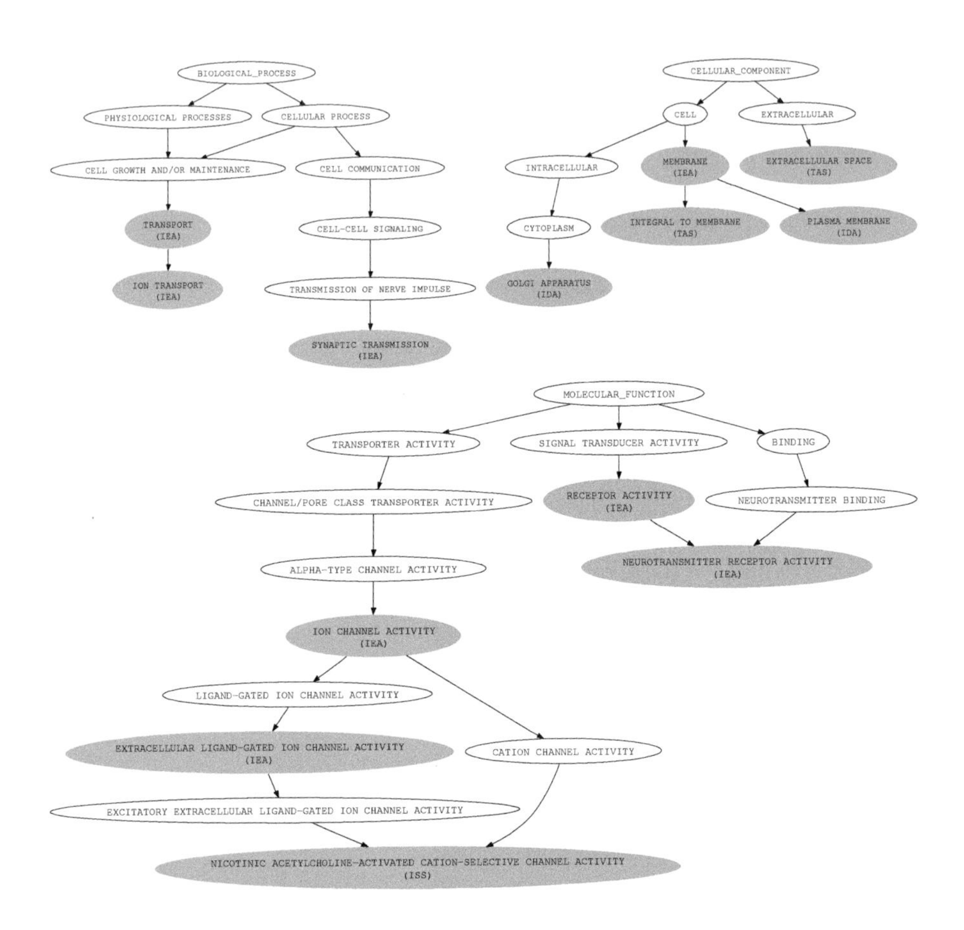

FIGURE 12.20: Gene ontology. GO terms associated with acetylcholine receptor protein, alpha chain precursor (Biozon 457-16). The terms are organized in three different trees: the top left one corresponds to the biological process, the top right tree corresponds to the cellular component and the bottom tree is the molecular function. The shaded ellipses mark GO terms that were associated directly with the protein by curators. The remaining terms in each hierarchy are the parent terms. Different curators may annotate the same protein at different "resolutions" (e.g., "transport" and "ion transport" in the top left tree). The three letter code in parentheses is the evidence code based on which an association is made. Evidence codes can be IDA (inferred from direct assay), TAS (traceable author statement), NAS (non-traceable author statement), ISS (inferred from sequence or structural similarity) and IEA (inferred from electronic annotation). The last one differs from ISS and NAS as it implies that an association is made without being reviewed by curators.

where N is the total number of proteins that are associated with GO terms. If two proteins share more than one GO term, then we can define the significance of their semantic similarity as the minimum probability over all common GO terms.

Unfortunately, this approach is not perfect either, as a larger set of proteins that is associated with a specific GO term does not necessarily mean a less significant functional similarity but can be merely a consequence of a historical bias since some protein families were more studied than others. Therefore, it cannot be used to "quantify" functional similarity. Furthermore, while the GO database is an excellent source of biological knowledge (that was manually curated for the most part), it is not ideal for performance evaluation. GO data is partial, and since it is derived from multiple sources, it is not necessarily coherent. Consequently, in many cases proteins that are very similar based on sequence, structure or other attributes are not necessarily associated with the same set of GO terms. This can pose a big problem when evaluating results (e.g., clusters), since true relationships can be wrongly labeled as false relations. The problem is especially pronounced when the query protein is semantically similar to only a few proteins according to GO but is similar (based on sequence or other attributes) to many proteins, most of which are associated with other, unrelated GO terms. A similar mismatch problem occurs when there are few similar proteins but many proteins with similar GO terms. Nevertheless, the GO database can be used to see whether a group of proteins is enriched with similar GO terms more than is expected by chance [647–649]. In that sense, it is a useful tool for assessing the validity of gene sets. For other GO-based measures of functional similarity and evaluation see [650,651].

12.5.3.2 Gene set enrichment

When analyzing expression data obtained from samples with different phenotypes, one of the main goals is to characterize the cellular processes that operate substantially differently between the samples.

In Section 12.3 we discussed differential analysis of gene expression data. In a typical setup, the samples belong to two different classes (e.g., diseased cells vs. healthy ones). The goal of differential analysis is to identify genes whose expression values differ markedly between the two classes. Differential analysis can reveal which genes play a significant role in a certain process or disease.

To identify the "meaningful" genes, one can sort the list of genes based on the ratio of their expression values between class one and class two. The simple approach would be to mark those genes at the top or the bottom of the list (above or below a certain threshold) as the characteristic genes. For example, if the logarithm of the expression ratio is used, then genes with highly positive values are strongly correlated with class one.

However, in some cases there are no genes (or only a few) whose differential expression is statistically significant. Indeed, often the differences between samples is the collective effect of multiple genes, each one only *slightly* differentially expressed. The collective effect can be more significant than the effect of a single gene whose expression has drastically changed. However, with the noise levels that are inherent to microarray data (see Section 12.2), it may be hard to identify these genes.

To discover these gene sets, we can use more advanced methods that utilize weak and subtle correlations between genes, from clustering to SVMs. Such methods are used for class discovery, as discussed in Section 12.5. Unlike differential analysis which is concerned with single genes, class discovery considers the collective properties of sets of genes. The coordinated behavior of a gene set in a specific class of samples suggests an association with a certain key biological process that operates fundamentally differently between classes. This kind of characterization can help to better understand the functional role of the genes that make up the set and link them with specific cellular processes (or unravel new pathways).

To identify the key process, one can inspect the annotations of the genes that are significantly differentially expressed (as individuals or as part of a set) and try to associate a certain common function. However, sometimes there are too many genes which are differentially expressed, without a clear association with one key process or a common related function. Moreover, the differences in the phenotypic traits between samples might be a combined effect of multiple genes involved in multiple processes. Crosstalk between cellular processes (see Section 14.3.1) may further obscure the common theme and complicate functional analysis of the data. In either case, characterizing the molecular processes that underly the differences in phenotypes can be a difficult task.

A complementary approach to functional annotation of microarray data is to start with a predefined gene set of common biological function, and assess its importance in the context of each class. For example, a gene set could be a group of genes that participate in the same cellular pathway, or a group of genes in the same GO category, or the same subcellular location.

In Section 12.5.2 we already discussed one such approach for identifying significant links between cellular pathways and clusters of co-expressed genes. A similar analysis can be done with respect to the list of differentially expressed genes. In this section we will describe another approach, called **gene set enrichment analysis (GSEA)** [652, 653] or gene set analysis [654]. Given as input a set of genes with a common biological function, the goal of GSEA is to assess the biological importance of the set with respect to the sorted list, and see whether the set can be significantly associated with one of the two classes. In other words, whether the collective activity level can be differentiated between the classes.

We are given a list L with N genes, ranked based on their expression ratio in the two classes. Let e_{ik} denote the expression of gene i in class k ($k = 1, 2$), and define $e_i = \log \frac{e_{i1}}{e_{i2}}$, such that genes that are highly expressed in class one are assigned a positive log-ratio value. We are also given a predefined set $\mathbf{S}$ with M genes. The goal is to determine whether the set $\mathbf{S}$ is distributed randomly throughout the list L or not. GSEA associates a score with $\mathbf{S}$ that assesses how well the set correlates with the extremes of the list. The score is a function of the position along the list, and reflects the correlation of the subset of genes up to that point with $\mathbf{S}$. Specifically, if e_i denotes the expression value of the gene that is ranked in the i-th place in the list, then the score at position l in the list is defined as the cumulative sum

$$S(l) = \sum_{i=1, i \in \mathbf{S}}^{l} |e_i|^p - \sum_{j=1, j \notin \mathbf{S}}^{l} |e_j|^p$$

The first sum is referred to as "hits" while the second is referred to as "misses". Thus, every gene that is in the set $\mathbf{S}$ increases the score S and every gene that is not decreases S. The exponent p controls the weight of each gene. The definition of the score is slightly different in [653], where the weight of each miss is always one, and the contributions of hits and misses is normalized by their total sums, such that

$$S(l) = \sum_{i=1, i \in \mathbf{S}}^{l} \frac{|e_i|^p}{Z} - \sum_{j=1, j \notin \mathbf{S}}^{l} \frac{1}{N - M}$$

where $Z = \sum_{i \in \mathbf{S}} |e_i|^p$. By default, p is set to one.

The **enrichment score** is defined as the maximum deviation from zero (or maximum absolute score). That is,

$$ES(\mathbf{S}) = \max_{l} |S(l)|$$

If the set $\mathbf{S}$ tends to concentrate around the top of the list, then the score S will tend to increase up to a certain point, as demonstrated in Figure 12.21 (the procedure can be reversed and start from the bottom of the list, to detect strong concentration at the bottom). The set of genes that belong to $\mathbf{S}$ and appear in the list before the maximum point is considered the core of that gene set, and is likely to underlie the differences in the phenotypic traits.

When can the ES score be considered as significant? One way to assess the significance is by using permutation-based sampling (as in [653]), which generates an empirical null distribution based on which the pvalue can be estimated. Specifically, the list is reordered randomly many times and an ES score is computed for each shuffled list. The distribution of ES scores for the shuffled lists is the null distribution. To account for multiple testing the pvalue has to be adjusted by the size of the set (Bonferroni correction), or by using the FDR method (see Section 3.12.8).

Each shuffled list is created by changing the class (label) of a subset of randomly selected genes. The reassignment of labels to gene i means that e_{i1}

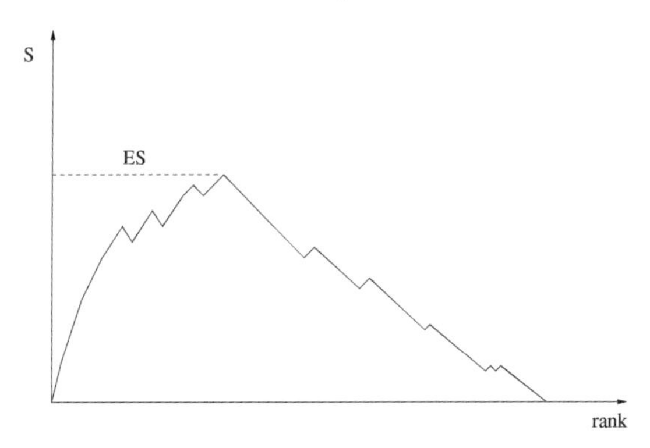

FIGURE 12.21: **Gene enrichment curve.** The score S as a function of the position along the ranked list. The score increases with every member of **S** (hit) and decreases with every gene that is not (miss). The increase depends on the expression value of a hit, but is fixed for every miss (which explains the constant downslope). The maximum score is defined as the enrichment score ES.

and e_{i2} are interchanged and translates to reversal of the sign of the log-ratio value. That is,

$$e_i' = \log \frac{e_{i1}'}{e_{i2}'} = \log \frac{e_{i2}}{e_{i1}} = -e_i$$

Thus the procedure places the gene at the other extreme of the list. This results in a random ordering of genes while preserving gene-gene correlations. Namely, if two genes are strongly expressed in one of the classes, then both will be in one extreme of the list. If one is relabeled in a shuffled dataset, then both will still be extreme points, but on opposing sides of the list. If they both belong to the set **S**, this procedure will break the association with the set when computing the enrichment score and provide a null distribution.

12.5.4 Limitations of mRNA arrays

While microarrays have made their mark on the field, it is important to note their limitations. We already pointed out that microarray data is very noisy as a result of multitude of experimental factors. To increase the quality and reliability of the results, it has become a common practice to repeat the experiment with multiple assays under the same conditions (referred to as **replication**) and average the replicated results.

However, expression data is not only noisy because of technical reasons, it also does not necessarily provide us with a true snapshot of the cell machinery. The biggest criticism regarding mRNA arrays is that they do not reflect the true quantities of genes in cell [655, 656]. As their name implies, mRNA microarrays measure mRNA expression levels. These are not necessarily correlated with the expression levels of proteins, which is really what we are after.

For most genes the mRNA is an intermediate product before being translated to proteins[10]. The quantities of the translated proteins might differ greatly from the mRNA levels, as a result of post-translational modifications such as phosphorylation, methylation and ubiquitination (see Section 13.3.2.4) and other factors that affect translation rate. Finally, the lack of standardization in experimental design and conditions and array fabrication methods makes the cross-comparison of microarray data especially difficult. Thus, despite the relative success with clustering and other data analysis algorithms, conclusions based on expression data should be made with caution. Nevertheless, there is substantial information in these arrays, and when exploited carefully, one can derive meaningful conclusions with high confidence.

12.6 Protein arrays

The limitations of mRNA arrays triggered the development of a different type of arrays to assess the gene content in samples. Protein arrays can be used to measure more precisely protein abundance levels in cell extracts. Protein arrays (also called capture arrays) use molecules with high affinity to other proteins, such as antibodies, to detect the molecular content of a sample and compare protein levels in different tissues and cells. As with mRNA microarrays, protein arrays affix proteins to a glass slide and rely on the binding properties of these proteins to analyze the protein content of a sample that is exposed to the slide. As with microarrays, detection is based on fluorescent markers. The samples are dyed with fluorescent dyes and the slides are usually hybridized with samples from two different sources (healthy vs. cancerous cells, different tissues, control vs. test). The outcome is very similar to that obtained with mRNA arrays, with spot intensities indicating the relative quantities of the corresponding molecules.

The first protein arrays were immune related arrays. They were designed to detect antibodies in blood samples taken from diseased individuals, or to detect antigens such as cytokines[11]. These arrays were instrumental in advancing our understanding of the immune system. The technique was later extended to detect a wider range of proteins by replacing the binding agents (antibodies) with enzymes or receptors and even single-stranded DNA segments.

[10]Some RNA molecules become functional in their own right, without being translated to proteins. Such genes are called **non-coding RNA genes**.

[11]Cytokines are signaling molecules secreted by certain cells to communicate with other cells and regulate immune response.

Unlike mRNA arrays, protein arrays are more sensitive to cross-reactivity since certain agents, like antibodies, can bind to multiple different proteins. Cross-reactivity affects the specificity, but can be controlled by using a combination of two or more arrays with different binding agents. Using multiple arrays in sequence improves specificity as well as sensitivity and is especially effective for antigens of low concentration or antigens that bind poorly to proteins.

Unfortunately, there is a limited number of capture agents, and none of them bind to all proteins. Partial specificity and limited sensitivity further complicate the use of protein arrays. However, protein arrays are becoming a major tool for genome-wide studies. Their most common use is to detect protein-protein interactions or identify the ligands of certain molecules, such as the substrates of enzymes or the targets of antibodies. We will get back to this topic in Chapter 13.

12.6.1 Mass-spectra data

Unlike mRNA arrays, probes in protein arrays are not necessarily exclusive to specific genes. Some capture agents can bind to multiple genes. Therefore, the content of the solution analyzed cannot be identified directly and it is necessary to further process the outcome of these experiments. This is usually done by means of mass spectrometry.

Mass Spectrometry (MS), also abbreviated mass-spec, is a method for determining the molecular content of a sample. The method works by ionizing molecules and running them through an electro-magnetic field that separates the sample into its building blocks (that is, amino acids or smaller molecules) based on the charge-to-mass ratio of each molecule. Each molecule is characterized by a specific mass-to-charge ratio, and by analyzing the observed spectrum of ratios it is possible to determine which molecules exist in the sample and in what quantity.

The mass-spectrometer apparatus consists of three parts. The **ionizer**, the **analyzer/separator** and the **detector**. The **ionizer** charges the molecules of the sample, using for example an electron beam. In some cases, the sample is vaporized and then bombarded with an electron beam to form ions. An electron beam that hits a molecule causes that molecule to lose electrons. For liquids and solids, the ionization is obtained by techniques called electrospray and matrix-assisted laser ionization, for example by exposing the sample to plasma that consists of charged (ionized) atoms, although neutral overall.

The **mass analyzer** accelerates the ions by applying an electromagnetic force. The beams (rays) are passed through a canal where the electromagnetic force is applied vertically so that the mass differences between the different components start to show up (the heavier ones move closer to one side of the magnet). The analyzer has both a magnetic field that alters the direction of the beam and an electric field (voltage) that alters the speed or changes the path of the ion beam. The change depends on the ratio of charge to mass. Some analyzers use either magnetic or electric, while other analyzers use both. In tandem mass spectrometry, multiple analyzers are used one after the other to gradually fragment a sample into its components, for example by first isolating a subset of molecules, and then passing the subset through a second analyzer.

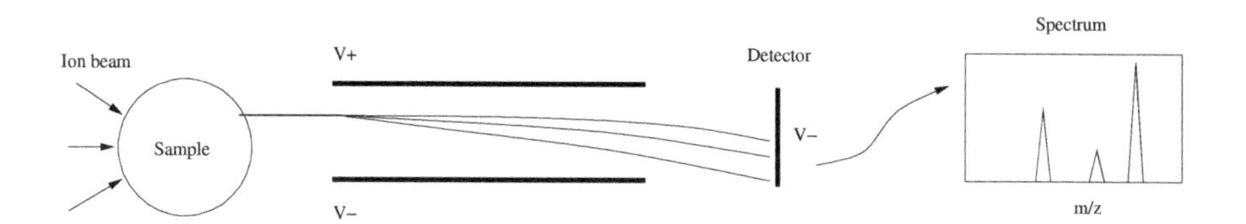

FIGURE 12.22: **Mass spectrometer.**

The **detector** records the position each ray hits and the time of impact. From this information it derives the velocity and the acceleration of each beam. The key to devising the mass-to-charge ratio is two classical equations that govern the motion of charged particles in electro-magnetic fields. According to Newton's laws, an atom with mass m and charge q that is in a force field F accelerates with acceleration a such that

$$F = ma$$

In addition, according to Lorentz' law

$$F = q(E + vB)$$

where E and B are the electric and magnetic fields and v is the velocity. Combining these equations we get that

$$\frac{m}{q}a = E + vB$$

and based on the acceleration and speed, it is possible to determine the mass-to-charge ratio of each component. If only an electric field is used then the separation between atoms depends on the time they hit the detector. If a magnetic field is used, then the separation is in the location on the detector. If both are used then both the time and the location differ from one atom to another.

Beyond identifying the molecular components of a sample, the detector can also estimate the quantity of each component, based on the strength of the current induced when the atoms in each beam hit the detector (the current is proportional to the total charge). The result is a spectrum of mass-to-charge ratios as shown in Figure 12.22 (referred to as **mass spectrum**), which plots the intensity as a function of the mass-to-charge ratio. The spectrum is recorded on a photographic plate or a graph, and hence the name spectrograph or spectrometer.

The spectrum can be given in terms of absolute numbers or relative numbers, based on the relative abundance with respect to the most abundant component. Relative spectrum is useful when comparing samples that were generated in different experiments, or using different approaches. The base reaction is a function of whether the spectrometer uses positive charge (protons) or negative charge (electrons), which differ substantially in mass. Depending on the source, the charge might be a single electron or multiple electrons per atom. If so, then it is better to use $z = q/e$, which is the charge per atom, and then report the ratio m/z instead of m/q. If the charge is a single particle, then the path depends simply on the mass. This information is essential in order to interpret the spectrum correctly. The purity of the data and the quality of the spectrum also depend on the method used to produce the sample. Therefore, it is important to calibrate the detector by operating the mass-spectrometer on a molecule whose ratio is known.

Finally, the spectrum can be analyzed by comparing it to a library of known ratios of mass to charge (reference spectra). However, since the sample might contain components that have not been analyzed yet (e.g., proteins without homologs), then this search is likely to produce incomplete results. Using fragment mass-spec library, which can be simulated in-silico (computationally), can help in deciphering the sequences of new proteins through enumeration of multiple combinations.

12.7 Further reading

The roots of microarray technology date back to the 1980s [657, 658]. Over the next two decades the technology was refined and improved, with a major milestone in [659] describing what would become the seed for the affymetrix chip. The first large-scale microarrays were introduced in [660]. A review of the different types of microarray platforms appears in [624].

There is an overwhelming number of studies which describe microarray assays and discuss analysis of expression data. Single gene analysis is discussed extensively in [625, 661]. Pairwise analysis is discussed in [29, 628]. Cluster analysis in particular received a lot of attention, starting with the seminal work of [662]. Absurdly, in the first few years since the introduction of the technology, papers on new clustering algorithms appeared at a pace that exceeded by far the number of expression datasets that were available for analysis. In Chapter 10 we discussed at length clustering algorithms. For a review on application of clustering algorithms to mRNA expression data see [663–666].

Validation of clustering algorithms is another important problem that was addressed by many. Several studies proposed and applied external, knowledge-driven indices to compare and evaluate clusterings of gene expression data. Some used the Gene Ontology [524,526,667] or paper abstracts from PubMed [668]. Internal indices are also common, and clustering solutions were evaluated by using measures of cluster separation [669], by projecting the data on a lower dimensional host space and testing the stability of clusters in the host space [670], by comparing cluster densities to the local background [671], or by using measures of cluster reproducibility [672]. Cross-validation techniques and related re-sampling methods were employed in [673–676]. For more information on these and other studies, see [641].

We only touched on SVD analysis, but a series of interesting papers on the application of this technique and other variations followed the original paper by [642] and for a review see [677]. Gene set enrichment is discussed in [640, 653, 654, 678]. A review on applications of microarrays to clinical practice appears in [679]. For additional surveys of expression data analysis see [680–682]. For updates and additional references see the book's website at `biozon.org/proteomics/`.

12.8 Conclusions

- Different genes are expressed at different quantities in each cell, and under different cellular or experimental conditions.
- The expression of genes is controlled mostly by Transcription Factors (TF).
- Microarray expression data measures the activity levels of genes in a sample.
- Differential analysis of expression data can link genes with diseases and identify genes that are essential to specific cellular processes.
- Comparison of expression profiles can help to identify functional links between genes.
- Genes that are controlled by the same TF are usually similarly expressed.
- Genes that interact or are part of the same pathway are often similarly expressed.
- Anti-correlated expression profiles indicate regulatory relationships between genes.
- Cluster analysis can discover groups of genes that are similarly expressed. Such groups often contain genes that are part of the same cellular process or regulatory module.
- Gene Ontology can help to annotate gene clusters.
- Microarray data does not necessarily reflect the true quantities of proteins in a sample.
- Mass spectrometry data can provide more accurate measurements of protein levels and protein arrays are becoming more common.

12.9 Problems

For updates, additional problems, files and datasets check the book's website at `biozon.org/proteomics/`.

1. **Normalization and pre-processing of microarray data**

   The first step in processing microarray expression data is to normalize the data to correct for experimental variation between the different microarrays and eliminate biases due to various factors that have nothing to do with the real expression values. You are given a data file `expression.data`, with measurements of gene expression values in cell samples taken from leukemia patients. Usually, microarray data is normalized such that each row or each column has zero mean and standard deviation of one, after normalization. Which should you normalize in this case: the rows (gene expression profiles) or columns (cell's activity profiles)? Can you think of situations where normalization may hurt the analysis and is not advised?

2. **Differential analysis of microarray data**

   Identify the genes which are the most differentially expressed in the dataset of the previous problem, either positively (over-expression) or negatively (under-expression). Since the dataset is composed of multiple arrays (samples), how would you assess the significance of the differential expression of each gene? Characterize the biological function of the 10 most significant genes.

3. **Pairwise comparison**

   Compare all pairs of genes in the dataset using the Euclidean metric, Pearson correlation, Spearman rank and the mass-distance. Note that some values are missing and normalize the distances accordingly. Compare the performance of each metric using ROC-like curves as in Section 12.4.5. A list of functional links is provided in the book's website.

4. **Missing values**

   Given the dataset of the previous problems, estimate missing values by using the following methods

   (a) The average value in the same experiment.

   (b) The value of the nearest neighbor.

   (c) The weighted contributions of all neighbors within a neighborhood of genes whose similarity is statistically significant.

Take special care when defining the nearest neighbors (metric and normalization wise). Test the impact of missing values on pairwise comparison (using each of the metrics of the previous problem) by computing the ROC curves for each method, and compare to the results without estimation of missing values.

5. **Cluster analysis**

   In this problem we will cluster genes based on microarray expression data, to identify groups of co-expressed genes that are potentially associated with specific cellular processes.

   (a) Implement the k-means clustering algorithm (see Problem 10.8).

   (b) Apply the algorithm to cluster the expression data provided in the book's website. Use $k = 100$ and the Euclidean distance function. Use only the available data (i.e., the distance between each pair of vectors is computed only based on the available features and has to be normalized accordingly).

   (c) Consider the ten best nontrivial clusters, with at least five members each (e.g. the clusters with the highest ratio of average inter-cluster distance to average intra-cluster distance). For each cluster:

   - Compute the average distance between its members.
   - Draw the collective expression profile of all genes in the cluster.
   - Based on the gene descriptions, does the cluster make sense? Can you associate a "theme" or function with the cluster?

   (d) Say we want to replace the Euclidean metric with the Pearson correlation metric. Specify precisely the changes you need to make in the algorithm such that it will work with the Pearson correlation metric. What can you say about the general properties of clusters that are generated with the Euclidean metric as opposed to the Pearson correlation measure?

6. **Annotating clusters with GO**

   Annotate the top 10 clusters of the previous problem using the gene ontology. For each cluster, collect the GO terms that are associated with its members and generate the hierarchy of terms (note that proteins may be annotated at different resolutions, or GO levels). Compute the probability to observe each GO category by chance. Repeat the analysis at different levels of the GO hierarchy. Does it help to produce more accurate or meaningful annotation of clusters?

Chapter 13

Protein-Protein Interactions

13.1 Introduction

Each cell of every living organism contains many components, from free atomic ions such as hydrogen or calcium, small molecules such as water or ATP[1], to large molecules including proteins and RNA. These components move around and in and out of the different compartments of the cell. To perform the various processes and functions that constitute a cell, they must communicate with each other, and they communicate by means of interactions.

Molecular interactions are at the core of numerous basic reactions that make up cellular processes. They enable transmission of information between molecules and underlie mechanisms of metabolism, control, repair and regulation. Of special interest are direct interactions between large molecules such as proteins, DNA and RNA. Interactions between proteins and DNA or RNA molecules is the main method of regulation in the cell (see Figure 12.1 and Figure 12.13) and underlie DNA replication, transcription and translation. We already discussed certain aspects of detecting such interactions. For example, in Chapter 5 we discussed algorithms for motif analysis, which can identify genes that have similar promoter regions and are therefore predicted to bind to the same transcription factors. In Chapter 12 we discussed methods for expression data analysis that can detect co-expressed genes, which are often co-regulated. Predicting physical interactions between proteins and DNA or RNA has been the subject of many studies (see Further reading).

Molecular interactions also occur between proteins and lipids. For example, membrane proteins such as receptors are integrated across the membrane,

[1] **ATP**, which stands for adenosine-triphosphate, is a small molecule that serves as energy carrier in cells. It is composed of the nucleotide adenosine with three phosphate groups. ATP is one of the products in photosynthesis, for example, and it acts as a cofactor for enzymes that take part in various processes in the cell. Two closely related molecules are **AMP** (adenosine-monophosphate) and **ADP** (adenosine-diphosphate), which are similar in structure to ATP, but with only one or two phosphate groups, respectively. Many processes and reactions in the cell involve conversion of ATP to ADP or AMP, which releases and supplies the energy needed to carry out these reactions. The reverse conversion is part of energy producing processes such as photosynthesis in plants or glycolysis in animals (see Figure 14.1).

which is made up mostly of lipids. They are important for functions such as signal transduction and transport of small molecules in and out of the cell. Another type of interaction is observed between proteins and polysaccharides (such as cellulose), for example during the process of cell wall synthesis or destruction (e.g., as part of the adaptive response of the immune system). These interactions are of special interest for drug design.

In this chapter we will focus on protein-protein interactions, which are ubiquitous in the cell and are essential for signal transduction, metabolism and molecular transport. Protein-protein interactions occur, for example, between hormones and receptors (such as insulin and insulin-receptor) to stimulate reactions to changes in other cells. They regulate synaptic transmission between neurons, underlie the numerous defense reactions of the immune system (between antibodies and antigens) and facilitate cell motility, to name a few examples. Figure 13.1 shows an interaction involved in regulation of signal transduction via post-translational modification of proteins[2].

Interactions between proteins are established by forming hydrogen bonds, electrostatic bridges and hydrophobic interactions between atoms, usually confined to what is called the **interaction site** (see Figure 13.1). An electrostatic interaction can occur between an amino acid that has a positive charge and one that is negatively charged. A hydrogen bond can form between two residues that are capable of sharing a hydrogen atom. One of the residues "donates" the atom while the other has the ability to "accept" it. A hydrophobic interaction occurs between two hydrophobic (water-repelling) residues, resulting in a spatial arrangement that reduces the number of water molecules surrounding the residues and is energetically favorable.

Direct interactions as in Figure 13.1 are referred to as **physical interactions**. Non-physical interactions, or **functional interactions**, refer to proteins that associate in the same complex or process without interacting directly. Most interactions occur between proteins in the same subcellular location. Some interactions are **transient** while others are more permanent and are maintained for relatively long periods of time. Analyzing the former is obviously more difficult, and most interactions that have been determined experimentally (e.g., through X-ray crystallography) are of the latter type.

---

[2]**Post-Translational Modifications (PTMs)** serve as an engine to monitor and regulate response to changes in cellular content (such as the levels of minerals, nutrients and ions) due to changes in the cell state and condition or stimuli from the surrounding environment. The changes affect certain proteins and are propagated to others to activate response. Signal transfer, processing and response are "implemented" by modifying molecules within the cell. Modifications can be methylation (see Figure 13.1), phosphorylation, ubiquitination and others. For example, **phosphorylation** is addition of a phosphate ion to a single amino acid. **Ubiquitination** is the attachment of a protein called ubiquitin to a target protein, usually marking the target protein for degradation. The modifications can act upon multiple sites along a protein molecule, and some of them are reversible. As such, they provide a powerful mechanism for recording and memorization of intermediate computations. For more information on PTMs and their functional roles (for example in facilitating long-lasting memory), see [683, 684].

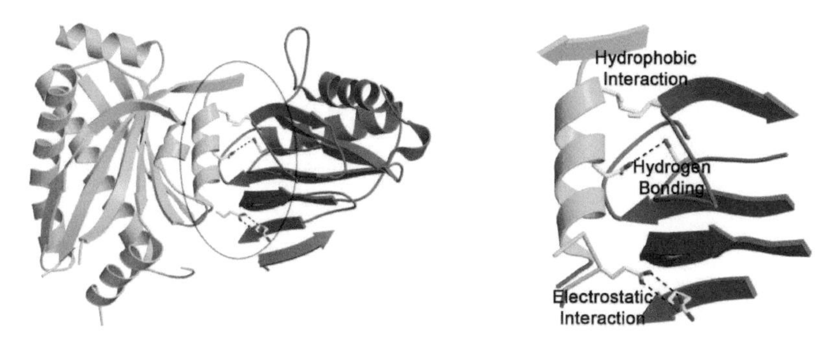

FIGURE 13.1: Protein-Protein interaction. A complex between two proteins, Che-C (green) and Che-D (brown), in the bacterium *Thermotoga maritima*. The interaction site is zoomed in on the right side. The interaction is formed through electrostatic, hydrophobic and hydrogen bonds between atoms of the two molecules. The residues that form the interactions are called **contact residues**. Che-D is a de-methylating protein that is part of a complex network of interactions that are involved in controlling methylated states of receptor proteins that bind chemicals on the cell surface and transmit signals into the cell. **Methylation**, which means an attachment of a small molecule called methyl to another molecule, is one of the molecular mechanisms in nature for tagging molecules. It can be perceived as a way of processing information and memorizing the outcome for subsequent "computations". Methylated and de-methylated states of receptors (at different sites) result in appropriate response of the bacterium to changing concentrations of chemicals in the outer environment. Che-C mimics the receptor structure and inhibits Che-D, adding an element of control to the cell.

The set of all interactions that exist in a certain genome is referred to as the **interactome** of that genome. Knowledge of the interactome is necessary if we want to understand the molecular machinery of the cell and can help to associate genes with pathways, elucidate their function, or infer their significance in certain diseases. It can also reveal cellular processes that have not been characterized so far. The interactome changes from one cell type to another within the same organism, and also as a function of the cellular state.

Traditionally, protein-protein interactions have been determined individually via conventional wet-lab experimental methods such as immunoprecipitation and other pull-down techniques. In the past decade, high-throughput technologies including the yeast two-hybrid system [685] and tandem affinity purification [686] combined with mass spectrometry [687, 688] enabled researchers to detect interactions on a genomic scale [689–691]. The mass of new data triggered many studies that analyzed protein interaction networks in search of distinct sub-networks, complexes and regular patterns and characterized their topological properties.

While groundbreaking, the high-throughput experimental methods proved to be noisy, and in an attempt to verify and complement the experimental data many computational methods for prediction of protein-protein interactions were developed. In this chapter we will discuss some of these methods. We will start with a brief description of the experimental methods. Later in this chapter we will discuss the properties of protein interaction networks and network motifs.

13.2 Experimental detection of protein interactions

There are many experimental methods for detection of protein-protein interactions (see Table 13.1). Here we describe briefly some of the more common ones. For more details see Further reading.

Method	No. of interactions	No. of interactions (unique)
yeast two-hybrid	47069	45857
immunoprecipitation	7717	5572
tandem affinity purification (tap)	4108	3375
affinity chromatography	1645	842
X-ray diffraction	688	394
in vitro binding	646	238
cross linking	363	159
gel filtration chromatography	351	69
copurification	327	112
biochemical	217	95
competition binding	214	87
in vivo kinase activity	198	176
immunoblotting	189	51
biophysical	188	79
gel retardation assays	144	52
cosedimentation	124	33
alanine scanning	119	44
three-dimensional structure	118	67
native gel electrophoresis	114	78
genetic	113	50
other	99	72
electron microscopy	86	18
experimental	71	54
density gradient sedimentation	66	12
surface plasmon resonance	62	22
interaction adhesion assay	61	33
filter overlay assay	58	11
calcium mobilization assay	55	15
autoradiography	51	4

TABLE 13.1: Main experimental methods for detection of protein interactions. Numbers are based on a dataset of about 60,000 unique protein-protein interactions that were gathered from BIND [692], DIP [693, 694] and HPRD [695, 696]. The unified and non-redundant dataset was retrieved from the Biozon database [494]. Many interactions are associated with multiple evidence codes (e.g., yeast two-hybrid and immunoprecipitation). More than 75% of the interactions were determined exclusively with the yeast two-hybrid test. The breakdown by method is given above. First column is the number of interactions that were verified by each method. The second is the number of interactions that were determined *only* by that method. (Table reproduced from [697] with permission.)

13.2.1 Traditional methods

Given a **target protein**, whose interactions we would like to determine, all methods work by isolating complexes in which it is involved. The target molecule (also referred to as the **binding agent**) could be for example an antibody, an enzyme or a receptor. Once isolated, the complexes can be disintegrated into their interacting partners using a **2D gel electrophoresis** technique called SDS-PAGE. This process breaks all molecular interactions (such as hydrophobic and electrostatic), separates the complex into the individual molecules and denatures each protein into its unfolded state. Each protein chain appears as a separate band in the gel, whose location depends on the mass of the chain (for more details see Section 13.7.2). The relevant gel bands that correspond to the interacting partners of the target molecule can then be cut and broken into smaller peptides using enzymes called proteases, and the peptides are identified using mass spectrometry, as described in Section 12.6.1. The binding agent can be identified from the peptide spectrum.

The difference between the methods is in the technique used to pull the complexes out of a solution containing a cell extract or a blood serum.

13.2.1.1 Affinity chromatography

Affinity chromatography is a technique that has been around for a while. It works by attaching a target molecule with affinity to selected proteins to a solid surface (made of tiny beads) and then exposing it to the cell sample. Most proteins in the sample pass through and are washed away, while interacting proteins remain attached. This process is often referred to as a **pull down** experiment.

The experiment is conducted in two steps. To bind the target protein to the beads (which are usually made of polysaccharides and are rich with nickel ions) the protein is expressed with a his-tag (six histidines in a row) and then passed through the beads. The nickel atoms in the beads bind to the his-tag, and hence to the target protein as well. Proteins that bind to the target protein will remain attached once the sample is washed away and can be analyzed using gel electrophoresis followed by mass-spectrometry, as described above. As a control, the process is repeated with the cell extract but without the target molecules. Non-specific interactions involving proteins that attach directly to the nickel beads are ruled out in the gel electrophoresis phase.

Affinity chromatography was first proposed for detection of protein-protein interactions by [698]. The method is sensitive to weak interactions and its main advantage is that interactions are kept in their natural state and concentration in the cell. On the other hand, it is limited to proteins that can bind to the target protein. In addition, interactions detected with this method might not be direct, if the binding protein is part of a complex.

13.2.1.2 Co-immunoprecipitation

Immunoprecipitation has been traditionally utilized to isolate and identify protein antigens that bind to specific antibodies out of a cell extract. Immunoprecipitation is a special case of a pull down experiment, where the target protein is an antibody. The antibodies are attached to a solid surface and then added to a protein mixture from a cell extract. In the mixture, the antibodies and the antigens form an immune complex. This process of pulling out the protein by forming a solid in a solution is called **precipitation**. A different approach, which is sometimes preferred with proteins of low concentration or low affinity, is to first mix the antibody with the protein mixture, letting them bind, and only then expose the mixture to the solid surface. In a second step, the complex is separated from the solid surface (for example, by competitively binding a small molecule to the solid surface, which will result in the release of the complex) and analyzed using gel electrophoresis and mass-spectrometry.

Co-immunoprecipitation (CO-IP) is similar and works by using antibodies that target antigens that are believed to be part of a complex. The antibody interacts with the antigen and precipitates both the antigen and the protein complex attached to it. As with other methods, the interactions detected are not necessarily direct.

The method may fail if the **epitope** (the short segment of the antigen based on which the antibody recognizes and binds to the antigen) is covered by other members of the complex or if the members of the complex are not tightly bound to each other. Therefore, it might require repeating the procedure with several antibodies that can work on the same protein, using other epitopes. By repeating this procedure with antibodies that target other proteins in the same complex, it is possible to verify complexes, increase the reliability of the method, and detect new complexes that involve these proteins.

13.2.2 High-throughput methods

Experimental techniques such as affinity chromatography and coimmunoprecipitation are quite accurate but are low-throughput methods. In addition, since the experiments are run in vitro, they might falsely report interactions between proteins that do not interact in vivo, for example when the two proteins are located in different subcellular locations. High-throughput methods that use model organisms report interactions that are likely to occur in vivo and can detect interactions on a large scale.

13.2.2.1 The two-hybrid system

Two-hybrid systems have become very popular in recent years, in particular the **yeast two-hybrid** described in [685]. The technique is based on manipulation of a transcription factor that consists of two domains: a binding domain,

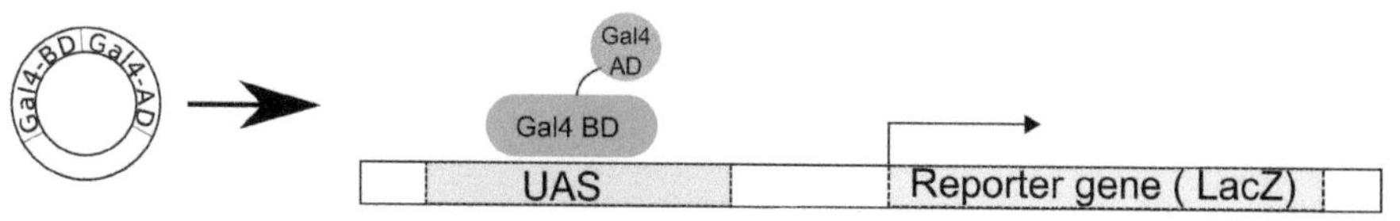

A. Regular transcription of the reporter gene

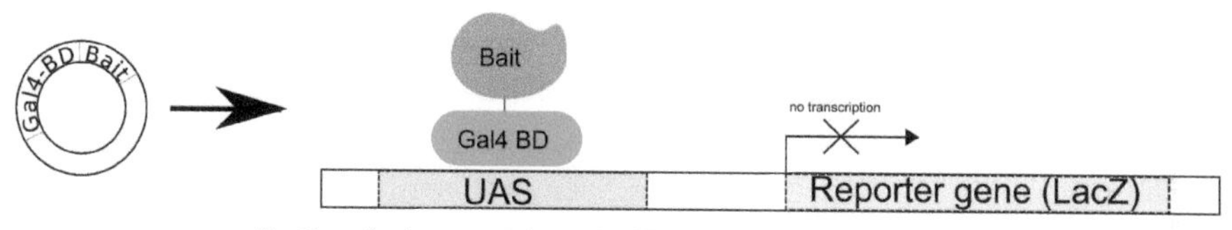

B. One fusion protein only (Gal4-BD + Bait) - no transcription

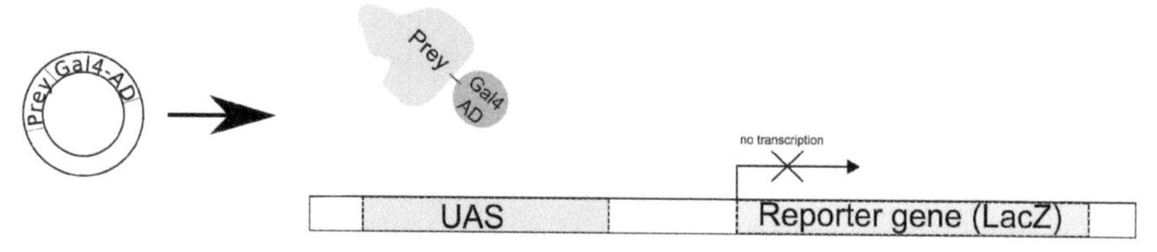

C. One fusion protein only (Gal4-AD + Prey) - no transcription

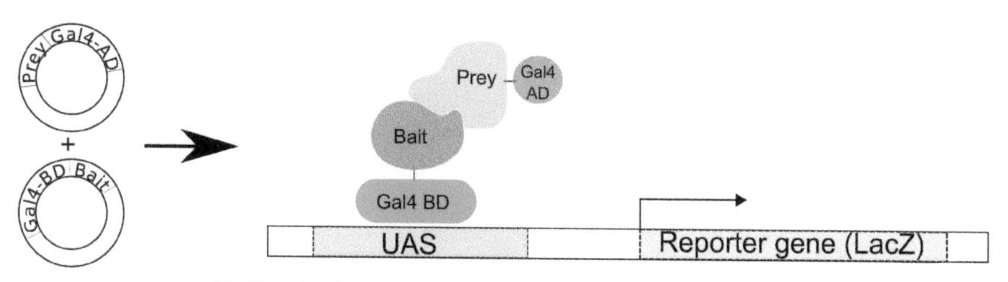

D. Two fusion proteins with interacting Bait and Prey

FIGURE 13.2: **The two-hybrid system.** (Figure reproduced from Wikipedia [699]. Original figure created by Anna K.) Two-hybrid assays check for interactions between two proteins, sometimes referred to as "Bait" (the target protein) and "Prey" (potential interacting protein). (A) The Gal4 transcription factor consists of two domains, a binding domain (BD) and an activation domain (AD); both are necessary for transcription of the reporter gene (LacZ). A fusion of the bait protein to the binding domain (B) or the prey protein with the activation domain (C) is not sufficient for transcription initiation of the reporter gene. (D) When both fusion proteins are synthesized and the bait protein binds to the prey protein, transcription of the reporter gene is initiated.

capable of binding to a DNA sequence, and an activation domain, capable of activating the transcription of DNA (see Figure 13.2). A *transcriptional activator* of a gene is a protein containing both a binding and an activation domain, where the binding domain is capable of binding to the gene's promoter region. The gene's transcription is initiated when its transcriptional activator binds to the promoter region.

This mechanism may be used for detection of protein-protein interactions by fusing the binding domain to a target protein and the activation domain to a potential interacting protein. Fusion is done by inserting the DNA sequences of the binding domain and the target protein to one plasmid and those of the activation domain and the potential interacting protein to another plasmid (more on plasmids in Section 13.7.1). In addition, a reporter gene is inserted to the host cell's DNA, following the promoter region to which the binding domain binds. A culture of host cells containing both types of plasmids is then created, and the interaction of the two proteins is checked through the reporter gene, which will be transcribed if the two proteins interact.

The yeast two-hybrid technique has several advantages, including its scale and relative speed compared to traditional methods and the fact that it is an in vivo technique, testing interactions under the actual conditions of the cell. On the other hand, the technique has a high false-positive rate, which may be as high as 50% [689, 700, 701]. Although the method is an in vivo technique, false positives may occur for example if the target protein and the potentially interacting protein are in physical proximity in the model system but are truly in different subcellular localizations in their natural setting.

The yeast two-hybrid method also has many false negatives. Not all proteins can bind to the activation or binding domains, although they might interact with each other. Even when binding to the activation and binding domains is possible, stereochemical collisions between the components or misfolding may still prevent the formation of the interacting complex. False negatives can also occur even if a complex is formed, owing to steric hindrances between the interacting complex and the DNA structure or RNA polymerase, which prevent transcription.

13.2.2.2 Tandem affinity purification

Tandem Affinity Purification (TAP [686]) is another method that uses a model system to detect interactions. It was initially developed for yeast but may be applied to other organisms as well. The method works by fusing in a plasmid a certain molecular tag, called TAP, at the C terminus of target proteins (by appending the tag's DNA sequence before the coding region of each target protein). The plasmid with the fused sequences is inserted into a host organism such as yeast, where the sequences can be expressed. After transcription and translation, the target proteins might interact with other proteins in the cell.

The TAP tag consists of a protein called protein A (a cell-wall protein that can bind well to certain antibodies), followed by a protease cleavage site and a calmodulin binding peptide. Being attached to the tag, any complex involving the target proteins can be pulled out from the cell extract using affinity chromatography, as described above, with the antibodies that act on protein A as the binding agent. The protein A is then cleaved out using proteases that target the cleavage site. In a second affinity step, the sample is washed through calmodulin binding agents to remove the protease. Lastly, the clean sample is analyzed by means of gel electrophoresis and mass spectrometry.

13.2.2.3 Protein arrays

We already mentioned arrays for protein identification in Section 12.6. These arrays are similar to mRNA arrays, but while DNA microarrays are mostly used to measure the dynamic aspect of cell's activity, protein arrays are mostly used to test interactions. Protein arrays are basically an extension of traditional methods such as affinity chromatography and co-immunoprecipitation, and they provide an in-vitro alternative to the yeast two-hybrid system. They are especially useful for interactions that cannot be detected by means of the transcriptional-activator system, such as interactions with integral membrane proteins, which are localized far away from the nucleus (where transcription occurs), or interactions between secreted (extra-cellular) proteins, which are excluded in yeast two-hybrid assays.

Variations over this technology can be used to detect a variety of other molecular interactions, including protein-DNA and protein-lipid interactions. For example, **ChIP-chip** (ChIP-on chip) [702, 703] combines Chromatin ImmunoPrecipitation (ChIP) with microarrays (chip). The first step involves immunoprecipitation of the protein-DNA complexes containing the protein of interest. In a second step, the DNAs are separated from the proteins and are identified by hybridizing them with a microarray of the genome under study. Another technology, called **ChIP-seq**, combines chromatin immunoprecipitation with fast sequencing platforms [704, 705], thus bypassing the need to use DNA chips and mitigating some of the problems associated with them (such as limited prob density and specificity).

Protein arrays are commonly applied to study the properties of receptor-ligand or enzyme-substrate compounds and can help to assess the binding capabilities of proteins to potential drugs.

13.3 Prediction of protein-protein interactions

Experimental methods for high-throughput determination of interactions are expensive and time consuming. They are also incomplete and inconsistent and have high error rates. Therefore, there is a growing interest in computational methods that can help verify inconclusive experimental results.

Furthermore, with the rapid increase in the number of interactions that were verified experimentally and the wealth of information on new genes, it is natural to assume that computational methods might be able to detect patterns that are common to interacting proteins and use these patterns to reliably predict *novel* protein-protein interactions.

Successful prediction of protein-protein interactions is instrumental to molecular design of therapeutic agents. It can help to assess the impact of mutations on interactions and possibly link them to diseases (such as the transmembrane conductance regulator gene CFTR and cystic fibrosis). It can also help in designing synthetic protein complexes with potential new functions.

The field of protein-protein interaction prediction via computational methods is very active, and many methods have been developed. In general, we distinguish between structure-based methods that use structural information on proteins and sequence-based methods that are based purely on sequence analysis. In the next sections we discuss the different methods in detail.

13.3.1 Structure-based prediction of interactions

Prediction of protein-protein interactions entails two tasks. Our first goal is to predict whether two proteins interact. The second is to determine and localize the interaction interface. If we are given that two proteins interact, then the problem of identifying the interaction sites is easier, although still challenging. However, in the general case we do not know whether the proteins interact, neither do we know whether an interaction site exists. Most structure-based methods address the first problem by tackling the second problem first, and look for possible interaction interfaces between the two input structures.

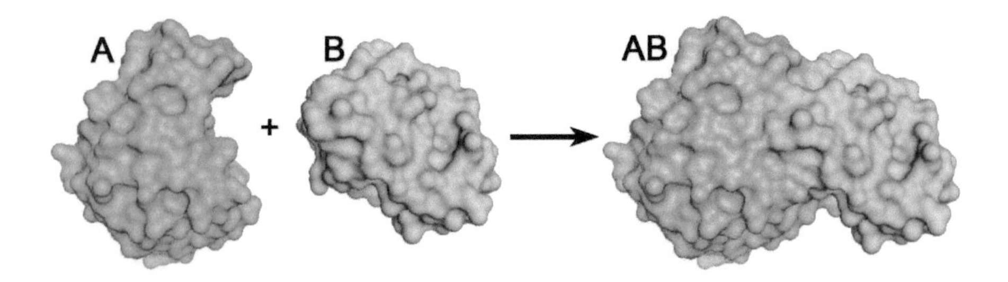

FIGURE 13.3: **Protein docking.** Molecular surface representation of PDB complex 1brs between a ribonuclease protein barnase (green, A) and its inhibitor barstar (blue, B). Barnase (for *Bacillus amyloliquefaciens* RNAse) is a toxic enzyme secreted by the bacterium *Bacillus amyloliquefaciens*, which breaks RNA indiscriminately. Its activity is controlled inside the bacterium's cell by the inhibitor, barstar. But once secreted it is lethal to neighboring cells.

Structure-based approaches that predict the interaction site are also useful for studying the properties of protein-ligand interactions[3] and are especially important for rational drug design. For example, if a certain protein is targeted, then a drug can be designed by studying interaction sites in the 3D structure of the target and identifying complementary molecules. We will discuss a few structure-based methods next.

13.3.1.1 Protein docking and prediction of interaction sites

Given the solved structures of two proteins, one approach to predict a possible interaction and the formation of a complex is to search for a combined conformation of both structures with minimal free energy that is lower than the sum of the free energies of the individual structures. However, the space of possible combined conformations is very large, and an exhaustive search is practically infeasible.

Identifying complementary shapes can greatly reduce the search space for configurations of protein complexes and increase the success of structure-based prediction [706]. To facilitate an interaction, two proteins should have compatible surfaces, both in shape and physico-chemical properties (e.g., residues with opposite charges or the potential to form hydrogen bonds). Based on this premise, **protein docking** methods analyze tertiary structures to see whether they are compatible with each other and predict the most probable binding site between the two (Figure 13.3).

The problem of identifying compatible shapes in 3D structures is actually NP hard [707], and existing methods rely on sampling the conformation space in one way or another. An efficient method for detecting compatible structures is to use a grid representation and compute the correlation between the positions of the structures in the grid, under different orientations, as described in [708]. Each structure **A** is projected on an $n \times n \times n$ grid and represented as a function $a(x, y, z)$ such that $a(x, y, z) = 1$ if position x, y, z of the grid is inside the molecule and 0 otherwise. A position is considered inside if there is least one atom within radius r from that position, where r is the average van der Waals radius of atoms[4]. In a second step, the function is redefined such that $a(x, y, z) = \rho$ for all grid points that are inside the molecule but not on the surface of the structure. A second structure **B** is processed in a similar way such that $b(x, y, z) = 1$ if x, y, z is on the surface of **B**, $b(x, y, z) = 0$ if outside the molecule and $b(x, y, z) = \delta$ if inside the molecule. The constant ρ

[3]**Ligand** is a general term referring usually to small molecules that bind to other molecules (such as receptors).

[4]Atoms are often modeled as spheres encompassing the nucleus and the cloud of electrons orbiting around the nucleus. Although atoms are somewhat compressible, they are usually modeled with hard spheres, and the van der Waals radius is the radius of the sphere. It changes from one atom to another and ranges between 1.2 Å (hydrogen) and 1.7 Å (carbon) for the atoms that make up amino acids. The minimal distance below which two atoms with radii r_1 and r_2 start to repulse each other is $r_1 + r_2$.

is set to a large positive value while the constant δ is set to a large negative value, such that their product results in a negative value, as will be explained next.

The complementariness of the structures is assessed by computing the correlation function between the two point sets for each possible translation vector $\vec{t} = (\alpha, \beta, \gamma)$

$$c(\alpha, \beta, \gamma) = \sum_{x=1}^{n} \sum_{y=1}^{n} \sum_{z=1}^{n} a(x, y, z) \cdot b(x + \alpha, y + \beta, z + \gamma) \tag{13.1}$$

If there is no contact between the two structures the correlation score will be zero, while if the structures penetrate each other then the total correlation score will be negative. The best scenario is when the two structures touch each other only at surface positions, and the higher the correlation score the larger the interface between the two. In practice, good mappings can be produced also when the contacts at the surface outweigh those that result in penetration.

A straightforward computation of the correlation score is expensive since there are n^3 possible translations, and computing the score for each one takes $O(n^3)$ operations. However, by using the Discrete Fourier Transform (DFT) it is possible to compute these much more efficiently. The DFT of a 3D function $f(x, y, z)$ is defined as

$$F(a, b, c) = \sum_{x=1}^{n} \sum_{y=1}^{n} \sum_{z=1}^{n} \exp[-2\pi i(ax + by + cz)/n] \cdot f(x, y, z)$$

where $i = \sqrt{-1}$. One of the useful properties of the DFT involves convolutions. A convolution of two discrete 1D functions, $f(x)$ and $g(x)$, is a transformation of the two functions into a third function $h(\alpha)$ defined as

$$h(\alpha) = (f * g)(\alpha) = \sum_{x} f(x)g(\alpha - x)$$

The notation $f * g$ denotes convolution rather than multiplication. For 3D functions, the definition is generalized by summing over y and z. According to the convolution theorem, the DFT of a convolution $f * g$ satisfies

$$F(f * g) = F(f) \cdot F(g)$$

That is, the Fourier transform of the convolution function can be computed by applying the Fourier transform to the two component functions and multiplying the values.

Note that the definition of the correlation function in Equation 13.1 is almost identical to the definition of the convolution of the functions a and b. That is,

$$h(\alpha, \beta, \gamma) = (f * g)(\alpha, \beta, \gamma) = \sum_{x=1}^{n} \sum_{y=1}^{n} \sum_{z=1}^{n} a(x, y, z) \cdot b(\alpha - x, \beta - y, \gamma - z)$$

and the only difference is in the sign. Similar to the convolution theorem, the DFT of the correlation function can be written as the product of the DFTs of the component functions

$$C(a, b, c) = A^*(a, b, c) \cdot B(a, b, c)$$

where C is the DFT of the correlation function c, A and B are the DFTs of the structure functions a and b defined above and

$$A^*(a, b, c) = \sum_{x=1}^{n} \sum_{y=1}^{n} \sum_{z=1}^{n} \exp[2\pi i(ax + by + cz)/n] \cdot a(x, y, z)$$

is the complex conjugate of the DFT of a, which is obtained by reversing the sign of the imaginary part in the complex number. As such, the DFT of the correlation score, $C(a, b, c)$, can be computed by computing the DFTs of the component functions and multiplying them.

Given $C(a, b, c)$ we can then apply the inverse Fourier transform to obtain the original correlation score

$$c(\alpha, \beta, \gamma) = \sum_{a=1}^{n} \sum_{b=1}^{n} \sum_{c=1}^{n} \exp[2\pi i(\alpha a + \beta b + \gamma c)/n] \cdot C(a, b, c)$$

The real speedup is obtained by applying the fast Fourier transform (FFT), which can compute the DFT values and their inverse for all translation vectors in one run, in only $O(n^3 \ln n^3)$ steps. If there is a strong correlation for some translation vector, it can be identified quickly by scanning the FFT scores and picking the shift with the highest score. This clearly depends on the relative orientation of the two structures, and an arbitrary orientation is unlikely to produce a significant match. To find the optimal rotation the rotation space is scanned in fixed intervals of α degrees, and the procedure is repeated $(360/\alpha)^3$ times for each rotation. Exclusions can be made if there is prior knowledge of binding sites in one of the proteins (which is the case for many antibodies). Other approaches for detecting complementary shapes include the geometric hashing method described in Chapter 8 [709], genetic algorithms [710] and others (see Further reading).

High scoring orientations are indicative of complementary structures, geometrically. But a geometrical match is insufficient to indicate a potential interaction. The complementary sites should also be compatible in their biochemical properties. The compatibility can be assessed energy wise, looking for matches of minimal energy. However, modeling the exact underpinnings of molecular interactions is difficult, since there are multiple forces at play and a cumulative effect of neighboring residues and surrounding molecules, which is harder to quantify and model. Knowledge-based force fields or mean force potentials are an alternative approach (referred to as **knowledge-based potentials**), based on frequencies of residue pairs in interaction sites [711]. They are derived from experimental data on solved structures of protein complexes and are based on the principle that frequent structures must be energetically favorable. This is established in the principle known as Boltzmann's principle, which relates probabilities with energy. An empirical mean force potential between amino acids a and b at distance r is defined as

$$E_{a,b}(r) = -kT \ln f_{a,b}(r)$$

where T is the temperature, k is Boltzmann's constant, and $f_{a,b}(r)$ is the relative frequency of residues a and b at that distance. If the residues are close in sequence, then another parameter that is often introduced is the sequence separation. This potential reflects all forces that act between the residues (including electrostatic and van der Waals interactions), as well as the influence of the surrounding medium (such as water molecules). Essential to the definition of a mean-force potential is the reference state, which is the average level of interaction in the system over all possible residue types

$$\bar{E}(r) = -kT \ln f(r)$$

where $f(r) = \sum_{a,b} f_{a,b}(r)$. Given the reference state, the energy associated with a specific interaction is defined as

$$\Delta E_{ab}(r) = E_{a,b}(r) - \bar{E}(r) = -kT \ln \left(\frac{f_{a,b}(r)}{f(r)} \right)$$

Over the years many variations of this idea were proposed. For example, one such variation is to define the empirical potentials based on the frequencies of all neighboring residues within a certain threshold distance, such that

$$s(a,b) = \ln \frac{f_{ab}}{e_{ab}}$$

where f_{ab} is the fraction of observed residue pairs of type a and b and e_{ab} is their expected frequency $e_{ab} = p_a p_b$, where p_a and p_b are estimated based on their frequency in interaction sites in general.

Protein docking methods may fail in the presence of structural deformations at the time of interaction. If the changes involve only side chains, then their orientation can be tuned to minimize the total free energy [712, 713]. More substantial deformations can be addressed by allowing flexible docking and searching for the optimal configuration, considering local adjustments (re-configurations) after binding [714, 715]. However, predicting the changes between the bound and unbound states remains one of the most difficult problems in structure prediction.

A different approach toward predicting the interface is based on analysis of residue conservation among surface residues [716, 717]. The assumption is that residues in interaction sites are more conserved than other surface residues (which usually tend to be more versatile and less conserved than core residues). Indeed, it has been shown that these residues, sometimes referred to as **hotspots**, are important for the stabilization of protein complexes and often contribute most of the binding energy for the interaction [718–720]. Conserved positions can be identified by analyzing MSA positions associated with protein structures using measures such as the entropy function or other methods similar to those discussed in Section 9.2.4. Point mutations that are known to affect the interaction can further help to determine functional residues that might be part of the interaction site. However, conserved residues and destructive point mutations appear also outside of the interaction sites, which makes the prediction just based on patches with highly conserved residues difficult and unreliable. Nevertheless, they can reduce the search space for docking methods by suggesting candidate interaction sites.

Interaction sites can be characterized in terms of other properties, determined by analysis of residue contacts in complexes whose structures are known. It has been observed that the frequencies of amino acids in interaction sites is distinct [721–723]. For example, proline and arginine residues are more likely in these sites, and many sites consist of hydrophobic segments. Characteristic residues can be predictive of interfaces, given protein structures [724, 725]. However, different studies are sometimes contradictory, and the results depend on the dataset used and on the definition of the interaction interface (a patch of surface residues vs. contact residues only). For example, some have found that charged pairs are under-represented in interaction sites [726], while others detected the opposite.

In summary, methods using protein tertiary structure have had some success and have the advantage that they can predict proteins that can physically interact. On the other hand, they are limited to proteins of known structures and are relatively slow. They might also fail to distinguish between two proteins that have the biochemical potential of interaction and two proteins that physiologically interact.

13.3.1.2 Extensions to sequences of unknown structures

All the methods described above can also be applied to sequences of unknown structures, if the structures can be predicted reliably. However, as we discussed in Chapter 8, tertiary structure prediction is a difficult problem and existing methods have limited success. Structure prediction is more successful if there is a close homolog with a solved structure and a model is created based on sequence similarity (homology modeling).

One of the more effective methods for predicting interactions based on predicted structures is by using solved structures of protein *complexes*. Homologs of the complex's components are expected to interact as well, depending on the extent and the degree of the similarity. Such predictions can "transfer" interactions from one organism, where the complex was solved, to other organisms, although the contact residues and the affinity of the interaction may be different. Quantifying the degree of confidence in such predictions is difficult since it is hard to evaluate the effect of mutations on the interaction site. The homologs can be modeled structurally based on their sequence similarity with the complex's components; however, modeling the loops in the interaction site and the conformational changes to the backbone can be challenging. The overall compatibility of the homologs in the interaction site can be evaluated using empirical potentials derived from known interaction sites, such as those discussed above. The total score of a hypothesized interaction site for a new pair of proteins is the sum of scores for all residues that are in contact, and its significance can be assessed in terms of a zscore based on the distribution of scores for randomized sites [727]. The method is more successful on close homologs (more than 30% identity) than it is on remote homologs.

Protein threading methods can further extend predictions to remote homologs with undetectable sequence similarity. For example, the "Multiprospector" algorithm [728] threads sequences in a library of monomer structures that are known to participate also as part of dimer structures. If two sequences have significant signal with respect to two chains that are part of the same complex, the sequences are re-aligned (considering the other sequence in the template structure) and the energy between the interacting residues is computed. If the zscore of the dimer is significant compared to that of the monomers, the sequences are predicted to interact. However, this method is limited to solved complexes of interacting proteins, of which only a few hundred are known in the protein data bank.

13.3.2 Sequence-based inference

Sequence-based methods utilize prior information on interactions to extrapolate known interactions to similar proteins and learn patterns that can be used to predict the existence of *novel* interactions. Sequence-based methods often predict functional interactions rather than physical interactions, but they are simple and among the most successful in the field of computational protein-protein interaction inference.

13.3.2.1 Gene preservation and locality

Gene preservation is the most natural way of extrapolating known interactions from one organism to another. The method predicts interactions by looking for homologs of interacting proteins in other organisms. The underlying assumption of gene preservation is that if two proteins interact to perform a vital biological function, then both proteins will be passed on during speciation. Indeed, many interactions are conserved across species, in particular interactions with functions such as protein translation, ribosomal structure, DNA binding and ATP metabolism. The confidence or reliability of prediction depends on the level of the similarity between the homologous genes.

Homologs that are not similar along the whole sequence might not form stable interactions, as the differences might exclude the interaction site or might result in different tertiary structures. Closely related genes with significant similarity along the whole sequence are good candidates, but as sequences diverge their function might change too.

The **phylogenetic-profile** method [729, 730] is an extension of the gene preservation idea and is based on the same assumption that interacting proteins are inherited together during speciation events due to strong selective pressure. Therefore, if one is present in a certain genome then the other one is expected to be present as well. In other words, these proteins are expected to have similar phylogenetic profiles, composed of those organisms where their homologs are present (see Figure 13.4). To build such a profile, we need a set of complete genomes. Therefore, this kind of analysis can predict interactions only in genomes that have been fully sequenced and have had all their genes identified

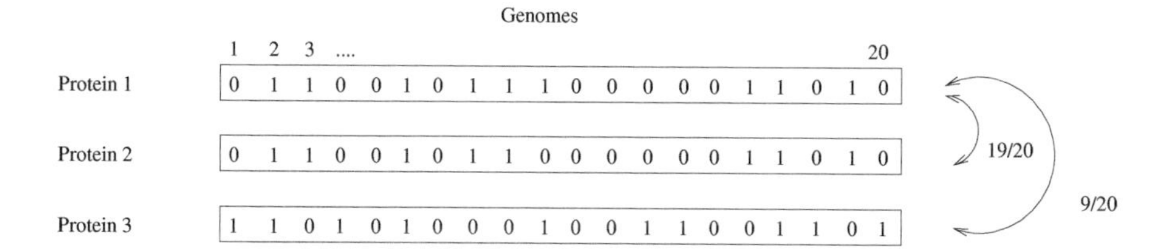

FIGURE 13.4: **Phylogenetic profiles.** In a phylogenetic profile each protein is represented as a binary vector, where each coordinate is associated with one organism and its value is 1 if the gene exists in that genome. In the example above, the phylogenetic profile of the first protein is very similar to that of the second protein (with 19 identical entries out of 20) but is quite different from that of the third protein (only 9 out 20 identical entries).

Since proteins are expected to exist in multiple genomes regardless of whether they interact or not, it is important to assess the prior probability that two proteins appear in the same subset of organisms. This probability can be estimated empirically based on the background distribution of profile similarity scores over the set of genomes used. Pairs that have a high score and low probability are candidates for interactions. However, certain proteins are crucial to sustain life and are therefore expected to occur in almost all organisms (e.g., certain transcription factors, or DNA and RNA polymerase proteins, which are essential for transcription). Clearly, not every protein in this subset interacts with every other protein, yet they all appear in coordination. Since it is difficult to reject the existence of an interaction, it is hard to assess the reliability of such predictions, and only a fraction were verified experimentally on well-studied genomes.

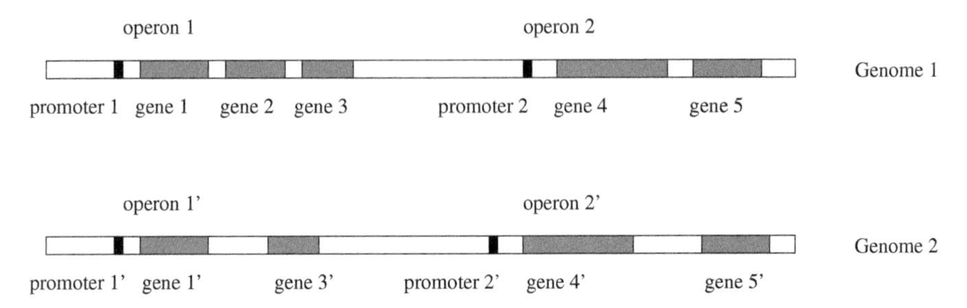

FIGURE 13.5: **Gene locality.** Gene 1 and gene 3 preserve their proximity in the second genome and are therefore predicted to interact. So are genes 4 and 5.

A related approach, which works only in prokaryotes (such as bacteria and archea) and simple eukaryotes (such as nematodes), is the **gene locality** method. In bacterial genomes, genes are often organized into **operons**; a group of genes that appear in sequence following a promoter region, such that they are all controlled and regulated by the same transcription factor (Figure 13.5). Once on, the transcription factor activates the transcription of all genes in the same operon, and hence they tend to be co-expressed. It has been hypothesized that co-expression is indicative of interaction since interacting proteins must be available at the same time to form an interaction (see also Section 13.3.3). The operon structure is an efficient mechanism for synchronizing expression, and examination of well-studied genomes confirmed that genes that are part of the same operon are very likely to interact or be involved in the same metabolic or signaling pathway [731, 732]. The gene locality method predicts interactions by looking for genes that are in close physical proximity along the DNA of one organism and that preserve that proximity in other organisms. This does not seem to hold in higher organisms, and interacting genes are no more localized than expected by chance.

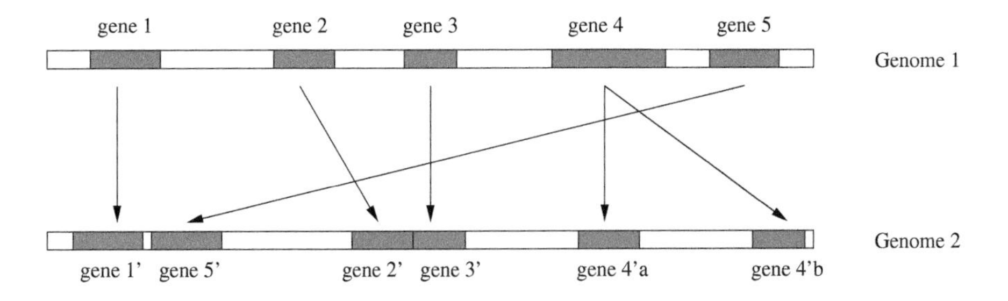

FIGURE 13.6: **Gene fusion.** Genes 1 and 5 fuse to a single gene in the second genome, and so are genes 2 and 3. Gene 4 is a composite protein that splits into two domains. All three instances are presumed to represent interacting proteins.

The relation between protein interactions and gene locality is also the basis of a technique termed **domain fusion** [733, 734]. That is, if two domains A and B exist in certain genomes as individual genes but are adjacent to each other or fused to a single gene $A'B'$ in another genome (Figure 13.6), then it is likely that the two are part of the same process (functional interaction) or are physically interacting. The method is sometimes referred to as the "Rosetta stone" approach, since a fused sequence $A'B'$ "deciphers" an interaction between A and B. The underlying assumption is that this scenario is more favorable thermodynamically, since the interacting partners do not need to find each other in the cell, which increases the chances of an interaction and reduces the reaction time and the quantities the cell needs to produce of each protein to carry out the interaction at the required rate. It has been hypothesized that this could be especially efficient for channeling in metabolic pathways or signal transduction pathways where multiple enzymes (e.g., kinase cascades) can form a complex with substrates passing from one enzyme to another directly rather than after diffusing in the cytosol at large. However, interacting proteins can also be the result of a "composite" protein evolving into "component" proteins [734], which is harder to justify based on thermodynamics alone.

Sequence-based methods based on gene preservation and locality can be successful, but only for interactions of a similar nature or when there is evidence of gene fusion/decomposition events in sequence databases. Such evolutionary circumstances are not the case for all protein-protein interactions. Note that all these methods rely on gene annotations and are therefore sensitive to the method used to identify genes along genomes. Determining the presence of an homologous gene can be difficult, especially when the similarity is weak or partial or when there are multiple similar genes in the same genome (paralogs). Moreover, as sequences diverge they might change their function as well. Sometimes even a single mutation can have a drastic impact on the function, for example if it occurs in the active site or the interaction site.

13.3.2.2 Co-evolution analysis

The basic assumption of the co-evolution principle is that two proteins that interact will tend to co-evolve in a coordinated manner resulting in a higher evolutionary correlation between their corresponding homologs. (We have already mentioned this principle in Section 9.2.2 when discussing methods for predicting protein domains, but in a local, position-specific context.) The intuition behind this premise is fairly simple; if one partner in an interaction mutates, then its counterpart will have to adapt in order to preserve the interaction. Therefore, given two query proteins and their homologs one can theoretically predict an interaction if there is evidence that the groups co-evolve.

The principle of co-evolution seems quite convincing and natural in the context of the interaction site (as will be discussed in the next section). However, the interaction site is an unknown parameter when the two query proteins are only hypothesized to interact. Even for known interactions, the interaction site is often unknown. Nonetheless, beyond adaptation in the interaction site, one would expect interacting proteins to have similar *overall rates* of mutation in the course of evolution [735]. Assuming that the general structure of each interacting protein has to be preserved to maintain a structurally and functionally active interaction site, it is less likely that the two proteins will evolve at significantly different paces. This is not true for different protein families in general, as different families have different **molecular clocks**, some exhibiting faster mutation rates than others[5], where rapidly changing proteins might undergo significant changes in conformation.

This is the motivation behind **mirror trees** methods, which compare homologs of interacting proteins (i.e., protein families) and their phylogenetic trees: if the two trees are very similar, then it is assumed that the proteins have co-evolved and possibly interact [741, 742]. Unlike the sequence-based methods of the previous section, which are based on some simple and ad hoc rules, the co-evolution principle implies a certain process that can be modeled mathematically. We will discuss this method in more length as this is an interesting exercise in translating a principle to a working model, which requires attention to detail.

The mirror tree method relies on knowledge of the underlying phylogenetic trees. However, in practice, the exact evolutionary path of different proteins is unknown, and reconstructing phylogenetic trees for a set of homologous proteins is an NP-complete problem [122]. Usually, the trees are approximated by applying hierarchical pairwise clustering algorithms (see Section 4.2.2) to the set of sequence similarities. An alternative to phylogenetic trees is the **correlated divergence** approach, which compares the distance matrices of the two protein families that are hypothesized to interact (Figure 13.7). This approach avoids the construction of trees by comparing the divergence rates between *all* pairs of homologs instead. If the matrices are strongly correlated, then it is presumed that members of the two families co-evolved and potentially interact with each other, either directly (physically) or as part of the same complex.

Correlated divergence methods are composed of the following three steps: given two query proteins (i) identify their homologs (family members) in a common set of n organisms O_{common}, (ii) construct distance matrices for the two families and (iii) measure the correlation between the two matrices.

[5]There are conflicting reports regarding the correlation between a gene's essentiality and its mutation rate [736–738]. However, it has been shown that the rate can depend on the function of the protein (e.g., proteins that are part of the immune system often exhibit rapid mutation rates, enabling fast adaptation to pathogens and attacks from ever-mutating viruses [736, 739]) as well as on the number of interactions the protein forms [740].

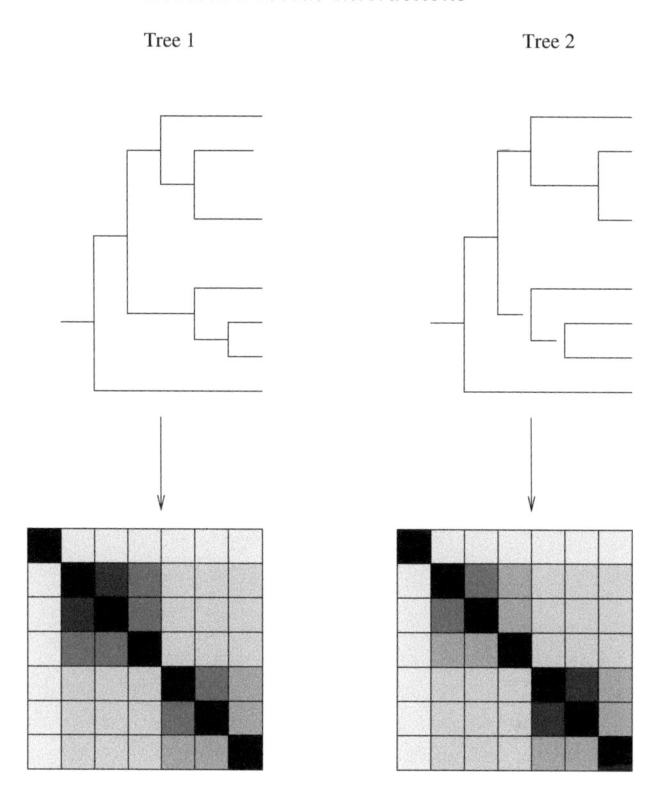

FIGURE 13.7: **Mirror trees and distance matrices.**

To compute the distance matrix we need a measure of divergence between proteins. One possible measure is the average number of mutations $d(p_1, p_2)$ in the sequence alignment of p_1 and p_2, normalized per 100 residues to prevent biases due to different lengths. While this measure is not as sensitive as the measures we discussed in Chapter 3, it is simple to implement and compute. Moreover, this measure qualifies as a distance metric, since it is symmetric, non-negative and it satisfies the triangle inequality. This property is necessary to accurately estimate distances between entities that are evolutionarily more distant. It is also a necessary condition for constructing a phylogenetic topology.

The correlated divergence is estimated by computing the Pearson correlation coefficient r between the mutation levels of organism pairs in family F_1

and family F_2:

$$r = \frac{\sum_{i=1}^{n-1} \sum_{j=i+1}^{n} (D_1(o_i, o_j) - \mu_1)(D_2(o_i, o_j) - \mu_2)}{\sqrt{\sum_{i=1}^{n-1} \sum_{j=i+1}^{n} (D_1(o_i, o_j) - \mu_1)^2} \sqrt{\sum_{i=1}^{n-1} \sum_{j=i+1}^{n} (D_2(o_i, o_j) - \mu_2)^2}} \quad (13.2)$$

where $n = |O_{common}|$ is the number of organisms in the common set, $D_k(o_1, o_2)$ is the divergence between protein pairs in organisms o_1 and o_2 in family F_k

$$D_k(o_1, o_2) = d(p_1, p_2) \qquad p_1 \in F_k(o_1) \quad p_2 \in F_k(o_2)$$

and

$$\mu_k = \frac{2 \sum_{i=1}^{n-1} \sum_{j=i+1}^{n} D_k(o_i, o_j)}{n(n-1)}$$

is the average divergence observed in family F_k. The time complexity of this method is approximately $O(mn^2)$, where m is the length of the longest pairwise alignment. The correlation coefficient ranges in value from -1 (anti-correlation) to 1 (perfect correlation). A value of 0 indicates no correlation. If two protein families co-evolve, these mutation levels should correlate, indicating similar protein clocks.

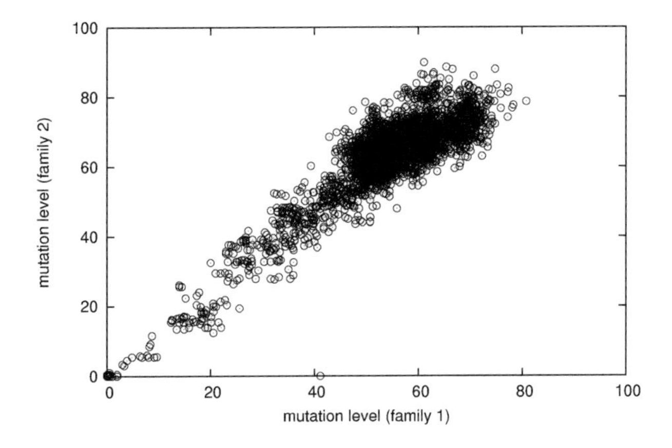

FIGURE 13.8: **Correlation plot between interacting proteins.** (Figure reproduced from [697] with permission.) Functional interaction between Biozon 269-168 (60S ribosomal protein L9, mitochondrial precursor YmL9) and Biozon 286-147 (60S ribosomal protein YmL6, mitochondrial precursor) shows strong signs of correlated divergence ($r = 0.90$). To view the Biozon profile page of a protein with an identifier x follow the URL biozon.org/db/id/x.

While this method is successful for some specific examples (see Figure 13.8), it fails on other cases where the correlation signal is too weak (Figure 13.9). False positives are abundant too, since the pearson correlation score can be sensitive to outliers (see Section 13.8). A more robust measure is the zscore of the correlation score. The zscore is obtained from the mean and standard deviation of the background distribution of scores (see Section 3.4.2.3)

$$zscore(r) = \frac{r - \mu}{\sigma}$$

where the background distribution can be generated by permuting the entries in one of the distance matrices and recomputing the correlation score for each permutation. The zscore is less sensitive to outliers, and true correlation signals over large enough dataset can result in significant zscores. However, it is computationally intensive and it does not resolve other issues with the mirror-tree method.

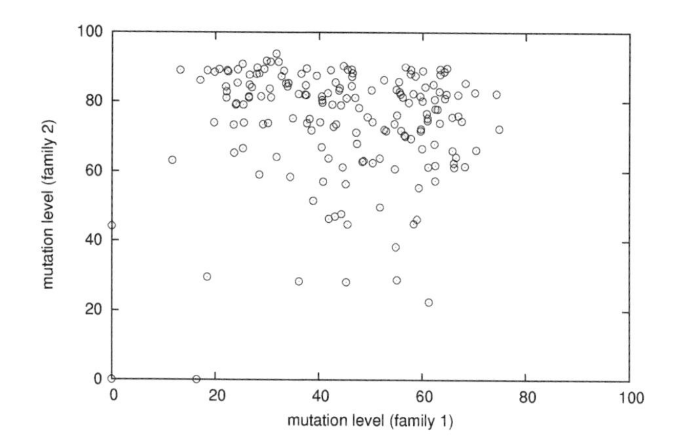

FIGURE 13.9: **Interacting proteins do not necessarily co-evolve.** Low correlated divergence score ($r = 0.03$) is assigned to the interaction between Biozon 318-2190 (TATA box binding protein) and Biozon 1230-9 (TBP-interacting protein 120A). (Figure reproduced from [697] with permission.)

One problem, which is actually common to many other sequence-based algorithms for prediction of protein-protein interactions, is the presence of paralogs. Often, there are multiple homologs of a protein family in the same organism, and a priori it is unclear which ones if any should be considered in the analysis. In the correlated divergence model we can circumvent this problem by estimating the family-specific divergence between o_1 and o_2 as the average over all protein pairs in the cross product of $F_k(o_1)$ and $F_k(o_2)$

$$D_k(o_1, o_2) = \frac{\sum_{p_1 \in F_k(o_1), p_2 \in F_k(o_2)} d(p_1, p_2)}{|F_k(o_1)||F_k(o_2)|}$$

With this definition, computing the correlated divergence between the two families implies a total of

$$N = \sum_{i=1}^{n} |F_1(o_i)||F_2(o_i)|$$

interactions. However, it is unlikely that in each organism all paralogs from one family interact with all paralogs of the other family. Rather it is more likely that each paralog is "tuned" to perform a different function (see also Section 14.3.2.2) and the maximal number of expected interactions is on the order of n. Moreover, after speciation an interacting pair might mutate and lose its ability to interact. Therefore, the interaction is not even guaranteed to exist in all organisms in the common set.

This protein multiplicity weakens co-evolutionary signals, since many of the pairs considered are not interacting and therefore are likely to evolve without explicit co-evolution constraints. If the total number of pairs considered in the analysis N is much larger than n, then the majority of pairs is only indirectly constrained, and co-evolution signals from interacting proteins can quite easily be overwhelmed by data from the numerous non-interacting pairs. Therefore it is important to minimize the dataset to the minimal set of truly interacting proteins.

Identifying which pairs actually interact is not a trivial problem. Assuming that interacting proteins are likely to mutate less (in order to preserve the interaction) than their paralogs that are not involved in an interaction, we can target the group of orthologous proteins that are most closely related in evolution. A possible simple solution is to pick in each organism the protein closest to the query protein (as in [742]). The Reciprocal Best BLAST Hit (RBH) algorithm [737, 743] uses a stricter approach to identify orthologous proteins. Given two proteomes A and B, RBH considers proteins $a \in A$ and $b \in B$ as orthologs if b is the first hit in a BLAST search in which a is the query and B is the database, and a is the first hit in the search of b against A. However, the algorithm may falsely reject true orthologs since both proteins must be first in each other's list in order to be considered as orthologs.

Both algorithms might produce suboptimal sets, depending on the query proteins. An alternative approach, which is applicable to small datasets, is to select the subset of proteins (one from each organism in O_{common}) that minimizes the total sum of pairwise distances

$$D(p_1 \in F_k(o_1), ..., p_n \in F_k(o_n)) = \sum_{i=1}^{n-1} \sum_{j=i+1}^{n} d(p_i, p_j)$$

by exhaustive enumeration of all possible combinations. However, for large datasets the number of possible combinations can be prohibitive, and exhaustive enumeration is impossible. An alternative solution is to employ a local hill-climbing search algorithm, starting from a random subset containing one protein from each organism and iteratively updating the chosen protein from one organism at a time based on its distance from the other members of the subset.

A more direct approach than the minimal distance is to identify the subsets of proteins (one from each family) that maximize the correlated divergence score. As with the minimal distance, the size of typical datasets precludes exhaustive search. An approximate locally maximal solution can be found by sampling the search space [744, 745] or by considering only subsets with similar tree topologies in both families [746]. Another possible solution is to use a hill-climbing procedure, similar to the one described above. Starting from a random subset, or a set that was produced with another method, the algorithm iterates through all organisms until the subset remains unchanged and the correlation scores stop improving. This procedure resembles expectation maximization procedures (Section 5.7), with the hidden variables being the indicator variables that specify whether a protein interacts with another protein. The maximization step selects from one organism at a time the pair of proteins that maximizes the co-evolution signal, given the remaining proteins in the subset. The expectation step is essentially the averaging over all pairs in all other genomes[6]. One can use a similar EM-like procedure to select the subset of organisms that maximizes the correlation score, under the assumption that the interaction is more likely to occur in these organisms.

The second and more major issue with mirror tree and correlated divergence methods is that they can confuse correlation due to interaction and co-evolution, with correlation due to similar evolutionary trees in general. Naturally, any two sets of homologous proteins over the same set of organisms have the same true phylogenetic tree structure with similar distance matrices. The basic correlation between these distance matrices may shadow the more subtle signals of co-evolution. The signal can be enhanced by subtracting information about the phylogenetic relations of the proteins represented in the distance matrix [747]. While determining the exact phylogeny is a difficult problem, we can approximate it with distance matrices of protein families that are present in all organisms, are fairly conserved and are presumed to be free of horizontal gene transfer (segments that were inserted into a host genome from bacterial or viral genomes). Ribosomal proteins are often the first choice, as they satisfy these conditions. To adjust for the different clocks in general, the matrices should be scaled before subtracting them from the distance matrices of the proteins under study, based on the ratio of distance

[6]Formally, this procedure is not exactly an expectation maximization algorithm; however it is inspired by the EM algorithm and is therefore referred to as an EM-like procedure.

values [748].

However, despite all these enhancements, signals of co-evolution are fairly weak when comparing the distributions of correlated divergence scores for interacting proteins vs. non-interacting proteins [697], and one should set the expectations accordingly.

13.3.2.3 Predicting the interaction interface

Predicting potential interaction interfaces from sequence information alone is a difficult problem. When structural information is available on the putatively interacting partners, the search for interaction sites can be confined to local (spatially close) patches of surface residues in each protein (as in Section 13.3.1.1). However, a sequence-based search for "hot spots" that have the characteristics of interaction sites has to consider all positions in both candidate proteins, since residues that are located close to each other in the 3D structure can be far apart in sequence. This is not only computationally hard but can also lead to false predictions, as sequence positions that contain residues characteristic of interaction sites are not necessarily part of an interaction site or in structural proximity to each other. In the next subsection we will discuss a few heuristics for identifying local sites that potentially can be interaction sites.

Charge and hydrogen-bonding potentials: Vital to any protein-protein interaction is a biochemically feasible binding site. In order for two proteins to interact, they must be able to physically bond. The types of *atomic* interactions that underlie interaction sites are electrostatic interactions (also called ionic or salt-bridges), hydrogen bonds and hydrophobic interactions (see Figure 13.1). Table 13.2 lists the properties and simple binding potentials for each of the 20 amino acids. Given two protein sequences $\mathbf{a}, \mathbf{b}$ of length L_1 and L_2 we can generate an $L_1 \times L_2$ matrix and quantify the binding potential $e(a_i, b_j)$ for each pair of positions in the matrix, using residue-residue potentials as in Table 13.2. There is no point in trying to compute the exact energy between two candidate sites, as the relative placement of residues is unknown. However, using these potentials we can identify sites that are rich with residues capable of forming an atomic interaction.

If homologs of the two sequences are known too, then an MSA of each family would be a better starting point. The binding potential between two positions in that case is a function of the residues observed in these positions in the MSAs and can be defined as the average

$$E(i,j) = \frac{1}{|O_{common}|} \sum_{o \epsilon O_{common}} e(a_{oi}, b_{oj})$$

where O_{common} is the set of organisms that are preserved in position i of family F_1 and position j of family F_2, and a_{oi} is the residue in position i of

Amino Acid	Charge	Hydrogen bonds	Hydrophobic
Alanine	0	None	Yes
Arginine	+	Acceptor	No
Asparagine	0	Acceptor	No
Aspartic acid	-	Donor	No
Cysteine	0	Acceptor, Donor	No
Glutamine	0	Donor	No
Glutamic acid	-	Donor	No
Glycine	0	None	Yes
Histidine	+	Acceptor, Donor	No
Isoleucine	0	None	Yes
Leucine	0	None	Yes
Lysine	+	Acceptor	No
Methionine	0	Donor	Yes
Phenylalanine	0	None	Yes
Proline	0	None	Yes
Serine	0	Acceptor, Donor	No
Threonine	0	Acceptor, Donor	No
Tryptophan	0	Acceptor	Yes
Tyrosine	0	Acceptor, Donor	Yes
Valine	0	None	Yes

TABLE 13.2: Charge, hydrophobic interactions and hydrogen bonds. A simple scoring function can be defined based on these properties of amino acids, where opposite charges, pairs of hydrophobic residues or donor/acceptor pairs are scored +1. Similarly charged or pairs of donor/donor and acceptor/acceptor are assigned a score of -1. So are pairs of hydrophobic and hydrophilic residues. Otherwise the score is set to 0. These scores are not energy potentials, since different pairings contribute differently to the binding energy (see Section 13.3.1.1). They merely serve to indicate the potential of binding.

the sequence in organism o. Similarly, b_{oj} is defined with respect to position j in family F_2. When the MSAs have multiple paralogs from the same organism then $e(a_i(o), b_j(o))$ can be computed as the average over all possible pairs within the same organism. This multiplicity however might weaken signals of binding potential, as discussed in Section 13.3.2.2.

As mentioned above, the interaction sites can contain residues that are far apart in sequence. Since the 3D structures of the proteins are unknown, it is hard to verify which pairs are in structural proximity to each other, a necessary condition to form an interaction site. However, for the most part it is residues along local secondary structures that make up the interaction interface (see Figure 13.1). These sites tend to take on the form of a "hot spot" in this matrix. Therefore, possible interaction sites can be identified by searching locally for sub-matrices (**bounding boxes**) with a high ratio of positions that can form bonds. The search can be implemented using a sliding box of size $l \times l$ with a shift of 1, where l is a user-defined parameter. The highest scoring boxes can be tuned and resized to maximize the ratio or the score, as discussed ahead. While this approach does not guarantee a global optimum or the most significant bounding box, it can identify good candidates.

Correlated mutations: The correlated mutations method is a variant of the co-evolution model. As opposed to correlated divergence methods that compare general divergence patterns across species in two protein families, this approach searches for correlated, compensating mutations in specific positions between two candidate interacting proteins. The assumption is that such mutations are characteristic of binding sites since binding sites have stronger selective pressure for adaptation and co-evolution. The general perception is that both phenomena (correlated divergence *and* correlated mutations) should happen to preserve the viability of an interacting protein pair.

We have already seen the correlated mutations model at work when we discussed the problem of protein domain prediction (Section 9.2.2). Recall that the correlation mutations score of two MSA columns i and j is defined as

$$r(i,j) = \frac{1}{n^2} \sum_{k=1}^{n} \sum_{l=1}^{n} \frac{(s(a_{ki}, a_{li}) - \bar{s}_i)(s(a_{kj}, a_{lj}) - \bar{s}_j)}{\sigma_i \cdot \sigma_j}$$

where $s(a,b)$ is the similarity score of amino acids a and b according to the scoring matrix used. The term $\bar{s}_i$ is the average similarity in position i, σ_i is the standard deviation and n is the number of sequences that are common to both columns.

To detect potential interaction sites, the correlation score is calculated for each pair of positions to create a correlation matrix of size $L_1 \times L_2$. Assuming that binding sites have significantly higher correlation scores than non-binding sites, regions that are rich with correlated mutations can be indicative of an interaction site and physical interaction between proteins. The significance of the bounding box can be estimated using similar methods to those applied to charge and hydrogen-bond potentials, as discussed in the next section.

Searching for potential binding sites entails comparison of all possible pairs of residues, one from each protein. However, the number of positions that actually interact is usually small, and most pairs are not expected to be correlated. To eliminate positions that are uninformative we can compile the background distribution of scores across the matrix and assess the significance of each position based on that distribution. Since many residue positions within a protein family are not well preserved and might not exist in certain organisms, a separate background distribution needs to be calculated for each position in the correlated mutation matrix. This can be done in the same way as in the correlated divergence method; the organism order is shuffled for one of the families, r is recomputed, and a zscore is associated with the raw correlation score based on the distribution of scores for the shuffled datasets. A significant position is defined as a position with a correlated mutations zscore $> T$ (a parameter that can be optimized for performance).

While intuitively appealing, the correlated mutations method is computationally intensive with a runtime of $O(m^2 n^2)$ where m = length of the longer of the aligned protein families and $n = |O_{common}|$. As such it is less suited for large-scale prediction of interaction sites. It also suffers from low sensitivity. When the compensating mutations are mediated by more than one of the neighboring positions, this model will not necessarily detect the signal. Similar to the correlated divergence method, the model might also fail to distinguish a true signal from background noise. This can be addressed to some extent by integrating into the model information about phylogenetic relationships between sequences [749–751].

Estimating the significance of bounding boxes: How can we score a bounding box and assess its significance? Clearly, not all positions in the bounding box co-evolve or form bonds. Stereo-chemical constraints allow each amino acid to be spatially close to only a very few other amino acids. Therefore, given a bounding box of size $l \times l$ it is likely that the interaction interface involves only $O(l)$ amino-acid pairs. In that view, averaging the correlated mutations score or the binding potentials over the whole bounding box could diminish real signals. Scoring functions based on a tally of the "surprising" events, such as the number of positive pairs or the number of significant positions in the box, are more likely to distinguish between interaction sites and other sites. Other alternatives are the average score or zscore of the l highest scoring positions in the box.

The significance of these measures can be assessed empirically by characterizing the background distribution of scores over a large population of randomly selected bounding boxes of the same size, located outside of the true interacting binding box. Given the background distribution of scores P_0 the significance of score S is estimated as the sum

$$pvalue(S) = P(S' \geq S) = \sum_{S' \geq S} P_0(S')$$

The significance of each measure can also be computed without having to generate the background distribution, using a simple model. Consider the correlated mutations matrix for example (the same model holds also for the binding potential matrices). We are given a matrix of size $N = L_1 \cdot L_2$ with a total of T significant correlated mutations and a putative binding box of size $n = l \cdot l$ with t significant correlated mutations. The null hypothesis is that the correlated mutations are randomly scattered in the matrix. Therefore, the probability that a position will be deemed significant by chance is $p = n/N$. The number of significant positions within a random box of size n is expected to follow the binomial distribution

$$P(t) = \binom{n}{t} p^t (1-p)^{T-t}$$

and the pvalue of a bounding box with t significant mutations is given by

$$pvalue(t) = P(t' \geq t) = \sum_{t'=t}^{n} P(t)$$

For a typical interaction site n is small enough that the pvalue can be computed directly from the sum. For n large the binomial distribution can be approximated by a normal distribution with $\mu = np$ and $\sigma = \sqrt{np(1-p)}$ (see Section 3.12.4). In that case, the pvalue can be computed from approximated closed-form solutions for the cumulative normal distribution[7].

13.3.2.4 Sequence signatures and domain-based prediction

Given the complexity of the problem and the many unknowns, predicting the interaction interface from sequence information alone produces only mediocre results. A more successful approach for predicting the interaction site (and protein-protein interactions in general) is based on analysis of over-represented sequence signatures, such as Prosite or InterPro motifs, in databases of interacting proteins [752]. Let $P(x, y)$ denote the relative frequency of motifs x and y in interacting protein pairs. That is, the number of protein pairs containing x in one protein and y in the second protein, divided by the number of protein pairs in the dataset. Let $P(x)$ denote the relative frequency of motif x in the dataset. Then the ratio

$$\frac{P(x, y)}{P(x)P(y)}$$

measures the likelihood ratio of the pair x, y under two hypotheses: the occurrence of x and y is coordinated vs. random. If the ratio > 1 then the signature pair x, y appears in interacting proteins more than expected by chance. These signatures characterize interaction sites and can be used for prediction by searching for similar patterns in other proteins. This idea is an extension of the conservation studies we discussed in Section 13.3.1.1 that linked sequence conservation to structural properties important to facilitate an interaction. The signatures need not be known motifs, and a similar analysis can be carried on interaction interfaces extracted from solved structures of protein complexes [753]. For example, the study by [754] works by identifying clusters of similar protein segments that are present in interaction sites. Pairs of clusters with many links between the source proteins (derived from protein interaction databases) are predicted as interacting motifs and can be searched against sequence data to predict interactions.

Protein-protein interactions can also be predicted by studying interactions between domains [755, 756]. As we discussed in Chapter 9, the majority of proteins are composed of multiple domains. Domains tend to form compact substructures and are assumed to be capable of folding independently. The interaction of two multi-domain proteins is almost always enabled by one of the domains in each protein. Some domains have multiple binding sites that can bind to different partners. Characterizing which domains actually interact can help to localize the interaction between multi-domain proteins and improve the accuracy of predictions. Furthermore, it can help to determine

[7]Procedures for computing the tail of the normal distribution are available in most programming languages.

which interactions can occur simultaneously and which are mutually exclusive[8].

According to the domain-based approach, the probability that two proteins q_1, q_2 interact depends on their domain content. Domain interactions can be determined based on inspection of 3D structures of protein complexes. For example, pairs of SCOP domains that are in contact, or are on the same PDB chain and adjacent to each other can be considered interacting domains. Domains can also be determined from sequence-based databases. For example, the work by [755] characterizes proteins based on their Pfam domains and computes their probability of interaction based on the probability that the component domains will interact. In this section we will describe this method in detail.

A simple way of assessing the probability that domains x and y interact is from the empirical counts in a training set of interacting proteins. Let F_x denote the set of proteins containing domain x, then the empirical probability of domains x and y to interact is defined as

$$P(x, y) = \frac{m_{xy}}{n_{xy}} \tag{13.3}$$

where n_{xy} is the number of protein pairs q_i, q_j in the dataset such that $q_i \in F_x$ and $q_j \in F_y$, and m_{xy} is the subset of protein pairs that interact. Given the domain probabilities, we can compute the probability that two multi-domain proteins will interact. Let I_{ij} be an indicator variable denoting whether proteins q_i and q_j interact or not. The model of [755] assumes independence between domains and considers all pairs of domains, one from each protein. The term

$$\prod_{x \in q_i, y \in q_j} (1 - P(x, y))$$

denotes the total probability that *none* of the domain pairs in the query proteins interact. For an interaction to occur it is enough that one of the pairs interacts, which is the complementary probability

$$P(I_{ij} = 1) = 1 - \prod_{x \in q_i, y \in q_j} (1 - P(x, y)) \tag{13.4}$$

[8]Interestingly, domain interactions play a special role in processing post-translational modifications (PTMs, see Section 13.1). Different domains recognize different modifications, based on short peptides in the target proteins that bind to the domain only once the peptide has been processed by one of the specific PTMs. Very much like antibodies, which can recognize multiple antigens by binding to different peptides with a common signature, it has been hypothesized that PTM domains also recognize multiple peptides with a specific PTM [757, 758]. In some cases, multiple PTMs are necessary to trigger a reaction, which can be recognized by duplicated domains, or multiple recognition "pockets" in the binding domain.

While this approach is simple and fast, it relies heavily on the empirical data, which can be sparse, and it ignores the possibility of other domains carrying out the interaction. For example, consider the following scenario: given two interacting proteins $q_1 = (a, x)$ and $q_2 = (b, y)$ with domains x and y being unique to proteins q_1 and q_2 in the whole dataset, we might conclude that $P(x, y) = 1$. However, the interaction might be facilitated by any of the four possible domain combinations and might not involve x and y at all.

More accurate estimates of domain interaction probabilities (and of protein interaction probabilities) can be obtained by computing the Maximum Likelihood (ML) parameters. The ML estimators are the probabilities that maximize the likelihood of the training set (the observed data). Let λ_{xy} denote the probability that domains x and y interact. Under this notation we rewrite Equation 13.4 such that

$$P(I_{ij} = 1) = 1 - \prod_{x \in q_i, y \in q_j} (1 - \lambda_{xy}) \tag{13.5}$$

Let X_{ij} denote the observed data with $X_{ij} = 1$ indicating that there is experimental evidence that i and j are interacting and 0 otherwise. Since the experimental information is often erroneous, an interaction might be reported even if the proteins do not interact. By the complete probability formula

$$\begin{aligned} P(X_{ij} = 1) &= P(X_{ij} = 1 | I_{ij} = 1) P(I_{ij} = 1) + P(X_{ij} = 1 | I_{ij} = 0) P(I_{ij} = 0) \\ &= P(X_{ij} = 1 | I_{ij} = 1) P(I_{ij} = 1) + P(X_{ij} = 1 | I_{ij} = 0)(1 - P(I_{ij} = 1)) \end{aligned}$$

Note that $P(X_{ij} = 1 | I_{ij} = 0)$ is the probability of a false positive (denoted p_1) while

$$P(X_{ij} = 1 | I_{ij} = 1) = 1 - P(X_{ij} = 0 | I_{ij} = 1) = 1 - p_2$$

where $p_2 = P(X_{ij} = 0 | I_{ij} = 1)$ is the probability of a false negative. Hence

$$P(X_{ij} = 1) = (1 - p_2) P(I_{ij} = 1) + p_1 (1 - P(I_{ij} = 1)) \tag{13.6}$$

The probability $P(X_{ij} = 0)$ is simply the complementary probability $P(X_{ij} = 0) = 1 - P(X_{ij} = 1)$. Thus, the likelihood that a pair of proteins q_i and q_j interacts depends on p_1, p_2 and $P(I_{ij} = 1)$, which in turn depends on the parameters of the model, the probabilities $\{\lambda_{xy}\}$.

Assuming that p_1 and p_2 are fixed (and a characteristic of the dataset), we are interested in determining the domain interaction probabilities $\{\lambda_{xy}\}$. The ML approach looks for the set that maximizes the likelihood of the observed data $\mathbf{X} = \{x_{ij}\}$, where x_{ij} is the specific realization (0 or 1) of X_{ij} in the experimental dataset. The likelihood of an individual pair of proteins can be written in an abbreviated form as

$$P(x_{ij}) = P(X_{ij} = x_{ij}) = \prod_{ij} P(X_{ij} = 1)^{x_{ij}} P(X_{ij} = 0)^{1 - x_{ij}} \tag{13.7}$$

and the likelihood of the whole set is simply the product over all protein pairs

$$\mathcal{L}(\mathbf{X}|\theta) = \prod_{ij} P(x_{ij}|\theta) = \prod_{i,j} P(X_{ij} = 1|\theta)^{x_{ij}} P(X_{ij} = 0|\theta)^{1-x_{ij}} \quad (13.8)$$

where the implicit dependency on the parameters is now introduced explicitly, with θ denoting the parameter vector consisting of all domain probabilities $\theta = \{\lambda_{xy}\}$. However, it is difficult to maximize this function (or its logarithm). Given the number of known domains, there are tens of thousands of parameters (which correspond to domain pairs), and exact solutions for the ML estimators are not attainable.

The problem is complicated by the fact that we do not know which domain pairs actually interact. We have already discussed similar problems in the context of motif detection (Section 5.3) and HMMs (Section 6.3). Similar to the setup of these problems, the problem of predicting protein interactions from domain interactions has to account for hidden data, since we do not know which domains facilitate the interaction for each pair. Let D^{ij}_{xy} be an indicator variable signifying whether domains x and y are interacting in proteins i and j. The set of missing data is

$$\mathbf{D} = \{d^{ij}_{xy} | i \in F_x \;\; j \in F_y \;\; x, y \in \mathcal{D}\}$$

where $\mathcal{D}$ is the domain space and F_x is the set of proteins containing domain x (lower case d^{ij}_{xy} again denotes a specific assignment of values $\in \{0, 1\}$ to the variables D^{ij}_{xy}). When computing the likelihood we now have to consider both the observed variables $\mathbf{X}$ (which characterizes the interactions between proteins) and the hidden data $\mathbf{D}$. Note that the likelihood of the complete data can be written as

$$\mathcal{L}(\mathbf{X}, \mathbf{D}|\theta) = \prod_{i,j} P(x_{ij}, \mathbf{d}_{ij}|\theta) = \prod_{i,j} P(x_{ij}|\mathbf{d}_{ij}, \theta) P(\mathbf{d}_{ij}|\theta)$$

where $\mathbf{d}_{ij} = \{d^{ij}_{xy} | x \in q_i \;\; y \in q_j\}$, and

$$P(\mathbf{d}_{ij}|\theta) = \prod_{x \in q_i, y \in q_j} P(d^{ij}_{xy}|\theta) = \prod_{x \in q_i, y \in q_j} \lambda_{xy}^{d^{ij}_{xy}} (1 - \lambda_{xy})^{1-d^{ij}_{xy}}$$

where the second term simply utilizes the fact that $P(d^{ij}_{xy} = 1|\theta) = \lambda_{xy}$ and $P(d^{ij}_{xy} = 0|\theta) = 1 - \lambda_{xy}$, and uses the same trick as in Equation 13.7. Note also that

$$P(x_{ij}|\mathbf{d}_{ij}, \theta) = \sum_{i_{ij} \in \{0,1\}} P(x_{ij}|i_{ij}, \mathbf{d}_{ij}, \theta) \cdot P(i_{ij}|\mathbf{d}_{ij}, \theta) \quad (13.9)$$

and both probabilities do not depend on θ. The first is only a function of p_1 and p_2, and the second is either zero or one. For maximization purposes we can treat the term in Equation 13.9 as a constant and denote it as C_{ij}. All together, we get that

$$\mathcal{L}(\mathbf{X}, \mathbf{D}|\theta) = \prod_{i,j} C_{ij} \prod_{x \in q_i, y \in q_j} \lambda_{xy}^{d_{xy}^{ij}} (1 - \lambda_{xy})^{1 - d_{xy}^{ij}}$$

$$= C \cdot \prod_{x,y \in \mathcal{D}} \lambda_{xy}^{\sum_{i \in F_x, j \in F_y} d_{xy}^{ij}} (1 - \lambda_{xy})^{n_{xy} - \sum_{i \in F_x, j \in F_y} d_{xy}^{ij}}$$

where $n_{xy} = |F_x||F_y|$ is the number of protein pairs over F_x and F_y, and $C = \prod_{i,j} C_{ij}$.

Our goal is to find the parameters that maximize this likelihood. However, we do not know the values of the hidden variables. As with HMMs, a possible solution is to use the available information with the expectation maximization (EM) algorithm to derive locally optimal parameters. Starting from some initial values for the parameters, the EM algorithm works in two steps. In the first step, the likelihood of the complete data $\mathcal{L}(\mathbf{X}, \mathbf{D})$ is averaged over all possible values of the hidden data, using the parameters from the previous iteration. In the second step, a new estimate for the parameters is obtained by looking for the values that maximize the average likelihood $E_{\mathbf{D}}(\mathcal{L}(\mathbf{X}, \mathbf{D}))$ or the average log likelihood $E_{\mathbf{D}}(\log \mathcal{L}(\mathbf{X}, \mathbf{D}))$.

We will start with the expectation step. First note that the logarithm of the likelihood is given by

$$\log \mathcal{L}(\mathbf{X}, \mathbf{D}|\theta) = \log(C) + \sum_{x,y} \log(\lambda_{xy}) \sum_{i \in F_x, j \in F_y} d_{xy}^{ij}$$
$$+ \sum_{x,y} \log(1 - \lambda_{xy}) \cdot \left(n_{xy} - \sum_{i \in F_x, j \in F_y} d_{xy}^{ij} \right) \quad (13.10)$$

In Section 5.7 we discussed in detail the EM algorithm. The expectation step (E-step) is described in Equation 5.10. The function we would like to maximize is the average log likelihood, as defined in Equation 5.10. Substituting d for h (hidden data) we obtain that

$$F(\theta|\theta^{t-1}) = E_{\mathbf{D}} \left[\log \mathcal{L}(\mathbf{X}, \mathbf{D}|\theta) \right] + H(\mathbf{D})$$

where $E_{\mathbf{D}}[]$ denotes averaging over the hidden variables and $H(\mathbf{D})$ does not depend on θ. Ignoring $H(\mathbf{D})$ and plugging in Equation 13.10 we get that

$$F(\theta|\theta^{t-1}) = \log(C) + \sum_{x,y} \log(\lambda_{xy}) E_{\mathbf{D}} \left[\sum_{i \in F_x, j \in F_y} d_{xy}^{ij} \right]$$
$$+ \sum_{x,y} \log(1 - \lambda_{xy}) \cdot \left(n_{xy} - E_{\mathbf{D}} \left[\sum_{i \in F_x, j \in F_y} d_{xy}^{ij} \right] \right) \quad (13.11)$$

Now we can proceed to the maximization step. When computing the derivative of Equation 13.11 with respect to θ we can focus on one parameter λ_{xy} at a time, assuming the domain interaction probabilities are independent of each other. In other words, we can ignore the sum over x, y in Equation 13.11. Therefore,

$$\frac{\partial F(\theta|\theta^{t-1})}{\lambda_{xy}} = \frac{1}{\lambda_{xy}} E_{\mathbf{D}}\left[\sum_{i \in F_x, j \in F_y} d_{xy}^{ij}\right] - n_{xy}\frac{1}{1-\lambda_{xy}} + \frac{1}{1-\lambda_{xy}} E_{\mathbf{D}}\left[\sum_{i \in F_x, j \in F_y} d_{xy}^{ij}\right]$$

and by equating the derivative to zero we can obtain the optimal parameter value

$$\lambda_{xy}^t = \frac{1}{n_{xy}} E_{\mathbf{D}}\left[\sum_{i \in F_x, j \in F_y} d_{xy}^{ij}\right]$$

Note that the averaging in $E_{\mathbf{D}}[]$ is done with the $P(\mathbf{D}|\mathbf{X}, \theta^{t-1})$ conditional probability (left integral in Equation 5.10). Therefore,

$$\lambda_{xy}^t = \frac{1}{n_{xy}} \sum_{i \in F_x, j \in F_y} \sum_{d_{xy}^{ij} \in \{0,1\}} d_{xy}^{ij} \cdot P(d_{xy}^{ij}|\mathbf{X}, \theta^{t-1})$$

The result is an intuitive one. The probability of domains x and y interacting in iteration t can be estimated from the average value of the indicator variables d_{xy}^{ij} in iteration $t-1$, over all protein pairs i, j that contain the domains x and y. Namely,

$$\lambda_{xy}^t = \frac{1}{n_{xy}} \sum_{i \in F_x, j \in F_y} 1 \cdot P(D_{xy}^{ij} = 1|\mathbf{X}, \theta^{t-1}) + 0 \cdot P(D_{xy}^{ij} = 0|\mathbf{X}, \theta^{t-1})$$

If $P(D_{xy}^{ij} = 1|\mathbf{X}, \theta^{t-1})$ for most pairs, then λ_{xy}^t is close to 1.

The probability that $D_{xy}^{ij} = 1$ in iteration $t-1$ depends on the observed data and the current parameters of the model. Note that the only observed data that is relevant to the pair i and j is x_{ij} (the rest is summarized in the parameter vector θ), hence

$$P(D_{xy}^{ij} = 1|\mathbf{X}, \theta^{t-1}) = P(D_{xy}^{ij} = 1|x_{ij}, \theta^{t-1})$$

Using Bayes' formula we interchange d and x

$$P(D_{xy}^{ij} = 1|x_{ij}, \theta^{t-1}) = \frac{P(x_{ij}|D_{xy}^{ij} = 1, \theta^{t-1})P(D_{xy}^{ij} = 1|\theta^{t-1})}{P(x_{ij}|\theta^{t-1})}$$

where the dependency on θ^{t-1} is unaffected (see Section 3.12.3). Note that $P(D_{xy}^{ij} = 1|\theta^{t-1})$ is the domain interaction probability λ_{xy}^{t-1} at time $t-1$, and $P(x_{ij}|\theta^{t-1})$ can be computed from Equation 13.7. If $D_{xy}^{ij} = 1$ then proteins i and j interact and $I_{ij} = 1$. Therefore, the first term is simply

$$P(x_{ij}|D_{xy}^{ij} = 1, \theta^{t-1}) = P(x_{ij}|I_{ij} = 1)$$

This can be written in an abbreviated form (similar to Equation 13.7) as

$$P(x_{ij}|I_{ij}=1) = (1-p_2)^{x_{ij}} p_2^{1-x_{ij}}$$

Therefore

$$P(D_{xy}^{ij}=1|x_{ij},\theta^{t-1}) = \frac{(1-p_2)^{x_{ij}} p_2^{1-x_{ij}} \cdot \lambda_{xy}^{t-1}}{P(x_{ij}|\theta^{t-1})}$$

In conclusion,

$$\lambda_{xy}^{t} = \frac{\lambda_{xy}^{t-1}}{n_{xy}} \sum_{i\in F_x \ j\in F_y} \frac{(1-p_2)^{x_{ij}} p_2^{1-x_{ij}}}{P(x_{ij}|\theta^{t-1})} \tag{13.12}$$

Hence λ_{xy} can be rewritten as a sum where protein pairs are weighted based on their probability of interacting. The weight of interactions that are supported by the parameters increases over time (and vice versa), while the weight of interactions that are not decreases. This leads to a new set of parameters, and the process repeats until convergence.

To recap, the algorithm starts with empirical probabilities, as in Equation 13.3 and Equation 13.4, derived from datasets of experimentally verified interactions. From these initial parameters, we can compute $P(I_{ij}=1)$ using Equation 13.5 and $P(x_{ij}|\theta)$ using Equation 13.7. From these probabilities we can recompute the parameters as in Equation 13.12 and the likelihood as in Equation 13.8, and repeat these steps until the likelihood value does not change. Note that this model considers only pairs of domains, while neighboring domains, which might be essential to stabilize the structure, are ignored. However, this framework can be extended to domain combinations, as in [759].

An alternative approach to predicting interactions from domains is to cast the problem as a classification problem and tackle it by training a classifier, as discussed in Chapter 7. For example, each protein can be represented as a sparse binary vector in the domain space [760]. Coordinates that correspond to domains that are present in the protein are assigned a value of 1. These vectors can then be analyzed with SVMs or decision trees to derive a decision rule that discriminates between interacting and non-interacting proteins.

To summarize, domain information can help to infer interactions between protein complexes and is reportedly more successful than most sequence-based approaches. On the down side, the rate of false positives can be high, as the domain approach ignores the fact that domains might not fold independently. As with other relational data mining methods, sequence signatures and domain-based prediction can only identify protein interactions similar to those found in a training dataset.

13.3.3 Gene co-expression

Gene expression data provides useful information that can help to predict protein-protein interactions. The paper by [761] was one of the first to study and verify the correlation between the interactome (interaction data) and the transcriptome (expression data) of the yeast genome, showing that the ratio of interacting proteins within clusters of co-expressed genes is higher than across clusters. Although the original study was somewhat biased due to the inclusion of self-interacting proteins [762], the results were later confirmed by other studies as well (e.g., [29, 763]). However, gene co-expression can be indicative of other types of functional links, as discussed in Section 12.4. Therefore, co-expression in itself is insufficient to predict interaction but could be useful when combined with other sources of information. The other direction is not necessarily true either, and interacting proteins are not always co-expressed (for further discussion, see Section 12.4).

13.3.4 Hybrid methods

Hybrid methods combine different sources of information or different methods for the task at hand. They often produce better results than the methods they combine. Hybrid methods are especially useful when the individual methods use different and complementary sources of information. For example, when predicting interactions, co-expression and same cellular location are two events that increase the likelihood of an interaction if both are true. That is, if A denotes the event that two proteins are co-expressed and B denotes the events that the two are in the same cellular compartment then

$$P(\texttt{interaction}|A \cap B) > \max\{P(\texttt{interaction}|A), P(\texttt{interaction}|B)\}$$

Each one of the measures and methods described in the previous sections is based on a different element providing some information about the potential of forming an interaction, but the formation of an interaction usually requires the combination of all these elements.

However, some of the information is redundant since different sources can overlap and depend on each other. For example, co-expression is more likely when proteins are in the same subcellular location. Similarly, solved structures of protein complexes and information extracted from databases of interacting proteins cannot be considered independent sources, since some of the interactions in interaction databases are deduced from structural information. Therefore, combining these sometimes redundant sources requires a careful design.

In general, assume we have k sources of information, each providing us with a certain evidence e regarding the likelihood of an interaction between two proteins p_1, p_2. This evidence could be the divergence correlation score, the expression similarity score, presence of sequence signatures typical of interaction sites, identical subcellular location, evidence of a domain fusion event, sequence similarity with proteins that are known to interact and so on. Other features we could consider include, for example, the binding potentials in the highest scoring bounding box (Section 13.3.2.3). We are interested in computing the posterior probability

$$P_1 = P(\texttt{interaction}|e_1, ..., e_k)$$

The probability of no interaction is the complementary probability

$$P_2 = P(\texttt{no interaction}|e_1, ..., e_k) = 1 - P_1$$

If $P_1 > P_2$ then we can predict that the proteins interact, minimizing the total error probability of false positives and false negatives (see Section 3.12.8). By Bayes' formula

$$P(\texttt{interaction}|e_1, ..., e_k) = \frac{P(e_1, ..., e_k|\texttt{interaction})P(\texttt{interaction})}{P(e_1, ..., e_k)}$$

but the amount of data available is usually insufficient to estimate reliably the multi-dimensional distribution $P(e_1, ..., e_k|\texttt{interaction})$. To simplify the analysis, we can make the naive assumption that the sources are independent or weakly dependent on each other and approximate

$$P(\texttt{interaction}|e_1, ..., e_k) = \frac{\prod_{i=1}^{k} P(e_i|\texttt{interaction})P(\texttt{interaction})}{\prod_{i=1}^{k} P(e_i)}$$

Under the assumption of independence between the variables, we can estimate $P(e_i|\texttt{interaction})$ from the empirical data. With continuous variables it is often more useful to define the event such that the cumulative distributions are used, since the probability of observing a specific value is very small. That is, instead of computing the probability of observing a specific value, we are interested in the total probability of observing outcomes that are as or more extreme than the outcome observed. For example, the probability of a specific correlated divergence score D is less informative than the probability of observing a correlated divergence score $D' \geq D$.

The term $P(\texttt{interaction})$ can be estimated from the relative number of known interactions in well-studied genomes, such as yeast and *E. coli*, where there is extensive information on interactions. The denominator can be estimated from the marginal probability distributions, again assuming independence, or ignored if we are only interested in comparing $P_1 > P_2$.

It should be noted that while Bayes' decision rule, which classifies instances based on comparison of P_1 and P_2, minimizes the total error probability, it is not the best decision rule in this case. The number of interacting proteins is orders of magnitude smaller than the number of non-interacting proteins, hence the relative number of false positives outnumbers the relative number of false negatives even if P_1 is much higher than P_2. Usually high specificity (smaller number of false positives) is considered to be a higher priority, and therefore adopting a stricter approach and setting a threshold T such that the proteins are predicted to interact only if P_1 exceeds the threshold T is recommended.

A more statistically meaningful approach is to assess the significance of the observations in terms of a null hypothesis of non-interaction. Establishing the distribution of each source (feature) in a population of non-interacting proteins can provide us with a measure of significance for a pair of putatively interacting proteins. Namely, for each source we can assess the probability of observing by chance a value as extreme as the one observed (the pvalue). If the pvalue is lower than a certain preset threshold we can reject the null hypothesis with confidence that depends on the threshold. If the distributions are uni-modal and normal-like then we can also assess the significance in terms of the zscore (see Section 3.4.2.3).

However, establishing the significance of multi-dimensional feature vectors is more complicated. Not only is there usually insufficient data to estimate the multi-dimensional distribution reliably; it is also not so obvious which feature vectors should be considered more extreme than the observed one (see discussion in Section 7.7). Furthermore, there might be intricate dependencies between the variables, as noted above.

Indeed, in practice some of the sources we mentioned before are dependent on each other, and the assumption of independence might result in a poor approximation. A proper model should take into account this dependency, and one such integrative approach is to use a Bayesian network (as in [764]). We will discuss Bayesian networks in Chapter 15. For other integrative approaches, see Further reading.

13.3.5 Training and testing models on interaction data

Any learning task requires the availability of positive and negative examples. In the context of protein-protein interactions the positives are interacting proteins and the negatives are non-interacting proteins. This is actually not as simple as it sounds. While there is plenty of information on experimentally verified interactions, little information is available on non-interacting proteins [765]. Since the experimental data is partial, a dataset that consists of protein pairs that are not documented as interacting or a dataset of randomly chosen pairs might contain many proteins that actually do interact. If information on the subcellular locations of proteins is available, the pairs can be chosen from different locations to reduce the chance of an interaction. However, subcellular location is available for a relatively small number of proteins.

Care should also be given to the preparation of the interacting dataset. The dataset should exclude interactions that were determined exclusively with the yeast two-hybrid test, since they are considered less reliable. To minimize redundancy, only one pair from all pairs of homologous interactions should be retained. Depending on the method used, additional filters might be necessary. For example, co-evolution methods can be applied only when there are enough homologs from different organisms [697]. Otherwise, the correlated divergence scores are insignificant. Furthermore, interactions between homologous proteins (including self-interaction) should be excluded, since they bias the correlated divergence score.

Most importantly, in order to prevent biases and false conclusions, the datasets of interacting and non-interacting proteins should have similar statistical properties. Otherwise, it is difficult to know whether the results obtained are reflective of a true signal (e.g., correlated divergence) or an artifact (e.g., high frequency of outliers or different number of entries in the distance matrices). To neutralize such irrelevant artifacts it is important to process the datasets using the same criteria and choose samples such that certain global properties of the datasets are similar.

It is convenient to think of the datasets in terms of graphs where the proteins can be viewed as nodes and each pair that is chosen introduces an edge in the graph. The degree of each node is the number of edges incident on that node, or the number of pairs this node participates in. A proper selection should result in two graphs (of interacting and of non-interacting proteins) over the *same* set of nodes where certain graph properties are preserved. These could be, for example, the degree distribution or the degree of each node. While having two graphs where every node has the same degree in both does not guarantee similar properties at the interaction/protein-pair level, it significantly reduces statistical biases that might affect the results. This can be verified by comparing some key properties of the interacting and non-interacting datasets that can influence the method (e.g. Pearson correlation), such as the distribution of $|O_{common}|$ and the distribution of the average number of paralogs for each organism in O_{common}.

Ultimately, to test the premise that interacting proteins or interaction sites are characterized with a certain property, we need to compare the scores observed for interacting proteins vs. those observed for non-interacting proteins. A good separation is indicative that the property is predictive of protein interactions.

13.4 Interaction networks

The set of all interactions that can occur in a cell (the interactome) forms a complex network with interesting properties. Some proteins tend to interact with single proteins, while others are hubs that interact with many others. Proteins with higher numbers of partners are often essential genes that are crucial for proper functioning of the cell. Multiple partners can be the result of multiple interfaces, some of which are **exclusive** (meaning they preclude interactions at other interfaces at the same time), while others are **inclusive** and can work in concert with each other.

Multiple partners can also be the result of complex expression patterns, with different proteins interacting with each other at different times. The combinatorial complexity of such networks results in different dynamics and outcomes, depending on the interacting partners at any given time. Analyzing the properties of such networks has been the subject of many studies. Characterizing the topological properties of these networks can lead to the detection of bottlenecks and central proteins and help in studying the effect of knockouts and mutations (such as SNPs) on individual proteins and their significance in various diseases. It is also instrumental for studies of the stability of biological systems and mechanisms of fault tolerance, such as redundancy. Detecting recurring patterns and clusters can be useful to the prediction of interactions and identification of functional modules or motifs. Finally, the dynamic of the network can shed light on the way information is being processed and transfered in the cell.

In this section we will discuss models of biological networks, their properties and their biological significance. The discussion is not limited to interaction networks, and is relevant to a broader class of networks, including regulatory networks and metabolic networks. We will elaborate on metabolic pathways in Chapter 14. Regulatory networks are of special interest, and we will discuss them in Section 14.4 and in Chapter 15.

13.4.1 Topological properties of interaction networks

A network is basically a graph with nodes and edges between nodes. The emergence of a structure in a graph depends on the edge placement. The number of nodes that node i is connected to is called the **degree** and is denoted by k_i. When characterizing the topology of a network we are mainly interested in the "connectedness" and the "orderness" of the graph. The connectedness can be quantified in terms of several properties, such as the average path length between nodes (the **diameter** of the graph) and the distribution of node degrees. These properties (graph indices) can differ markedly between graphs, depending on the process that generated the graph.

One way to assess the order in a graph is the **clustering coefficient** [766]. For each node, the clustering coefficient measures the "tightness" or "cohesiveness" of a node and its immediate neighbors in terms of how close they are to being a clique. If a node of degree k is part of a clique, then all k nodes it is connected to must be also connected to each other for a total of $k(k-1)/2$ edges. The clustering coefficient of a node is the ratio between the number of edges that do exist between the nodes, e_i, and the maximal number possible. That is,

$$c_i = \frac{e_i}{k_i(k_i-1)/2}$$

as illustrated in Figure 13.10. The clustering coefficient of a graph is the average coefficient of its nodes

$$C = \frac{1}{N}\sum_{i=1}^{N} c_i$$

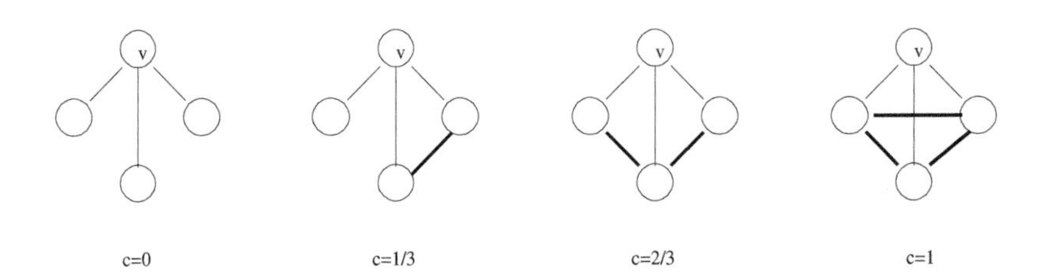

FIGURE 13.10: Clustering coefficient. The clustering coefficient of a node v depends on the connectivity between the nodes it is connected to. In the graph on the left there are no edges between the three nodes that v is connected to; therefore the clustering coefficient is 0. In the graph on the right there are three edges, which is the maximal number of possible edges between the three nodes; therefore the clustering coefficient is 1.

When characterizing graphs and networks, we can think of two extremes. Highly ordered graphs and random graphs. In **random graphs** edges are drawn randomly between nodes, according to a certain probability distribution. For example, in the traditional Erdos-Renyi model [767], edges are drawn according to a uniform distribution and each edge has the same fixed probability of occurring, independent of other edges. There are other models of random graphs that satisfy other properties, but this model is simple and widely used.

Given a graph with N nodes, the Erdos-Renyi (ER) model (in one of its variants) connects every two nodes with probability p and rejects a connection with probability $1-p$. The higher the probability, the more connected the graph is. There is a total of $\binom{N}{2}$ potential edges, and the average number of edges in the graph is $\binom{N}{2}p$. On average, the clustering coefficient of a node is $c_i = p$, and the coefficient for the whole graph is

$$\frac{1}{N}\sum_{i=1}^{N} c_i = \frac{1}{N}Np = p$$

Since the edge selection events are independent, one can think of this process as a set of $N-1$ experiments (connection attempts) for every node, with p being the probability of success. The outcome of this process is a graph where the node degrees are distributed like the binomial distribution. That is, the probability that a node is connected to k other nodes in a random graph with

N nodes is given by

$$P(k) = \binom{N-1}{k} p^k (1-p)^{N-1-k}$$

The binomial distribution can be approximated with the normal distribution for large N and $Np > 5$, such that

$$P(k) = \frac{1}{\sqrt{2\pi\sigma^2}} e^{\frac{-(k-\mu)^2}{2\sigma^2}}$$

with $\mu = Np$ and $\sigma = \sqrt{np(1-p)}$. As we can see from the normal approximation[9], the probability of observing a node with degree k decreases exponentially with k. Random graphs are also characterized by a large diameter.

The main limitation of the ER model is that it does not describe realistic networks, where edges are not equally probable and not necessarily independent of each other. The model precludes the possibility of graphs with locally dense subgraphs or clusters. In other words, random graphs based on the ER model tend to have low clustering coefficient. Moreover, due to the exponential decay, nodes that are highly connected (**hubs**) are very unlikely in this model, although they are often observed in real-life networks.

In **highly ordered graphs**, the edges are not placed randomly, but according to some directed process. For example, a square lattice is a highly structured graph, where every node is associated with a position in a grid (Figure 13.11) and an edge exists between every two adjacent points. A ring lattice is another example of an ordered graph where the N nodes are numbered by integers $1..N$ and organized in a ring such that node 1 is adjacent to node N (see Figure 13.11). Each node i in this ring is connected to its k ($k < N$) nearest neighbors. That is, a node i is connected to all nodes j such that $\min\{|i-j|, N-|i-j|\} \leq k/2$. Since every node is connected to k other nodes, the total number of edges in this graph is $N\frac{k}{2}$, where every edge is counted only once. The diameter of the graph is relatively high, and so is the clustering coefficient.

Most real-life networks happen to be neither of the completely random type nor the highly ordered one, but rather in between. These include biological networks (such as interaction networks), man-made networks (such as the Internet with links between webpages, or the network of papers and citations in scientific literature), and social networks (with nodes representing individuals and edges formed between friends, colleagues or acquaintances).

[9]In the literature, the Poisson approximation to the binomial distribution is often used, such that

$$P(k) = \frac{\lambda^k}{k!} e^{-\lambda}$$

with $\lambda = N \cdot p$. This approximation holds for large N and small p. However, the exponential decay is not as obvious.

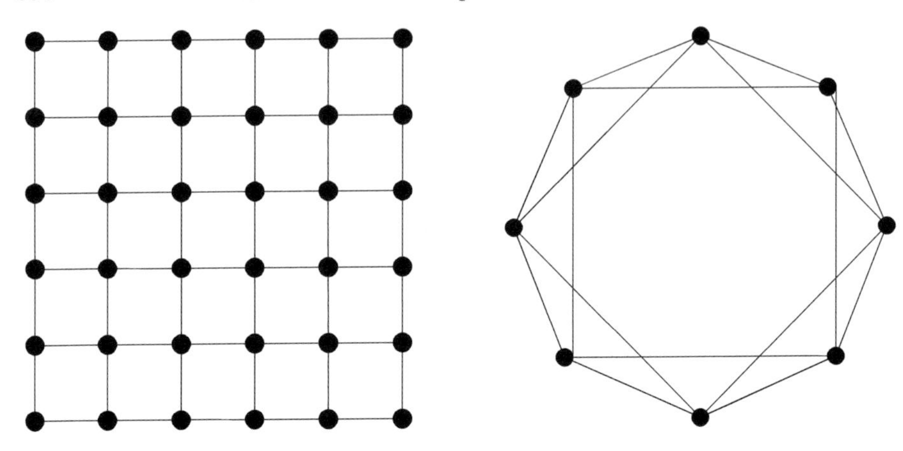

FIGURE 13.11: **Left:** square lattice. **Right:** ring lattice with $k = 4$.

Over the past decade two different models were introduced in an attempt to explain the properties of real-life networks. The **small-world model** of Watts and Strogatz [766] starts with a ring lattice with N nodes (as described above) and then reconnects every edge with probability q. That is, for every node $i \in 1..N$ and for every edge connecting i to a node $j > i$, replace (i, j) with an edge (i, k) with probability q. The node k is selected uniformly from the set of all nodes, excluding i (to prevent loops) and nodes that are already connected to i (to prevent duplication of edges). The total number of edges is fixed and equals $N\frac{k}{2}$, as in the lattice. Of these edges, about $qN\frac{k}{2}$ are rewired. With $q = 0$ the graph remains a regular lattice while for $q = 1$ the graph becomes essentially a random ER. By equating the average number of edges in a random graph $\binom{N}{2}p$ and the number of edges in the randomized lattice $N\frac{k}{2}$, we conclude that for $q = 1$ the outcome is a random ER graph with

$$p = \frac{Nk}{2}\frac{1}{\binom{N}{2}} = \frac{k}{N-1}$$

For any other $0 < q < 1$, this process forms long-distance connections between nodes that tend to reduce the diameter of the graph significantly, leading to the phenomenon called "small-world". Usually in the context of social networks, the term refers to the phenomenon where random people are often connected by a small number of acquaintances (the famous "six degrees of separation" [768]).

Small-world graphs are characterized with (i) high clustering coefficient, reflecting the formation of local clusters, and (ii) a smaller average path between nodes. This is often the result of a few hub nodes that are connected to many other nodes and can cross "quadrants" of the graph (Figure 13.12). By studying the distribution of the degrees we can identify these nodes that have exceptionally high degree and can be qualified as hubs. If the number of such nodes is higher than expected by chance then the graph is a small-world graph.

The degree distribution changes with the probability of redirection q. For $q = 0$ the structure is governed by the initial lattice construction and the degree is fixed and equals k. For $q > 0$ the degree distribution tends to be similar to the binomial distribution of a random graph with a peak at k. Thus the probability of observing nodes that are highly connected (large k) decreases exponentially, as in random graphs, and hub nodes are not very likely to occur, although they are more probable than in random ER graphs (depending on the ratio between Np and k).

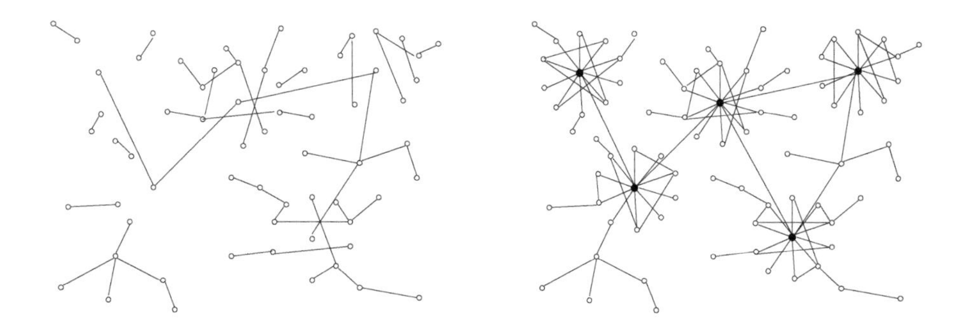

FIGURE 13.12: **Left:** random graph. **Right:** small-world graph.

The emergence of small-world networks depends also on the network content. Some networks can be defined such that in a certain context they are small-world. For example, social networks are small-world only if limited to the group of people within a certain age group. They also depend on the model of edge drawing. When the edges are added not randomly but rather based on a certain proximity function then the outcome is not necessarily a small-world network.

A different network model is the **scale-free model** of [769], which postulates that real-life networks tend to evolve according to certain principles that are generic and independent of the type of the network and reach a state of scale-free expansion.

A scale-free network is one of the types of small-world networks per the two criteria mentioned above, but the model is characterized by two unique elements: growth and preferential attachment. Unlike the other models, which assume a fixed size network, the scale-free model adopts a more realistic approach that enables gradual growth, which is typical of real-life systems that expand over time by adding new entities and connecting them to existing entities. For example, the Internet expands constantly by adding webpages, the scientific literature expands with new papers, biological systems evolve by adding and eliminating genes.

The second element, the preferential attachment, differs from the random placement of edges that characterizes the other models. The assumption is that new entities are more likely to connect to existing entities with high degree. Indeed, this is characteristic of real-life networks. For example, new webpages are likely to have links to established and highly linked websites, and new publications are more likely to cite papers that are already highly cited. It has been shown that interaction networks also grow by means of preferential attachment [770].

The outcome of this model is the **power law** distribution of degrees. That is, the probability that a node is connected to k other nodes follows a decaying function of the form

$$P(k) \sim k^{-\gamma}$$

where γ is a certain constant that is characteristic of the network. Nodes of high degree are assigned higher probabilities with the power-law model than with the other models, which makes them more abundant. This function is scale-free, meaning it is independent of the size of the system, unlike the other models. Empirical distributions of node degrees in real-life networks (such as social, Internet and biological networks) strongly support this model, with γ ranging from 2 to 4.

Formally, the model starts with a certain number of nodes m_0 and in every time step adds one node to the network. Every new node is connected with $m \leq m_0$ edges to m nodes of the network. The probability of attaching a new node to an existing node i depends on the degree of that node and is defined as

$$p_i = \frac{k_i}{\sum_{j=1}^{n} k_j}$$

where n is the number of nodes in the network at that time. The probability assignment function is independent of time and the size of the system. Note that $\sum_{i=1}^{n} p_i = 1$ as required. With this function, the probability of connecting a node to an existing node of degree k_i increases with the connectivity of the node. This results in the **"rich gets richer"** phenomenon.

This model generates a network with $m_0 + t$ nodes and tm edges at time t. The degree of a node is clearly a function of time $k_i(t)$. Since in every time step we add m edges, distributed according to p_i, then the degree of node i increases by mp_i at time t. That is,

$$\frac{\partial k_i(t)}{\partial t} = mp_i = m \frac{k_i(t)}{\sum_{j=1}^{n(t)} k_j(t)}$$

By time t we added tm edges to the network; therefore $\sum_{j=1}^{n(t)} k_j(t) = 2tm$, since every edge (i, j) is counted twice in this sum. Hence,

$$\frac{\partial k_i(t)}{\partial t} = \frac{k_i(t)}{2t}$$

Note that a function $f(t)$ whose derivative is $f'(t) = f(t)/2t$ must be of the form $f(t) = c\sqrt{t}$ where c is a constant. The constant can be determined from the requirement that at time t_i, which is the time the node i was added to the network, the degree was m (the initial degree of every new node). That is, $m = c\sqrt{t_i}$ and therefore

$$k_i(t) = m\sqrt{\frac{t}{t_i}} \tag{13.13}$$

This function, the degree as a function of time, indicates that earlier nodes increase their degree faster than newer nodes.

In their seminal paper from 2000, Barbási and Albert showed that both growth and preferential attachment are essential to the development of power-law scale-free network. To understand the power law distribution of degrees we need to track the changes in node degrees. Consider the probability that a node has a degree smaller than a certain constant k. From Equation 13.13 it follows that

$$P(k_i(t) < k) = P\left(t_i > \frac{m^2 t}{k^2}\right) = 1 - P\left(t_i \leq \frac{m^2 t}{k^2}\right) \tag{13.14}$$

Although it is a bit unintuitive to think about t_i (the time the node i was first introduced to the network) as a random variable, one can think of the following event: picking a node at random, what is the probability $P(t_i)$ that the node was introduced at time t_i? At time t the network has $m_0 + t$ nodes, and since the nodes are added at fixed time intervals, one at a time, then each $0 < t_i \leq t$ is associated with a single node. The probability of picking this node at random is $\frac{1}{m_0+t}$. The probability of selecting any of the nodes that were introduced in the initial step is $\frac{m_0}{m_0+t}$. Therefore,

$$\begin{aligned} P\left(t_i \leq \frac{m^2 t}{k^2}\right) &= \sum_{t_i=0}^{m^2 t/k^2} P(t_i) = \frac{m_0}{m_0 + t} + \frac{m^2 t}{k^2} \cdot \frac{1}{m_0 + t} \\ &\sim \frac{m^2 t}{k^2(m_0 + t)} \end{aligned} \tag{13.15}$$

where in the last step we ignored the first term, assuming $m_0 << t$ and $k^2 << t$ (in other words, we are interested in the asymptotic behavior of the system for large t). From Equation 13.14 and Equation 13.15 we get

$$P(k_i(t) < k) = 1 - \frac{m^2 t}{k^2(m_0 + t)} \tag{13.16}$$

By definition

$$P(k_i < k) = \int_0^k P(k_i)dk_i$$

where we omitted the dependency on time for clarity. According to one of the fundamental theorems of calculus, the derivative of a function $F(c) = \int_0^c f(x)dx$ is

$$\frac{\partial F(c)}{\partial c} = \frac{\partial}{\partial c}\int_0^c f(x)dx = f(c)$$

Therefore

$$\frac{\partial P(k_i(t) < k)}{\partial k} = P(k)$$

The derivative of $P(k_i(t) < k)$ (Equation 13.16) is

$$\frac{\partial P(k_i(t) < k)}{\partial k} = \frac{2m^2t}{k^3(m_0 + t)}$$

hence the probability to observe a certain degree $P(k)$ is

$$P(k) = \frac{2m^2t}{k^3(m_0 + t)} \sim k^{-3}$$

In conclusion, the distribution of degrees ends up being a power law with $\gamma = 3$. The exact gamma differs from one system to another depending on the specific properties of the system. For example, some systems evolve not just by adding new nodes but also by removal of existing nodes. The system could also evolve by establishing connections between existing nodes, and not just between new nodes and existing nodes. The relative pace of these events can affect the exponent γ.

Despite their similar properties, it has been argued that the scale-free model does not describe biological systems accurately. For one, the graph properties that we mentioned earlier, such as the average path length or the distribution of degrees, reflect the global properties of the network but do not necessarily characterize graphs uniquely. That is, two different graphs with very different local structures can give rise to the same global properties. A different approach to describe the local structure is to enumerate all possible *induced* subgraphs (called **graphlets**) in the graph [771]. Graphlets can be viewed as network signatures. Unlike subgraphs or partial graphs (referred to as network motifs, as discussed in the next section), an induced graph over a subset of n nodes contains *all* edges in the graph G between the n nodes. That is, a projection of G on a subset of n nodes results in a single induced graph, which can give rise to many partial graphs. Given the complexity of graphlets for large n, topological analysis of a network usually considers only graphlets of five nodes at most. There is only one possible graphlet on two nodes, but two graphlets over three nodes, six graphlets over four nodes (see Figure 13.13) and 21 graphlets over five nodes. In total, there are 30 graphlets of size 2-5, and graphs can be characterized and compared based on the distributions of their graphlets, which better reflect their local structure.

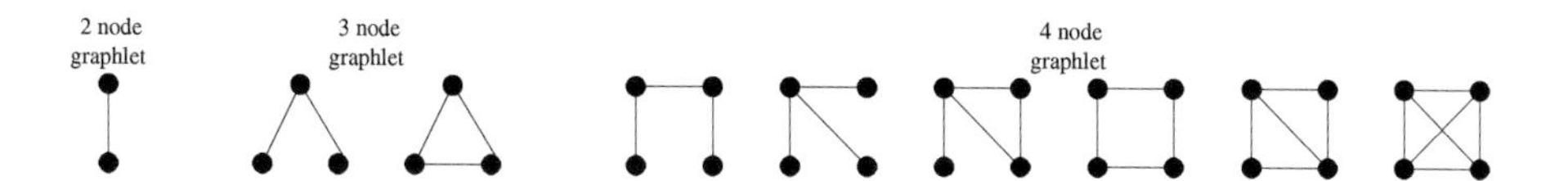

FIGURE 13.13: **Graphlets.**

Using this distribution, it has been suggested that a different class of random graphs, called **random geometric graphs**, describe biological systems more accurately. A random geometric graph $G_d(n, r)$ is generated by placing n random points in the d-dimensional unit square and connecting every two nodes whose distance is less than r. For more details see [772, 773].

13.4.2 Applications

The scale-free model has been applied to many types of networks. This includes protein interaction networks [774, 775], regulatory networks (of transcription factors and the genes they regulate) [776,777], metabolic networks (of substrates that are connected by reactions, catalyzed by enzymes) [778] and even networks of residue contacts in interaction sites (with residues connected if their distance is below a certain threshold distance) [779, 780] or domain co-occurrence networks (where nodes correspond to individual domains and edges are formed between domains that co-occur in the same protein) [781].

Beyond characterizing the general properties of the network and the complexity of the host system, topological analysis can provide insight into fundamental properties of biological systems. For example, the small-world property has been linked to the bi-stability of neural networks, which tend to switch between two stable states [777]. Comparative network analysis can help to understand differences across organisms from a topological point of view and provide deeper insight into the underlying evolutionary mechanisms. Topological analysis is especially useful for identification of central nodes, which correspond to key genes in interaction networks [774], essential substrates in metabolic networks [778], or key residues in protein complexes that are hypothesized to contribute most of the binding energy (hot spots) [780].

Network analysis can also lead to prediction of novel features, from detection of clusters of genes that correspond to complexes and pathways, or prediction of interaction interfaces based on the presence of key residues. Similarly, the small-world properties of interaction networks can be utilized to predict interactions, or assess the reliability of interactions that were determined in error-prone experiments such as yeast two-hybrid. For example, in [782] the probability that each edge in the graph is a true interaction is estimated based on the neighborhood "cohesiveness" around the edge, measured by counting how many triangles it is a side of. The mutual neighbors that form the other sides of the triangles serve to corroborate the edge, as it is less likely to happen by chance in a random graph. This method is mostly effective for complexes or **"central" genes** (often disease-related genes) that have dominated interaction databases so far. The cohesiveness measure seems to correlate well with the reliability of the interaction.

It has been suggested that the significance of scale-free networks is in their robustness, since their topology is more resistant to random perturbations. For example, mutations that knock out nodes have minimal effect on the average path length and the topology in general, unless the node deleted is one of the hub nodes. In that case the impact can be drastic, or even fatal in the case of biological systems. However, the number of hub nodes is insignificant compared to the network size, and hence the probability of that happening by chance is very small. In contrast, a deletion of a node or an edge in a random graph will have a non-negligible effect on the path length on average (and hence on the ability of nodes to communicate); therefore random graphs are more vulnerable to mutations. On the other hand, since they do not have hubs, they are less likely to completely fail if one node is deleted.

Thus, in some respects scale-free networks might have a certain evolutionary advantage over other types of networks. Indeed, biological systems show high tolerance for errors, and knockout mutations often show no phenotypic differences. This is attributed to the scale-free topology, since communication between genes across the network is less likely to be affected, as well as to fail-safe mechanisms [783]. One such mechanism is **redundancy**. Redundancy arises due to gene duplication events that result in multiple genes with similar functions. The extent of this phenomenon is non-negligible: for example, almost 60% of yeast genes have at least one other gene with similar sequence and an overlapping function [784]. Other studies suggest that less than 20% of the genes in prokaryotes and eukaryotes genomes are essential [785–787].

On the other hand, the presence of hub genes that are irreplaceable makes scale-free systems fragile to targeted attacks. Indeed, certain viruses target key genes, like p53 (an important transcription factor that regulates cell cycle and functions as a tumor suppressor by preventing uncontrollable cell division [788, 789]), recruiting the whole cell replication machinery to their benefit, with drastic results. However, connectivity is not necessarily correlated with lethality, and hub genes may be knocked out without a significant effect on the phenotype [790]. For example, it has been shown that cells can survive mutations even in regulatory hubs [791]. This has been explained by the presence of **distributed backup systems**. With distributed systems, the loss of function of a major regulator or even a cell-wide process (such as a regulation program, see Section 14.4) is handled by a collective of multiple correlated genes [791] or by using alternative pathways [792].

13.4.3 Network motifs and the modular organization of networks

The models discussed in the previous section address the formation and growth of complex networks from a purely mathematical point of view. However, biological networks evolve according to somewhat different principles. The addition of new nodes (genes) is mostly done through gene duplication events, which result in the addition of genes that resemble existing genes in the network. Second, genes are not stationary. Once introduced, genes undergo a gradual process of divergence. Mutations, insertions and deletions result in variations on existing genes, which explains why interacting proteins can acquire multiple new partners, not just paralogs of their interacting partners[10]. Similarly, in regulatory networks, new relations can be formed between genes and TFs. These interactions and relations correspond to new edges in the network. Third, the success of an addition depends on how beneficial the addition is to the network. Therefore, there are additional constraints that over time determine whether a new gene or a new relation will be accepted or not.

One of the most interesting aspects of biological systems is their modularity. Consider regulatory networks. Owing to duplications of promoter sites and genes (including transcription factors), a gene may acquire additional regulators and a transcription factor may become a regulator of additional genes over time. However, the process is not completely random, and certain patterns emerge, in the form of small **modules** (subnetworks consisting of several genes) with similar topologies. The set of relations (edges) that make up each module determines its topology. Modules with different functional roles may have similar or identical topologies. Small topologies that occur more frequently than one would expect by chance are referred to as **network motifs**. Network motifs are believed to be the building blocks of larger networks, implementing basic mechanisms of cellular computations. Network motifs can be analyzed in regulatory networks (where edges correspond to regulatory relationships between genes) or interaction networks (where edges represent interactions). The meaning of these motifs clearly depends on the context. Even within the same type of network, the functional meaning of the same motif in different parts of the network may vary.

An interesting study of the regulatory network of *E. coli* revealed several such motifs [793]. In a graph where nodes represent individual genes or operons and directed edges are drawn between transcription factors and the operons they regulate, network motifs are frequent partial graphs. They can be identified by searching the graph connectivity matrix for frequent submatrices of size $n \times n$ that represent the set of connections between n connected

[10]Multiple partners may indicate a basic function that can act upon many other proteins. For example, PTMs (see Section 13.1) are "implemented" through interactions involving short sequence motifs, which are common to many proteins.

vertices. The significance of motifs can be estimated using sampling methods, for example, by counting their frequency in a large population of simulated random networks with the same characteristic as the *E. coli* network. That is, simulated networks should have the same components and the same connectivity (i.e., same rank per node) as the *E. coli* network, as discussed in Section 12.4.5.1 and Section 13.3.5.

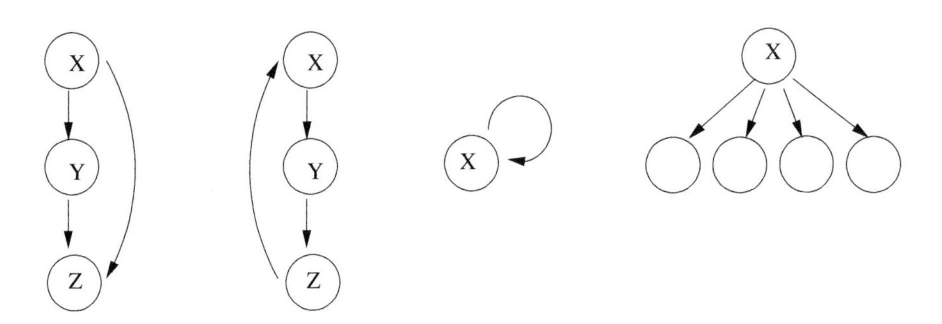

FIGURE 13.14: **Network motifs.** Left: feedforward loop. There are eight total variations over this motif, depending on the nature of the relation (positive or negative) between every pair of genes. The more common variations are +++ and +-+ [794]. The outcome at Z depends on how the signals from X and Y are combined (e.g., AND or OR). Left middle: feedback loop. Right middle: auto-regulation. Right: single input modules (a motif which is common in regulation programs, Section 14.4).

For example, a frequent motif with three nodes is a **feedforward loop** (see Figure 13.14). There are several variations over this motif, depending on whether the regulatory relation between genes is positive (one gene activates another) or negative (one gene inhibits the other). Feedforward loops play an important role in systems that require rapid response times (immediate shutdown) and yet are robust to transient random signals. Another interesting motif is the **feedback loop**. Feedback loops are regulatory motifs that help to adjust and stabilize systems under changes in environmental and cellular conditions (such as concentrations of minerals, co-factors or other molecules which are essential for certain functions). In a feedback loop the output is connected back to the input, and changes in the quantity of the output affect the input. Thus the input can be perceived as a sensor element (e.g. a receptor) which monitors a specific molecule. A change in the quantity of that molecule often results in a conformational change of the sensor. The modified sensor then activates or deactivates another molecule to reduce (or increase) the amount of the molecular output, in order to establish balance. The exact rate or quantity of the molecular output at any given moment depends on the ratio between the modified sensor and the unbound one at that time. Positive feedback loops accelerate the rate of reactions (or transcription), while negative feedback loops slow them down. A special case of a feedback loop is **auto-regulation**, where a protein has the capacity to control and adjust its own activity level, for example, by controlling its own transcription. Auto-regulation can be positive or negative, repressing or enhancing the gene's activity level. Another significant motif is the **single-input module**, where a set of genes or operons is controlled by a single transcription factor. Other motifs implement more complex modes of regulations where operons are regulated by two or more transcription factors. For example, the **overlapping regulation motif** consists of four nodes, where two operons are regulated by the same two transcription factors. Such motifs result in strong co-expression patterns, more so than when different genes are controlled by a single common TF [795].

The high frequency of such network motifs, despite no homology between the constituent genes, suggests that these topologies were re-selected again and again in the course of evolution through combinations of regulatory elements across the network. Indeed, random duplication events may add or remove a regulatory motif (e.g., a promoter) before gene sequences along the DNA, or a binding site in a protein (e.g. a transcription factor), gradually introducing new regulatory relationships between existing elements of the regulatory network. Considered as "easier" and faster than the evolution of new genes capable of regulating other genes, stochastic network modifications provide an efficient mechanism to adapt, modulate and enhance the regulatory network of an organism.

This evolutionary process of network expansion gradually leads to the formation of larger modules from basic network motifs. For example, **regulatory modules** are subnetworks consisting of several genes that are regulated by the same transcription factor(s) and are co-expressed. Not every set of co-expressed genes is a module, and there are other conditions (from cellular location, to the presence of interaction interfaces) that must be met before a new module emerges. This lends more control to the evolution of biological systems. Many of the new modules share genes or regulators with other modules, due to co-inheritance and duplication events. Some of the links (interactions or regulatory relations) between members of an "old" module may be carried over as well to a "new" module, leading gradually to the formation of a more complex and dense network, with overlapping modules. Similar to the evolution of multi-domain proteins, this process resulted in a modular organization of cellular networks from simpler functional elements. This modular organization is not only more efficient, it also increases fault-tolerance, since functional modules often have related backup modules in the same network. Thus, if an essential module is disrupted, another module or pathway can compensate for the loss of function. We will elaborate on regulatory modules in Section 14.4.

13.5 Further reading

The emerging field of systems biology triggered a strong interest in characterizing the interactome of different organisms and under different cellular conditions. The reviews by [796–798] provide excellent summaries of experimental methods for determining interactions. The accumulated knowledge on protein-protein interactions was compiled into several databases, including DIP [693,694], BIND [692], MINT [799], IntAct [800,801] and others [802–805].

The first methods to predict interactions utilized structural data. A review of structure-based methods appears in [797]. Docking algorithms are reviewed in [806–809]. Extensions to multi-unit complexes are discussed in [810]. Because of the wide availability of sequence data, sequence-based methods (including the gene preservation approach and the phylogenetic-profile method) attracted the most attention in the past decade. For reviews of sequence-based methods for prediction of protein-protein interactions see [811–813].

The concept of co-evolution dates back to Darwin. Molecular co-evolution is discussed in [814] and in many publications that followed. It was first applied to the prediction of interactions in [429]. Mirror tree methods gained popularity in recent years, and the reported results on domains and families that are known to interact were promising [741, 815, 816]; however, recent studies [697, 817] indicate that signals of co-evolution are often too weak.

Several studies use hybrid methods to predict interactions. For example, [818] combines information from sources such as DIP, PDB complexes and domain-fusion analysis. Integrative algorithms that are based on SVMs were proposed in [819] and [820], using features such as charge, hydrophobicity and surface area [819] or sequence similarity, homology to other interacting pairs and GO terms [820]. In [818], a simple scoring function is used to combine the different sources, while in [820] a different kernel function is constructed for each information source and then combined together in a single classifier that separates interacting from non-interacting proteins. Other integrative approaches include the study of [760], who use random forests (see Section 7.4.3) combined by majority voting, the study of [764], who use Bayesian networks, and others [821–823].

Properties of interaction networks have been analyzed in many studies (e.g., [691, 824, 825]). In Section 13.4.2 we mentioned several applications of topological analysis. For a review of network analysis, see [826]. Network motifs are discussed in [793,827–832]. A thorough introduction to the subject appears in [833].

This chapter focuses on protein-protein interactions. There are many papers describing methods for prediction of protein-DNA and protein-RNA interactions. For protein-DNA interactions see [378, 834–837]. For protein-RNA interactions see [838–842]. For updates and additional references see the book's website at `biozon.org/proteomics/`.

13.6 Conclusions

- Interactions between macromolecules such as proteins, DNA and RNA are the basis of many reactions that underlie signal transduction, transcriptional regulation and metabolism, among many other cellular processes.
- Interactions between proteins are established through hydrogen bonds, electrostatic and hydrophobic interactions between atoms in the interaction sites.
- Knowledge of protein-protein interactions can help to characterize the function of genes and the cellular processes they participate in. It is especially important for drug design and protein engineering.
- High-throughput experimental methods can detect interactions on a genomic scale, but are noisy and somewhat unreliable.
- Prediction of protein-protein interactions can be done based on protein sequence, protein structure, gene expression data, subcellular location data or a combination of all of the above.
- Most structure-based prediction methods rely on identification of complementary shapes (docking).
- Other structure-based methods search for conserved surface residues which are presumed to contribute most of the binding energy (hotspots).
- Sequence-based methods such as gene preservation, phylogenetic-profiles, and domain fusion utilize information on homologous genes across different organisms to predict interactions.
- Co-evolution methods predict interactions based on correlated patterns of evolution between protein families.
- Predictions based on sequence signatures and domain interactions are among the most successful methods.
- The graph of all interactions in a genome forms a complex network with interesting properties.
- Interaction networks are scale-free networks, where the probability to observe a highly connected node (high degree) decreases with the degree of that node.
- Network analysis can identify motifs and central genes which are crucial to sustain normal function of cells.

13.7 Appendix - DNA amplification and protein expression

13.7.1 Plasmids

Plasmids are usually circular DNA molecules that can replicate in a host cell and naturally occur in several bacteria. One of the uses of plasmids in molecular biology is for gene amplification. This is done by inserting the gene of interest into the plasmid using recombinant DNA technology and then utilizing the bacterial resources to over-express the protein encoded by that gene. The bacterial cells (usually *E. coli*) first undergo a process that makes their cell walls permeable, after which they become "competent" and are able to take up DNA molecules from their surroundings. The bacterial cells are very weak during this stage and are stored frozen at -80°C. When one is ready to do a transformation (insertion of plasmid DNA into the cell), the frozen cells are thawed on ice. Plasmid DNA is added to a small amount of competent cells in a plastic tube on ice. The cell-DNA mixture is then given a heat shock by incubating the tube at a higher temperature (usually 42°C) for 30-40 seconds. This heat shock enables the DNA to penetrate into already weak cell walls. The cells are then rapidly cooled on ice. Once cool, nutrients are added and the cells are allowed to grow and recover for about an hour. These cells are then spread on a surface that again contains nutrients for the cells to be able to grow. Most plasmids contain a known gene that provides the host cell with an antibiotic resistance. This property is then used to select for only those cells that have the plasmid of interest by having an antibiotic molecule in the same plate. All cells that have no plasmid do not survive. This results in formation of colonies of cells that are now ready to over-express the protein of interest, as they contain the necessary plasmid.

13.7.2 SDS-PAGE

SDS-PAGE (Sodium Dodecyl Sulfate Poly-Acrylamide Gel Electrophoresis) is a common technique for verifying the presence of a protein of an expected molecular weight in a given sample. Addition of SDS to a protein sample results in the protein becoming negatively charged, with the net charge on the protein proportional to the molecular weight of the protein. When the charged protein sample is introduced into the poly-acrylamide gel and an electric field is applied, the proteins move in the gel matrix according to their mass, independent of their native charge. The molecular weight of a sample of interest is estimated by using known molecular weight markers. Figure 13.15 shows the markers with indicated molecular weights in the left lane and a purified protein with a molecular weight of 45 kDa (with a cleavable tag) in the middle lane. Minor impurities of varying sizes can be seen in the same lane. The right lane shows the same protein after the tag has been removed by site-specific cleavage, which results in the protein becoming slightly smaller in size. This lane is loaded with more protein than is necessary, resulting in a thick band.

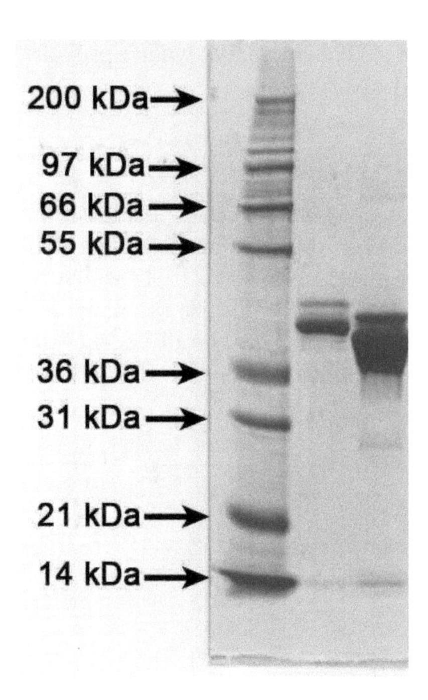

FIGURE 13.15: SDS-PAGE.

13.8 Appendix - the Pearson correlation

Although widely used in mirror tree-based algorithms, the Pearson correlation coefficient is not a very robust test statistic, and it has several properties that should be taken into consideration when applying it for the specific task of detecting correlated divergence:

13.8.1 Uneven divergence rates

The Pearson correlation coefficient assigns high scores for data points with linear correlation, regardless of the slope of the line. This may lead to undesirable situations as presented in Figure 13.16, in which two families are assigned a high correlated divergence score despite the fact that the divergence rate of one family is about twice as fast as the divergence rate of the other family. One possible solution is to measure how well the data fits the line $y = x$.

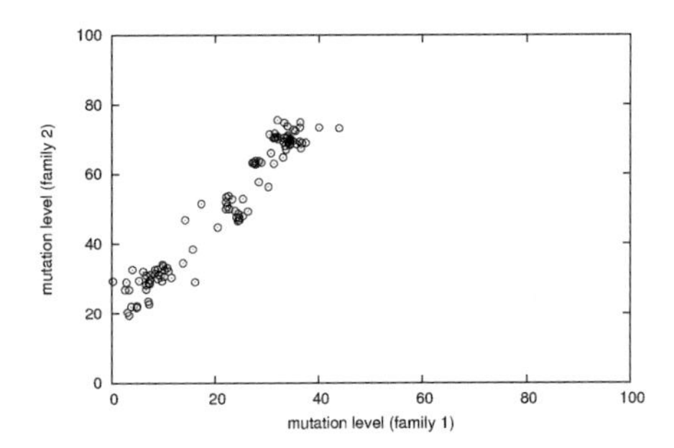

FIGURE 13.16: High Pearson correlation coefficient does not necessarily indicate correlated divergence. The interaction between Biozon 214-1532 (ribosomal protein L10) and Biozon 467-93 (presenilin 1) is assigned a high correlation score ($r = 0.97$), but the divergence rates are different in the two families, with the second diverging twice as fast as the first one. We would expect the data points of two families whose divergence rates are in correlation due to interaction (and not because of similar evolutionary trees in general) to be concentrated around the $y = x$ diagonal. (Figure reproduced from [697] with permission.)

13.8.2 Insensitivity to the size of the dataset

In general, we would consider correlation detected in a larger set of data points more reliable than the same correlation patterns found in a smaller set. However, the Pearson correlation measure assigns the same values for different sets showing similar signs of correlation, regardless of their size.

To address this problem the Pearson score should be normalized with respect to a background distribution of correlation scores of non-corresponding proteins. This is done by randomly permuting the organism order of one of the families and recomputing r as defined above. From this distribution, the mean Pearson correlation score, $\bar{r}$, and standard deviation, $\bar{\sigma}$, are calculated. Along with the true correlation score, r^*, the normalized zscore is calculated as

$$z = \frac{r^* - \bar{r}}{\sigma}$$

The zscore based normalization mitigates the aforementioned problem, since it assigns higher scores to larger sets. However, on the other hand, the size of the dataset tends to dominate this measure, thus creating a bias that can mask differences between signals of correlation, either positive or negative, between the two distance matrices.

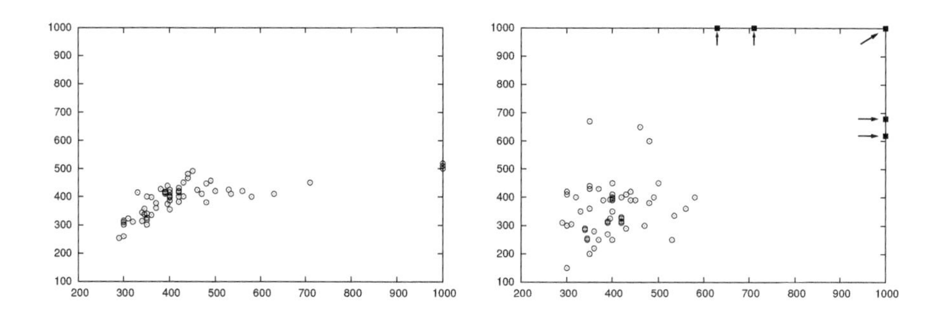

FIGURE 13.17: **Instability of the Pearson correlation coefficient.** Scatterplots of artificial data. **Left:** a dataset showing signs of a real correlation ($r = 0.69$). **Right:** a dataset heavily influenced by a few outliers, marked with arrows ($r = 0.67$ with outliers, $r = 0.17$ without outliers). (Figure reproduced from [697] with permission.)

13.8.3 The effect of outliers

Another issue with the Pearson score is that it can be heavily influenced by a few outliers. Consider the scatterplots of artificial data points given in Figure 13.17. The plot on the left shows signs of correlation, as the data points fit the linear regression model quite well. The plot on the right, however, is heavily influenced by five outliers marked by black squares. When computing the Pearson correlation coefficient for both examples the results are similar ($r = 0.69$ and $r = 0.67$ for the left and right plots, respectively). When the five outliers are removed from the right plot, however, the correlation coefficient decreases to an insignificant score of $r = 0.17$. The case of a few outliers is common when using the correlated divergence model and is usually caused by proteins that are either very similar (almost identical) to the query proteins or by remote homologs that are weakly similar to the query proteins.

There are several approaches one can take to handle these kinds of situations. One possibility is to weigh the entries based on their relative similarity, to decrease the contributions of highly similar or highly dissimilar proteins. There are quite a few weighting schemes that are commonly used when constructing profiles or HMMs from MSAs (for example, see Section 4.3.3.3). However, these methods underweight either the most similar or the most dissimilar proteins, but none of them is designed to underweight both.

Another possibility is to exclude outliers (above a certain threshold) from the distance matrices. Very high values in the distance matrix of one family are usually the result of comparing two distant organisms. In such cases the real co-evolution signal, if it exists, is likely to be masked by noise because of the large number of mutations overall, making it almost impossible to detect signs of correlated divergence. The opposite situation, in which mutation levels are too low, is undesirable as well: the proteins of two close organisms probably did not diverge much, again making it difficult to detect co-evolution.

13.9 Problems

For updates, additional problems, files and datasets check the book's website at `biozon.org/proteomics/`.

1. **Docking - identifying complementary shapes**

   (a) You are given two protein structures. Apply the DFT-based method of Section 13.3.1.1 to search for complementary substructures.

   (b) Assess the biochemical compatibility of the top 10 conformations detected with DFT, using the knowledge-based potential provided in the book's website.

   (c) Compare your top prediction to the known interaction site.

2. **Phylogenetic profiles**

   (a) You are given a dataset with phylogenetic profiles of a set of proteins. Identify the most similar pairs of proteins. Use GO to associate a biological function with each pair. Are the results biologically meaningful?

   (b) Suggest a theoretical method for estimating the statistical significance of the similarity scores between phylogenetic profiles.

   (c) Estimate the statistical significance of similarity scores based on the background distribution. Is it consistent with theoretical estimates? If not, what is the reason?

3. **Mirror trees**

   (a) Given two MSAs of two protein families, assess their correlated divergence using the correlation score as in Section 13.3.2.2. Apply the method to the pairs of MSAs provided in the website. Based on the score, which pairs are likely to interact?

   (b) Compute the correlated divergence using the zscore instead of the raw correlation score. Apply it to the same set of pairs. Are the results different?

4. **Topology of interaction networks**

   You are given a file with all pairs of interacting proteins in yeast.

   (a) Construct the interaction network graph from this file. Compute the clustering coefficient and the diameter of the graph. Are they characteristic of random graphs of the same size?

   (b) Compute the distribution of degrees and show that it follows a scale-free distribution.

5. **Network motifs**

   Given the graph of the previous problem, identify common sub-matrices over 3, 4 and 5 nodes. Assess the significance of each one and characterize the topology of the motifs which correspond to the 10 most significant sub-matrices. Can you associate a function with each motif?

Chapter 14

Cellular Pathways

14.1 Introduction

How does a cell utilize and orchestrate all the individual components it contains to work together and form the complex entity it is? Earlier we mentioned transcription factors (Chapter 12) and promoters (Chapter 5) as the main players involved in monitoring and controlling gene expression. In Chapter 13 we discussed protein-protein and protein-RNA interactions as mechanisms of signal transduction. However, it is the "compilation" of the various molecules into an intricate network of **pathways** that truly underlies the complexity of living organisms[1]. The collective set of pathways and their crosstalk is what makes a cell a highly elaborate and almost autonomous entity, equivalent in a way to a sophisticated operating system.

Pathways are cellular procedures that carry out specific functions in the cell, such as amino acid synthesis and degradation, energy metabolism, signal transduction and more. Consider for example glycolysis, the first pathway that was discovered in the 1920s. This pathway converts glucose (and other sugar molecules, such as fructose and galactose) into another molecule, called pyruvate, for the purpose of storage (Figure 14.1). During the conversion energy is produced in the form of ATP molecules. The pathway is an important component of the catabolic network that generates the energy necessary to maintain cells.

There are numerous other pathways with various functions, many of which have not been characterized yet. Pathway discovery and analysis have been garnering much attention following recent breakthroughs in DNA sequencing technologies, as the number of genomes that are completely sequenced is increasing steadily. One of the main challenges when analyzing this data is to decipher the complicated network of cellular pathways that is coded in each genome. By analyzing this network we wish to map the functional procedures of the cell, understand the functional role of individual genes and identify genes that potentially play a significant role in certain diseases or genetic disorders.

[1] This network is different from the protein interaction network we discussed in Chapter 13, although they are closely related.

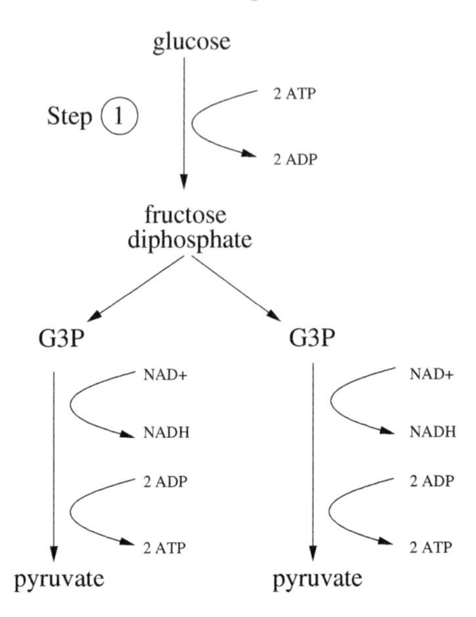

FIGURE 14.1: **The glycolysis pathway.** Glycolysis is a metabolic pathway which processes glucose while releasing energy in the form of ATP molecules. To prevent the glucose from leaving the cell (and reduce glucose levels in the cell, thus keeping the influx of glucose through the cell's membrane), the first reaction in the pathway utilizes ATP to phosphorylate the glucose molecule (as a result the ATP molecule is converted to ADP, which has one less phosphate group). Another ATP molecule is utilized, donating its phosphate group, to form fructose-diphosphate. Subsequent steps split fructose-diphosphate into two molecules of glyceraldehyde 3-phosphate (G3P), each one is further processed and converted into pyruvate while donating phosphate groups to two ADP molecules, thus resulting in a net gain of two ATP molecules (two ATP per one molecule of G3P, minus the two ATP used in step 1 of the pathway).

Each pathway usually involves several reactions and interactions, starting from a certain molecular "input" and producing a certain "output" or "action". The input can be small molecules of a certain type, and the output can be a compound that is made of these molecules and other molecules that are present in the cytoplasm at large. Abundance (or lack thereof) of the input molecules could trigger the pathway or increase its activity. The process of generating the output molecule from the input molecules usually proceeds through a sequence of one or more intermediate steps, each one mediated by RNA molecules or proteins, such as enzymes. The series of steps is not necessarily linear, and often a pathway will split into multiple branches (depending for example on the subcellular location) or work in a cycle to amplify a certain product. A common way of representing pathways is as a diagram such as the ones in Figure 14.1 and Figure 14.2.

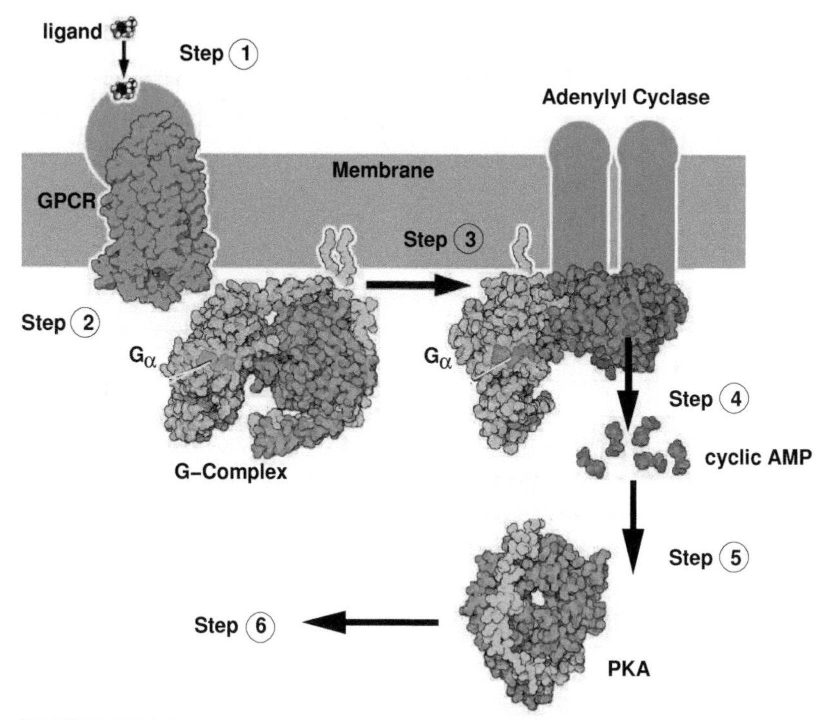

FIGURE 14.2: The cAMP-dependent pathway. (Figure modified from Wikipedia [843]. Original figure created by David S. Goodsell.) The cAMP-dependent pathway is a signal transduction pathway, which transfers extracellular signals into the cell through a signaling cascade. It is an important element of intercellular communication. The pathway involves three main components. The first is a G-protein coupled receptor (GPCR), a membrane protein that traverses the membrane and enables the cell to communicate with other cells and respond to changes in extracellular conditions. The GPCR interacts with a G-protein complex inside the cell. The binding of an extracellular agent (step 1) affects the conformation of the GPCR, and results in detachment of one protein, called G_s alpha subunit, from the G-complex (step 2). There are many different GPCRs, which bind to a wide variety of ligands (such as adrenaline or caffeine). Different ligands result in unique conformational changes in the GPCR, and as a result, each GPCR can interact with different G-protein complexes inside the cell. The different G-protein complexes can further pass the signal to other proteins, activating different pathways. In the cAMP pathway, the released G_s subunit binds to an enzyme called adenylyl cyclase (step 3). As a result the enzyme is activated and catalyzes a reaction that converts ATP to cyclic adenosine monophosphate (cAMP) in step 4. Abundance of cAMP may then result in activation of several other proteins (step 5), such as Protein Kinase A (PKA). PKA works by phosphorylating other proteins (see PTMs, page 582) and is involved in multiple cellular processes (step 6). These include, for example, regulation of fat breakdown (lipolysis), cell-aging processes [844], and regulation of cell adhesion and migration [845]. The pathway is deactivated when GDP levels in the cell are high, as the G-protein complex is locked in the inactive state (step 2). When uncontrolled, the cAMP-dependent pathway has been implicated in uncontrolled growth (cancer) [846, 847].

A cell's functional capacity depends on its internal procedures. Some pathways are ubiquitous and exist in cells of all organisms from all kingdoms of life, but other pathways exist only in certain organisms (e.g., photosynthesis is unique to plants and certain aqueous organisms). Larger genomes typically encode not only for a larger number of genes but also for more pathways than smaller genomes, diversifying the functions they can execute. Furthermore, owing to the genetic variation across organisms, organisms of comparable size have different pathways.

The goal of pathway research is to chart the complete network of pathways in each organism. Hundreds of pathways have already been determined and characterized over the years. These pathways were compiled into publicly available databases such as MetaCyc [848] and KEGG [849]. Beyond storing and analyzing the biochemical information on cellular processes, pathway research also strives to extend and improve the known data by mapping pathways in newly sequenced genomes, identifying variations to pathways in different organisms and discovering novel pathways. Identifying the active pathways in different organisms is important for identifying the key processes that are essential to sustain life and the pathways that are unique to each organism. This knowledge can also shed light on the mechanisms a cell uses to acquire its functional role.

In the next sections we will discuss methods for pathway analysis and prediction. Specifically, we will focus on metabolic pathways. As we will see, expression data plays an important role in this type of analysis.

14.2 Metabolic pathways

Metabolic pathways are pathways which consist mostly of enzymatic reactions. Metabolic processes make up a substantial part of the cell's activity and are essential to maintain internal equilibrium (called **homeostasis**) under different physiological and environmental conditions. For example, metabolic pathways are involved in synthesis of various components of the cell, from amino acids to cell walls. They are also an essential part of the energy-producing machinery in cells, generating ATP through breakdown of complex molecules into simpler ones (a process called **catabolism**) or recycling of stored ATP derivatives (such as ADP or AMP). Many metabolic pathways have been studied and documented in the literature, usually in specific model organisms such as yeast or *E. coli.*

The enzymatic reactions of a metabolic pathway are typically organized in a sequence, where the product of one reaction is the input for the subsequent reaction. Two examples are shown in Figure 14.3. Each reaction involves a certain molecular compound (called **metabolite** or **substrate**) and modifies it through synthesis or breakdown. To be completed, some reactions require minerals or vitamins (called enzyme **co-factors**). Co-factors can also inhibit or activate enzymes. The whole pathway results in a transformation of a metabolite (which can be an intermediate compound of another pathway) to another output compound (called **product**). The output molecule may be utilized by other molecules, it may trigger another set of reactions and actions, it may be stored in the cell for future needs or it may be the input for another pathway. Either way, the number of possibilities to manipulate the molecular content of the cell is enormous, resulting in sophisticated mechanisms of cell monitoring and regulation, ensuring stability under a wide range of conditions, and producing complex procedures with diverse functions.

Each reaction in a metabolic pathway is catalyzed by a specific type of enzyme. There are many types of enzymes, depending on the molecules (substrates) they act upon and the reaction they catalyze. To distinguish between their many functions, enzymes are usually characterized using **Enzyme Classification (EC) numbers**. The enzyme classification system is a hierarchical scheme devised by the IUB Enzyme Commission [851] to organize enzymes into functional classes. The EC number of an enzyme is of the form A.B.C.D, where the first digit A indicates the enzyme's major functional family: oxidoreductase, transferase, hydrolase, lyasase, isomerase, or ligase. The second digit B places the enzyme in a functional subfamily, the third into a sub-subfamily and so on. For example, the EC number 1.2.3.4 designates enzymes that are oxidoreductase[2] (the first digit) that act on compounds containing aldehyde or oxo groups as donors (the second digit) and have oxygen as an acceptor (the third digit). In this case, the last digit specifies the particular reaction that is catalyzed by this family of enzymes ($C_2O_4^{2-}$ + oxygen $\leftrightarrow H_2O_2 + 2CO_2$). Enzymes whose functions are poorly understood have incomplete EC numbers, i.e., they have only been assigned the first few digits of an EC number, such as 1.2.3.- or 1.2.-.-. Interestingly, some enzymes can work on multiple substrates. Furthermore, an enzyme can be associated with more than one EC number if it has multiple sites or domains that can catalyze different reactions.

Accurate EC annotation of proteins is an important step toward pathway prediction and functional annotation of genes in general (protein classification and function prediction are the goals of the models we discussed in Chapters 4 to 7).

[2]Oxidoreductases are enzymes that catalyze reactions in which one molecule is oxidized (loses electrons) while the other is reduced (accepts electrons). The first is called a donor while the latter is called an acceptor.

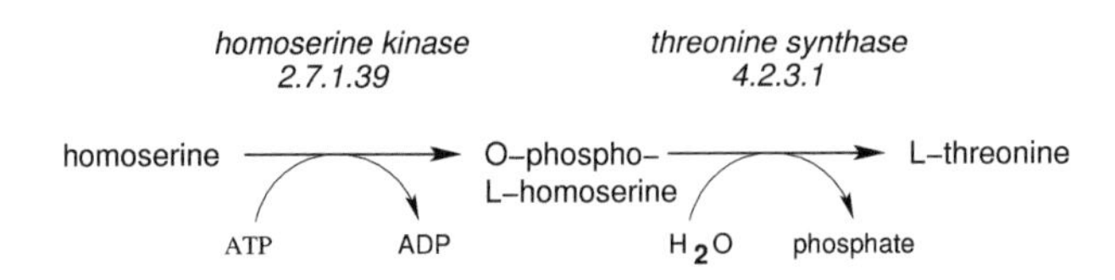

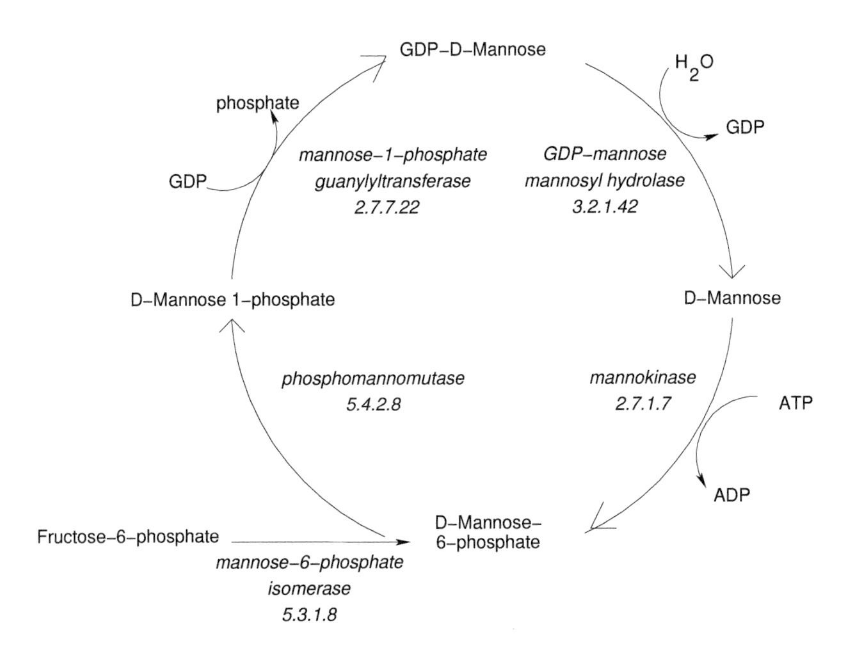

FIGURE 14.3: **Top: Glycine biosynthesis pathway. Bottom: I/GDP-mannose metabolism.** (Pathway layouts were recreated with permission from diagrams in the MetaCyc database [850]). Metabolic pathways consist of sequences of consecutive enzymatic reactions. Each reaction starts with a certain metabolite and produces another metabolite through synthesis, breakdown or other transformations. The glycine biosynthesis pathway produces the amino acid glycine from the amino acid serine. The GDP-mannose metabolism is one of the pathways that produce the compounds necessary to build cellular structures and organelles, such as cell walls.

14.3 Pathway prediction

Experimental reconstruction of pathways is a long and costly process. In fact, the known pathways have been verified and studied extensively only in a few organisms. Therefore, there is a great interest in computational methods that can extend the existing knowledge about pathways to newly sequenced genomes and predict novel pathways. Given the number of molecules involved and the number of possible combinations, computational pathway analysis can greatly speed up experimental verification of pathways. As an example, we will discuss the case of metabolic pathways.

14.3.1 Metabolic pathway prediction

The collection of all metabolic pathways forms a complex network with many inter-connections between pathways, referred to as **crosstalk**[3]. Over the years, a significant amount of effort was devoted to the reconstruction of *cell-wide* metabolic networks, rather than just individual pathways. Metabolic networks were constructed manually from a variety of data sources and literature for selected organisms, including *Escherichia coli* [853–855], *Haemophilus influenzae* [856] and *Saccharomyces cerevisiae* [857, 858]. Manually reconstructed networks are more likely to be biologically plausible; however, they require close human intervention and are naturally labor intensive.

Automated network reconstruction can be divided into two main categories: unsupervised and supervised. Unsupervised methods based on models such as boolean networks [859] and Bayesian networks [621] were applied for cellular network inference from high-throughput datasets such as expression data and protein-protein interaction data. These models were applied to infer mostly regulatory networks, but can also be utilized to infer other types of networks, including metabolic. We will elaborate on unsupervised learning of gene networks based on Bayesian networks in Chapter 15. In this chapter we will focus mostly on supervised methods.

The supervised approach to automated metabolic pathway reconstruction is to create a metabolic network graph from known enzymatic reactions and chemical rules. Chemical rules are constraints characterizing consecutive reactions, where compatibility can be assessed for example by checking the

[3]Crosstalk between pathways plays an important role in pathway regulation. One interesting example is the mitogen activated protein kinase (MAPK) pathway which regulates a diverse set of cellular functions such as cell proliferation, differentiation and motility [852]. This is achieved with a topology which traverses multiple subcellular locations, feedback loops (see Section 13.4.3) and crosstalk with other pathways. Since some reactions can be processed by multiple proteins, crosstalk provides multiple ways to interface, intervene and regulate the pathway.

EC numbers of the reactions (see Problem 14.2). Edges are introduced between chemically compatible reactions, or the corresponding enzymes. The resulting graph can then be analyzed and delineated into local subnetworks, which can be considered as pathways. However, chemical compatibility is not sufficient when constructing such graphs, and spurious edges may be common [860]. To improve prediction, a search for pathways can be directed by considering also the substrates and products that can lead from one reaction to another [861, 862]. Other related studies attempt to find the shortest path between enzymes (referred to as their **metabolic distance**) [732, 863]. To eliminate false edges, edges can be further verified with other datasets such as protein-protein interactions, subcellular location information, gene locality data or expression data (see Further reading). For example, an edge between two genes that physically interact or are co-expressed is more likely to be valid (Problem 14.3). Such automated procedures are capable of detecting novel pathways. On the other hand, given the size of the search space, these procedures may result in predictions which are inconsistent with known pathways.

Here we will focus on prediction based on **pathway blueprints**. Most pathways exist in multiple organisms, with slight variations. The different variations of the same pathway are "homologous" to each other in the sense that they have the same overall structure. In each organism, different genes fill the various roles in the pathway. Hence, a pathway can be viewed as a generic graph diagram or a blueprint, where the nodes of the graph are the locations in the pathway, annotated with just the function associated with each location (see Figure 14.3). This is the way pathways are represented in pathway databases such as KEGG and MetaCyc[4].

The most popular approach for pathway prediction is based on extrapolation from one organism to another. Procedures developed by KEGG, MetaCyc and others use blueprints of pathways collected either from biochemical charts or from actually observed pathways in specific organisms. These procedures map genes from a newly sequenced genome onto these diagrams based on homology between genes across organisms, as well as database annotations (e.g., [864–866]).

With pathway blueprints, the problem of pathway prediction is essentially a problem of gene assignment (that is, assigning genes to the nodes of the pathway diagram). Clearly, accurate assignment of genes to pathways is crucial in order to correctly map the existing pathways in a given genome. By assigning genes to pathways one can better understand the exact functional role of these genes and identify key genes whose presence is crucial to ensure normal cell functionality. Most of the existing methods for metabolic pathway prediction are based on pathway blueprints, assigning genes to pathways simply

[4]The KEGG representation is slightly different from that of MetaCyc. In KEGG nodes correspond to enzymes and edges are introduced between successive reactions in a pathway, such that each is associated with a certain compound [849]. In MetaCyc, nodes correspond to compounds and edges are associated with enzymes [848].

based on their EC designation. For example, to map the glycine biosynthesis pathway of Figure 14.3 in a specific genome, all we need to do is to identify genes which can catalyze the reactions 2.7.1.39 and 4.2.3.1.

14.3.2 Pathway prediction from blueprints

While simple and straightforward, automated pathway reconstruction from EC numbers does not necessarily produce biologically valid metabolic networks, and requires manual inspection to verify the correctness of these predictions. In the next sections we will discuss some of the issues with automatic pathway prediction and the role of expression data in addressing these problems. Specifically, we will address the following problems:

14.3.2.1 The problem of pathway holes

To extrapolate metabolic pathways we need a functional association of genes (and their protein products) with enzymatic reactions or enzyme families. However, many genes have unknown function and therefore the cellular processes in which they participate remain largely unknown. We refer to a node in a pathway diagram as a **pathway hole** in a particular organism, if the enzyme corresponding to this node is unknown in that organism. Only proteins from that organism with the proper enzymatic activity can be candidates for filling the hole.

To identify the set of candidate enzymes for each pathway hole we can compare the genome against related enzymes (that is, enzymes with the correct EC number from other genomes), using BLAST or other sequence-based methods. The most similar gene can be marked as a potential member of the pathway. However, enzymes with similar enzymatic activities are not necessarily similar in sequence (often referred to as **enzyme recruitment** [867, 868]) and therefore it is necessary to consider also other genes as candidates. Clearly, identifying related enzymes without sequence similarity is a difficult problem[5]. Furthermore, in some pathways, the enzymatic activity of certain nodes has not been fully characterized and is associated with a partial EC number such as 1.2.-.-. In such cases, there is a broader set of candidates that should be evaluated consisting of all enzymes that match the partial EC number. Identifying the correct enzyme in this case is even more difficult. The larger the candidate set, the higher the chances we pick the wrong one [870].

14.3.2.2 The problem of ambiguity

The problem of pathway assignment is further complicated by the multiplicity of reactions and enzymes. First, certain reactions appear in multiple pathways. Second, in most genomes some reactions can be associated with

[5] In some cases related enzymes can be detected even without sequence similarity, using for example SVMs or the decision tree model we discussed in Section 7.4 (see also [264]). If the structure is known then enzymatic function can be sometimes predicted from structure [869].

multiple genes (that is, there are multiple enzymes with the same EC number). Therefore, a straightforward assignment of all enzymes that can catalyze a reaction (termed **isozymes**) to all pathways that contain that reaction leads to nondiscriminatory assignments, where certain enzymes are associated with multiple pathways (see Problem 14.5). In other words, pathway prediction based only on EC numbers results in *many-to-many* ambiguous mapping between genes and pathways.

The extent of this problem is not negligible. Of the 469 pathways in release 7.0 of MetaCyc, 336 have at least one reaction in common with another pathway. For instance, the oxidization reaction associated with EC number 1.1.1.37 is part of several different pathways, including mixed acid fermentation, gluconeogenesis, superpathway of fatty acid oxidation and glyoxylate cycle, respiration, and more. In yeast there are three genes that can catalyze this reaction (referred to as malate dehydrogenases). In theory, each of the three isozymes can be involved in all these pathways.

However, it is unlikely that *all* genes with the same EC designation are used in *all* pathways that contain the corresponding reaction. Rather, it is more likely that different genes are used in different pathways. It has been suggested [732,871] that different isozymes are independently co-regulated with different groups of genes, helping to regulate reactions that are shared by different pathways or switch between alternate routes. In other cases, the multiplicity is an evidence of backup systems (see Section 13.4.2). Unfortunately, this information is sparse and without additional experiments it is hard to make this type of functional differentiation.

Another source of ambiguity is that an EC number might specify not a single reaction but rather a class of reactions having common characteristics. One such example is the family of alcohol dehydrogenases which oxidize a variety of alcohols. All alcohol dehydrogenases are associated with the EC number 1.1.1.1 although they act on different substrates. In yeast, for instance, there are six enzymes annotated with 1.1.1.1. Therefore, EC numbers might not be specific enough, and even database annotations may be insufficient to differentiate between the different functions of these enzymes.

14.3.3 Expression data and pathway analysis

Addressing the problems of ambiguity and pathway holes mentioned above and narrowing down the candidate set of genes to the right gene(s) requires utilizing additional information. One approach is to consider only enzymes that have phylogenetic profiles (see Section 13.3.2.1) similar to the profiles of nearby enzymes in the pathway [872]. Proteins that participate in the same pathway are expected to be conserved through evolution and exist in all genomes where the pathway is present. However, filtering based on phylogenetic profiles may not be selective enough since many genes are conserved across multiple organisms. Another approach is to utilize protein interaction data, or datasets indicative of protein interactions such as gene fusion data, protein localization and gene expression data (see Further reading). Proteins that are known to interact with other members of the pathway are more likely candidates to be part of the pathway. Note however that lack of interaction does not exclude the possibility that a protein participates in a pathway.

Expression data is an important source of information on the activity of genes, which can help to assess the relevance of different genes in each pathway. By analyzing expression data we can identify groups of genes that are co-expressed, i.e., that are likely to exist in the cell at the same time or under the same set of conditions. Although the sequence of reactions in a pathway does not take place simultaneously, these reactions can be considered to occur instantly and simultaneously for all practical purposes, given the time resolution of mRNA expression measurements. Therefore, it is expected that genes that participate in the same pathway will have similar expression profiles, i.e., they will co-exist and will be concurrently available at the cell's disposal to complete the pathway (see Figure 14.4). Furthermore, expression data can also reveal control mechanisms and pathway switches. For example, when a pathway has a fork, isozymes might be used to switch between the alternate routes, resulting in anti-correlated expression profiles [871].

Indeed, it has been shown that genes that are co-expressed are likely to participate in the same pathway [29,873]. Furthermore, it has been suggested that correlation in expression profiles is an indication of linear pathways that consist of sequences of reactions [871]. Based on this premise many studies analyzed expression data in search of biologically significant sets of co-expressed genes (see Section 12.5). Searching for co-expressed subnetworks is also the methodology of studies that seek functional modules, as discussed in Section 14.4. However, co-expressed subnetworks do not necessarily correspond or overlap with known pathways and when utilized regardless of pathway blueprints, expression data does not necessarily improve pathway prediction.

Nonetheless, pathway analysis can definitely benefit from expression data, especially when combined with pathway blueprints and metabolic information. We already mentioned the role of expression data in studies that reconstruct metabolic networks from basic chemical reactions, both with unsupervised methods and supervised methods (Section 14.3.1). Other studies utilize metabolic information when clustering genes based on expression data, to obtain clusters that better correlate with pathways (see Further reading). Expression data can also be used to score known pathways, verify their existence in a genome and assess their activity level, in search for "interesting pathways" (for example, those whose aggregated transcription is most affected by certain knockout experiments or under certain conditions, see Problem 14.7). Pathway expression analysis is especially important in pharmaceutical studies, for example, when evaluating the toxicity of drugs on pathways (see Section 12.5).

In the next sections we will focus on the role of expression data in prediction based on pathway blueprints. We will discuss two algorithms that use expression data when assigning genes to metabolic pathways, while addressing the problems we mentioned above.

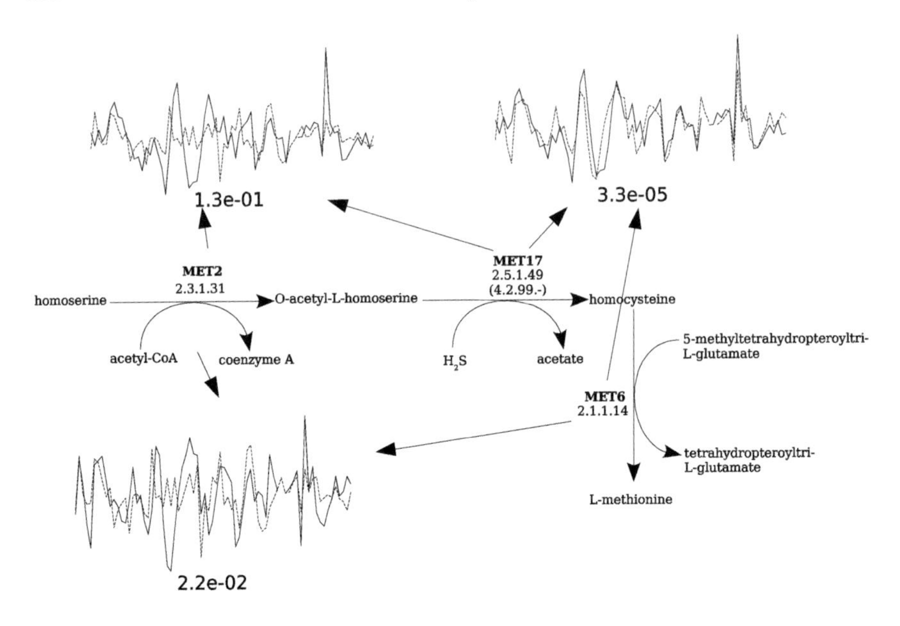

FIGURE 14.4: The homoserine methionine biosynthesis pathway in yeast. The diagram was obtained with permission from the MetaCyc database and augmented with genes that are experimentally verified to participate in the pathway and their expression profiles. There is a strong correlation between the genes catalyzing the last two reactions while the first gene is less correlated. (Profiles shown are for the yeast cell-cycle dataset). (Figure reproduced from [870] with permission from Imperial College Press.)

14.3.3.1 Deterministic gene assignments

One way to assign genes to pathways based on expression data is by looking for assignments that maximize the co-expression of genes that are assigned to the same pathway.

A search for pathways starts by mapping *probable pathways*. Given a genome $\mathbf{G}$ with N genes, a set of enzyme families $\mathbf{F} = \{F_1, F_2, ..., F_M\}$ and pathways $\mathbf{P} = \{P_1, P_2, ..., P_K\}$, the straightforward approach we mentioned before is to match each pathway $P_k \in \mathbf{P}$ with enzymes that can catalyze the reactions that make up the pathway. Denote by $\mathbf{F}(P_k) = F_{k1}, ..., F_{km} \quad F_{kj} \in \mathbf{F}$ the set of protein families that are associated with the reactions of pathway P_k and let $\mathbf{G}(F_j)$ represent the set of genes that can be assigned to enzyme family F_j based on their database records (or based on their similarity with known enzymes of family F_j). Pathways that are only partly covered (e.g., less than half of their reactions can be associated with genes in $\mathbf{G}$) may be eliminated, since it is less likely that they exist in the given genome[6].

Given a probable pathway P_k with m reactions, there are $|F_{k1}| \times |F_{k2}| \times \cdots |F_{km}|$ possible assignments of genes to reactions, consisting of all possible combinations of genes in the pathway families $F_{k1}, F_{k2}, ..., F_{km} \in \mathbf{F}(P_k)$. A simple approach to assessing the biological significance of each one is to compute the total correlation score between genes, as in [874], i.e., the score of assignment $\mathbf{A} = (g_1, g_2, ..., g_m)$ s.t. $g_i \in F_{ki}$ can be defined as the average co-expression score

$$score(\mathbf{A}(P_k)) = \frac{2}{m(m-1)} \sum_{i=1}^{m-1} \sum_{j=i+1}^{m} w_i w_j \cdot sim(\mathbf{e}_i, \mathbf{e}_j) \tag{14.1}$$

where $sim(\mathbf{e}_i, \mathbf{e}_j)$ is the expression similarity of genes g_i and g_j using any of the measures we discussed in Section 12.4.1 and w_i is a weight coefficient that represents the likelihood that gene g_i belongs to family F_{ki}. The weight can be defined based on the similarity score of gene g_i with a model (such as a profile, HMM or PST) representing the enzyme family[7].

An initial assignment of genes to pathways is obtained by selecting the assignment that maximizes the average correlation score out of all possible assignments to pathway P_k and repeating the procedure for each pathway. This initial set of assignments is likely to produce a good mapping between genes and pathways. However, since each pathway is analyzed independently it might happen that the same gene is assigned to the same reaction in multiple pathways. Each such assignment is considered a *conflict*. In some cases, the same gene may play the same or similar functional roles in different pathways. However, if a certain reaction is part of multiple pathways and there are multiple enzymes in the same genome that can catalyze the reaction, then it is more probable that each enzyme is "specialized" to catalyze this reaction in a different pathway.

To eliminate the conflicts we need to revisit the assignments and consider alternative genes, whenever possible. To find the best global assignment without conflicts, all pathways have to be considered at once. We will represent the set of pathways using a relation graph where each pathway is a node, and two nodes are connected by an edge if the two pathways represented by the nodes share a reaction (see Figure 14.5a). One edge is introduced for each

[6]The accuracy of this initial screening depends on the accuracy and completeness of the functional annotation of genes in the genome $\mathbf{G}$.

[7]One simple way to assign weights is to define

$$w_i = \frac{\log evalue(i)}{\min_{i' \in F_i} \{\log evalue(i')\}}$$

where $evalue(i)$ is the significance of the match between gene i and the model of family F_i. For example, assume the best match with family F_i is observed for an annotated gene with evalue of 10^{-20}. Then a gene that is classified to that family with evalue of 10^{-10} is assigned a weight of 0.5.

such reaction (i.e., there might be multiple edges connecting the same two nodes). The *pathway conflict graph* is derived from this graph: an edge is marked as a conflict if the corresponding reaction is associated with the same gene in both pathways, based on the initial assignments (see Figure 14.5b).

The pathway conflict graph can be split into connected components, each one containing several pathways connected by edges (reactions), some of which are marked and indicate possible conflicts. If several genes are associated with a marked reaction, it might be possible to resolve this conflict. Clearly, the assignments in one connected component have no effect on the other connected components, and therefore we can revisit these assignments independently for each connected component.

To find the best assignment of genes to pathways in a connected component we can generalize the scoring function of Equation 14.1. That is, the score of an assignment is defined as the sum of the scores of the assignments to pathways contained in the component

$$score(\mathbf{A}(\mathbf{P})) = \sum_i score(\mathbf{A}(P_i))$$

where $\mathbf{A}(P_i)$ is the subset of genes assigned to pathway P_i and $score(\mathbf{A}(P_i))$ is as defined previously. Inter-pathway expression data correlations are ignored, assuming different cellular processes are independent of each other (in reality, the situation is more complex, as crosstalk between pathways is common). As before, we have to consider all possible combinations of genes in the Cartesian product of all families associated with reactions in the connected component. However, to eliminate conflicts we also introduce the restriction that no enzyme can be used twice to catalyze the same reaction in different pathways.

It should be noted that not all conflicts can be resolved. If in the given genome there is only a single gene that can be associated with a specific reaction, then clearly it is not possible to avoid conflicts associated with that reaction[8]. Even when the algorithm produces a non-conflicting assignment, it may score significantly lower than the highest scoring assignment with conflicts. Indeed, some shared reactions are central and are best catalyzed by the same enzyme. In such cases, the other genes that can catalyze the reaction may be considered as alternative ("backup") genes which may be invoked only under specific cellular conditions.

Another interesting observation is that multi-domain enzymes are more likely to contribute all their domains to the same pathway [874]. This is not

[8]In some cases there might be other genes that can catalyze the reaction, but they are not annotated as such. Considering a broader set of genes for each family can sometimes help to identify additional candidate enzymes, based on the correlation of their expression profile with other members of the pathway. However, correlation in itself is not enough to indicate catalytic activity.

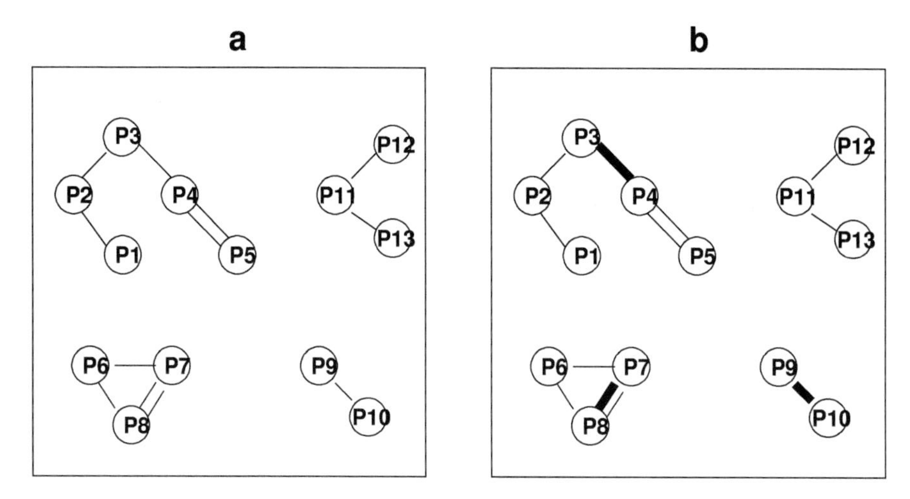

FIGURE 14.5: **Pathway graphs.** (a) The pathway relation graph. Each pathway is represented as a node, and an edge is drawn between two pathways for each reaction that they share in common. (b) The pathway conflict graph. Thick edges represent conflicts (i.e., the same gene was assigned to catalyze the same reaction in both pathways connected by the edge). (Figure reproduced from [874].)

surprising, as multi-functional proteins would be thermodynamically favorable in pathways. If two reactions in a pathway can be catalyzed by the same protein, the efficiency of the reaction can significantly increase, since it saves the need to localize and control the expression of multiple proteins. If the two reactions are consecutive, it is quite likely that the output of one reaction is immediately transferred as an input to the second reaction catalyzed by the second domain. To account for this scenario and create a natural bias toward multi-domain proteins we can use the self-similarity score in Equation 14.1 when assigning these genes to two (or more) different reactions within the same pathway. With that bias, multi-domain proteins will be preferred whenever some or all their domains can be used in the same pathway.

14.3.3.2 Fuzzy assignments

The algorithm described in the previous section assigns a single gene to each reaction. However, in reality this is not always the case. Different genes might catalyze the same reaction in the same pathway, but with different affinities. In some cases multiple isozymes might participate in the same pathway in response to slightly different conditions and substrates. Other enzymes have low specificity and can accept diverse substrates and therefore catalyze several different reactions. Therefore, a more reasonable approach would be to assign a probabilistic measure to indicate the level of association of each gene with a given pathway.

A probabilistic approach seems also more appropriate since pathways are not well-defined entities, and many of them are tightly related. Since cellular processes form a complex and highly connected network it is difficult to delineate the boundaries of individual pathways. Indeed, there has been disagreement about the definition of certain pathways (see [874]). This further motivates a probabilistic approach that assigns each gene with a certain probability to each pathway.

Formally, given a genome $\mathbf{G}$ with N genes, enzyme families $F_1, F_2, ..., F_M$ and pathways $P_1, P_2, ..., P_K$, the goal is to predict which genes take part in each pathway and determine their level of association with each pathway (pathway assignment probabilities), now assuming that the same gene can participate in multiple pathways. In other words, we need to compute the probability $p(i|P_k)$ of gene i participating in pathway P_k as well as the posterior probability $p(P_k|i)$, which can be referred to as the **affinity of gene i with pathway P_k**.

How should we define the probabilities $p(i|P_k)$ and $p(P_k|i)$? Naturally, we can utilize measurements associated with genes, such as their expression levels under different conditions. We will assume that each cellular process (in this case a metabolic pathway) can be modeled as a statistical source generating measurable observations over genes. That is, each gene i is associated with a feature vector $\vec{x}_i$, and the conditional probability $p(\vec{x}_i|P_k)$ denotes the probability that the k-th source emits $\vec{x}_i$. In other words, we estimate $p(i|P_k) \sim p(\vec{x}_i|P_k)$, assuming that genes participating in the same biological process have similar features (e.g., are similarly expressed).

The observations can be a combination of different types of measurements. Assume for now we have only expression profiles that are generated from multiple experiments, i.e., we approximate $p(\vec{x}_i|P_k) \sim p(\vec{e}_i|P_k)$ where $p(\vec{e}|P_k)$ is the probability of observing expression vector $\vec{e}$ in pathway P_k. In other words, we assume that each pathway can be modeled as a probabilistic source with a *characteristic* expression profile, generating expression profiles for the pathway genes with different probabilities. For simplicity we will assume that $p(\vec{e}|P_k)$ follows a Gaussian distribution $N(\vec{\mu}_k, \mathbf{\Sigma}_k)$ or a mixture of Gaussian sources (if there are several underlying processes or branches, intermingled together). In addition, each pathway is associated with a prior probability $p(P_k)$.

Expression profiles are typically composed of measurements taken from a set of independent experiments. For example, in time-series datasets (see Section 12.2.1) the measurements are collected at different time points usually spaced at relatively large time intervals during which the cell has undergone significant changes and the correlation between consecutive time points is relatively weak. Other datasets (e.g., Rosetta) are generated from experiments that are conducted practically independently under different conditions. Therefore, it is usually safe to assume that the microarray experiments are independent of each other such that the expression vector $\vec{e}$ is composed of d *independent* measurements $\{e_1, e_2, ..., e_d\}$, i.e., $p(\vec{e}|P_k) = \prod_{l=1}^{d} p(e_l|P_k)$.

Hence, the multi-dimensional normal distribution can be replaced with a product of one-dimensional normal distributions $N(\mu_{kl}, \sigma_{kl})$, one per experiment, and the covariance matrix $\mathbf{\Sigma}_k$ is reduced to a diagonal matrix whose non-zero elements are denoted by $\vec{\sigma}_k$.

In order to obtain a useful model for prediction we first have to learn the parameters of the model $\vec{\theta} = \{(\vec{\mu}_1, \vec{\sigma}_1), (\vec{\mu}_2, \vec{\sigma}_2), ..., (\vec{\mu}_K, \vec{\sigma}_K)\}$. We can employ an Expectation-Maximization (EM) algorithm, similar to the one described in Section 10.3.2, to find the parameters that maximize (locally) the likelihood of the data

$$p(\mathcal{D}|\vec{\theta}) = \prod_{i=1}^{N} p(\vec{e}_i|\vec{\theta}) = \prod_i \sum_k p(\vec{e}_i|P_k)p(P_k)$$

The EM algorithm consists of two steps: the E-step and the M-step. The parameters of the pathway models $\vec{\mu}_k$, $\vec{\sigma}_k$ and $p(P_k)$ can be initialized based on prior knowledge of metabolic reactions, database annotations and similarity data, as described in Section 14.3.4. These initial parameters as well as the probabilities $p(P_k|\vec{e}_i)$ are then revisited and recomputed based on experimental observations until convergence to maximum likelihood solutions. Denote the parameters in iteration t by $\vec{\mu}_k^t$, $\vec{\sigma}_k^t$ and the likelihood and posterior probabilities in iteration t by $p(\vec{e}_i|P_k)^t$ and $p(P_k|\vec{e}_i)^t$, respectively.

Expectation step: For each gene i and pathway k compute the posterior probability (the pathway affinity)

$$p(P_k|\vec{e}_i)^t = \frac{p(\vec{e}_i|P_k)^t \cdot p(P_k)^t}{\sum_{k'} p(\vec{e}_i|P_{k'})^t \cdot p(P_{k'})^t} \tag{14.2}$$

where

$$p(\vec{e}|P_k)^t = \prod_{l=1}^{d} p(e_l|P_k)^t = \prod_{l=1}^{d} \frac{1}{\sqrt{2\pi}\sigma_{kl}^t} \exp\left[-\frac{(e_l - \mu_{kl}^t)^2}{2(\sigma_{kl}^t)^2}\right] \tag{14.3}$$

Maximization step: Maximize the likelihood by recomputing the parameters for every pathway P_k.

$$\vec{\mu}_k^{t+1} = \frac{\sum_{i=1}^{N} p(P_k|\vec{e}_i)^t \vec{e}_i}{\sum_{i=1}^{N} p(P_k|\vec{e}_i)^t}$$

$$\vec{\sigma}_k^{t+1} = \frac{\sum_{i=1}^{N} p(P_k|\vec{e}_i)^t (\vec{e}_i - \vec{\mu}_k^{t+1})^2}{\sum_{i=1}^{N} p(P_k|\vec{e}_i)^t}$$

$$p(P_k)^{t+1} = \frac{\sum_{i=1}^{N} p(P_k|\vec{e}_i)^t}{\sum_{k'=1}^{K} \sum_{i=1}^{N} p(P_k'|\vec{e}_i)^t}$$

Termination: Exit if $|\mu_k^{t+1} - \mu_k^t| < \epsilon$ for every k or after repeating for a maximum number of iterations.

Upon convergence, this iterative procedure results in steady state pathway assignment probabilities. Such fuzzy assignments, based on measurable observations, can also help to fill pathway holes and assign genes to partially characterized reactions.

14.3.4 From model to practice

As with any model describing a biological system, real data is usually more complicated and requires adaptation, exceptions and human intervention. This is typical of any learning task involving biological data. Specifically, we will focus on the model of the previous section.

Handling missing data. One of the first issues we have to address when applying this model to real expression data is the problem of missing values. This is a major hurdle when analyzing microarray data (see Section 12.4.2). Incomplete profiles pose a problem in Equation 14.3, when computing the probability of a profile with respect to the probabilistic model of a pathway. We can solve this problem by either inferring the missing data or by ignoring it. The first approach (as discussed in Section 12.4.2) is the easier one, since we do not need to change anything in the equations, but it may introduce more noise to the data, which is already noisy in nature. On the other hand, the second approach has to deal with profiles of different lengths, which is a matter of concern when computing the likelihood and posterior probabilities $p(\vec{e}|P_k)$ and when recomputing the means and variances. However, this can be addressed with a simple solution utilizing the independence between experiments (see Problem 14.6). If the entire expression profile of a gene is missing from the dataset, we can use the prior information to determine its probability to belong to a pathway, as explained next.

Incorporating prior knowledge. Utilization of prior knowledge is another important aspect of tailoring a model to the problem at hand. For example, to improve the performance of the classical EM algorithm described above we can initialize it based on prior biological knowledge as embodied in database annotations and similarity data (see [870]). Specifically, we can define

$$p(i|P_k)^0 = \sum_{F_{kj} \in \mathbf{F}(P_k)} p(i|F_{kj})$$

as the initial probability that gene i participates in pathway P_k, where $p(i|F_{kj})$ is the probability that gene i belongs to family F_{kj} and is associated with the corresponding enzymatic function[9]. (Note that we use i instead of $\vec{x}_i$ in these definitions because this probability is based on the functional knowledge of genes rather than on measurable observations, as explained in Section 14.3.3.2.) This initialization ensures that every gene that can evidently catalyze one or more of the reactions in a certain pathway has a non-zero probability to be assigned to that pathway. Note also that this definition assigns higher initial probability to multi-functional (multi-domain) genes that can catalyze multiple reactions in the same pathway. This is a desirable effect, as multi-functional enzymes are likely to be favorable in such cases (Section 14.3.3.1).

Having computed these probabilities, we can then initialize $p(\vec{e}_i|P_k)^0 = p(i|P_k)^0$ to account for the prior information about the functional assignment of genes. Priors over pathways can be initialized using uniform[10] prior $p(P_k)^0 = 1/K$ and using Bayes' formula we can derive the initial posterior probabilities

$$p(P_k|\vec{e}_i)^0 = \frac{p(i|P_k)^0 \cdot p(P_k)^0}{\sum_{k'} p(i|P_{k'})^0 \cdot p(P_{k'})^0}$$

From these probabilities we can now compute the initial parameters of each Gaussian, and continue as before by iterating through the M-step and the E-step until convergence.

Problem-specific modifications. To ensure the validity of the model, it is also necessary to verify its consistency with the existing knowledge of metabolic pathways. Specifically, all reactions in a pathway must take place in order for that pathway to exist. To enforce this constraint we can require for each reaction that the total affinity of genes associated with it be greater than a certain threshold and guide the clustering so as to meet this requirement (if possible, given the data). These constraints can be formalized as:

$$\sum_{i \in \mathbf{G}(F_{kj})} p(P_k|\vec{e}_i)^t > \theta \qquad \text{for each } F_{kj} \in \mathbf{F}(P_k)$$

[9]This probability can be approximated heuristically based on similarity data. For example,

$$p(i|F_{kj}) = \begin{cases} \frac{\log(evalue(i,j))}{\sum_{i' \in \mathbf{G}(F_{kj})} \log(evalue(i',j))} & \text{if } i \in \mathbf{G}(F_{kj}) \\ 0 & \text{otherwise} \end{cases}$$

where $evalue(i,j)$ is the statistical significance of the match between gene i and the model of family F_{kj}. In order to show our confidence in the database annotations we can set the $evalue(i,j)$ to the maximum observed evalue if gene i is annotated to belong to family j.

[10]The prior can be set based on the number of reactions in each pathway. Most pathways are of similar size, and therefore a uniform prior is a reasonable approach.

where θ is a global threshold (e.g., 0.5). Constraints can be enforced by checking whether they are met after each iteration of the EM algorithm. If they fail at the $t+1$ iteration we can reset $\vec{\mu}_k$ and $\vec{\sigma}_k$ such that $\vec{\mu}_k^{t+1} = \vec{\mu}_k^t$ and $\vec{\sigma}_k^{t+1} = \vec{\sigma}_k^t$.

This approach assumes that no constraints failed in the previous iteration. However, this does not guarantee that the model has no violated constraints since a constraint can fail as early as the first iteration. Furthermore, even if we freeze the mean and variance for one pathway the other pathways continue to change and might affect further the posterior of the genes involved in the constraint for the frozen pathway. However, this approach can reduce the number of violated constraints.

Alternatively, we can remove the constraints altogether. Thus, although the algorithm starts with pathway diagrams that were extracted from the literature, the procedure gradually alters and redefines pathway boundaries so as to correlate better with expression data and regulatory modules (see Section 14.4). Depending on the goal, this may be perceived as an advantage of the probabilistic approach.

Another type of problem-specific modification is the integration of metrics that are more suited to expression profiles, such as the mass-distance measure of Section 12.4.1. While the Gaussian mixture model can work quite well on many real-world problems, it is not necessarily well-suited for modeling pathways based on expression data. First, sometimes the sources generate samples that are not distributed normally. Second, Gaussians are defined based on the Euclidean metric. However, the Euclidean metric detects a significantly smaller number of functional links between genes based on expression data compared to other metrics (see Section 12.4.5).

To use the mass-distance measure with the EM algorithm described above we can replace the multi-dimensional Gaussian with a product of one-dimensional radial basis functions, defined for each experiment based on the distribution of MASS values in that experiment. Under this new setup the conditional probability is defined as

$$
\begin{aligned}
p(\vec{e}_i|P_k)^t &= \prod_{l=1}^{d} p(e_l|P_k)^t \\
&= \prod_{l=1}^{d} \frac{1}{\sqrt{2\pi}\sigma_{kl}^t} \exp\left(-\frac{MASS(e_l,\mu_{kl}^t)^2}{2(\sigma_{kl}^t)^2}\right)
\end{aligned}
$$

where μ is defined as before and the variance is given by

$$
\sigma_{kl}^{t+1} = \frac{\sum_{i=1}^{N} p(P_k|\vec{e}_i)^t MASS(e_{il},\mu_{kl}^{t+1})^2}{\sum_{i=1}^{N} p(P_k|\vec{e}_i)^t}
$$

Even with the constraints and inclusion of prior knowledge, clustering based on expression data alone cannot recover and discover all pathways. Therefore, it is necessary to integrate additional high-throughput datasets such as protein-protein interaction data and subcellular location data. The framework discussed in the previous section can be extended to include such datasets, each one providing information on different aspects of the same cellular process. For example, we can represent each gene with a feature vector $\vec{x}_i = \{\vec{e}_i, \vec{i}_i, ...\}$ where $\vec{e}_i$ is the expression profile of gene i, $\vec{i}_i$ is the interaction profile and so on. Assuming independence between these features we can decompose $p(\vec{x}_i|P_k) = p(\vec{e}_i|P_k)p(\vec{i}_i|P_k)\cdots$ and determine the parameters using the same equations we derived in Section 14.3.3.2.

14.4 Regulatory networks: modules and regulation programs

Earlier we mentioned that clustering genes based on expression data can identify co-expressed subnetworks, which often correspond to functionally related sets of genes. However, as pointed out in Section 14.3.3, clusters of co-expressed genes do not necessarily correspond to pathways. Furthermore, crosstalk between cellular processes makes it difficult to partition genes into *distinct* groups of co-expressed genes.

A different approach to expression data analysis, which is closely related to pathway prediction based on clustering of expression data, is **module discovery** in regulatory networks [622]. Modules are defined as groups of co-expressed genes. A cellular process may be comprised of several modules, each one involving multiple genes that are co-regulated and hence are co-expressed. However, the same gene may participate in more than one module and therefore each module accounts for a fraction of the gene's expression in a particular experiment. The idea is similar to the concept of fuzzy clustering we discussed in Section 14.3.3.2.

What differentiates module discovery from fuzzy clustering in transcriptional regulatory networks is the interplay between clusters of co-expressed (co-regulated) genes and their regulators. **Regulatory modules** are sets of genes whose expression is controlled by the *same* group of control genes. The set of control genes (regulators) is referred to as a **regulation program**. A regulation program can "implement" quite complex regulation patterns over a set of regulators.

One way to represent a regulation program is as a tree, similar to the decision tree model we discussed in Section 7.4. Figure 14.6 illustrates a regulation program over four regulators, where each node corresponds to one regulator, each branch specifies its activity level and each leaf node corresponds to a specific regulation mode (**context**). Each regulation program regulates a specific set of genes (a regulatory module). In each context, the module exhibits similar expression patterns, which may differ from one context to another. For example, the module's genes may be down-expressed in one context while over-expressed in another.

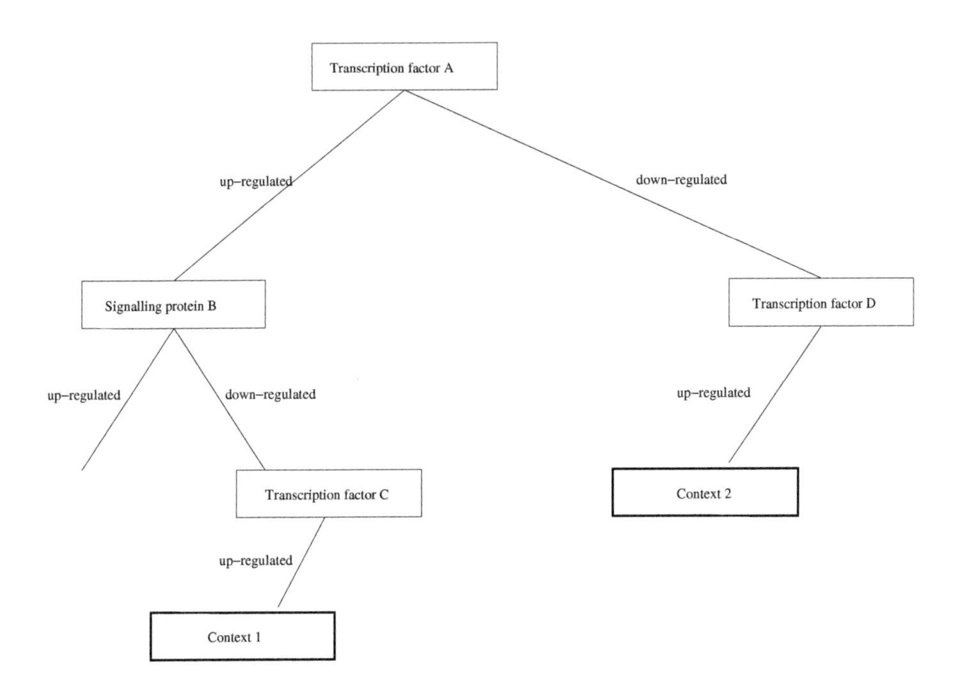

FIGURE 14.6: A regulation program. A tree over four regulators. The top node corresponds to the question "Is transcription factor A up-regulated?". The left branch corresponds to the answer yes, while the right branch corresponds to the answer no. The second node in the left branch corresponds to the question "Is signaling protein B up-regulated" and so on. Each path from the root node to the leaf node corresponds to a specific **context** where certain regulators are active while others are repressed. Given a set of arrays (experiments), each one is associated with one context based on the value of the regulators in that array. Each context may be associated with several arrays.

Due to the interplay between modules and their regulation programs, module discovery is a challenging problem. The procedure we describe in this section follows the general outline[11] of the EM algorithm presented in [622].

[11]The model of [622] consists of a Bayesian network combined with regression trees. We will defer the discussion on Bayesian networks to the next chapter and will focus here on the other elements of the model.

The input data is a set of genes and their expression under multiple experiments. Since the algorithm has to learn two different entities, modules and programs, the procedure switches iteratively between two steps. In one step, regulation programs are learned from the set of modules. In the second step, the modules are optimized given the set of regulation programs from the previous iteration.

The procedure starts by clustering genes into a fixed number of clusters using one of the hierarchical pairwise clustering algorithms we described in Section 10.4.1. The resulting clusters are the initial modules for the EM procedure. We will use the term cluster as opposed to module to focus on the algorithmic aspect of the method rather than the interpretation.

To initialize the regulation programs, a set of potential regulators is first collected based on database annotations (in [622], the authors include in the set all genes that are documented as transcription factors, as well as signaling proteins with potential impact on transcription, and genes that are similar to these regulators). The expression of the set of regulators is discretized in each experiment and classified to one of three categories: down-regulated (under-expression), up-regulated (over-expression) and no-change.

The EM procedure works iteratively, refining and redefining clusters as well as their regulation programs, until convergence. In the M-step, the best regulation program over the set of potential regulators is learned for each cluster (module). This is done by learning a regression tree, using a procedure very similar to the one we discussed in Section 7.4.2 on decision trees. Regression trees are similar in structure to decision trees but are used to learn continuous functions rather than for classification. When training decision trees the goal is to build a tree that splits the dataset perfectly into cluster members and non-members. Here, the goal is to build a tree that best predicts the expression values of the cluster's genes.

Recall that a typical decision tree learning algorithm selects at each node the attribute that most reduces the impurity (the class mix-up) if splitting samples based on the possible values of the attribute. In our case, the setting is slightly different. Each data sample is an array (which corresponds to one experiment over all genes), a regulator's expression value is an attribute of an array, and the possible values of the attribute are the three expression categories: up-regulated, down-regulated and no-change. During tree learning, the dataset is partitioned based on the attributes' values, such that each node is associated with a subset of the data samples (arrays). The continuous function we attempt to learn is a function that characterizes the distribution of expression values over genes in these arrays. Specifically, we are interested in the distribution of expression values of the cluster's genes.

Learning a regulation program for cluster C means finding a set of regulators that can predict the expression values of the cluster's genes. The regulator at the top of the tree can be selected by testing each regulator in the global pool of regulators and picking the one that best "separates" cluster members from non-members in all arrays with the right context. That is, we are looking for a regulator that when up-regulated (or alternatively, when down-regulated) produces the most distinctive expression patterns among cluster members in the arrays in which it is up-regulated. The process continues recursively, and at each node of the tree another regulator is selected (from the pool of regulators which have not been used already in the tree), the one that reduces the most the mix-up of cluster members and non-members in that node[12]. Thus, a regulation program for the cluster C is gradually learned.

While several different regulators may affect the expression values of cluster members, not all regulators are equally "important". Some result in more distinct expression patterns than others, with cluster members having either strongly positive or strongly negative expression values. To account for such differences and select the regulators which maximize the separation between expression values of members and non-members, the algorithm of [622] uses a score that measures how distinct are the two distributions of expression values when samples (arrays) are split based on the regulator's value (similar to the criterion function of Figure 7.10). Specifically, say we are considering adding a node that corresponds to a regulator T and a branch that corresponds to that regulator being up-regulated. The arrays are partitioned based on the value of the regulator and the distribution of expression values of the cluster's genes is collected in all arrays in which the regulator is up-regulated. We denote this distribution by N_T^C. Similarly, we generate the distribution of expression values of all genes that are not part of the cluster, and denote this distribution by N_T^{NC}. The distance $D(N_T^C, N_T^{NC})$ between the two distributions (assessed using measures as those discussed in Section 4.3.5) indicates how strongly T regulates C. If the regulator T plays no role in regulating members of C, then the distribution of expression values in C will be similar to the distribution of expression values of the other genes. However, if T regulates members of C then we expect the distributions to differ. For example, members of C may be down-regulated when T is up-regulated, resulting in a distinct distribution of expression values, where negative expression values are more frequent[13]. Thus, the value of T can be predictive of the expression values of cluster members. The higher the distance $D(N_T^C, N_T^{NC})$, the stronger is the regulatory relationship between T and the cluster C.

With a distance-based scoring function, the regulator selected at each node of the regression tree is the one which maximizes the distance between the distributions[14]. For each context, a normal distribution is fitted to the distri-

[12]The term mix-up is used here with respect to the distribution of expression values, as explained next.

[13]Expression values are usually logarithms of ratios of expression values (see Section 12.2). Hence down-regulation translates to negative values.

[14]It should be noted that not every value of a regulator needs to be considered. For example, the value no-change is ignored and only one of the values up-regulated and down-regulated may be used in the tree in one of the contexts (the other one might be ignored if it is not very common or if it results in weak separation between cluster members and non-members). Therefore, not every inner node splits into two child nodes.

bution of expression values of cluster members and the parameters μ and Σ are estimated (see Section 3.12.4). Since each context (a leaf node) may be associated with one or more arrays, the data is combined from multiple arrays when estimating the parameters. The learning process terminates when the error rate over all leaf nodes is below a certain preset threshold.

The M-step ends when regulation programs were learned for all clusters. In the E-step, the algorithm re-assesses the classification of genes to clusters, considering the regulation programs from the M-step. For each gene g we identify the regulation program that best predicts the expression value of that gene in all experiments. The gene is then assigned to the cluster associated with that regulation program. Specifically, let R_j denote the regulation program associated with cluster C_j, and let $P_{jk}(x) = N_{jk}(\mu_{jk}, \Sigma_{jk})$ denote the distribution of expression values induced by regulation program R_j in array k. Then the probability that gene g is regulated by the regulation program R_j is estimated by

$$P(i|j) = \prod_{k=1}^{d} P_{jk}(e_{ik})$$

where e_{ik} is the expression value of gene i in array k and d is the number of arrays. (Note that several arrays may be associated with the same context in regulation program R_j and hence with the same normal distribution.) This calculation is repeated for every cluster and the gene is classified to the cluster whose regulation program best predicts its expression value. That is,

$$class(i) = \arg\max_j P(i|j)$$

To avoid overfitting, self assignments are avoided. That is, regulators which are part of regulation program R_j cannot be assigned to cluster C_j. The whole process iterates between M-step and E-step until convergence.

As mentioned before, module discovery is important to the analysis of regulatory networks, but is not necessarily geared toward pathway prediction. Genes in a module are assumed to have a common function. It is also commonly assumed that genes in the same pathway are co-regulated, thus there is an overlap between a pathway and a regulatory module. However, they are not identical and the relationships between pathways and modules can be one of several types as illustrated in Figure 14.7:

1. One to one: a pathway overlaps with a regulatory module (Figure 14.7a), i.e., the genes participating in the pathway are co-regulated (e.g., *Homoserine methionine biosynthesis*).
2. Many to one (module sharing): a module is shared by several pathways (Figure 14.7b), i.e., the genes participating in several pathways are co-regulated (e.g., *valine biosynthesis* and *isoleucine biosynthesis*).

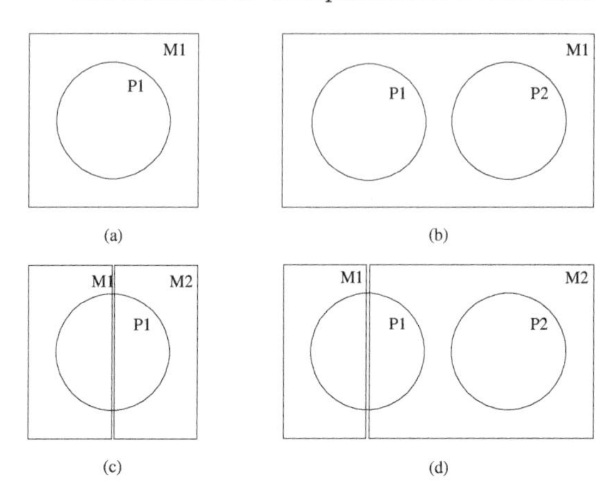

FIGURE 14.7: Relationships between pathways and modules. (a) One to one: the pathway P_1 overlaps module M_1. (b) Many to one (module sharing): the pathways P_1 and P_2 share the module M_1. (c) One to many: the pathway P_1 overlaps both modules M_1 and M_2. (d) Mixed: the pathways P_1 and P_2 share the module M_2 while P_1 overlaps with module M_2 as well. (Figure reproduced from [870] with permission from Imperial College Press.)

3. One to many: a pathway overlaps several modules (Figure 14.7c), i.e., not all the genes participating in a pathway are co-regulated but they can be grouped in a few co-regulated groups.
4. Mixed: a pathway overlaps several modules and shares some of them with other pathways, as in Figure 14.7d (e.g., *folic acid biosynthesis*).

For more information on module discovery see Further reading.

14.5 Pathway networks and the minimal cell

The collective set of cellular pathways forms a very interesting system. Consider for example the metabolic network that is the collective graph of all metabolic pathways. In this graph there are hundreds of compounds that are associated with hundreds of reactions. Some compounds are involved in multiple reactions. Furthermore, some reactions can be catalyzed by multiple genes (the multiplicity is an evidence of backup systems or complex regulation mechanisms, as discussed in Section 13.4.2 and Section 14.3.2.2) and some reactions are used in multiple pathways. The result is a complex and tightly connected graph, with many links (crosstalk) between different pathways.

Knowledge of this network enables us to ask and answer questions regarding components of the network. For example, by understanding the topology of a pathway we can envision ways to perturb or interfere with its normal course (usually by knocking out one of the genes) and verify them experimentally. Such mechanisms, which can drastically change the function of a cell, are referred to as **switches**. We can identify all occurrences of a certain metabolite and find all reactions that operate on it and all pathways it is involved in, providing insights into mechanisms of metabolic computations. We can study the effect of changing the concentration of a certain metabolite on the cell, or the effect of targeting a certain reaction. Knowing for example that a certain reaction in a certain pathway can be catalyzed by multiple genes is important for drug design, as targeting these genes in order to halt the reaction would be ineffective due to the presence of backup genes[15].

Pathway network analysis can also help to characterize topological properties of the network. It has been observed that the topology of metabolic networks is scale-free with small-world properties (see Section 13.4.1). In such networks the number of nodes which are highly connected is small. These so-called "hub" nodes shorten the average distance between nodes of the graph and between metabolites. However, mutations or knockouts of such genes can be lethal. It should be noted though that some studies challenge this finding, and argue that the small-world property is a result of shortcuts through common metabolites which are shared by many reactions and become the hubs in the scale-free network [863].

Beyond these immediate observations, pathway networks are capable of analyzing and verifying theorems regarding cellular computations and checking their consistency with experimental data. A theory or a model can be, for example, a set of specific relationships describing transcription regulation. The data can be measurements of the expression of genes. A pathway network can provide a qualitative evaluation of such theories and assess their plausibility. Quantitative analysis can be carried out too, and features such as flux rates, equilibrium concentrations, and composition can be computed from initial concentrations of metabolites and a list of reactions.

By further extending the pathway network, connecting genes to their protein products and proteins to the reactions they catalyze, it is possible to simulate cellular computations in a specific genome by following the edges of the network. This is important to the understanding of systems as complex as cells and can help to identify different types of regulation, such as feedback loops and auto-regulation (see Section 13.4.3). Whole-cell pathway simulations have been the subject of multiple studies, including [875–879]. A pathway graph can also be used to check whether a set of pathways is complete. For example, MetaCyc uses a protocol involving a *set of rules* regarding transformations of metabolites, and a *working memory* [880]. Initially, the working memory is loaded with the metabolites available in the cytoplasm at

[15] In actuality, the system may be more complicated, because in some cases the same reaction appears in different pathways and is catalyzed by different genes in different pathways.

large (assuming a minimal growth medium). Every "rule" or process that uses metabolites in the working memory is fired and the products are added to the working memory. The simulation continues iteratively until convergence. If all essential compounds are created during the simulation then the model is theoretically correct. Otherwise, this simulation can help to detect pathways that are missing from the model, missing precursor substrates or unknown synthesis rules.

The concept of the **minimal cell** follows [881–884]. What is the minimal genome which can synthesize all the compounds necessary for its normal functioning and catalyze all essential reactions? It has been shown in multiple studies that most genes in large genomes are non-essential [784–787]. Small organisms, such as the bacteria *Mycoplasma capricolum* (with about 500 genes) and *Mycoplasma genitalium* (which contains about 480 genes), are often used as model organisms for the minimal cell [885]. Identifying the subset of genes which constitute the minimal cell has been the subject of many studies as well [886–888], with estimates ranging between 200 and 300 genes [889]. In [632], the authors conclude that only 382 of the 482 genes of *Mycoplasma genitalium* are actually essential to life. More recent estimates set the number at around 150 [890]. Indeed, this is in correlation with the discovery of even smaller organisms such as the bacterium *Carsonella* with 182 genes [891]. It should be noted though that such small genomes might not be free-living organisms and can be sustained only through a symbiotic relationship with a host organism, which provides some of the bacteria's needs. It has been hypothesized that in the course of evolution such small bacteria sometimes became an integral part of the host genome, stored in separate organelles inside the cell in some cases. For example, this is the common perception regarding the origin of mitochondria [892, 893].

14.6 Further reading

Computational pathway analysis is a relatively new field, but a very active one. Several pathway databases were established over the years, in addition to KEGG [849, 894] and MetaCyc [848, 850], including ERGO [895] and its previous incarnations [896–898], UM-BBD [899], aMAZE [900] and others [901–905].

Metabolic pathway reconstruction from enzymatic reactions and chemical rules is discussed in multiple studies (see Section 14.3.1). For recent reviews see [906–908]. For a review on reconstruction of transcriptional regulatory networks see [909]. Supervised approaches for inference of interaction and metabolic networks are described in [821, 910–912], integrating phylogenetic profiles, protein interactions, protein localization, promoter data and gene expression data.

Pathway analysis and prediction based on expression data are among the most popular approaches, and there are many studies in this category. Some predict pathways from clusters of co-expressed genes, others extract active pathways and assess changes in their response to varying conditions by analyzing their aggregated activity profile [913–919]. Statistical significance is assessed with respect to the null hypothesis, for example by estimating the probability to obtain such changes by chance. The methodology can also be used to evaluate new pathways or assess other gene sets. Other studies focus on projecting and visualizing expression data (as well as other sources of information such as GO), on top of pathway diagrams, enabling detailed analysis of pathways [860, 865, 920–922].

Some studies integrate metabolic information to tweak clusters of co-expressed genes so as to correspond better with pathways. For example, in [923] the distance function between genes is a combination of the distance between the two reactions they catalyze in the pathway reaction graph, and a correlation-based distance between their expression profiles. The method of [924] compares graphs of co-expressed genes and interacting proteins (or metabolic networks with graphs indicating physical proximity along the genome) in search of overlapping gene clusters. A variation of the method is described in [918], where expression data and information on enzymatic reactions are encoded into two kernel functions, representing a metabolic network and a network of expression similarities. To identify active pathways, the method compares the kernels and computes their correlation. The approach is extended in [925], by considering also the physical position of the genes on the DNA. Another approach which combines interaction data and expression data is described in [873], using a Bayesian network model.

The concept of regulatory modules is explored in depth in a series of papers by Segal, Koller and colleagues, where several probabilistic models and inference algorithms are presented [622, 926–928]. Studying regulatory networks from expression data is also the subject of [929], where expression data is used in conjunction with DNA binding data on transcription factors to generate a regulatory network and identify modules and motifs. An excellent review of modules in regulatory networks appears in [776].

The intricate relationship between metabolic networks and regulatory networks is discussed in depth in [871, 930–932]. For example, in [871] several important observations are made regarding principles of gene regulation in metabolic networks, such as co-expression as an engine to ensure metabolic flow, the role of isozymes in regulating crosstalk between pathways and the presence of co-expressed pathways. The study by [932] focuses on metabolites as the driving force behind mechanisms of metabolic regulation. They search for **reporter** metabolites around which the most significant transcriptional changes occur (as measured by the expression data of the genes that catalyze reactions in which this metabolite is involved). Changes in cellular conditions or content which perturb the concentration of reporter metabolites may have a more global effect on transcriptional responses across the network, coordinating the activity of multiple metabolic pathways.

Pathway comparison is another emerging topic. The study by [933] is concerned with comparison of paths in interaction and pathway networks, seeking common pathways across different organisms. Pathways can also be compared and organized into a hierarchy. For example, [934] align pathways based on the EC hierarchy of enzymes. In [871], the authors construct a hierarchy of pathways based on their mutual correlation as measured by expression data.

The problem of finding missing enzymes has been addressed in several studies, by looking for similar proteins based on sequence similarity [935,936], using machine learning models [252], considering alternative reactions [937,938], utilizing expression data [870,874,939], or by using protein interaction and gene fusion data [940].

For more information on these and other related studies see the appendix in [874]. For reviews of pathway analysis see also [941–943]. For updates and additional references see the book's website at `biozon.org/proteomics/`.

14.7 Conclusions

- Pathways are cellular procedures that carry out numerous functions in cells.
- Each pathway has an input and an output.
- Signaling pathways transfer signals into the cell and between cells, through signal cascades.
- Metabolic pathways are composed of enzymatic reactions and are involved in energy metabolism, synthesis and degradation of various molecules and other processes.
- Many reactions in the metabolic network are shared by multiple pathways and facilitate crosstalk between pathways.
- Pathways in new genomes can often be predicted based on homology with pathways which were already characterized in other genomes.
- Some reactions can be catalyzed by multiple enzymes in the same genome (isozymes).
- Different isozymes can have different affinities to different pathways.
- Gene expression data can help to assess the association of genes with pathways and resolve pathway holes.
- A regulatory module is a group of co-expressed genes that have the same regulation program.
- Regulatory modules are not identical to pathways although there is overlap in some cases.
- The minimal cell is composed of the minimal set of pathways that are necessary to sustain life and produce all the essential compounds.

14.8 Problems

For updates, additional problems, files and datasets check the book's website at `biozon.org/proteomics/`.

1. **Enzyme families**

   You are given a set of several enzyme families. Compute the sequence similarity of all pairs of enzymes within each family. What is the average and minimal sequence identity in each family? Can you establish the relationship between functionally similar enzymes based on sequence similarity?

2. **Creating a network from compatible enzymes**

   You are given a detailed description of enzymatic reactions with their substrates and products. Establish a set of rules that specify which reactions can follow each other. Construct a complete graph (that is, nodes and the set of edges) between compatible enzymes. Assess the correlation of connected components in this graph with pathways.

3. **Pathway reconstruction from correlated subgraphs**

   You are given a dataset of protein-protein interactions, and a list of similarly expressed genes. Starting from the graph of the previous problem, increase the weight of each edge that appears in these datasets. Apply hierarchical clustering to group the genes into clusters. How would you delineate the graph into pathways? Assess the correlation of each cluster with known pathways. How would you assess the significance of each cluster?

4. **Clustering with hybrid distance functions**

   Use the graph of Problem 14.2 to define a distance function between genes as a weighted combination of their metabolic distance and their expression similarity. Apply hierarchical clustering to group the genes into clusters and assess their overlap with pathways. Test different weights in the combined distance function to obtain the best mapping between pathways and clusters.

5. **The many-to-many mapping problem**

   (a) List all enzymes that can catalyze reactions in the gluconeogenesis pathway in the yeast genome (pathway blueprint and list of enzymes are available at the book's website).

   (b) Identify additional genes which can potentially be part of the pathway based on similarity in expression profiles.

(c) How many reactions of the pathway are shared with other pathways?

(d) How many possible assignments of genes to the gluconeogenesis pathway can you generate?

6. **Fuzzy assignments with missing values**

   In Section 14.3.3.2 we discussed an EM algorithm for fuzzy assignment of genes to pathways, given gene expression data. Rewrite the E-step and the M-step, considering missing data. Explain why Equation 14.2 can be computed while ignoring missing data. Rewrite the mean and and variance of pathway P_k such that each component (experiment) is computed considering only the genes whose expression is defined in that experiment.

7. **Scoring pathways**

   Given the expression data of genes in yeast under different conditions, and a list of pathway diagrams, score each pathway based on the average expression similarity in each experiment. Can you conclude which pathways are active and which are not in each experiment?

Chapter 15

Learning Gene Networks with Bayesian Networks

15.1 Introduction

Biological systems are often complex and involve many variables with intricate dependencies. For example, gene networks (like signaling or regulatory networks) consist of hundreds or thousands of genes that interact with each other. These networks are designed to react to changes in the molecular content of the cell, the state of the cell or outside stimuli. The elaborate dependencies between genes determine which genes are active or inactive at any given moment. Deducing these dependencies from experimental data is a major goal of systems biology and is instrumental to understanding the genotypic factors of certain diseases. Consider cancer for example; cancer is triggered when genes that control replication malfunction and cells start to proliferate without control. There are many genes that take part in the replication process, directly or indirectly, and different kinds of cancers in different tissues are linked to different genes. However, our knowledge of these networks is very limited. Therefore, figuring out which gene(s) trigger a certain cancer is a challenge.

Bayesian Belief Networks (BBNs) is a family of models that can be trained from observations (i.e., experimental data) to learn the structure and dynamics of a complex network of entities and unravel the dependencies between them. The field of Bayesian networks emerged as a very promising one in computational systems biology soon after the introduction of microarray technology, as the technology provided the data needed to determine the existence and type of dependencies between genes in gene networks. Bayesian networks are also called **Bayesian nets** or **causal networks**. These models can use other types of data and be applied to other problems where dependencies have to be learned from data, such as data fusion and various classification and decision problems. For example, an integrative approach to prediction of protein-protein interactions would utilize the multiple types of data that are indicative of interactions (see Section 13.3.4). This includes subcellular location, structural compatibility, experimental evidence for interaction, domain fusion events and others. However, the different sources of data are often related and dependent on each other and cannot be simply integrated or summed without taking into account this dependency. Similar problems arise in the context of domain prediction (Chapter 9) and function prediction. In this chapter we will discuss a certain type of BBNs that can be applied to these problems and others, with the main motivation being the study of gene networks. Bayesian networks can learn and recover the structure of gene networks from experimental data, such as expression data and interaction data. For example, a typical experiment might study the effect of a mutant or a knocked out gene on the activity levels of other genes. Dependencies can be learned from the results of these experiments, measured in terms of the expression levels of genes (see Chapter 12), revealing intricate regulation mechanisms and signaling pathways.

15.1.1 The basics of Bayesian networks

Given a system with n variables $X_1, X_2, ..., X_n$, a Bayesian network characterizes their joint probability distribution $P(X_1, X_2, ..., X_n)$ by modeling the statistical dependencies between variables, rather than modeling the underlying source distribution explicitly, as in Chapter 6. This kind of model is appropriate when some of the variables are dependent on each other while others are independent.

BBNs can model complex systems that otherwise would be too difficult to model using one of the common probability distributions. For example, a source-based approach might assume that the data can be modeled by a multivariate normal distribution with an $n \times n$ covariance matrix Σ. The non-diagonal elements of the covariance matrix Σ characterize the dependency between variables. However, if the number of variables n is large then estimating Σ reliably requires very large datasets. Furthermore, often the individual variables do not follow a normal distribution, which deems this common distribution irrelevant. Even if we assume that the variables are discrete and binary (e.g., a gene can be "on" or "off"), it is practically impossible to obtain accurate estimates of the joint probability distribution for networks with more than a few tens of genes, since there are 2^n possible combinations of values, which is usually far more than the amount of available experimental data.

The BBN model provides an appealing alternative to the source-based approach. Bayesian networks are useful when most variables are independent of each other, in other words, when the dependency graph is sparse. By focusing on the dependencies, the BBN model reduces the dimensionality of the model (i.e., the number of parameters). BBNs can also tolerate missing data and noise that are typical of biological systems. Most importantly, the model provides an insightful description of the data that reflects its inherent structure.

To represent the dependencies between the elements of a system, a Bayesian network uses a graph consisting of nodes, edges and conditional probability distributions. The nodes correspond to random variables. Usually the variables are discrete, but BBN can be generalized to non-discrete variables. With discrete variables, each variable (node) X can take on a value from a set of possible values $\{\mathtt{x}_1, \mathtt{x}_2, ...\}$ with probability $P(\mathtt{x}_\mathtt{i})$ such that $\sum_i P(\mathtt{x}_\mathtt{i}) = 1$. For example, in gene networks the nodes correspond to genes and the variables can indicate whether the genes are active or inactive or can take on a numeric value corresponding to their expression level.

The edges of the graph represent dependencies between the variables. For example, the functional relation between a regulator (e.g., a transcription factor) and a regulatee (the gene it regulates) introduces a strong dependency between their expression values. In general, we say that two variables X, Y are **statistically independent** if

$$P(X,Y) = P(X) \cdot P(Y)$$

If the variables are dependent on each other then the probability that Y takes on a certain value depends on the specific value of X and

$$P(X,Y) = P(X)P(Y|X) \neq P(X) \cdot P(Y)$$

Given observations this can be easily verified by computing the empirical distributions $\hat{P}(X,Y), \hat{P}(X), \hat{P}(Y)$ and comparing the joint distribution $\hat{P}(X,Y)$ to the product $\hat{P}(X)\hat{P}(Y)$. In some cases even a simple correlation test between X and Y can indicate statistical dependency[1].

If the graph is directed, then its structure also entails certain causal relationships. For example, an edge from X to Y, denoted by $X \rightarrow Y$, signifies that X affects the value of Y. We usually assume that the graph is directed and acyclic, meaning there are no edges going both ways between the same two nodes[2]. The immediate predecessors of a node X are referred to as the **parents** of X, and the **children** of X are its immediate successors (see Figure 15.1). Node Y is called a **descendant** of X if there is a directed path from X to Y. Unlike statistical dependency, however, we cannot necessarily infer causal relationships from experimental data. For example, both $X \rightarrow Y$ and $Y \rightarrow X$ result in the same statistical dependency, and without additional information we cannot determine which one of the causal relations is the correct one. We will get back to this issue later in this chapter.

[1]There are several different correlation measures, including the Pearson correlation and the Spearman rank correlation (see Section 12.4.1). The Pearson correlation coefficient measures linear correlation between two variables. If it does not equal zero then we can infer statistical dependency with confidence that depends on the value of the coefficient and the size of the datasets (see Section 13.3.2.2 and Section 13.8). However, lack of correlation does not necessarily imply independence (see Problem 15.1).

[2]The model can be generalized to include cycles although it increases its complexity and mathematical tractability.

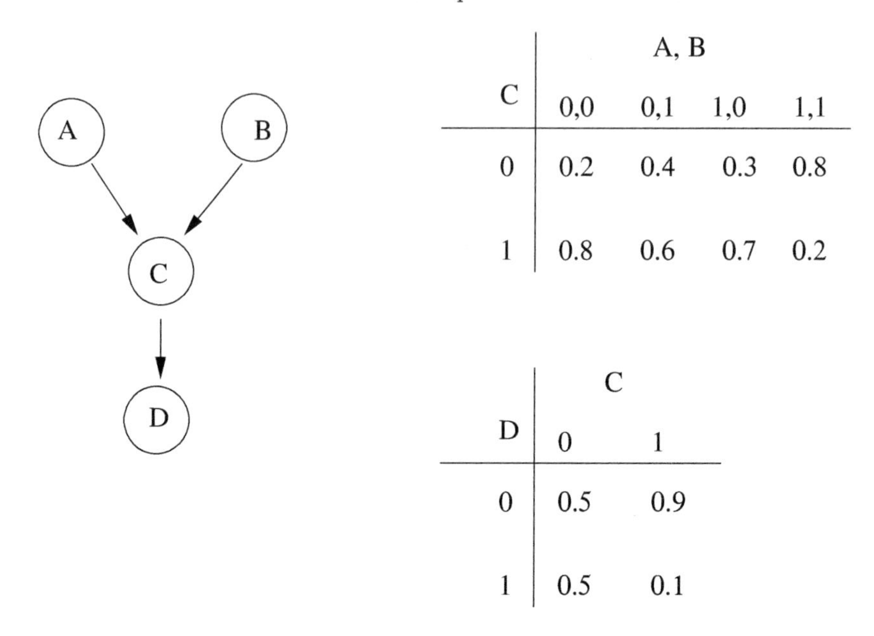

C	A, B: 0,0	0,1	1,0	1,1
0	0.2	0.4	0.3	0.8
1	0.8	0.6	0.7	0.2

D	C: 0	1
0	0.5	0.9
1	0.5	0.1

FIGURE 15.1: A simple Bayesian network. The nodes A and B are parents of C and D is a child of C. Node D is a descendant of A as well as of B. The probability tables associated with C and D are shown on the right. For example, the probability that $C = 1$ given that $A = 1$ and $B = 0$ is 0.7. The complete probability of $C = 1$ is given by the sum $P(C = 1) = \sum_{a,b} P(C = 1|A = a, B = b)P(A = a)P(B = b)$.

An important notion of dependency is **conditional dependency** or **conditional independence**. We say that Z is conditionally independent of X if

$$P(Z|Y, X) = P(Z|Y)$$

For example, given three variables X, Y, Z such that $X \rightarrow Y \rightarrow Z$ then knowledge of the value X does not affect our belief regarding the value of Z if we already know the value of Y. In this case, knowledge of Y *unlinks* X and Z, which are indirectly related. A different scenario is observed if the three variables X, Y, Z are linked as in the right graph of Figure 15.2, where Y depends on X and Z but there is no direct dependency between X and Z. This structure implies that X and Z are conditionally dependent given Y, although they are independent when Y is unknown. In other words, because X and Z affect the value of Y, knowledge of Y *creates* a dependency between X and Y. Inheritance is a fine example of such dependency (see [944]), where X and Z represent the copies of a certain gene from both parents, Y denotes the child gene, and their value is one if a mutation is observed and 0 otherwise. Knowing that there is a mutation in Y affects the probability of mutation in X or Z given that there is no mutation in the other. Hence, X and Y are statistically dependent on each other *given* Y, although they are independent entities in their nature. With the concept of conditional independence, a Bayesian network can learn which dependencies are direct and which are indirect.

Each node in a Bayesian network is associated with a conditional probability distribution that specifies its exact dependency on the values of its parents. Consider for example node D in Figure 15.1. The edge linking node C to node D is directional and represents the dependency of the value of D on the value of C. The node D is associated with the conditional probability distribution $P(D|C)$. Assuming that the variables are discrete, then $P(D|C)$ is the set of conditional probabilities $\{p(\mathtt{d_j}|\mathtt{c_i})\}$, where $p(\mathtt{d_j}|\mathtt{c_i})$ is the probability that D takes on the value $\mathtt{d_j}$ of all possible values for D given that C's value is $\mathtt{c_i}$. These probabilities are the **parameters** of the network. The sum of the probabilities must satisfy $\sum_j p(\mathtt{d_j}|\mathtt{c_i}) = 1 \ \ \forall i$. The conditional probabilities are summarized in tables (one per node), as illustrated in Figure 15.1. With multiple parents, the tables are generalized to multi-dimensional ones. For example, node C depends on both A and B. The conditional probability table of node C defines $P(C|A, B)$ and consists of the probabilities $\{p(\mathtt{c_k}|\mathtt{a_i}, \mathtt{b_j})\}$ for all possible combinations of i, j, k.

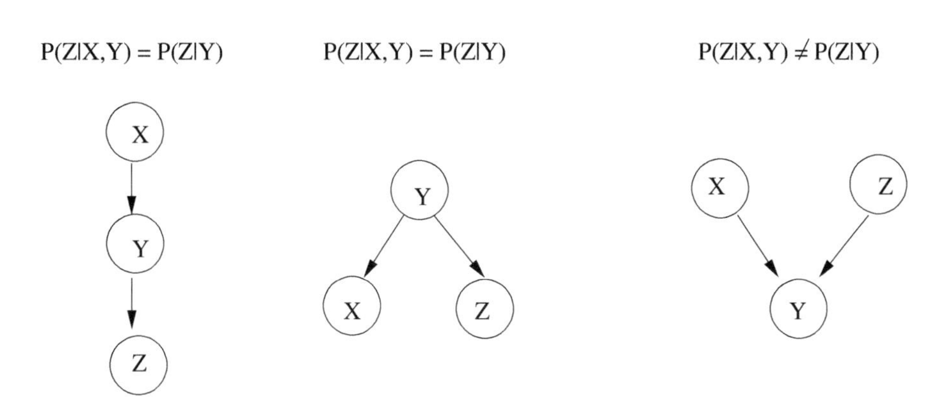

FIGURE 15.2: Conditional dependency vs. conditional independence. Three graphs with different causal structures and statistical dependencies over X, Y and Z. In the graph on the left, Y unlinks X and Z. That is, X and Z are conditionally independent of each other given Y. This can be tested by comparing $\hat{P}(Z|X, Y)$ to $\hat{P}(Z|Y)$. The graph in the middle has a different causal structure but implies the same statistical dependencies as the graph on the left. Therefore, given observations (e.g., three genes whose expression values are correlated) we cannot tell which of the two graphs describes the true causal relationships between them. In the graph on the right Y creates a statistical dependency between X and Z. That is, X and Z are conditionally dependent on each other given Y. Note that X and Z are not necessarily correlated in this case, and therefore the structure can be distinguished from the other two.

The structure of the network and the parameters (the conditional probabilities) determine the joint probability distribution over the network's variables. Assuming the network structure is acyclic then there are two rules (called **network assertions**) that characterize the joint probability distribution implied by a BBN:

1. The value of a node depends only on the value of its parents.
2. The value of a node is conditionally independent of its non-descendants, given its parents.

The first assertion is the essence of the *causal relation* encoded in the network structure. The parents of X *determine* the value of X with certain conditional probability, and X affects the values of its children. However, as the second assertion emphasizes, there is a *statistical dependency* between the value of a node and its descendants. In other words, when *inferring* the most likely values across the network, knowledge of the children is also relevant. We will elaborate on this problem in Section 15.3.1.

In the next sections we will discuss several common problems and applications of Bayesian networks. The main inference and learning tasks involve the probabilistic inference of unknown variables given evidence, and learning parameters and structure from data. Given the structure and the parameters of the network, we can compute the likelihood of a set of observations or infer the most likely values of unknown variables in the network, as discussed in Section 15.3. The remaining sections will discuss the problem of learning BBNs from data instances.

15.2 Computing the likelihood of observations

How do we compute the likelihood of a set of observations for a complex network with many nodes and edges? Standard factorization expresses the joint probability in terms of a product of conditional probabilities

$$P(x,y) = P(x)P(y|x)$$

which stands for

$$P(X = x, Y = y) = P(X = x)P(Y = y|X = x)$$

With multiple variables this is generalized to

$$P(\vec{x}) = P(x_1, x_2, ..., x_n) = P(x_1)P(x_2|x_1)P(x_3|x_1,x_2)... = \prod_{i=1}^{n} P(x_i|x_1,...,x_{i-1})$$

However, this general formula ignores the causal structure of the network and introduces unnecessary dependencies that complicate the computation. Based on the first network assertion we can eliminate the unnecessary dependencies from the factorization. For example, $P(x|y,z) = p(x|y)$ if Z is not a parent of X. Therefore, given a network with n nodes the probability of an individual assignment of values $x_1, x_2, ..., x_n$ to the variables $X_1, X_2, ..., X_n$ is

$$P(x_1, x_2, ..., x_n) = \prod_{i=1}^{n} P(x_i|parents(X_i)) \tag{15.1}$$

where $P(x_i|parents(X_i))$ is given by the conditional probability table associated with the variable X_i. We use the short notation $pa(X_i)$ to denote $parents(X_i)$.

Determining the ordering of variables that reflects the true causal structure with minimal conditional dependencies can be done efficiently using **topological ordering**. In topological ordering of nodes in a directed acyclic graph, each node appears before all its children, and this guarantees that the number of conditional dependencies is minimal. The algorithm starts the expansion with nodes that have no parents. The nodes are removed and the search continues for nodes that have no parents in the remaining set. These nodes are next in order. This process is repeated until we exhausted all nodes, and the factorization is determined based on the order nodes were removed from the set.

For example, the probability of an assignment a, b, c, d to the variables A, B, C, D of Figure 15.1 using the standard factorization over an arbitrarily chosen ordering D, C, B, A is

$$P(a, b, c, d) = P(d)P(c|d)P(b|c, d)P(a|b, c, d)$$

while using topological ordering and removing unnecessary dependencies, the probability can rewritten as

$$P(a, b, c, d) = P(a)P(b)P(c|ab)P(d|c)$$

15.3 Probabilistic inference

A Bayesian network reflects our belief (either prior or learned) regarding the relations and dependencies between the variables of a complex system. Probabilistic inference with Bayesian networks uses the prior knowledge together with evidence regarding some of the network's variables to produce probabilistic statements regarding the value of other variables in the network.

15.3.1 Inferring the values of variables in a network

To understand how a Bayesian network can be used for inference, we will start with a relatively simple task. Assume we are given a model where the network structure and all the parameters (the conditional probabilities) are known. Assume we also know the values of some of the variables in the network. We want to find the most probable values for the *unknown* variables. The simplest case is when only one variable X is unknown.

Note that the conditional probabilities $p(x|pa(X))$ do not give us all the information we need to determine the most likely value of X. They just tell us the (prior) probability X takes on a certain value, given the values of its parents. If we have other information relevant to X, i.e., if there is a variable Y such that Y is a descendant of X and the value of Y is known, then the most probable assignment to X depends also on the value of Y since X influences Y. In other words, knowing $pa(X)$ and the value of Y is more informative than knowing just $pa(X)$ and

$$P(X = x|Y, Pa(X)) \neq P(X = x|Pa(X))$$

While knowledge of descendants of X matters for inference, information about non-descendants is irrelevant because either they do not influence X, or if they influence X indirectly then only through its parents, which are already assumed to be known.

The distinction between inference based on posterior and prior knowledge is identical to the distinction between prior and posterior probabilities we discussed early on in Section 3.12.7. For example, the prior probability of rain depends on the presence of clouds. If the sky is cloudy then we can infer with a certain probability that it rained in the past couple of hours. However, certain observations can increase or decrease this probability. For example, if we see puddles of water outside then it tells us something about the probability it rained recently. The posterior probability of rain, given the observations, revises our prior belief and affects our decision.

To compute the posterior probability of $X = x$ given an assignment $(y_1, y_2, ..., y_n)$ to all other variables[3] we can use Bayes' formula

$$P(X = x|y_1, ..., y_n) = \frac{P(x, y_1, ..., y_n)}{P(y_1, ..., y_n)} = \frac{P(x, y_1, ..., y_n)}{\sum_{x'} P(x', y_1, ..., y_n)}$$

This is usually impractical for large n because we need many observations to estimate the joint probability distributions $P(x, y_1, ..., y_n)$ reliably. However, using the network structure we can reduce this term to a more tractable one as in Equation 15.1

$$P(X = x|y_1, ..., y_n) = \frac{P(x|pa(X)) \cdot \prod_{i=1}^{n} P(y_i|pa(Y_i))}{\sum_{x'} [P(x'|pa(X)) \cdot \prod_{i=1}^{n} P(y_i|pa(Y_i))]} \tag{15.2}$$

[3]Our network in this example contains $n+1$ variables. To make a distinction between them, notation wise, we refer to them as X (the unknown) and $Y_1, ..., Y_n$ (known variables).

Note that the product over Y_i cannot be just canceled, because $P(y_i|pa(Y_i))$ depends on the exact value of x' if X is one of the parents of Y_i. However, we can further reduce Equation 15.2 by considering only the nodes that are directly connected to X. Per the second network assertion[4], the value of X is conditionally independent of all the other nodes in the network given the values of the parents $pa(X)$ and the children $ch(X)$. Indeed, the probabilities $P(y_i|pa(Y_i))$ cancel each other in the numerator and denominator for variables that are conditionally independent of X. Therefore,

$$P(X = x|y_1, ..., y_n) = \frac{P(x|pa(X)) \cdot \prod_{Y_i \in ch(X)} P(y_i|x, pa(Y_i))}{\sum_{x'} \left[P(x'|pa(X)) \cdot \prod_{Y_i \in ch(X)} P(y_i|x', pa(Y_i))\right]} \quad (15.3)$$

Note that if $Y_i \in ch(X)$ then $X \in pa(Y_i)$ and $P(y_i|x', pa(Y_i)) = P(y_i|pa(Y_i))$ but the notation $P(y_i|x', pa(Y_i))$ emphasizes the dependency on the value of X.

For example, consider the network of Figure 15.3 that describes causes and symptoms of the flu and the common cold. The variable of interest is I. If we can determine the value of I based on the other variables then we can diagnose the illness. If the values of A, S, F and R are known then we can easily compute $P(I = 1|a, s, f, r)$ as

$$P(I = 1|a, s, f, r) = \frac{P(I = 1, a, s, f, r)}{P(I = 1, a, s, f, r) + P(I = 2, a, s, f, r)} \quad (15.4)$$

Based on the network structure the joint probability of the five variables is

$$P(I, A, S, F, R) = P(A)P(S)P(I|A, S)P(F|I)P(R|I) \quad (15.5)$$

Therefore, each term on the right side of Equation 15.4 can be computed using the expansion of Equation 15.5 and the conditional probability tables. For example, given the assignment $A = 1, S = 4, R = 2, F = 1$ the joint probability is

$$\begin{aligned} P_1 &= P(I = 1, A = 1, S = 4, R = 2, F = 1) \\ &= P(A = 1)P(S = 4)P(I = 1|A = 1, S = 4)P(R = 2|I = 1)P(F = 1|I = 1) \\ &= 0.5 \cdot 0.25 \cdot 0.8 \cdot 0.4 \cdot 0.8 = 0.32 \end{aligned}$$

For simplicity, we assumed a uniform prior over the sample space of each variable, hence $P(A = 1) = 0.5$ and $P(S = 4) = 0.25$. Similarly we can compute $P_2 = P(I = 2|A, S, R, F)$ and decide $I = 1$ if $P_1 > P_2$ (see decision rules, Section 3.12.8).

[4]The second assertion is stated more generally, and while eliminating non-descendants it leaves all descendants (direct and indirect) as part of the equation. However, assuming all variables other than X are known then we can ignore the non-direct descendants as well, since their value is determined based on their parents, which are known.

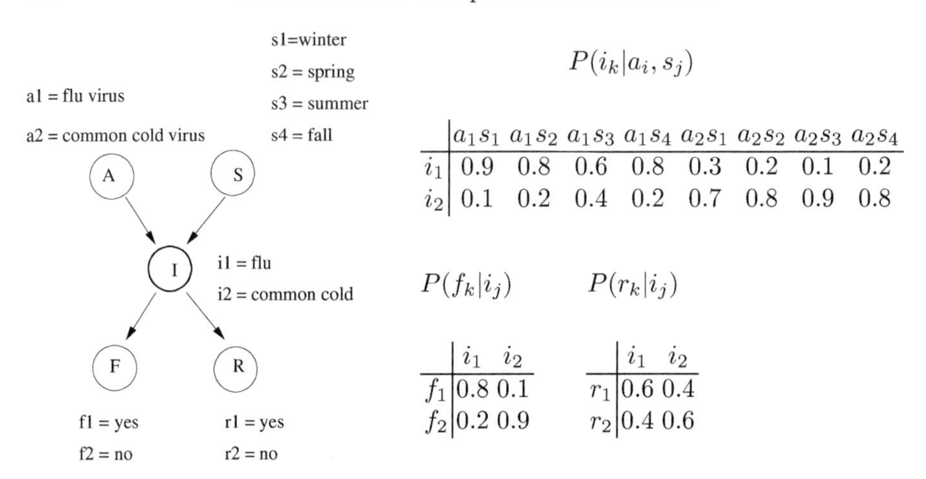

$P(i_k|a_i, s_j)$

	a_1s_1	a_1s_2	a_1s_3	a_1s_4	a_2s_1	a_2s_2	a_2s_3	a_2s_4
i_1	0.9	0.8	0.6	0.8	0.3	0.2	0.1	0.2
i_2	0.1	0.2	0.4	0.2	0.7	0.8	0.9	0.8

$P(f_k|i_j)$

	i_1	i_2
f_1	0.8	0.1
f_2	0.2	0.9

$P(r_k|i_j)$

	i_1	i_2
r_1	0.6	0.4
r_2	0.4	0.6

FIGURE 15.3: A Bayesian network that represents the dependencies between illnesses (flu and common cold), causes and symptoms. The network has the following nodes: viral **a**gent, **S**eason, **I**llness, **F**ever and **R**espiratory tract infection. The conditional probability tables are listed on the right and reflect known dependencies between the variables. For example, if a person was infected with the flu virus then there is a high probability that he will develop the flu (top table). However, being infected with the flu virus does not necessarily lead to flu, depending on the person's immune system and whether he is already immune to the specific viral strain. Since the flu virus is seasonal, the probability of developing the flu is higher during the winter. On the other hand, the common cold virus usually prevails throughout the year. Even if the common cold virus is not present, the patient might have the flu due to the presence of a different strain. The table is artificial but simulates statistics of patients with flu-like symptoms. The conditional table $P(f_k|i_j)$ reflects the fact that flu is usually accompanied with fever and higher temperatures are more probable. Similarly, $P(r_k|i_j)$ reflects the fact that cough is one of the symptoms of flu but less so of the common cold.

Usually we are given certain observations or measurements regarding some of the network variables and not necessarily their exact values. This is referred to as **evidence**. The evidence provides us with prior knowledge on the network variables, which translates to prior probabilities over their values. For example, we might have the result of a swab test, which could help us to determine whether the cause of the illness is an influenza (flu) virus as opposed to one of the common cold viruses. These kind of tests are not 100% accurate, and therefore they only give us a certain probability that the viral agent is a flu virus. Similarly, we can measure the temperature of the patient. However, depending on the temperature, we cannot always determine whether the patient has fever or not and we can only associate a certain probability with each possible value of F.

Computing the probability that an unknown variable X takes on a certain value, given the evidence $\vec{e} = (e_1, e_2, ..., e_n)$ is very similar to the case we described above. The posterior probability $P(X = x|\vec{e})$ can be computed using Bayes' formula as before

$$P(X = x|e_1, ..., e_n) = \frac{P(x, e_1, ..., e_n)}{P(e_1, ..., e_n)} = \frac{P(x, e_1, ..., e_n)}{\sum_{x'} P(x', e_1, ..., e_n)}$$

and utilizing the network structure this is reduced to

$$P(X = x|e_1, ..., e_n) = \frac{P(x|e_p(X)) \cdot \prod_{Y_i \in ch(X)} P(e_i|x, pa(Y_i))}{\sum_{x'} \left[P(x'|e_p(X)) \cdot \prod_{Y_i \in ch(X)} P(e_i|x', pa(Y_i))\right]}$$

where $e_p(X)$ denotes the evidence regarding the parents of X. In the toy example of Figure 15.3

$$P(I = 1|e_a, e_s, e_r, e_f) = \frac{P(I = 1, e_a, e_s, e_r, e_f)}{P(I = 1, e_a, e_s, e_r, e_f) + P(I = 2, e_a, e_s, e_r, e_f)}$$

and the joint probability can be written as

$$P(I = i, e_a, e_s, e_r, e_f) = P(I = i|e_a, e_s)P(e_r|I = i)P(e_f|I = i) \quad (15.6)$$

Let's look at the components of Equation 15.6. The first term $P(I = 1|e_a, e_s)$ can be generalized from $P(I = 1|A, S)$ by using the complete probability rule and accounting for the prior knowledge. That is,

$$P(I = 1|e_a, e_s) = \sum_{a,s} P(I = 1|A = a, S = s)P(A = a|e_a)P(S = s|e_s)$$

The term $P(e_f|I = 1)$ assesses the probability of the evidence regarding a descendant of I and is formulated according the causal structure of the network

$$P(e_f|I = 1) = \sum_{f} P(e_f|F = f)P(F = f|I = 1)$$

Similarly, we can compute $P(e_r|I = 1)$.

Consider the following case, where a patient complains about flu symptoms to his doctor. The doctor is presented with the following evidence $\vec{e}$:

1. **Season:** It is winter time, therefore $P(s_1|e_s) = 1$ while $P(S = s_j|e_s) = 0$ for $j \neq 1$.
2. **Viral agent:** Swab test suggests that the patient has the flu virus, but the test is only 70% accurate; therefore $P(a_1|e_a) = 0.7$ and $P(a_2|e_a) = 0.3$ (for simplicity we assume that the cause of the illness can only be one of the two; hence the total probability must be 1).
3. **Fever:** The patient has mildly high temperature of 98.8°F. Based on statistics accumulated over the years the doctor assigns $P(e_f|f_1) = 0.3$ and $P(e_f|f_2) = 0.5$. Note that unlike the evidence over the parents, $P(e_f|f_1)$ and $P(e_f|f_2)$ need not sum to 1, since f_1 and f_2 are different events. It is the distribution of temperatures in each case that should sum to 1. That is, $\sum_{e_f} P(e_f|f_1) = 1$ and similarly $\sum_{e_f} P(e_f|f_2) = 1$.

4. **Respiratory tract infection:** The patient complains about mild coughing. This could be indicative of a respiratory tract infection, but since it is mild the doctor assigns $P(e_r|r_1) = 0.6$ and $P(e_r|r_2) = 0.6$.

Given the evidence we can now compute $P(I = 1|\vec{e}) = P(i_1|\vec{e})$ and $P(i_2|\vec{e})$

$$\begin{aligned}
P(i_1|e_a, e_s) &= \sum_{i,j} P(i_1|a_i, s_j) \cdot P(a_i|e_a) \cdot P(s_j|e_s) \\
&= \sum_{i} P(i_1|a_i, s_1) \cdot P(a_i|e_a) \\
&= P(i_1|a_1 s_1) \cdot P(a_1|e_a) + P(i_1|a_2 s_1) \cdot P(a_2|e_a) \\
&= 0.9 \cdot 0.7 + 0.3 \cdot 0.3 = 0.72 \\
P(e_f, e_r|i_1) &= P(e_f|i_1) \cdot P(e_r|i_1) \\
&= [P(e_f|f_1) \cdot P(f_1|i_1) + P(e_f|f_2) \cdot P(f_2|i_1)] \cdot \\
&\quad [P(e_r|r_1) \cdot P(r_1|i_1) + P(e_r|r_2) \cdot P(r_2|i_1)] \\
&= (0.3 \cdot 0.8 + 0.5 \cdot 0.2)(0.6 \cdot 0.6 + 0.6 \cdot 0.4) = 0.204
\end{aligned}$$

Similarly, we derive that $P(i_2|e_a, e_s) = 0.28$ and $P(e_f, e_r|i_2) = 0.288$. Therefore,

$$\begin{aligned}
P(i_1|\vec{e}) &= \frac{P(i_1|e_a, e_s) \cdot P(e_f, e_r|i_1)}{P(i_1|e_a, e_s) \cdot P(e_f, e_r|i_1) + P(i_2|e_a, e_s) \cdot P(e_f, e_r|i_2)} \\
&= \frac{0.72 \cdot 0.204}{0.72 \cdot 0.204 + 0.28 \cdot 0.288} \approx 0.65
\end{aligned}$$

while $P(i_2|\vec{e}) \approx 0.35$ and since $P(i_1|e) > P(i_2|e)$ we conclude that the patient is more likely to have the flu.

15.3.2 Inference of multiple unknown variables

Determining the most probable values is much more difficult if there are multiple variables that are unknown. In such cases we can use the following iterative sampling procedure to assign values to variables

- **Initialize:** Assign values at random to all unknown variables.
- **Iterate:** (1) Pick one of the variables, Z. (2) Assign Z the most probable value, given all other values, using the procedure we described in the previous section.

The value we just estimated for Z is used when assigning a value to the next variable we pick. The process is iterated until we exhaust all variables and then repeated until convergence. Note however that the algorithm is not guaranteed to converge to the most probable assignment overall, and it might end up with a locally optimal assignment.

15.4 Learning the parameters of a Bayesian network

In the previous section we discussed inference with Bayesian networks, assuming the model is completely known. However, usually we do not know the structure of the network (meaning the set of direct dependencies between variables), as well as the parameters (the conditional probabilities associated with every variable). In this section we discuss the problem of learning the parameters $\vec{\theta}$ from a sample of N observations $D = \{\vec{x}^1, \vec{x}^2, ..., \vec{x}^N\}$, where every observation $\vec{x}$ is a vector of measurements over all n variables in the network

$$\vec{x} = (x_1, x_2, ..., x_n)$$

For now we will assume that the model structure S is given (in the next section we will generalize the discussion to the case of unknown structure). We will also assume that the dataset D is complete.

The Bayesian approach to solving this problem would be to compute the set of parameters that maximize the posterior probability

$$P(\vec{\theta}|D, S) = \frac{P(D|\vec{\theta}, S)P(\vec{\theta}|S)}{P(D|S)}$$

These parameters are called the MAP estimators (see Section 3.12.7). Since we assume the structure is fixed, we will remove the dependency on S to simplify the notation and will keep in mind that $\vec{\theta}$ is a parameter vector for the specific structure S. Our goal then is to find the parameter vector that optimizes

$$P(\vec{\theta}|D) = \frac{P(D|\vec{\theta})P(\vec{\theta})}{P(D)}$$

If the prior $P(\vec{\theta})$ is uniform we can pick the parameter vector $\vec{\theta}$ that maximizes the likelihood of the observations $P(D|\vec{\theta})$, but we will discuss the Bayesian approach in the general case of non-uniform prior, using both $P(D|\vec{\theta})$ and $P(\vec{\theta})$.

When computing the likelihood of the data D given a model we assume the samples are i.i.d.

$$P(D|\vec{\theta}) = \prod_{l=1}^{N} P(\vec{x}^l|\vec{\theta})$$

and for each individual sample $\vec{x} = \vec{x}^l$ the probability is given by the expansion

$$P(\vec{x}|\vec{\theta}) = \prod_{i=1}^{n} P(x_i|pa(X_i), \vec{\theta}_i)$$

where $\vec{\theta}_i$ is a subset of $\vec{\theta}$, with the parameters relevant to the variable X_i (that is, the parameters of the conditional probability distributions associated with the variable X_i as in Figure 15.4).

There are many ways to model the conditional probability distributions. For simplicity, we will assume a collection of multinomial distributions, one for each possible configuration of values for the parents of X_i. If k_i is the number of possible values for X_i, then the number of possible configurations m_i of the parents is

$$m_i = \prod_{X_r \in pa(X_i)} k_r$$

For each parent configuration $pa(X_i)_j$ we define a multinomial distribution

$$P(X_i = \mathrm{x_k} | pa(X_i)_j) = \theta_{ijk}$$

such that $0 < \theta_{ijk} < 1$ and $\sum_{k=1}^{k_i} \theta_{ijk} = 1$. The parameter θ_{ijk} is the probability that X_i takes on the k-th possible value, given that the parents of X_i are in configuration j (see Figure 15.4). We denote the parameter vector associated with X_i and configuration j as

$$\vec{\theta}_{ij} = (\theta_{ij1}, \theta_{ij2}, ..., \theta_{ijk_i})$$

and the complete parameter vector for a network with a structure S is the combined vector of all the conditional probability distributions

$$\vec{\theta} = (\vec{\theta}_{11}, \vec{\theta}_{12}, \vec{\theta}_{1m_1}, \vec{\theta}_{21}, ..., \vec{\theta}_{nm_n})$$

We will assume that the parameter vectors $\vec{\theta}_{ij}$ are mutually independent.

Given a dataset with observations over all variables in the network, let's focus on the subset of measurements over X_i for which the parents are in configuration j. If N_{ijk} is the number of times we observe the variable X_i take on the k-th value when its parents are in configuration j, then the probability of observing the subset $D_{ij} = \{N_{ij1}, N_{ij2}, ..., N_{ijk_i}\}$ is given by the multinomial distribution with the parameter vector $\vec{\theta}_{ij}$

$$P(D_{ij}|\vec{\theta}) = P(N_{ij1}, ..., N_{ijk_i}|\vec{\theta}_{ij}) = \binom{N_{ij}}{N_{ij1}N_{ij2}...N_{ijk_i}} \theta_{ij1}^{N_{ij1}} \theta_{ij2}^{N_{ij2}} \cdots \theta_{ijk_i}^{N_{ijk_i}}$$

where $N_{ij} = \sum_k N_{ijk}$ is the number of instances associated with configuration j and $\sum_j N_{ij} = N_i$ is the number of observations over X_i. Assuming the data is complete then $N_i = N$.

The Maximum Likelihood (ML) estimator of the multinomial distribution is

$$\theta_{ijk} = \frac{N_{ijk}}{N_{ij}} \tag{15.7}$$

(see Section 5.2.2.1). The ML estimator is a good approximation to the MAP estimator when there is plenty of data, because the prior plays a less significant role. Since it is simpler to compute, it is often the default choice. However, the ML approach may overfit the model to the observations and is likely to result in some parameters with zero probabilities, especially if the size of the dataset is smaller than the number of possible combinations ijk (which is the case for even small networks). The MAP approach uses also the prior information and is more effective in avoiding overfitting. The simplest prior is a uniform prior. The uniform prior solves the zero-probability problem but is considered an uninformative prior, since it does not provide any new information. In the general case we might have a certain informative prior $P(\vec{\theta}|S)$ that we would like to incorporate.

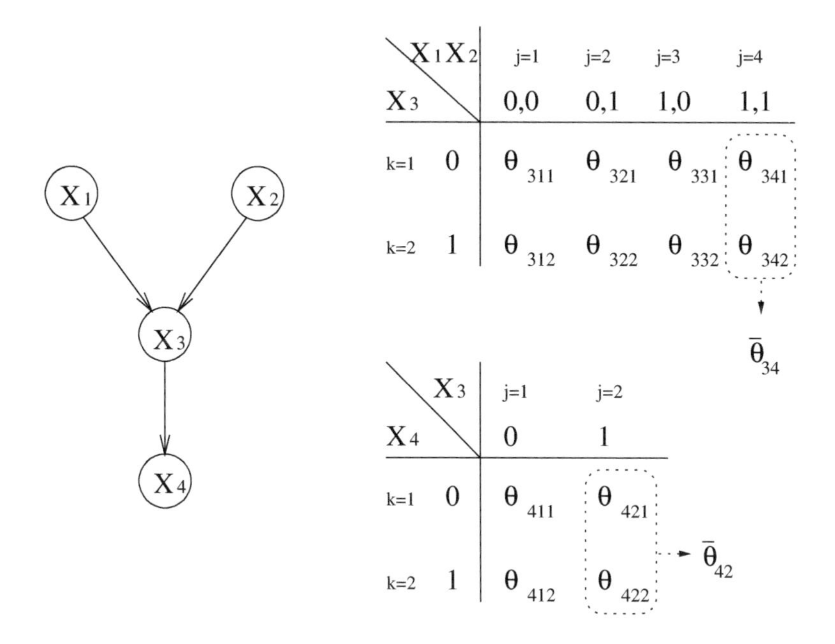

FIGURE 15.4: Parameter vectors in Bayesian networks. The network in this example has four binary variables. Each variable is associated with a conditional probability distribution table. In this table, each column corresponds to one configuration of the parents. For example, variable X_3 has two parents, X_1 and X_2. There are four possible configurations for the two parents, depending on their values. The first configuration ($j = 1$) corresponds to the assignment $X_1 = 0$ and $X_2 = 0$; the second configuration corresponds to the assignment $X_1 = 0$ and $X_2 = 1$; and so on. The variable X_3 can take one of two possible values, 0 or 1, and their probabilities depend on the configuration of the parents. The parameters θ_{ijk} denote these probabilities, and for each variable X_i and parent configuration j the sum over the sample space of X_i should satisfy $\sum_k \theta_{ijk} = 1$. For binary variables $\theta_{ij2} = 1 - \theta_{ij1}$.

Since we assume that the parameters are mutually independent, the prior can be decomposed into

$$P(\vec{\theta}|S) = P(\vec{\theta}) = \prod_{i=1}^{n} \prod_{j=1}^{m_i} P(\vec{\theta}_{ij}) \tag{15.8}$$

It is convenient to assume that the prior for each multinomial distribution follows a Dirichlet distribution (see Section 4.3.3.2), such that

$$P(\vec{\theta}_{ij}) = Dir(\vec{\theta}_{ij}|\alpha_{ij1}, \alpha_{ij2}, ..., \alpha_{ijk_i}) = Dir(\vec{\theta}_{ij}|\vec{\alpha}_{ij}) \tag{15.9}$$

The choice of the Dirichlet distribution as the functional form of the prior is motivated by the fact that it can be decomposed into local likelihood functions for each variable X_i and parent configuration j, as outlined above. The second reason is that the Dirichlet distribution is the **conjugate distribution** to the multinomial distribution. We say that a certain prior distribution $p(\vec{\theta})$ is a conjugate distribution of a sample generating distribution $p(x|\vec{\theta})$ if $p(\vec{\theta})$ and $p(\vec{\theta}|\vec{x}) \sim p(\vec{x}|\vec{\theta})p(\vec{\theta})$ both belong to the same family of distributions. For example, the conjugate distribution of a normal distribution is also a normal distribution. The conjugate distribution of a binomial distribution is the beta distribution, and the conjugate distribution of the multinomial distribution is the Dirichlet distribution.

Recall (Section 4.3.3.2) that the Dirichlet distribution gives a prior over K-dimensional parameter vectors $\vec{\theta}$ that satisfy $0 < \theta_k < 1$ and $\sum_k \theta_k = 1$ (for clarity, we ignore for a minute the subscripts i, j). The distribution has a hyper-parameter vector of counts $\vec{\alpha} = (\alpha_1, \alpha_2, ..., \alpha_K)$ such that $\alpha_k > 0 \quad \forall k$, and is defined

$$Dir(\vec{\theta}|\vec{\alpha}) = \frac{\prod_k \theta_k^{\alpha_k - 1}}{Z(\vec{\alpha})} \tag{15.10}$$

where $Z(\vec{\alpha})$ is a normalization factor over all possible parameters (the numbers θ_k being exponentiated in Equation 15.10) and not the counts (the exponents), given by the K-dimensional integral

$$Z(\vec{\alpha}) = \int \prod_{k=1}^{K} \theta_k^{\alpha_k - 1} d\vec{\theta} = \frac{\prod_k \Gamma(\alpha_k)}{\Gamma(\sum_k \alpha_k)} \tag{15.11}$$

The Γ function is defined recursively such that $\Gamma(x+1) = x\Gamma(x)$ and $\Gamma(n) = (n-1)!$ for integers. The Dirichlet function reflects our belief or prior knowledge we have on the value of the parameters. The mean of the Dirichlet distribution is a vector where $\theta'_k = \frac{\alpha_k}{\sum_k \alpha_k}$, and the distribution obtains its peak at the mean.

With all the components in place we can now write an explicit term for the posterior probability

$$P(\vec{\theta}|D) = \frac{P(D|\vec{\theta}) \cdot P(\vec{\theta})}{P(D)} \tag{15.12}$$

Since we assumed independence between the parameters of the distributions of different variables and different parent configurations, we can decompose the likelihood into the following product

$$P(D|\vec{\theta}) = \prod_{i=1}^{n} \prod_{j=1}^{m_i} P(D_{ij}|\vec{\theta}_{ij})$$

(where D_{ij} already groups the statistics from the different observations) and

$$P(\vec{\theta}) = \prod_{i=1}^{n} \prod_{j=1}^{m_i} P(\vec{\theta}_{ij})$$

Therefore, we can consider and optimize each set of parameters independently

$$\begin{aligned} P(D_{ij}|\vec{\theta}_{ij})P(\vec{\theta}_{ij}) &= \binom{N_{ij}}{N_{ij1}...N_{ijk_i}} \theta_{ij1}^{N_{ij1}} \theta_{ij2}^{N_{ij2}} \cdots \theta_{ijk_i}^{N_{ijk_i}} \cdot Dir(\vec{\theta}_{ij}|\vec{\alpha}_{ij}) \\ &= \binom{N_{ij}}{N_{ij1}...N_{ijk_i}} Z^{-1}(\vec{\alpha}_{ij}) \theta_{ij1}^{N_{ij1}+\alpha_{ij1}-1} \cdots \theta_{ijk_i}^{N_{ijk_i}+\alpha_{ijk_i}-1} \\ &= \binom{N_{ij}}{N_{ij1}...N_{ijk_i}} Z^{-1}(\vec{\alpha}_{ij}) \prod_{k=1}^{k_i} \theta_{ijk}^{N_{ijk}+\alpha_{ijk}-1} \end{aligned} \tag{15.13}$$

The marginal likelihood $P(D_{ij})$ is given by the integral over all possible parameters

$$\begin{aligned} P(D_{ij}) &= \int_{\vec{\theta}_{ij}} P(D_{ij}|\vec{\theta}_{ij}) P(\vec{\theta}_{ij}) d\vec{\theta}_{ij} \qquad (15.14) \\ &= \binom{N_{ij}}{N_{ij1}...N_{ijk_i}} Z^{-1}(\vec{\alpha}_{ij}) \int \prod_{k=1}^{k_i} \theta_{ijk}^{N_{ijk}+\alpha_{ijk}-1} d\vec{\theta}_{ij} \end{aligned}$$

The integral has a closed-form solution as in Equation 15.11, such that

$$\begin{aligned} P(D_{ij}) &= \binom{N_{ij}}{N_{ij1}...N_{ijk_i}} Z^{-1}(\vec{\alpha}_{ij}) \frac{\prod_k \Gamma(N_{ijk} + \alpha_{ijk})}{\Gamma(\sum_k N_{ijk} + \alpha_{ijk})} \\ &= \binom{N_{ij}}{N_{ij1}...N_{ijk_i}} Z^{-1}(\vec{\alpha}_{ij}) Z(\vec{\alpha}_{ij} + \vec{N}_{ij}) \end{aligned} \tag{15.15}$$

Therefore, from Equation 15.12, Equation 15.13 and Equation 15.15

$$\begin{aligned} P(\vec{\theta}_{ij}|D_{ij}) &= \frac{P(D_{ij}|\vec{\theta}_{ij}) \cdot P(\vec{\theta}_{ij})}{P(D_{ij})} = \frac{\prod_{k=1}^{k_i} \theta_{ijk}^{N_{ijk}+\alpha_{ijk}-1}}{Z(\vec{\alpha}_{ij} + \vec{N}_{ij})} \qquad (15.16) \\ &= Dir(\vec{\theta}_{ij}|\alpha_{ij1} + N_{ij1}, \alpha_{ij2} + N_{ij2}, ..., \alpha_{ijk_i} + N_{ijk_i}) \end{aligned}$$

and we have a closed-form solution for the posterior probability distribution over the parameters. As we mentioned earlier, the Dirichlet distribution peaks at the mean and hence the MAP solution is given by

$$\theta_{ijk} = \frac{N_{ijk} + \alpha_{ijk}}{\sum_k N_{ijk} + \alpha_{ijk}} = \frac{N_{ijk} + \alpha_{ijk}}{N_{ij} + \alpha_{ij}} \tag{15.17}$$

Thus the hyper-parameters of the Dirichlet distributions essentially serve as pseudo-counts. All that is left to do is to set the prior $\vec{\alpha}_{ij}$ over parameters. A simple approach for estimating the hyper-parameters of the Dirichlet distributions is to ignore the value of the parents such that the hyper-parameters for each variable X_i are set based on the relative number of times the k-th value was observed. That is,

$$\alpha_{ijk} = \frac{N_{i*k}}{N} \cdot \alpha \tag{15.18}$$

for all $1 \leq j \leq m_i$. The constant α is a user-defined parameter. As a rule of thumb we can set $\alpha = \sqrt{N}$ (see Section 4.3.3.2). For other alternatives, see [945].

15.4.1 Computing the probability of new instances

Using the formalism of the previous section we can also compute the probability of new instances, which is useful for prediction. For example, if we have different networks that were trained from data observed in different cellular states or different tissues, we can compute the likelihood of a new instance according to each and classify it to the network that assigns the maximum likelihood.

The likelihood can be computed using the ML or MAP estimators of each network. The Bayesian approach does not commit to a single parameter vector but rather averages the contributions of all possible parameter vectors

$$P(\vec{x}_{new}|D) = \int_{\vec{\theta}} P(\vec{x}_{new}|\vec{\theta}) \cdot P(\vec{\theta}|D) \cdot d\vec{\theta}$$

where the likelihood $P(\vec{x}_{new}|\vec{\theta})$ is weighted with the posterior probability $P(\vec{\theta}|D)$ of each parameter vector. By averaging over the posterior probabilities the marginal likelihood accounts for the uncertainty in the parameter value.

Usually, the ML or MAP parameter vectors will contribute the most and the integral can be approximated by keeping just the one term that corresponds to that parameter vector. However, under the assumptions of the previous section we can derive a closed-form solution for the marginal likelihood. For each variable i in the network the sample $\vec{x}_{new}$ assigns a certain value x_i out of k_i possible values. Denote the index of that value as k_x and the parent configuration of X_i that is set by the sample $\vec{x}_{new}$ as j_x, then the probability of $\vec{x}_{new}$ depends only on the parameters associated with these configurations

$$P(\vec{x}_{new}|\vec{\theta}) = \prod_{i=1}^{n} \theta_{ij_x k_x} \tag{15.19}$$

Hence

$$\begin{aligned} P(\vec{x}_{new}|D) &= \int_{\vec{\theta}} \prod_{i=1}^{n} \theta_{ij_x k_x} \cdot P(\vec{\theta}|D) d\vec{\theta} \\ &= \prod_{i=1}^{n} \int_{\vec{\theta}_{ij_x k_x}} \theta_{ij_x k_x} \cdot P(\theta_{ij_x k_x}|D) d\theta_{ij_x k_x} = \prod_{i=1}^{n} E(\theta_{ij_x k_x}) \\ &= \prod_{i=1}^{n} \frac{\alpha_{ij_x k_x} + N_{ij_x k_x}}{\sum_k \alpha_{ijk} + N_{ijk}} = \prod_{i=1}^{n} \frac{\alpha_{ij_x k_x} + N_{ij_x k_x}}{\alpha_{ij} + N_{ij}} \end{aligned} \tag{15.20}$$

The transition from the first step to the second reduces the integral to the subset of the relevant parameters, since the parameters are independent of each other. Note that each integral in the second line is basically the average over the k-th component of the distribution. The mean of the Dirichlet distribution $Dir(\vec{\theta}|\vec{\alpha})$ is $\theta_i = \alpha_i / \sum_i \alpha_i$, which explains the fourth step.

15.4.2 Learning from incomplete data

So far we assumed that we have a complete dataset to learn the model from. However, this is almost never the case. For example, expression data, which is usually the type of data used to train Bayesian networks for gene networks, is often incomplete due to various experimental and technical problems as discussed in Section 12.2. Another type of missing data is due to hidden variables. These could be unknown genes, or genes whose expression we cannot measure or other control elements for which we do not have measurement data.

Consider the latter type first. Our goal is to approximate $P(\vec{\theta}|D)$ given the network structure and an incomplete dataset, assuming the parameters are mutually independent. For each individual instance $\vec{y} = (y_1, ..., y_n)$ denote by $X \subset Y$ the set of observed variables and $Z \subset Y$ the set of unobserved variables, such that $\vec{y} = (\vec{x}, \vec{z})$. Note that $\vec{\theta}$ characterizes both the hidden and the observed data and therefore we cannot just apply the solutions of the previous section to the observed data. One possible roundabout is to apply an EM algorithm, similar to the one we described in Chapter 5, and compute the posterior probability of $\vec{\theta}$ given the observed data by marginalizing over the hidden data

$$P(\vec{\theta}_{ij}|\vec{x}) = \sum_{\vec{z}} P(\vec{\theta}_{ij}|\vec{x}, \vec{z}) P(\vec{z}|\vec{x})$$

The algorithm starts from an initial guess for the parameters and then iterates through the E-step and the M-step until convergence. In the E-step, the probability is marginalized over the missing variables (computing their expectation values using the current estimators of the model's parameters). In the M-step, we look for the parameter vector that maximizes the marginalized likelihood. That is, the expectation values substitute for the real values and together with the observed data they form a complete dataset. The complete dataset is then used as in the first part of Section 15.4 to find a new locally optimal parameter vector.

A different approach is to use sampling methods such as Gibbs sampling. Such methods are less susceptible to be trapped in local maxima and are also better suited for handling missing data of the first type. We are given an incomplete dataset with N instances $D = \{\vec{x}^1, ..., \vec{x}^N\}$ each one is missing a different subset. In Chapter 5 we described the Gibbs sampling technique in the context of motif detection. The procedure in this case is similar:

1. Initialize the states of the unobserved variables in each instance randomly, to generate a complete set D_c.

2. Choose a variable X_i^l that is missing from the original dataset (that is, variable i in the l instance). Denote by D_c^{li} the set D_c excluding X_i^l. For every possible value of X_i^l generate a different complete dataset and compute the probabilities

$$P(X_i^l = \mathrm{x_k} | D_c^{li}) = \frac{P(X_i^l = \mathrm{x_k}, D_c^{li})}{\sum_{\mathrm{x}'_\mathrm{k}} P(X_i^l = \mathrm{x}'_\mathrm{k}, D_c^{li})}$$

   The probability $P(D'_c) = P(X_i^l = \mathrm{x}'_\mathrm{k}, D_c^{li})$ can be computed as in Equation 15.14.

3. Pick a value for X_i^l according to this probability distribution.

4. Repeat steps 2-3 for every unobserved variable in D. Denote the new complete dataset D_c.

5. Compute $P(\vec{\theta}|D_c)$.

6. Repeat steps 2-5 many times and take the average of $P(\vec{\theta}|D_c)$ as the approximation for $\vec{\theta}$.

If this process is repeated many times then we eventually sample every possible configuration of values to the unknown variables and converge to a locally optimal assignment of values, as in Section 15.3.2.

15.5 Learning the structure of a Bayesian network

We now proceed to the more general case where the parameters as well as the structure of the network S are unknown. Learning the structure means learning which variables are dependent on each other and which are independent.

When learning structure from data, we usually cannot fully recover the true causal structure, since there may be many structures that can explain the set of observed dependencies equally well. The simplest example is $X \to Y$ and $Y \to X$, which imply the same statistical dependency between X and Y. Unfortunately, empirical data cannot resolve the directionality and distinguish between the two. That is, if we observe a correlation in the values of X and Y, we still cannot determine which one affects the other. Thus, both structures are good candidates.

However, we can discern between structures that imply that certain variables are conditionally independent from other structures that imply they are not. For example, the two left structures in Figure 15.2 are distinguishable from the right structure, since both imply that X and Z are independent given Y, while the right structure implies exactly the opposite. Therefore, when learning structure we actually learn an **equivalence class** of structures. Each equivalence class consists of a set of graphs that describe the same set of dependencies (see Problem 15.2). In Figure 15.2 the left and middle graphs belong to the same class (together with the structure $Z \to Y \to X$), while the graph on the right belongs to a different class.

The members of each equivalence class "agree" on some edges, while "disagreeing" on others. Hence, the directionality of some edges can be determined, and the equivalence class can be described in terms of a partially directed **consensus graph**. A direct edge between two variables in the consensus graph signifies that all members of the equivalence class share this edge in the same direction, while an undirected edge indicates that the edge appears in both directions.

Further complicating the task of learning structure from data is the fact that the data observations are a noisy sample of the source distribution, with missing values and possibly corrupted measurements. Therefore, even determining the exact set of dependencies reliably is difficult. Our goal is to approximate the source distribution and search for a structure that best describes the observed dependencies. The best we can do is to learn an equivalence class of graphs or an instance of the equivalence class that is equivalent to the real structure in terms of the dependencies it induces. As before, the Bayesian approach looks for the model that maximizes the posterior probability

$$P(\vec{\theta}_S, S|D)$$

where both $\vec{\theta}_S$ and S are unknown. The notation $\vec{\theta}_S$ emphasizes the dependency of the parameter vector on the structure S. The joint probability distribution can be written as

$$P(\vec{\theta}_S, S|D) = P(S|D) \cdot P(\vec{\theta}_S|D, S)$$

and maximizing $P(\vec{\theta}_S, S|D)$ is equivalent to maximizing the logarithm

$$\log P(\vec{\theta}_S, S|D) = \log P(S|D) + \log P(\vec{\theta}_S|D, S)$$

We already know how to estimate $P(\vec{\theta}_S|D, S)$ for a specific structure. This is $P(\vec{\theta}|D)$ we computed in the previous section under certain assumptions on the conditional probability distributions (Equation 15.16). The second term, $P(S|D)$, is given by

$$P(S|D) = \frac{P(D|S) \cdot P(S)}{P(D)}$$

where we assume there is a certain prior distribution over structures $P(S)$. With the logarithm transformation

$$\log P(S|D) = \log P(D|S) + \log P(S) - \log P(D)$$

and when maximizing over S and $\vec{\theta}_S$ we can ignore the last term, since it is constant for the dataset D. Therefore,

$$S_{MAP}, \vec{\theta}_{MAP} = \arg\max_{S,\vec{\theta}} \left[\log P(\vec{\theta}|D, S) + \log P(D|S) + \log P(S)\right]$$

The second and third terms are independent of $\vec{\theta}$, and since $P(\vec{\theta}|D, S)$ is maximal for $\vec{\theta}_{MAP}$ we can reduce this problem to a simpler one and optimize over structures while using the MAP estimator for the parameters of each structure. That is,

$$S_{MAP} = \arg\max_{S} \left[\log P(\vec{\theta}_{MAP}|D, S) + \log P(D|S) + \log P(S)\right] \quad (15.21)$$

To find the best structure according to Equation 15.21 we need to compute $P(D|S)$ and $P(S)$. The marginal likelihood $P(D|S)$ can be computed by integrating over all parameter vectors, as before

$$\begin{aligned}
P(D|S) &= \int_{\vec{\theta}_S} P(D|\vec{\theta}_S, S)P(\vec{\theta}_S|S)d\vec{\theta}_S \\
&= \prod_{i=1}^{n}\prod_{j=1}^{m_i}\int_{\vec{\theta}_{ij}} P(D_{ij}|\vec{\theta}_{ij})P(\vec{\theta}_{ij}|S)d\vec{\theta}_{ij} \\
&= \prod_{i=1}^{n}\prod_{j=1}^{m_i}\binom{N_{ij}}{N_{ij1}...N_{ijk_i}} Z^{-1}(\vec{\alpha}_{ij})Z(\vec{\alpha}_{ij}+\vec{N}_{ij}) \\
&= \prod_{i=1}^{n}\prod_{j=1}^{m_i}\binom{N_{ij}}{N_{ij1}...N_{ijk_i}} \frac{\Gamma(\sum_k \alpha_{ijk})}{\prod_k \Gamma(\alpha_{ijk})}\frac{\prod_k \Gamma(N_{ijk}+\alpha_{ijk})}{\Gamma(\sum_k N_{ijk}+\alpha_{ijk})} \\
&= \prod_{i=1}^{n}\prod_{j=1}^{m_i}\binom{N_{ij}}{N_{ij1}...N_{ijk_i}} \frac{\Gamma(\alpha_{ij})}{\Gamma(N_{ij}+\alpha_{ij})}\prod_k \frac{\Gamma(N_{ijk}+\alpha_{ijk})}{\Gamma(\alpha_{ijk})} \qquad (15.22)
\end{aligned}$$

where $P(\vec{\theta}_{ij}|S)$ is $P(\vec{\theta}_{ij})$ as in Equation 15.8 and Equation 15.9, and the transition from the second to the third step is based on Equation 15.14 and Equation 15.15.

Note that for each structure S, a different set of variables constitutes the parents of X_i, and hence m_i, the number of parent configurations, is structure dependent. Therefore, the parameter θ_{ijk} might be part of completely different conditional probability tables in two different structures. Similarly, the counts N_{ijk} correspond to different combinations of observations and are dependent on the structure.

Computing the prior over structures $P(S)$ is more complicated. The simplest approach would be to use a uniform prior. However, this would also create a preference toward larger networks with a higher number of dependencies, since by adding edges (dependencies) we add parameters and we can better fit the data. Therefore, if we were to explore the space of all possible networks just based on the likelihood, then the best model would be the fully connected one (see Problem 15.3). Note that this is a similar situation to the one we had in clustering where the maximal likelihood is obtained when every data point is in its own cluster, since the likelihood is one in that case (Section 10.5.2.1).

A more informative prior can be defined by excluding unreasonable structures (for example, structures that contain invalid dependencies) or using a network derived from prior biological knowledge (such as protein-protein interaction networks or pathway blueprints). The prior $P(S)$ can be estimated based on the deviation of S from that prior network. For example, let $P(e)$ denote the prior probability of an edge e, then $P(S)$ can be defined as the product over all edges

$$P(S) = \prod_{e \in S} P(e) \qquad (15.23)$$

The prior over edges can be defined based on the confidence in the dependencies between the corresponding variables. For example, in [946] the support to edges comes from promoter analysis of genes and their descendants. Protein interactions can also provide support for edges. However, all these approaches require user input. Alternatively, we can assign priors based on the complexity of the networks, such that more complex networks (with a larger number of edges) are assigned a lower probability. This is the basis for one of the alternatives described in the next section.

It is sometimes convenient to use a slightly different formulation of the posterior probability (with identical results)

$$P(\vec{\theta}, S|D) = \frac{P(D|\vec{\theta}, S)P(\vec{\theta}, S)}{P(D)} = \frac{P(D|\vec{\theta}, S)P(\vec{\theta}|S)P(S)}{P(D)}$$

and maximizing $P(\vec{\theta}, S|D)$ is equivalent to maximizing

$$\log\left(P(D|\vec{\theta}, S)P(\vec{\theta}|S)\right) + \log P(S)$$

once again ignoring $P(D)$, which is independent of S and $\vec{\theta}$. We already have closed-form solutions for $P(D|\vec{\theta}, S)$ and $P(\vec{\theta}|S) = P(\vec{\theta})$, as well as for their product (Equation 15.13)[5]. Therefore, we only need to compute $P(S)$, as before.

Some studies maximize the marginalized posterior probability of the structure given the data by averaging over all possible parameters

$$P(S|D) = \int_{\vec{\theta}} P(\vec{\theta}, S|D)d\vec{\theta} = \frac{P(D|S)P(S)}{P(D)} \tag{15.24}$$

as this function is easier to compute and maximize.

15.5.1 Alternative score functions

Maximizing the posterior probability is a rigorous approach to structure learning. However, estimating the prior is not simple and can bias the search toward a small subset of the space of all possible structures. The score-based approach is an alternative to the Bayesian approach. Instead of maximizing the posterior probability, we define and maximize a score function that measures the quality of the structure in terms of its ability to explain the observations as well as the plausibility of the structure.

One way to assess the quality of the model is to test whether pairs of variables are correlated or independent as implied by the model. For example,

[5] Note that the solution is identical to the one obtained by multiplying the first two terms of Equation 15.21, $P(\vec{\theta}_{MAP}|D, S)$ and $P(D|S)$, which are given in Equation 15.16 and Equation 15.22, respectively.

in gene networks the expression profiles of all pairs can be compared using measures of statistical similarity or independence, such as correlation or mutual information (Problem 15.1). Every pair of nodes that is connected by a path should show signs of correlation or anti-correlation (Section 12.4), while no correlation is expected if there is no connection (this is partly true, see Footnote 15.1 on page 651). The overall quality of the model can be defined as a function of the correspondence between the graph dependencies and the dependencies that are observed in the experimental data (Problem 15.4). The higher it is, the better fit the model is. If testing all possible pairs becomes prohibitive, it can be circumvented by using a random sample of all pairs.

Another popular scoring function is the MDL inspired **Bayesian Information Criterion (BIC)** [550]. Recall the discussion on the MDL (minimum description length) principle in Section 10.5.2.1. The total description length of a model (e.g., network structure) is the sum of the model description length and the description of the data given the model. The BIC score is defined as

$$score(S) = \log P(D|\vec{\theta}_{MAP}, S) - \frac{d}{2} \ln N \tag{15.25}$$

where the first term $\log P(D|S)$ is computed with the $\vec{\theta}$ that is optimal for that structure. The second term is related to the structure complexity and penalizes complex models with a larger number of parameters d. For a network with n variables, each one with m_i possible parent configurations, the number of parameters is

$$d = \sum_{i=1}^{n} \sum_{j=1}^{m_i} (k_i - 1)$$

The MDL approach is strongly linked to the Bayesian approach for model selection. Given a hypothesis space $\mathcal{S}$, the Bayesian approach selects the hypothesis $S \in \mathcal{S}$ such that the posterior probability is maximal. That is, per Equation 15.24, S is the structure that maximizes $\log P(S|D)$ or

$$\log P(D|S) + \log P(S) = \log P(D|S) - (-\log P(S)) \tag{15.26}$$

The first term is the log-likelihood, which can be viewed as a measure of the description of the data given the model, while the second is related to the description of the model and is a function of the uncertainty we have about the model. The BIC score is an approximation of Equation 15.26 where $\log P(D|S)$ is approximated by $P(D|\vec{\theta}_{MAP}, S)$, assuming the probability $P(D|\vec{\theta}, S)$ peaks sharply around $\vec{\theta}_{MAP}$, and $-\log P(S)$ is approximated by $\frac{d}{2} \ln N$, where N is the number of data instances, assuming a fixed penalty (description length) for each parameter.

We can actually estimate the uncertainty linked with $-\log P(S)$ (and the description length) more accurately, following the method we described in Section 10.5.2.1. The structure S determines the number of parameters of the model. The model parameters are estimated from the data, and since the data is finite there is uncertainty associated with each one. The total uncertainty is the sum of uncertainties over all parameters.

Specifically, assume each variable is a binary variable and consider the conditional probability distribution associated with one variable X_i and a specific configuration of parents j. Assuming a binomial distribution with parameter $p = p_{ij}$ and $N = N_{ij}$ samples with that specific configuration of parents, then the mean of the distribution is $\mu_{ij} = Np$ and the standard deviation $\sigma_{ij} = \sqrt{Np(1-p)}$.

The ML estimator of the binomial distribution is the empirical mean $\hat{\mu}_{ij}$. Note that the empirical mean can be written as the sum of N i.i.d. random variables $\hat{\mu}_{ij} = \frac{1}{N}\sum_{l=1}^{N} X_i^l$. According to the central limit theorem the sum is distributed normally with mean μ_{ij} and standard deviation $\sigma_{ij}^N = \frac{\sigma_{ij}}{N}$. Therefore, as the number of instances N increases, $\hat{\mu}_{ij}$ will converge to the true mean.

For finite datasets the standard deviation σ_{ij}^N is a measure of the uncertainty in the value of the parameter $\hat{\mu}_{ij}$. This can be converted to bits by taking $\log_2$, and the total uncertainty is the sum over all parameters

$$-\log P(S) \sim \sum_{i=1}^{n}\sum_{j=1}^{m_i} \log_2 \sigma_{ij}^N = \sum_{i=1}^{n}\sum_{j=1}^{m_i} \log_2 \left(\frac{\sigma_{ij}}{N_{ij}}\right) = \frac{1}{2}\sum_{i=1}^{n}\sum_{j=1}^{m_i} \log_2 \left(\frac{p_{ij}(1-p_{ij})}{N_{ij}}\right)$$

Note that this term increases with the number of dependencies as N_{ij} is smaller for each possible configuration, increasing the uncertainty in each parameter as well as adding additional terms to the total uncertainty.

15.5.2 Searching for optimal structures

To find the structure S that results in maximal score (or posterior probability), we need to compute the derivative of the score. Unfortunately, the structures are discrete and equations like Equation 15.22 are bitwise functions whose derivatives are undefined. The same is true for the prior, if using non-uniform prior. Therefore, to search for the best network we have to enumerate all possible structures, compute the score for each one and pick the one that maximizes the score. Since there are $n(n-1)/2$ possible undirected edges between n nodes, the number of possible networks over n nodes is $2^{n(n-1)/2}$ and this approach is not practical without even considering directionality. When adding directionality, there are 2^m possible configurations of m edges, which further increases the number of networks we need to consider. The problem is NP-hard even if we limit the number of parents of each node to k at most.

The practical solution is to explore a subset of all possible networks. Existing approaches are heuristics that work by sampling the network space or using a hill-climbing procedure to find locally optimal structures.

15.5.2.1 Greedy search

The greedy search algorithm searches the network space stochastically, by introducing random changes to the initial network structure. The search is done locally, focusing on one variable at a time, using the fact that the global likelihood function can be decomposed into local likelihood functions.

Greedy search can start with an empty network, a fully connected network or a random one. It can also start from a network based on prior knowledge, such as protein-protein interaction networks or known pathways. The algorithm changes one element of the network at a time, keeping the structure if its score improves. The changes are local perturbations that affect edges of the graph. The outline of the greedy search algorithm with the posterior probability as the score function is as follows:

1. Choose an initial structure.
2. Estimate $P(\vec{\theta}_S, S|D)$.
3. Make a single local change in the network structure, such as adding, removing or reversing an edge (subject to the constraint that no direct cycles are introduced). Denote the new structure by S'.
4. Compute $P(\vec{\theta}_{S'}, S'|D)$ and accept change if $P(\vec{\theta}_{S'}, S'|D) > P(\vec{\theta}_S, S|D)$.
5. Repeat steps 3-4 for r iterations or until no change produces improvement.

Note that if the criterion function is separable (as is the case for the posterior probability with multinomial distributions and Dirichlet priors) then $P(\vec{\theta}_{S'}, S'|D)$ can be computed from $P(\vec{\theta}_S, S|D)$ efficiently without recomputing all components. Specifically, as pointed out earlier, maximizing $P(\vec{\theta}_S, S|D)$ is equivalent to maximizing

$$\log P(\vec{\theta}_S|D, S) + \log P(D|S) + \log P(S)$$

as in Equation 15.21. The probability $P(\vec{\theta}_S|D, S)$ is separable and given by a product of Dirichlet functions over the local parameters, while $P(D|S)$ is given by a product of Gamma functions, and $\log P(S)$ is a sum of uncertainties. Therefore, introducing or removing an edge going into X_i requires recomputing only the components associated with X_i to determine the relative change in the total score. If an edge between X_i and X_j is reversed, all the components associated with both X_i and X_j need to be recomputed.

15.5.2.2 Sampling techniques

The greedy search algorithm can get easily trapped in local maximum. There are several variations that can help to obtain better models. For example, the network structure can be randomly perturbed upon reaching a local maximum. To avoid exponential expansion in the size of the network, a restriction is usually introduced such that no node can have more than k parents. To direct the search toward the gradient (largest improvement) we can select a set of candidate changes, test each one, and pick the one with the maximal improvement. Another approach is to consider only edges between pairs of variables that are strongly correlated [947].

Sampling techniques such as **simulated annealing** explore more of the search space and hence are less likely to be trapped in local maxima, although they are slower to converge. In simulated annealing[6] each candidate change is associated with a score

$$p = \exp\left(\frac{\Delta S}{T}\right)$$

where ΔS is the change in the score and T is a user defined parameter, representing the "temperature". Note that if ΔS is positive (the change improves the score) then $p > 1$ while if ΔS is negative then $0 < p < 1$ and therefore can be viewed as a probability. The simulated annealing algorithm accepts all positive changes but it also accepts some negative changes, probabilistically. That is, if $p > 1$ the change is approved, otherwise, it is approved with probability p. The more negative ΔS is, the lower the probability of acceptance. After r changes or evaluations, the temperature is lowered and the process is repeated until $T = 0$. The temperature parameter allows users to govern the rate of acceptance. As T decreases the probability of accepting negative changes decreases too. To ensure we converge to a good structure we can switch to the greedy approach after a certain number of random changes and accept only positive changes before reducing the temperature and starting the stochastic search again.

15.5.2.3 Model averaging

Since the available data is usually sparse, it is impossible to learn the structure of a network with high certainty. In fact, often there are many possible networks (in different equivalence classes) that can perform equally well, with almost the same posterior probability or score. Each one might contain some coincidental edges, and a priori it is unclear which network we should believe.

The model averaging approach looks for the consensus structure among many different structures, since edges that are common to many networks are more likely to reflect a true dependency in the data. Each edge e_{XY} is associated with a confidence value that reflects the agreement on that edge

[6] In physics, annealing refers to the process of slowly cooling a metal that was initially heated to a very high temperature. As the metal cools down the atoms are reorganized in a lattice-like structure, affecting the properties of the metal such as its elasticity and strength.

over all structures S, given by the weighted sum

$$P(e_{XY}|D) = \sum_{S} I_S(e_{XY})P(S|D)$$

where $I_S(e)$ is an indicator variable such that $I_S(e) = 1$ if the edge exists in S and 0 otherwise, and $P(S|D)$ is the posterior probability (weight) of the structure S. The higher the probability the more confident we are in the presence of the edge and the existence of dependency between the corresponding variables.

Computing this probability over the space of all possible networks is impossible for the same reasons mentioned before (super-exponential search space). A practical solution is to sample the network space and compute the probability based on that sample set $\mathbf{S}$

$$P(e_{XY}|D) = \frac{\sum_{S \in \mathbf{S}} I_S(e_{XY})P(S|D)}{\sum_{S \in \mathbf{S}} P(S|D)} \tag{15.27}$$

where the division by the sum of posterior probabilities ensures that the total probability of the sample is one.

The quality of this approximation depends on how well we sample the space of all possible networks. Naturally, if we have to select a subset then we are better off with a subset of "good" models that all can explain the data equally well. However, collecting such models is not trivial. The greedy search procedure samples the search space only locally and is therefore inadequate. A better procedure would be one with a stochastic element, like the simulated annealing procedure. Another alternative is to introduce perturbations in the data by using bootstrapping (sampling with replacements, as in Section 7.4.2.3) and learn a locally optimal network for each dataset. Each one is considered a good model, and to speed up the computation of the confidence we can approximate Equation 15.27 with

$$P(e_{XY}|D) = \frac{1}{|\mathbf{S}|} \sum_{S \in \mathbf{S}} I_S(e_{XY})$$

since the posterior probability of each is assumed to be more or less the same. The consensus structure can be computed by including the edges with highest confidence (above a certain threshold, which can be set by studying the distribution of confidence values) and then learning the optimal parameters given this structure.

The model averaging approach can also help to infer causal relationships in the network. As we mentioned earlier, we cannot guarantee that the structure learned is the correct structure, since there may be many other networks that imply the same set of dependencies but with different causal structures. However, by learning multiple graphs from bootstrapped datasets and computing their consensus we can identify directed edges that are in high agreement and suggest a causal relationship. This can be further helped by experimentally studying interventions, such as knockout experiments, and their effect on the network. For example, if knocking out gene X affects the distribution of values in Y but not the other way around then we can conclude that X affects Y.

15.5.3 Computing the probability of new instances

When both the structure and the parameters are unknown, we need to average over all possible structures *and* parameters when classifying a new instance. The total probability of $\vec{x}$ is given by

$$P(\vec{x}|D) = \sum_S \int_{\vec{\theta}_S} P(\vec{x}|\vec{\theta}_S, S)P(\vec{\theta}_S, S|D)d\vec{\theta}_S$$

and assuming that the structures are mutually exclusive

$$= \sum_S P(S|D) \int_{\vec{\theta}_S} P(\vec{x}|\vec{\theta}_S, S)P(\vec{\theta}_S|D, S)d\vec{\theta}_S \tag{15.28}$$

The integral is $p(\vec{x}|D, S)$ for which we already have a closed-form solution (Equation 15.20) under the assumptions of Section 15.4.1. However, the sum over all possible networks in Equation 15.28 makes this computation prohibitive. One possible solution is to pick a high-scoring model and treat it as if it were the correct model. Another approach is to average over a subset of models, as described in [621].

15.6 Learning Bayesian networks from microarray data

To wrap up the discussion on Bayesian networks, we will review the process of learning regulatory networks from expression data. As discussed in Chapter 12, expression data is one of the most useful sources of information on functional links between genes. Co-expression of a gene pair may indicate an interaction or a regulatory relationship between an activator TF and the gene it regulates. It might also indicate that both genes are regulated by the same TF. The degree of the correlated expression under different experimental conditions determines the extent of the co-dependency of a gene pair. Anti-correlation between expression profiles can also indicate a functional link such as a regulatory relationship between a repressor TF and the gene it regulates.

When learning a network from expression data we first have to define the set of entities that constitute the graph nodes. In the case of regulatory networks, the nodes correspond to genes and our goal is to learn the edge structure, i.e., which edges exist in the network. Each edge represents a regulatory relationship. The degree of confidence in the existence of an edge depends on the evidence we have. Some edges can be introduced based on prior knowledge collected from the literature or from databases of TFs and the genes they regulate (e.g. [948, 949]). Other edges can be inferred from prior knowledge, even if they were not verified experimentally. For example, certain TFs were associated with unique promoter sequences [950, 951]. All genes that share a sequence signature that is associated with a specific TF may be regulated by the same TF. Therefore, we can introduce an edge between the corresponding nodes, and assign the prior probability based on our confidence in the inference. One can also introduce edges based on interaction data and subcellular location data. However, that might affect the nature of the network learned[7].

A relatively small number of relations can be introduced based on prior knowledge, and the remaining relations have to be learned from the data. The next step would be to utilize the experimental data to identify dependencies and learn all other edges in the network. This entails learning the edges, as well as the specific statistical dependencies they imply.

As discussed in Section 15.5, searching for the "best" structure which explains the observations is a hard problem. However, we can use heuristic search algorithms to learn a locally optimal structure. For example, starting from a network based on prior knowledge we can modify the structure one edge at a time and accept the modification if it improves the posterior probability of the structure given the data, as outlined in Section 15.5.2.1. The posterior probability of a structure depends on the parameters of the network, but as we showed in Section 15.5 it is maximized when using the MAP estimators for the parameters (Equation 15.21). So the first step in learning the structure is actually to learn the optimal parameters *given* a structure. Learning the parameters means learning the conditional probability distributions associated with every node. That is, we have to learn the probability that a certain gene (node) will take on a certain expression value, given the specific expression values of its parents.

The expression value of a gene is a continuous variable. However, the typical number of microarray experiments available for analysis is usually too small to obtain reliable estimates of the conditional probability distributions for continuous variables. When the number of measurements is limited and data is sparse, it is advised to first discretize the data, since it is easier to work with discrete variables. That is, we can reduce the sample space of each variable

[7]Different networks consist of different elements and relations, as discussed in Chapter 13 and Chapter 14 (see also note Footnote 12.3 in page 506). The type of prior one should use depends on the context. For example, when learning metabolic networks, information on enzymes and the interactions they catalyze can be utilized. Interaction networks can be learned from multiple sources on protein-protein interactions. BBNs can also be applied to learn a broader class of gene networks that encompass different types of relations, from enzymatic to interactions and regulatory. This is especially useful, for example, when learning signaling pathways.

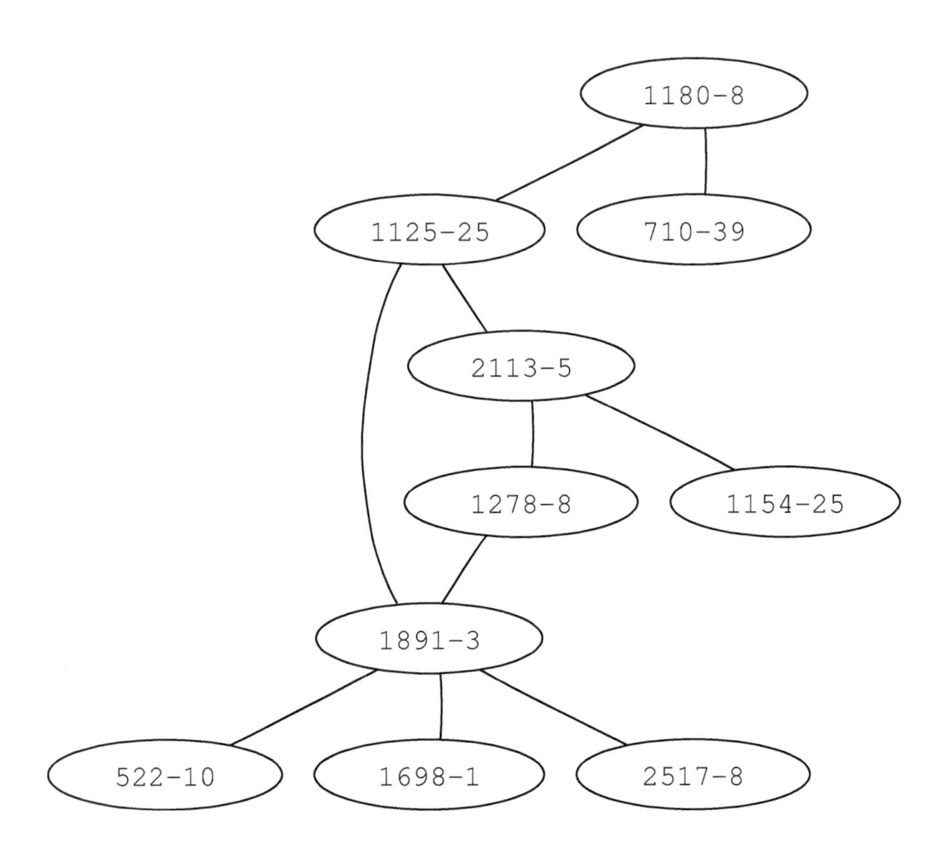

FIGURE 15.5: A gene subnetwork. A network over several genes, possibly involved in axon guidance (a component of nerve cells that is responsible for transmitting signals to other nerve cells) and processing of information in networks of nerve cells (neurons). A few of the genes were linked with down syndrome. Data is obtained from Biozon [494]. Identifiers in nodes are the Biozon IDs (to view the profile page of a protein with Biozon ID x follow the link `www.biozon.org/db/id/`x). Biozon 1180-8 is a transmembrane protein with immunoglobulin domains, 1125-25 is a membrane-associated kinase protein, 2113-5 is a down syndrome cell adhesion protein, 1278-8 is a synaptic scaffolding protein which interacts with atrophin (an important transcriptional regulator protein in neurons), 1891-3 is a receptor protein called plexin involved in axon growth, 1698-1 is a protein called cullin that is involved in ubiquitination (degradation) of other proteins and 2517-8 is a p53-associated protein.

(gene) to two possible values: on (1) and off (0). If the expression value is above a certain value, then the gene is considered on and otherwise off.

Throughout this chapter we assumed that the conditional probability distributions can be modeled as multinomial distributions. To determine the parameters, we can use the ML estimators as in Equation 15.7. Consider the toy example of Figure 15.5. To estimate the conditional probabilities associated with the node 1891-3 we have to build a two-dimensional table for each possible combination (configuration) of values of its parent nodes: 1125-25 and 1278-8. Consider the configuration (0,0), which means that both genes are off. We need to determine the probability that 1891-3 is off (the probability that it is on is the complementary probability). We denote this probability by the parameter $\theta_{1891-3,1,1}$. The first subscript refers to the variable (node) number, the second to the parent configuration and the third to the index of the specific value out of all possible values (see Figure 15.4). To estimate the parameter we first count in how many arrays both 1125-25 and 1278-8 are off (denote this number by $N_{1891-3,1}$). Second, we count in how many of these arrays 1891-3 is off (denote this number by $N_{1891-3,1,1}$). The ML estimator of $\theta_{1891-3,1,1}$ is simply $N_{1891-3,1,1}/N_{1891-3,1}$. The MAP estimator is given by considering also prior over parameters (Equation 15.17). Assuming Dirichlet prior, the MAP estimator is

$$\theta_{1891-3,1,1} = \frac{N_{1891-3,1,1} + \alpha_{1891-3,1,1}}{N_{1891-3,1} + \alpha_{1891-3,1}}$$

The pseudo-counts $\alpha_{1891-3,1,1}$ can be set as in Equation 15.18:

$$\alpha_{1891-3,1,1} = \frac{N_{1891-3,*,1}}{\sqrt{N}}$$

where $N_{1891-3,*,1}$ is the total number of arrays in which 1891-3 is off and $\alpha_{1891-3,1} = \sum_k \alpha_{1891-3,1,k}$.

The second step in learning a structure is to estimate the prior probability of the structure. As discussed in Section 15.5, the prior can be approximated based on prior knowledge of relations between genes, as in Equation 15.23. Alternatively, we can approximate the prior based on the complexity of the network, decreasing its probability with the addition of more edges and parameters (Section 15.5.1). The middle term in Equation 15.21 is given in Equation 15.22 and can be computed from the values we computed previously when estimating the parameters. Using one of the search techniques of Section 15.5.2 we can gradually learn a locally optimal network. By repeating this process multiple times with bootstrap datasets, we can obtain multiple "good" models and use the model averaging approach of Section 15.5.2.3 to highlight the stronger dependencies and expose causal relationships between genes.

15.7 Further reading

Bayesian networks were first introduced by Judea Pearl [952]. Excellent introductions to the subject appear in [445, 945, 953]. BBNs emerged as a powerful model for modeling causality in cells soon after genome-wide expression datasets became available, starting with the seminal paper by [621]. Many papers followed, discussing improvements, scoring functions and other variations. For reviews see [944, 954, 955].

The Bayesian network model we discussed in this chapter is acyclic. This kind of model cannot capture dynamic aspects of networks or patterns such as feedback loops. These, however, can be modeled with dynamic BBNs [956, 957]. Applications of BBNs include prediction of protein-protein interactions [764], pathway prediction [622,873] and function prediction [958]. For updates and additional references see the book's website at `biozon.org/proteomics/`.

15.8 Conclusions

- Bayesian networks can model complex dependencies between sets of variables.
- BBNs are useful when the dependency graph is relatively sparse.
- They are especially effective for learning gene networks from high throughput experimental data.
- The graph structure of a BBN implies certain causal relationships and statistical dependencies between variables.
- Given a BBN with a specific edge structure, the conditional probability distributions associated with each node (variable) and the values of *some* of the variables, one can infer the most likely values of all the unknown variables.
- The structure and the parameters of a BBN can be learned from empirical observations (such as gene expression data).
- BBNs can be applied also to protein function prediction, prediction of protein-protein interactions and other problems where there is dependency between observations.

15.9 Problems

For updates, additional problems, files and datasets check the book's website at `biozon.org/proteomics/`.

1. **Statistical independence**

   You are given sample datasets of two variables X and Y, where the i-th measurement of each is taken at the same time. Compare the following measures of dependency between the variables.

   - Pearson correlation between X and Y.
   - Mutual information of X and Y.

   Can you explain the difference between the two measures? Can you infer dependency based on either one? Can you associate a confidence value with each one?

2. **Equivalence classes**

   Specify all possible equivalence classes over three variables. Given a dataset over three genes, determine which equivalence class the dependency graph belongs to. Can you infer the direction of some of the edges?

3. **Model complexity and data likelihood**

   Explain why a fully connected model would result in higher likelihood of the observed data.

4. **Assessing model quality**

   Asses the quality of a model based on the correspondence between the dependencies implied from the graph structure and the statistical dependencies observed in the input dataset. How would you penalize lack of dependency? Should indirect dependencies be considered as well?

5. **Learning Bayesian networks from expression data**

   Train a Bayesian network from gene expression data to discover causal links between genes. We will use the yeast cell-cycle data available from the *Saccharomyces cerevisiae* (yeast) site.

   (a) In this assignment we will assume all values are given. Complete missing values in the data by using one of the methods of Section 12.4.2 (using a simple solution, such as the average in each experiment, should be good enough for starters).

   (b) Discretize the data, such that values between [-0.5,0.5] are mapped to zero, [0.5,] are mapped to 1 and [,-0.5] are mapped to -1.

(c) Train a network from this dataset. Consider each one of the measurements as a sample when computing the conditional probability tables.

(d) Initialize the network with a random structure, such that the maximal number of parents for each node is $k = 10$. Maintain this restriction throughout the learning process.

(e) Apply the greedy search algorithm to find a locally optimal network, using the posterior probability as the criterion function.

(f) Generate a graph with the posterior probability as a function of the iteration.

(g) Report the top two most connected genes and their local neighborhood, up to two hops away. Do they correspond to known biological processes? (Gene definitions are in `yeast.def`)

References

[1] Wilkins, M. R., Pasquali, C., Appel, R. D., Ou, K., Golaz, O., Sanchez, J. C., Yan, J. X., Gooley, A. A., Hughes, G., Humphery-Smith, I., Williams, K. L. & Hochstrasser, D. F. (1996). From proteins to proteomes: large scale protein identification by two-dimensional electrophoresis and amino acid analysis. *Nat. Biotechnol.* **14**, 61-65.

[2] Humphery-Smith, I. & Blackstock, W. (1997). Proteome analysis: genomics via the output rather than the input code. *J. Protein Chem.* **16**, 537-544.

[3] James, P. (1997). Protein identification in the post-genome era: the rapid rise of proteomics. *Q. Rev. Biophys.* **30**, 279-331.

[4] Persidis, A. (1998). Proteomics. *Nat. Biotechnol.* **16**, 393-394.

[5] Figure retrieved from http://en.wikipedia.org/wiki/File:DNA_Overview.png, December 2006.

[6] Figure retrieved from http://en.wikipedia.org/wiki/File:Aa.svg, July 2010.

[7] Figure retrieved from http://en.wikipedia.org/wiki/File:TRNA-Phe_yeast_1ehz.png, December 2006.

[8] Freeland, S. J., Knight, R. D. & Landweber, L. F. (1999). Do proteins predate DNA? *Science* **286**, 690-692.

[9] Yona, G. & Brenner, S. E. (2000). Comparison of protein sequences and practical database searching. In *Bioinformatics: Sequence, Structure and Databanks*, Oxford University Press.

[10] Pearson, W. R. (1996). Effective protein sequence comparison. *Methods Enzymol.* **266**, 227-258.

[11] Bellman, R. (1957). "Dynamic programming." Princeton University Press, Princeton, NJ.

[12] Needleman, S. B. & Wunsch, C. D. (1970). A general method applicable to the search for similarities in the amino acid sequence of two proteins. *J. Mol. Biol.* **48**, 443-453.

[13] Gonnet, G. H., Cohen, M. A. & Benner, S. A. (1992). Exhaustive matching of the entire protein sequence database. *Science* **256**, 1443-1445.

[14] Smith, T. F. & Waterman, M. S. (1981). Comparison of biosequences. *Adv. App. Math.* **2**, 482-489.

[15] Benson, D. A., Boguski, M. S., Lipman, D. J., Ostell, J., Ouellette, B. F., Rapp, B. A. & Wheeler, D. L. (1999). GenBank. *Nucl. Acids Res.* **27**, 12-17.

[16] Bairoch, A. & Apweiler, R. (1999). The SWISS-PROT protein sequence data bank and its supplement TrEMBL in 1999. *Nucl. Acids Res.* **27**, 49-54.

[17] Pearson, W. R. & Lipman, D. J. (1988). Improved tools for biological sequence comparison. *Proc. Natl. Acad. Sci. USA* **85**, 2444-2448.

[18] Altschul, S. F., Gish, W., Miller, W., Myers, E. W. & Lipman, D. J. (1990). Basic local alignment search tool. *J. Mol. Biol.* **215**, 403-410.

[19] Compugen LTD. BIOCCELERATOR Manual. http://www.compugen.co.il.

[20] Schatz, M. C., Trapnell, C., Delcher, A. L. & Varshney, A. (2007). High-throughput sequence alignment using graphics processing units. *BMC Bioinformatics* **8**, 474.

[21] Manavski, S. A. & Valle, G. (2008). CUDA compatible GPU cards as efficient hardware accelerators for Smith-Waterman sequence alignment. *BMC Bioinformatics* **9** Suppl 2, S10.

[22] Chothia, C. & Lesk, A. M. (1986). The relation between the divergence of sequence and structure in proteins. *EMBO J.* **5**, 823-826.

[23] Murzin, A. G., Brenner, S. E., Hubbard, T. & Chothia, C. (1995). SCOP: a structural classification of proteins database for the investigation of sequences and structures. *J. Mol. Biol.* **247**, 536-540.

[24] Levitt, M. & Gerstein, M. (1998). A unified statistical framework for sequence comparison and structure comparison. *Proc. Natl. Acad. Sci. USA* **95**, 5913-5920.

[25] Yona, G. & Kedem, K. (2004). The URMS-RMS hybrid algorithm for fast and sensitive local protein structure alignment. *J. Comp. Biol.* **12**, 12-32.

[26] Gribskov, M., Mclachlen, A. D. & Eisenberg, D. (1987). Profile analysis: detection of distantly related proteins. *Proc. Natl. Acad. Sci. USA* **84**, 4355-4358.

[27] Sadreyev R. & Grishin, N. (2003). COMPASS: a tool for comparison of multiple protein alignments with assessment of statistical significance. *J. Mol. Biol.* **326**, 317-336.

[28] Qian, J., Dolled-Filhart, M., Lin, J., Yu, H. & Gerstein, M. (2001). Beyond synexpression relationships: local clustering of time-shifted and inverted gene expression profiles identifies new, biologically relevant interactions. *J. Mol. Biol.* **312**, 1053-1066.

[29] Yona, G., Dirks, W., Rahman, R. & Lin, M. (2006). Effective similarity measures for expression profiles. *Bioinformatics* **22**, 1616-1622.

[30] Waterman, M. S. (1995). "Introduction to computational biology." Chapman & Hall, London.

[31] Erdös, P. & Renyi, A. (1970). On a new law of large numbers. *J. Anal. Math.* **22**, 103-111.

[32] Karlin, S. & Altschul, S. F. (1990). Methods for assessing the statistical significance of molecular sequence features by using general scoring schemes. *Proc. Natl. Acad. Sci. USA* **87**, 2264-2268.

[33] Dembo, A. & Karlin, S. (1991). Strong limit theorems of empirical functionals for large exceedances of partial sums of i.i.d. variables. *Ann. Prob.* **19**, 1737-1755.

[34] Dembo, A., Karlin, S. & Zeitouni, O. (1994). Limit distribution of maximal non-aligned two-sequence segmental score. *Ann. Prob.* **22**, 2022-2039.

[35] Gumbel, E. J. (1958). "Statistics of extremes." Columbia University Press, New York.

[36] Karlin, S. & Altschul, S. F. (1993). Applications and statistics for multiple high-scoring segments in molecular sequences. *Proc. Natl. Acad. Sci. USA* **90**, 5873-5877.

[37] Arratia, R. & Waterman, M. S. (1994). A phase transition for the score in matching random sequences allowing deletions. *Ann. App. Prob.* **4**, 200-225.

[38] Smith, T. F., Waterman, M. S. & Burks, C. (1985). The statistical distribution of nucleic acid similarities. *Nucl. Acids Res.* **13**, 645-656.

[39] Waterman, M. S. & Vingron, M. (1994). Rapid and accurate estimates of statistical significance for sequence data base searches. *Proc. Natl. Acad. Sci. USA* **91**, 4625-4628.

[40] Altschul, S. F. & Gish, W. (1996). Local alignment statistics. *Methods Enzymol.* **266**, 460-480.

[41] Drasdo, D., Hwa, T. & Lässig, M. (1998). A statistical theory of sequence alignment with gaps. In *The Proceedings of ISMB*, 52-58.

[42] Mott, R. & Tribe, R. (1999). Approximate statistics of gapped alignments. *J. Comp. Biol.* **6**, 91-112.

[43] Pearson, W. R. (1998). Empirical statistical estimates for sequence similarity searches. *J. Mol. Biol.* **276**, 71-84.

[44] Brenner, S. E., Chothia, C. & Hubbard, T. J. P. (1998). Assessing sequence comparison methods with reliable structurally identified distant evolutionary relationships. *Proc. Natl. Acad. Sci. USA* **95**, 6073-6078.

[45] Yona, G., Linial, N. & Linial, M. (1999). ProtoMap: automatic classification of protein sequences, a hierarchy of protein families, and local maps of the protein space. *Proteins* **37**, 360-378.

[46] Olsen, R., Bundschuh, R. & Hwa, T. (1999). Rapid assessment of extremal statistics for gapped local alignment. In *The Proceedings of ISMB*, 211-222.

[47] Mott, R. (2000). Accurate formula for P-values of gapped local sequence and profile alignments. *J. Mol. Biol.* **300**, 649-659.

[48] Altschul, S. F., Bundschuh, R., Olsen, R. & Hwa, T. (2001). The estimation of statistical parameters for local alignment score distributions. *Nucl. Acids Res.* **29**, 351-361.

[49] Sheetlin, S., Park, Y. & Spouge, J. L. (2005). The Gumbel pre-factor k for gapped local alignment can be estimated from simulations of global alignment. *Nucl. Acids Res.* **33**, 4987-4994.

[50] Promponas, V. J., Enright, A. J., Tsoka, S. T., Kreil, D. P., Leroy, C., Hamodrakas, S., Sander, C. & Ouzounis, C. A. (2000). CAST: an iterative algorithm for the complexity analysis of sequence tracts. *Bioinformatics* **16**, 915-922.

[51] Romero, P., Obradovic, Z., Li, X., Garner, E. C., Brown, C. J. & Dunker, A. K. (2001). Sequence complexity of disordered protein. *Proteins* **42**, 38-48.

[52] Alba, M. M., Laskowski, R. A. & Hancock, J. M. (2002). Detecting cryptically simple protein sequences using the SIMPLE algorithm. *Bioinformatics* **18**, 672-678.

[53] Wootton, J. C. & Federhen, S. (1993). Statistics of local complexity in amino acid sequences and sequence databases. *Comp. Chem.* **17**, 149-163.

[54] Claverie, J. M. & States, D. J. (1993). Information enhancement methods for large scale sequence analysis. *Comp. Chem.* **17**, 191-201.

[55] Wootton, J. C. (1994). Sequences with "unusual" amino acid compositions. *Curr. Opin. Struct. Biol.* **4**, 413-421.

[56] Yona, G. & Levitt, M. (2000). A unified sequence-structure classification of proteins: combining sequence and structure in a map of protein space. In *The Proceedings of RECOMB*, 308-317.

[57] Karplus, K., Karchin, R. & Hughey, R. (2003). Calibrating e-values for hidden Markov models with reverse-sequence null models.

[58] Sharon, I., Birkland, A., Chang, K., El-Yaniv, R. & Yona, G. (2005). Correcting BLAST and PSI-BLAST evalues for low-complexity sequences. *J. Comp. Biol.* **12**, 980-1003.

[59] Schaffer, A. A., Aravind, L., Madden, T. L., Shavirin, S., Spouge, J. L., Wolf, Y. I., Koonin, E. V. & Altschul, S. F. (2001). Improving the accuracy of PSI-BLAST protein database searches with composition-based statistics and other refinements. *Nucl. Acids Res.* **29**, 2994-3005.

[60] Yu, Y. K., Gertz, E. M., Agarwala, R., Schäffer, A. A. & Altschul, S. F. (2006). Retrieval accuracy, statistical significance and compositional similarity in protein sequence database searches. *Nucl. Acids Res.* **34**, 5966-5973.

[61] Sander, C. & Schneider, R. (1991). Database of homology-derived protein structures and the structural meaning of sequence alignment. *Proteins* **9**, 56-68.

[62] Flores, T. P., Orengo, C. A., Moss, D. & Thoronton, J. M. (1993). Comparison of conformational characteristics in structurally similar protein pairs. *Protein Sci.* **2**, 1811-1826.

[63] Hilbert, M., Bohm, G. & Jaenicke, R. (1993). Structural relationships of homologous proteins as a fundamental principle in homology modeling. *Proteins* **17**, 138-151.

[64] Murzin, A. G. (1993). OB(oligonucleotide/oligosaccharide binding)-fold: common structural and functional solution for non-homologous sequences. *EMBO J.* **12**, 861-867.

[65] Pearson, W. R. (1997). Identifying distantly related protein sequences. *Comp. App. Biosci.* **13**, 325-332.

[66] Doolittle, R. F. (1992). Reconstructing history with amino acid sequences. *Protein Sci.* **1**, 191-200.

[67] Harris, N. L., Hunter, L. & States, D. J. (1992). Mega-classification: Discovering motifs in massive datastreams. In *The Proceedings of Nat. Conf. on AI*, 837-842.

[68] Sonnhammer, E. L. L. & Kahn, D. (1994). Modular arrangement of proteins as inferred from analysis of homology. *Protein Sci.* **3**, 482-492.

[69] Watanabe, H. & Otsuka, J. (1995). A comprehensive representation of extensive similarity linkage between large numbers of proteins. *Comp. App. Biosci.* **11**, 159-166.

[70] Koonin, E. V., Tatusov, R. L. & Rudd, K. E. (1996). Protein sequence comparison at genome scale. *Methods Enzymol.* **266**, 295-321.

[71] Neuwald, A. F., Liu, J. S., Lipman, D. J. & Lawrence, C. E. (1997). Extracting protein alignment models from the sequence database. *Nucl. Acids Res.* **25**, 1665-1677.

[72] Park, J., Teichmann, S. A., Hubbard, T. & Chothia, C. (1997). Intermediate sequences increase the detection of homology between sequences. *J. Mol. Biol.* **273**, 349-354.

[73] Agrawal, A. & Huang, X. (2008). DNAlignTT: pairwise DNA alignment with sequence specific transition-transversion ratio. In *The Proceedings of IEEE Int. Conf. on EIT*, 457-459.

[74] Feng, D. F., Johnson, M. S. & Doolittle, R. F. (1985). Aligning amino acid sequences: comparison of commonly used methods. *J. Mol. Evol.* **21**, 112-125.

[75] Johnson, M. S. & Overington, J. P. (1993). A structural basis for sequence comparisons. An evaluation of scoring methodologies. *J. Mol. Biol.* **233**, 716-738.

[76] Fitch, W. M. & Margoliash, E. (1967). Construction of phylogenetic trees. *Science* **155**, 279-284.

[77] Benner, S. A. & Cohen, M. A. & Gonnet, G. H. (1994). Amino acid substitution during functionally constrained divergent evolution of protein sequences. *Protein Eng.* **7**, 1323-1332.

[78] Grantham, R. (1974). Amino acid difference formula to help explain protein evolution. *Science* **185**, 862-864.

[79] Miyata, T., Miyazawa, S. & Yasunaga, T. (1979). Two types of amino acid substitutions in protein evolution. *J. Mol. Evol.* **12**, 219-236.

[80] Dayhoff, M. O., Schwartz, R. M. & Orcutt, B. C. (1978). A model of evolutionary change in proteins. In *Atlas of Protein Sequence and Structure* **5** Suppl 3, 345-352, National Biomedical Research Foundation, Silver Spring, MD.

[81] Jones, D. T., Taylor, W. R. & Thornton, J. M. (1992). The rapid generation of mutation data matrices from protein sequences. *Comp. App. Biosci.* **8**, 275-282.

[82] Henikoff, S. & Henikoff, J. G. (1992). Amino acid substitution matrices from protein blocks. *Proc. Natl. Acad. Sci. USA* **89**, 10915-10919.

[83] McLachlan, A. D. (1971). Tests for comparing related amino-acid sequences. Cytochrome c and cytochrome c551. *J. Mol. Biol.* **61**, 409-424.

[84] Levin, J. M., Robson, B. & Garnier, J. (1986). An algorithm for secondary structure determination in proteins based on sequence similarity. *FEBS Lett.* **205**, 303-308.

[85] Mohana Rao, J. K. (1987). New scoring matrix for amino acid residue exchanges based on residue characteristic physical parameters. *Int. J. Pept. Protein. Res.* **29**, 276-281.

[86] Risler, J. L., Delorme, M. O., Delacroix, H. & Henaut, A. (1988). Amino acid substitutions in structurally related proteins. A pattern recognition approach. Determination of a new and efficient scoring matrix. *J. Mol. Biol.* **204**, 1019-1029.

[87] Setubal, J. C. & Meidanis, J. (1996). "Introduction to computational molecular biology." PWS Publishing Co., Boston.

[88] Henikoff, J. G., Henikoff, S. & Pietrokovski, S. (1999). New features of the blocks database servers. *Nucl. Acids Res.* **27**, 226-228.

[89] Altschul, S. F. (1991). Amino acid substitution matrices from an information theoretic perspective. *J. Mol. Biol.* **219**, 555-565.

[90] Henikoff, S. & Henikoff, J. G. (1993). Performance evaluation of amino acid substitution matrices. *Proteins* **17**, 49-61.

[91] Pearson, W. R. (1995). Comparison of methods for searching protein sequence databases. *Protein Sci.* **4**, 1145-1160.

[92] Flory, P. J. (1953). "Principles of polymer chemistry." Cornell University Press, Ithaca, NY.

[93] Durbin, R., Eddy, S. R., Krogh, A. & Mitchison, G. (1998). "Biological sequence analysis." Cambridge University Press.

[94] Miklos, I., Lunter, G. A. & Holmes, I. (2004). A long indel model for evolutionary sequence alignment. *Mol. Biol. Evol.* **21**, 529-540.

[95] Vingron, M. & Waterman, M. S. (1994). Sequence alignment and penalty choice: review of concepts, case studies and implications. *J. Mol. Biol.* **235**, 1-12.

[96] Sellers, P. H. (1974). On the theory and computation of evolutionary distances. *SIAM J. App. Math.* **26**, 787-793.

[97] Xu, W. & Miranker, D. P. (2004). A metric model of amino acid substitution. *Bioinformatics* **20**, 1-8.

[98] Feng, D. & Doolittle, R. F. (1987). Progressive sequence alignment as a prerequisite to correct phylogenetic trees. *J. Mol. Evol.* **60**, 351-360.

[99] Linial, M., Linial, N., Tishby, N. & Yona, G. (1997). Global self organization of all known protein sequences reveals inherent biological signatures. *J. Mol. Biol.* **268**, 539-556.

[100] Altschul, S. F., Madden, T. L., Schaffer, A. A., Zhang, J., Zhang, Z., Miller, W. & Lipman, D. J. (1997). Gapped BLAST and PSI-BLAST: a new generation of protein database search programs. *Nucl. Acids Res.* **25**, 3389-3402.

[101] Apostolico, A. & Giancarlo, R. (1998). Sequence alignment in molecular biology. *J. Comp. Biol.* **5**, 173-196.

[102] Pearson, W. R. & Sierk, M. L. (2005). The limits of protein sequence comparison? *Curr. Opin. Struct. Biol.* **15**, 254-260.

[103] Gusfield, D. (1997). "Algorithms on strings, trees, and sequences: computer science and computational biology." Cambridge University Press.

[104] Mount, D. W. (2001). "Bioinformatics: sequence and genome analysis." Cold Spring Harbor Laboratory Press.

[105] Lesk, A. (2002). "Introduction to bioinformatics." Oxford University Press.

[106] Jones, N. C. & Pevzner, P. A. (2004). "An introduction to bioinformatics algorithms." MIT Press, Cambridge, MA.

[107] Mitrophanov, A. Y. & Borodovsky, M. (2006). Statistical significance in biological sequence analysis. *Brief. Bioinform.* **7**, 2-24.

[108] Altschul, S. F., Boguski, M. S., Gish, W. G. & Wootton, J. C. (1994). Issues in searching molecular sequence databases. *Nat. Genet.* **6**, 119-129.

[109] Egan, J. P. (1975). "Signal detection theory and ROC analysis." Academic Press, New York.

[110] Hanley, J. A. & McNeil, B. J. (1982). The meaning and use of the area under the Receiver Operating Characteristic (ROC) curve. *Radiology* **143**, 29-36.

[111] Ponting, C. P., Schultz, J., Milpetz, F. & Bork, P. (1999). SMART: identification and annotation of domains from signalling and extracellular protein sequences. *Nucl. Acids Res.* **27**, 229-232.

[112] Ku, C. & Yona, G. (2005). The distance-profile representation and its application to detection of distantly related protein families. *BMC Bioinformatics* **6**, 282.

[113] Benjamini, Y. & Hochberg, Y. (1995). Controlling the false discovery rate: a practical and powerful approach to multiple testing. *J. Roy. Stat. Soc. B* **57**, 289-300.

[114] Benjamini, Y. & Yekutieli, D. (2001). The control of the false discovery rate in multiple testing under dependency. *Ann. Stat.* **29**, 1165-1188.

[115] Bairoch, A. & Boeckmann, B. (1991). The SWISS-PROT protein sequence data bank. *Nucl. Acids Res.* **19**, 2247-2249.

[116] UniProt Consortium. (2009). The Universal Protein Resource (UniProt) 2009. *Nucl. Acids Res.* **37**, D169-174.

[117] Carillo, H. & Lipman, D. (1988). The multiple sequence alignment problem in biology. *SIAM J. App. Math.* **48**, 1073-1082.

[118] Altschul, S. F., Carrol, R. J. & Lipman, D. J. (1989). Weights for data related by a tree. *J. Mol. Biol.* **207**, 647-653.

[119] Gupta, S. K., Kececioglu, J. D. & Schaffer, A. A. (1995). Improving the practical space and time efficiency of the shortest-paths approach to sum-of-pairs multiple sequence alignment. *J. Comp. Biol.* **2**, 459-472.

[120] Reinert, K., Stoye, J. & Will, T. (2000). An iterative method for faster sum-of-pairs multiple sequence alignment. *Bioinformatics* **16**, 808-814.

[121] Wang, L. & Jiang, T. (1994). On the complexity of multiple sequence alignment. *J. Comp. Biol.* **1**, 337-348.

[122] Day, W. H. E., Johnson, D. S. & Sankoff, D. (1986). The computational complexity of inferring phylogenies by parsimony. *Mathematical Biosciences* **81**, 33-42.

[123] Saitou, N. & Nei, M. (1987). The neighbor-joining method: a new method for reconstructing phylogenetic trees. *Mol. Biol. Evol.* **4**, 406-425.

[124] Hofmann, K., Bucher, P., Falquet, L. & Bairoch, A. (1999). The PROSITE database, its status in 1999. *Nucl. Acids Res.* **27**, 215-219.

[125] Henikoff, S. & Henikoff, J. G. (1994). Position-based sequence weights. *J. Mol. Biol.* **243**, 574-578.

[126] Henikoff, J. G. & Henikoff, S. (1996). Using substitution probabilities to improve position-specific scoring matrices. *Comp. App. Biosci.* **12**, 135-143.

[127] Sjölander, K., Karplus, K., Brown, M., Hughey, R., Krogh, A., Mian, I. S. & Haussler, D. (1996). Dirichlet mixtures: a method for improved detection of weak but significant protein sequence homology. *Comp. App. Biosci.* **12**, 327-345.

[128] Thompson, J. D., Higgins, D. G. & Gibson, T. J. (1994). Improved sensitivity of profile searches through the use of sequence weights and gap excision. *Comp. App. Biosci.* **10**, 19-29.

[129] Vingron, M. & Argos, P. (1989). A fast and sensitive multiple sequence alignment algorithm. *Comp. App. Biosci.* **5**, 115-121.

[130] Luthy, R., Xenarios, I. & Bucher, P. (1994). Improving the sensitivity of the sequence profile method. *Protein Sci.* **3**, 139-146.

[131] Fleissner, R., Metzler, D. & von Haeseler, A. (2005). Simultaneous statistical multiple alignment and phylogeny reconstruction. *Syst. Biol.* **54**, 548-561.

[132] Lunter, G., Miklós, I., Drummond, A., Jensen, J. L. & Hein, J. (2005). Bayesian coestimation of phylogeny and sequence alignment. *BMC Bioinformatics* **6**, 83.

[133] Yue, F., Shi, J. & Tang, J. (2009). Simultaneous phylogeny reconstruction and multiple sequence alignment. *BMC Bioinformatics* **10** Suppl 1, S11.

[134] Wang, T. & Stormo, G. D. (2003). Combining phylogenetic data with co-regulated genes to identify regulatory motifs. *Bioinformatics* **19**, 2369-2380.

[135] El-Yaniv, R., Fine, S. & Tishby, N. (1997). Agnostic classification of Markovian sequences. *NIPS* **10**, 465-471.

[136] Fuglede, B. & Topse, F. (2004). Jensen-Shannon divergence and Hilbert space embedding. In *The Proceedings of IEEE Int. Sym. on Information Theory.*

[137] Chung, R. & Yona, G. (2004). Protein family comparison using statistical models and predicted structural information. *BMC Bioinformatics* **5**, 183.

[138] Yona, G. & Levitt, M. (2002). Within the twilight zone: a sensitive profile-profile comparison tool based on information theory. *J. Mol. Biol.* **315**, 1257-1275.

[139] Edgar, R. C. & Sjölander, K. (2004). A comparison of scoring functions for protein sequence profile alignment. *Bioinformatics* **20**, 1301-1308.

[140] Söding, J. (2005). Protein homology detection by HMM-HMM comparison. *Bioinformatics* **21**, 951-960.

[141] Higgins, D. G., Thompson, J. D. & Gibson, T. J. (1996). Using CLUSTAL for multiple sequence alignments. *Methods Enzymol.* **266**, 383-402.

[142] Notredame, C., Higgins, D. G. & Heringa, J. (2000). T-Coffee: a novel method for fast and accurate multiple sequence alignment. *J. Mol. Biol.* **302**, 205-217.

[143] Heger, A., Lappe, M. & Holm, L. (2004). Accurate detection of very sparse sequence motifs. *J. Comp. Biol.* **11**, 843-857.

[144] Do, C. B., Mahabhashyam, M. S., Brudno, M. & Batzoglou, S. (2005). ProbCons: probabilistic consistency-based multiple sequence alignment. *Genome Res.* **15**, 330-340.

[145] Rausch, T., Emde, A. K., Weese, D., Döring, A., Notredame, C. & Reinert, K. (2008). Segment-based multiple sequence alignment. *Bioinformatics* **24**, i187-192.

[146] Lee, C., Grasso, C. & Sharlow, M. F. (2002). Multiple sequence alignment using partial order graphs. *Bioinformatics* **18**, 452-464.

[147] Morgenstern, B., Dress, A. & Werner, T. (1996). Multiple DNA and protein sequence alignment based on segment-to-segment comparison. *Proc. Natl. Acad. Sci. USA* **93**, 12098-12103.

[148] Shastry, B. S. (2002). SNP alleles in human disease and evolution. *J. Hum. Genet.* **47**, 561-566.

[149] Suh, Y. & Vijg, J. (2005). SNP discovery in associating genetic variation with human disease phenotypes. *Mutat. Res.* **573**, 41-53.

[150] De Gobbi, M., Viprakasit, V., Hughes, J. R., Fisher, C., Buckle, V. J., Ayyub, H., Gibbons, R. J., Vernimmen, D., Yoshinaga, Y., de Jong, P., Cheng, J. F., Rubin, E. M., Wood, W. G., Bowden, D. & Higgs, D. R. (2006). A regulatory SNP causes a human genetic disease by creating a new transcriptional promoter. *Science* **312**, 1215-1217.

[151] Vasmatzis, G., Essand, M., Brinkmann, U., Lee, B. & Pastan, I. (1998). Discovery of three genes specifically expressed in human prostate by expressed sequence tag database analysis. *Proc. Natl. Acad. Sci. USA* **95**, 300-304.

[152] Bera, T. K., Iavarone, C., Kumar, V., Lee, S., Lee, B. & Pastan, I. (2002). MRP9, an unusual truncated member of the ABC transporter superfamily, is highly expressed in breast cancer. *Proc. Natl. Acad. Sci. USA* **99**, 6997-7002.

[153] Katoh, K., Misawa, K., Kuma, K. & Miyata, T. (2002). MAFFT: a novel method for rapid multiple sequence alignment based on fast Fourier transform. *Nucl. Acids Res.* **30**, 3059-3066.

[154] Pei, J., Sadreyev, R. & Grishin, N. V. (2003). PCMA: fast and accurate multiple sequence alignment based on profile consistency. *Bioinformatics* **19**, 427-428.

[155] Van Walle, I., Lasters, I. & Wyns, L. (2004). Align-m: a new algorithm for multiple alignment of highly divergent sequences. *Bioinformatics* **20**, 1428-1435.

[156] Zhang, X. & Kahveci, T. (2007). QOMA: quasi-optimal multiple alignment of protein sequences. *Bioinformatics* **23**, 162-168.

[157] Notredame, C. (2007). Recent evolutions of multiple sequence alignment algorithms. *PLoS Comp. Biol.* **3**, e123.

[158] Pirovano, W. & Heringa, J. (2008). Multiple sequence alignment. *Methods Mol. Biol.* **452**, 143-161.

[159] Do, C. B. & Katoh, K. (2008). Protein multiple sequence alignment. *Methods Mol. Biol.* **484**, 379-413.

[160] Heringa, J. (2002). Local weighting schemes for protein multiple sequence alignment. *Comp. Chem.* **26**, 459-477.

[161] Vogt, G., Etzold, T. & Argos, P. (1995). An assessment of amino acid exchange matrices in aligning protein sequences: the twilight zone revisited. *J. Mol. Biol.* **249**, 816-831.

[162] Thompson, J. D., Plewniak, F. & Poch, O. (1999). BAliBASE: a benchmark alignment database for the evaluation of multiple alignment programs. *Bioinformatics* **15**, 87-88.

[163] Thompson, J. D., Koehl, P., Ripp, R. & Poch, O. (2005). BAliBASE 3.0: latest developments of the multiple sequence alignment benchmark. *Proteins* **61**, 127-136.

[164] Lassmann, T. & Sonnhammer, E. L. (2002). Quality assessment of multiple alignment programs. *FEBS Lett.* **529**, 126-130.

[165] Antonarakis, S. E. & McKusick, V. A. (2000). OMIM passes the 1,000-disease-gene mark. *Nat. Genet.* **25**, 11.

[166] Amberger, J., Bocchini, C. A., Scott, A. F. & Hamosh, A. (2009). McKusick's Online Mendelian Inheritance in Man (OMIM). *Nucl. Acids Res.* **37**, D793-796.

[167] Sherry, S. T., Ward, M. & Sirotkin, K. (1999). dbSNP-database for single nucleotide polymorphisms and other classes of minor genetic variation. *Genome Res.* **9**, 677-679.

[168] Barash, Y., Elidan, G., Friedman, N. & Kaplan, T. (2003). Modeling dependencies in protein-DNA binding sites. In *The Proceedings of RECOMB*, 28-37.

[169] Gorodkin, J., Stricklin, S. L. & Stormo, G. D. (2001). Discovering common stem-loop motifs in unaligned RNA sequences. *Nucl. Acids Res.* **29**, 2135-2144.

[170] Lawrence, C. E., Altschul, S. F., Boguski, M. S., Liu, J. S., Neuwald, A. F. & Wootton, J. C. (1993). Detecting subtle sequence signals: a Gibbs sampling strategy for multiple alignment. *Science* **262**, 208-214.

[171] Bailey, T. L. & Elkan, C. (1995). Unsupervised learning of multiple motifs in biopolymers using expectation maximization. *Machine Learning* **21**, 51-83.

[172] Pevzner, P. A. & Sze, S. H. (2000). Combinatorial approaches to finding subtle signals in DNA sequences. In *The Proceedings of ISMB*, 269-278.

[173] Stormo, G. D. & Hartzell, G. W. 3rd. (1989). Identifying protein-binding sites from unaligned DNA fragments. *Proc. Natl. Acad. Sci. USA* **86**, 1183-1187.

[174] Maier, D. (1978). The complexity of some problems on subsequences and supersequences. *J. ACM* **25**, 322-336.

[175] Li, M., Ma, B. & Wang, L. (2002). Finding similar regions in many strings. *J. Comp. Sys. Sci.* **65**, 73-96.

[176] Buhler, J. & Tompa, M. (2002). Finding motifs using random projections. *J. Comp. Biol.* **9**, 225-242.

[177] Smith, H. O., Annau. T. M. & Chandrasegaran, S. (1990). Finding sequence motifs in groups of functionally related proteins. *Proc. Natl. Acad. Sci. USA* **87**, 826-830.

[178] Neuwald, A. F. & Green, P. (1994). Detecting patterns in protein sequences. *J. Mol. Biol.* **239**, 698-712.

[179] Suyama, M., Nishioka, T. & Oda, J. (1995). Searching for common sequence patterns among distantly related proteins. *Protein Eng.* **8**, 1075-1080.

[180] Hanke, J., Beckmann, G., Bork, P. & Reich, J. G. (1996). Self-organizing hierarchic networks for pattern recognition in protein sequence. *Protein Sci.* **5**, 72-82.

[181] Jonassen, I., Collins, J. F. & Higgins, D. G. (1995). Finding flexible patterns in unaligned protein sequences. *Protein Sci.* **4**, 1587-1595.

[182] Keich, U. & Pevzner, P. A. (2002). Finding motifs in the twilight zone. *Bioinformatics* **18**, 1374-1381.

[183] Wang, L. & Dong L. (2005). Randomized algorithms for motif detection. *J. Bioinform. Comp. Biol.* **3**, 1039-1052.

[184] Sze, S. H., Gelfand, M. S. & Pevzner, P. A. (2002). Finding weak motifs in DNA sequences. In *The Proceedings of PSB*, 235-246.

[185] Styczynski, M. P., Jensen, K. L., Rigoutsos, I. & Stephanopoulos, G. N. (2004). An extension and novel solution to the (l,d)-motif challenge problem. *Genome Inform.* **15**, 63-71.

[186] Romer, K. A., Kayombya, G. R. & Fraenkel, E. (2007). WebMOTIFS: automated discovery, filtering and scoring of DNA sequence motifs using multiple programs and Bayesian approaches. *Nucl. Acids Res.* **35**, W217-220.

[187] Dopazo, J., Rodriguez, A., Saiz, J. C. & Sobrino, F. (1993). Design of primers for PCR amplification of highly variable genomes. *Comp. App. Biosci.* **9**, 123-125.

[188] Lucas, K., Busch, M., Mossinger, S. & Thompson, J. A. (1991). An improved microcomputer program for finding gene- or gene family-specific oligonucleotides suitable as primers for polymerase chain reactions or as probes. *Comp. App. Biosci.* **7**, 525-529.

[189] Hertz, G. Z., Hartzell, G. W. 3rd & Stormo, G. D. (1990). Identification of consensus patterns in unaligned DNA sequences known to be functionally related. *Comp. App. Biosci.* **6**, 81-92.

[190] Bejerano, G., Pheasant, M., Makunin, I., Stephen, S., Kent, W. J., Mattick, J. S. & Haussler, D. (2004). Ultraconserved elements in the human genome. *Science* **304**, 1321-1325.

[191] Duda, R. O., Hart, P. E. & Stork, D. G. (2000). "Pattern classification." John Wiley & Sons, New York.

[192] Rabiner, L. R. & Juang, B. H. (1986). An introduction to hidden Markov models. *IEEE ASSP Mag.* **3**, 4-15.

[193] Baum, L. E., Petrie, T., Soules, G. & Weiss, N. (1970). A maximization technique occurring in the statistical analysis of probabilistic functions of Markov chains. *Ann. Math. Statist.* **41**, 164-171.

[194] Hughey, R. & Krogh, A. (1996). Hidden Markov models for sequence analysis: Extension and analysis of the basic method. *Comp. App. Biosci.* **12**, 95-107.

[195] Eddy, S. R. (1998). HMMER user's guide: biological sequence analysis using profile hidden Markov models. http://hmmer.wustl.edu/

[196] Krogh, A., Brown, M., Mian, I. S., Sjölander, K. & Haussler, D. (1994). Hidden Markov models in computational biology: Application to protein modeling. *J. Mol. Biol.* **235**, 1501-1531.

[197] Eddy, S. R. (1998). Profile hidden Markov models. *Bioinformatics* **14**, 755-763.

[198] Sammut, S. J., Finn, R. D. & Bateman, A. (2008). Pfam 10 years on: 10,000 families and still growing. *Brief. Bioinform.* **9**, 210-219.

[199] Bell, T., Cleary, J. & Witten, I. (1990). "Text compression." Prentice Hall, New York.

[200] Feder, M. & Merhav, N. (1994). Relations between entropy and error probability. *IEEE Trans. on Information Theory* **40**, 259-266.

[201] Huffman, D. A. (1952). A method for the construction of minimum-redundancy codes. In *The Proceedings of the Institute of Radio Engineers* **40**, 1098-1102.

[202] Shannon, C. E. (1948). A mathematical theory of communication. *Bell System Technical Journal* **27**, 379-423, 623-656.

[203] Cover, T. M. & Thomas, J. A. (1991). "Elements of information theory." John Wiley & Sons, New York.

[204] Rissanen, J. & Langdon, G. (1979). Arithmetic coding. *IBM J. Research and Development* **23**, 149-162.

[205] Witten, I. H., Neal, R. M. & Cleary, J. G. (1987). Arithmetic coding for data compression. *Communications of the ACM* **30**, 520-540.

[206] McMillan, B. (1956). Two inequalities implied by unique decipherability. *IEEE Trans. on Information Theory* **2**, 115-116.

[207] Willems, F. M. J., Shtarkov, Y. M. & Tjalkens, T. J. (1995). The context-tree weighting method: basic properties. *IEEE Trans. on Information Theory* **41**, 653-664.

[208] Willems, F. M. J., Shtarkov, Y. M. & Tjalkens, T. J. (1996). Context weighting for general finite-context sources. *IEEE Trans. on Information Theory* **42**, 1514-1520.

[209] Begleiter, R., El-Yaniv, R. & Yona, G. (2004). On prediction using variable order markov models. *J. Artif. Intell. Res.* **22**, 385-421.

[210] Ziv, J. & Lempel, A. (1977). A universal algorithm for sequential data compression. *IEEE Trans. on Information Theory* **23**, 337-343.

[211] Ziv, J. & Lempel, A. (1978). Compression of individual sequences via variable-rate coding. *IEEE Trans. on Information Theory* **24**, 530-536.

[212] Langdon, G. (1983). A note on the Ziv-Lempel model for compressing individual sequences. *IEEE Trans. on Information Theory* **29**, 284-287.

[213] Rissanen, J. (1983). A universal data compression system. *IEEE Trans. on Information Theory* **29**, 656-664.

[214] Nisenson, M., Yariv, I., El-Yaniv, R. & Meir, R. (2003). Towards behaviometric security systems: Learning to identify a typist. In *The Proceedings of Conf. on Principles and Practice of Knowledge Discovery in Databases.*

[215] Ron, D., Singer, Y. & Tishby, N. (1996). The power of amnesia: learning probabilistic automata with variable memory length. *Machine Learning* **25**, 117-149.

[216] Bejerano, G. & Yona, G. (2001). Variations on probabilistic suffix trees: statistical modeling and prediction of protein families. *Bioinformatics* **17**, 23-43.

[217] Eskin, E., Grundy, W. N. & Singer, Y. (2000). Protein family classification using sparse markov transducers. In *The Proceedings of ISMB*, 20-23.

[218] Cleary, J. G. & Witten, I. H. (1984). Data compression using adaptive coding and partial string matching. *IEEE Trans. on Communications* **32**, 396-402.

[219] Rabiner, L. R. (1989). A tutorial on Hidden Markov Models and selected applications in speech recognition. *Proc. IEEE* **77**, 257-286.

[220] Kundu, A., He, Y. & Bahl, P. (1989). Recognition of handwritten words: first and second order hidden Markov model based approach. *Pattern Recognition* **22**, 283-297.

[221] Vlontzos, A. & Kung. S. Y. (1992). Hidden Markov models for character recognition. *IEEE Trans. on Image Processing* **1**, 539-543.

[222] Najmi, J. L. A. & Gray, R. M. (2000). Image classification by a two dimensional hidden Markov model. *IEEE Trans. on Signal Processing* **48**, 517-533.

[223] Wilson, A. D. & Bobick, A. F. (2001). Hidden Markov models for modeling and recognizing gesture under variation. *J. Pattern Recognition and AI* **15**, 123-160.

[224] Ourston, D., Matzner, S., Stump, W. & Hopkins, B. (2003). Applications of hidden Markov models to detecting multi-stage network attacks. In *The Proceedings of Int. Conf. on System Sciences.*

[225] Qian, B. & Goldstein, R. A. (2003). Detecting distant homologs using phylogenetic tree-based HMMs. *Proteins* **52**, 446-453.

[226] Siepel, A. & Haussler, D. (2004). Combining phylogenetic and hidden Markov models in biosequence analysis. *J. Comp. Biol.* **11**, 413-428.

[227] Baldi, P. & Brunak, S. (2001). "Bioinformatics: the machine learning approach." MIT Press, Cambridge, MA.

[228] Lukashin, A. V. & Borodovsky, M. (1998). GeneMark.hmm: new solutions for gene finding. *Nucl. Acids Res.* **26**, 1107-1115.

[229] Stanke, M. & Waack, S. (2003). Gene prediction with a hidden Markov model and a new intron submodel. *Bioinformatics* **19** Suppl 2, 215-225.

[230] Allen, J. E., Pertea, M. & Salzberg, S. L. (2004). Computational gene prediction using multiple sources of evidence. *Genome Res.* **14**, 142-148.

[231] Nielsen, P. & Krogh, A. (2005). Large-scale prokaryotic gene prediction and comparison to genome annotation. *Bioinformatics* **21**, 4322-4329.

[232] Majoros, H. (2007). "Methods for computational gene prediction." Cambridge University Press.

[233] Rosenblatt, F. (1958). The perceptron: a probabilistic model for information storage and organization in the brain. *Psychol Rev.* **65**, 386-408.

[234] Novikoff, A. B. (1962). On convergence proofs for perceptrons. In *The Proceedings of Sym. on Mathematical Theory of Automata* **12**, 615-622.

[235] Vapnik, V. N. (1982). "Estimation of dependences based on empirical data." Springer-Verlag, New York.

[236] Cortes, C. & Vapnik, V. (1995). Support-vector networks. *Machine Learning* **20**, 273-297.

[237] Joachims, T. (1999). Making large-scale SVM learning practical. In *Advances in Kernel Methods - Support Vector Learning*, MIT Press, Cambridge, MA.

[238] Fan, R. E., Chen, P. H. & Lin, C. J. (2005). Working set selection using the second order information for training SVM. *J. Machine Learning Res.* **6**, 1889-1918.

[239] Bryan, J. G. (1951). The generalized discriminant function: mathematical foundation and computational routine. *Harvard Educ. Rev.* **21**, 90-95.

[240] Boser, B. E., Guyon, I. M. & Vapnik, V. N. (1992). A training algorithm for optimal margin classifiers. In *The Proceedings of COLT*, 144-152.

[241] Aizerman, M., Braverman, E. & Rozonoer, L. (1964). Theoretical foundations of the potential function method in pattern recognition learning. *Automation and Remote Control* **25**, 821-837.

[242] Vapnik, V. N. (1995). "The nature of statistical learning theory." Springer-Verlag, New York.

[243] Salton, G. (1989). "Automatic text processing." Addison-Wesley, Reading, MA.

[244] Leslie, C. S., Eskin, E. & Noble, W. S. (2002). The spectrum kernel: a string kernel for SVM protein classification. In *The Proceedings of PSB*, 564-575.

[245] Leslie, C. S., Eskin, E., Cohen, A., Weston, J. & Noble, W. S. (2004). Mismatch string kernels for discriminative protein classification. *Bioinformatics* **20**, 467-476.

[246] Liao, L. & Noble, W. S. (2003). Combining pairwise sequence similarity and support vector machines for detecting remote protein evolutionary and structural relationships. *J. Comp. Biol.* **10**, 857-868.

[247] Jaakkola, T., Diekhans, M. & Haussler, D. (1999). Using the Fisher kernel method to detect remote protein homologies. In *The Proceedings of ISMB*, 149-158.

[248] Jaakkola, T., Diekhans, M. & Haussler, D. (2000). A discriminative framework for detecting remote protein homologies. *J. Comp. Biol.* **7**, 95-114.

[249] Seeger, M. (2001). Covariance kernels from Bayesian generative models. *NIPS* **14**, 905-912.

[250] Cuturi, M. & Vert, J. P. (2004). A mutual information kernel for sequences. In *The Proceedings of Int. Conf. on Neural Networks* **3**, 1905-1910.

[251] Breiman, L., Friedman, J. H., Olshen, R. A. & Stone, C. J. (1993). "Classification and regression trees." Chapman & Hall, New York.

[252] Syed, U. & Yona, G. (2009). Enzyme function prediction with interpretable models. *Methods Mol. Biol.* **541**, 373-420.

[253] Hyafil, L. & Rivest, R. L. (1976). Constructing optimal binary decision trees is NP-complete. *Information Processing Letters* **5**, 15-17.

[254] Mitchell, T. M. (1997). "Machine learning." McGraw-Hill.

[255] Quinlan, J. R. (1986). Induction of decision trees. *Machine Learning* **1**, 81-106.

[256] Mantaras, R. L. (1991). A distance-based attribute selection measure for decision tree induction. *Machine Learning* **6**, 81-92.

[257] Fayyad, U. M. & Irani, K. B. (1992). The attribute selection problem in decision tree generation. In *The Proceedings of Int. Conf. on AI*, 104-110.

[258] Hjorth, J. S. U. (1994). "Computer intensive statistical methods validation, model selection, and bootstrap." Chapman & Hall, London.

[259] Breiman, L. (1996). Bagging predictors. *Machine Learning* **24**, 123-140.

[260] Shakhnarovich, G., El-Yaniv, R. & Baram, Y. (2001). Smoothed bootstrap and statistical data cloning for classifier evaluation. In *The Proceedings of ICML*, 521-528.

[261] Jain, A. K., Dubes, R. C. & Chen, C. (1998). Bootstrap techniques for error estimation. *IEEE Trans. on Pattern Analysis and Applications* **9**, 628-633.

[262] Rissanen, J. (1978). Modeling by shortest data description. *Automatica* **14**, 465-471.

[263] Quinlan, J. R. & Rivest, R. L. (1989). Inferring decision trees using the minimum description length principle. *Information and Computation* **80**, 227-248.

[264] Syed, U. & Yona, G. (2003). Using a mixture of probabilistic decision trees for direct prediction of protein function. In *The Proceedings of RECOMB*, 289-300.

[265] Dietterich, T. G. (2000). An experimental comparison of three methods for constructing ensembles of decision trees: bagging, boosting, and randomization. *Machine Learning* **40**, 139-157.

[266] Ho, T. K. (1998). The random subspace method for constructing decision forests. *IEEE Trans. on Pattern Analysis and Machine Intelligence* **20**, 832-844.

[267] Breiman, L. (2001). Random forests. *Machine Learning* **45**, 5-32.

[268] Berman, H., Henrick, K., Nakamura, H. & Markley, J. L. (2007). The worldwide Protein Data Bank (wwPDB): ensuring a single, uniform archive of PDB data. *Nucl. Acids Res.* **35**, D301-D303.

[269] Fayyad, U. M. & Irani, K. B. (1993). Multi-interval discretization of continuous-valued attributes for classification learning. In *The Proceedings of Int. Conf. on AI*, 1022-1027.

[270] Kohavi, R. & Sahami, M. (1996). Error-based and entropy-based discretization of continuous features. In *The Proceedings of Int. Conf. on Knowledge Discovery and Data Mining*, 114-119.

[271] Breiman, L., Friedman, J. H., Olshen, R. A. & Stone, C. J. (1993). "Classification and regression trees." Wadsworth Int. Group, Belmont, California,

[272] Freund, Y. & Schapire, R. E. (1999). Large margin classification using the perceptron algorithm. *Machine Learning* **37**, 277-296.

[273] Vapnik, V. N. (1998). "Statistical learning theory." John Wiley & Sons, New York.

[274] Cristianini, N. & Shawe-Taylor, J. (2000). "An introduction to support vector machines and other kernel-based learning methods." Cambridge University Press.

[275] Schölkopf, B. & Smola, A. J. (2002). "Learning with kernels." MIT Press, Cambridge, MA.

[276] Burges, C. J. (1998). A tutorial on support vector machines for pattern recognition. *Data Mining and Knowledge Discovery* **2**, 121-167.

[277] Müller, K. R., Mika, S., Rätsch, G., Tsuda K. & Schölkopf, B. (2001). An introduction to kernel-based learning algorithms. *IEEE Trans. on Neural Networks* **12**, 181-201.

[278] Lecun, Y., Jackel, L. D., Bottou, L., Cortes, C., Denker, J. S., Drucker, H., Guyon, I., Müller, U. A., Säckinger, E., Simard, P. Y. & Vapnik, V. (1995). Learning algorithms for classification: a comparison on handwritten digit recognition. *Neural Networks* 261-276.

[279] Joachims, T. (1998). Text categorization with support vector machines: learning with many relevant features. In *The Proceedings of ECML*, 137-142.

[280] Phillips, P. J. (1999). Support vector machines applied to face recognition. *NIPS* **12**, 803-809.

[281] Müller, K. R., Smola, A. J., Rätsch, G., Schölkopf, B, Kohlmorgen, J. & Vapnik, V. (1997). Predicting time series with support vector machines. *Lecture Notes in Computer Science* **1327**, 999-1004.

[282] Chapelle, O., Weston, J. & Schölkopf, B. (2002). Cluster Kernels for Semi-Supervised Learning. *NIPS* **15**, 585-592.

[283] Mika, S., Schölkopf, B., Smola, A., Müller, K.R., Scholz, M. & Räthsch, G. (1998). Kernel PCA and de-noising in feature space. *NIPS* **11**, 536-542.

[284] Brown, M. P., Grundy, W. N., Lin, D., Cristianini, N., Sugnet, C. W., Furey, T. S., Ares, M. & Haussler, D. (2000). Knowledge-based analysis of microarray gene expression data by using support vector machines. *Proc. Natl. Acad. Sci. USA* **97**, 262-267.

[285] Furey, T. S., Cristianini, N., Duffy, N., Bednarski, D. W., Schummer, M. & Haussler, D. (2000). Support vector machine classification and validation of cancer tissue samples using microarray expression data. *Bioinformatics* **16**, 906-914.

[286] Lee, Y. & Lee C. K. (2003). Classification of multiple cancer types by multicategory support vector machines using gene expression data. *Bioinformatics* **19**, 1132-1139.

[287] Hua, S. & Sun, Z. (2001). Support vector machine approach for protein subcellular localization prediction. *Bioinformatics* **17**, 721-728.

[288] Chen, L., Wang, W., Ling, S., Jia, C. & Wang, F. (2006) KemaDom: a web server for domain prediction using kernel machine with local context. *Nucl. Acids Res.* **34**, W158-163.

[289] Baten, A. K., Chang, B. C., Halgamuge, S. K. & Li, J. (2006). Splice site identification using probabilistic parameters and SVM classification. *BMC Bioinformatics* **7** Suppl 5, S15.

[290] Shen, J., Zhang, J., Luo, X., Zhu, W., Yu, K., Chen, K., Li, Y. & Jiang, H. (2007). Predicting protein-protein interactions based only on sequences information. *Proc. Natl. Acad. Sci. USA* **104**, 4337-4341.

[291] Ben-Hur, A. & Brutlag, D. (2003). Remote homology detection: a motif based approach. *Bioinformatics* **19** Suppl 1, i26-33.

[292] Rangwala, H. & Karypis, G. (2005). Profile-based direct kernels for remote homology detection and fold recognition. *Bioinformatics* **21**, 4239-4247.

[293] Kuang, R., Ie, E., Wang, K., Wang, K., Siddiqi, M., Freund, Y. & Leslie, C. (2005). Profile-based string kernels for remote homology detection and motif extraction. *J. Bioinform. Comp. Biol.* **3**, 527-550.

[294] Eskin, E. & Snir, S. (2007). Incorporating homologues into sequence embeddings for protein analysis. *J. Bioinform. Comp. Biol.* **5**, 717-738.

[295] Lanckriet, G. R., De Bie, T., Cristianini, N., Jordan, M. I. & Noble, W. S. (2004). A statistical framework for genomic data fusion. *Bioinformatics* **20**, 2626-2635.

[296] Bejerano, G., Friedman, N. & Tishby, N. (2004). Efficient exact p-value computation for small sample, sparse and surprising categorical data. *J. Comp. Biol.* **11**, 867-886.

[297] Agresti, A. (2001). Exact inference for categorical data: recent advances and continuing controversies. *Stat. Med.* **20**, 2709-2722.

[298] Keich, U. & Nagarajan, N. (2006). A fast and numerically robust method for exact multinomial goodness-of-fit test. *J. Comp. Graph. Stat.* **15**, 779-802.

[299] Figure retrieved from http://en.wikipedia.org/wiki/File:Chromatin_Structures.png, October 2008.

[300] Cairns, B. R. (2009). The logic of chromatin architecture and remodelling at promoters. *Nature* **461**, 193-198.

[301] Zuker, M. (2000). Calculating nucleic acid secondary structure. *Curr. Opin. Struct. Biol.* **10**, 303-310.

[302] Mathews, D. H. & Turner, D. H. (2006). Prediction of RNA secondary structure by free energy minimization. *Curr. Opin. Struct. Biol.* **16**, 270-278.

[303] Levinthal, C. (1968). Are there pathways for protein folding? *Journal de Chimie Physique et de Physico-Chimie Biologique* **65**, 44-45.

[304] Shirts, M. R. & Pande, V. S. (2000). Screen savers of the world, unite! *Science* **290**, 1903-1904.

[305] Bradley, P., Misura, K. M. & Baker, D. (2005). Toward high-resolution de novo structure prediction for small proteins. *Science* **309**, 1868-1871.

[306] Bujnicki, J. M. (2006). Protein-structure prediction by recombination of fragments. *Chembiochem* **7**, 19-27.

[307] Kolodny, R., Koehl, P., Guibas, L. & Levitt, M. (2002). Small libraries of protein fragments model native protein structures accurately. *J. Mol. Biol.* **323**, 297-307.

[308] Bowie, J. U., Luthy, R. & Eisenberg, D. (1991). A method to identify protein sequences that fold into a known three-dimensional structure. *Science* **253**, 164-170.

[309] Jones, D. T., Taylor, W. R. & Thornton, J. M. (1992). A new approach to protein fold recognition. *Nature* **358**, 86-89.

[310] Jones, D. T. & Thornton, J. M. (1996). Potential energy functions for threading. *Curr. Opin. Struct. Biol.* **6**, 210-216.

[311] Shen, M. Y. & Sali, A. (2006). Statistical potential for assessment and prediction of protein structures. *Protein Sci.* **15**, 2507-2524.

[312] Gibrat, J. F., Madej, T. & Bryant, S. H. (1996). Surprising similarities in structure comparison. *Curr. Opin. Struct. Biol.* **6**, 377-385.

[313] Madej, T., Panchenko, A. R., Chen, J. & Bryant, S. H. (2007). Protein homologous cores and loops: important clues to evolutionary relationships between structurally similar proteins. *BMC Struct. Biol.* **7**, 23.

[314] Donate, L. E., Rufino, S. D., Canard, L. H. & Blundell, T. L. (1996). Conformational analysis and clustering of short and medium size loops connecting regular secondary structures: a database for modeling and prediction. *Protein Sci.* **5**, 2600-2616.

[315] Zhou, X., Alber, F., Folkers, G., Gonnet. G. H. & Chelvanayagam, G. (2000). An analysis of the helix-to-strand transition between peptides with identical sequence. *Proteins* **41**, 248-256.

[316] Figure retrieved from http://en.wikipedia.org/wiki/File:Peptide_angles.png, October 2008.

[317] Ramachandran, G. N., Ramakrishnan, C. & Sasisekharan, V. (1963). Stereochemistry of polypeptide chain configurations. *J. Mol. Biol.* **7**, 95-99.

[318] Frishman, D. & Argos, P. (1995). Knowledge-based secondary structure assignment. *Proteins* **23**, 566-579.

[319] Kabsch, W. & Sander, C. (1983). Dictionary of protein secondary structure: pattern recognition of hydrogen-bonded and geometrical features. *Biopolymers* **22**, 2577-2637.

[320] Chou, P. Y. & Fasman, G. (1978). Prediction of secondary structure of proteins from their amino-acid sequence. *Adv. Enzymol.* **47**, 45-148.

[321] Holley, L. H. & Karplus, M. (1989). Protein secondary structure prediction with a neural network. *Proc. Natl. Acad. Sci. USA* **86**, 152-156.

[322] Zhang, X., Mesirov, J. P. & Waltz, D. L. (1992). Hybrid system for protein secondary structure prediction. *J. Mol. Biol.* **225**, 1049-1063.

[323] Geourjon, C. & Deléage, G. (1995). SOPMA: significant improvement in protein secondary structure prediction by consensus prediction from multiple alignments. *Comp. App. Biosci.* **11**, 681-684.

[324] Jones, D. T. (1999). Protein secondary structure prediction based on position-specific scoring matrices. *J. Mol. Biol.* **292**, 195-202.

[325] Garnier, J., Osguthorpe, D. J. & Robson, B. (1978). Analysis of the accuracy and implications of simple methods for predicting the secondary structure of globular proteins. *J. Mol. Biol.* **120**, 97-120.

[326] Holmes, K. C., Sander, C. & Valencia, A. (1993). A new ATP-binding fold in actin, hexokinase and Hsc70. *Trends Cell Biol.* **3**, 53-59.

[327] Kosloff, M. & Kolodny, R. (2008). Sequence-similar, structure-dissimilar protein pairs in the PDB. *Proteins* **71**, 891-902.

[328] Alexander, P. A., He, Y., Chen, Y., Orban, J. & Bryan, P. N. (2009). A minimal sequence code for switching protein structure and function. *Proc. Natl. Acad. Sci. USA* **106**, 21149-21154.

[329] de la Cruz, X., Sillitoe, I. & Orengo C. (2002). Use of structure comparison methods for the refinement of protein structure predictions. I. Identifying the structural family of a protein from low-resolution models. *Proteins* **46**, 72-84.

[330] Kim, D. E., Chivian, D. & Baker, D. (2004). Protein structure prediction and analysis using the Robetta server. *Nucl. Acids Res.* **32**, W526-531.

[331] Skolnick, J. (2006). In quest of an empirical potential for protein structure prediction. *Curr. Opin. Struct. Biol.* **16**, 166-171.

[332] Orengo, C. A., Michie, A. D., Jones, S., Jones, D. T., Swindells, M. B. & Thornton, J. M. (1997). CATH - a hierarchic classification of protein domain structures. *Structure* **5**, 1093-1108.

[333] Holm, L., Ouzounis, C., Sander, C., Tuparev, G. & Vriend, G. (1992). A database of protein structure families with common folding motifs. *Protein Sci.* **12**, 1691-1698.

[334] Yona, G. & Levitt, M. (2000). Towards a complete map of the protein space based on a unified sequence and structure analysis of all known proteins. In *The Proceedings of ISMB*, 395-406.

[335] Levitt, M., Hirshberg, M., Sharon, R. & Daggett, V. (1995). Potential energy function and parameters for simulations of the molecular dynamics of proteins and nucleic acids in solution. *Comp. Phys. Comm.* **91**, 215-231.

[336] Kabsch, W. (1976). A solution for the best rotation to relate two sets of vectors. *Acta Crystallographica* **32**, 922-923.

[337] Coutsias, E. A., Seok, C. & Dill, K. A. (2004). Using quaternions to calculate RMSD. *J. Comp. Chem.* **25**, 1849-1857.

[338] Chew, L. P., Kedem, K., Huttenlocher, D. P., & Kleinberg, J. (1999). Fast detection of geometric substructure in proteins. *J. Comp. Biol.* **6**, 313-325.

[339] Shatsky, M., Nussinov, R. & Wolfson, H. J. (2002). Flexible protein alignment and hinge detection. *Proteins* **48**, 242-256.

[340] Yuzhen, Y. & Godzik, A. (2003). Flexible structure alignment by chaining aligned fragment pairs allowing twists. *Bioinformatics* **19**, 246-255.

[341] Rocha, J., Segura, J., Wilson, R. C. & Dasgupta, S. (2009). Flexible structural protein alignment by a sequence of local transformations. *Bioinformatics* **25**, 1625-1631.

[342] Vriend, G. & Sander, C. (1991). Detection of common three-dimensional substructures in proteins. *Proteins* **11**, 52-58.

[343] Griendley, H. M., Artymiuk, P. J., Rice D. W. & Willett, P. (1993). Identification of tertiary structure resemblance in proteins using a maximal common subgraph isomorphism algorithm. *J. Mol. Biol.* **229**, 707-721.

[344] Yee, D. & Dill, K. (1993). Families and the structural relatedness among globular proteins. *Protein Sci.* **2**, 884-899.

[345] Holm, L. & Sander, C. (1993). Protein structure comparison by alignment of distance matrices. *J. Mol. Biol.* **233**, 123-138.

[346] Dubnov, S., El-Yaniv, R., Gdalyahu, Y., Schneidman, E., Tishby, N. & Yona, G. (2002). A new nonparametric pairwise clustering algorithm based on iterative estimation of distance profiles. *Machine Learning* **47**, 35-61.

[347] Kolodny, R. & Linial, N. (2004). Approximate protein structural alignment in polynomial time. *Proc. Natl. Acad. Sci. USA* **101**, 12201-12206.

[348] Holm, L. & Sander, C. (1996). Mapping the protein universe. *Science* **273**, 595-602.

[349] Shindyalov, I. N. & Bourne, P. E. (1998). Protein structure alignment by incremental combinatorial extension (CE) of the optimal path. *Protein Eng.* **11**, 739-747.

[350] Fischer, D., Nussinov, R. & Wolfson, H. (1992). 3D substructure matching in protein molecules. In *The Proceedings of Int. Sym. on Combinatorial Pattern Matching* **644**, 136-150.

[351] Pennec, X. & Ayache, N. (1998). A geometric algorithm to find small but highly similar 3D substructures in proteins. *Bioinformatics* **14**, 516-522.

[352] Andreeva, A., Howorth, D., Chandonia, J. M., Brenner, S. E., Hubbard, T. J., Chothia, C. & Murzin, A. G. (2008). Data growth and its impact on the SCOP database: new developments. *Nucl. Acids Res.* **36**, D419-425.

[353] Wrabl, J. O. & Grishin, N. V. (2008). Statistics of random protein superpositions: p-values for pairwise structure alignment. *J. Comp. Biol.* **15**, 317-355.

[354] Nagarajan, N. & Yona, G. (2004). Automatic prediction of protein domains from sequence information using a hybrid learning system. *Bioinformatics* **20**, 1335-1360.

[355] Koehl, P. & Levitt, M. (2002). Protein topology and stability define the space of allowed sequences. *Proc. Natl. Acad. Sci. USA* **99**, 1280-1285.

[356] Yu, L., White, J. V. & Smith, T. F. (1998). A homology identification method that combines protein sequence and structure information. *Protein Sci.* **7**, 2499-2510.

[357] Hedman, M., Deloof, H., Von Heijne, G., & Elofsson, A. (2001). Improved detection of homologous membrane proteins by inclusion of information from topology predictions. *Protein Sci.* **11**, 652-658.

[358] Ginalski, K., Pas, J., Wyrwicz, L. S., von Grotthuss, M., Bujnicki, J. M. & Rychlewski, L. (2003). ORFeus: detection of distant homology using sequence profiles and predicted secondary structure. *Nucl. Acids Res.* **31**, 3804-3807.

[359] Teodorescu, O., Galor, T., Pillardy, J. & Elber, R. (2004). Enriching the sequence substitution matrix by structural information. *Proteins* **54**, 41-48.

[360] Creighton, T. E. (1992). "Proteins: structures and molecular properties." WH Freeman, New York.

[361] Sternberg, M. J. E., ed. (1996) "Protein structure prediction: a practical approach." Oxford University Press.

[362] Webster, D. M. ed. (2000). "Protein structure prediction: methods and protocols." *Methods Mol. Biol.* **143**, Humana Press.

[363] IBM. (2005). Blue gene. *IBM J. Research and Development.* **49**, 191-489.

[364] Shaw, D. E., Deneroff, M. M., Dror, R. O., Kuskin, J. S., Larson, R. H. et al. (2007). Anton: A special-purpose machine for molecular dynamics simulation. In *The Proceedings of Int. Sym. on Computer Architecture.*

[365] Zhang, Y. (2008). Progress and challenges in protein structure prediction. *Curr. Opin. Struct. Biol.* **18**, 342-348.

[366] Shen, Y., Lange, O., Delaglio, F., Rossi, P., Aramini, J. M., Liu, G., Eletsky, A., Wu, Y., Singarapu, K. K., Lemak, A., Ignatchenko, A., Arrowsmith, C. H., Szyperski, T., Montelione, G. T., Baker, D. & Bax, A. (2008). Consistent blind protein structure generation from NMR chemical shift data. *Proc. Natl. Acad. Sci. USA* **105**, 4685-4690.

[367] David, R., Korenberg, M. J. & Hunter, I. W. (2000). 3D-1D threading methods for protein fold recognition. *Pharmacogenomics* **1**, 445-455.

[368] Xu, J., Jiao, F. & Yu, L. (2008). Protein structure prediction using threading. *Methods Mol. Biol.* **413**, 91-121.

[369] Smith, T. F., Lo Conte, L., Bienkowska, J., Gaitatzes, C., Rogers, R. G. Jr & Lathrop, R. (1997). Current limitations to protein threading approaches. *J. Comp. Biol.* **4**, 217-225.

[370] Albrecht, M., Hanisch, D., Zimmer, R. & Lengauer, T. (2002). Improving fold recognition of protein threading by experimental distance constraints. *In Silico Biol.* **2**, 325-337.

[371] Dukka, B. K., Tomita, E., Suzuki, J., Horimoto, K. & Akutsu, T. (2006). Protein threading with profiles and distance constraints using clique based algorithms. *J. Bioinform. Comp. Biol.* **4**, 19-42.

[372] Taylor, W. R., Lin, K., Klose, D., Fraternali, F. & Jonassen, I. (2006). Dynamic domain threading. *Proteins* **64**, 601-614.

[373] Ellrott, K., Guo, J. T., Olman, V. & Xu, Y. (2006). A generalized threading model using integer programming with secondary structure element deletion. *Genome Inform.* **17**, 248-258.

[374] Yang, Y. D., Park, C. & Kihara, D. (2008). Threading without optimizing weighting factors for scoring function. *Proteins* **73**, 581-596.

[375] Pulim, V., Bienkowska, J. & Berger, B. (2008). LTHREADER: prediction of extracellular ligand-receptor interactions in cytokines using localized threading. *Protein Sci.* **17**, 279-292.

[376] Xu, B., Yang, Y., Liang, H. & Zhou, Y. (2009). An all-atom knowledge-based energy function for protein-DNA threading, docking decoy discrimination, and prediction of transcription-factor binding profiles. *Proteins* **76**, 718-730.

[377] Brylinski, M. & Skolnick, J. (2009). FINDSITE: a threading-based approach to ligand homology modeling. *PLoS Comp. Biol.* **5**, e1000405.

[378] Gao, M. & Skolnick, J. (2009). A threading-based method for the prediction of DNA-binding proteins with application to the human genome. *PLoS Comp. Biol.* **5**, e1000567.

[379] Rackovsky, S. & Scheraga, H. A. (1978). Differential geometry and polymer conformations 2. Development of a conformational distance function. *Macromolecules* **13**, 1440-1453.

[380] Rackovsky, S. & Goldstein, D. A. (1998). Protein comparison and classification: A differential geometry approach. *Proc. Natl. Acad. Sci. USA* **85**, 777-781.

[381] Taylor, W. R. & Orengo, C. A. (1989). Protein structure alignment. *J. Mol. Biol.* **208**, 1-22.

[382] Orengo, C. A., Brown, N. P. & Taylor, W. R. (1992). Fast structure alignment for protein databank searching. *Proteins* **14**, 139-167.

[383] Taylor, W. R. (1999). Protein stucture alignment using iterated double dynamic programming. *Protein Sci.* **8**, 654-665.

[384] Holm, L. & Park, J. (2000). DaliLite workbench for protein structure comparison. *Bioinformatics* **16**, 566-567.

[385] Zhu, J. & Weng, Z. (2005). FAST: a novel protein structure alignment algorithm. *Proteins* **58**, 618-627.

[386] Rogen, P. & Fain, B. (2003). Automatic classification of protein structure by using Gauss integrals. *Proc. Natl. Acad. Sci. USA* **100**, 119-124.

[387] Lotan, I. & Schwarzer, F. (2004). Approximation of protein structure for fast similarity measures. *J. Comp. Biol.* **11**, 299-317.

[388] Roach, J., Sharma, S., Kapustina, M. & Carter, C. W. Jr. (2005). Structure alignment via Delaunay tetrahedralization. *Proteins* **60**, 66-81.

[389] Pelta, D. A., Gonzalez, J. R. & Moreno Vega, M. (2008). A simple and fast heuristic for protein structure comparison. *BMC Bioinformatics* **9**, 161.

[390] Mavridis, L. & Ritchie, D. W. (2010). 3d-blast: 3d protein structure alignment, comparison, and classification using spherical polar fourier correlations. In *The Proceedings of PSB*, 281-292.

[391] Budowski-Tal, I., Nov, Y. & Kolodny, R. (2010). FragBag, an accurate representation of protein structure, retrieves structural neighbors from the entire PDB quickly and accurately. *Proc. Natl. Acad. Sci. USA* **107**, 3481-3486.

[392] Pearson, W. R. & Sierk, M. L. (2004). Sensitivity and selectivity in protein structure comparison. *Protein Sci.* **13**, 773-785.

[393] Kolodny, R., Koehl, P. & Levitt, M. (2005). Comprehensive evaluation of protein structure alignment methods: scoring by geometric measures. *J. Mol. Biol.* **346**, 1173-1188.

[394] Taylor, W. R., Flores, T. P. & Orengo, C. A. (1994). Multiple protein structure alignment. *Protein Sci.* **3**, 1858-1870.

[395] Chew, L. P. & Kedem, K. (2002). Finding the consensus shape for a protein family. In *The Proceedings of ACM Sym. on Computational Geometry*, 64-73.

[396] Dror, O., Benyamini, H., Nussinov, R. & Wolfson, H. J. (2003). Multiple structural alignment by secondary structures: algorithm and applications. *Protein Sci.* **12**, 2492-2507.

[397] Guda, C., Lu, S., Scheeff, E. D., Bourne, P. E. & Shindyalov, I. N. (2004). CE-MC: a multiple protein structure alignment server. *Nucl. Acids Res.* **32**, W100-103.

[398] Shatsky, M., Nussinov, R. & Wolfson, H. J. (2004). A method for simultaneous alignment of multiple protein structures. *Proteins* **56**, 143-156.

[399] Konagurthu, A. S., Whisstock, J. C., Stuckey, P. J. & Lesk, A. M. (2006). MUSTANG: a multiple structural alignment algorithm. *Proteins* **64**, 559-574.

[400] Ebert, J. & Brutlag, D. (2006). Development and validation of a consistency based multiple structure alignment algorithm. *Bioinformatics* **22**, 1080-1087.

[401] Menke, M., Berger, B. & Cowen, L. (2008). Matt: local flexibility aids protein multiple structure alignment. *PLoS Comp. Biol.* **4**, e10.

[402] Gerstein, M., Lesk, A. M. & Chothia, C. (1994). Structural mechanisms for domain movements in proteins. *Biochemistry* **33**, 6739-6749.

[403] Ye, Y. & Godzik, A. (2003). Flexible structure alignment by chaining aligned fragment pairs allowing twists. *Bioinformatics* **19** Suppl 2, i246-255.

[404] Chen, Y. & Crippen, G. M. (2005). A novel approach to structural alignment using realistic structural and environmental information. *Protein Sci.* **14**, 2935-2946.

[405] Mosca, R., Brannetti, B. & Schneider, T. R. (2008). Alignment of protein structures in the presence of domain motions. *BMC Bioinformatics* **9**, 352.

[406] Salem, S., Zaki, M. J. & Bystroff, C. (2010). FlexSnap: flexible non-sequential protein structure alignment. *Algorithms Mol. Biol.* **5**, 12.

[407] Ye, Y. & Godzik, A. (2005). Multiple flexible structure alignment using partial order graphs. *Bioinformatics* **21**, 2362-2369.

[408] Kempner, E. S. (1993). Movable lobes and flexible loops in proteins. Structural deformations that control biochemical activity. *FEBS Lett.* **326**, 4-10.

[409] Cunningham, B. A., Gottlieb, P. D., Pflumm, M. N. & Edelman, G. M. (1971). Immunoglobulin structure: diversity, gene duplication, and domains. In *Progress in Immunology*, Academic Press, New York.

[410] Wetlaufer, D. B. (1973). Nucleation, rapid folding, and globular intrachain regions in proteins. *Proc. Natl. Acad. Sci. USA* **70**, 697-701.

[411] Touriol, C., Bornes, S., Bonnal, S., Audigier, S., Prats, H., Prats, A. C. & Vagner, S. (2003). Generation of protein isoform diversity by alternative initiation of translation at non-AUG codons. *Biology of the Cell* **95**, 169-178.

[412] Ingolfsson, H. & Yona, G. (2008). Protein domain prediction. *Methods Mol. Biol.* **426**, 117-143.

[413] Liu, J. & Rost, B. (2004). CHOP proteins into structural domain-like fragments. *Proteins* **55**, 678-688.

[414] Levitt, M. (2009). Nature of the protein universe. *Proc. Natl. Acad. Sci. USA* **106**, 11079-11084.

[415] Branden, C. & Tooze, J. (1999). "Introduction to protein structure." Garland Publishing, Inc. New York, NY.

[416] Bornberg-Bauer, E., Beaussart, F., Kummerfeld, S. K., Teichmann, S. A. & Weiner III, J. (2005). The evolution of domain arrangements in proteins and interaction networks. *Cellular and Molecular Life Sciences* **62**, 435-445.

[417] Kim, D. E., Chivian, D., Malmström, L. & Baker, D. (2005). Automated prediction of domain boundaries in CASP6 targets using Ginzu and RosettaDOM. *Proteins* **61**, 193-200.

[418] Tai, C. H., Lee, W. J., Vincent, J. J. & Lee, B. (2005). Evaluation of domain prediction in CASP6. *Proteins* **61**, 183-192.

[419] Rose, G. D. (1979). Hierarchic organization of domains in globular proteins. *J. Mol. Biol.* **134**, 447-470.

[420] Lesk, A. M. & Rose, G. D. (1981). Folding units in globular proteins. *Proc. Natl. Acad. Sci. USA* **78**, 4304-4308.

[421] Holm, L. & Sander, C. (1994). Parser for protein folding units. *Proteins* **19**, 256-268.

[422] Hubbard, S. J. (1998). The structural aspects of limited proteolysis of native proteins. *Biochim. Biophys. Acta.* **1382**, 191–206.

[423] Crippen, G. M. (1978). The tree structural organization of proteins. *J. Mol. Biol.* **126**, 315-332.

[424] Xu, Y., Xu, D. & Gabow, H. N. (2000). Protein domain decomposition using a graph-theoretic approach. *Bioinformatics* **16**, 1091-1104

[425] Ford, L. R. & Fulkerson, D. R. (1962). "Flows in networks." Princeton University Press, Princeton, NJ.

[426] Pugalenthi, G., Archunan, G. & Sowdhamini, R. (2005). DIAL: a web-based server for the automatic identification of structural domains in proteins. *Nucl. Acids Res.* **33**, W130-132.

[427] Gelly, J. C., de Brevern, A. G. & Hazout, S. (2006). "Protein Peeling": an approach for splitting a 3D protein structure into compact fragments. *Bioinformatics* **22**, 129-133.

[428] Rigden, D. J. (2002). Use of covariance analysis for the prediction of structural domain boundaries from multiple protein sequence alignments. *Protein Eng.* **15**, 65-77.

[429] Gobel, U., Sander, C., Schneider, R. & Valencia, A. (1994). Correlated mutations and residue contacts in proteins. *Proteins* **18**, 309-317.

[430] Gouzy, J., Corpet, F. & Kahn, D. (1999). Whole genome protein domain analysis using a new method for domain clustering. *Comp. Chem.* **23**, 333-340.

[431] Gracy, J. & Argos, P. (1998). Automated protein sequence database classification. I. Integration of copositional similarity search, local similarity search and multiple sequence alignment. II. Delineation of domain boundaries from sequence similarity. *Bioinformatics* **14**, 164-187.

[432] George, R. A. & Heringa, J. (2002). Protein domain identification and improved sequence similarity searching using PSI-BLAST. *Proteins* **48**, 672-681.

[433] Kuroda, Y., Tani, K., Matsuo, Y. & Yokoyama, S. (2000). Automated search of natively folded protein fragments for high-throughput structure determination in structural genomics. *Protein Sci.* **9**, 2313-2321.

[434] Heger, A. & Holm, L. (2003). Exhaustive enumeration of protein domain families. *J. Mol. Biol.* **328**, 749-767.

[435] Gilbert, W. & Glynias, M. (1993). On the ancient nature of introns. *Gene* **135**, 137-144.

[436] Gilbert, W., de Souza, S. J. & Long, M. (1997). Origin of genes. *Proc. Natl. Acad. Sci. USA* **94**, 7698-7703.

[437] Saxonov, S., Daizadeh, I., Fedorov, A. & Gilbert, W. (2000). EID: the Exon-Intron Database - an exhaustive database of protein-coding intron-containing genes. *Nucl. Acids Res.* **28**, 185-190.

[438] Enright, A. J. & Ouzounis, C. A. (2000). GeneRAGE: a robust algorithm for sequence clustering and domain detection. *Bioinformatics* **16**, 451-457.

[439] Ferran, E. A., Pflugfelder, B. & Ferrara P. (1994). Self-organized neural maps of human protein sequences. *Protein Sci.* **3**, 507-521.

[440] Miyazaki, S., Kuroda, Y. & Yokoyama, S. (2002). Characterization and prediction of linker sequences of multi-domain proteins by a neural network. *J. Struct. Funct. Genomics* **15**, 37-51.

[441] Miyazaki, S., Kuroda, Y. & Yokoyama, S. (2006). Identification of putative domain linkers by a neural network - application to a large sequence database. *BMC Bioinformatics* **7**, 323.

[442] Suyama, M. & Ohara, O. (2003). DomCut: prediction of inter-domain linker regions in amino acid sequences. *Bioinformatics* **19**, 673-674.

[443] Saini, H. K. & Fischer, D. (2005) Meta-DP: domain prediction meta-server. *Bioinformatics* **21**, 2917-2920.

[444] Ireland, C. T. & Kullback, S. (1968). Contingency tables with given marginals. *Biometrika* **55**, 179-189.

[445] Pearl, J. (1997). "Probabilistic reasoning in intelligent systems: networks of plausible inference." Morgan Kaufmann Publishers Inc., San Mateo, CA.

[446] Apic, G., Gough, J. & Teichmann, S. A. (2001). Domain combinations in archaeal, eubacterial and eukaryotic proteomes. *J. Mol. Biol.* **310**, 311-325.

[447] Bashton, M. & Chothia, C. (2002). The geometry of domain combination in proteins. *J. Mol. Biol.* **315**, 927-939.

[448] Vogel, C., Teichmann, S. A. & Pereira-Leal, J. (2005). The relationship between domain duplication and recombination. *J. Mol. Biol.* **346**, 355-365.

[449] Apic, G., Huber, W. & Teichmann, S. A. (2003). Multi-domain protein families and domain pairs: comparison with known structures and a random model of domain recombination. *J. Struct. Funct. Genomics* **4**, 67-78.

[450] Przytycka, T., Davis, G., Song, N. & Durand, D. (2006). Graph theoretical insights into evolution of multidomain proteins. *J. Comp. Biol.* **13**, 351-363.

[451] Hegyi, H. & Gerstein, M. (2001). Annotation transfer for genomics: measuring functional divergence in multi-domain proteins. *Genome Res.* **11**, 1632-1640.

[452] Ku, C. & Yona, G. (2006). Domain-based hierarchy of multi-domain proteins and characterization of semantically significant domain architectures. `html://www.biozon.org/people/golan/papers/`.

[453] Lord, P. W., Stevens, R. D., Brass, A. & Goble, C. A. (2003). Investigating semantic similarity measures across the Gene Ontology: the relationship between sequence and annotation. *Bioinformatics* **19**, 1275-1283.

[454] Ashburner, M., Ball, C. A., Blake, J. A., Botstein, D., Butler, H., Cherry, J. M., Davis, A. P., Dolinski, K., Dwight, S. S., Eppig, J. T., Harris, M. A., Hill, D. P., Issel-Tarver, L., Kasarskis, A., Lewis, S., Matese, J. C., Richardson, J. E., Ringwald, M., Rubin, G. M. & Sherlock, G. (2000). Gene ontology: tool for the unification of biology. The Gene Ontology Consortium. *Nat. Genet.* **25**, 25-29.

[455] Gross, M. (1997). The construction of local grammars. In *Finite-State Language Processing*, MIT Press, Cambridge, MA.

[456] Klein, D. & Manning, C. D. (2001). Natural language grammar induction using a constituent-context model. *NIPS* **14**.

[457] Solan, Z., Horn, D., Ruppin, E. & Edelman, S. (2003). Unsupervised context sensitive language acquisition from a large corpus. *NIPS* **16**.

[458] Phillips, D. C. (1966). The three-dimensional structure of an enzyme molecule. *Sci. Am.* **215**, 78-90.

[459] Schulz, G. E. (1981). Protein differentiation: emergence of novel proteins during evolution. *Angew. Chem. Int. Edit.* **20**, 143-151.

[460] Richardson, J. S. (1981). The anatomy and taxonomy of protein structure. *Adv. Protein Chem.* **34**, 167-339.

[461] Swindells, M. B. (1995). A procedure for detecting structural domains in proteins. *Protein Sci.* **4**, 103-112.

[462] Siddiqui, A. S. & Barton, G. J. (1995). Continuous and discontinuous domains: an algorithm for the automatic generation of reliable protein domain definitions. *Protein Sci.* **4**, 872-884.

[463] Taylor, W. R. (1999). Protein structural domain identification. *Protein Eng.* **12**, 203-216.

[464] Alexandrov, N. & Shindyalov, I. (2003). PDP: protein domain parser. *Bioinformatics* **19**, 429-430.

[465] George, R. A. & Heringa, J. (2002). SnapDRAGON: a method to delineate protein structural domains from sequence data. *J. Mol. Biol.* **316**, 839-851.

[466] Sowdhamini, R. & Blundell, T. L. (1995). An automatic method involving cluster analysis of secondary structures for the identification of domains in proteins. *Protein Sci.* **4**, 506-520.

[467] Linding, R., Russell, R. B., Neduva, V. & Gibson, T. J. (2003). GlobPlot: exploring protein sequences for globularity and disorder. *Nucl. Acids Res.* **31**, 3701-3708.

[468] Marsden, R. L., McGuffin, L. J. & Jones, D. T. (2002). Rapid protein domain assignment from amino acid sequence using predicted secondary structure. *Protein Sci.* **11**, 2814-2824.

[469] Park, J. & Teichmann, S. A. (1998). DIVCLUS: an automatic method in the GEANFAMMER package that finds homologous domains in single- and multi-domain proteins. *Bioinformatics* **14**, 144-150.

[470] Portugaly, E., Harel, A., Linial, N. & Linial, M. (2006). EVEREST: automatic identification and classification of protein domains in all protein sequences. *BMC Bioinformatics* **7**, 277.

[471] Coin, L., Bateman, A. & Durbin, R. (2003). Enhanced protein domain discovery by using language modeling techniques from speech recognition. *Proc. Natl. Acad. Sci. USA* **100**, 4516-4520.

[472] Przytycka, T., Srinivasan, R. & Rose, G. D. (2002). Recursive domains in proteins. *Protein Sci.* **11**, 409-417.

[473] Vogel, C., Bashton, M., Kerrison, N. D., Chothia, C. & Teichmann, S. A. (2004). Structure, function and evolution of multidomain proteins. *Curr. Opin. Struct. Biol.* **14**, 208-216.

[474] Marsh, J. A. & Teichmann, S. A. (2010). How do proteins gain new domains? *Genome Biol.* **11**, 126.

[475] Krishnamurthy, N., Brown, D. & Sjölander, K. (2007). FlowerPower: clustering proteins into domain architecture classes for phylogenomic inference of protein function. *BMC Evol. Biol.* **7** Suppl 1, S12.

[476] Song, N., Joseph, J. M., Davis, G. B. & Durand, D. (2008). Sequence similarity network reveals common ancestry of multidomain proteins. *PLoS Comp. Biol.* **4**, e1000063.

[477] Bairoch, A. (1991). PROSITE: a dictionary of sites and patterns in proteins. *Nucl. Acids Res.* **19**, 2241-2245.

[478] Hulo, N., Bairoch, A., Bulliard, V., Cerutti, L., Cuche, B., De Castro, E., Lachaize, C., Langendijk-Genevaux, P. S. & Sigrist, C. J. A. (2008). The 20 years of PROSITE. *Nucl. Acids Res.* **36**, D245-249.

[479] Attwood, T. K. & Beck, M. E. (1994). PRINTS - a protein motif fingerprint database. *Protein Eng.* **7**, 841-848.

[480] Attwood, T. K., Bradley, P., Flower, D. R., Gaulton, A., Maudling, N., Mitchell, A. L., Moulton, G., Nordle, A., Paine, K., Taylor, P., Uddin, A. & Zygouri, C. (2003). PRINTS and its automatic supplement, prePRINTS. *Nucl. Acids Res.* **31**, 400-402.

[481] Henikoff, S. & Henikoff, J. G. (1991). Automated assembly of protein blocks for database searching. *Nucl. Acids Res.* **19**, 6565-6572.

[482] Henikoff, J. G., Greene, E. A., Pietrokovski, S. & Henikoff, S. (2000). Increased coverage of protein families with the blocks database servers. *Nucl. Acids Res.* **28**, 228-230.

[483] Servant, F., Bru, C., Carrere, S., Courcelle, E., Gouzy, J., Peyruc, D. & Kahn, D. (2002). ProDom: automated clustering of homologous domains. *Brief. Bioinform.* **3**, 246-251.

[484] Sonnhammer, E. L., Eddy, S. R. & Durbin, R. (1997). Pfam: a comprehensive database of protein domain families based on seed alignments. *Proteins* **28**, 405-420.

[485] Finn, R. D., Tate, J., Mistry, J., Coggill, P. C., Sammut, J. S., Hotz, H. R., Ceric, G., Forslund, K., Eddy, S. R., Sonnhammer, E. L. & Bateman, A. (2008). The Pfam protein families database. *Nucl. Acids Res.* **36**, D281-D288.

[486] Schultz, J., Milpetz, F., Bork, P. & Ponting, C. P. (1998). SMART, a simple modular architecture research tool: identification of signaling domains. *Proc. Natl. Acad. Sci. USA* **95**, 5857-5864.

[487] Letunic, I., Doerks, T. & Bork P. (2009). SMART 6: recent updates and new developments. *Nucl. Acids Res.* **37**, D229-232.

[488] Haft, D. H., Loftus, B. J., Richardson, D. L., Yang, F., Eisen, J. A., Paulsen, I. T. & White, O. (2001). TIGRFAMs: a protein family resource for the functional identification of proteins. *Nucl. Acids Res.* **29**, 41-43.

[489] Cuff, A. L., Sillitoe, I., Lewis, T., Redfern, O. C., Garratt, R., Thornton, J. & Orengo, C. A. (2009). The CATH classification revisited - architectures reviewed and new ways to characterize structural divergence in superfamilies. *Nucl. Acids Res.* **37**, D310-314.

[490] Siddiqui, A. S., Dengler, U. & Barton, G. J. (2001). 3Dee: a database of protein structural domains. *Bioinformatics* **17**, 200-201.

[491] Vinayagam, A., Shi, J., Pugalenthi, G., Meenakshi, B., Blundell, T. L. & Sowdhamini, R. (2003). DDBASE2.0: updated domain database with improved identification of structural domains. *Bioinformatics* **19**, 1760-1764.

[492] Marchler-Bauer, A., Anderson, J. B., Cherukuri, P. F., DeWeese-Scott, C., Geer, L. Y., Gwadz, M., He, S., Hurwitz, D. I., Jackson, J. D., Ke, Z., Lanczycki, C. J., Liebert, C. A., Liu, C., Lu, F., Marchler, G. H., Mullokandov, M., Shoemaker, B. A., Simonyan, V., Song, J. S., Thiessen, P. A., Yamashita, R. A., Yin, J. J., Zhang, D. & Bryant, S. H. (2005). CDD: a Conserved Domain Database for protein classification. *Nucl. Acids Res.* **33**, D192-196.

[493] Hunter, S., Apweiler, R., Attwood, T. K., Bairoch, A., Bateman, A., Binns, D., Bork, P., Das, U., Daugherty, L., Duquenne, L., Finn, R. D., Gough, J., Haft, D., Hulo, N., Kahn, D., Kelly, E., Laugraud, A., Letunic, I., Lonsdale, D., Lopez, R., Madera, M., Maslen, J., McAnulla, C., McDowall, J., Mistry, J., Mitchell, A., Mulder, N., Natale, D., Orengo, C., Quinn, A. F., Selengut, J. D., Sigrist, C. J., Thimma, M., Thomas, P. D., Valentin, F., Wilson, D., Wu, C. H. & Yeats, C. (2009). InterPro: the integrative protein signature database. *Nucl. Acids Res.* **37**, D224-228.

[494] Birkland, A. & Yona, G. (2006). The BIOZON database: a hub of heterogeneous biological data. *Nucl. Acids Res.* **34**, D235-D242.

[495] Vlahovicek, K., Kajan, L., Agoston, V. & Pongor, S. (2005). The SBASE domain sequence resource, release 12: prediction of protein domain-architecture using support vector machines. *Nucl. Acids Res.* **33**, D223-225.

[496] Apweiler, R., Attwood, T. K., Bairoch, A., Bateman, A., Birney, E., Biswas, M., Bucher, P., Cerutti, L., Corpet, F., Croning, M. D., Durbin, R., Falquet, L., Fleischmann, W., Gouzy, J., Hermjakob, H., Hulo, N., Jonassen, I., Kahn, D., Kanapin, A., Karavidopoulou, Y., Lopez, R., Marx, B., Mulder, N. J., Oinn, T. M., Pagni, M., Servant, F., Sigrist, C. J. & Zdobnov, E. M. (2001). The InterPro database, an integrated documentation resource for protein families, domains and functional sites. *Nucl. Acids Res.* **29**, 37-40.

[497] Geer, L. Y., Domrachev, M., Lipman, D. J. & Bryant, S. H. (2002). CDART: protein homology by domain architecture. *Genome Res.* **12**, 1619-1623.

[498] Kohonen, T. (1982). Self-organized formation of topologically correct feature maps. *Biological Cybernetics* **43**, 59-69.

[499] Rose, K., Gurewitz, E. & Fox, G. (1990). A deterministic annealing approach to clustering. *Pattern Recognition Letters* **11**, 589-594.

[500] Smith, P. S. (1993). Threshold validity for mutual neighborhood clustering. *IEEE Trans. on Pattern Analysis and Machine Intelligence* **15**, 89-92.

[501] Shafer, P., Isganitis, T. & Yona, G. (2006). Hubs of knowledge: using the functional link structure in Biozon to mine for biologically significant entities. *BMC Bioinformatics* **7**, 71.

[502] Cormen, T. H., Leiserson, C. E. & Rivest, R. L. (1990). "Introduction to algorithms." MIT press/McGraw-Hill Book Company, New York, NY.

[503] Wu, Z. & Leahy, R. (1993). An optimal graph theoretic approach to data clustering: theory and its application to image segmentation. *IEEE Trans. on Pattern Analysis and Machine Intelligence* **15**, 1101-1113.

[504] Karger, D. R., Klein, P. N. & Tarjan, R. E. (1995). A randomized linear-time algorithm to find minimum spanning trees. *J. ACM* **42**, 321-328.

[505] Gomory, R. E. & Hu, T. C. (1961). Multi-terminal network flows. *J. Soc. Indust. App. Math.* **9**, 551-570.

[506] Hotelling, H. (1933). Analysis of a complex of statistical variables into principal components. *J. Educ. Psych.* **24**, 417-441, 498-520.

[507] Heath, M. T. (2002). "Scientific computing." McGraw-Hill, New York.

[508] Golub, G. H. & Van Loan, C. F. (1996). "Matrix computations (3rd edition)." Johns Hopkins University Press, Baltimore, MD.

[509] Perona, P. & Freeman, W. (1998). A factorization approach to grouping. In *The Proceedings of ECCV*, 655-670.

[510] Shi, J. & Malik, J. (1997). Normalized cuts and image segmentation. In *The Proceedings of CVPR*, 731-737.

[511] Weiss, Y. (1999). Segmentation using eigenvectors: a unifying view. In *The Proceedings of ICCV*, 975-988.

[512] Meila, M. & Shi, J. (2000). Learning segmentation by random walks. *NIPS* **13**, 873-879.

[513] Ng, A. Y, Jordan, M. & Weiss, Y. (2001). On spectral clustering: analysis and an algorithm. *NIPS* **14**.

[514] Kannan, R., Vempala, S. & Vetta, A. (2004). On clusterings - good, bad and spectral. *J. ACM* **51**, 497-515.

[515] Enright, A. J., Van Dongen, S. & Ouzounis, C. A. (2001). An efficient algorithm for large-scale detection of protein families. *Nucl. Acids Res.* **30**, 1575-1584.

[516] Camoglu, O., Can, T. & Singh, A. K. (2006). Integrating multi-attribute similarity networks for robust representation of the protein space. *Bioinformatics* **22**, 1585-1592.

[517] Blatt, M., Wiseman, S. & Domany, E. (1997). Data clustering using a model granular magnet. *Neural Computation* **9**, 1805-1842.

[518] Getz, G., Gal, H., Notterman, D. A. & Domany, E. (2003). Coupled two-way clustering analysis of breast cancer and colon cancer gene expression data. *Bioinformatics* **19**, 1079-1089.

[519] Tetko, I. V., Facius, A., Ruepp, A. & Mewes, H. W. (2005). Super paramagnetic clustering of protein sequences. *BMC Bioinformatics* **6**, 82.

[520] Radjiman, S., Lianyi, H., Jian-Sheng, W. & Zong, C. Y. (2006). Super paramagnetic clustering of DNA sequences. *J. Biol. Phys.* **32**, 11-25.

[521] Jain, A. K. & Dubes, R. C. (1988). "Algorithms for clustering data." Prentice Hall, New Jersey.

[522] Halkidi, M., Batistakis, Y. & Vazirgiannis, M. (2002). Cluster validity methods: part I. *ACM SIGMOD Record* **31**, 40-45.

[523] Ben-Dor, A., Friedman, N. & Yakhini, Z. (2001). Class discovery in gene expression data. In *The Proceedings of RECOMB*, 31-38.

[524] Gat-Viks, I., Sharan, R. & Shamir, R. (2003). Scoring clustering solutions by their biological relevance. *Bioinformatics* **19**, 2381-2389.

[525] Handl, J., Knowles, J. & Kell, D. B. (2005). Computational cluster validation in post-genomic data analysis. *Bioinformatics* **21**, 3201-3212.

[526] Bolshakova, N., Azuaje, F. & Cunningham, P. (2005). A knowledge-driven approach to cluster validity assessment. *Bioinformatics* **21**, 2546-2547.

[527] Levine, E. & Domany, E. (2001). Resampling method for unsupervised estimation of cluster validity. *Neural Computation* **13**, 2573-2593.

[528] Roth, V., Lange, T., Braun, M. & Buhmann, J. (2002). A resampling approach to cluster validation. In *The Proceedings of Int. Conf. on Computational Statistics*, 123-128.

[529] Pal, N. R. & Biswas, J. (1997). Cluster validation using graph theoretic concepts. *Pattern Recognition* **30**, 847-857.

[530] Bezdek, J. C. & Pal, N. R. (1998). Some new indexes of cluster validity. *IEEE Trans. Syst. Man. Cybern. B Cybern.* **28**, 301-315.

[531] Langan, D. A., Modestino, J. W. & Zhang, J. (1998). Cluster validation for unsupervised stochastic model-based image segmentation. *IEEE Trans. on Image Processing* **7**, 180-195.

[532] Güter, S. & Bunke, H. (2003). Validation indices for graph clustering. *Pattern Recognition Letters* **24**, 1107-1113.

[533] Bouguessa, M., Wang, S. & Sun, H. (2006). An objective approach to cluster validation. *Pattern Recognition Letters* **27**, 1419-1430.

[534] Rissanen, J. (1989). Stochastic complexity in statistical inquiry. *World Scientific.*

[535] Grünwald, P. (2007). "The minimum description length principle." MIT Press, Cambridge, MA.

[536] Barron, A., Rissanen, J. & Yu, B. (1998). The minimum description length principle in coding and modeling. *IEEE Trans. on Information Theory* **44**, 2743-2760.

[537] Wolpert, D. (1996). The lack of a priori distinctions between learning algorithms. *Neural Computation* **8**, 1341-1390.

[538] Tatusov, R. L., Eugene, V. K. & David, J. L. (1997). A genomic perspective on protein families. *Science* **278**, 631-637.

[539] Barker, W. C., Pfeiffer, F. & George, D. G. (1996). Superfamily classification in PIR-international protein sequence database. *Methods Enzymol.* **266**, 59-71.

[540] Krause, A. & Vingron, M. (1998). A set-theoretic approach to database searching and clustering. *Bioinformatics* **14**, 430-438.

[541] Yooseph, S., Sutton, G., Rusch, D. B., Halpern, A. L., et al. (2007). The Sorcerer II global ocean sampling expedition: expanding the universe of protein families. *PLoS Biol.* **5**, e16.

[542] Jain, A., Murty, M. & Flynn, P. (1999). Data clustering: a review. *ACM Computing Surveys* **31**, 264-323.

[543] Kotsiantis, S. & Pintelas, P. (2004). Recent advances in clustering: a brief survey. *WSEAS Transactions on Information Science and Applications* **1**, 73-81.

[544] Barbakh, W. A., Wu, Y. & Fyfe, C. (2009). Review of clustering algorithms. *Studies in Computational Intelligence* **249**, 7-28.

[545] Meila, M. & Shi, J. (2001). A random walks view of spectral segmentation. In *The Proceedings of Int. Conf. on AI and Statistics.*

[546] Tishby, N., Pereira, F. C. & Bialek, W. (1999). The information bottleneck method. In *The Proceedings of Ann. Conf. on Communication, Control and Computing*, 368-377

[547] Gdalyahu, Y., Weinshall, D. & Werman, M. (1998). A randomized algorithm for pairwise clustering. *NIPS* **11**.

[548] Rissanen, J. (1983). A universal prior for integers and estimation by minimum description length. *Ann. Stat.* **11**, 416-431.

[549] Hansen, M. H. & Yu, B. (2001). Model selection and the principle of minimum description length. *J. Am. Stat. Assoc.* **96**, 746-774.

[550] Schwarz, G. E. (1978). Estimating the dimension of a model. *Ann. Stat.* **6**, 461-464.

[551] Akaike, H. (1974). A new look at the statistical model identification. *IEEE Trans. on Automatic Control* **19**, 716-723.

[552] Kearns, M., Mansour, Y., Ng, A. Y. & Ron, D. (1997). An experimental and theoretical comparison of model selection methods. *Machine Learning* **27**, 7-50.

[553] Kyrgyzov, I. O., Kyrgyzov, O. O., Maître, H. & Campedel, M. (2007). Kernel MDL to determine the number of clusters. *Lecture Notes in Computer Science* **4571**, 203-217.

[554] Krause, A., Stoye, J. & Vingron, M. (2005). Large scale hierarchical clustering of protein sequences. *BMC Bioinformatics* **6**, 15.

[555] Paccanaro, A., Casbon, J. A.& Saqi, M. A. (2006). Spectral clustering of protein sequences. *Nucl. Acids Res.* **34**, 1571-1580.

[556] Wittkop, T., Baumbach, J., Lobo, F. P. & Rahmann, S. (2007). Large scale clustering of protein sequences with FORCE - A layout based heuristic for weighted cluster editing. *BMC Bioinformatics* **8**, 396.

[557] Joseph, J. M. & Durand, D. (2009). Family classification without domain chaining. *Bioinformatics* **25**, i45-53.

[558] Haykin, S. (1998). "Neural networks: a comprehensive foundation (2nd edition)." Prentice Hall, New Jersey.

[559] Francis, J. G. F. (1961). The QR transformation I. *The Computer Journal* **4**, 265-271.

[560] Press, W. H., Teukolsky, S. A., Vetterling, W. T. & Flannery, B. P. (1993). "Numerical recipes in C: the art of scientific computing (2nd edition)." Cambridge University Press.

[561] Cox, T. F. & Cox, M. A. A. (2001). "Multidimensional scaling (2nd edition)." Chapman & Hall CRC, New York.

[562] Sammon, J. W. (1969). A nonlinear mapping for data structure analysis. *IEEE Trans. on Computers* **18**, 401-409.

[563] Klock, H. & Buhmann, J. M. (1997). Multidimensional scaling by deterministic annealing. In *The Proceedings of Int. Workshop on Energy Minimization Methods in Computer Vision and Pattern Recognition*, 245-260.

[564] Apostol, I. & Szpankowski, W. (1999). Indexing and mapping of proteins using a modified nonlinear Sammon projection. *J. Comp. Chem.* **20**, 1049-1059.

[565] Basalaj, W. (1999). Incremental multidimensional scaling method for database visualization. In *The Proceedings of Visual Data Exploration and Analysis VI* **3643**, 149-158.

[566] Birkland, A. & Yona, G. (2006). BIOZON: a system for unification, management and analysis of heterogeneous biological data. *BMC Bioinformatics* **7**, 70.

[567] Johnson, W. B. & Lindenstrauss, J. (1984). Extensions of Lipschitz mappings into a Hilbert space. *Contemporary Mathematics* **26**, 189-206.

[568] Frankl, P. & Maehara, H. (1988). The Johnson-Lindenstrauss lemma and the sphericity of some graphs. *J. Comb. Th. B* **44**, 355-362.

[569] Linial, N., London, E. & Rabinovich, Y. (1995). The geometry of graphs and some of its algorithmic applications. *Combinatorica* **15**, 215-245.

[570] Kleinberg, J. (1997). Two algorithms for nearest-neighbor search in high dimensions. In *The Proceedings of Ann. ACM Sym. on Theory of Computing*, 599-608.

[571] Figure retrieved from http://en.wikipedia.org/wiki/File:M%C3%B6bius_strip.jpg, August 2010.

[572] Tenenbaum, J. B., de Silva, V. & Langford, J. C. (2000). A global geometric framework for nonlinear dimensionality reduction. *Science* **290**, 2319-2323.

[573] Agrafiotis, D. K. & Xu, H. (2002). A self-organizing principle for learning nonlinear manifolds. *Proc. Natl. Acad. Sci. USA* **99**, 15869-15872.

[574] Roweis, S. T. & Saul, L. K. (2000). Nonlinear dimensionality reduction by locally linear embedding. *Science* **290**, 2323-2326.

[575] Saul, L. K. & Roweis, S. T. (2003). Think globally, fit locally: unsupervised learning of low dimensional manifolds. *J. Machine Learning Res.* **4**, 119-155.

[576] Quist, M. & Yona, G. (2004). Distributional scaling: an algorithm for structure-preserving embedding of metric and nonmetric spaces. *J. Machine Learning Res.* **5**, 399-420.

[577] Camastra, F. (2003). Data dimensionality estimation methods: a survey. *Pattern Recognition* **36**, 2945-2954.

[578] Casari, G., Sander, C. & Valencia, A. (1995). A method to predict functional residues in proteins. *Nat. Struct. Biol.* **2**, 171-178.

[579] van Heel, M. (1991). A new family of powerful multivariate statistical sequence analysis techniques. *J. Mol. Biol.* **220**, 877-887.

[580] Wu, C., Whitson, G., Mclarty, J., Ermongkonchai A. & Chang, T. (1992). Protein classification artificial neural system. *Protein Sci.* **1**, 667-677.

[581] Hobohm, U. & Sander, C. (1995). A sequence property approach to searching protein database. *J. Mol. Biol.* **251**, 390-399.

[582] Shi, J., Blundell, T. L. & Mizuguchi, K. (2001). FUGUE: sequence-structure homology recognition using environment-specific substitution tables and structure-dependent gap penalties. *J. Mol. Biol.* **310**, 243-257.

[583] Freund, Y. & Schapire, R. E. (1997). A decision-theoretic generalization of on-line learning and an application to boosting. *J. Comp. Sys. Sciences* **55**, 119-139.

[584] Valentini, G. & Masulli, F. (2002). Ensembles of learning machines. In *Series Lecture Notes in Computer Science* **2486**, 3-19, Springer-Verlag, Heidelberg, Germany.

[585] Ali, K. M. & Pazzani, M. J. (1996). Error reduction through learning multiple descriptions. *Machine Learning* **24**, 173-202.

[586] Young, G. & Householder, A. S. (1938). Discussion of a set of points in terms of their mutual distances. *Psychometrika* **3**, 19-22.

[587] Gower, J. C. (1966). Some distance properties of latent root and vector methods in multivariate analysis. *Biometrika* **53**, 325-338.

[588] Kouropteva, O., Okun, O. & Pietikäinen, M. (2005). Incremental locally linear embedding algorithm. *Pattern Recognition* **38**, 1764-1767.

[589] Belkin, M. & Niyogi, P. (2001). Laplacian eigenmaps and spectral eigenmaps for embedding and clustering. *NIPS* **14**, 585-591.

[590] Belkin, M. & Niyogi, P. (2003). Laplacian Eigenmaps for dimensionality reduction and data representation. *Neural Computation* **15**, 1373-1396.

[591] Donoho, D. L. & Grimes, C. (2003). Hessian eigenmaps: locally linear embedding techniques for high-dimensional data. *Proc. Natl. Acad. Sci. USA* **100**, 5591-5596.

[592] Havel, T. F., Kuntz, I. D. & Crippen, G. M. (1983). The theory and practice of distance geometry. *Bull. Math. Biol.* **45**, 665-720.

[593] Whitney, H. (1944). The self-intersections of a smooth n-manifold in 2n-space. *Ann. Math.* **45**, 220-246.

[594] Nash, J. (1954). C^1-isometric imbeddings. *Ann. Math.* **60**, 383-396.

[595] Broomhead, D. S. & Kirby, M. (2000). A new approach to dimensionality reduction: theory and algorithms. *SIAM J. App. Math.* **60**, 2114-2142.

[596] Fukunaga, K. & Olsen, D. R. (1971). An algorithm for finding intrinsic dimensionality of data. *IEEE Trans. on Computers* **20**, 176-183.

[597] Pettis, K. W., Bailey, T. A., Jain, A. K. & Dubes, R. C. (1979). An intrinsic dimensionality estimator from near-neighbor information. *IEEE Trans. on Pattern Analysis and Machine Intelligence* **1**, 25-37.

[598] Bruske, J. & Sommer, G. (1998). Intrinsic dimensionality estimation with optimally topology preserving maps. *IEEE Trans. on Pattern Analysis and Machine Intelligence* **20**, 572-575.

[599] Camastra, F. & Vinciarelli, A. (2002). Estimating the intrinsic dimension of data with a fractal-based method. *IEEE Trans. on Pattern Analysis and Machine Intelligence* **24**, 1404-1407.

[600] Kegl, B. (2002). Intrinsic dimension estimation using packing numbers. *NIPS* **15**.

[601] Costa, J. A. & Hero, A. O. (2004). Geodesic entropic graphs for dimension and entropy estimation in manifold learning. *IEEE Trans. on Signal Processing* **52**, 2210-2221.

[602] Levina, E. & Bickel, P. (2004). Maximum likelihood estimation of intrinsic dimension. *NIPS* **17**.

[603] Hein, M. & Audibert, J. Y. (2005). Intrinsic dimensionality estimation of submanifolds in $\mathbb{R}R^d$. In *The Proceedings of ICML*, 289-296.

[604] Fan, M., Qiao, H. & Zhang, B. (2009). Intrinsic dimension estimation of manifolds by incising balls. *Pattern Recognition* **42**, 780-787.

[605] Verveer, P. J. & Duin, R. P. W. (1995). An evaluation of intrinsic dimensionality estimators. *IEEE Trans. on Pattern Analysis and Machine Intelligence* **17**, 81-86.

[606] Stawiski, E. W. , Baucom, A. E. , Lohr, S. C. & Gregoret, L. M. (2000). Predicting protein function from structure: unique structural features of proteases. *Proc. Natl. Acad. Sci. USA* **97**, 3954-3958.

[607] Spellman, P. T., Sherlock, G., Zhang, M., Iyer, V., Eisen, M., Brown, P., Botstein, D. & Futcher, B. (1998). Comprehensive identification of cell cycle-regulated genes of the yeast Saccharomyces cerevisiae by microarray hybridization. *Mol. Biol. Cell* **9**, 3273-3297.

[608] Shapira, M., Segal, E. & Botstein, D. (2004). Disruption of yeast forkhead-associated cell cycle transcription by oxidative stress. *Mol. Biol. Cell* **15**, 5659-5669.

[609] Bammer, G. & Fostel, J. (2000). Genome-wide expression patterns in Saccharomyces cerevisiae: comparison of drug treatments and genetic alterations affecting biosynthesis of ergosterol. *Antimicrobial Agents and Chemotherapy* **44**, 1255-1265.

[610] McCormick, S. M., Frye S. R., Eskin, S. G., Teng, C. L., Lu, C. M., Russell, C. G., Chittur, K. K. & McIntire L. V. (2003). Microarray analysis of shear stressed endothelial cells. *Biorheology* **40**, 5-11.

[611] Yoo, M. S., Chun, H. S., Son, J. J., DeGiorgio, L. A., Kim, D. J., Peng, C. & Son J. H. (2003). Oxidative stress regulated genes in nigral dopaminergic neuronal cells: correlation with the known pathology in Parkinson's disease. *Brain Res. Mol. Brain Res.* **110**, 76-84.

[612] Diffee, G. M., Seversen E. A., Stein, T. D.& Johnson, J. A. (2003). Microarray expression analysis of effects of exercise training: increase in atrial MLC-1 in rat ventricles. *American Journal of Physiology Heart and Circulatory Physiology* **284**, 830-837.

[613] Lopez, I. P., Marti, A., Milagro, F. I., Zulet, Md Mde L., Moreno-Aliaga, M. J., Martinez, J. A. & De Miguel, C. (2003). DNA microarray analysis of genes differentially expressed in diet-induced (cafeteria) obese rats. *Obesity Research* **11**, 188-194.

[614] Alter, O., Brown, P. O. & Botstein, D. (2003). Generalized singular value decomposition for comparative analysis of genome-scale expression data sets of two different organisms. *Proc. Natl. Acad. Sci. USA* **100**, 3351-3356.

[615] Hughes, T., Marton, M., Jones, A., Roberts, C., Stoughton, R., Armour, C., Bennett, H., Coffey, E., Dai, H., He, Y., Kidd, M., King, A., Meyer, M., Slade, D., Lum, P., Stepaniants, S., Shoemaker, D., Gachotte, D., Chakraburtty, K., Simon, J., Bard, M. & Friend, S. (2000). Functional discovery via a compendium of expression profiles. *Cell* **102**, 109-126.

[616] Mei, R., Galipeau, P. C., Prass, C., Berno, A., Ghandour, G., Patil, N., Wolff, R. K., Chee, M. S., Reid, B. J. & Lockhart, D. J. (2000). Genome-wide detection of allelic imbalance using human SNPs and high-density DNA arrays. *Genome Res.* **10**, 1126-1137.

[617] Masters, J. R. & Lakhani, S. R. (2000). How diagnosis with microarrays can help cancer patients. *Nature* **404**, 921.

[618] Yeatman, T. J. (2003). The future of clinical cancer management: one tumor, one chip. *The American Surgeon* **69**, 41-44.

[619] Liu, E. T. (2003). Classification of cancers by expression profiling. *Curr. Opin. Genet. Devel.* **13**, 97-103.

[620] Perez-Diez, A., Morgun, A. & Shulzhenko, N. (2007). Microarrays for cancer diagnosis and classification. *Adv. Exp. Med. Biol.* **593**, 74-85.

[621] Friedman, N., Linial, M., Nachman, I. & Pe'er, D. (2000). Using Bayesian network to analyze expression data. *J. Comp. Biol.* **7**, 601-620.

[622] Segal, E., Shapira, M., Regev, A., Pe'er, D., Botstein, D., Koller, D. & Friedman, N. (2003). Module networks: identifying regulatory modules and their condition-specific regulators from gene expression data. *Nat. Genet.* **34**, 166-176.

[623] Figure retrieved from http://en.wikipedia.org/wiki/File:Microarray2.gif, October 2008.

[624] Ahmed, F. E. (2006). Microarray RNA transcriptional profiling: part I. Platforms, experimental design and standardization. *Expert. Rev. Mol. Diagn.* **6**, 535-550.

[625] Speed, T. P. (2003). "Statistical analysis of gene expression microarray data." CRC Press, Chapman & Hall, New York.

[626] Ramnarain, D. B., Park, S., Lee, D. Y., Hatanpaa, K. J., Scoggin, S. O., Otu, H., Libermann, T. A., Raisanen, J. M., Ashfaq, R., Wong, E. T., Wu, J., Elliott, R. & Habib, A. A. (2006). Differential gene expression analysis reveals generation of an autocrine loop by a mutant epidermal growth factor receptor in glioma cells. *Cancer Res.* **66**, 867-74.

[627] Herbert, J. M., Stekel, D., Sanderson, S., Heath, V. L. & Bicknell, R. (2008). A novel method of differential gene expression analysis using multiple cDNA libraries applied to the identification of tumour endothelial genes. *BMC Genomics* **9**, 153.

[628] Heyer, L. J., Kruglyak, S. & Yooseph, S. (1999). Exploring expression data: identification and analysis of coexpressed genes. *Genome Res.* **9**, 1106-1115.

[629] Tusher, V. G., Tibshirani, R. & Chu, G. (2001). Significance analysis of microarrays applied to the ionizing radiation response. *Proc. Natl. Acad. Sci. USA* **98**, 5116-5121.

[630] Kooperberg, C., Sipione, S., LeBlanc, M., Strand, A. D., Cattaneo, E. & Olson, J. M. (2002). Evaluating test statistics to select interesting genes in microarray experiments. *Hum. Mol. Genet.* **11**, 2223-2232.

[631] Gasch, A. P., Spellman, P. T., Kao, C. M, Carmel-Harel, O., Eisen, M. B., Storz, G., Botstein, D. & Brown, P. O. (2000). Genomic expression programs in the response of yeast cells to environmental changes. *Mol. Biol. Cell* **11**, 4241-4257.

[632] Glass, J. I., Assad-Garcia, N., Alperovich, N., Yooseph, S., Lewis, M. R., Maruf, M., Hutchison, C. A. 3rd, Smith, H. O. & Venter, J. C. (2006). Essential genes of a minimal bacterium. *Proc. Natl. Acad. Sci. USA* **103**, 425-430.

[633] Miklos, G. & Rubin, G. (1996). The role of the Genome Project in determining gene function: insights from model organisms. *Cell* **86**, 521-529.

[634] D'haeseleer P., Wen, X., Fuhrman, S. & Somogyi, R. (1998). Mining the gene expression matrix: inferring gene relationships from large scale gene expression data. In *Information Processing in Cells and Tissues*, Plenum Publishing.

[635] Alizadeh, A., Eisen, M., Davis, R. E., Ma, C. A., Lossos, I., Rosenwald, A., Boldrick, J., Sabet, H., Tran, T., Yu, X., Powell, J., Yang, L., Marti, G., Moore, T., Hudson, J., Chan, W. C., Greiner, T. C., Weissenberger, D. D., Armitage, J. O., Levy, R., Grever, M. R., Byrd, J. C., Botstein, D., Brown, P. O. & Staudt, L. M. (2000). Distinct types of diffuse large B-cell lymphoma identified by gene expression profiling. *Nature* **403**, 503-511.

[636] Troyanskaya, O., Cantor, M., Sherlock, G., Brown, P., Hastie, T., Tibshirani, R., Botstein, D. & Altman, R. (2001). Missing value estimation methods for DNA microarrays. *Bioinformatics* **17**, 520-525.

[637] Dudoit, S., Yang, Y. H., Callow, M. J. & Speed, T. P. (2002). Statistical methods for identifying differentially expressed genes in replicated cDNA microarray experiments. *Statistica Sinica* **12**, 111-139.

[638] Algorithm AS 89. (1975). Tail probabilities for Spearman's rho. *App. Stat. Alg.* **24**, 377.

[639] Chiaretti, S., Li, X., Gentleman, R., Vitale, A., Wang, K. S., Mandelli, F., Foa, R. & Ritz, J. (2005). Gene expression profile of adult B-lineage Adult Acute Lymphocytic Leukemia Reveal Genetic Patterns that Identify Lineage Derivation and Distinct Mechanisms of Transformation. *Clin. Cancer Res.* **11**, 7209-7219.

[640] Oron, A. P., Jiang, Z. & Gentleman, R. (2008). Gene set enrichment analysis using linear models and diagnostics. *Bioinformatics* **24**, 2586-2591.

[641] Yona, G., Dirks, W. & Rahman, S. (2009). Comparing algorithms for clustering of expression data: how to assess gene clusters. *Methods Mol. Biol.* **541**, 479-509.

[642] Alter, O., Brown, P. O. & Botstein, D. (2000). Singular value decomposition for genome-wide expression data processing and modeling. *Proc. Natl. Acad. Sci. USA* **97**, 10101-10106.

[643] Golub, T. R., Slonim, D. K., Tamayo, P., Huard, C., Gaasenbeek, M., Mesirov, H. P., Coller , H. Loh, M., Downing, J. R., Caligiuri, M. A., Bloomfield, C. D. & Lander, E. S. (1999). Molecular classification of cancer: class discovery and class prediciton by gene expression monitoring. *Science* **286**, 531-537.

[644] von Heydebreck, A., Huber, W., Poustka, A. & Vingron, M. (2001). Identifying splits with clear separation: a new class discovery method for gene expression data. *Bioinformatics* **17** Suppl 1, S107-114.

[645] Steinfeld, I., Navon, R., Ardigó, D., Zavaroni, I. & Yakhini, Z. (2008). Clinically driven semi-supervised class discovery in gene expression data. *Bioinformatics* **24**, i90-97.

[646] Gray, R. M., Kieffer, J. C., & Linde, Y. (1980). Locally optimal block quantizier design. *Information and Control* **45**, 178-198.

[647] Wren, J. D. & Garner, H. R. (2004). Shared relationship analysis: ranking set cohesion and commonalities within a literature-derived relationship network. *Bioinformatics* **20**, 191-198.

[648] Cora, D., Di Cunto, F., Provero, P., Silengo, L. & Caselle, M. (2004). Computational identification of transcription factor binding sites by functional analysis of sets of genes sharing overrep-resented upstream motifs. *BMC Bioinformatics* **5**, 57.

[649] Bejerano, G. (2003). Efficient exact p-value computation and applications to biosequence analysis. In *The Proceedings of RECOMB*, 38-47.

[650] Xu, T., Du, L. & Zhou, Y. (2008). Evaluation of GO-based functional similarity measures using S. cerevisiae protein interaction and expression profile data. *BMC Bioinformatics* **9**, 472.

[651] Pesquita, C., Faria, D., Bastos, H., Ferreira, A. E., Falcão, A. O. & Couto, F. M. (2008). Metrics for GO based protein semantic similarity: a systematic evaluation. *BMC Bioinformatics* **9** Suppl 5, S4.

[652] Mootha, V. K., Lindgren, C. M., Eriksson, K. F., Subramanian, A., Sihag, S., Lehar, J., Puigserver, P., Carlsson, E., Ridderstrale, M., Laurila, E., Houstis, N., Daly, M. J., Patterson, N., Mesirov, J. P., Golub, T. R., Tamayo, P., Spiegelman, B., Lander, E. S., Hirschhorn, J. N., Altshuler, D. & Groop, L. C. (2003). PGC-1alpha-responsive genes involved in oxidative phosphorylation are coordinately downregulated in human diabetes. *Nat. Genet.* **34**, 267-273.

[653] Subramanian, A., Tamayo, P., Mootha, V. K., Mukherjee, S., Ebert, B. L., Gillette, M. A., Paulovich, A., Pomeroy, S. L., Golub, T. R., Lander, E. S. & Mesirov, J. P. (2005). Gene set enrichment analysis: a knowledge-based approach for interpreting genome-wide expression profiles. *Proc. Natl. Acad. Sci. USA* **102**, 15545-15550.

[654] Efron, B. & Tibshirani, R. (2007). On testing the significance of sets of genes. *Ann. App. Stat.* **1**, 107-129.

[655] Schweitzer, B. & Kingsmore, S. F. (2002). Measuring proteins on microarrays. *Curr. Opin. Biotechnol.* **13**, 14-19.

[656] Gygi, S. P., Rochon, Y., Franza, B. R. & Aebersold, R. (1999). Correlation between protein and mRNA abundance in yeast. *Mol. Cell Biol.* **19**, 1720-1730.

[657] Kulesh, D. A., Clive, D. R., Zarlenga, D. S. & Greene, J. J. (1987). Identification of interferon-modulated proliferation-related cDNA sequences. *Proc. Natl. Acad. Sci. USA* **84**, 8453-8457.

[658] Bains, W. & Smith, G. C. (1988). A novel method for nucleic acid sequence determination. *J. Theor. Biol.* **135**, 303-307.

[659] Fodor, S. P., Read, J. L., Pirrung, M. C., Stryer, L., Lu, A. T. & Solas, D. (1991). Light-directed, spatially addressable parallel chemical synthesis. *Science* **251**, 767-773.

[660] Schena, M., Shalon, D., Davis, R. W. & Brown, P. O. (1995). Quantitative monitoring of gene expression patterns with a complementary DNA microarray. *Science* **270**, 467-470.

[661] Durinck, S. (2008). Pre-processing of microarray data and analysis of differential expression. *Methods Mol. Biol.* **452**, 89-110.

[662] Eisen, M. B., Spellman, P. T., Brown, P. O. & Botstein, D. (1998). Cluster analysis and display of genome-wide expression patterns. *Proc. Natl. Acad. Sci. USA* **95**, 14863-14868.

[663] Valafar, F. (2002). Pattern recognition techniques in microarray data analysis: a survey. *Ann. NY Acad. Sci.* **980**, 41-64.

[664] Jiang, D., Tang, C. & Zhang, A. (2004). Cluster analysis for gene expression data: a survey. *IEEE Trans. on Knowledge and Data Engineering* **16**, 1370-1386.

[665] Boutros, P. C. & Okey, A. B. (2005). Unsupervised pattern recognition: an introduction to the whys and wherefores of clustering microarray data. *Brief. Bioinform.* **6**, 331-343.

[666] D'haeseleer, P. (2005). How does gene expression clustering work? *Nat. Biotechnol.* **23**, 1499-1501.

[667] Speer, N., Spieth, C. & Zell, A. (2004). A memetic clustering algorithm for the functional partition of genes based on the gene ontology. In *The Proceedings of IEEE Sym. on Computational Intelligence in Bioinformatics and Computational Biology*, 252-259.

[668] Raychaudhuri, S., Schutze, H. & Altman, R. B. (2002). Using text analysis to identify functionally coherent gene groups. *Genome Res.* **12**, 1582-1590.

[669] Bolshakova, N. & Azuaje F. (2003). Machaon CVE: cluster validation for gene expression data. *Bioinformatics* **19**, 2494-2495.

[670] Bertoni, A. & Valentini, G. (2006). Randomized maps for assessing the reliability of patients clusters in DNA microarray data analyses. *Artif. Intell. Med.* **37**, 85-109.

[671] Olman, V., Xu, D. & Xu, Y. (2003). CUBIC: identification of regulatory binding sites through data clustering. *J. Bioinform. Comp. Biol.* **1**, 21-40.

[672] McShane, L. M., Radmacher, M. D., Freidlin, B., Yu, R., Li, M. C. & Simon, R. (2002). Methods for assessing reproducibility of clustering patterns observed in analyses of microarray data. *Bioinformatics* **18**, 1462-1469.

[673] Zhang, K. & Zhao, H. (2000). Assessing reliability of gene clusters from gene expression data. *Funct. Integr. Genomics* **1**, 156-173.

[674] Yeung, K. Y., Haynor, D. R. & Ruzzo, W. L. (2001). Validating clustering for gene expression data. *Bioinformatics* **17**, 309-318.

[675] Smolkin, M. & Ghosh, D. (2003). Cluster stability scores for microarray data in cancer studies. *BMC Bioinformatics* **4**, 36.

[676] Dudoit, S. & Fridlyand, J. (2003). Bagging to improve the accuracy of a clustering procedure. *Bioinformatics* **19**, 1090-1099

[677] Alter O. (2007). Genomic signal processing: from matrix algebra to genetic networks. *Methods Mol. Biol.* **377**, 17-60.

[678] Kong, S. W., Pu, W. T. & Park, P. J. (2006). A multivariate approach for integrating genome-wide expression data and biological knowledge. *Bioinformatics* **22**, 2373-2380.

[679] Abdullah-Sayani, A., Bueno-de-Mesquita, J. M. & van de Vijver, M. J. (2006). Technology insight: tuning into the genetic orchestra using microarrays–limitations of DNA microarrays in clinical practice. *Nat. Clin. Pract. Oncol.* **3**, 501-516.

[680] Allison, D. B., Cui, X., Page, G. P. & Sabripour, M. (2006). Microarray data analysis: from disarray to consolidation and consensus. *Nat. Rev. Genet.* **7**, 55-65.

[681] Wheelan, S. J., Martínez Murillo, F. & Boeke, J. D. (2008). The incredible shrinking world of DNA microarrays. *Mol. Biosyst.* **4**, 726-732.

[682] Roberts, P. C. (2008). Gene expression microarray data analysis demystified. *Biotechnol. Ann. Rev.* **14**, 29-61.

[683] Routtenberg, A. & Rekart, J. L. (2005). Post-translational protein modification as the substrate for long-lasting memory. *Trends Neurosci.* **28**, 12-19.

[684] Walsh, G. & Jefferis, R. (2006). Post-translational modifications in the context of therapeutic proteins. *Nat. Biotechnol.* **24**, 1241-1252.

[685] Fields, S. & Song, O. (1989). A novel genetic system to detect protein-protein interactions. *Nature* **340**, 245-246.

[686] Rigaut, G., Shevchenko, A., Rutz, B., Wilm, M., Mann, M. & Seraphin, B. (1999). A generic protein purification method for protein complex characterization and proteome exploration. *Nat. Biotechnol.* **17**, 1030-1032.

[687] Cooks, R. G., Busch, K. L. & Glish, G. L. (1983). Mass spectrometry: analytical capabilities and potentials. *Science* **222**, 273-291.

[688] Fenn, J. B., Mann, M., Meng, C. K., Wong, S. F. & Whitehouse, C. M. (1989). Electrospray ionization for mass spectrometry of large biomolecules. *Science* **246**, 64-71.

[689] Uetz, P., Giot, L., Cagney, G. et al. (2000). A comprehensive analysis of protein-protein interactions in Saccharomyces cerevisiae. *Nature* **403**, 623-627.

[690] Ito, T., Chiba, T., Ozawa, R., Yoshida, M., Hattori, M. & Sakaki, Y. (2001). A comprehensive two-hybrid analysis to explore the yeast protein interactome. *Proc. Natl. Acad. Sci. USA* **98**, 4569-4574.

[691] Ho, Y., Gruhler, A., Heilbut, A., et al. (2002). Systematic identification of protein complexes in Saccharomyces cerevisiae by mass spectrometry. *Nature* **415**, 180-183.

[692] Bader, G. D., Donaldson, I., Wolting, C., Ouellette, B. F., Pawson, T. & Hogue, C. W. (2001). BIND - The Biomolecular Interaction Network Database. *Nucl. Acids Res.* **29**, 242-245.

[693] Xenarios, I., Rice, D. W., Salwinski, L., Baron, M. K., Marcotte, E. M. & Eisenberg, D. (2000). DIP: The Database of Interacting Proteins. *Nucl. Acids Res.* **28**, 289-291.

[694] Salwinski, L., Miller, C. S., Smith, A. J., Pettit, F. K., Bowie, J. U. & Eisenberg, D. (2004). The database of interacting proteins: 2004 update. *Nucl. Acids Res.* **32**, D449-451.

[695] Peri, S., Navarro J. D., Amanchy, R. et al. (2003). Development of human protein reference database as an initial platform for approaching systems biology in humans. *Genome Res.* **13**, 2363-2371.

[696] Prasad, T. S. K., Goel, R., Kandasamy, K. et al. (2009). Human protein reference database - 2009 update. *Nucl. Acids Res.* **37**, D767-772.

[697] Sharon, I., Davis, J. V. & Yona, G. (2009). Prediction of protein-protein interactions: a study of the co-evolution model. *Methods Mol. Biol.* **541**, 61-88.

[698] Ratner, D. (1974). The interaction of bacterial and phage proteins with immobilized Escherichia coli RNA polymerase. *J. Mol. Biol.* **88**, 373-383.

[699] Figure retrieved from http://en.wikipedia.org/wiki/File:Two_hybrid_assay.svg, October 2008.

[700] von Mering, C., Krause, R., Snel, B., Cornell, M., Oliver, S. G., Fields, S. & Bork, P. (2002). Comparative assessment of large-scale data sets of protein-protein interactions. *Nature* **417**, 399-403.

[701] Sprinzak, E., Sattath, S. & Margalit, H. (2003). How reliable are experimental protein-protein interaction data? *J. Mol. Biol.* **327**, 919-923.

[702] Ren, B., Robert, F., Wyrick, J. J., Aparicio, O., Jennings, E. G., Simon, I., Zeitlinger, J., Schreiber, J., Hannett, N., Kanin, E., Volkert, T. L., Wilson, C. J., Bell, S. P. & Young, R. A. (2000). Genome-wide location and function of DNA binding proteins. *Science* **290**, 2306-2309.

[703] Wu, J., Smith, L. T., Plass, C. & Huang, T. H. (2006). ChIP-chip comes of age for genome-wide functional analysis. *Cancer Res.* **66**, 6899-6902.

[704] Kharchenko, P. V., Tolstorukov, M. Y. & Park, P. J. (2008). Design and analysis of ChIP-seq experiments for DNA-binding proteins. *Nat. Biotechnol.* **26**, 1351-1359.

[705] Jothi, R., Cuddapah, S., Barski, A., Cui, K. & Zhao, K. (2008). Genome-wide identification of in vivo protein-DNA binding sites from ChIP-Seq data. *Nucl. Acids Res.* **36**, 5221-5231.

[706] Janin, J., Henrick, K., Moult, J., Eyck, L. T., Sternberg, M. J., Vajda, S., Vakser, I. & Wodak, S. J. (2003). CAPRI: a Critical Assessment of PRedicted Interactions. *Proteins* **52**, 2-9.

[707] Garey, M. R. & Johnson, D. S. (1978). Strong NP-completeness results: motivation, examples and implications. *J. ACM* **25**, 499-508.

[708] Katchalski-Katzir, E., Shariv, I., Eisenstein, M., Friesem, A. A., Aflalo, C. & Vakser, I. A. (1992). Molecular surface recognition: determination of geometric fit between proteins and their ligands by correlation techniques. *Proc. Natl. Acad. Sci. USA* **89**, 2195-2199.

[709] Sandak, B., Nussinov, R. & Wolfson, H. J. (1998). A method for biomolecular structural recognition and docking allowing conformational flexibility. *J. Comp. Biol.* **5**, 631-654.

[710] Gardiner, E. J., Willett, P. & Artymiuk, P. J. (2001). Protein docking using a genetic algorithm. *Proteins* **44**, 44-56.

[711] Sippl M. J. (1995). Knowledge-based potentials for proteins. *Curr. Opin. Struct. Biol.* **5**, 229-235.

[712] Zacharias, M. (2003). Protein-protein docking with a reduced protein model accounting for side-chain flexibility. *Protein Sci.* **12**, 1271-1282.

[713] Wang, C., Schueler-Furman, O. & Baker, D. (2005). Improved side-chain modeling for protein-protein docking. *Protein Sci.* **14**, 1328-1339.

[714] Teodoro, M., Phillips, G. & Kavraki, L. (2001). Molecular docking: a problem with thousands of degrees of freedom. In *The Proceedings of IEEE Int. Conf. on Robotics and Automation.*

[715] Andrusier, N., Mashiach, E., Nussinov, R. & Wolfson, H. J. (2008). Principles of flexible protein-protein docking. *Proteins* **73**, 271-289.

[716] Lichtarge, O., Bourne, H. R. & Cohen, F. E. (1996). An evolutionary trace method defines binding surfaces common to protein families. *J. Mol. Biol.* **257**, 342-358.

[717] Jones, S. & Thornton, J. M. (1997). Prediction of protein-protein interaction sites using patch analysis. *J. Mol. Biol.* **272**, 133-143.

[718] Clackson, T. & Wells, J. A. (1995). A hot spot of binding energy in a hormonereceptor interface. *Science* **267**, 383-386.

[719] Hu, Z., Ma, B., Wolfson, H. & Nussinov, R. (2000). Conservation of polar residues as hot spots at protein interfaces. *Proteins* **39**, 331-342.

[720] Lise, S., Archambeau, C., Pontil, M. & Jones, D. T. (2009). Prediction of hot spot residues at protein-protein interfaces by combining machine learning and energy-based methods. *BMC Bioinformatics* **10**, 365.

[721] Kini, R. M. & Evans, H. J. (1995). A hypothetical structural role for proline residues in the flanking segments of protein-protein interaction sites. *Biochem. Biophys. Res. Commun.* **212**, 1115-1124.

[722] Lo Conte, L., Chothia, C. & Janin, J. (1999). The atomic structure of protein-protein recognition sites. *J. Mol. Biol.* **285**, 2177-2198.

[723] Gallet, X., Charloteaux, B., Thomas, A. & Brasseur, R. (2000). A fast method to predict protein interaction sites from sequences. *J. Mol. Biol.* **302**, 917-926.

[724] Espadaler, J., Romero-Isart, O., Jackson, R. M. & Oliva, B. (2005). Prediction of protein-protein interactions using distant conservation of sequence patterns and structure relationships. *Bioinformatics* **21**, 3360-3368.

[725] Aytuna, A. S., Gursoy, A. & Keskin, O. (2005). Prediction of protein-protein interactions by combining structure and sequence conservation in protein interfaces. *Bioinformatics* **21**, 2850-2855.

[726] Halperin, I., Wolfson, H. & Nussinov, R. (2004). Protein-protein interactions; coupling of structurally conserved residues and of hot spots across interfaces. Implications for docking. *Structure* **12**, 1027-1038.

[727] Aloy, P. & Russell, R. (2002). Interrogating protein interaction networks through structural biology. *Proc. Natl. Acad. Sci. USA* **99**, 5896-5901.

[728] Lu, L., Lu, H. & Skolnick, J. (2002). MULTIPROSPECTOR: an algorithm for the prediction of protein-protein interactions by multimeric threading. *Proteins* **49**, 350-364.

[729] Pellegrini, M., Marcotte, E. M., Thompson, M. J., Eisenberg, D. & Yeates, T. O. (1999). Assigning protein functions by comparative genome analysis: protein phylogenetic profiles. *Proc. Natl. Acad. Sci. USA* **96**, 4285-4288.

[730] Sun, J., Xu, J., Liu, Z., Liu, Q., Zhao, A., Shi, T. & Li, Y. (2005). Refined phylogenetic profiles method for predicting protein-protein interactions. *Bioinformatics* **21**, 3409-3415.

[731] Dandekar, T., Snel, B., Huynen, M. & Bork, P. (1998). Conservation of gene order: a fingerprint of proteins that physically interact. *Trends Biochem. Sci.* **23**, 324-328.

[732] Rison, S. C., Teichmann, S. A. & Thornton, J. M. (2002). Homology, pathway distance and chromosomal localization of the small molecule metabolism enzymes in Escherichia coli. *J. Mol. Biol.* **318**, 911-932.

[733] Marcotte, E. M., Pellegrini, M., Ng, H. L., Rice, D. W., Yeates, T. O. & Eisenberg, D. (1999). Detecting protein function and protein-protein interactions from genome sequences. *Science* **285**, 751-753.

[734] Enright, A. J., Iliopoulos, I., Kyrpides, N. C. & Ouzounis, C. A. (1999). Protein interaction maps for complete genomes based on gene fusion events. *Nature* **402**, 86-90.

[735] Williams, E. J. & Hurst, L. D. (2000). The proteins of linked genes evolve at similar rates. *Nature* **407**, 900-903.

[736] Hurst, L. D. & Smith, N. G. (1999). Do essential genes evolve slowly? *Curr. Biol.* **9**, 747-750.

[737] Hirsh, A. E. & Fraser, H. B. (2001). Protein dispensability and rate of evolution. *Nature* **411**, 1046-1049.

[738] Zhang, J. & He, X. (2005). Significant impact of protein dispensability on the instantaneous rate of protein evolution. *Mol. Biol. Evol.* **22**, 1147-1155.

[739] Kuma, K., Iwabe, N. & Miyata, T. (1995). Functional constraints against variations on molecules from the tissue level: slowly evolving brain-specific genes demonstrated by protein kinase and immunoglobulin supergene families. *Mol. Biol. Evol.* **12**, 123-130.

[740] Fraser, H. B. & Hirsh, A. E. (2004). Evolutionary rate depends on number of protein-protein interactions independently of gene expression level. *BMC Evol. Biol.* **4**, 13.

[741] Goh, C., Bogan, A., Joachimiak, M., Walther, D. & Cohen, F. (2000). Co-evolution of proteins with their interaction partners. *J. Mol. Biol.* **299**, 283-293.

[742] Pazos, F. & Valencia, A. (2001). Similarity of phylogenetic trees as indicator of protein-protein interaction. *Protein Eng.* **14**, 609-614.

[743] Jordan, I. K., Rogozin, I. B., Wolf, Y. I. & Koonin, E. V. (2002). Essential genes are more evolutionarily conserved than are nonessential genes in bacteria *Genome Res.* **12**, 962-968.

[744] Ramani, A. K. & Marcotte, E. M. (2003). Exploiting the co-evolution of interacting proteins to discover interaction specificity. *J. Mol. Biol.* **327**, 273-284.

[745] Gertz, J., Elfond, G., Shustrova, A., Weisinger, M., Pellegrini, M., Cokus, S. & Rothschild, B. (2003). Inferring protein interactions from phylogenetic distance matrices. *Bioinformatics* **19**, 2039-2045.

[746] Jothi, R., Kann, M. G. & Przytycka, T. M. (2005). Predicting protein-protein interaction by searching evolutionary tree automorphism space. *Bioinformatics* **21**, i241-i250.

[747] Sato, T., Yamanishi, Y., Kanehisa, M. & Toh, H. (2005). The inference of protein-protein interactions by co-evolutionary analysis is improved by excluding the information about the phylogenetic relationships. *Bioinformatics* **21**, 3482-3489.

[748] Pazos, F., Ranea, J. A., Juan, D. & Sternberg, M. J. (2005). Assessing protein co-evolution in the context of the tree of life assists in the prediction of the interactome. *J. Mol. Biol.* **352**, 1002-1015.

[749] Pollock, D., Taylor, W. & Goldman, N. (1999). Coevolving protein residues: maximum likelihood identification and relationship to structure. *J. Mol. Biol.* **287**, 187-198.

[750] Dimmic, M. W., Hubisz, M. J., Bustamante, C. D. & Nielsen, R. (2005). Detecting coevolving amino acid sites using Bayesian mutational mapping. *Bioinformatics* **21** Suppl 1, i126-135.

[751] Dutheil, J. & Galtier, N. (2007). Detecting groups of coevolving positions in a molecule: a clustering approach. *BMC Evol. Biol.* **7**, 242.

[752] Sprinzak, E. & Margalit, H. (2001). Correlated sequence-signatures as markers of protein-protein interaction. *J. Mol. Biol.* **311**, 681-692.

[753] Aloy, P. & Russell, R. (2003). InterPreTS: protein interaction prediction through tertiary structure. *Bioinformatics* **19**, 161-162.

[754] Wojcik, J. & Schächter, V. (2001). Protein-protein interaction map inference using interacting domain profile pairs. *Bioinformatics* **17**, S296-305.

[755] Deng, M., Mehta, S., Sun, F. & Chen, T. (2002). Inferring domain-domain interactions from protein-protein interactions. *Genome Res.* **12**, 1540-1548.

[756] Liu, Y., Liu, N. & Zhao, H. (2005). Inferring protein-protein interactions through high-throughput interaction data from diverse organisms. *Bioinformatics* **21**, 3279-3285.

[757] Yang, X. J. (2005). Multisite protein modification and intramolecular signaling. *Oncogene* **24**, 1653-1662.

[758] Seet, B. T., Dikic, I., Zhou, M. M. & Pawson, T. (2006). Reading protein modifications with interaction domains. *Nat. Rev. Mol. Cell. Biol.* **7**, 473-483.

[759] Han, D. S., Kim, H. S., Jang, W. H., Lee, S. D. & Suh, J. K. (2004). PreSPI: a domain combination based prediction system for protein-protein interaction. *Nucl. Acids Res.* **32**, 6312-6320.

[760] Chen, X. W. & Liu, M. (2005). Prediction of protein-protein interactions using random decision forest framework. *Bioinformatics* **21**, 4394-4400.

[761] Ge, H., Liu, Z., Church, G. & Vidal, M. (2001). Correlation between transcriptome and interactome mapping data from Saccharomyces cerevisiae. *Nat. Genet.* **29**, 482-486.

[762] Mrowka, R., Liebermeister, W. & Holste, D. (2003). Does mapping reveal correlation between gene expression and protein-protein interaction? *Nat. Genet.* **33**, 15-16.

[763] Jansen, R., Greenbaum, D. & Gerstein, M. (2002). Relating whole-genome expression data with protein-protein interactions. *Genome Res.* **12**, 37-46.

[764] Jansen, R., Yu, H., Greenbaum, D., Kluger, Y., Krogan, N.J., Chung, S., Emili, A., Snyder, M., Greenblatt, J. F. & Gerstein, M. (2003). A Bayesian networks approach for predicting protein-protein interactions from genomic data. *Science* **302**, 449-453.

[765] Smialowski, P., Pagel, P., Wong, P., Brauner, B., Dunger, I., Fobo, G., Frishman, G., Montrone, C., Rattei, T., Frishman, D. & Ruepp, A. (2010). The Negatome database: a reference set of non-interacting protein pairs. *Nucl. Acids Res.* **38**, D540-544.

[766] Watts, D. J. & Strogatz, S. (1998). Collective dynamics of small-world networks. *Nature* **393**, 440-442.

[767] Erdös, P. & Renyi, A. (1959). On random graphs I. *Publicationes Mathematicae* **6**, 290-297.

[768] Travers, J. & Milgram, S. (1969). An experimental study of the small world problem. *Sociometry* **32**, 425-443.

[769] Barabási, A. L. & Réka, A. (1999). Emergence of scaling in random networks. *Science* **286**, 509-512.

[770] Davids, W. & Zhang, Z. (2008). The impact of horizontal gene transfer in shaping operons and protein interaction networks - direct evidence of preferential attachment. *BMC Evol. Biol.* **8**, 23.

[771] Przulj, N. (2007). Biological network comparison using graphlet degree distribution. *Bioinformatics* **23**, e177-183.

[772] Penrose, M. (2003). "Random geometric graphs." Oxford University Press.

[773] Przulj, N., Corneil, D. G. & Jurisica, I. (2004). Modeling interactome: scale-free or geometric? *Bioinformatics* **20**, 3508-3515.

[774] Jeong, H., Mason, S. P., Barabási, A. L. & Oltvai, Z. N. (2001). Lethality and centrality in protein networks. *Nature* **411**, 41-42.

[775] Park, D., Lee, S., Bolser, D., Schroeder, M., Lappe, M., Oh, D. & Bhak, J. (2005). Comparative interactomics analysis of protein family interaction networks using PSIMAP (protein structural interactome map). *Bioinformatics* **21**, 3234-3240.

[776] Babu, M. M., Luscombe, N. M., Aravind, L., Gerstein, M. & Teichmann, S. A. (2004). Structure and evolution of transcriptional regulatory networks. *Curr. Opin. Struct. Biol.* **14**, 283-291.

[777] Roxin, A., Riecke, H. & Solla, S. A. (2004). Self-sustained activity in a small-world network of excitable neurons. *Phys. Rev. Lett.* **92**, 198101.

[778] Jeong, H., Tombor, B., Albert, R., Oltvai, Z. N. & Barabási, A. L. (2000). The large-scale organization of metabolic networks. *Nature* **407**, 651-654.

[779] del Sol, A., Fujihashi, H. & O'Meara, P. (2005). Topology of small-world networks of protein-protein complex structures. *Bioinformatics* **21**, 1311-1315.

[780] Huang, J., Kawashima, S. & Kanehisa, M. (2007). New amino acid indices based on residue network topology. *Genome Inform.* **18**, 152-161.

[781] Wuchty, S. & Almaas, E. (2005). Evolutionary cores of domain co-occurrence networks. *BMC Evol. Biol.* **5**, 24.

[782] Goldberg, D. S. & Roth, F. P. (2003). Assessing experimentally derived interactions in a small world. *Proc. Natl. Acad. Sci. USA* **100**, 4372-4376.

[783] Albert, R., Jeong, H. & Barabasi, A. L. (2000). Error and attack tolerance of complex networks. *Nature* **406**, 378-382.

[784] Winzeler, E. A., Shoemaker, D. D., Astromoff, A., Liang, H., Anderson, K., Andre, B., Bangham, R. et al. (1999). Functional characterization of the S. cerevisiae genome by gene deletion and parallel analysis. *Science* **285**, 901-906.

[785] Giaever, G., Chu, A. M., Ni, L., Connelly, C., Riles, L. et al. (2002). Functional profiling of the Saccharomyces cerevisiae genome. *Nature* **418**, 387-391.

[786] Kobayashi, K., Ehrlich, S. D., Albertini, A., Amati, G., Andersen, K. K. et al. (2003). Essential Bacillus subtilis genes. *Proc. Natl. Acad. Sci. USA* **100**, 4678-4683.

[787] Boutros, M., Kiger, A. A., Armknecht, S., Kerr, K., Hild, M., Koch, B., Haas, S. A., Paro, R., Perrimon, N. & Heidelberg Fly Array Consortium. (2004). Genome-wide RNAi analysis of growth and viability in Drosophila cells. *Science* **303**, 832-835.

[788] Vogelstein, B., Lane, D. & Levine, A. J. (2000). Surfing the p53 network. *Nature* **408**, 307-310.

[789] Harris, S. L. & Levine, A. J. (2005). The p53 pathway: positive and negative feedback loops. *Oncogene* **24**, 2899-2908.

[790] Siegal, M. L., Promislow, D. E., Bergman, A. (2006). Functional and evolutionary inference in gene networks: does topology matter? *Genetica.* **129**, 83-103.

[791] Balaji, S., Iyer, L. M., Aravind, L. & Babu, M. M. (2006). Uncovering a hidden distributed architecture behind scale-free transcriptional regulatory networks. *J. Mol. Biol.* **360**, 204-212.

[792] Wagner, A. (2000). Robustness against mutations in genetic networks of yeast. *Nat. Genet.* **24**, 355-361.

[793] Shen-Orr, S. S., Milo, R., Mangan, S. & Alon U. (2002). Network motifs in the transcriptional regulation network of Escherichia coli. *Nat. Genet.* **31**, 64-68.

[794] Mangan, S. & Alon, U. (2003). Structure and function of the feed-forward loop network motif. *Proc. Natl. Acad. Sci. USA* **100**, 11980-11985.

[795] Yu, H., Luscombe, N. M., Qian, J. & Gerstein, M. (2003). Genomic analysis of gene expression relationships in transcriptional regulatory networks. *Trends Genet.* **19**, 422-427.

[796] Phizicky, E. M. & Fields, S. (1995). Protein-protein interactions: methods for detection and analysis. *Microbiological Reviews* **95**, 94-123.

[797] Szilágyi, A., Grimm, V., Arakaki, A. K. & Skolnick, J. (2005). Prediction of physical protein-protein interactions. *Phys. Biol.* **2**, S1-16.

[798] Shoemaker, B. A. & Panchenko, A. R. (2007). Deciphering protein-protein interactions. Part I. Experimental techniques and databases. *PLoS Comp. Biol.* **3**, e42.

[799] Ceol, A., Chatr Aryamontri, A., Licata, L., Peluso, D., Briganti, L., Perfetto, L., Castagnoli, L. & Cesareni G. (2010). MINT, the molecular interaction database: 2009 update. *Nucl. Acids Res.* **38**, D532-539.

[800] Hermjakob, H., Montecchi-Palazzi, L., Lewington, C., Mudali, S., Kerrien, S., Orchard, S., Vingron, M., Roechert, B., Roepstorff, P., Valencia, A., Margalit, H., Armstrong, J., Bairoch, A., Cesareni, G., Sherman, D. & Apweiler, R. (2004). IntAct: an open source molecular interaction database. *Nucl. Acids Res.* **32**, D452-455.

[801] Aranda, B., Achuthan, P., Alam-Faruque, Y., Armean, I., Bridge, A., Derow, C., Feuermann, M., Ghanbarian, A. T., Kerrien, S., Khadake, J., Kerssemakers, J., Leroy, C., Menden, M., Michaut, M., Montecchi-Palazzi, L., Neuhauser, S. N., Orchard, S., Perreau, V., Roechert, B., van Eijk, K. & Hermjakob, H. (2010). The IntAct molecular interaction database in 2010. *Nucl. Acids Res.* **38**, D525-531.

[802] Huang, T., Tien, A., Huang, W., Lee, Y. G., Peng, C., Tseng, H., Kao, C. & Huang, C. F. (2004). POINT: a database for the prediction of protein-protein interactions based on the orthologous interactome. *Bioinformatics* **20**, 3273-3276.

[803] Dou, Y., Baisnée, P. F., Pollastri, G., Pécout, Y., Nowick, J. & Baldi, P. (2004). ICBS: a database of interactions between protein chains mediated by beta-sheet formation. *Bioinformatics* **20**, 2767-2677.

[804] Raghavachari, B., Tasneem, A., Przytycka, T. M. & Jothi, R. (2008). DOMINE: a database of protein domain interactions. *Nucl. Acids Res.* **36**, D656-661.

[805] Chautard, E., Ballut, L., Thierry-Micg, N. & Ricard-Blum, S. (2009). MatrixDB, a database focused on extracellular protein-protein and protein-carbohydrate interactions. *Bioinformatics* **25**, 690-691.

[806] Sternberg, M. J., Gabb, H. A. & Jackson, R. M. (1998). Predictive docking of protein-protein and protein-DNA complexes. *Curr. Opin. Struct. Biol.* **8**, 250-256.

[807] Halperin, I., Ma, B., Wolfson, H. & Nussinov, R. (2002). Principles of docking: an overview of search algorithms and a guide to scoring functions. *Proteins* **47**, 409-443.

[808] Ritchie, D. W. (2008). Recent progress and future directions in protein-protein docking. *Curr. Protein Pept. Sci.* **9**, 1-15.

[809] Vajda, S. & Kozakov, D. (2009). Convergence and combination of methods in protein-protein docking. *Curr. Opin. Struct. Biol.* **19**, 164-170.

[810] Inbar, Y., Benyamini, H., Nussinov, R. & Wolfson, H. J. (2005). Combinatorial docking approach for structure prediction of large proteins and multi-molecular assemblies. *Phys. Biol.* **2**, S156-165.

[811] Shoemaker, B. A. & Panchenko, A. R. (2007). Deciphering protein-protein interactions. Part II. Computational methods to predict protein and domain interaction partners. *PLoS Comp. Biol.* **3**, e43.

[812] Skrabanek, L., Saini, H. K., Bader, G. D. & Enright, A. J. (2008). Computational prediction of protein-protein interactions. *Mol. Biotechnol.* **38**, 1-17.

[813] Pitre, S., Alamgir, M., Green, J. R., Dumontier, M., Dehne, F. & Golshani, A. (2008). Computational methods for predicting protein-protein interactions. *Adv. Biochem. Eng. Biotechnol.* **110**, 247-267.

[814] Dover, G. A. & Flavell, R. B. (1984). Molecular coevolution: DNA divergence and the maintenance of function. *Cell* **38**, 622-623.

[815] Pazos, F., Helmer-Citterich, M., Ausiello, G. & Valencia, A. (1997). Correlated mutations contain information about protein-protein interaction. *J. Mol. Biol.* **271**, 511-523.

[816] Pazos, F. & Valencia, A. (2002). In silico two-hybrid system for the selection of physically interacting protein pairs. *Proteins* **47**, 219-227.

[817] Kann, M. G., Shoemaker, B. A., Panchenko, A. R & Przytycka, T. M. (2009). Correlated evolution of interacting proteins: looking behind the mirrortree. *J. Mol. Biol.* **385**, 91-98.

[818] Ng, S. K., Zhang, Z. & Tan, S. H. (2003). Integrative approach for computationally inferring protein domain interactions. *Bioinformatics* **19**, 923-929.

[819] Bock, J. R. & Gough, D. A. (2001). Predicting protein–protein interactions from primary structure. *Bioinformatics* **17**, 455-460.

[820] Ben-Hur, A. & Noble, W. S. (2005). Kernel methods for predicting protein-protein interactions. *Bioinformatics* **21** Suppl 1, i38-46.

[821] Yamanishi, Y., Vert, J. P. & Kanehisa, M. (2004). Protein network inference from multiple genomic data: a supervised approach. *Bioinformatics* **20** Suppl 1, i363-370.

[822] Sun, J., Sun, Y., Ding, G., Liu, Q., Wang, C., He, Y., Shi, T., Li, Y. & Zhao, Z. (2005). InPrePPI: an integrated evaluation method based on genomic context for predicting protein-protein interactions in prokaryotic genomes. *BMC Bioinformatics* **8**, 414.

[823] Schelhorn, S. E., Lengauer, T. & Albrecht, M. (2008). An integrative approach for predicting interactions of protein regions. *Bioinformatics* **24**, i35-41.

[824] Schwikowski, B., Uetz, P. & Fields, S. (2000). A network of protein-protein interactions in yeast. *Nat. Biotechnol.* **18**, 1257-1261.

[825] Gavin, A. C., Bosche, M., Krause, R., Grandi, P. et al. (2002). Functional organization of the yeast proteome by systematic analysis of protein complexes. *Nature* **415**, 141-147.

[826] Albert, R. (2005). Scale-free networks in cell biology. *J. Cell Sci.* **118**, 4947-4957.

[827] Milo, R., Shen-Orr, S., Itzkovitz, S., Kashtan, N., Chklovskii, D. & Alon, U. (2002) Network motifs: simple building blocks of complex networks. *Science* **298**, 824-827.

[828] Yeger-Lotem, E., Sattath, S., Kashtan, N., Itzkovitz, S., Milo, R., Pinter, R. Y., Alon, U. & Margalit, H. (2004). Network motifs in integrated cellular networks of transcription-regulation and protein-protein interaction. *Proc. Natl. Acad. Sci. USA* **101**, 5934-5939.

[829] Wang, E. & Purisima, E. (2005). Network motifs are enriched with transcription factors whose transcripts have short half-lives. *Trends Genet.* **21**, 492-495.

[830] Mazurie, A., Bottani, S. & Vergassola, M. (2005). An evolutionary and functional assessment of regulatory network motifs. *Genome Biol.* **6**, R35.

[831] Ward, J. J. & Thornton, J. M. (2007). Evolutionary models for formation of network motifs and modularity in the Saccharomyces transcription factor network. *PLoS Comp. Biol.* **3**, 1993-2002.

[832] Brandman, O. & Meyer, T. (2008). Feedback loops shape cellular signals in space and time. *Science* **322**, 390-395.

[833] Alon, U. (2007). Network motifs: theory and experimental approaches. *Nat. Rev. Genet.* **8**, 450-461.

[834] Jones, S. & Thornton, J. M. (2003). Protein-DNA interactions: the story so far and a new method for prediction. *Comp. Funct. Genomics.* **4**, 428-431.

[835] Ahmad, S., Gromiha, M. M. & Sarai, A. (2004). Analysis and prediction of DNA-binding proteins and their binding residues based on composition, sequence and structural information. *Bioinformatics* **20**, 477-486.

[836] Bhardwaj, N., Langlois, R. E., Zhao, G. & Lu, H. (2005). Kernel-based machine learning protocol for predicting DNA-binding proteins. *Nucl. Acids Res.* **33**, 6486-6493.

[837] Sathyapriya, R., Vijayabaskar, M. S. & Vishveshwara, S. (2008). Insights into protein-DNA interactions through structure network analysis. *PLoS Comp. Biol.* **4**, e1000170.

[838] Jones, S., Daley, D. T., Luscombe, N. M., Berman, H. M. & Thornton, J. M. (2001). Protein-RNA interactions: a structural analysis. *Nucl. Acids Res.* **29**, 943-954.

[839] Morozova, N., Allers, J., Myers, J. & Shamoo, Y. (2006). Protein-RNA interactions: exploring binding patterns with a three-dimensional superposition analysis of high resolution structures. *Bioinformatics* **22**, 2746-2752.

[840] Ellis, J. J., Broom, M. & Jones, S. (2007). Protein-RNA interactions: structural analysis and functional classes. *Proteins* **66**, 903-911.

[841] Shazman, S. & Mandel-Gutfreund, Y. (2008). Classifying RNA-binding proteins based on electrostatic properties. *PLoS Comp. Biol.* **4**, e1000146.

[842] Kumar, M., Gromiha, M. M. & Raghava, G. P. (2008). Prediction of RNA binding sites in a protein using SVM and PSSM profile. *Proteins* **71**, 189-194.

[843] Figure retrieved from http://en.wikipedia.org/wiki/File:G_protein_signal_transduction_(epinephrin_pathway).png, May 2010.

[844] Guarente, L. & Kenyon, C. (2000). Genetic pathways that regulate ageing in model organisms. *Nature* **408**, 255-262.

[845] Imhof, B. A. & Aurrand-Lions, M. (2004). Adhesion mechanisms regulating the migration of monocytes. *Nat. Rev. Immunol.* **4**, 432-444.

[846] Prasad, K. N., Cole, W. C., Yan, X. D., Nahreini, P., Kumar, B., Hanson, A. & Prasad, J. E. (2003). Defects in cAMP-pathway may initiate carcinogenesis in dividing nerve cells: a review. *Apoptosis* **8**, 579-586.

[847] Naviglio, S., Spina, A., Marra, M., Sorrentino, A., Chiosi, E., Romano, M., Improta, S., Budillon, A., Illiano, G., Abbruzzese, A. & Caraglia, M. (2007). Adenylate cyclase/cAMP pathway downmodulation counteracts apoptosis induced by IFN-alpha in human epidermoid cancer cells. *J. Interferon Cytokine Res.* **27**, 129-136.

[848] Karp, P. D., Riley, M., Saier, M., Paulsen, I. T., Paley, S. M. & Pellegrini-Toole, A. (2000). The EcoCyc and MetaCyc databases. *Nucl. Acids Res.* **28**, 56-59.

[849] Kanehisa, M. (1996). Toward pathway engineering: a new database of genetic and molecular pathways. *Science & Technology Japan* **59**, 34-38.

[850] Caspi, R., Altman, T., Dale, J. M., Dreher, K., Fulcher, C. A., Gilham, F., Kaipa, P., Karthikeyan, A. S., Kothari, A., Krummenacker, M., Latendresse, M., Mueller, L. A., Paley, S., Popescu, L., Pujar, A., Shearer, A. G., Zhang, P. & Karp, P. D. (2010). The MetaCyc database of metabolic pathways and enzymes and the BioCyc collection of pathway/genome databases. *Nucl. Acids Res.* **38**, D473-479.

[851] http://www.chem.qmw.ac.uk/iubmb/enzyme/

[852] Kolch, W., Calder, M. & Gilbert, D. (2005). When kinases meet mathematics: the systems biology of MAPK signalling. *FEBS Lett.* **579**, 1891-1895.

[853] Pramanik, J. & Keasling, J. 1997. Stoichiometric model of Escherichia coli metabolism: incorporation of growth-rate dependent biomass composition and mechanistic energy requirements. *Biotechnol. Bioeng.* **56**, 398-421.

[854] Edwards, J. S. & Palsson, B. O. (2000). The Escherichia coli MG1655 in silico metabolic genotype: its definition, characteristics, and capabilities. *Proc. Natl. Acad. Sci. USA* **97**, 5528-5533.

[855] Reed, J., Vo, T., Schilling, C. & Palsson, B. (2003). An expanded genome-scale model of Escherichia coli K-12 (iJR904 GSM/GPR). *Genome Biol.* **4**, R54.

[856] Edwards, J. S. & Palsson, B. O. (1999). Systems Properties of the Haemophilus influenzae Rd Metabolic Genotype. *J. Biol. Chem.* **274**, 17410-17416.

[857] Forster, J., Famili, I., Fu, P., Palsson, B. O. & Nielsen, J. (2003). Genome-scale reconstruction of the saccharomyces cerevisiae metabolic network. *Genome Res.* **13**, 244-253.

[858] Duarte, N. C., Herrgard, M. J. & Palsson, B. O. (2004). Reconstruction and validation of saccharomyces cerevisiae iND750, a fully compartmentalized genome-scale metabolic model. *Genome Res.* **14**, 1298-1309.

[859] Akutsu, T., Miyano, S. & Kuhara, S. (2000). Algorithms for identifying boolean networks and related biological networks based on matrix multiplication and fingerprint function. *J. Comp. Biol.* **7**, 331-343.

[860] Kuffner, R., Zimmer, R. & Lengauer, T. (2000). Pathway analysis in metabolic databases via differential metabolic display (DMD). *Bioinformatics* **16**, 825-836.

[861] McShan, D., Rao, S. & Shah, I. (2003). PathMiner: predicting metabolic pathways by heuristic search. *Bioinformatics* **19**, 1692-1698.

[862] Boyer, F. & Viari, A. (2003). Ab initio reconstruction of metabolic pathways. *Bioinformatics* **19** Suppl 2, i26-34.

[863] Croes, D., Couche, F., Wodak, S. J. & van Helden, J. (2006). Inferring meaningful pathways in weighted metabolic networks. *J. Mol. Biol.* **356**, 222-236.

[864] Bono, H., Ogata, H., Goto, S, & Kanehisa, M. (1998). Reconstruction of amino acid biosynthesis pathways from the complete genome sequence. *Genome Res.* **8**, 203-210.

[865] Sirava, M., Schafer, T., Eiglsperger, M., Kaufmann, M., Kohlbacher, O., Bornberg-Bauer, E. & Lenhof, H. P. (2002). BioMiner - modeling, analyzing, and visualizing biochemical pathways and networks. *Bioinformatics* **18** Suppl 2, S219-230.

[866] Paley, S. M. & Karp, P. D. (2002). Evaluation of computational metabolic-pathway predictions for Helicobacter pylori. *Bioinformatics* **18**, 715-724.

[867] Jensen, R. A. (1976). Enzyme recruitment in evolution of new function. *Ann. Rev. Microbiol.* **30**, 409-425.

[868] Galperin, M. Y., Walker, D. R. & Koonin, E. V. (1998). Analogous enzymes: independent inventions in enzyme evolution. *Genome Res.* **8**, 779-790.

[869] von Grotthuss, M., Plewczynski, D., Vriend, G. & Rychlewski, L. (2008). 3D-Fun: predicting enzyme function from structure. *Nucl. Acids Res.* **36**, W303-307.

[870] Popescu, L. & Yona, G. (2006). Expectation-maximization algorithms for fuzzy assignment of genes to cellular pathways. In *The Proceedings of CSB*, 281-291.

[871] Ihmels, J., Levy, R. & Barkai, N. (2004). Principles of transcriptional control in the metabolic network of Saccharomyces cerevisiae. *Nat. Biotechnol.* **22**, 86-92.

[872] Chen, L. & Vitkup, D. (2006). Predicting genes for orphan metabolic activities using phylogenetic profiles. *Genome Biol.* **7**, R17.

[873] Segal, E., Wang, H. & Koller, D. (2003). Discovering molecular pathways from protein interaction and gene expression data. *Bioinformatics* **19** Suppl 1, i264-272.

[874] Popescu, L. & Yona, G. (2005). Automation of gene assignments to metabolic pathways using high-throughput expression data. *BMC Bioinformatics* **6**, 217.

[875] Tomita, M., Hashimoto, K., Takahashi, K., Shimizu, T. S., Matsuzaki, Y., Miyoshi, F., Saito, K., Tanida, S., Yugi, K., Venter, J. C. & Hutchison, C. A. 3rd. (1999). E-CELL: software environment for whole-cell simulation. *Bioinformatics* **15**, 72-84.

[876] Tomita, M. (2001). Whole-cell simulation: a grand challenge of the 21st century. *Trends Biotechnol.* **19**, 205-210.

[877] Steuer, R., Gross, T., Selbig, J. & Blasius, B. (2006). Structural kinetic modeling of metabolic networks. *Proc. Natl. Acad. Sci. USA* **103**, 11868-11873.

[878] Azuma, Y. & Ota, M. (2009). An evaluation of minimal cellular functions to sustain a bacterial cell. *BMC Sys. Biol.* **3**, 111.

[879] Helikar, T. & Rogers, J. A. (2009). ChemChains: a platform for simulation and analysis of biochemical networks aimed to laboratory scientists. *BMC Sys. Biol.* **3**, 58.

[880] Karp, P. D. (2001). Pathway databases: a case study in computational symbolic theories. *Science* **293**, 2040-2044.

[881] Bork, P., Ouzounis, C., Casari, G., Schneider, R., Sander, C., Dolan, M., Gilbert, W. & Gillevet, P. M. (1995). Exploring the Mycoplasma capricolum genome: a minimal cell reveals its physiology. *Mol. Microbiol.* **16**, 955-967.

[882] Browning, S. T. & Shuler, M. L. (2001). Towards the development of a minimal cell model by generalization of a model of Escherichia coli: use of dimensionless rate parameters. *Biotechnol. Bioeng.* **76**, 187-192.

[883] Castellanos, M., Wilson, D. B. & Shuler, M. L. (2004). A modular minimal cell model: purine and pyrimidine transport and metabolism. *Proc. Natl. Acad. Sci. USA* **101**, 6681-6686.

[884] Murtas, G. (2009). Artificial assembly of a minimal cell. *Mol. Biosyst.* **5**, 1292-1297.

[885] Fraser, C. M., Gocayne, J. D., White, O., Adams, M. D., Clayton, R. A., Fleischmann, R. D., Bult, C. J., Kerlavage, A. R., Sutton, G., Kelley, J. M., Fritchman, R. D., Weidman, J. F., Small, K. V., Sandusky, M., Fuhrmann, J., Nguyen, D., Utterback, T. R., Saudek, D. M., Phillips, C. A., Merrick, J. M., Tomb, J. F., Dougherty, B. A., Bott, K. F., Hu, P. C., Lucier, T. S., Peterson, S. N., Smith, H. O., Hutchison, C. A. 3rd & Venter, J. C. (1995). The minimal gene complement of Mycoplasma genitalium. *Science* **270**, 397-403.

[886] Maniloff, J. (1996). The minimal cell genome: "on being the right size". *Proc. Natl. Acad. Sci. USA* **93**, 10004-10006.

[887] Gil, R., Silva, F. J., Peretó, J. & Moya, A. (2004). Determination of the core of a minimal bacterial gene set. *Microbiol. Mol. Biol. Rev.* **68**, 518-537.

[888] Hutchison, C. A., Peterson, S. N., Gill, S. R., Cline, R. T., White, O., Fraser, C. M., Smith, H. O. & Venter, J. C. (1999). Global transposon mutagenesis and a minimal Mycoplasma genome. *Science* **286**, 2165-2169.

[889] Luisi, P. L., Ferri, F. & Stano, P. (2006). Approaches to semi-synthetic minimal cells: a review. *Naturwissenschaften* **93**, 1-13.

[890] Forster, A. C. & Church, G. M. (2006). Towards synthesis of a minimal cell. *Mol. Syst. Biol.* **2**, 45.

[891] Nakabachi, A., Yamashita, A., Toh, H., Ishikawa, H., Dunbar, H. E., Moran, N. A. & Hattori, M. (2006). The 160-kilobase genome of the bacterial endosymbiont Carsonella. *Science* **314**, 267.

[892] Sagan, L. (1967). On the origin of mitosing cells. *J. Theor. Biol.* **14**, 255-274.

[893] Gray, M. W., Burger, G. & Lang, B. F. (2001). The origin and early evolution of mitochondria. *Genome Biol.* **2**, reviews1018.

[894] Kanehisa, M., Goto, S., Furumichi, M., Tanabe, M. & Hirakawa, M. (2010). KEGG for representation and analysis of molecular networks involving diseases and drugs. *Nucl. Acids Res.* **38**, D355-D360.

[895] Overbeek, R., Larsen, N., Walunas, T., D'Souza, M., Pusch, G., Selkov, E., Liolios, K., Joukov, V., Kaznadzey, D., Anderson, I., Bhattacharyya, A., Burd, H., Gardner, W., Hanke, P., Kapatral, V., Mikhailova, N., Vasieva, O., Osterman, A., Vonstein, V., Fonstein, M., Ivanova, N. & Kyrpides, N. (2003). The ERGO genome analysis and discovery system. *Nucl. Acids Res.* **31**, 164-171.

[896] Selkov, E., Basmanova, S., Gaasterland, T., Goryanin, I., Gretchkin, Y., Maltsev, N., Nenashev, V., Overbeek, R., Panyushkina, E., Pronevitch, L., Selkov, E. Jr. & Yunus, I. (1996). The metabolic pathway collection from EMP: the enzymes and metabolic pathways database. *Nucl. Acids Res.* **24**, 26-28.

[897] Selkov, E. Jr., Grechkin, Y., Mikhailova, N. & Selkov E. (1998). MPW: the Metabolic Pathways Database. *Nucl. Acids Res.* **26**, 43-45.

[898] Overbeek, R., Larsen, N., Pusch, G. D., D'Souza, M., Selkov, E. Jr, Kyrpides, N., Fonstein, M., Maltsev, N. & Selkov, E. (2000). WIT: integrated system for high-throughput genome sequence analysis and metabolic reconstruction. *Nucl. Acids Res.* **28**, 123-125.

[899] Ellis, L. B. M., Hou, B. K., Kang, W. & Wackett, L. P. (2003). The University of Minnesota Biocatalysis/Biodegradation Database: post-genomic data mining. *Nucl. Acids Res.* **31**, 262-265.

[900] Lemer, C., Antezana, E., Couche, F., Fays, F., Santolaria, X., Janky, R., Deville, Y., Richelle, J. & Wodak, S. J. (2004). The aMAZE LightBench: a web interface to a relational database of cellular processes. *Nucl. Acids Res.* **32**, D443-448.

[901] Vastrik, I., D'Eustachio, P., Schmidt, E., Gopinath, G., Croft, D., de Bono, B., Gillespie, M., Jassal, B., Lewis, S., Matthews, L., Wu, G., Birney, E. & Stein, L. (2007). Reactome: a knowledge base of biologic pathways and processes. *Genome Biol.* **8**, R39.

[902] Chowbina, S. R., Wu, X., Zhang, F., Li, P. M., Pandey, R., Kasamsetty, H. N. & Chen, J. Y. (2009). HPD: an online integrated human pathway database enabling systems biology studies. *BMC Bioinformatics* **10** Suppl 11, S5.

[903] Geer, L. Y., Marchler-Bauer, A., Geer, R. C., Han, L., He, J., He, S., Liu, C., Shi, W. & Bryant, S. H. (2010). The NCBI BioSystems database. *Nucl. Acids Res.* **38**, D492-496.

[904] Frolkis, A., Knox, C., Lim, E., Jewison, T., Law, V., Hau, D. D., Liu, P., Gautam, B., Ly, S., Guo, A. C., Xia, J., Liang, Y., Shrivastava, S. & Wishart, D. S. (2010). SMPDB: The Small Molecule Pathway Database. *Nucl. Acids Res.* **38**, D480-487.

[905] Schellenberger, J, Park, J. O., Conrad, T. M. & Palsson B. O. (2010). BiGG: a Biochemical Genetic and Genomic knowledgebase of large scale metabolic reconstructions. *BMC Bioinformatics* **11**, 213.

[906] Durot, M., Bourguignon, P. Y. & Schachter, V. (2009). Genome-scale models of bacterial metabolism: reconstruction and applications. *FEMS Microbiol. Rev.* **33**, 164-190.

[907] Oberhardt, M. A., Palsson, B. O. & Papin, J. A. (2009). Applications of genome-scale metabolic reconstructions. *Mol. Syst. Biol.* **5**, 320.

[908] Feist, A. M., Herrgard, M. J., Thiele, I., Reed, J. L. & Palsson, B. O. (2008). Reconstruction of biochemical networks in microorganisms. *Nat. Rev. Microbiol.* **7**, 129-143.

[909] Herrgård, M. J., Covert, M. W. & Palsson, B. O. (2004). Reconstruction of microbial transcriptional regulatory networks. *Curr. Opin. Biotechnol.* **15**, 70-77.

[910] Yamanishi, Y., Vert, J. P. & Kanehisa, M. (2005). Supervised enzyme network inference from the integration of genomic data and chemical information. *Bioinformatics* **21** Suppl 1, i468-477.

[911] Myers, C. L., Robson, D., Wible, A., Hibbs, M. A., Chiriac, C., Theesfeld, C. L., Dolinski, K. & Troyanskaya, O. G. (2005). Discovery of biological networks from diverse functional genomic data. *Genome Biol.* **6**, R114.

[912] Bleakley, K., Biau, G. & Vert, J. P. (2007). Supervised reconstruction of biological networks with local models. *Bioinformatics* **23**, i57-65.

[913] Nakao, M., Bono, H., Kawashima, S., Kamiya, T., Sato, K., Goto, S. & Kanehisa M. (1999). Genome-scale gene expression analysis and pathway reconstruction in KEGG. *Genome Inform.* **10**, 94-103.

[914] Zien, A., Küffner, R., Zimmer, R. & Lengauer, T. (2000). Analysis of gene expression data with pathway scores. In *The Proceedings of ISMB*, 407-417.

[915] Kurhekar, M., Adak, S., Jhunjhunwala, S. & Raghupathy, K. (2002). Genome-wide pathway analysis and visualization using gene expression data. In *The Proceedings of PSB*, 462-473.

[916] Grosu, P., Townsend, J. P., Hartl, D. L. & Cavalieri, D. (2002). Pathway processor: a tool for integrating whole-genome expression results into metabolic networks. *Genome Res.* **12**, 1121-1126.

[917] Pavlidis, P., Lewis, D. P. & Noble, W. S. (2002). Exploring gene expression data with class scores. In *The Proceedings of PSB*, 474-485.

[918] Vert, J. P. & Kanehisa, M. (2003). Extracting active pathways from gene expression data. *Bioinformatics* **19**, i238-i244.

[919] Rahnenfuhrer, J., Domingues, F. S., Maydt, J. & Lengauer, T. (2004). Calculating the statistical significance of changes in pathway activity from gene expression data. *Statistical Applications in Genetics and Molecular Biology* **3**, 16.

[920] Goesmann, A., Haubrock, M., Meyer, F., Kalinowski, J. & Giegerich, R. (2002). PathFinder: reconstruction and dynamic visualization of metabolic pathways. *Bioinformatics* **18**, 124-129.

[921] Pan, D., Sun, N., Cheung, K. H., Guan, Z., Ma, L., Holford, M., Deng, X. & Zhao, H. (2003). PathMAPA: a tool for displaying gene expression and performing statistical tests on metabolic pathways at multiple levels for Arabidopsis. *BMC Bioinformatics* **4**, 56.

[922] Doniger, S., Salomonis, N., Dahlquist, K., Vranizan, K., Lawlor, S. & Conklin B. (2003). MAPPFinder: using Gene Ontology and GenMAPP to create a global gene-expression profile from microarray data. *Genome Biol.* **4**, R7.

[923] Hanisch, D., Zien, A., Zimmer, R. & Lengauer, T. (2002). Co-clustering of biological networks and gene expression data. *Bioinformatics* **18**, S145-S154.

[924] Nakaya, A., Goto, S. & Kanehisa, M. (2001). Extraction of correlated gene clusters by multiple graph comparison. *Genome Inform.* **12**, 44-53.

[925] Yamanishi, Y., Vert, J. P., Nakaya, A. & Kanehisa, M. (2003). Extraction of correlated gene clusters from multiple genomic data by generalized kernel canonical correlation analysis. *Bioinformatics* **19**, i323-i330.

[926] Segal, E., Battle, A. & Koller, D. (2003). Decomposing gene expression into cellular processes. In *The Proceedings of PSB*, 89-100.

[927] Segal, E., Yelensky, R. & Koller, D. (2003). Genome-wide discovery of transcriptional modules from DNA sequence and gene expression. *Bioinformatics* **19** Suppl 1, i273-282.

[928] Battle, A., Segal, E. & Koller, D. (2004). Probabilistic discovery of overlapping cellular processes and their regulation. In *The Proceedings of RECOMB*, 167-176.

[929] Bar-Joseph, Z., Gerber, G. K., Lee, T. I., Rinaldi, N. J., Yoo, J. Y., Robert, F., Gordon, D. B., Fraenkel, E., Jaakkola, T. S., Young, R. A. & Gifford, D. K. (2003). Computational discovery of gene modules and regulatory networks. *Nat. Biotechnol.* **21**, 1337-1342.

[930] DeRisi, J. L., Iyer, V. R. & Brown, P. O. (1997). Exploring the metabolic and genetic control of gene expression on a genomic scale. *Science* **278**, 680-686.

[931] Ideker, T., Thorsson, V., Ranish, J. A., Christmas, R., Buhler, J., Eng, J. K., Bumgarner, R., Goodlett, D. R., Aebersold, R. & Hood, L. (2001). Integrated genomic and proteomic analyses of a systematically perturbed metabolic network. *Science* **292**, 929-934.

[932] Patil, K. R. & Nielsen, J. (2005). Uncovering transcriptional regulation of metabolism by using metabolic network topology. *Proc. Natl. Acad. Sci. USA* **102**, 2685-2689.

[933] Yang, Q. & Sze, S. H. (2007). Path matching and graph matching in biological networks. *J. Comp. Biol.* **14**, 56-67.

[934] Tohsato, Y., Matsuda, H & Hashimoto, A. (2000). A multiple alignment algorithm for metabolic pathway analysis using enzyme hierarchy. In *The Proceedings of ISMB*, 376-383.

[935] Bono, H., Goto, S., Fujibuchi, W., Ogata, H. & Kanehisa, M. (1998). Systematic prediction of orthologous units of genes in the complete genomes. *Genome Inform.* **9**, 32-40.

[936] Green, M. & Karp, P. (2004). A Bayesian method for identifying missing enzymes in predicted metabolic pathway databases. *BMC Bioinformatics* **5**, 76.

[937] Omelchenko, M. V., Galperin, M. Y., Wolf, Y. I. & Koonin, E. V. (2010). Non-homologous isofunctional enzymes: A systematic analysis of alternative solutions in enzyme evolution. *Biol. Direct.* **5**, 31.

[938] Saigo, H., Hattori, M., Kashima, H. & Tsuda, K. (2010). Reaction graph kernels predict EC numbers of unknown enzymatic reactions in plant secondary metabolism. *BMC Bioinformatics* **11** Suppl 1, S31.

[939] Kharchenko, P., Vitkup, D. & Church, G. M. (2004). Filling gaps in a metabolic network using expression information. *Bioinformatics* **20** Suppl 1, i178-185.

[940] Kharchenko, P., Chen, L., Freund, Y., Vitkup, D. & Church, G. M. (2006). Identifying metabolic enzymes with multiple types of association evidence. *BMC Bioinformatics* **7**, 177.

[941] van Baarlen, P., van Esse, H. P., Siezen, R. J. & Thomma, B. P. (2008). Challenges in plant cellular pathway reconstruction based on gene expression profiling. *Trends Plant Sci.* **13**, 44-50.

[942] Werner, T. (2008). Bioinformatics applications for pathway analysis of microarray data. *Curr. Opin. Biotechnol.* **19**, 50-54.

[943] Viswanathan, G. A., Seto, J., Patil, S., Nudelman, G. & Sealfon, S. C. (2008). Getting started in biological pathway construction and analysis. *PLoS Comp. Biol.* **4**, e16.

[944] Pe'er, D. (2005). Bayesian network analysis of signaling networks: a primer. *Sci. STKE* **281**, pl4.

[945] Heckerman, D. (1998). A tutorial on learning with Bayesian networks. In *Learning in Graphical Models*, MIT Press, Cambridge, MA.

[946] Tamada, Y., Kim, S., Bannai, H., Imoto, S., Tashiro, K., Kuhara, S. & Miyano, S. (2003). Estimating gene networks from gene expression data by combining Bayesian network model with promoter element detection. *Bioinformatics* **19** Suppl 2, ii227-236.

[947] Friedman, N., Nachman, I. & Pe'er, D. (1999). Learning Bayesian network structure from massive datasets: the sparse candidate algorithm. In *The Proceedings of Conf. on Uncertainty in AI*, 206-221.

[948] Guo, A. Y., Chen, X., Gao, G., Zhang, H., Zhu, Q. H., Liu, X. C., Zhong, Y. F., Gu, X., He, K. & Luo, J. (2008). PlantTFDB: a comprehensive plant transcription factor database. *Nucl. Acids Res.* **36**, D966-969.

[949] Wingender, E. (2008). The TRANSFAC project as an example of framework technology that supports the analysis of genomic regulation. *Brief. Bioinform.* **9**, 326-332.

[950] Sandelin, A., Alkema, W., Engström, P., Wasserman, W. W. & Lenhard, B. (2004). JASPAR: an open-access database for eukaryotic transcription factor binding profiles. *Nucl. Acids Res.* **32**, D91-94.

[951] Loots, G. & Ovcharenko, I. (2007). ECRbase: database of evolutionary conserved regions, promoters, and transcription factor binding sites in vertebrate genomes. *Bioinformatics* **23**, 122-124.

[952] Pearl, J. (1988). "Probabilistic reasoning in intelligent systems." Morgan Kaufmann Publishers Inc., San Francisco, CA.

[953] Cowell, R. (1998). Introduction to inference for Bayesian networks. In *Learning in Graphical Models*, MIT Press, Cambridge, MA, USA.

[954] Friedman, N. (2004). Inferring cellular networks using probabilistic graphical models *Science* **303**, 799-805.

[955] Needham, C. J., Bradford, J. R., Bulpitt, A. J. & Westhead, D. R. (2007). A primer on learning in Bayesian networks for computational biology. *PLoS Comp. Biol.* **3**, e129.

[956] Kim, S. Y., Imoto, S. & Miyano, S. (2003). Inferring gene networks from time series microarray data using dynamic Bayesian networks. *Brief. Bioinform.* **4**, 228-235.

[957] Zou, M. & Conzen, S. D. (2005). A new dynamic Bayesian network (DBN) approach for identifying gene regulatory networks from time course microarray data. *Bioinformatics* **21**, 71-79.

[958] Troyanskaya, O. G., Dolinski, K., Owen, A. B., Altman, R. B. & Botstein, D. (2003). A Bayesian framework for combining heterogeneous data sources for gene function prediction (in Saccharomyces cerevisiae). *Proc. Natl. Acad. Sci. USA.* **100**, 8348-8353.

Conference Abbreviations

- COLT: Computational Learning Theory
- CSB: Computational Systems Bioinformatics
- CVPR: Computer Vision and Pattern Recognition
- ECCV: European Conference on Computer Vision
- ECML: European Conference on Machine Learning
- ICCV: International Conference on Computer Vision
- ICML: International Conference on Machine Learning
- ISIT: International Symposium on Information Theory
- ISMB: Intelligent Systems for Molecular Biology
- NIPS: Neural Information Processing Systems
- PAMI: Pattern Analysis and Machine Intelligence
- PSB: Pacific Symposium on Biocomputing
- RECOMB: Research in Computational Molecular Biology

Acronyms

- AI - Artificial Intelligence
- DP - Dynamic Programming
- EC - Enzyme Classification
- EM - Expectation Maximization
- GO - Gene Ontology
- HMM - Hidden Markov Model
- MAP - Maximum a Posteriori
- ML - Maximum Likelihood
- MSA - Multiple Sequence Alignment
- NN - Neural Network
- PST - Probabilistic Suffix Tree
- PTM - Post Translational Modification
- SNP - Single Nucleotide Polymorphism
- SVM - Support Vector Machine
- VMM - Variable Order Markov Model
- Y2H - Yeast Two-Hybrid

Index

σ-distance, 452

accuracy, 76, 272
activator, 155
activity profile, 506
adenine, 11
adjacency matrix, 416
ADP, 553
affinity chromatography, 557
alignment, 24
    global, 25
    local, 30
    semi-global, 30
allele, 146
alpha helix, *see* secondary structure
alternative splicing, 18, 359
amino acid, 13
AMP, 553
analogs, 23
anti-codon, 17
anti-correlated genes, 523
antibody, 13, 545, 558, 583
antigen, 545, 558, 583
apoptosis, 315
arithmetic coding, 215
atomic interaction, 350, 554, 578
ATP, 553
auto-regulation, 605
average linkage clustering, 414
Azuma-Hoeffding lemma, 35

backbone, 304
background distribution, *see* null distribution
backup gene, 516
backup system, 602
bag-of-words, 256, 493
bagging, 270
base pair, 11
base vector, 474
baseline distribution, *see* null distribution
baseline performance, 528
Bayes' decision rule, 95
Bayes' formula, 88
Bayesian information criterion, *see* BIC
Bayesian network, 651
    assertions, 656
    consensus graph, 671
    equivalence class, 671
    evidence, 660
    inference, 657
    model averaging, 678
    parameter learning, 663
    structure learning, 671
beta strand, *see* secondary structure
BIC, 449, 675
binding agent, 557
binomial distribution, 88
BLAST, 31, 72
    gapped, 46, 74
BLOCKS, 57, 173
BLOSUM, 57
Bonferroni correction, 96
bootstrapping, 270
branch-and-bound, 286

cancer, 651
CART algorithm, 278
CASP, 296
catabolism, 620
categorical data, 281
CATH, 391
causal network, 651
cDNA, *see* DNA
cDNA array, 509
CE algorithm, 322
cell, 9
cell cycle, 512
cell extract, 509
cell lysate, 509
center-of-gravity score, *see* CG-score
central dogma, 15
central limit theorem, 89
centroid, 402
CG-score, 108
chaining effect, 413
chaperone, 19, 292, 293
chi-square distribution, 90
chi-square test, 282
ChIP-chip, 561
ChIP-seq, 561
chromatin, 292
chromosome, 11, 292

circular dichroism, 302
classification, 235
classifier, 237
CLUSTALW, 137
cluster validation, 428, 534
clustering, 349, 397
  agglomerative, 408
  batch, 400
  configuration, 401
  divisive, 408
  fuzzy, 400, 404
  hard, 400
  hierarchical, 400
  online, 400
  partitional, 400
  random, 433
clustering coefficient, 593
co-evolution, 354, 571
co-expression, 515, 589, 627
co-factor, 621
code, 214
coding sequence, 12, 184
codon, 15
  start, 15
  stop, 15
coil, *see* secondary structure
collaborative clustering, 415
comparative modeling, *see* structure prediction
complete linkage clustering, 414
composition, 492
composition based statistics, 46
compression, 217
conditional data, 437
conditional independence, 654
conjugate distribution, 666
conjugate prior, 94
connected component, 412
connectivity matrix, 416
consensus profile, 532
consensus sequence, 113
conservation measure, 361
contact profile, 352
contact residues, 555
convergent evolution, 21
convex hull, 483
core structure, 19
correlated divergence, 354, 572
correlated mutations, 353, 579
cross-validation, 83, 269, 439, 451
cryo electron microscopy, 293
curse of dimensionality, 460
cytochrome, 320
cytokine, 347, 545
cytosine, 11
dali algorithm, 319, 391
decision boundary, 235
decision rule, 267
decision surface, 235, 239
decision tree, 262
deformation, 317
dendrogram, 532
deterministic annealing, 409
dihedral angle, 298
Dirichlet distribution, 120, 666
discriminant function, 237
  linear, 238
  non-linear, 249
discriminative model, 237
distance function, 98
distance matrix, 318
distance profile, 318, 421, 493
distortion
  in clustering, 402
  in embedding, 468, 474
distribution
  binomial, 88
  chi-square, 90
  exponential, 90
  extreme value, 39
  multinomial, 90
  normal, 89
  Poisson, 90
  uniform, 85
distributional scaling, 484
DNA, 10
  cDNA, 509
  junk DNA, 184
DNA binding, 507
DNA chip, *see* microarray
DNA polymerase, 510
docking, 563
domain, 20, 345
domain fusion, 571
domain architecture, 380
domain length, 378
domain structure, 346
dynamic programming, 26, 518
  matrix, 26
  trace, 27

earth-mover's distance, 486
EC number, 621
edit distance, 62
EGF domain, 224
eigenarray, 533
eigenfunction, 254
eigengene, 533
eigenvalue, 464
eigenvector, 464
electrostatic interaction, 554

EM algorithm, 162, 176, 198, 669
embedding, 459
    distance preserving, 467
    manifold learning, 478
    maximal variance, 461
    random projection, 474
    structure preserving, 461
enhancer, 155
entropy, 91, 265
    relative, 92
enzyme, 12, 262
enzyme classification, *see* EC number
enzyme recruitment, 625
epitope, 558
equivalence number, 78, 273
ergodic model, 202
error rate, 77
EST, 147
Euclidean distance, 99, 517
eukaryotes, 10
evalue, 42, 86
event, *see* random variable
exclusive interaction, 592
exon, 17, 359
expectation value, *see* evalue
expectation-maximization, *see* EM algorithm
exponential distribution, 90
expression level, 505
expression profile, 510
extrapolation, 521
extreme value distribution, 39

factorization, 88, 656
false discovery rate, 97
false negative, 76
false positive, 76
farthest neighbor algorithm, 414
FASTA, 31, 72
feature space, 249
feature vector, 249
feedback loop, 604
feedforward loop, 604
ferredoxin, 304
Fischer kernel, 258
fluorescent dye, 510
fold, 19, 21
fold recognition, 295
Fourier transform, 564
Frobenius distance, 313
FSSP, 391
functional interaction, 554

gain ratio, 267
Gamma function, 121, 666
gap function, 27, 61
    linear, 28
    sub-additive, 28
Gaussian distribution, 89
gel electrophoresis, 557, 609
gene, 12, 185
gene chip, *see* microarray
gene duplication, 156, 379, 602, 603
gene expression profiling, 506
gene locality, 570
gene ontology, *see* GO
gene prediction, 184
gene preservation, 568
generalization, 398, 429
generalized profile, 332
generative model, 183, 236
genetic code, 15
genome, 11
genotype, 516
geodesic distance, 479
geometric hashing, 324
geometry, 467
Gibbs sampling, 161, 670
global alignment, *see* alignment
glycolysis, 618
GO, 82, 444, 538
GO term, 538
    common level, 539
    depth, 539
    semantic similarity, 383, 539
gradient descent procedure, 242
gradient vector, 242
Gram matrix, *see* kernel
graph, 410
    cut, 418
    cut capacity, 418
    diameter, 593
    hub, 595
    lattice, 595
graph-based clustering, 400, 410
graphlet, 600
guanine, 11
guide tree, 112
guilt by association, 397

haplotype, 146
harvesting, 509
hash function, 324
heatmap, 532
hemoglobin, 292
hidden Markov model, *see* HMM
hierarchical clustering, 408
Hilbert space, 255
histone, 292
HMM, 188
    backward algorithm, 194
    Baum-Welch algorithm, 198

decoding problem, 196
evaluation problem, 192
forward algorithm, 194
learning problem, 198
parameters, 188
Viterbi algorithm, 197
homeostasis, 620
homologs, 20, 23
homology, 20, 23, 32, 49
homology modeling, *see* structure prediction
horizon effect, 269
host space, 461
HSP, 32
Huffman code, 214
hybridization, 509
hydrogen bond, 554
hydrophobic interaction, 554
hyper-parameter, 94
hypothesis testing, 95

i.i.d. variables, 89
ID3 algorithm, 266
image point, 461
impurity, 265
Gini, 265
information, 265
in silico, 262
in vitro, 262, 558
in vivo, 262, 558
inclusive interaction, 592
independent variables, *see* statistical independence
information content, 91
information gain, 265
information impurity, *see* impurity
inner-product kernel, *see* kernel
inter-atomic, 305
inter-cluster, 399
interaction network, 593
interaction site, 554
interactome, 555
intra-atomic, 305
intra-cluster, 399
intrinsic dimensionality, 466, 488
intron, 17, 359
isomap algorithm, 479
isometry, 467
isozyme, 626, 627
iterative alignment, 132

jackknife test, 83, 514
Jensen-Shannon divergence, *see* JS-divergence
Johnson-Lindenstrauss theorem, 455, 474
JS-divergence, 130, 274, 456, 485

k-means, 402
fuzzy, 407
hierarchical, 409
kernel, 250
Gaussian, 252
inner-product, 251
linear, 251
matrix, 253
Mercer, 253
polynomial, 252
RBF, 252
kernel trick, 251
kinesin, 292
KL-divergence, 92, 129, 274, 285, 419, 456, 485
knn, 236
knockout experiment, 512, 643
knowledge-based potential, 565
Kolmogorov complexity, 435
Kullback-Leibler divergence, *see* KL-divergence

learning rate, 243
leave-one-out, 83
Lempel-Ziv algorithm, 218
Levinthal's paradox, 19, 293
ligand, 563
likelihood, 85, 93
likelihood ratio, 52, 284
log-likelihood ratio, 129
log-odds, 52, 59
loop, *see* secondary structure
low-complexity sequence, 45

Mahalanobis distance, 408
majority voting, 373
Manhattan distance, 98
manifold, 478
MAP estimator, 94, 663
margin, 244
marginal likelihood, 94
Markov model, 184
context, 212
fixed-order, 211
memory length, 211
order, 189
variable-order, 213
Markovian clustering, 425
mass spectrometry, 546
mass-distance, 497, 518
maximum a posteriori, *see* MAP
maximum likelihood, *see* ML
MDL, 276, 434, 489, 534, 675
MDS, 468, 484
meiosis, 105
MEME algorithm, 162
Mercer kernel, *see* kernel

metabolic distance, 624
metabolic network, 601
metabolic pathway, 620
metabolite, 621
methylation, 555
metric space, 98
microarray, 509
  class discovery, 533
  cluster analysis, 529
  differential analysis, 513
  enrichment analysis, 538
  gene set enrichment analysis, 541
  missing values, 521
  pairwise analysis, 515
  prob, 509
  replication, 544
minimal cell, 644
minimal cut algorithm, 418
minimum description length, *see* MDL
mirror tree, 572
mitosis, 105
mixture density, 404
mixture model, 209, 270, 500
ML estimator, 93, 406, 664
model averaging, 270
molecular clock, 20, 572
moment, 86
motif, 20
mRNA, *see* RNA
MSP, 32
multi-category classifier, 248
multi-domain protein, 346, 380
multidimensional scaling, *see* MDS
multinomial distribution, 90
multipartite graph, 168
multiple testing, 96
mutual information, 92
mutual information kernel, 259

native structure, 293
nearest neighbor algorithm, 111, 412
Needleman-Wunsch algorithm, 26
negatives, 75
neighbor-joining, 112
network analysis, 593
  redundancy, 602
  robustness, 602
network motif, 603
network switch, 643
neural network, 249
NMR, 293
no-free-lunch theorem, 439
non-parametric model, 411
norm, 98
normal distribution, 89
normalized cut algorithm, 422
NP-hard problem, 109
nucleic acid, 10
nucleosome, 291
nucleotide, 11
null distribution, 37, 39, 95, 433
null hypothesis, 95
null probability, 281

Occam's razor, 435
odds ratio, 52
operon, 570, 603
overfitting, 83, 269, 429

p53 (gene), 602
pair statistic, 377
pairwise clustering, 111, 400, 411
pairwise kernel, 257
PAM, 52
paralogs, 23
partial order alignment, *see* POA
partial order graph, *see* POG
pathway, 617
  blueprint, 624
  crosstalk, 623, 642
  hole, 625
pattern matrix, *see* vector data
pattern recognition, 398
PCA, 454, 462, 533
PCR, 510
Pearson correlation, 517, 611
Pearson distance, 129
peptide, 345
peptide bond, 16
percent identity, 48
perceptron algorithm, 241
Pfam, 390, 444
phenotype, 516
phosphorylation, 554
phylogenetic profile, 569
physical interaction, 554
physico-chemical property, 13, 363
planted motif problem, 167
plasmid, 609
POA, 141
POG, 142
point mutation, 24
Poisson distribution, 90
Poisson process, 90
position-specific scoring matrix, *see* PSSM
positives, 75
post-translational modification, *see* PTM
power iteration method, 421
power-law distribution, 598
PPM, 229
precipitation, 558
precision, 77

prefix code, 214
primary structure, 18
primer, 510
principal component analysis, *see* PCA
prior
    Dirichlet, 119
    Dirichlet mixture, 120
    fixed, 118
    position-dependent, 119
    uniform, 118
probabilistic suffix tree, *see* PST
probability
    conditional, 87
    density function, 85
    distribution, 85
    joint distribution, 87
    marginal distribution, 87
    mass function, 85
    posterior, 94
    prior, 94
    value, 85
ProDom, 390
profile, 116
profile-HMM, 204
profile-profile comparison, 128
progressive alignment, 136
projection, 463
prokaryotes, 10
promiscuous domain, 380, 443
promoter, 155, 515, 603
protease, 557
protein, 12
protein family, 20
protein superfamily, 20
protein array, 545, 561
protein folding, 19, 293
protein interactions, 554
    hotspots, 567
    prediction, 561
proximity data, 400
proximity matrix, *see* proximity data
pseudo-counts, 117, 222, 668
pseudo-distance, 64
pseudo-gene, 185
PSIBLAST, 132
PSIPRED, 301
PSSM, 126
PST, 219
PTM, 18, 554, 583, 603
pull down experiment, 557
purine, 10, 50
pvalue, 41, 96, 283
pyrimidine, 10, 50

quaternary structure, 19

Ramachandran plot, 298
random geometric graph, 601
random forest, 271
random graph, 594
random projection, 170
random sequence, 33
random variable, 85
    continuous, 85
    discrete, 85
    event, 85
    sample space, 85
real-normed space, 99
recall, 76
reference set, 75, 493
regular expression, 114
regulation program, 637
regulatory element, 155, 184
regulatory module, 605, 637
regulatory network, 506, 601
relative entropy, *see* entropy
repressor, 155
reverse transcription, 509
ribose, 11
ribosome, 17, 514
rigid transformation, 306
RMSd, 305
RNA, 16
    mRNA, 16, 506, 509
    siRNA, 512
    tRNA, 17
RNA binding, 523
RNA gene, 545
RNA labeling, 509
RNA polymerase, 16
RNA world, 18
RNase, 17
ROC, 79, 273
rotation matrix, 306

salt bridge, *see* electrostatic interaction
sampling, 270
scaffold, 296
scale-free model, 597
scaling, 366
SCOP, 82, 330, 390, 444
scoring matrix, 50
SDS-PAGE, *see* gel electrophoresis
search space, 38
secondary structure, 18, 296, 352
secondary structure prediction, 299
selectivity, 77, 273
semantic similarity, *see* GO term
sensitivity, 76, 273
sequence space, 20
sequence termination analysis, 355
sequence to structure alignment, 295

sequence weighting, 122
signal transduction, 347, 554, 617
signaling network, 506
signaling pathway, 619
simulated annealing, 678
single linkage clustering, 412
single nucleotide polymorphism, *see* SNP
single-input module, 605
singular value decomposition, *see* SVD
siRNA, *see* RNA
slack variable, 246
small-world model, 596
SMART, 390
Smith-Waterman algorithm, 30
smoothing, 118
SNP, 146
    nonsynonymous, 146
    synonymous, 146
soft margin, 246
SP-score, 107, 362
spanning tree, 412
sparse Markov transducer, 227
Spearman rank correlation, 517
spectral clustering, 421
splicing, 17, 22
split information, 267
squared error
    in clustering, 401
    in embedding, 469
standard deviation, 86
star alignment, 109
statistic, 89
statistical independence, 87, 653
statistical significance, 32, 96
Steiner string, 114
stochastic decision tree, 271
stress function, 469
string kernel, 256
    mismatch spectrum, 257
    spectrum, 256
structal algorithm, 309
structure comparison, 303
structure prediction, 293
    ab initio, 294
    comparative modeling, 294
    homology modeling, 294
structure space, 20
substrate, 621
sum-of-pairs score, *see* SP-score
super-paramagnetic clustering, 427
supervised learning, 235
support vector, 245
support vector machine, *see* SVM
SVD, 467, 533
SVM, 246
systems biology, 8, 651

T-coffee, 139
tandem affinity purification, 560
target frequencies, 58
tertiary structure, 19
test set, 83
threading, 295
thymine, 11
TigrFam, 390
time-series expression data, 512
    shifts, 520
topological ordering, 657
topology, 467
torsion angle, *see* dihedral angle
training set, 83, 269
transcription, 16
transcription factor, 155, 505, 515
transcription regulation, 155, 320, 505
transient interaction, 554
transitive alignment, 138
transitive chaining, 360, 441
transitive closure, 417
transitive relation, 49
transitivity test, 360
translation, 17
tree alignment, 110
tree of life, 20
tree pruning, 269
triangle inequality, 98
troponin, 315
true negative, 76
trypsin, 292
twilight zone, 43, 48, 361

ubiquitin, 304
ubiquitination, 554
uniform distribution, 85
unit vector model, 311
unsupervised learning, 397
uracil, 16
URMS, 311

validation set, 269
variance, 86
vector quantization, 398
vector data, 400
vector space clustering, 400
vector space model, 256
vectorial representation, 490
    external, 493
    internal, 492

weighted sum, 128
whitening transformation, 37, 366

yeast two-hybrid, 558

zinc finger, 115
zscore, 35, 514